AF342669

Quantum Coherence in Mesoscopic Systems

NATO ASI Series

Advanced Science Institutes Series

*A series presenting the results of activities sponsored by the NATO Science Committee,
which aims at the dissemination of advanced scientific and technological knowledge,
with a view to strengthening links between scientific communities.*

The series is published by an international board of publishers in conjunction with the
NATO Scientific Affairs Division

A	Life Sciences	Plenum Publishing Corporation
B	Physics	New York and London
C	Mathematical and Physical Sciences	Kluwer Academic Publishers
D	Behavioral and Social Sciences	Dordrecht, Boston, and London
E	Applied Sciences	
F	Computer and Systems Sciences	Springer-Verlag
G	Ecological Sciences	Berlin, Heidelberg, New York, London,
H	Cell Biology	Paris, Tokyo, Hong Kong, and Barcelona
I	Global Environmental Change	

Recent Volumes in this Series

Volume 248—Lower-Dimensional Systems and Molecular Electronics
  edited by R. M. Metzger, P. Day, and G. C. Payavassiliou

Volume 249—Advances in Nonradiative Processes in Solids
  edited by B. Di Bartolo

Volume 250—The Application of Charge Density Research
  to Chemistry and Drug Design
  edited by George A. Jeffrey and Juan F. Piniella

Volume 251—Granular Nanoelectronics
  edited by David K. Ferry, John R. Barker, and Carlo Jacoboni

Volume 252—Laser Systems for Photobiology and Photomedicine
  edited by A. N. Chester, S. Martellucci, and A. M. Scheggi

Volume 253—Condensed Systems of Low Dimensionality
  edited by J. L. Beeby

Volume 254—Quantum Coherence in Mesoscopic Systems
  edited by B. Kramer

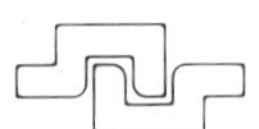

Series B: Physics

Quantum Coherence in Mesoscopic Systems

Edited by

B. Kramer

Physikalisch–Technische Bundesanstalt Braunschweig und Berlin
Braunschweig, Germany

Plenum Press
New York and London
Published in cooperation with NATO Scientific Affairs Division

Proceedings of a NATO Advanced Study Institute on
Quantum Coherence in Mesoscopic Systems,
held April 2–13, 1990,
in Les Arcs, France

Library of Congress Cataloging-in-Publication Data

NATO Advanced Study Institute on Quantum Coherence in Mesoscopic
  Systems (1990 : Les Arcs, Savoie, France)
    Quantum coherence in mesoscopic systems / edited by B. Kramer.
        p.   cm. -- (NATO ASI series. Series B, Physics ; v. 254)
    "Proceedings of a NATO Advanced Study Institute on Quantum
  Coherence in Mesoscopic Systems, held April 2-13, 1990, in Les Arcs,
  France"--T.p. verso.
    "Published in cooperation with NATO Scientific Affairs Division."
    "Held within the program of activities of the NATO Special Program
  on Condensed Systems of Low Dimensionality, running from 1983 to
  1988 as part of the activities of the NATO Science Committee--P.
    Includes bibliographical references and index.
    ISBN 0-306-43889-5
    1. Mesoscopic phenomena (Physics)--Congresses.   I. Kramer, B.
  (Bernhard), 1942-   . II. Special Program on Condensed Systems of
  Low Dimensionality (NATO) III. Title. IV. Series.
  QC176.8.M46N38   1990
  530.4'1--dc20   91-10837
                                                             CIP

© 1991 Plenum Press, New York
A Division of Plenum Publishing Corporation
233 Spring Street, New York, N.Y. 10013

SPECIAL PROGRAM ON CONDENSED SYSTEMS OF LOW DIMENSIONALITY

This book contains the proceedings of a NATO Advanced Research
Workshop held within the program of activities of the NATO
Special Program on Condensed Systems of Low Dimensionality,
running from 1985 to 1990 as part of the activities of the NATO
Science Committee.

Other books previously published as a result of the activities of
the Special Program are:

SPECIAL PROGRAM ON CONDENSED SYSTEMS OF LOW DIMENSIONALITY

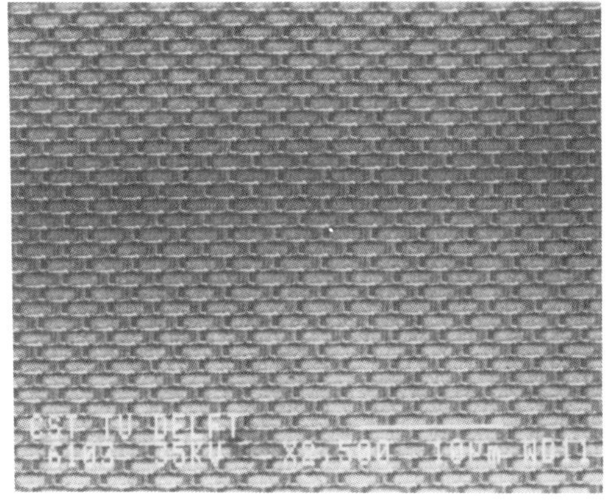

Scanning electron microscope photograph of a man-made two dimensional Aluminum array. Tunnel junctions of the approximate size 100x100nm^2 are situated within the narrow lines between the broad H-like islands. The distances between the broad islands are about 3μm and 1μm in horizontal and vertical direction, respectively. The structure was fabricated by L.J. Geerligs, H.S.J van der Zant, and J.E. Mooij at the Delft Centre for Submicron Technology/DIMES.

Deep-mesa-etched quantum dots in a Al-GaAs/GaAs heterostructure. Diameter of the dots is 400 nm. The lattice distance is about 1μm. The number of electrons per dot can be varied from 25 to 200 by using the persistent photo-conductivity effect. The structure was prepared by T. Demel, P. Grambow and D. Heitmann at the Max-Planck Institut für Festkörperphysik in Stuttgart.

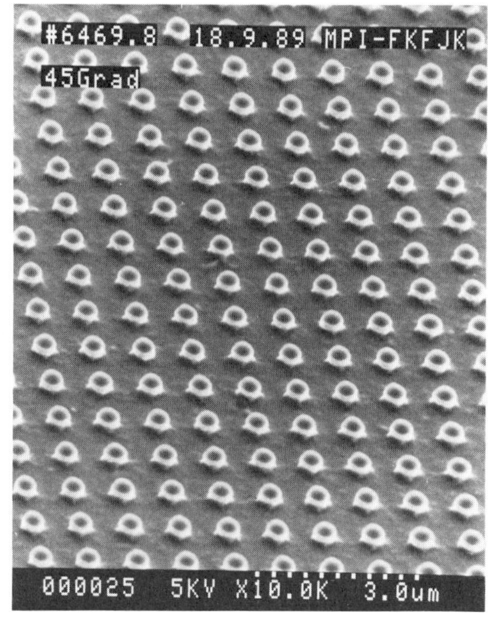

Aluminum network having the shape of a Sierpinski Gasket. The smallest wire is 3.2μm long. The linear size is increased by a factor of two at each iteration step whereas the wire length increased by a factor of three. The fractal dimension is 1.585. The sample was realized by B. Pannetier at the Centre de Recherches sur les Très Bas-ses Temperatures (CRTBT) from a lift-off stencil electron beam written in Centre National d'Etudes des Télécommunications (CNET) in Grenoble.

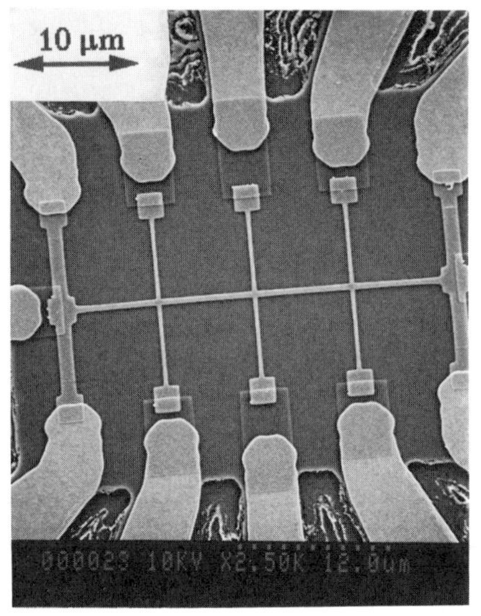

Photograph of the surface of a AlGaAs/GaAs heterostructure patterned to generate a quasi-one dimensional electron gas underneath. Shown are the gate (thin wire structure in the center) and the voltage probes. The sample was prepared by T.J. Thornton, M.L. Roukes, A. Scherer, B.P. Van der Gaag at Bellcore in Red Bank, New Jersey.

Scanning transmission electron micrograph of an Sb "spider" (0.999999 starting material). The thickness of the wires is 80nm. The narrowest lines are 95-98nm wide. The distance along the edge of the image between a pair of large dark areas is about 1μm. The sample was made by C. P. Umbach at IBM Watson Research Center in Yorktown Heights, New York.

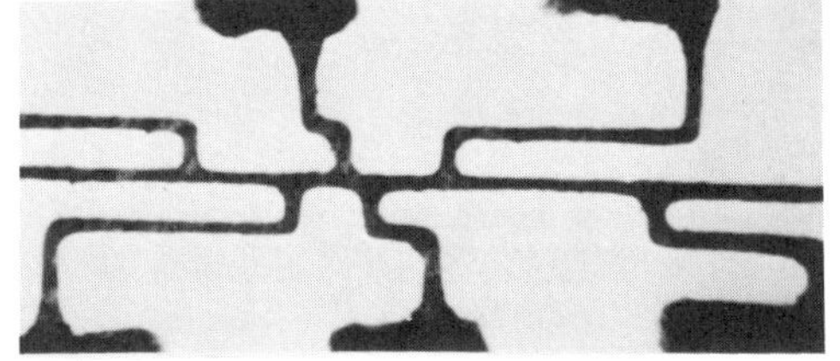

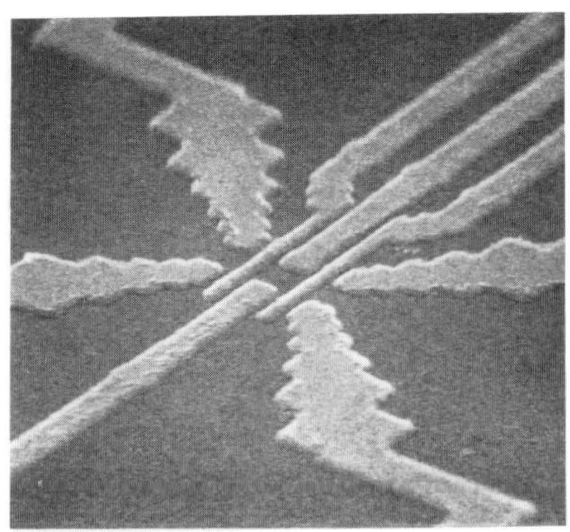

An electron micrograph of three split-gate electrodes in series separated by barriers. The structure was fabricated on a high mobility heterostructure. The distance between the outer arms of the electrodes (upper left and lower right corners of the figure) is about 1μm. The gaps between the gates in the center are about 300nm wide. The sample was made by G. Timp, R.E. Behringer, E.H. Westerwick, and J.E. Cunningham at AT&T Bell Laboratories in Holmdel, New Jersey.

PREFACE

Mesoscopic condensed matter systems have small spatial dimensions ($< 1\,\mu$m) and are characterized by the presence of a certain amount of disorder, for instance due to the incorporation of impurities. At very low temperatures (< 1 K), these systems show a variety of interesting transport phenomena that are related to quantization, and to the fact that the coherence of the quantum states is not destroyed by inelastic and phase randomizing processes such as electron-electron, and electron-phonon collisions. One of the main aims of the school has been to bring together experienced researchers working in different areas of this new and rapidly developing field, and to provide a broad survey to those who just started or plan to enter.

The most important mesoscopic phenomena which were almost exclusively discovered during the past decade are

- oscillations of the magneto-resistance of small normal metallic cylinders and rings,

- the quantized Hall effect in MOSFETs and heterostructures subject to high magnetic field,

- reproducible statistical fluctuations of the conductance of quasi-one dimensional inversion layers in MOSFETs when the gate voltage is changed,

- universal fluctuations of the magnetoconductance in small metallic wires and rings,

- quantization of the DC conductance of two-dimensional point contacts,

- non-local and time dependent transport effects in small metallic structures,

- existence of a critical resistance for the superconducting transition in granular superconductors and Josephson-junction networks,

- single electron effects in Josephson tunnel junctions,

- persistent currents in normal metallic rings.

Among the effects that were theoretically predicted but are not yet experimentally verified are

- a logarithmic decay of the time-dependent response to an applied electric field,

- non-linear properties of the AC transport due to impurity induced asymmetries,

- quantum oscillations of the AC conductance of ballistic systems of finite width and length due to electron-hole interference,

- frequency dependent mesoscopic fluctuations of the AC conductance.

The lectures and seminar talks at this NATO Advanced Study Institute reviewed theory and experimental work concerning these transport phenomena, and the closely related quantum mechanics such as the statistical properties of the energy spectra, and the shape of the corresponding (fractal) wave functions. Historically, the field of mesoscopic research emerged from the study of the localization properties of the states in random systems. There seems to exist a close relation between mesoscopic fluctuations and chaotic behavior in non-linear systems. Therefore, the respective lectures and seminar talk were organized as a symposium in order to emphasize also the topical nature of these questions.

Among the highlights were the reports on the recent first observation of the above mentioned persistent currents, and the first realization of a current standard by applying an alternating voltage to an ultra-small tunnel junction such that single electron effects dominate.

Although many of the phenomena are now reasonably well understood there are a number of as yet not fully explained effects such as the quantized Hall effect, and the mesoscopic fluctuations in the regime of localized states. The field is certainly only at the beginning of its development, and new surprises may be expected at any time especially when looking into the somewhat remote corners of the subject.

Braunschweig, October 30, 1990   Bernhard Kramer

ACKNOWLEDGEMENTS

It is a pleasure to acknowledge the financial support by NATO and by the Office of Naval Research (ONR), as well as the contributions of other institutions which made the organization of this school possible.

I want to thank Frau Elisabeth Meyer for her devoted help during the preparation of the event, and for preparing the manuscripts for publication. The assistance of Bodo Huckestein in setting up the $\TeX$ formats is especially appreciated.

CONTENTS

CHAPTER 1

PREPARATION AND CHARACTERIZATION

FAR-INFRARED SPECTROSCOPY OF TWO-DIMENSIONAL ELECTRONIC SYSTEMS WITH TUNABLE CHARGE DENSITY

Detlef Heitmann and Klaus Ensslin

Max-Planck-Institut für Festkörperforschung
Heisenbergstraße 1
W-7000 Stuttgart 80, F. R. Germany

1 Introduction

Mesoscopic electronic systems, i.e. systems in the intermediate regime between totally classical and totally quantized behavior are very often studied in two-dimensional electronic systems (2 DES). 2 DES can be realized in semiconductor heterostructures or in metal-oxide-semiconductor (MOS) systems. They have attracted a great interest in the last decade [1]. This interest arose from many proposed and realized technical applications and – in an interacting way – also from fundamental physical investigations. In these systems many of the interesting physical properties can be tailored during the fabrication process and, for a particular sample, varied over a wide range via electric or magnetic fields, stress, temperature, etc. This allows very detailed studies of fundamental physical properties and of interactions that are unique to the 2 D system. For the investigation of mesoscopic effects it is advantageous that laterally confined electron systems in 2 D samples can be tuned, e.g. via a gate voltage, from a macroscopic systems through the mesoscopic regime into the quantized regime. Further, a very sensitive and reproducible adjustment of e.g. the Fermi energy or other parameters is easily possible.

In our series of lectures on basic facts we would like to explain the basic properties of 2 DES by describing spectroscopic experiments on these systems. Since typical energies in the 2 DES, e.g., subband spacing, 2 D-plasmon and cyclotron resonance (CR) energies, are in the regime of 1-100 meV, far-infrared (FIR) spectroscopy [1, 2, 3, 4] is widely used to study the 2 DES. We will first give a short introduction into the experimental techniques. Then we discuss elementary excitations in 2 DES, e.g., 2 D plasmons, inter-subband resonances (ISR), and CR. It is not the goal of these notes to present a complete survey, but they are rather an attempt to provide a general understanding by describing of some selected examples. Most of these will be devoted to $Al_xGa_{1-x}As/GaAs$ heterostructures. We will, however, also report about experiments on the Si-MOS system. Due to the special surface bandstructure of Si, which arises from the indirect band

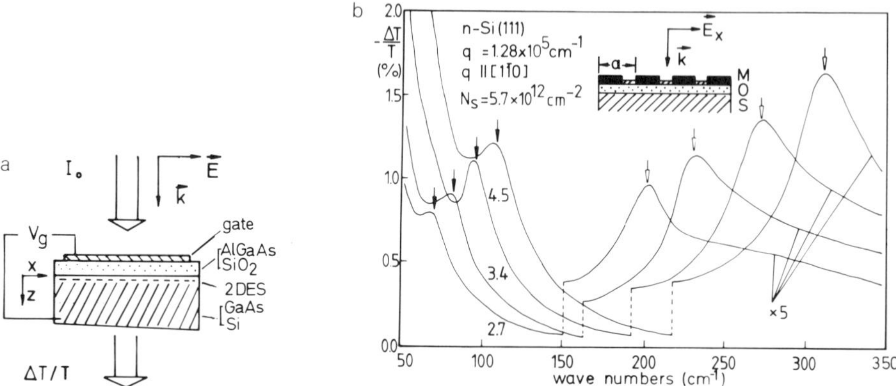

Figure 1. (a) Schematics for the transmission of FIR radiation through MOS- or heterostructure samples, (b) relative change of transmission $-\Delta T/T$ as a function of the wavenumber in 2 D electron accumulation layers of $n-$Si (111) at different charge densities N_S. Full and open arrows indicate, respectively, 2 D plasmon- and inter-subband resonance excitations [3].

gap, unique excitations can be observed which are not easily accessible in the GaAs system. They are, however, important for an understanding of spectroscopy in 2 DES. In these notes we address particularly CR experiments in gated heterostructures with field effect tunable charge density [5, 6, 7, 8, 9, 10]. These are very useful examples for a tutorial introduction to CR itself. They are also a valuable tool for the study of bandstructure, occupation and other properties of 2 DES. Currently there is an increasing interest in electronic systems with further reduced dimensionality, 1 D quantum wires and 0 D quantum dots. They are prepared by imposing a lateral potential onto the original 2 DES. FIR spectroscopy has proved to be a very useful tool to study these systems (e.g. [11, 12, 13, 14]). Spectroscopy on these systems has been reviewed recently in [15, 16, 17, 18].

2   Experimental Techniques

The most common arrangement for FIR spectroscopy is shown schematically in Fig. 1a. FIR radiation is incident normally onto the sample and the transmitted radiation is detected. The samples are usually wedged ($2° - 3°$) to avoid interference effects which would arise from multi-reflections in plane-parallel samples. If the signal is large enough as, e.g. in CR of high-mobility GaAs samples, the transmission can be recorded directly and the resonance position and linewidth can be analyzed. In many cases, however, differential techniques are necessary to resolve even a very weak response with a sufficient signal-to-noise ratio. Here it is most convenient if one can tune the 2 D charge density, N_S, via a gate voltage, V_g, which is applied to a semitransparent gate. Then the differential transmission $\Delta T/T = (T(N_S) - T(N_S = 0))/T(N_S = 0)$ can be evaluated. This is possible for Si-MOS structures and gated heterostructures where the gate voltage can be switched between a certain value, V_g, and the threshold voltage, V_t, with $N_S(V_t) = 0$.

Many spectroscopic investigations are performed with FIR lasers. In a laser experiment the magnetic field or, if possible, N_S is varied. A drawback of laser experiments is that by sweeping N_S or B the physical states of the 2 DES are changed, e.g. the subband spacing, the filling factor etc. Thus, it is highly desirable to investigate the sample at fixed N_S and B in the frequency domain. Since broadband light sources (Hg-lamps, Globar-lamps) have a weak spectral emission in the FIR, the technique of Fourier transform spectroscopy is applied, where all spectral information is measured simultaneously. The principle of the operation is that one of the mirrors of a Michelson-type interferometer is moved with a constant velocity v in the x-direction. Thus, each spectral component S_ν of wavenumber ν is modulated with $\cos(\nu x)$. At the output of the interferometer the signal $I(x) \propto \sum S_\nu \cdot \cos(\nu x)$ is measured. $I(x)$, the so-called interferogram, is digitized and the spectral components S_ν are computed from a numerical Fourier transform $S_\nu = FT[I(x)]$. In a rapid scan Fourier transform spectrometer a large number of interferograms, measured in different succeeding scans, are collected ("co-added") to improve the signal-to-noise ratio. In samples where one can switch the gate voltage one can apply differential techniques. Alternately small numbers of interferograms are taken at a gate voltage V_g and at threshold voltage V_t. The interferograms are co-added in different files and Fourier transformed separately at the end of the experiments. In this way, the effective delay between measuring the sample spectrum $I(V_g)$ and the reference spectrum $I(V_t)$ is small (1s to 10s) and very weak excitations (as low as 0.01% change in relative transmission) can be resolved without problems from long time drifts. These techniques are described in more detail in [3]. The typical spectral range of commercial FT-spectrometers is $10\,\mathrm{cm}^{-1}$ to $4000\,\mathrm{cm}^{-1}$. The highest spectral resolution, which depends on the inverse optical path length of the moving mirror, is typically $0.03\,\mathrm{cm}^{-1}$. Normally all experiments are performed at low temperatures $T < 10K$.

As an example for differential spectroscopy on 2 D systems we show in Fig. 1b spectra measured on electron-accumulation layers in an n-type Si-(111)-MOS system. The carrier density N_S has been varied via the gate voltage. The background of the spectra, which decreases monotonically with frequency, is caused by the non-resonant intra-band absorption ($=$ Drude absorption) of the 2 DES. The resonances in the frequency regime $30\text{-}150\,\mathrm{cm}^{-1}$ are 2 D plasmon resonances (PR) which shift with increasing N_S to higher frequencies. The excitation of these resonances is possible due to the grating coupler effect of the gate which consists of periodic stripes of high and low conductivity materials. At high frequencies ISR are observed, i.e. resonant transitions from the lowest subband E_0 to the first excited subband E_1, which shift with increasing surface electric field and corresponding N_S to higher frequencies. These ISR can be excited with normally incident radiation due to the special subband structure of Si(111) (see below) [1]. 2 D PR, ISR and various other kinds of excitations will be the topic of this paper.

3 FIR Transmission Spectroscopy of 2 DES

2 DES are defined here in the sense that a confinement of electrons in a narrow potential well quantizes the motion of the electrons in the z-direction perpendicular to the layered systems. In combination with the free motion of the electrons parallel to the interface, this leads to an electron energy spectrum that consists of different subbands (index i) with dispersion

$$E_i(K) = E_i + \frac{\hbar^2(K_x^2 + K_y^2)}{2m^*}. \tag{1}$$

Here K_x and K_y are quasi-momenta in the 2 D plane, and m^* is the effective mass of the parabolic system. An additional static magnetic field B applied normally to the

2 D-plane leads to a totally quantized energy spectrum where each subband forms a set of discrete Landau levels with energies

$$E_{in} = E_i + \hbar\omega_c(n + 1/2) \tag{2}$$

$\omega_c = eB/m^*$ is the cycloton frequency. An important quantity in a magnetic field is the filling factor $\nu = N_S h/eB$. It denotes the number and fraction of occupied spin resolved Landau levels. We use in the following also $n = 0.5\nu$ for the number of filled Landau levels if the spin is not important.

The dynamical response of a 2 DES is strongly anisotropic. Excitations with electric fields in the plane are collective PR and CR in a magnetic field, i.e. the transitions between the Landau levels. Excitations with electric fields perpendicular to the 2 D plane are ISR, i.e. transitions between the different subbands of the 2 DES. There are also various combined resonances, e.g., plasmon-ISR [19], plasmon-CR [20], CR-ISR [21, 22, 23, 7] and others, some examples will be discussed later in this paper.

The simplest microscopic model for the in-plane response of a 2 DES to electromagnetic fields is the Drude model.

$$\sigma(\omega) = \sigma_0(1 - i\omega\tau)^{-1}; \quad \sigma_0 = N_S e^2 \tau/m^*. \tag{3}$$

Here σ_0 is the DC conductivity and the damping of the electrons is characterized by a scattering time τ. The real part of this Drude conductivity, $\sigma_r = \sigma_0(1+\omega^2\tau^2)^{-1}$, decreases with frequency and drops to $\sigma_o/2$ at a frequency $\omega' = \tau^{-1}$. There are more sophisticated models for the in-plane excitations and also formulations for the perpendicular response. Some of these models will be explained later. For further information we refer to a tutorial introduction in [4] and the references therein.

Based on such models, the aim of FIR spectroscopy is to extract information on the microscopic properties of the investigated system. To demonstrate this for transmission spectroscopy let us consider the transmission of normally incident FIR radiation through a sample as shown in Fig. 1a. We normalize the conductivity to the reciprocal vacuum impedance, $\epsilon_0 c = 1/377\Omega$, by defining $s = \sigma/\epsilon_0 c$. The dielectric function of the semiconductor is ϵ. The conductivity of the gate film, σ_g, is characterized by the normalized value $s_g = 377\Omega/R_g$, where R_g is the gate resistivity. The very small distance between the gate and the 2 DES ($\approx 50 - 100\,\text{nm}$) can be neglected for the long wavelengths FIR radiation. Using Maxwells equations we find at normal incidence for the transmission coefficient of the amplitudes (e.g. [4])

$$t_p = 2\sqrt{\epsilon}/(1 + \sqrt{\epsilon} + s_g + s). \tag{4}$$

The transmitted intensity is $T = |t_p|^2$. In the experiment one often measures the differential transmission $\Delta T/T = (T(s) - T(s = 0))/T(s = 0)$. We find from eq. (4)

$$\frac{\Delta T}{T} = -\frac{2(1 + \sqrt{\epsilon} + s_g)s_r + s_r^2 + s_i^2}{(1 + \sqrt{\epsilon} + s_g + s_r)^2 + s_i^2}. \tag{5}$$

For small signals, if both the real and imaginary part, s_r and s_i, respectively, are small compared with $1 + \sqrt{\epsilon} + s_g$, we obtain the well-known result

$$\Delta T/T = -2\sigma_r(\omega)/c\epsilon_0(1 + \sqrt{\epsilon} + s_g) \quad \propto \quad Re(\sigma(\omega)). \tag{6}$$

The relative change of transmission is directly proportional to the real part of the dynamic conductivity. This approximation can be used to extract $\sigma_r(\omega)$ if $\Delta T/T$ is small.

However, one has to be careful in the case of large signals, e.g. CR in high mobility GaAs-heterostructures, then eq. (5) should be used to extract $\sigma(\omega)$ from $\Delta T/T$. One should also keep in mind that sometimes, e.g., in the Reststrahlen regime of polar materials, $\epsilon = \epsilon(\omega)$ depends strongly on the frequency and can influence the lineshape of a resonance (see e.g. [24]). If we insert into eq. (6) the Drude conductivity eq. (3) we find that $\Delta T/T$ decreases monotonically with increasing ω, in particular for large ω with ω^{-2}. This is the background in the spectra of Fig. 1b.

4 Experimental Results

In the experimental part of these lecture notes we discuss some typical examples for spectroscopic investigations. Within the available space we cannot explain all the details and did not attempt to give a complete list of references. The goal is to provide the general ideas.

4.1 Cyclotron Resonance

By far the most FIR investigations of heterostructures are devoted to CR. To demonstrate some characteristic features of CR we discuss some measurements on $Al_xGa_{1-x}As/GaAs$ heterostructures with front gates [5]. We have prepared heterostructures with a 50 nm Si-doped AlGaAs-layer and with 20 nm spacer layer. On top of this structure a semi-transparent NiCr gate was prepared and contacts were made to the channel. Via a gate voltage that was applied between the gate and the channel, we were able to tune N_S from $5 \cdot 10^{11} cm^{-2}$ to nearly zero. The charge density could be determined in situ under experimental conditions from magnetocapacitance measurements. Experimental CR measurements are shown in Fig. 2a for different densities N_S. The CR excitation leads to a resonant decrease of the transmission at frequencies about $112\,cm^{-1}$ and we see from Fig. 2a that not only the intensity of the signal increases with N_S but also the resonance position and halfwidth depend on N_S.

To extract from these curves information on the 2 DES one usually applies the Drude model to describe the 2 D conductivity in a magnetic field:

$$\sigma_{xx}(\omega, B) = \sigma_0[(1 - i(\omega - \omega_c)\tau)^{-1} + (1 - i(\omega + \omega_c)\tau)^{-1}]. \tag{7}$$

For linearly polarized radiation σ_{xx} includes the active $(\omega - \omega_c)$ and the inactive mode $(\omega + \omega_c)$. Inserting eq. (7) into the transmission formula eq. (5) we obtain that the maximum change in transmission of the active mode at ω_c for circular polarized radiation is given as follows.

$$(-\Delta T/T)_{\mathrm{max}} = 1 - (1 + b \cdot N_S \cdot \tau)^{-2} \tag{8}$$

and the halfwidth $\Delta\omega$ (FWHM) is given by

$$\Delta\omega = \frac{1}{\tau} + b \cdot N_S \tag{9}$$

where $b = e^2/(\epsilon_o c m^*(1 + \sqrt{\epsilon} + 377/R_g))$. We see that for small N_S and τ the resonance amplitude increases linearly with N_S and τ, $((-\Delta T/T)_{\mathrm{max}} \approx 2 \cdot b \cdot N_S \cdot \tau)$. The halfwidth of the resonance is directly $\Delta\omega = 1/\tau$. For large N_S and τ, however, the relative change of transmission $\Delta T/T$ cannot exceed 100% (50%) for circularly (lincarly) polarized radiation. Therefore, with increasing N_S or τ, the amplitude $(\Delta T/T)_{\mathrm{max}}$ saturates and the

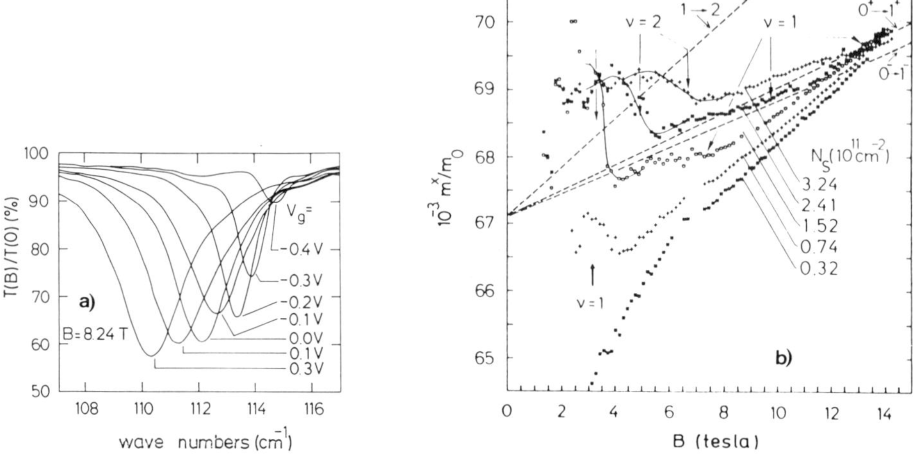

Figure 2. (a) Experimental CR excitation in front-gated $Al_x Ga_{1-x}As$-GaAs heterostructures. The transmission $T(B)$ is measured at fixed $B = 8.24T$ and is normalized to the transmission $T(0)$ at $B = 0$. From magnetocapacitance we find that the gate voltages V_g (V)=0.3, 0.2, 0.0, -0.1, -0.2, -0.3 correspond to densities $N_S(10^{11}\mathrm{cm}^{-2}) = 5.35$, 3.94, 3.16, 2.37, 1.74, and 0.80, respectively. (b) Experimental CR positions expressed in terms of an effective mass $m^* = eB/\omega_r$ versus B for five different N_S. For some densities experimental data have been connected by full lines for clarity. Broken lines represent calculated m^* for $0^+ \rightarrow 1^+$, $0^- \rightarrow 1^-$, and $1 \rightarrow 2$ transitions at $N_S = 1.1$ x 10^{11} cm^{-2} and $N_{depl} = 5 \cdot 10^{10}$ cm^{-2} from [26]. Positions of integer filling factors ν are indicated [5].

resonance broadens with respect to $\Delta\omega = 1/\tau$. Actually, this signal saturation occurs very often in high mobility $Al_x Ga_{1-x}As$/GaAs samples and makes it difficult to extract the microscopic scattering time τ. There is a trick which diminishes this problem. If we use a gate with a high conductivity then we make the parameter b and thus the second term in eq. (9) small. Then it is easier to determine τ. This impedance matching has recently been demonstrated experimentally [25].

For the spectra shown in Fig. 2a we have determined N_S by fitting the theoretical transmission eq. (5) and eq. (7) to the experimental resonance profile and we have obtained excellent agreement with N_S values determined from magnetocapacitance. From the same fits we also can determine an effective CR mass, $m_c = eB/\omega_c$, and τ. The experimental resonance position expressed in terms of this effective cyclotron mass is plotted in Fig. 2b. For all densities we evaluate well pronounced jumps in m_c near filling factors $\nu = 1$ and $\nu = 2$. If we concentrate for a moment on the curve for $N_S = 1.52 \cdot 10^{11}$ cm^{-2} in Fig. 2b, we find that for large B (i.e. thus that $\nu < 2$) m_c increases due to the non-parabolic bandstructure of GaAs in good agreement with self-consistent $k \cdot p$ subband calculations [26]. For smaller B, i.e. $\nu > 2$, where the $n = 1$ Landau level is partially filled, one expects in a one-particle model two transitions, one from the $n = 0$ Landau level to $n = 1$ and another from $n = 1$ to $n = 2$. Due to the non-parabolicity they should have different transition energies as shown by the calculated one-particle mass in Fig. 2b. Experimentally one observes that m_c jumps abruptly near $\nu = 2$ and that, in particular for small N_S, this jump is even larger than expected from non-parabolicity. A similar behaviour has been observed by several authors [26], even on very high mobility samples

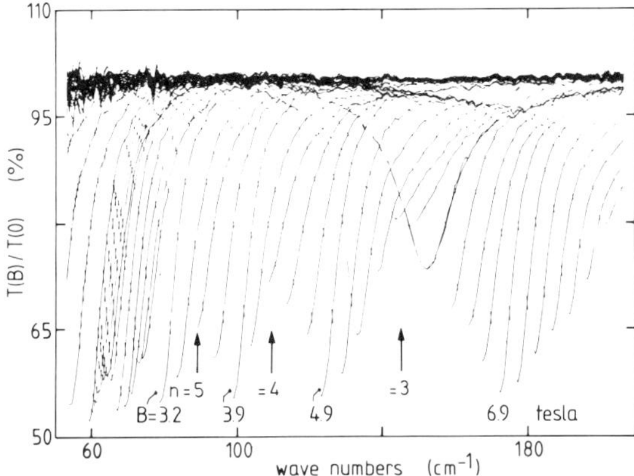

Figure 3. CR excitation in InAs QW [31]. Original experimental spectra of the transmission $T(B)$ are normalized to the transmission $T(0)$ at $B = 0$. The figure shows 42 different spectra measured at different B which have been increased in small - sometimes different - increments from $B \approx 2.1T$ to $B \approx 8.2T$. Some values of B are indicated. Temperature is 1.8K, spectral resolution is 1cm^{-1}.

[27]. This puzzling behaviour, which does not occur for a 3 DES where different Landau level transitions have well separated resonances in full accordance with non-parabolicity [28], indicates that CR in 2 DES cannot be explained in a one-particle model. This, in a certain sense, collective behaviour was tentatively explained in [5] by energy renormalization and interaction effects. In recent calculations some of the experimentally aspects, i.e. interacting CR, could be explained by the influence of non-parabolicity [29].

Besides the resonance position it is found that also the CR linewidth depends on the filling factor [5, 30, 31]. This is beautifully demonstrated in the spectra in Fig. 3 for an InAs quantum well (QW) sandwiched between GaSb layers [31]. The linewidth and correspondingly the amplitude show well pronounced oscillations up to Landau levels $n > 10$. Broad resonances occur if the highest Landau level is completely filled, i.e. $n = 3, 4, 5 \ldots$ and sharp resonances are observed for half-filled Landau levels, i.e. for $n = 2.5, 3.5, \ldots$. This behaviour can be explained by considering filling factor depending self-consistent screening and corresponding Landau level widths. For half-filled Landau levels the density of states (DOS) at the Fermi energy E_F is large. The system behaves "metal like", impurities are very effectively screened which results in sharp Landau levels. For completely filled Landau levels the DOS at E_F is small and the screening is poor. This results in broad Landau levels and broad CR. Experiments seem to indicate that an oscillatory behaviour of the CR line widths is particularly pronounced in relatively dirty samples. In high mobility samples linewidth oscillations are weak but nevertheless still observable. They seem to be related to the oscillation of the resonance position via some type of Kramers-Kronig relation [5, 27]. There is a general agreement that all the anomalies, even in high mobility samples, are related or induced via residual impurities. Thus a much better understanding came from experiments where intentionally small, well defined numbers of impurities were built into the 2 D channel of the heterostructure during the MBE growth [32, 33]. It was found that there is a characteristically different

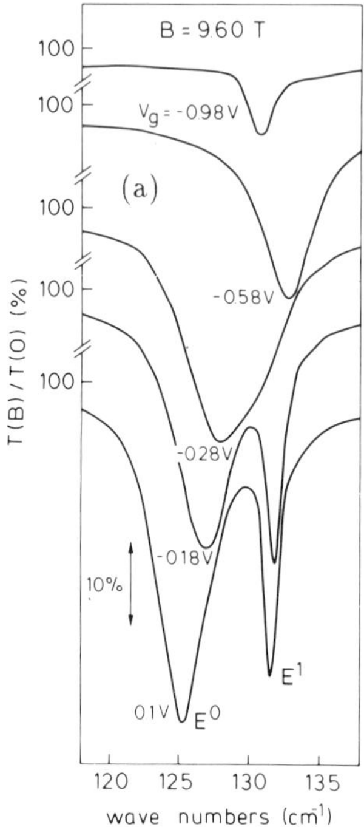
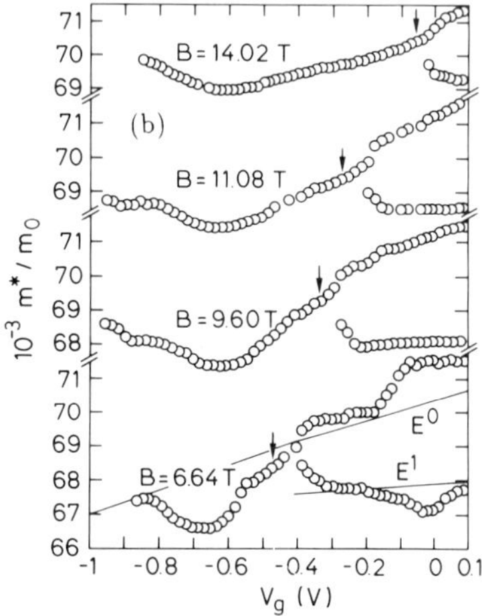

Figure 4. (a) Normalized transmission $T(B)/T(B=0)$ of unpolarized FIR radiation measured at fixed $B = 9.6T$ for various gate voltages V_g. For $V_g > -0.25V$ two subbands E^0 and E^1 are occupied, which show separated CR excitation. (b) Experimental resonance positions in terms of an effective mass m^* vs V_g at various magnetic fields B. For $B = 6.6T$ the results of a self-consistent calculation are indicated by solid lines. The arrows indicate the position of $\nu^0 = 2$. For $\nu^0 > 2$ two subbands are occupied, where the energetically higher E^1 subband has the lower effective mass m^* [6].

behaviour of the CR depending on the charge of the impurities. 2 DES with positively charged repulsive impurities exhibit for $\nu > 2$ a more or less normal CR. For $\nu < 2$ the linewidth decreases drastically. The CR position ω_r shifts to higher frequencies according to $\omega_r^2 = \omega_c^2 + \omega_o^2$. This behaviour seems to indicate some kind of confinement effect on the electrons [34] where the confinement energy ω_o, surprisingly, increases with increasing number of impurities. This anomalous behaviour for $\nu < 2$ is not found for samples with negatively charged repulsive scatterers in the channel. These samples, however, exhibit in general strong filling-factor depending oscillations of linewidths and resonance positions.

Although there are many open questions concerning the CR in 2 DES it is nevertheless a very useful tool to study numerous effects in these systems. Figure 4a shows CR spectra measured at a fixed magnetic field $B = 9.6T$ for different gate voltages [6]. With increasing V_g suddenly a second resonance appears at higher frequencies. This resonance represents a CR in the first excited subband E^1 which becomes populated for $V_g > -0.25V$. The resonance in this E^1 subband is very sharp which indicates a higher dynamic mobility. On a first glance it seems to be surprising that the higher subband has a higher CR frequency and thus smaller mass. This effect can be well explained from bandstructure calculations [6]. One can use the following hand waving argument: the energy difference between the 2 D energy levels and the conduction band that determines via the non-parabolicity the effective mass must not be taken at z=0 at the interface, but should rather be taken at the position z_{max} where the 2 D wavefunction has its maximum. In the nearly triangular potential well of a heterostructure this energy difference is smaller for higher subbands. Figure 4b shows resonance positions for different B and indicates that both, the magnetic field and the gate voltage, can be applied to populate or depopulate the second subband.

As a further example we show results obtained from a modulation-doped double QW system [8]. The separation of the two wells is 80 nm. They are electronically connected due to common contacts. Self-consistent calculations of the potential and the wave functions are shown in Fig. 5c. Figure 5a indicates that at low gate voltages only the back QW is occupied. Occupation of the front well starts at $V_g > -0.6V$. From the strength of the CR we can see directly in Fig. 5 that beyond this value of V_g the density in the back QW stays constant and all additional electrons are transferred with increasing V_g into the front well. This behaviour is also seen in magnetocapacitance and is confirmed very well by self-consistent subband calculations.

4.2   2 D Plasmons

So far we have considered excitations where the wave vector component in the 2 D plane, q, is zero. 2 D plasmons, the elementary collective charge density oscillations of the 2 D electron plasma are excitations with $q > \omega/c$. 2 D plasmons have, in contrast to 3 DES, a strong characteristic $\omega - q$ dispersion (e.g. [1, 35, 36, 37, 38])

$$\omega_p^2 = \frac{N_S e^2}{2\epsilon_{\mathrm{eff}}\epsilon_o m^*} \cdot q \tag{10}$$

where $\epsilon_{\mathrm{eff}}(\omega, q)$ is the effective dielectric function which takes account of the surrounding dielectric semiconductor and the screening of a gate electrode. q is the wave vector in the 2 D plane. 2 D plasmons were first observed for electrons on the surface of liquid He [39] and little later also in semiconductor systems [40, 41, 42]. The first calculations of the 2 D plasmon dispersion, eq. (10), were performed by Ritchie [46], Stern [35] and Chaplik [36]. Extended reviews on 2 D plasmons are given in [43, 37] (experiments) and [38] (theory).

4.3   Grating Coupler

2 D plasmons have wave vectors $q > \omega/c$. Thus, no direct coupling with radiation is possible. An effective way to excite these non-radiative modes are grating couplers [40, 41, 44, 45]. A grating coupler may consist of periodic stripes (periodicity a) of high and low conductivity materials (Fig. 6a). [1] Then the incident electric field, which has a homogeneous E_x component only, is short-circuited in the high conductivity regime. Thus, in the near field of the grating, the electromagnetic fields are spatially modulated and consist of a series of Fourier components

$$\vec{E}(x,\omega) = \sum_{n=-\infty}^{+\infty} (E_x^n, 0, E_z^n) \cdot exp(i(k_x^n \cdot x - \omega t)) \tag{11}$$

---

[1]In principle, any periodic variation of the dielectric properties in the vicinity of, or in the 2 DES itself, causes a grating coupler effect.

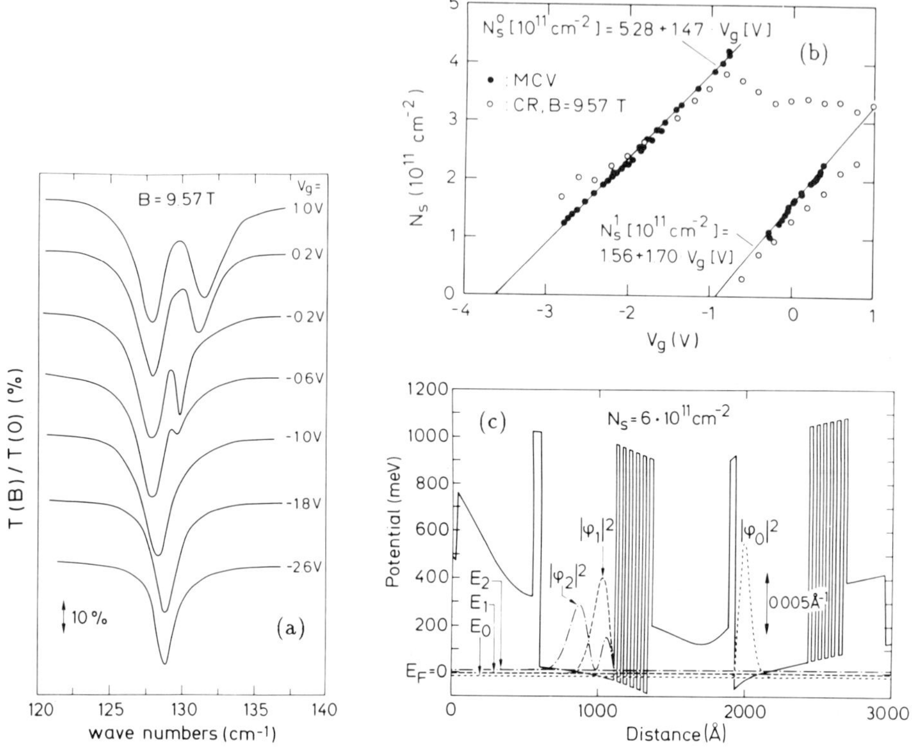

Figure 5. (a) Normalized transmission of unpolarized far-infrared radiation for fixed $B = 9.57T$ and various gate voltages for a double quantum well system [8]. The higher (lower) energy resonance corresponds to the CR of the electron in the front (back) QW. (b) Carriers densities vs V_g of both QW's evaluated from CV($\bullet$) and CR($\circ$) measurements. The straight lines are the result of a linear regression to the experimental points ($\bullet$) of the CV measurements. In the voltage regime $V_g > V_{t1}$ the results from the CR measurements ($\circ$) show that the carrier density N_S^0 is constant within the experimental accuracy. V_{t1} is the threshold voltage for the occupation of the front QW. (c) Potential of the double QW obtained from a self-consistent calculation. The surface of the sample is on the left-hand side of the figure. The dotted, dashed, and dash-dotted lines mark the wave functions and subband energies of the lowest three quantum states. The vertical arrow gives the scale for the squared wave functions.

with $k_x^n = n \cdot 2\pi/a$. Fourier components with wavevector k_x^n will couple to 2D plasmons if $q = k_x^n$ and ω satisfy the dispersion relation (Fig. 6b). It is also important that in the near field of the grating E_z components of the electric field are induced. These field components can be used to excite inter-subband resonances [44, 45] as will be discussed below. A full general calculation of the dielectric response in the presence of a grating is a complicated numerical problem. Approximate treatments for the conditions here are given in [40, 41, 43]. The efficiency of the grating coupler depends on its design. Since the n-th Fourier component of the field decays with $exp(-|k_x^n| \cdot z)$ for $|k_x| \gg \omega/c$, the distance d between the 2 DES and the grating should be small compared with the periodicity ($d < 0.1a/n$). 2D PR in MOS structures for Si(111) are observed in the spectra

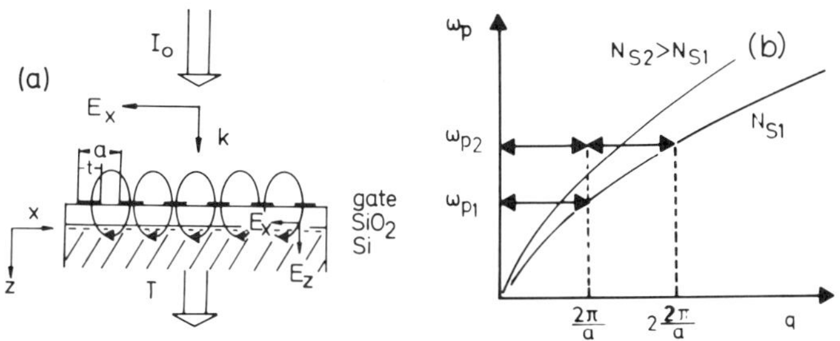

Figure 6. (a) grating coupler effect of a periodical structure. For normally incident FIR radiation spatially modulated parallel (E_x) and perpendicular (E_z) electric field components are induced. (b) Coupling of FIR radiation to 2 D plasmons of wavevectors $q_1=2\pi/a$ and $q_2=2\cdot2\pi/a$.

of Fig. 1b in the frequency regime 50-150 cm^{-1}. As we expect from the 2 D plasmon dispersion eq. (10) the resonances shift with increasing N_S to higher frequencies. Extended studies on 2 D plasmons for Si(100), (111), and (110) show that the experimental dispersion, in general, agrees with the theoretical predictions [44, 47]. For Si(110), e.g., the plasmon dispersion is found to be anisotropic with respect to e.g. the [100] direction in the surface [47]. This directly reflects the anisotropic bandstructure $E(\vec{K})$. Also 2 D plasmons in hole space charge layers of Si(110) have been investigated [48]. Here the plasmon dispersion indicates the non-parabolicity of the hole subband structure and the anisotropy of the Si(110) surface. A more detailed analysis of the plasmon excitation shows that, particularly at low N_S and, for Si(100) also at high N_S, there are characteristic deviations, in general a decrease, of the experimental plasmon dispersion with respect to the calculated dispersion. Most likely these deviations are caused by interaction of 2 D plasmons with different excitations in the system, e.g. plasmons of different $\vec{q}$ and single particle excitations. A theoretical treatment of this coupling via extrinsic potential fluctuations for Si-MOS structures within a memory function approach is given in [49].

For the high mobility Al$_x$Ga$_{1-x}$As/GaAs heterostructure system extrinsic coupling is not so important and intrinsic effects on the 2 D plasmon frequency can be studied [50]. Figure 7a shows the relative change in transmission $\Delta T/T$ for a GaAs heterostructure with a grating coupler of periodicity $a = 880$ nm. Two resonances are observed which are excited via the first and second Fourier component of the grating and correspond to wavevectors $q_1 = 2\pi/a$ and $q_2 = 2 \cdot 2\pi/a$, respectively. In the differential spectrum in Fig. 7b three PR are resolved. This allows us to study in detail the q-dependence of the dispersion, as shown in Fig. 7c. In eq. (10) the plasmon dispersion is calculated in a local and strictly 2 D limit. In higher orders of q, several effects become important [43, 37, 38]. Two of these effects should be mentioned. Non-local effects, the so-called Fermi pressure which arises from the finite compressibility of the 2 DES, increase the plasmon frequency [35, 36] $\omega_p^2 = \omega_L^2 + 3/4(qv_F)^2$. Here v_F is the Fermi velocity and ω_L is the "local" plasmon frequency (10) considered so far. For high q the finite thickness of the 2 D system also becomes important [51, 52]. The plasmon frequency may then be written as

$$\omega_p^2 = \omega_L^2 \left(1 + \frac{1}{g_v} \cdot \frac{3}{4} a^* \cdot q - F(q)d_c q\right). \tag{12}$$

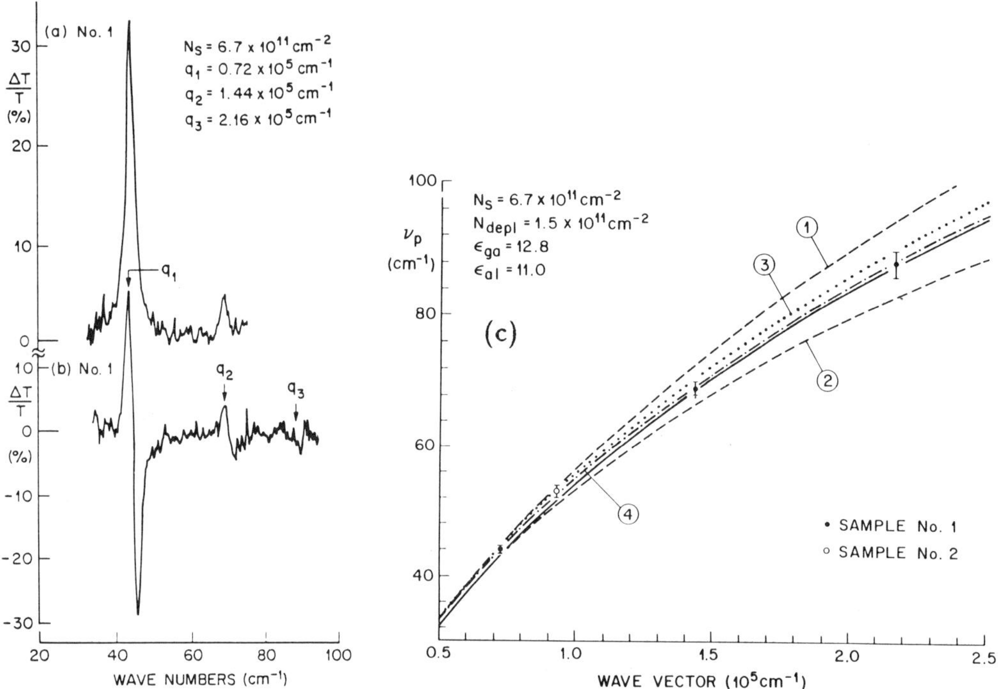

Figure 7. Experimental PR for a space-charge layer in GaAs with $N_S = 6.7 \cdot 10^{11}\,cm^{-2}$. The arrows mark plasmon resonance positions. For the upper trace (a) transmissions at $B = 0T$ and $B \geq 8T$ have been ratioed to determine $\Delta T/T$. The lower trace (b), qualitatively the derivative of the upper trace, is obtained by changing N_S slightly via the persistent photo-effect. (c) Theoretical and experimental plasmon dispersion. The solid line is the classical local plasmon dispersion. The curves marked 1-4 are defined as follows: curve 1, plasmon dispersion including non-local correction; curve 2, plasmon dispersion including finite-thickness effect; curve 3, plasmon dispersion including non-local and finite-thickness corrections combined; and curve 4, plasmon dispersion

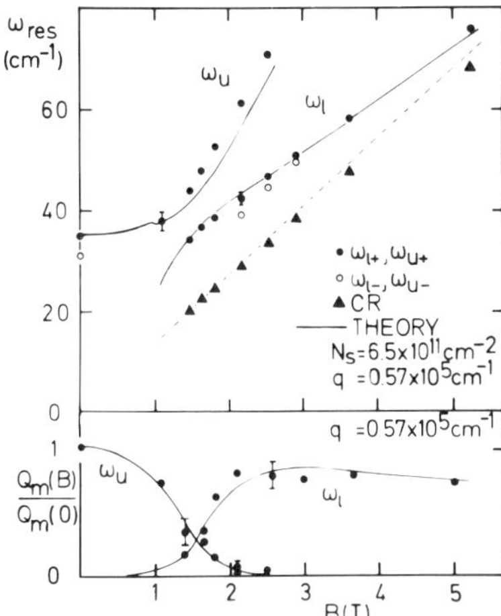

Figure 8. Splitting of the magneto-plasmon dispersion due to the non-local interaction with the harmonic cyclotron frequency [20]. The solid line is the theoretical dispersion [55] calculated for the parameters of the experiment. In the lower part, experimental (filled circles) and theoretical (full lines) normalized excitation amplitudes are compared. $Q_m(B)$ is the peak absorption of the plasmon excitation in an ω sweep at fixed B.

The first correction term, i.e. the term $3/4(q \cdot v_F)^2$, has been rewritten in terms of the effective Bohr radius $a^* = a_0 \epsilon m_0/m^*$ and the valley degeneracy g_v. Both $g_v = 1$ for GaAs instead of $g_v = 2$ for Si(100) and the larger effective Bohr radius of GaAs which arises from the small effective mass of GaAs ($a^*_{GaAs} \approx 10\,\mathrm{nm}, a^*_{Si(100)} \approx 2\,\mathrm{nm}$) make the non-local effect much more important for GaAs than for Si. The finite thickness becomes important if q approaches $1/d_e$, the inverse spatial extent of the electron wavefunction in the z-direction. The formfactor $F(q)$ has a value of about 1 and is positive and hence the finite thickness effect lowers the plasmon frequency. In Fig. 7c we compare the experimental plasmon dispersion with the calculations. Taking into account only the non-local corrections we would expect a significantly stronger increase of ω_p with q as it is observed in the experiment. The finite thickness effect alone would decrease the plasmon frequency below the observed values. However, both effects together nearly cancel each other in agreement with the experimental results. In the next paragraph we will see that the non-local effect can be observed directly in a resonance experiment. Thus, the experimental results (Fig. 7c) implicitly demonstrate the presence of finite thickness effects.

A magnetic field B perpendicular to the 2 D plane shifts the plasmon frequency. In a classical local approach [36]

$$\omega^2_{mp} = \omega^2_p(B = 0) + \omega^2_c. \tag{13}$$

This equation shows that the magneto-plasmon frequency approaches ω_c with increasing B. Evaluations of the magneto-plasmon dispersion and relative amplitudes from experiments [20] on an $Al_xGa_{1-x}As/GaAs$ heterostructure with a grating coupler of $a = 1150\,\mathrm{nm}$ are shown in Fig. 8. The magneto-plasmon dispersion approximately follows the classical dispersion eq. (13). However, in addition to this, there is obviously an interaction of the magneto-plasmon with the first harmonic of the CR, $2\omega_c$. This resonant anti-crossing arises from non-local interaction, i.e. the Fermi pressure. In a homogeneous system CR excitation is only allowed for $\omega_c = eB/m^*$, corresponding to transitions between adjacent Landau levels ($\Delta n = 1$). For the dynamic plasma oscillation at large q and corresponding short plasmon wavelengths the Fermi pressure becomes effective and leads to a strong resonantly enhanced interaction of magneto-plasmons with harmonics of the CR and causes a splitting of the magneto-plasmon dispersion at the crossing with $2\omega_c$.

The strength of this non-local interaction, represented by the amount of the splitting, is governed by the parameter [53, 54, 55]

$$(qv_F/\omega_c)^2 = 3a^*q/g_v. \tag{14}$$

For GaAs, due to the small mass and large a^*, this effect is strongly pronounced and can be observed very well in an experiment. As shown in Fig. 8, the observed splitting and the experimental dependence of the amplitude on B are in excellent agreement with calculations [55] of magneto-plasmon excitations. More details about magneto-plasmon-CR coupling, also for higher harmonics ($3\omega_c$) and for larger q, are described in [50]. Very recently it has been shown that similar non-local effects lead also to very interesting interactions in the dynamic response of very small 0 D quantum dot structures [14].

4.4   Inter-Subband Resonances

Besides 2 D plasmons, inter-subband resonances (ISR) are characteristic excitations in heterostructures and QW. ISR represent oscillations of the carriers perpendicular to the interface. Thus in highly symmetrical systems, e.g. electrons in Si(100) MOS systems or in GaAs-heterostructures, an E_z-component of the exciting electric field is necessary to couple to these transitions. Strip-line and prism coupler arrangements have been used to study ISR on Si(100) [56, 57]. Another very powerful method is the grating coupler technique [44, 45]. As we have discussed above, a grating coupler excites in the near field E_z components of the electric field (Fig. 6a). Experimental spectra measured on Si(100) samples with homogeneous charge density are shown in Fig. 9a. The grating periodicity is $a = 1800\,\text{nm}$. For $N_S = 3.3 \cdot 10^{12}\,\text{cm}^{-2}$ two resonances, $\tilde{E}_{01}$ and $\tilde{E}_{02}$, are observed which correspond to resonant transitions from the lowest subband to the first and second excited subbands, respectively. With increasing N_S the resonances shift to higher frequencies, corresponding to a larger subband separation in the steeper potential well at larger surface electric fields. For $N_S > 8 \cdot 10^{12}\,\text{cm}^{-2}$ additional resonances, $\tilde{E}'_{01}$, are observed, which can be attributed to resonant transitions in the primed subband system [1]. The primed subband system arises from the projection of four energy ellipsoids of the bulk Si onto the Si(100) surface and is separated in K-space by $0.86 \cdot 2\pi/A$ ($A=$ crystal lattice constant) in [001] and equivalent directions. It is known that this subband system is occupied for $N_S > 7.5 \cdot 10^{12}\,\text{cm}^{-2}$ [1].

The resonance energy measured in an ISR spectrum is not directly the subband spacing $E_{01} = E_1 - E_0$. Two effects shift the observed resonance with respect to the subband spacing. The first, the so-called exciton shift [58], results from the energy renormalization when an electron is transferred to the first excited subband E_1 leaving a "hole" in the E_0 subband. The exciton effect, characterized in a two-band model by a parameter β, shifts the resonance energy to smaller energies. A second effect, which increases the resonance energy, is the depolarization shift. In a microscopic one-particle

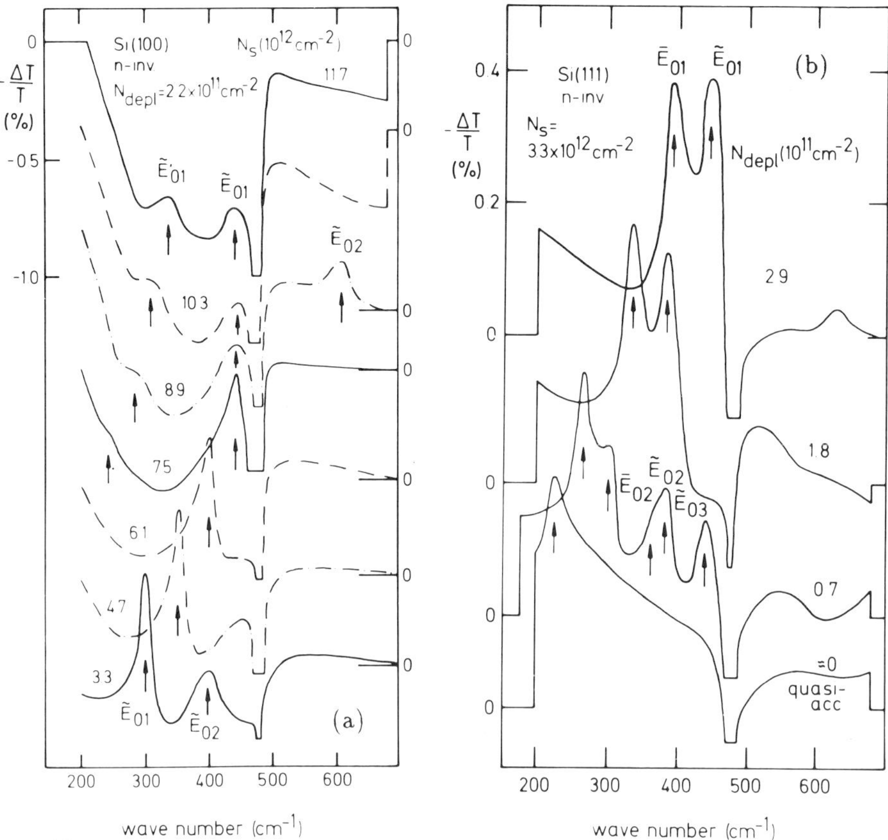

Figure 9. (a) grating coupler induced ISR for Si(100) at different charge densities N_S. Resonances $\tilde{E}_{01}$ and $\tilde{E}_{02}$ in the lower subband system and $\tilde{E}'_{01}$ in the second subband system are observed. In the regime of the optical phonon frequency of SiO$_2$ (about $480\,\mathrm{cm^{-1}}$), a resonant coupling to polaritons is measured, which is not fully shown here for clarity. (b) Excitation of ISR on a Si(111) sample with a grating coupler for different N_{depl}. Directly parallel excited $\bar{E}_{01}$ resonances and grating coupler induced depolarization shifted $\tilde{E}_{01}$ resonances are present [45].

model the inter-subband transition is a dipole oscillation at frequency ω_{01}. Macroscopically all the other electrons in the channel screen this dipole oscillation which shifts the resonance frequency according to $\tilde{E}_{01} = \sqrt{E_{01}^2 + \hbar^2 \omega_d^2}$ where $\hbar \omega_d = \sqrt{\alpha} E_{01}$ is an effective 3 D plasma frequency [59, 60].

In Si(111) and Si(110), the surface subband results from the projection of bulk energy ellipsoids which are tilted with respect to the surface [1]. Due to the anisotropic energy contours, the j_x and j_z component of the surface current are coupled and thus ISR can be excited with a parallel field component E_x [61]. Whereas parallel excited ISR (labelled $\bar{E}_{01}$) is not affected by the depolarization shift, $\bar{E}_{01} = E_{01} \cdot \sqrt{1 - \beta}$, the perpendicular excited resonance ($\tilde{E}_{01}$) is affected by both effects, $\tilde{E}_{01} = E_{01} \cdot \sqrt{1 + \alpha - \beta}$. Thus, if a grating coupler is used on Si(111) (Fig. 9b), both resonances, directly parallel excited resonances and perpendicular grating coupler excited, and thus depolarization shifted resonances, are observed. We see from the spectra in Fig. 9b that the depolarization shift slightly increases with the depletion field that is characterized by N_{depl}. It becomes much smaller for transitions to higher subbands (E_{02} and E_{03}). Very detailed investigations of grating coupler-induced ISR for different surface orientation on Si are reported in [45]. Very interesting effects are also observed for ISR in hole inversion layers of Si, in particular the intrinsic spin splitting due to spin-orbit interaction in the asymmetric potential of a Si-MOS system [62].

In principle it is possible to induce an electric field component in z-direction also by using non-normal incidence. However, due to the high index of refraction of semiconductor materials, this component is small and the signal is very weak. This is however not a problem if one uses QW superlattices with many 2 D layers. Such structures have also an interesting potential as FIR detectors where the sensitive wavelength can be tailored during the MBE growth via the width of the QW which determines the ISR frequency [63, 64].

Another method to study ISR in GaAs systems is resonant subband Landau level coupling (RSLC) in tilted magnetic fields [21, 22, 23, 7]. If a magnetic field is tilted slightly by an angle α with respect to the surface normal, the Landau levels of different subbands couple, e.g. the n = 1-Landau level of the E_0 subband and the n = 0-Landau level of the E_1 subband. In resonance anti-level crossing occurs leading to a splitting of the CR resonance near the energy E_{01}. Thus, from the position of the splitting the information on the subband separation can be extracted. The amount of the splitting is, for small α, $c_{01} \cdot sin\alpha$. From the matrix element c_{01}, information on the wavefunctions and the shape of the potential can be obtained [22].

We demonstrate RSLC for a gated $Al_xGa_{1-x}As/GaAs$ heterostructure [7]. Figure 10a shows experimental spectra for a magnetic field of $B = 13.2T$ which is tilted by $\alpha = 5°$ with respect to the surface normal. The strengths of the CR increases with increasing gate voltage. For $V_g = -1.1V$ resonant interaction occurs which arises from RSLC. In this experiment we shift the position of the ISR via the gate voltage until it comes into resonance with the CR which has a frequency of $\nu = 175\,cm^{-1}$. The experimental resonance positions $\tilde{E}_{10}$ and amplitudes are depicted in Fig. 10b. They allow a very accurate comparison with self-consistent subband calculations. From measurements at different B we can determine $\tilde{E}_{10}$ for a wide range of N_S. This allows us to address an interesting question. Although it has been theoretically suggested [65, 66] it was so far not experimentally confirmed that the resonance in a RSLC experiment occurs at the depolarization shifted value $\tilde{E}_{10}$ and not at the bare value of E_{01}. A comparison of experimentally observed resonance positions with self-consistent subband calculations alone is not enough since normally the depletion charge, N_{depl}, of a heterostructure, which depends on the residual doping in the GaAs and which strongly influences the subband spacing, is not known accurately enough. In our experiments on front-gated heterostructures we can vary N_S and thus determine $\tilde{E}_{10}$ in its dependence on N_S without changing N_{depl}. From the experimental slope $\tilde{E}_{10}$ (N_S) we can uniquely determine N_{depl} and demonstrate that RSLC occurs at the depolarization shifted position, $\tilde{E}_{10}$. This has recently also been found in samples with parabolic potentials [67, 68]. In these samples the depolarization shift is strongly pronounced and can be clearly extracted. In contrast to these observations and to the theory it was concluded in [69] by comparing

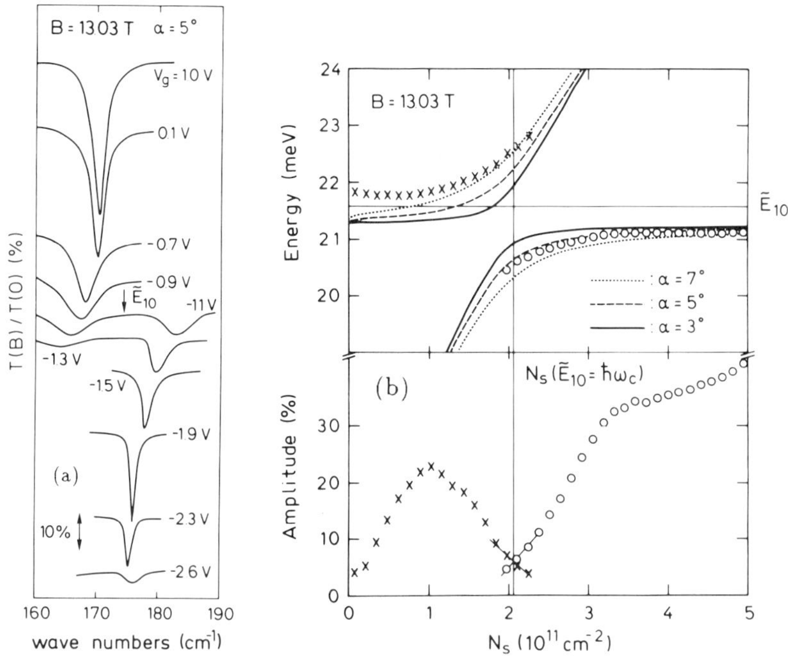

Figure 10. (a) RSLC in a tilted magnetic field. Normalized transmission $T(B)/T(B=0)$ of unpolarized FIR radiation measured at fixed $B = 13.03T$ for various gate voltages V_g. For $V_g \approx -1.1V$ the two resonances have approximately the same amplitude and linewidth and thus also the same oscillator strength. The arrow indicates the position of $\tilde{E}_{10}$. (b) Amplitude and resonance position from Fig. 1 plotted vs carrier density N_S. Circles (o) and crosses (x) mark lower and higher energy resonances, respectively. The straight lines show how the inter-subband resonance $\tilde{E}_{10}$ and the corresponding N_S are obtained from the measurement. The curved lines in the upper part of the figure are the result of the coupling calculation for different tilt angles as indicated. The depletion charge was extracted to be $N_{depl} = 6.3 \cdot 10^{10} \, \mathrm{cm}^{-2}$.

grating coupler experiments with RSLC that the ISR in a RSCL experiment does not occur at the depolarization shifted position.

We find, both for the Si system (Fig. 9) and for the GaAs heterostructure, that the depolarization shift is about 10% of the resonance frequency. Thus the 2D ISR is dominantly a one-particle-resonance with a small collective contribution. In recent experiments on 1 D quantum confined systems it has been found [11, 12, 15, 16] that the depolarization shift is very strong and exceeds in general the one-particle energy. Thus the resonances in these 1 DES have a strong collective, plasmon like character.

5 Conclusions

The basic properties of 2 DES can be studied by investigating the dynamic response in the FIR regime. The possibility to tune the electronic properties reproducibly over wide regimes via electric and magnetic fields make 2 DES ideally suited to explore many directions of solid state physics, in particular to study the elementary excitations, i.e. plasmons, subband transitions, CR and various combined resonances.

6   Acknowledgement

In these lecture notes we have reported on various kinds of experiments. We would like to thank all our colleagues, as listed in the references, who have been working with us on the different subjects. We also acknowledge support from the BMFT.

References

[1] T. Ando, A.B. Fowler, and F. Stern, Rev. Mod. Phys. **54**, 437(1982)

[2] D. Heitmann, in: Festkörperprobleme/Advances in Sol. St. Phys. **25**, P. Grosse, ed., p. 429, Vieweg, Braunschweig (1985)

[3] E. Batke and D. Heitmann, Infrared Phys. **24**, 189 (1984)

[4] D. Heitmann, in: Physics and Applications of Quantum Wells and Superlattices, E.E. Mendez and K. v. Klitzing, eds., p. 317, Plenum Press (1987)

[5] K. Ensslin, D. Heitmann, H. Sigg, and K. Ploog, Phys. Rev. **B 36**, 8177 (1987)

[6] K. Ensslin, D. Heitmann, and K. Ploog, Phys. Rev. **B 37**, 10150 (1988)

[7] K. Ensslin, D. Heitmann, and K. Ploog, Phys. Rev. **B 39**, 10879 (1989)

[8] K. Ensslin, D. Heitmann, M. Dobers, K. v. Klitzing, and K. Ploog, Phys. Rev. **B 39**, 11179 (1989)

[9] K. Ensslin, D. Heitmann, R.R. Gerhardts, and K. Ploog, Phys. Rev. **B 39**, 12993 (1989)

[10] K. Ensslin, D. Heitmann, and K. Ploog, Appl. Phys. Lett. **55**, 368 (1989)

[11] W. Hansen, M. Horst, J. P. Kotthaus, U. Merkt, Ch. Sikorski, and K. Ploog, Phys. Rev. Lett. **58**, 2586 (1987)

[12] T. Demel, D. Heitmann, P. Grambow, and K. Ploog, Phys. Rev. **B 38**, 12732 (1988)

[13] Ch. Sikorski and U. Merkt, Phys. Rev. Lett. **62**, 2164 (1989)

[14] T. Demel, D. Heitmann, P. Grambow, and K. Ploog, Phys. Rev. Lett. **64**, 788 (1990)

[15] T. Demel, D. Heitmann, and P. Grambow in: Spectroscopy of Semiconductor Microstructures, G. Fasol, A. Fasolino, and P. Lugli, eds., p. 75, Plenum Press (1989)

[16] D. Heitmann, T. Demel, P. Grambow, and K. Ploog, in: Festkörperprobleme, Advances in Solid State Physics **29**, U. Rössler, ed., p. 285, Vieweg, Braunschweig (1989)

[17] T. Demel, D. Heitmann, P. Grambow, and K. Ploog, Proc. of the Mauterndorf Winterschool 1990, G. Bauer, H. Heinrich, and F. Kuchar, eds., Springer-Verlag

[18] W. Hansen, this volume

[19] S. Oelting, D. Heitmann, and J.P. Kotthaus, Phys. Rev. Lett. **56**, 1846 (1986)

[20] E. Batke, D. Heitmann, J.P. Kotthaus, and K. Ploog, Phys. Rev. Lett. **54**, 2367 (1985)

[21] Z. Schlesinger, J.C.M. Hwang, and S.J. Allen, Phys. Rev. Lett. **50**, 2098 (1983)

[22] G.L.J.A. Rikken, H. Sigg, G.J.G.M. Langerak, H.W. Myron, and J.A.A.J. Perenboom, Phys. Rev. **B 34**, 5590 (1986)

[23] A.D. Wieck, J.C. Maan, U. Merkt, J.P. Kotthaus, K. Ploog, and G. Weimann, Phys. Rev. **B 35**, 4145 (1987)

[24] M. Ziesmann, D. Heitmann, and L.L. Chang, Phys. Rev. **B 35**, 4541 (1987)

[25] M. Watts, R.J. Nicholas, N.J. Pulsford, and J.J. Harris, C.T. Foxon, Solid State Physics Conference Warwick, England (1989)

[26] F. Thiele, U. Merkt, J.P. Kotthaus, G. Lommer, F. Malcher, U. Rössler, and G. Weimann, Sol. St. Commun. **62**, 841 (1987)

[27] R.J. Nicholas, M.A. Hopkins, D.J. Barnes, M.A. Brummell, H. Sigg, D. Heitmann, K. Ensslin, J.J. Harris, C.T. Foxon, and G. Weimann, Phys. Rev. **B 39**, 10955 (1989)

[28] G. Lindemann, R. Lassnig, W. Seidenbusch, and E. Gornik, Phys. Rev. **B 28**, 4693 (1983)

[29] A.M. MacDonald and C. Kallin, Phys. Rev. **B 40**, 5795 (1989)

[30] Th. Englert, J. C. Maan, Ch. Uihlein, D.C. Tsui, and A.C. Gossard, Sol. St. Commun. **46**, 545 (1983)

[31] D. Heitmann, M. Ziesmann, and L.L. Chang, Phys. Rev. **B 34**, 7463 (1986)

[32] J. Richter, H. Sigg, K. v. Klitzing, and K. Ploog, Phys. Rev. **B 39**, 6268 (1989)

[33] H. Sigg, D. Weiss, and K. v. Klitzing, Surf. Sc. **196**, 293 (1988)

[34] H.J. Mikeska and H. Schmid, Z. Phys. **20**, 43 (1975)

[35] F. Stern, Phys. Rev. Lett. **18**, 546 (1967)

[36] A.V. Chaplik, Soviet Phys. JETP **35**, 395 (1972)

[37] D. Heitmann, Surf. Sc. **170**, 332 (1986)

[38] A.V. Chaplik, Surf. Science Reports **5**, 289(1985)

[39] C.C. Grimes and G. Adams, Phys. Rev. Lett. **36**, 145 (1976)

[40] S.J. Allen, jr., D.C. Tsui, and R.A. Logan, Phys. Rev. Lett. **38**, 980 (1977)

[41] T.N. Theis, J.P. Kotthaus, and P.J. Stiles, Sol. St. Commun. **24**, 273 (1977)

[42] T.N. Theis, J.P. Kotthaus, and J.P. Stiles, Sol. St. Commun. **26**, 603 (1978)

[43] T.N. Theis, Surf. Sc. **98**, 515 (1980)

[44] D. Heitmann, J.P. Kotthaus, and E.G. Mohr, Sol. St. Commun. **44**, 715 (1982)

[45] D. Heitmann and U. Mackens, Phys. Rev. **B 33**, 8269 (1986)

[46] R.H. Ritchie, Phys. Rev. **B 106**, 874 (1957)

[47] E. Batke and D. Heitmann, Sol. St. Commun. **47**, 819 (1983)

[48] E. Batke, D. Heitmann, A.D. Wieck, and J.P. Kotthaus, Sol. St. Commun. **46**, 269 (1983)

[49] A. Gold, Phys. Rev. **B 32**, 4014 (1985)

[50] E. Batke, D. Heitmann, and C.W. Tu, Phys. Rev. **B 34**, 6951 (1986)

[51] T.K. Lee, C.S. Ting, and J.J. Quinn, Sol. St. Commun. **16**, 1309 (1975)

[52] R.Z. Vitlina and A.V. Chaplik, Soviet Phys. JETP **54**, 536 (1981)

[53] T.K. Lee and J.J. Quinn, Phys. Rev. **B 11**, 2144 (1975)

[54] N.J.M. Horing and M.M. Yildiz, Ann. Physik **97**, 216 (1976)

[55] A.V. Chaplik and D. Heitmann, J. Phys. C: Solid State Physics **18**, 3357 (1985)

[56] P. Kneschaurek, A. Kamgar, and J.F. Koch, Phys. Rev. **14**, 1610 (1976)

[57] B.D. McCombe, R.T. Holm, and D.E. Schafer, Sol. St. Commun. **32**, 603 (1979)

[58] T. Ando, Z. Phys. **B 26**, 263 (1977)

[59] W.P. Chen, Y.J. Chen, and E. Burstein, Surf. Sc. **58**, 263 (1976)

[60] S.J. Allen, D.C. Tsui, and B. Vinter, Sol. St. Commun. **20**, 425 (1976)

[61] T. Ando, T. Eda, and M. Nakayama, Sol. St. Commun. **23**, 751 (1977)

[62] A.D. Wieck, E. Batke, D. Heitmann, J.P. Kotthaus, and E. Bangert, Phys. Rev. Lett. **53**, 493 (1984)

[63] L.C. West and S.J. Eglash, Appl. Phys. Lett. **46**, 1156 (1985)

[64] R. Heinrich, R. Zachai, M. Besson, T. Egeler, G. Abstreiter, W. Schlapp, and G. Weimann Surf. Sc. (1990), in press

[65] M. Zaluzny, Sol. St. Commun. **56**, 235 (1985)

[66] M. Zaluzny, Phys. Rev. **40**, 8495 (1989)

[67] K. Karrai, H. D. Drew, M.W. Lee, and M. Shayegan, Phys. Rev. **39**, 1426 (1989)

[68] L. Brey, N.F. Johnson, and B.I. Halperin, Phys. Rev. **40**, 10647 (1989)

[69] J. Pillath, E. Batke, G. Weimann, and W. Schlapp, Phys. Rev. **40**, 5879 (1989)

SPECTROSCOPY ON LATERALLY CONFINED ELECTRON SYSTEMS

Wolfgang Hansen

Sektion Physik, Universität München
Geschwister-Scholl-Platz 1
W-8000 München 22, F. R. Germany

1 Introduction

Progressive decrease of the size of mesoscopic devices enables us to enter the area of quantum confined electron systems in which the system extensions are comparable to the Fermi-wavelength. It is intriguing to investigate how spatial confinement influences system properties when the system size drops from macroscopic dimensions ($W \gg \lambda_F$) to extents comparable to the Fermi wavelength ($W \simeq \lambda_F$). In such systems not only phase coherence phenomena but also confinement induced quantization of the conduction band energies into subbands or discrete levels start to influence system properties. Such devices are preferentially realized on semiconductors because of the relatively large Fermi wavelengths and small effective masses in these materials. Presently, the fabrication of mesoscopic devices in semiconductors starts in most cases from the two-dimensional (2 D) electron system in metal-oxide-semiconductor (MOS) structures or epitaxially grown heterojunctions [1]. In these systems the electrons are strongly bound at the interface so that the system is quantum confined in the direction normal to the interface. Free motion is possible only in the remaining lateral dimensions. Most advantageous is that very high mobilities are achievable (up to $10^7 \mathrm{cm}^2/\mathrm{Vs}$ in present GaAs heterojunctions), correspondingly the elastic mean free path can be as high as several microns, and the carrier density can be tuned over a wide range (up to several $10^{12}\mathrm{cm}^{-2}$ in Si-MOS structures). The lateral confinement of the 2 D electron system to quasi one-dimensional (1 D) wires or quasi zero-dimensional (0 D) electron dots is provided by micro-structuring processes that pattern the device surface [2].

In the following the influence of lateral confinement on two different system properties, namely the far-infrared conductivity and the device capacitance, will be discussed. For experimental reasons both properties have to be measured on large arrays of many parallel wires or matrices of many dots. In the experimental systems discussed here the electron confinement is defined by a purely electrostatic potential. The potential can be varied via voltages applied on patterned gate electrodes, so that the geometry of the electron system can be changed within certain limits.

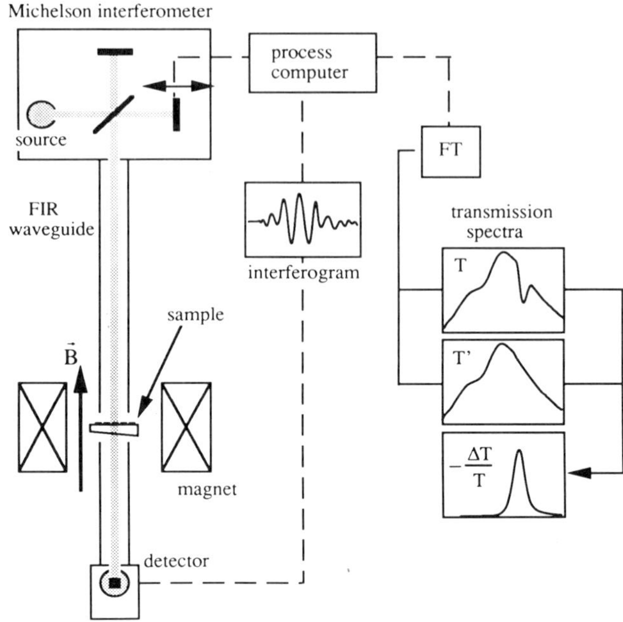

Figure 1. Sketch of the experimental setup for far-infrared
transmission measurements. Sample, magnet and detector
are usually cooled to liquid Helium temperatures.

2 Far-Infrared Spectroscopy

High frequency intra conduction band excitations in low-dimensional electron systems of
semiconductor devices exhibit their prominent features in the microwave and far-infrared
(FIR) regime. Cyclotron resonances measured in 2 D electron systems at amenable magnetic fields (of the order of 10 T) as well as plasmon excitations with wave vectors corresponding to feasible grating coupler periods (of the order of 0.5μm) are found in this
frequency regime. Since the presently feasible extensions of laterally confined electron
systems (about 100 nm) are of the same order as the cyclotron radius at moderate magnetic fields and only little smaller than present grating couplers, it is not surprising to
find the conductivity resonances of laterally confined systems at very similar frequencies.
Yet, we will see that there are drastic differences in the responses of systems of different
dimensionality.

Usually the FIR conductivity of the electron system $\sigma(\omega)$ is measured in transmission geometry. In the experiments discussed in the following the radiation polarization
is always parallel to the device surface, i.e. in the lateral direction. In Fig. 1 a typical
experimental setup employing Fourier transform spectroscopy is sketched. The sample
covers the cross section of a FIR wave guide in the center of a superconducting magnet.
The transmitted radiation is measured as a function of the position of the moving mirror
in a Michelson interferometer with a bolometer located sufficiently far from the magnet
to avoid field dependent detector response. The Fourier-transform of the detector signal
is the transmission spectrum $T(\omega)$ of the whole experimental setup including the characteristics of the radiation source, optical components and the spectral sensitivity of the
detector. The conductivity of the electron system can be controlled by the application

of either a gate bias V_g or a magnetic field B perpendicular to the device surface. A reference spectrum $T'(\omega)$ recorded at parameters V_g' and B' such that the conductivity of the electron system is essentially zero in the frequency domain of interest allows one to extract the conductivity at V_g and B from the difference $\Delta T = T' - T$. In the small signal limit and assuming the radiation transmitted through the sample does not change the polarization, the relative change of transmission $\Delta T/T$ is directly proportional to the real part of the conductivity

$$-\frac{\Delta T}{T} = \frac{T' - T}{T'} \cong \frac{2}{(1 + \sqrt{\epsilon_s})Y_0 + \sigma_G} Re[\sigma(\omega)]. \tag{1}$$

Here ϵ_s is the dielectric constant of the semiconductor material, $Y_0 = (\mu_0/\epsilon_0)^{1/2}$ the vacuum admittance, and σ_G accounts for an effective areal conductivity of the metal gate configuration. Application of eq. (1) is straightforward only in the case of a homogeneous electron system with a uniform metal gate. If the conductivities of the electron system or the gate configuration are periodically modulated averaged areal conductivities $\bar{\sigma}_g$ and $\bar{\sigma}$ may be defined since the FIR radiation wavelength λ is much longer than the superlattice period a. However, the averaged conductivities can no longer be regarded as independent. For instance, the grating coupler spatially modulates the radiation field sensed by the electron system and, in addition, polarizations of the electron system are screened by the gate. This is well known from plasmon resonance experiments [3]. Furthermore, the condition of an unaltered polarization after transmission becomes non-trivial in a magnetic field, if for example a linearly polarizing grating coupler is combined with a circularly polarizing electron system [4]. This becomes obvious in cyclotron resonance experiments with high mobility heterojunctions that are covered with a grating coupler [5,6]. Relative transmission changes close to 100% are recorded in contrast to maximum transmission changes of 50% in homogeneous systems and linearly polarized radiation.

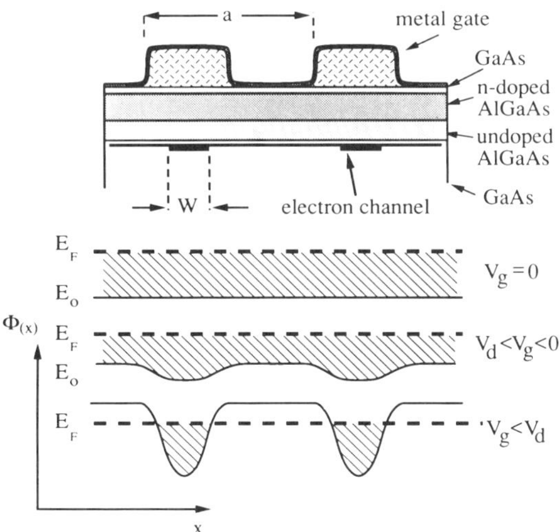

Figure 2. Cross section of an $Al_xGa_{1-x}As/GaAs$ heterojunction with periodically corrugated gate. In the lower part of the figure the effect of negative biases applied between gate and electron system on the spatial density dependence is sketched.

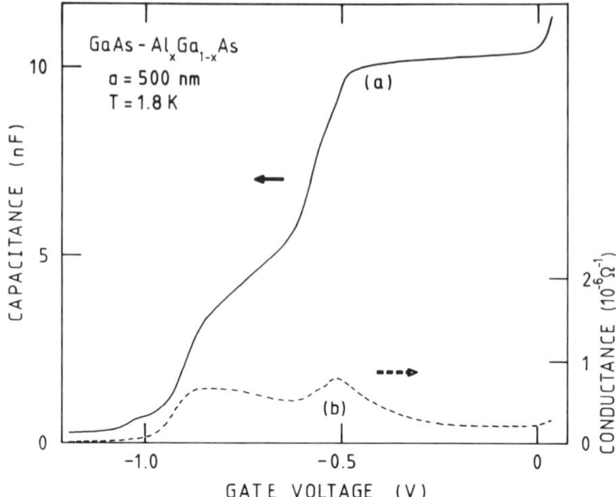

Figure 3. Gate voltage dependence of the capacitance measured on an $Al_xGa_{1-x}As/GaAs$ heterojunction with modulated gate as depicted in Fig. 2. The dashed line denotes the conductance signal measured out of phase to the capacitance signal [13].

It is instructive first to discuss the far-infrared excitations in an experimental system, which can be transformed from a 2 D electron sheet into an array of 1 D wires in situ by simply changing a gate bias V_g. Such devices have been realized in dual gate Si-devices [7-9] as well as in $Al_xGa_{1-x}As/GaAs$ heterojunctions with periodically corrugated gates [10]. A cross section of the $Al_xGa_{1-x}As/GaAs$ heterojunction device as well as the effect of a negative bias applied at the corrugated gate are sketched in Fig. 2. In devices with an unstructured GaAs cap layer the surface potential modulation which is caused by the periodically varying interface between GaAs cap layer and Schottky gate or photoresist spacer is usually small. Thus at the gate voltage $V_g = 0$ the electron system is 2 D and essentially has a homogeneous carrier density. Negative gate biases preferentially deplete those areas that are beneath the closer parts of the gate, so that a 1 D density modulation is induced. Below a certain gate voltage ($V_g \leq V_d$) the density modulation is sufficiently large to isolate the electron channels that remain beneath the photoresist pattern.

The operation of the device is easily examined by recording the gate voltage dependence of the capacitance as depicted for an $Al_xGa_{1-x}As/GaAs$ heterojunction device with a grating period $a = 500$ nm in Fig. 3. The capacitance is essentially constant at gate voltages $V_d < V_g \leq 0$ and decreases abruptly below $V_g = V_d$. The abrupt decrease of the capacitance reflects the reduction of the effective area contributing to the capacitance once gaps open in the electron sheet and an array of isolated wires is formed. Note that in this measurement the charge modulation of the electron system is built up by currents flowing from a 2 D contact area through the 1 D channels. In this respect the

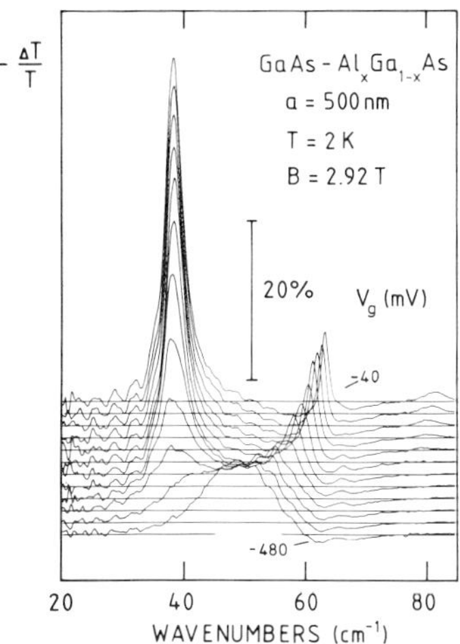

Figure 4. Far-infrared transmission spectra recorded on the sample of Fig. 3 with a magnetic field applied normally to the sample surface. The polarization of the radiation is perpendicular to the grating stripes of the corrugated gate. The gate voltage at which the bottom trace was recorded is $V_g = -0.48\,\text{V}$. It is increased by $0.04\,\text{V}$ between the subsequent traces plotted with a zero offset above the first trace to enhance visibility [13].

measurement differs significantly from capacitance spectroscopy experiments described later where carriers are injected from an especially designed back contact.

The traces of Fig. 4 show FIR spectra in the gate voltage regime $-0.5V < V_g \leq 0V$, where the electron system is essentially 2 D and has a 1 D density modulation. The radiation is chosen to be linearly polarized with the electric field perpendicular to the stripes of the corrugated gate. A magnetic field of $B = 2.9\,\text{T}$ is applied perpendicular to the sample surface, so that the cyclotron resonance is observed at $\bar{\nu} \approx 38\,\text{cm}^{-1}$. With decreasing gate bias the cyclotron resonance energy essentially remains constant whereas the oscillator strength decreases reflecting reduction of the average areal density. The excitations at higher frequencies are magneto-plasmon resonances [3]. They can be excited because the corrugated gate acts as a grating coupler. Because of the periodicity of the gate, the wave vectors of the excited plasmons are in general multiples of $q_0 = 2\pi/a$. In the sample of Fig. 4, however, the higher order excitations at $q = n \cdot q_0$ with $n > 1$ are very weak. The resonances observed at frequencies $\bar{\nu} = 63\,\text{cm}^{-1}$ and $\bar{\nu} = 80\,\text{cm}^{-1}$ at gate voltage $V_g = -0.04\,\text{V}$ are excited at wave vector $q = q_0$. The fact that two resonances are observed originates from non-local interaction of the magneto-plasmon with the first harmonic of the cyclotron resonance [5,11]. The dynamic charge density modulation of the plasmon results in the excitation of harmonics of the cyclotron resonance, which is forbidden in the case of a homogeneous system [12]. Surprisingly

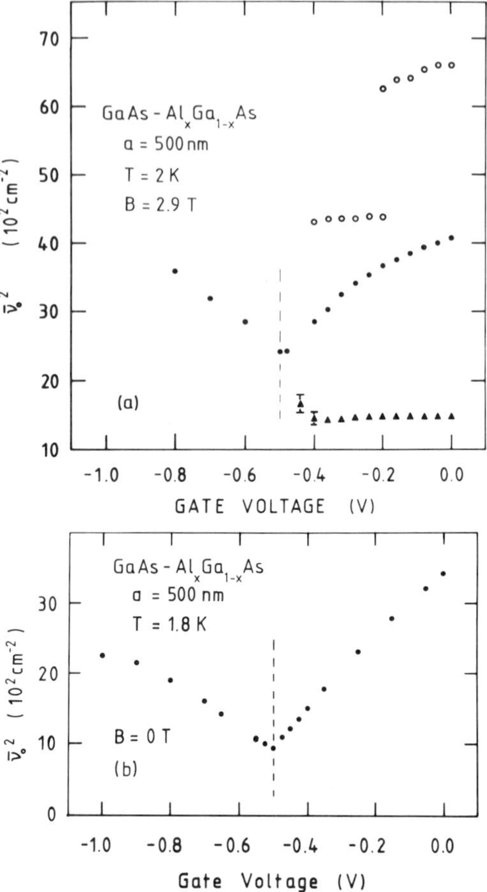

Figure 5. Squared resonance positions plotted versus the gate voltage at (a) magnetic field $B = 2.9\,\mathrm{T}$ and (b) zero magnetic field. The dashed line indicates the gate voltage V_d at which the array of electron wires transforms into a density modulated 2 D gas. Filled triangles in the gate voltage regime of a 2 D electron system in (a) denote the cyclotron resonance position, filled and open circles strong and weak plasmon resonances, respectively.

the splitting does not become larger when the static density modulation is increased by a negative gate bias. Contrary to the behavior of the cyclotron resonance the plasmon resonance frequencies decrease with the gate voltage and the oscillator strength increases. Close to the threshold voltage at which the transition into an array of 1 D electron channels occurs both resonances merge into one resonance that is significantly broadened. At gate voltages below the threshold voltage ($V_g < -0.5\,\mathrm{V}$) the resonances of an array of electron wires are observed. A single resonance occurs at a frequency that increases with decreasing gate voltage [10] as illustrated in Fig. 5a. The resonance linewidth is smaller than the resonance linewidth at the transition point ($V_g - -0.5\,\mathrm{V}$) [10].

The corresponding behavior of the resonances at zero magnetic field is shown in Fig. 5b. At zero gate voltage a single first order plasmon is observed at $\bar{\nu} = 59\,\mathrm{cm}^{-1}$. The second order resonance is much weaker [13]. The resonance position is well described by [3]

$$\omega_p^2 = \frac{e^2 n_s}{2\bar{\epsilon}\epsilon_0 m^*} q_0 \tag{2}$$

with an effective dielectric constant $\bar{\epsilon} = 13.9$, an electron density of $n_s = 6 \cdot 10^{11}\,\mathrm{cm}^{-2}$ derived from Shubnikov-de Haas (SdH) oscillations in the magnetoresistance, and a period of the metal gate $a = 500\,\mathrm{nm}$. The effective dielectric constant accounts for the polarization of the medium in which the electron system is embedded as well as for the screening by the metal gate. Again, the resonance frequency decreases with decreasing gate voltage. For relatively small absolute values of the gate voltage in Fig. 5b we expect a situation similar to the density modulated system in Si-MOS devices with corrugated gate [14]. Plasmon excitations in such devices have been well described within a perturbational approach by Krasheninnikov and Chaplik [15]. A charge density modulation given by

$$n(x) = \sum_{m=0}^{\infty} n_m \cos\left(m\frac{2\pi}{a}\right) \tag{3}$$

results in gaps in the plasmon dispersion at wave vectors $q_m = m\pi/a$ with gap size

$$\frac{\omega_+^2(q_m) - \omega_-^2(q_m)}{\omega_p^2(q_m)} = 2\frac{n_m}{\bar{n}_s}. \tag{4}$$

Here $\omega_p = [e^2 \bar{n}_s q_0/2\bar{\epsilon}\epsilon_0 m^*]^{1/2}$ is the center frequency of the gap determined by the average electron density $\bar{n}_s = n_0$ in the system. The fact that only one resonance is observed indicates a small $m = 2$ harmonic content of the density modulation. The linear decrease of the squared resonance position is well described by a linear decrease of the average electron density according to a rate $\Delta n_s/\Delta V_g = C/eA = 7.8 \cdot 10^{11} V^{-1}\mathrm{cm}^{-2}$ which agrees well with the almost constant capacitance of $C \approx 10\,nF$ (see Fig. 2) and the gate area of about $A = 0.08\mathrm{cm}^2$. The perturbational approach is justified at relatively small absolute values of the gate voltage in Fig. 5b, where the charge density modulation is sufficiently low. In fact, the resonance positions depart from eq. (4) to lower frequencies at gate voltages close to the threshold voltage V_d. Lower resonance frequencies compared to those derived from eq. (2) with $n_s = \bar{n}_s$ are predicted by classical approaches that remain valid at stronger charge density modulations [16,17]. Recently, quantum mechanical calculations of the excitations in the transition regime have been performed for zero magnetic field [18] as well as for finite fields [19]. These calculations reflect the experimental fact that at moderate magnetic fields the plasmon resonance energy remains finite at the threshold voltage $V_g = V_d$. Furthermore, the zero field calculations of excitation spectra reflect qualitatively the behavior of the linewidth at the threshold voltage [18]. The calculated spectra demonstrate that actually the notion of a discrete plasmon resonance becomes invalid close to the threshold, where isolated channels are formed. Single particle excitations that are well separated from the plasmon branch in homogeneous systems are folded back in strongly modulated systems and thus cause Landau damping of the plasmon excitations. Hence, the calculations predict a considerable broadening of the excitation spectra close to the transition point. Damping is reduced when the gate voltage is decreased beyond the transition point, because 1 D subbands are formed with no dispersion along the direction of the potential modulation.

The physical nature of the excitations in the array of 1 D electron channels has been discussed in several publications [20-25]. Generally, it is expected, that with decreasing channel width a collective type geometric resonance [26-28] (also referred to as dimensional resonance [8,29] or confined plasmon [30]) turns into a collective intersubband resonance (also referred to as intersubband plasmon [21,22] or depolarization shifted intersubband resonance [23]) once the subband separation of the single-particle energies contributes significantly to the excitation energies. Collective geometric resonances arise from the charge density polarization induced by the radiation field at the system boundary. In a classical model the resonance frequency is described by [28,16,13]

$$\omega_d^2 = \alpha \frac{e^2 n_s}{\epsilon_0 \bar{\epsilon} m^* W}. \tag{5}$$

Here a constant 2 D charge density is assumed along the channel width W. The constant α is of order of unity and accounts for corrections due to the actual charge density profile within the channel as well as the Coulomb interaction between adjacent channels.

In quantum confined electron channels the collective charge density polarization perpendicular to the wires is caused by Coulomb coupled virtual intersubband transitions. In a simplifying two-level model the subband separation $\hbar\Omega$ contributes to the resonance frequency ω via

$$\omega^2 = \Omega^2 + \omega_d^2. \tag{6}$$

In analogy to the terminology used for intersubband resonances between 2 D subbands [1] the frequency ω_d is often called the depolarization shift.

In more sophisticated models the random phase approximation is used taking into account the presence of many occupied as well as unoccupied subbands [22]. Magnetotransport measurements on a sample similar to the one of Fig. 3 to 5 show that, in the gate voltage regime investigated, between 6 and 9 subbands are occupied [31]. Que and Kirczenow find in calculations that consider 6 occupied as well as 3 unoccupied subbands 6 different modes that correspond to transitions between adjacent subbands. If the confinement potential is nearly parabolic, 5 of these modes are very close to the energy separation of the subbands and one mode has a considerably larger energy. Inspection of the potentials induced by the different modes shows that only the latter mode has a large dipole moment and thus is excited with large oscillator strength in the FIR resonance experiment. It has been pointed out by several authors, that under special conditions the resulting resonance may be equivalent to the excitation of a system with only one occupied and one empty subband [22,23,25]. For this case formulas for the resonance frequency are derived, that are equivalent to eq. (6) up to a factor of order unity multiplied with the depolarization frequency ω_d.

Magnetotransport measurements allow one to determine values for energy separations of the 1 D subbands within simplifying models for the Hartree potential of the channels [31-33]. Values measured for subband spacings in $Al_x Ga_{1-x} As/GaAs$ heterojunctions with corrugated gate are a factor of about 3 to 4 smaller than the observed FIR resonance energies [31]. Within the two-level model this means that the depolarization frequency in eq. (6) is large and the resonances are dominantly of collective

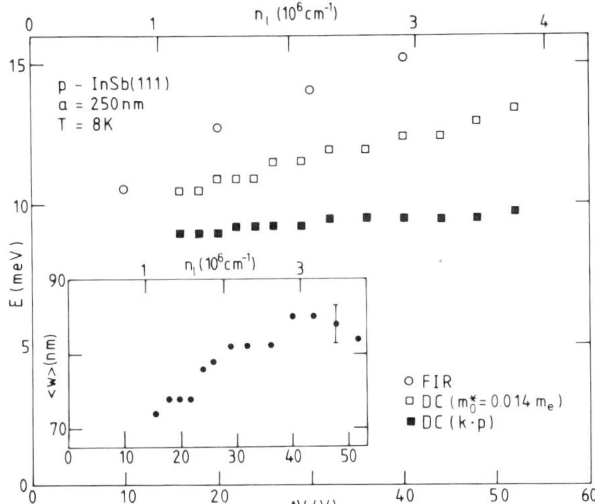

Figure 6. Far-infrared resonance energies and subband spacings measured in 1 D electron channels in an InSb MOS structure as a function of the gate voltage ΔV_g above the inversion threshold. Far-infrared resonance energies are marked with open circles, the subband spacings obtained from magnetotransport experiments by squares. The upper scale which denotes 1 D electron densities as well as the effective channel widths of the inset are also obtained from the magnetotransport experiments [36].

character [31,34]. A different situation is met in InSb MOS structures with 1 D electron channels defined by a grating type microstructured Schottky gate [10,33]. As in $Al_xGa_{1-x}As/GaAs$ heterojunctions a single FIR resonance is observed that qualitatively behaves like the excitation in the GaAs devices at gate voltages below the threshold voltage. However, the data of Fig. 6 demonstrate that in these devices the quantization into 1 D subbands contributes significantly to the resonance frequency. Open circles denote FIR resonance energies. Squares denote subband spacings deduced from magnetotransport experiments within a constant effective mass approximation (open squares) and a Fermi energy dependent mass (filled squares) which is more appropriate in InSb. At high electron densities ($\Delta V_g = 40\,\mathrm{V}$) the depolarization frequency is slightly larger than the subband spacing. However, the FIR resonance energy decreases with decreasing electron density whereas the subband spacings deduced from $k \cdot p$ approximation remain almost constant. Note that all three sets of data approach the same energy of $\hbar\Omega = 9\,\mathrm{meV}$ if extrapolated to zero electron density. Thus the collective contributions become smaller than the quantization energies and eventually vanish at vanishing electron density in correspondence with the density dependence of the depolarization shift eq. (5). The fact that the FIR resonances of quantum confined channels in InSb are much more influenced by single-particle quantization energies has been attributed to the smaller effective mass compared to the GaAs effective mass as well as to strong screening of the collective depolarization by the gate configuration in close vicinity to the electron channels [22,35,36].

Comparison of the above data obtained on InSb and GaAs shows that the transition from a predominantly collective geometric resonance to an intersubband plasmon dominated by the quantization into 1 D subbands may take place without noticeable change in the behavior of the resonances. The quantum nature of the system is verified only by magnetotransport measurements. This is clearly reflected within a simplifying model that considers the excitations of electrons confined in a parabolic bare potential of the form $V(x) = m^*\Omega_0^2 x^2/2$. Here the bare potential is understood to be the potential induced by the external charges, e.g. charges on the electrodes, excluding those carriers confined in the channel. In close analogy to Kohns theorem [12] for the cyclotron resonance in homogeneous electron systems it is shown that a homogeneous radiation field excites a single resonance at the characteristic frequency Ω_0 of the potential independently of the number of electrons confined in the channel and thus of the number of occupied subbands. In other words in a harmonically bound plasma electron-electron interactions do not influence the resonance frequency. The above argument has been applied recently to discuss excitations in quasi 3 D quantum wells, in which by means of epitaxial composition the conduction band has been designed to form a parabolic confinement potential [37]. However, it can be applied to 1 D or 0 D systems as well, if the assumptions of a harmonic lateral bare potential and a homogeneous excitation field are reasonable. The most intriguing consequence of above interpretation is that the FIR resonance position directly reflects the characteristic frequency Ω_0 of the bare potential when a parabola represents a good approximation to the form of the potential. The fact that so far only a single resonance has been resolved in the FIR conductivity of quantum confined electron systems may indicate that the above model describes the experiments appropriately. Indeed, Que and Kirczenow find that excitation spectra calculated for square-well confinement disagree with experiments [22].

Several higher order modes of dimensional resonances have been observed in Si inversion channels, where the above assumptions can be definitely violated. In the so-called subgrating mode electron channels are induced beneath a grating gate of a dual stacked gate configuration [8] as indicated in the inset of Fig. 7a. The homogeneous top gate biased relative to the substrate to $V_{gt} = -8\,\text{V}$ creates a high potential barrier between adjacent channels. The distance $d_1 = 50\,\text{nm}$ between the gate electrode and the inversion channel is 30 times smaller than the width $W = 1.5\mu\text{m}$ of the metal stripes of the bottom gate. From the above dimensions it is clear that the bare potential is not parabolic and the radiation field also may have considerable modulation. The fundamental mode and up to 4 higher order modes are observed as shown by the FIR spectra in Fig. 7a. The squared resonance frequencies plotted in Fig. 8 versus index n are readily described as higher order modes of the fundamental eq. (5) with $\alpha = 1$ and an effective dielectric constant $\bar{\epsilon}(n)$ dependent on the order n of the excitation [8]

$$\omega_{dn}^2 = \frac{e^2 n_s}{\epsilon_n \bar{\epsilon} m^* W} n. \tag{7}$$

Figure 7b demonstrates that the strength of the higher order modes decreases if the top gate voltage is chosen to be $V_{gt} = 0\,\text{V}$ and thus the confinement is less steep. The same devices can also be operated in the so-called subgap inversion mode, where electron channels are created by a positive top gate between the metal stripes of the grating gate. The negatively biased bottom gate provides isolation of adjacent channels. In this mode electron channels much narrower than the gap between the metal stripes can be created and the FIR spectra again show only a single resonance [8]. Experiments on

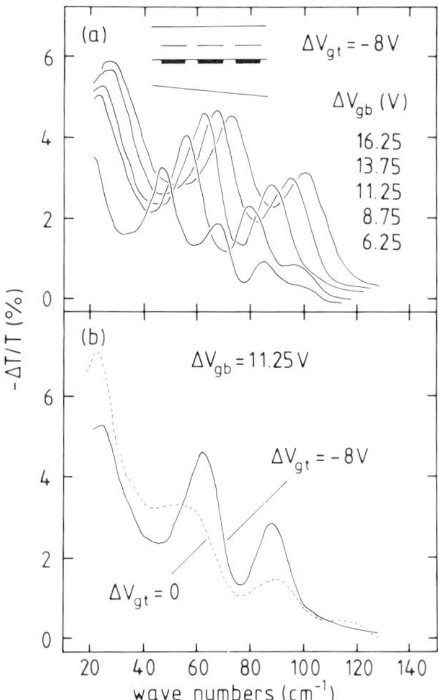

Figure 7. Far-infrared spectra of a Si-MOS structure with dual stacked gate configuration at zero magnetic field and temperature $T = 2\,\mathrm{K}$. The gate geometry and the location of the electron channels in the subgrating mode is sketched in the inset of (a). The stripes of the grating gate are $1.5\,\mu$m wide, the period is $a = 2\,\mu$m. In (a) the top gate voltage is constant, in (b) spectra obtained with two different top gate voltages are compared [8].

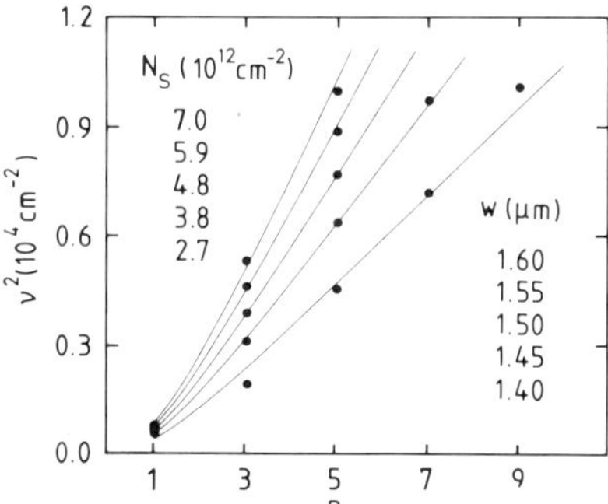

Figure 8. Squared resonance positions of the sample of Fig. 7 plotted versus a mode index n for different areal densities N_s. The solid lines are calculated from eq. (5) assuming a continuous mode index n and an effective dielectric constant $\bar{\epsilon}$ that depends on n and the channel width W. Channel widths obtained for the best fits are indicated for the different densities N_s [8].

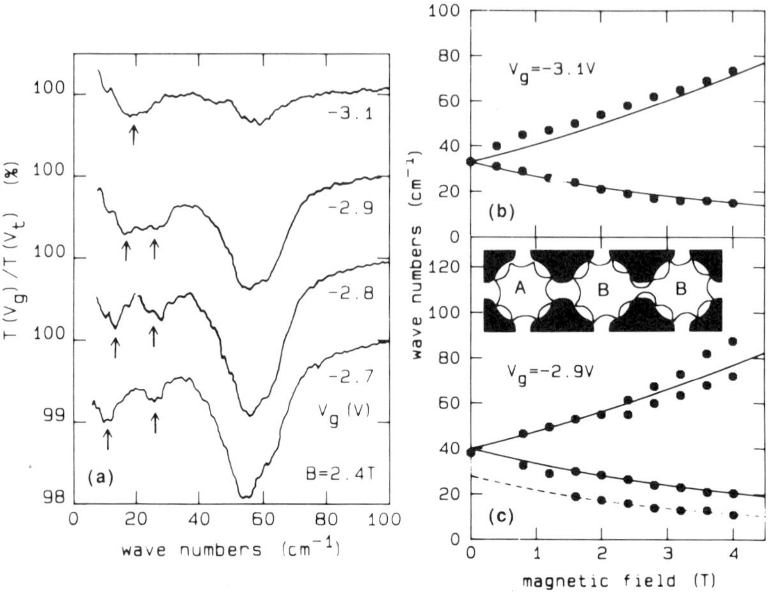

Figure 9. (a) Far-infrared transmission spectra of an $Al_xGa_{1-x}As$/GaAs heterojunction with different voltages applied at a front gate periodically modulated in both lateral directions. A magnetic field of $B = 2.4\,\mathrm{T}$ is applied perpendicular to the sample surface. Measured resonance positions as function of the magnetic field are depicted in (b) for isolated and in (c) for coupled electron dots. Solid lines are calculated with eq. (6). The dashed line indicates the magnetic field dispersion of the ω_- mode calculated with a doubled perimeter. The inset sketches classical trajectories of single- and double-perimeter modes [39].

electron channels quantum confined in a bare potential of square-well shape remain to be performed in the future.

An array of spatially completely confined electron systems, so-called electron dots, can be easily defined with a photo resist mask periodically modulated in both lateral dimensions. Such a mask can for instance be generated by means of holographic lithography [36] or electron beam writing [38]. Thus in $Al_xGa_{1-x}As$/GaAs heterojunctions with a front gate modulated in the two lateral dimensions a matrix of electron dots can be formed out of a density modulated 2D electron system similarly to the generation of 1D wires discussed in Fig. 2. FIR spectra of an $Al_xGa_{1-x}As$/GaAs heterojunction with a photoresist matrix of period $a = 450\,\mathrm{nm}$ in both lateral dimensions are depicted in Fig. 9 [39]. Inspection of the gate voltage dependent capacitance shows that at gate voltage $V_g = -2.7\,\mathrm{V}$ the electron system represents a mesh of electrically connected dots, whereas at $V_g = -3.1\,\mathrm{V}$ a matrix of isolated dots is formed. Thus in the spectra shown in Fig. 9 the electron dots become more and more electrically decoupled with decreasing gate voltage. Figure 9b shows the resonance frequencies of the device at the gate voltage $V_g = -3.1\,\mathrm{V}$ where the electron dots are isolated as a function of a magnetic field applied perpendicularly to the sample surface. A striking difference to the excitation in 1D channels is that the zero field resonance of electron dots splits into two branches, one approaching at high fields the cyclotron resonance energy whereas the other branch tends to zero. Whereas the resonance frequency in 1 D wires is described by $\omega^2 = \omega_c^2 + \omega_0^2$ the magnetic field dependences of the high frequency mode at ω_+ and the low frequency mode at ω_- are approximately given by [29,40,41]

$$\omega_\pm = \sqrt{\omega_0^2 + \left(\frac{\omega_c}{2}\right)^2} \pm \frac{\omega_c}{2}. \tag{8}$$

Here we have the cyclotron frequency ω_c and the zero field resonance position ω_0. In the limit of high magnetic fields the high frequency mode may be regarded as a "bulk like" mode that hardly senses the spatial confinement, whereas the low frequency resonance corresponds to a charge density wave more localized at the boundary of the dot. It is related to the edge magneto-plasmons observed in 2 D electron systems on liquid Helium [26,27,41] and at the boundaries of high mobility 2 D devices [42,43]. The number of electrons contained at $V_g = -3.1\,\mathrm{V}$ in each dot is estimated to about 50 and the excitations are of collective type. Similar resonances have previously been observed in electron dots fabricated on InSb [44]. Although the energy quantization is expected to be of more importance in InSb and the average number of electrons per dot could be reduced to only three, no qualitative changes in the resonance behavior were observed analogous to the behavior of 1 D quantum channels.

At gate voltages above $V_g = -3.1\,\mathrm{V}$ the electron dots become electrically connected and two new features arise. At the high energy side of the "bulk like" resonance a shoulder arises that becomes the magneto-plasmon resonance at higher gate voltages, where a 2 D density modulated system is formed. This is analogous to the formation of the magneto-plasmon in Fig. 4 when the 1 D electron channels become connected. The second most striking feature observed in coupled quantum dots is the splitting of the low field resonance indicated by the arrows in Fig. 9a. The magnetic field dependence of the resonance position for gate voltage $V_g = -2.9\,\mathrm{V}$ is depicted in Fig. 9c. The full line indicates the magnetic field dependence of resonance positions calculated with eq. (8). The dashed line is calculated with eq. (8) and a zero field frequency that is a factor $1/\sqrt{2}$ smaller than the one of the full line. In a classical picture the squared zero field frequency is inversely proportional to the perimeter of the electron dot. A virtually doubled dot perimeter is achieved if two adjacent dots are electrically connected as sketched by means of trajectories in the inset of Fig. 9c. Thus from the frequency ratio, the magnetic field dependence and the gate voltage dependence it is inferred that the splitting directly reflects the charge exchange between edge type modes of adjacent dots [39].

3   Capacitance Spectroscopy

The front gate employed in many semiconductor devices allows one to measure the device capacitance by a periodic modulation of the gate voltage and phase sensitive measurement of the AC current through the sample. In the previous section it was pointed out that capacitance measurements can be employed to independently determine the gate voltage at which large areas of the device are depleted in order to form isolated electron wires or dots. Actually, besides the geometry the density of states in the electron system also influences the capacitance. For instance, the changes of the density of states at the Fermi energy have been studied by capacitance measurements in 2 D systems with a magnetic field applied perpendicular to the sample surface [45,46,47]. Often a specially designed back contact is employed to avoid contributions to the signal arising from changes in the inevitable series resistance through which the electron system is charged [45,46]. Such low resistive back contacts essentially consist of an n^+-doped GaAs substrate separated from the $Al_xGa_{1-x}As$/GaAs interface by a shallow barrier formed by a layer of undoped GaAs. The electron channel at the interface is charged via tunneling processes from the degenerate substrate. The thickness of the undoped GaAs spacer (typically 50 nm to 80 nm) is chosen to avoid deterioration of the electron mobility at the interface on the one hand and to warrant low contact resistances on the other. The capacitance measured between the gate and the back contact can be separated into a geometric contribution C_{geom} in series with a second capacitance C_e that describes the rise of the Fermi energy with increasing channel occupation [45,46,47]

$$\frac{1}{C_{tot}} = \frac{1}{C_{geom}} + \frac{1}{C_e}. \tag{9}$$

The geometric contribution C_{geom} depends on the area of the electron system, the distance and geometry of the front gate, the insulator material, and the extent of the electron system in the direction normal to the interface. The second contribution is essentially proportional to the density of states at the Fermi energy $D(E_F)$, if many particle contributions and band non-parabolicity are neglected: $C_e = 1/e^2 D(E_F)$. In 2 D systems C_e is generally very large so that the total capacitance is dominated by the geometric contribution. Furthermore, the density of states within each 2 D subband is constant and usually the 2 D electron system is in the quantum limit, i.e. only the lowest 2 D subband is occupied. However, the density of states changes drastically if a magnetic field is applied perpendicular to the sample surface and can drop significantly if the Fermi energy lies between Landau levels. Thus strong Shubnikov-de Haas oscillations can be observed in the magnetocapacitance [45,46].

The situation changes if the device is microstructured so that the electron system consists of an array of 1 D electron wires. The density of states within a 1 D subband decreases with increasing energy according to $1/\sqrt{E}$ and thus changes rapidly each time a new 1 D subband is occupied. Since the lateral confinement is weaker, the 1 D subband spacing is small compared to the energy separation between 2 D subbands, so that within the experimentally accessible range of electron densities several 1 D subbands are occupied. Gate voltage dependent oscillations of the low temperature capacitance

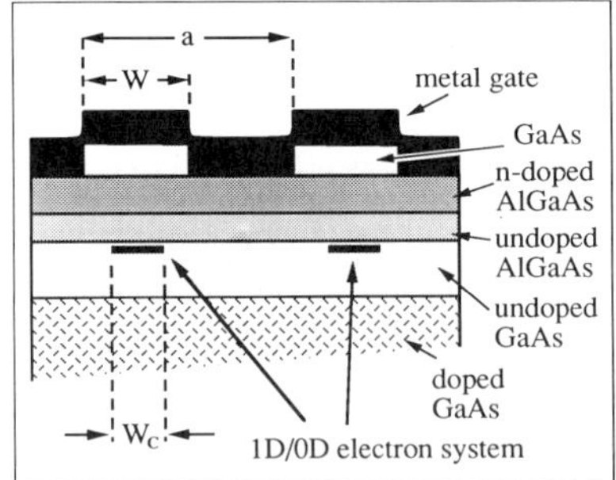

Figure 10. Cross section of an $Al_xGa_{1-x}As$/GaAs heterojunction as has been employed for capacitance spectroscopy. The front gate is periodically corrugated by a patterned GaAs cap layer. Contact to the electron system is provided by the doped GaAs substrate.

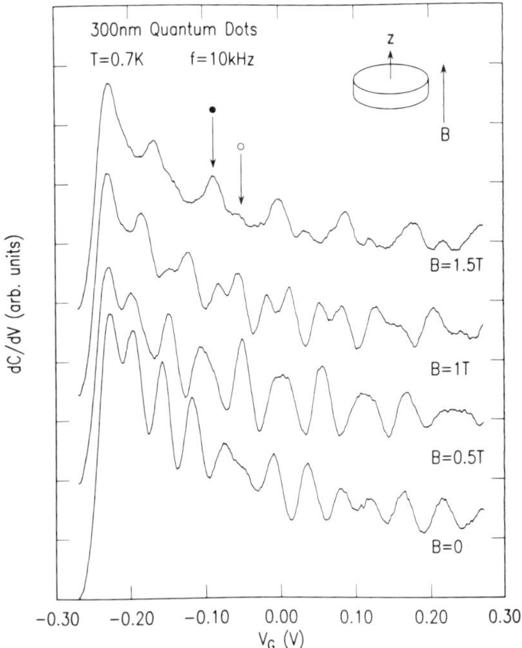

Figure 11. Gate voltage derivative of the capacitance measured on an $Al_xGa_{1-x}As/GaAs$ heterojunction like the one sketched in cross section in Fig. 10. The front gate is periodically microstructured in both lateral directions with a matrix of 300 nm wide GaAs squares. The traces are recorded at temperature $T = 0.7$ K and at four different magnetic field strengths. The direction of the magnetic field with respect to the dot geometry is indicated in the inset [51].

reflecting the gate voltage dependent density of states at the Fermi energy have been observed in the capacitance in microstructured $Al_xGa_{1-x}As/GaAs$ heterojunctions. Figure 10 illustrates the cross section of an $Al_xGa_{1-x}As/GaAs$ heterojunction employed for these measurements. Here the GaAs cap layer is partially removed by a selective etching process to form a GaAs grating that modulates the distance between gate and electron system similar to the device in Fig. 2. Typical values for the thickness and the width W of the GaAs stripes, the thickness of the $Al_xGa_{1-x}As$ layer, and the period a are 30 nm, $W = 300$ nm, 40 nm, and a = 600 nm, respectively. Unlike in the device of Fig. 2, the potential modulation induced in the device of Fig. 10 by the etched heterojunction surface is typically so strong that the areas beneath the removed GaAs cap remain depleted at all gate voltages applied in the experiment so that no transformation into a density modulated 2 D system is observed. Furthermore, the static potential distribution induced by a gate bias is different, since here the gate voltage is applied with respect to the doped back contact instead of the electron system itself. In order to analyze the experiments numerical model calculations were performed solving the Schrödinger and Poisson equations self-consistently in the Hartree approximation with boundary conditions derived from the device geometry [48,49]. Calculated gate voltage separations at which the Fermi level crosses subsequent subbands agree well with the gate voltage separations of the measured capacitance oscillations. Calculations performed for low electron densities and geometrical widths W between 0.21μm and 0.42μm result in typical level separations at the Fermi energy between 2 and 5 meV and channel widths W_c a factor two to three smaller than the geometrical line width W of the GaAs stripes [49]. Since the period in the devices is much larger than the widths of the electron channels charge exchange between adjacent channels can take place only via tunneling processes into and from the back contact.

Corresponding measurements have been performed on matrices of 0 D electron systems in similar heterojunction devices [50,51]. In the device of Fig. 10 a matrix of 0 D electron systems is generated if the GaAs cap layer is patterned in both lateral dimensions such that it forms a matrix of GaAs squares. Figure 11 depicts experimental traces recorded on a device with GaAs squares of width $W = 300$ nm and a period of 600 nm. In order to get a sufficiently large capacitance signal very many (of the order of 10^5) electron dots are measured in parallel. The derivative of the capacitance with respect to the gate voltage is recorded to enhance the capacitance oscillations against the background capacitance. As in the case of devices containing arrays of 1 D channels, throughout the whole gate voltage range recorded, the electron system consists of isolated electron dots in the sense that charge exchange between adjacent dots can take place only through the back contact. The traces are recorded at different magnetic fields applied perpendicular to the sample surface as indicated in the inset of Fig. 11. Electron dots start to be generated in the potential minima at the $Al_xGa_{1-x}As/GaAs$ interface at the gate voltages $V_g = -0.27$ mV resulting in a strong onset of the derivative signal. At higher gate voltages magnetic field dependent oscillations of the signal are observed. Fig. 12 depicts the gate voltage positions of oscillation maxima in the device of Fig. 11 versus the magnetic field. At high gate voltages and low magnetic fields a fairly complex behavior is observed. A pronounced splitting and anti-crossing like behavior is observed at fields around $B = 1$ T. At high magnetic fields ($B > 2$ T) the gate voltage positions increase linearly with the magnetic field as is expected for SdH oscillations of the capacitance signal in a 2 D system with areal electron density increasing linearly with the gate voltage. The behavior of the oscillation maxima in devices with different dot sizes is similar at magnetic fields that roughly scale with the inverse dot diameter. At magnetic fields above 4 T an additional splitting similar to the spin splitting of Landau levels is observed.

A numerical model to calculate the electron states in the 0 D electron dots is far more elaborate than in the case of 1 D channels [52]. The Schrödinger and Poisson equation now have to be solved self-consistently for wave functions and charge densities that vary in all three spatial dimensions. For an analysis of the gate voltage dependence it is important to properly include the back contact in the calculation, since charging of the small electron systems significantly alters the energy levels with respect to the Fermi energies in the back contact and front gate [52,53]. When an electron is injected into the 0 D system the Coulomb interaction with the electrodes will basically lift the energy levels by the so-called Coulomb gap energy, which may be described as e^2/C. Here we have the capacitance C of the 0 D electron system with respect to the electrodes. The value of this paramcter is very sensitive to the width and the background doping of the nominally undoped GaAs layer. Numerical model calculations in Hartree approximation similar to those performed for 1 D channels [48] have been performed for the 0 D electron

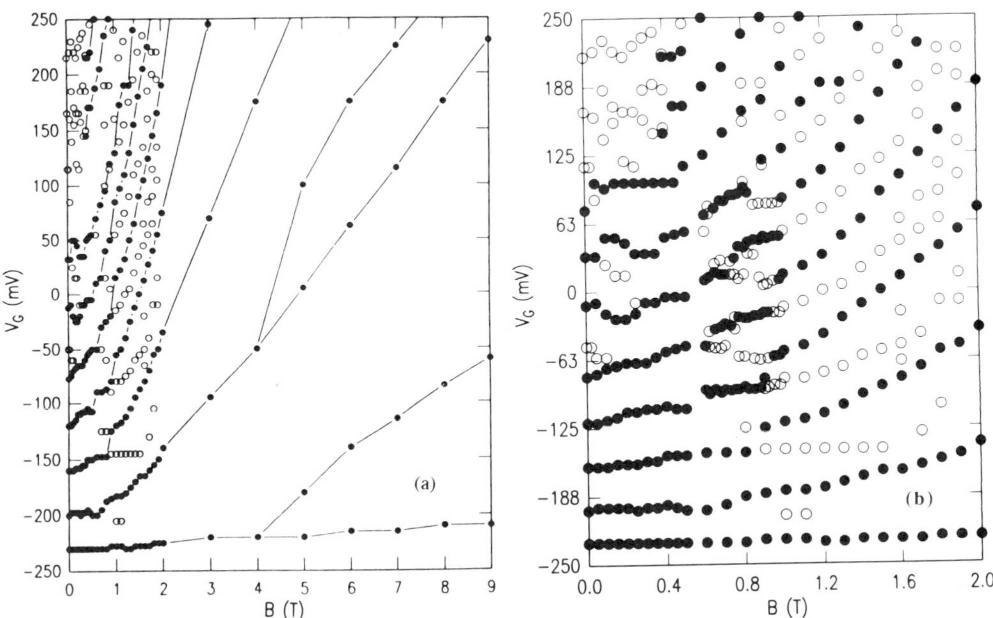

Figure 12. (a) Magnetic field dependence of the maxima in the capacitance derivative measured as function of the gate voltage on the device of Fig. 11; (b) extended low field range of (a) [51].

dots by Kumar et al. [52]. The results show that like in the case of 1 D channels the size of the electron system is considerably smaller than the width of the etched GaAs pattern. The shape of the confinement potential is found to be almost circular although the GaAs cap consists of squares. Apart from small splittings arising from the imperfect circular symmetry the energy levels calculated by Kumar et al. [52] have the same degeneracies as the ones in a single-particle harmonic oscillator. The calculated gate voltages at which electrons are injected into the lowest quantum levels have typical separations of about $10\,\mathrm{mV}$ at zero magnetic field. In contrast the first four capacitance maxima above the turn-on voltage $V_g = -0.27\,\mathrm{V}$ in Fig. 11 have a separation of about $40\,\mathrm{mV}$. This value is similar to the calculated gate voltage separations at which almost degenerate levels start to be filled. Kumar et al. [52] have also modelled the effect of a magnetic field applied perpendicular to the device. The calculations show that the qualitative similarity of the energy level structure in the self-consistent Hartree approximation to the level structure in a single-particle harmonic oscillator model remains in finite magnetic fields. The magnetic field dependence of the single particle energy levels in a 2 D harmonic oscillator potential $V(x) = m^{*}\Omega^{2}x^{2}/2$ is described by [54]

$$E_{n,l} = \hbar(2n + |l| + 1)\sqrt{\Omega^{2} + \frac{1}{4}\omega_{c}^{2}} + \frac{1}{2}l\hbar\omega_{c}^{2}. \tag{10}$$

At low magnetic fields the energy levels are dominantly determined by the electrostatic confinement via the characteristic harmonic oscillator frequency Ω. The level degeneracy at zero magnetic field is lifted by the magnetic field and the energy levels rearrange to form in the high field limit levels close to the Landau energies $[n + (|l| + l)/2 + 1/2]\hbar\omega_{c}$. Physically l is the angular momentum quantum number and $N = n + (|l| - l)/2$ is the

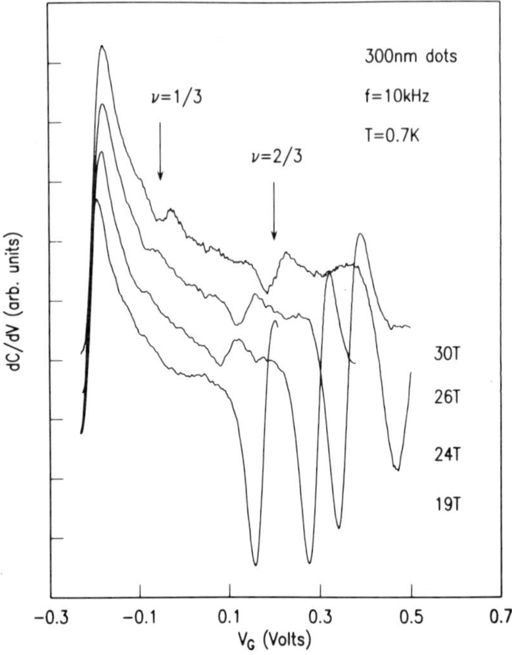

Figure 13. Gate voltage derivative of the capacitance recorded on 0 D electron systems in an $Al_x Ga_{1-x}As/GaAs$ heterojunction with 300 nm square size at high magnetic fields. The strong structure at the high gate voltage end of the traces corresponds to filling factor $\nu = 1$. The additional structures at approximate filling factors $\nu = 1/3$ and $\nu = 2/3$ are marked with arrows [59].

Landau level index. Independently of details in the shape of the confinement potential, the crossover from electrostatic to magnetic confinement is always accompanied by a lifting of the relatively low level degeneracy at zero magnetic field and pronounced level crossings at higher fields in order to form at highest fields a level structure similar to the highly degenerate Landau levels. This behavior is qualitatively reflected in the magnetic field dependence of the capacitance data in Fig. 12 [51].

Model calculations performed for 0 D electron systems in quantum well boxes of sizes comparable to the present experimental systems demonstrate that exchange correlation has a large effect on the level spectrum in few electron systems [55]. Corresponding model calculations for finite magnetic fields have been performed recently by Maksym and Chakraborty [56]. The results of the calculations demonstrate that at high magnetic fields the total angular momentum of the energetically lowest lying many particle states can become much larger than the momentum that would be derived from the corresponding sum of the momenta in the single-particle level structure. Accordingly, with the ratio of the above momenta a filling factor may be defined that is linked to the Landau level filling factor in 2 D electron systems [56].

It is indeed intriguing to investigate whether in quantum dots containing only few electrons energy gaps at fractional filling factors are still observable. Energy gaps at

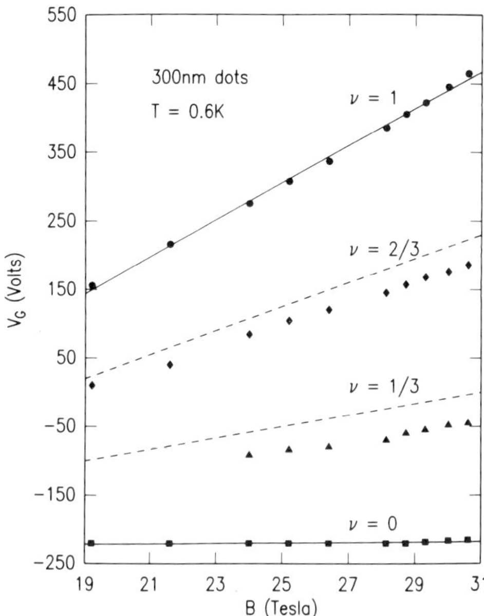

Figure 14. Gate voltage positions of the minima measured in the capacitance derivative of the sample in Fig. 13 vs magnetic field. The dashed lines indicate the interpolated gate voltages at which the filling factor is $\nu = 1/3$ and 2/3 [59].

the Fermi energy arising at fractional Landau level filling factors cause in high mobility 2 D electron systems additional plateaus in the Hall voltage [57]. Fractional states have been also observed in capacitance measurements in 2 D as well as 1 D systems [46,58]. Figure 13 shows traces of the capacitance derivative recorded at high magnetic fields on dots with 300nm square size [59]. The turn-on voltage at which carriers start to be trapped in the 0 D dots is close to the zero field value $V_g = -0.27\,\mathrm{V}$. The capacitance traces are recorded up to gate voltages at which from occupation of the first spin split Landau level a strong structure in the capacitance signal arises. At lower filling factors additional structures are observed that are marked with arrows in Fig. 13. The structures become more pronounced with increasing magnetic field. Furthermore they have a strong temperature dependence. They vanish if at $B = 30\,\mathrm{T}$ the temperature is increased up to $T = 3\,\mathrm{K}$. Figure 14 shows the magnetic field dependence of the gate voltages at which the additional structures are observed. The gate voltage positions increase linearly with the magnetic field indicating, that the filling factor at which they are observed is constant. For comparison, the dashed lines mark the positions of filling factors 1/3 and 2/3 estimated with the gate voltage positions of the two lowest spin split Landau levels and a constant capacitance. The magnetic field dependence, the temperature dependence as well as the gate voltage positions indicate that the observed additional structures are indeed caused by fractional states in the 0 D electron systems. At filling factor 1/3 and magnetic field $B = 24\,\mathrm{T}$ only about 22 electrons are contained within a dot of 120 nm diameter. These results demonstrate that in the future, numerical few-particle calculations of fractional quantization may be directly compared to experiments.

4 Concluding Remarks

$Al_xGa_{1-x}As$/GaAs heterojunctions with microstructured front gate make it possible to investigate the transformation of systems with higher dimensionality into lower dimensional systems by controlled depletion of geometric patterns in the electron system. lateral confinement of 2 D electron systems to 1 D wires or 0 D electron dots causes characteristic changes in the FIR conductivity. A strong resonance occurs at finite frequencies if the polarization of the radiation is in the direction of the confinement. The resonance is caused by Coulomb coupled virtual intersubband or interlevel transitions in 1 D or 0 D systems, respectively. Present experiments can be described by a model employing a harmonic bare potential of the system. Besides its amazing simplicity it also demonstrates how a collective type plasma resonance can smoothly transform into an intersubband resonance with decreasing electron density or effective channel width.

Quantum dots in semiconductor devices represent a link from artificial mesoscopic devices to atomic like impurities in semiconductors. Besides their technological importance they are also of physical interest, since the number of electrons confined in the dot as well as the shape of the potential can be controlled. This enables one to study how a few electron system transforms into a highly correlated electron fluid in mesoscopic devices. For experimental investigations of 0 D electron systems the contacts become a crucial part of the experiment, since with decreasing size of the few electron systems the properties of the contact increasingly contributes to the experimental results. So far carrier injection through tunnel barriers has been employed for transport [60] as well as capacitance measurements [50,51]. Capacitance spectra recorded on arrays of 1 D channels and 0 D electron dots exhibit at zero magnetic field gate voltage dependent oscillations in contrast to the capacitance of 2 D systems. In 1 D electron systems the oscillations directly reflect the quantization into 1 D subbands. The magnetic field dependence of capacitance oscillations in 0 D electron dots is similar to the field dependence of single-particle energies in simple confinement models. However, the question of the level resolution in the capacitance measurements on 0 D electron systems remains to be solved. At high magnetic fields fractional Landau states are observed in electron dots containing as few as 22 electrons.

5 Acknowledgements

The experiments discussed in this article where performed partly at the University of Hamburg, West-Germany, partly at IBM- Yorktown Heights, USA. The author has particularly benefitted from many discussions with J. Alsmeier, A. Lorke, J. A. Brum, C. Dahl, D. Heitmann, J. P. Kotthaus, U. Merkt and T. P. Smith. Financial support by Volkswagenstiftung is gratefully acknowledged.

References

[1] T. Ando, A.B. Fowler, and F. Stern, Rev. Mod. Phys. **54**, 437 (1982)

[2] "Proc. of the 33rd International Symposium on Electron, Ion and Photon Beams", Monterey 1989, published in J. Vac. Sci. Technol. **B 7**, 1373 (1989)

[3] T.N. Theis, Surf. Sci. **98**, 515 (1980)

[4] L. Zheng, W.L. Schaich, and A.H. MacDonald, Phys. Rev. **B 41**, 8493 (1990)

[5] E. Batke, D. Heitmann, and C.W. Tu, Phys. Rev. **B 34**, 6951 (1986)

[6] M. Tewordt, E. Batke, J.P. Kotthaus, G. Weimann, and W. Schlapp, in: "High Magnetic Fields in Semiconductor Physics II", G. Landwehr, ed., Springer Series in Sol. St. Sc. **87**, 297, Springer Verlag, Berlin (1988)

[7] A.C. Warren, D.A. Antoniadis, and H.I. Smith, Phys. Rev. Lett. **56**, 1858 (1986)

[8] J. Alsmeier, E. Batke, and J.P. Kotthaus, Phys. Rev. **B 40**, 12574 (1989)

[9] J. Alsmeier, E. Batke, and J. P. Kotthaus, Phys. Rev. **B 41**, 1699 (1990)

[10] W. Hansen, M. Horst, J.P. Kotthaus, U. Merkt, Ch. Sikorski, and K. Ploog, Phys. Rev. Lett. **58**, 2586 (1987)

[11] A.V. Chaplik and D. Heitmann, J. Phys. **C 18**, 3357 (1985)

[12] W. Kohn, Phys. Rev. **123**, 1242 (1961)

[13] W. Hansen, in: Festkörperprobleme, Advances in Sol. St. Phys. **28**, U. Rössler, ed., 121, Vieweg, Braunschweig (1988)

[14] U. Mackens, D. Heitmann, L. Prager, J. P. Kotthaus, and W. Beinvogl, Phys. Rev. Lett. **53**, 1485 (1984)

[15] M.V. Krasheninnikov and A.V. Chaplik, Fiz. Tekh. Poluprovodn. **15**, 32 (1981) [Sov. Phys. Semicond. **15**, 19 (1981)]

[16] G. Eliasson, J.-W. Wu, P. Hawrylak, and J.J. Quinn, Sol. St. Commun. **60**, 41 (1986)

[17] V. Cataudella and V.M. Ramaglia, Phys. Rev. **B 38**, 1828 (1988)

[18] C. Dahl, Phys. Rev. **B 41**, 5763 (1990)

[19] U. Wulf, E. Zeeb, P. Gies, R.R. Gerhardts, and W. Hanke, preprint

[20] V. Cataudella, Phys. Rev. **B 38**, 7828 (1988)

[21] W. Que and G. Kirczenow, Phys. Rev. **B 37**, 7153 (1988)

[22] W. Que and G. Kirczenow, Phys. Rev. **B 39**, 5998 (1989)

[23] A. V. Chaplik, Superlattices and Microstructures **6**, 329 (1989)

[24] V. Shikin, T. Demel, and D. Heitmann, Surf. Sc. **229**, 276 (1990)

[25] Q. Li and S. DasSarma, Phys. Rev. **B 40**, 5860 (1989)

[26] D.B. Mast, A.J. Dahm, and A. L. Fetter, Phys. Rev. Lett. **54**, 1706 (1985)

[27] D.C. Glattli, E.Y. Andrei, G. Deville, J. Poitrenaud, and F.I.B. Williams, Phys. Rev. **54**, 1710 (1985)

[28] S.J. Allen, F. DeRosa, G.J. Dolan, and C.W. Tu, Proc. of the 17th Int. Conf. Physics of Semiconductors, J.D. Chadi and W.A. Harrison, eds., 313, Springer, New York (1985)

[29] S.J. Allen, Jr. ,H.L. Störmer, and J.C. M. Hwang, Phys. Rev. **B 28**, 4875 (1983)

[30] T. Demel, D. Heitmann, P. Grambow, and K. Ploog, Phys. Rev. **B 38**, 12732 (1988)

[31] F. Brinkop, W. Hansen, J.P. Kotthaus, and K. Ploog, Phys. Rev. **B 37**, 6547 (1988)

[32] K.-F. Berggren, T.J. Thornton, D.J. Newson, and M. Pepper, Phys. Rev. Lett. **57**, 1769 (1986)

[33] J. Alsmeier, Ch. Sikorski, and U. Merkt, Phys. Rev. **B 37**, 4314 (1988)

[34] T. Demel, D. Heitmann, P. Grambow, and K. Ploog, Superlattices and Microstructures **5**, 287 (1989)

[35] U. Merkt, Superlattices and Microstructures **6**, 341 (1989)

[36] U. Merkt, Ch. Sikorski, and J. Alsmeier, in: "Spectroscopy of Semiconductor Microstructures", G. Fasol, A. Fasolino, and P. Lugli, eds., 89, Plenum Press, New York (1989)

[37] L. Brey, N.F. Johnson, and B.I. Halperin, Phys. Rev. **B 40**, 10647 (1989)

[38] K.Y. Lee, T.P. Smith, III, H. Arnot, C.M. Knoedler, J.M. Hong, D.P. Kern, and S.E. Laux, J. Vac. Sci. Technol. **B 6**, 1856 (1988)

[39] A. Lorke, J.P. Kotthaus, and K. Ploog, Phys. Rev. Lett. **64**, 2559 (1990)

[40] B.A. Wilson, S.J. Allen, Jr., and D.S. Tsui, Phys. Rev. **B 24**, 5887 (1981)

[41] A.L. Fetter, Phys. Rev. **B 32**, 7676 (1986)

[42] S.A. Govorkov, M.I. Reznikov, A.P. Senichkin, and V.I. Talyanskii, Pisma Zh. Eksp. Teor. Fiz. **44**, 380 (1986) [JETP Lett. **44**, 487 (1986)]

[43] V.I. Talyanskii, M. Wassermeier, A. Wixforth, J. Oshinowo, J.P. Kotthaus, I.E. Batov, H. Nickel, and W. Schlapp, Surf. Sc. **229**, 40 (1990)

[44] Ch. Sikorski and U. Merkt, Phys. Rev. Lett. **62**, 2164 (1989)

[45] T.P. Smith, B.B. Goldberg, P.J. Stiles, and M. Heiblum, Phys. Rev. **B 32**, 2696 (1985)

[46] T.P. Smith, III, W.I. Wang, and P.J. Stiles, Phys. Rev. **B 34**, 2995 (1986)

[47] D. Weiss and K. v. Klitzing, in: "High Magnetic Fields in Semiconductor Physics", G. Landwehr, ed., Springer Series in Sol. St. Sc., **71**, 57, Springer, Heidelberg (1987)

[48] S.E. Laux, D.J. Frank, and F. Stern, Surf. Sc. **196**, 101 (1988)

[49] T.P. Smith, III, H. Arnot, J.M. Hong, C.M. Knoedler, S.E. Laux, and H. Schmid, Phys. Rev. Lett. **59**, 2802 (1987)

[50] T.P. Smith, III, K.Y. Lee, C.M. Knoedler, J.M. Hong, and D.P. Kern, Phys. Rev. **B 38**, 2172 (1988)

[51] W. Hansen, T.P. Smith,III, K.Y. Lee, J.A. Brum, C.M. Knoedler, J.M. Hong, and D.P. Kern, Phys. Rev. Lett. **62**, 2168 (1989)

[52] A. Kumar, S.E. Laux, and F. Stern, preprint (1990)

[53] R. H. Silsbee and R.C. Ashoori, Phys. Rev. Lett. **64**, 1991 (1990)

[54] V. Fock, Z. Phys. **47**, 446 (1928)

[55] G.W. Bryant, Phys. Rev. Lett. **59**, 1140 (1987)

[56] P.A. Maksym and T. Chakraborty, Phys. Rev. Lett. **65**, 108 (1990)

[57] for a review, see T. Chakraborty and P. Pietilinen, "The Fractional Quantum Hall Effect", Springer Series in Sol. St. Sc. **85**, Springer, New York, (1988)

[58] T.P. Smith, III, K.Y. Lee, J.M. Hong, C.M. Knoedler, C.H. Arnot, and D.P. Kern, Phys. Rev. **B 38**, 1558 (1988)

[59] W. Hansen, T.P. Smith, III, K.Y. Lee, J.M. Hong, and C.M. Knoedler, Appl. Phys. Lett. **56**, 168 (1990)

[60] M.A. Reed, J.N. Randall, R.J. Aggarwal, R.J. Matyi, T.M. Moore, and A.E. Wetsel, Phys. Rev. Lett. **60**, 535 (1988)

ELECTRONIC STRUCTURE OF SMALL SYSTEMS

Ulrich Rössler

Institut für Theoretische Physik
Universität Regensburg
W-8400 Regensburg, F. R. Germany

1 Quantum Confinement and Physical Length Scales

When talking about small systems in the context of quantum coherence in mesoscopic systems, we mean all those artificially prepared metal and semiconductor structures, in which transport properties exhibit the coherence of the electronic states between the leads [1]. A large variety of such systems has been realized over the past few years [2] and new techniques are currently developed [3,4] (Fig. 1).

Most of these small systems, known as quantum wires, are fabricated by imposing a lateral structure on otherwise quasi-2 D semiconductor systems like single or multiple quantum well [5,6,7], MIS (metal-insulator-semiconductor) [8,9], or modulation doped heterostructures [9,10,11,12]. The lateral structure is realized by photolithography and etching techniques [5], by interdiffusion caused by focused-ion bombardment [6], by inhomogeneous strain patterns [7], by grating of metal gates [8,9], by split gates [11,12], or in-plane gates [3]. Moreover, quantum well wires can be obtained also from epitaxial growth on tilted superlattices [13].

The geometry of these systems is characterized by three lengths (Fig. 2):

1. The width L_z of the quantum well or inversion channel which defines the quasi-2 D system (taken to be in the $x - y$ plane); L_z can be varied in a wide range by epitaxial growth (QW) or changing the gate voltage (MIS), but is usually about 10 nm.

2. The width L_x of the quantum wire, which is determined by the lateral structure and is of the order of 100-1000 nm.

3. Finally, the length L_y of the quasi-1 D system, defined by the separation of the leads for static transport in this direction; it can be of the order of 1 μm. For quantum dots or quasi-0 D systems $L_y \simeq L_x$.

Comparison of these geometrical lengths with fundamental physical length scales for electrons [14] allows already to draw conclusions on the electronic structure and related properties of these systems (Fig. 2).

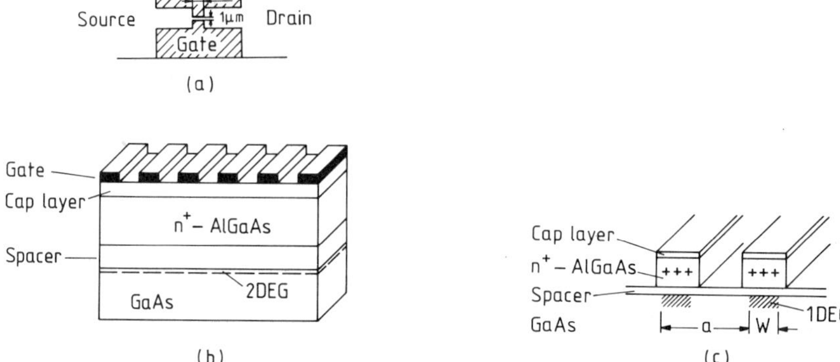

Figure 1. Examples of quasi-1 D systems: (a) split gate, after [21], (b) grating of metal gates, after [5], (c) mesa-etching, after [5].

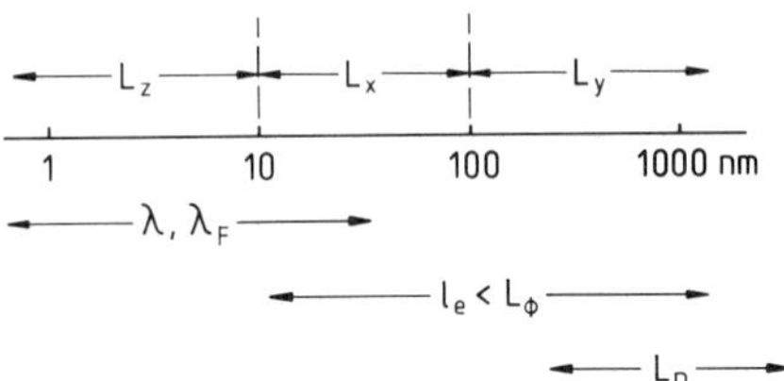

Figure 2. Comparison of geometrical lengths of small structures with physical length scales.

Starting from below, the de Broglie wavelength

$$\lambda = 2\pi \sqrt{\frac{\hbar^2}{2m^*E}} \tag{1}$$

of a free particle with effective mass m^* and energy E marks the length scale on which in a microscopic description quantum size effects show up. For electrons in semiconductors close to the conduction band edge, $E \lesssim 100$ meV and $m^* \lesssim 0.1 m_0$, λ is of the order of 100-1000 Å. This length rules the quantization of electron states by confinement in the z and x directions. Due to the relation $L_z \ll L_x$ the spacing of the quantized energy levels is much larger due to the confinement in z than in x direction. It is larger also for a system with smaller effective mass m^* [15].

For a degenerate electron system the Fermi energy defines via eq. (1) the Fermi wave length $\lambda_F = \lambda\left(E_F\right)$, which gives the extension of an electron state at E_F. If λ_F is of the order of L_z or L_x these states are bound and do not contribute to static transport in these directions. However, in this case dynamical transport measurements provide informations on the structure of the confined electron system [16,17].

The mean free path

$$\ell_e = v_{th} \, \tau_{coll} \tag{2}$$

is defined by the thermal velocity v_{th} (for low temperatures likewise the Fermi velocity) and the collision time τ_{coll}. These quantities are mean values and depend on the statistics of the electron system. They are used to describe the macroscopic transport properties. If τ_{coll} refers to the mean time between elastic collisions, e.g. scattering with impurities which dominates at low temperatures, than ℓ_e is called elastic mean free path. It defines the length on which the transport is ballistic, i.e. without any scattering process. Because elastic collisions do not destroy the phase memory of an electron state, a particle can undergo several elastic collisions before the phase coherence of the state is broken by an inelastic scattering process. The distance covered by a carrier between two inelastic collisions is, therefore, called coherence length L_ϕ (ϕ reminds of the phase of the particle wave function). In static transport experiments universal conductance fluctuations show up if the distance between the leads is comparable with L_ϕ, which in high-mobility semiconductor heterostructures can be of the order of 1 μm as in the y-direction of quantum wires. The coherence length L_ϕ defines the limit, at which transport is not characterized any more by a macroscopic material property (conductivity) but by the quantum mechanics and geometry of the sample and its leads (conductance) [1]. This specifies a mesoscopic system.

In semiconductor systems with two kinds of carriers, electrons and holes in quantum well structures, the $e - h$ recombination time τ_{rec} defines another length scale, the diffusion length

$$L_D = (D\,\tau_{rec})^{1/2} \tag{3}$$

where D is the diffusion constant. It is the length covered by a carrier during its lifetime irrespective of collision processes and can be of the order of several 100 mean free paths. Finally, the presence of a magnetic field introduces with the magnetic length

$$\lambda_c = (\hbar c/eB) \tag{4}$$

a physical length scale, which can be changed continuously from the outside.

Coincidence of the confinement length with one of the physical length scales signalizes a change of the physical properties connected with this length scale. If the confinement length or the physical length can be changed, e.g. by changing the gate voltage or the magnetic field, respectively, one can observe crossover from 2 D to 1 D [9,17,18,19] or even to 0 D behavior [20].

When looking on these length scales, different interesting questions arise. One group is related with the verification of the quantum size effect due to the lateral confinement in z or x : oscillations of conductance [4,8] or capacitance [10], deviations from 2 D behavior of magnetoconductance oscillations [21,22], finestructure in luminescence spectra [6,23], or FIR excitations observed in the dynamical conductivity [17,18,19,20]. The other group is devoted to the mesoscopic aspects due to quantum coherence in y direction: universal conductance fluctuations, quenching of the Hall effect, Aharonov-Bohm oscillations and non-local effects [1]. In this lecture we shall look only into the electronic structure defined by the lateral confinement and into the physical phenomena connected with this structure.

2   Electronic Structure of Quasi-1 D Systems

With the arguments of section 1 in mind the electronic structure of a quasi-1 D system can be calculated by solving the Schrödinger equation

$$\left[-\frac{\hbar^2}{2m^*}\left(\partial_x^2 + \partial_y^2 + \partial_z^2\right) + V(x,z)\right]\phi(x,y,z) = E\phi(x,y,z) \tag{5}$$

with the eigenfunctions

$$\phi_{\nu\mu k_y}(x, y, z) = \frac{1}{\sqrt{L_y}} e^{ik_y y} \varphi_{\nu\mu}(z, x). \tag{6}$$

Here we use effective-mass approximation, which for semiconductor structures is justified as the electron states are formed from near band edges states of the Bloch bands [24]. The simplest model for the confining potential $V(x, z)$ is the rectangular square-well potential in x and z, which is separable only for infinite barrier height and yields the eigenvalues

$$E_{\nu\mu}(k_y) = \frac{\hbar^2}{2m^*} \left[\left(\frac{\nu\pi}{L_z} \right)^2 + \left(\frac{\mu\pi}{L_x} \right)^2 + k_y^2 \right]. \tag{7}$$

Corresponding to $L_z \ll L_x$ the spacing of confined states at $k_y = 0$ is much larger due to quantization in z (of the order of several 10 meV) than in x direction (usually several meV). The density of states

$$z_{1D}(E) = \frac{1}{\pi} \sqrt{\frac{m^*}{2\hbar^2}} \sum_{\mu\nu} \theta(E - E_{\nu\mu})(E - E_{\nu\mu})^{-1/2} \tag{8}$$

exhibits the characteristic 1/square-root singularities which – even if they are broadened by disorder or finite temperature – are indicative in experimental data of quasi-1 D nature (Fig. 3). The optical absorption due to direct intersubband transitions has been calculated for this model [25]. The subband spacing has been detected in inelastic resonant light scattering [26] and the corresponding quantization of exciton states in photoluminescence experiments [6,26]. The 1 D character of exciton transitions is reflected also in a strong polarization dependence of the photoluminescence [13,27].

While the epitaxial growth of quantum wells, heterostructures, and superlattices leads to well defined confinement potentials in z direction, the methods of lateral confinement in x direction do not result in a clear-cut situation. Therefore, the proper modelling of the potential $V(x, z)$ is a central problem in a quantitative theory of these systems. Here, in particular, the systems in which confinement is achieved by gate structures or well defined doping profiles have been investigated theoretically [28,29,30,31]. These studies provide some principal insight into the electronic structure of 1 D systems.

Shik [28] considers the case of an n-channel in a MIS structure with additional confinement in x direction by two $p - n$ junctions (Fig. 4 a). The potential in the n-type region follows from Poisson's equation

$$\left(\partial_x^2 + \partial_z^2 \right) V(x, z) = -\frac{4\pi e}{\varepsilon_s} \rho(x, z) \tag{9}$$

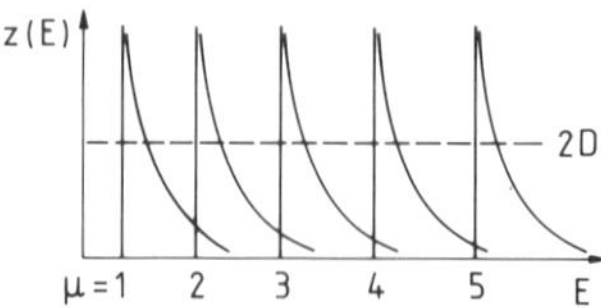

Figure 3. Electronic density of states of a quasi-1 D structure.

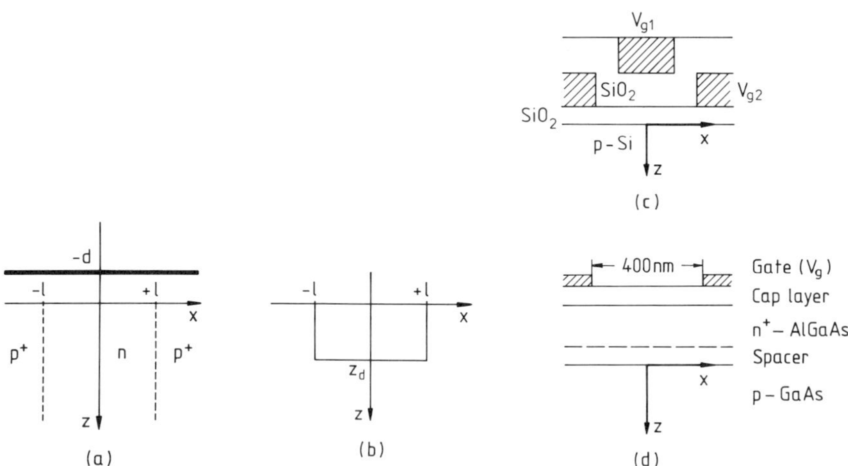

Figure 4. Modelling of gate structures used for calculations of electronic states: (a) after [28], (b) after [29], (c) after [30], (d) after [31].

where $\rho(x, z) = \rho_0$ is the constant volume density of negatively charged acceptors in the n region. The z dependence of the potential is obtained by assuming for a gate distance much smaller than the channel width ($d \ll l$, Fig. 4 a) a constant slope in the insulator starting at the gate voltage V_1

$$V(x, z) \simeq \left(1 + \frac{z}{d}\right) V(x, 0) - \frac{z}{d} V_1 \qquad -d < z < 0 \tag{10}$$

and by the condition of continuity for the dielectric displacement at $z = 0$

$$\varepsilon_s \partial_z V(x, z > 0) - \varepsilon_I \partial_z V(x, z < 0) = 4\pi e N_s \tag{11}$$

where N_s is the areal density of electrons at the interface ($z = 0$). In x direction the potential has to fulfill the boundary condition $V(x = \pm l, z) = V_2$, where V_2 is the (negative) potential of the p-side of the $p - n$ junctions. For $N_s = 0$ one obtains

$$V_0(x, z) = \frac{2\pi e \rho_0}{\varepsilon_s}(\ell^2 - x^2) + V_2 + \sum_{k=0}^{\infty} A_k \cos\left(\frac{\pi x}{2\ell}(2k + 1)\right) \exp\left(-\frac{\pi z}{2\ell}(2k + 1)\right) \tag{12}$$

with A_k being defined through the boundary conditions by $V_1, V_2, d, \ell, \rho_0, \varepsilon_s$ and ε_I. Using the limiting condition $V_0(x = 0, z = 0) = 0$ to form an n-channel at $z = 0 (N_s \simeq 0)$ Shik finds the approximate solution

$$V(x, z) \simeq \frac{\varepsilon_I}{\varepsilon_s} V_1 \frac{z}{d} - \frac{2\pi e \rho_0}{c_s}(1 + \beta)x^2 \tag{13}$$

with β being determined by the system parameters. The essential result of this study is, that at low electron concentrations N_s the confinement potential is separable and parabolic in x.

Lai and Das Sarma [29] performed an essentially analytical variational calculation for the ground state and charge density of a quantum wire without restriction to small electron concentration. They solved eqs. (5) and (9) self-consistently for the charge density $\rho(x,z) = \rho_d(x,z) + \rho_s(x,z)$, where

$$\rho_d(x,z) = -e(N_A - N_D)\,\theta(\ell - |x|)\,\theta(-z)\,\theta(z_d - z) \tag{14}$$

is the depletion charge to mimic a narrow-gate structure (Fig. 4 b) and

$$\rho_s(x,z) = N_s|\varphi_{00}(z,x)|^2 \tag{15}$$

is the electron density for occupation of the lowest quantized state only. The Hartree potential has to vanish at the boundary of the depletion charge region. In order to accomplish a mostly analytical treatment the authors make use of the formal solution of Poisson's equation with a Green's function

$$G(x,z;x',z') \simeq \frac{2}{\pi} \frac{\sin\dfrac{\pi x}{\ell}\sin\dfrac{\pi x'}{\ell}}{\sinh\dfrac{\pi z_d}{\ell}} \begin{cases} \sinh\dfrac{\pi(z_d - z')}{\ell}\sinh\dfrac{\pi z}{\ell} & z < z' \\[2ex] \sinh\dfrac{\pi z'}{\ell}\sinh\dfrac{\pi(z_d - z)}{\ell} & z' < z \end{cases} \tag{16}$$

which is the leading term of a highly convergent series. Moreover the variational wave function for the lowest occupied subband

$$\varphi_{00}(z,x) = A\,Z e^{-b_0 z/2}\,e^{-b_0 b_1 |x|z/2}\,F(x) \tag{17}$$

with

$$F(x) = \begin{cases} \cos\dfrac{\pi x}{2\ell} + \dfrac{\pi}{2\ell\gamma} & |x| \leq \ell \\[2ex] \dfrac{\pi}{2\ell\gamma}\,e^{-\gamma(|x|-\ell)} & |x| \geq \ell \end{cases} \tag{18}$$

(b_0, b_1 and γ being variational parameters) allows to give an analytical expression for the potential $V(x,z)$. It is important to note, that the potential (due to the form of the Green's function eq. (16)) is not separable, and as a consequence the wave function eq. (17) is not written as a product of functions in x and z. In addition, expansion of the second exponential factor in eq. (17) shows, that this form models a combination of ground and excited state wave functions of the Fang-Howard-Stern type [24] for MIS structures. The expression for the ground state energy $E(b_0, b_1, \gamma)$ is obtained analytically, but minimized numerically. Calculations of the envelope function for different carrier and depletion concentrations representing narrow inversion layers at Si-SiO$_2$ interfaces demonstrate the influence of coupling between x and z dependence due to the lateral confinement, which results in a larger distance of the maximum charge density from the interface at the center of the channel ($x = 0$) than close to its edges ($x \simeq \ell$) ("bending effect", Fig. 5). Although this study does not discuss the quantitative dependence of the ground state energy on this coupling (e.g. by comparison with the result for $b_1 = 0$, corresponding to neglect of this coupling), its effect is expected to be stronger for narrower channels, i.e. if the characteristic confinement lengths in x and z direction are closer to each other.

The results of [28,29] are consistent with complete self-consistent numerical calculations for quasi-1 D channels at Si-SiO$_2$ interfaces [30] and for a Al$_x$Ga$_{1-x}$As/GaAs heterostructure [31] with more realistic gate models (see Fig. 4 c,d). The calculations for a 80 nm slit for the Si-SiO$_2$ system gives occupation of only the lowest quantized subband E_{00} for electron concentrations $N_L \lesssim 10^6 cm^{-1}$ (Fig. 6 a) which is considered to be the

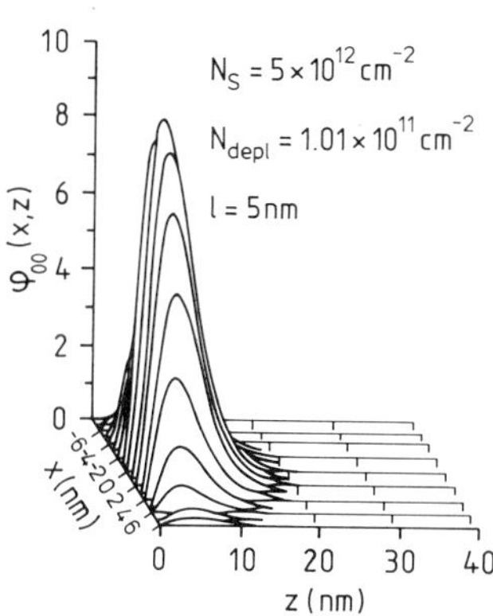

Figure 5. "Bending effect" in the electronic wave function, demonstrating the coupling between x and z dependence (after [29]).

strict 1 D region. The results for a 400 nm wide split gate on an $Al_x Ga_{1-x} As/GaAs$ heterostructure demonstrate in particular the change of the lateral potential profile from almost parabolic for very low electron concentration in the channel to a truncated parabola with flat bottom due to screening effects by increasing electron densities (Fig. 6 c) [32]. It can be read from these profiles that the width of the electron distribution in the channel is considerably smaller than the geometrical width of the gate structure. The strict 1 D regime with occupation of only the lowest quantized state E_{00} is obtained here also for charge densities below $10^6 cm^{-1}$ (see Fig. 6 b). Calculations for narrow semiconducting wires near threshold, i.e. for gate voltages at which the channel starts to be populated, can be performed analytically [33] and are consistent with the corresponding results of [31].

3 Crossover from Two– to One Dimension

An experimental investigation of the electron states considered in these theoretical papers is feasible only for multiple quantum wire systems with gratings of metal gates, which at the same time open the possibility to study the crossover from 2 D to 1 D behavior [9,17,18,19,] by changing the gate voltage and/or a magnetic field applied perpendicular to the system.

Theoretically, the situation in periodic multiple wire systems (Fig. 1 b) has been investigated in self-consistent calculations by Wulf and Gerhardts [34]. They solve Poisson's equation for a periodic electron density

$$\rho_s(x, z) = \sum_r \rho_{s,r}(z) \exp(iq_r x), \quad q_r = \frac{2\pi}{a} \tag{19}$$

where a is the period of the grating and corresponding Hartree-potential

$$V_H(x, z) = \sum_r V_{H,r}(z) \exp(iq_r x). \tag{20}$$

The boundary condition at the average position, $z = -d$, of the gate is simplified to

$$V_H(x, -d) = V_G + V_C \cos \frac{2\pi x}{a} \tag{21}$$

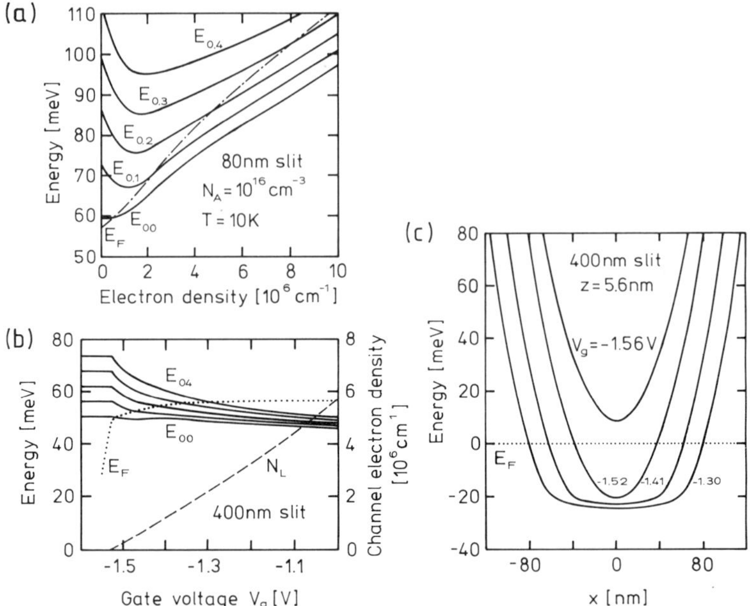

Figure 6. (a) Subband energies from confinement in x and Fermi energy vs. electron density for the gate configuration of Fig. 4c, after [30], (b) subband energies, Fermi energy and channel electron density vs. gate voltage for the gate configuration of Fig. 4d, after [31], (c) self-consistent potential profile in x for different gate voltages, after [31].

and the assumption of only one occupied subband in the inversion channel (energy E_0), caused by the confinement potential for $r = 0$ allows to factorize the wave function $\varphi_{0,\mu}(z,x) = \xi_0(z)\chi_\mu(x)$ (see eq. (6)) and to obtain the one-dimensional Schrödinger equation

$$\left[-\frac{\hbar^2}{2m^*}\partial_x^2 + V(x) - E_{0,\mu}\right]\chi_\mu(x) = 0 \tag{22}$$

where

$$V(x) = \sum_r \exp(iq_r x)\int_0^\infty dz\, V_{H,r}(z)\xi_0^2(z). \tag{23}$$

Due to the periodicity of the potential $V(x)$ the solutions of eq. (22) form minibands. Self-consistent calculations using parameters typical for an inversion layer at a Si-SiO$_2$ interface are shown in Fig. 7. The minimum energy of the two lowest subbands, $E_{\nu,0}$ for $\nu = 0, 1$, decreases only slowly with increasing amplitude V_C of the external potential up to about 40 mV, above which the decrease becomes much steeper. This crossover is not obtained in the linear screening approximation (dash-dotted line in Fig. 7). The insert in Fig. 7 demonstrates that this behavior is correlated with a change of the x-dependent electron density from a slightly modulated one to a strongly localized one if V_C is increased from below to above 40 mV, i.e. the disconnection of the 2 D electron system into separate 1 D channels takes place in a small interval of V_C and indicates the importance of non-linear screening.

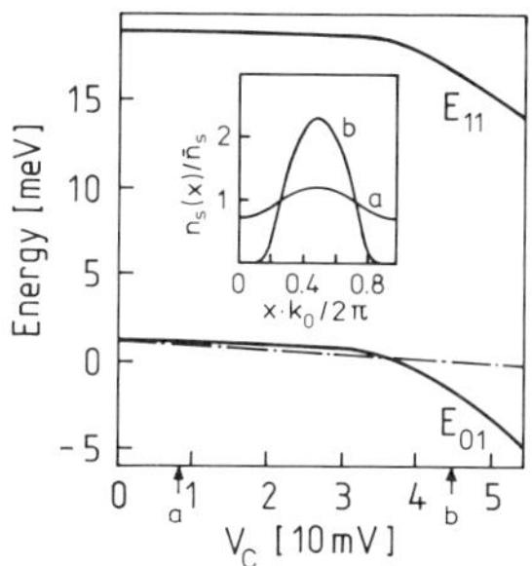

Figure 7. Lowest subband energies $E_{\nu 1}(\nu = 0, 1)$ vs. modulation amplitude of external potential (solid lines) and linear screening result for E_{01} (dash-dotted line) for a Si-SiO_2 system similar to Fig. 1 b, after [34].

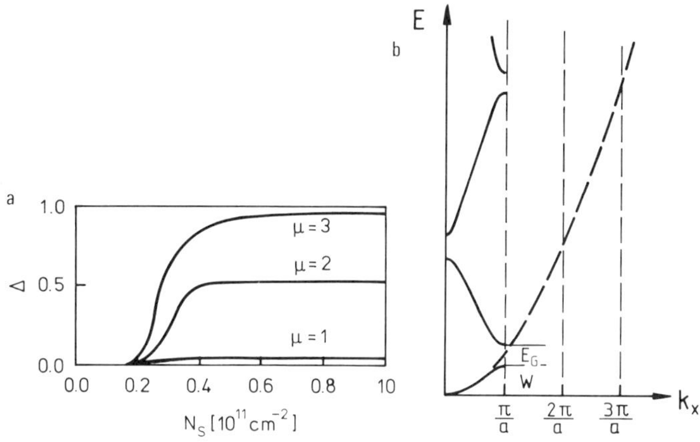

Figure 8. (a) Relative width $\Delta_\mu = W_\mu/(W_\mu + E_{G,\mu})$ for lowest subbands vs. areal electron density, after [34], (b) formation of subbands due to periodic potential in x. W denotes the bandwidth for $\mu = 1$, E_G the gap between the two lowest subbands.

For the lowest minibands ($\mu = 1, 2, 3$) connected with the lowest subband E_0 of Fig. 7 the relative width $\Delta = W/(E_G + W)$ (W being the width of a miniband, E_G the energy gap which separates it from the next higher miniband) is shown in Fig. 8 a as a function of the electron concentration N_s in the inversion channel at fixed V_C. For small N_s the minibands degenerate into discrete levels if the wave function overlap between neighboring 1 D channels vanishes. With increasing N_s $\Delta(\mu = 3)$ approaches 1 as expected for free electrons ($E_G \ll W$), whereas it stays small for the lowest miniband which corresponds to the tight-binding situation, as illustrated in Fig. 8 b. A similar change of Δ is expected at fixed N_s for changing V_C.

FIR transmission experiments on $Al_x Ga_{1-x} As/GaAs$ heterostructures and Si-MOS structures decorated with a periodic grating gate give strong evidence for a crossover from 2 D to 1 D behavior [9,17,18]. Spectra of the relative transmission change $\Delta T/T$ induced by the gate voltage V_G for light polarized in x-direction, i.e. perpendicular to the grating gate are shown in Fig. 9. In a simple description $\Delta T/T$ is about proportional to the real part of the dynamical conductivity [35]

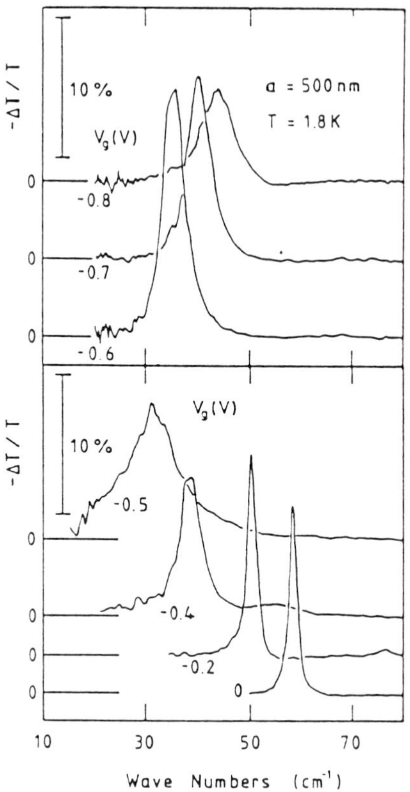

Figure 9. Relative transmission change due to gate voltage
V_G vs. wave number of FIR excitation, after [17, 18].

$$\sigma_{xx}(\omega) = i\omega \frac{e^2 N_s}{m^*} \left(\omega^2 - \omega_p^2 - i\frac{\omega}{\tau} \right)^{-1} \tag{24}$$

where N_s is the electron density in the channel and τ a phenomenological scattering
time. For the 2 D limit (lower part of Fig. 9), when the electron density at the interface is
modulated only slightly and the gate is used mainly to break the translational symmetry
in x-direction (grating-coupler effect [16]),

$$\omega_p^2 = \frac{N_s e^2 q}{2\bar{\varepsilon}\varepsilon_0 m^*} \tag{25}$$

is the 2 D plasmon frequency [36] observed here for $q = 2\pi/a$. Upon lowering the gate
voltage the average areal density of electrons N_s is reduced and the resonance shifts to
smaller wave numbers. Starting at $V_G = -0.5$ V the shift is reversed and the line shows
all characteristics of a depolarization shifted intersubband resonance [24,37] (upper part
of Fig. 9) at

$$\tilde{\omega}_\mu^2 = \omega_\mu^2 + \omega_d^2 \tag{26}$$

which in eq. (24) then has to replace ω_p^2. $\hbar\omega_\mu = E_{\nu,\mu+1} - E_{\nu,\mu}$ is the subband splitting
due to confinement in x, and the depolarization shift due to collective excitations of the

Q 1 D-plasma in x-direction is approximated by $\omega_d^2 = e^2 N_s / \bar{\varepsilon}\varepsilon_0 w m^*$, where w is the width of the Q 1 D-channel. The bare intersubband spacings have been found in oscillations of the magnetoresistance in y-direction [19].

In spite of this qualitative understanding of the FIR experiments and the correspondence with the calculations by Wulf and Gerhardts a thorough theoretical concept of these phenomena is still missing. For the quasi-2 D system it is known [38], that the spatial inhomogeneity allows for correlations between charge and current densities, which are responsible for the depolarization shift. For a multiple quantum wire system a calculation of the dynamical conductivity is even more complicated because excitations within the individual wires and between wires are coupled by Coulomb interaction. Some recent theoretical work, which considers the weak tunneling situation for a periodic wire system [39] or the modification of the classical depolarization shift due to dynamical screening [40], does not provide an explanation of the 2 D-1 D crossover, including the continuous change of the 2 D plasmon resonance into the depolarization shifted 1 D resonance. Instead, a more rigorous treatment of the collective excitations in periodically modulated inversion layers stresses the importance of Landau damping for this crossover behavior [41].

As mentioned in section 1 the crossover from 2 D to 1 D can be realized also by changing the magnetic length. This situation has been used to demonstrate the formation of 1 D channels in magneto resistance [5,21,22] and FIR transmission [5] experiments with multiple quantum-wire systems and a magnetic field in z-direction. A surprisingly simple model is available to describe the electron states in this situation [42] if the confinement potential in x-direction is assumed to be parabolic, as theoretical results (section 2) suggest for low electron density. The Hamiltonian for the electron motion in the $x - y$ plane

$$H = \frac{1}{2m^*}\left(\vec{p} - \frac{e}{c}\vec{A}\right)^2 + \frac{1}{2}m^*\omega_0^2 x^2 \tag{27}$$

with $\vec{A} = (0, Bx, 0)$ is reduced to a one-dimensional problem if we consider the translational invariance in y-direction. The last term in eq. (27) breaks the translational symmetry in x and removes the degeneracy of the Landau levels. Its parabolic form however does not change the harmonic cyclotron motion but only its frequency. The eigenvalues of eq. (27) are

$$E_n(k_y) = \hbar\omega(n + \frac{1}{2}) + \frac{\hbar^2 k_y^2}{2m^*}\frac{\omega_0^2}{\omega^2} \tag{28}$$

where $\omega^2 = \omega_c^2 + \omega_0^2$ defines the hybrid frequency of this harmonic oscillator ($\omega_c = eB/m^*c$). The second term of eq. (28) accounts for the dependence of the oscillator energy on the centercoordinate defined by k_y. For $B = 0$ the spectrum (Fig. 10 a) is that of a system confined in x-direction by a harmonic potential with subband spacings $\hbar\omega_0$ and free-particle behavior in y-direction (1 D limit). For $B \to \infty$ the oscillator levels approach the Landau levels of the 2 D system and the second term in eq. (28) vanishes (2 D limit). Changing the magnetic field for a system with fixed electron concentration shifts oscillator levels across the Fermi energy and gives rise to SdH oscillations. In contrast to the 2 D (or 3 D) case, however, where for $B \to 0$ the Fermi number crosses an infinite number of Landau levels – causing an infinite number of oscillations – this

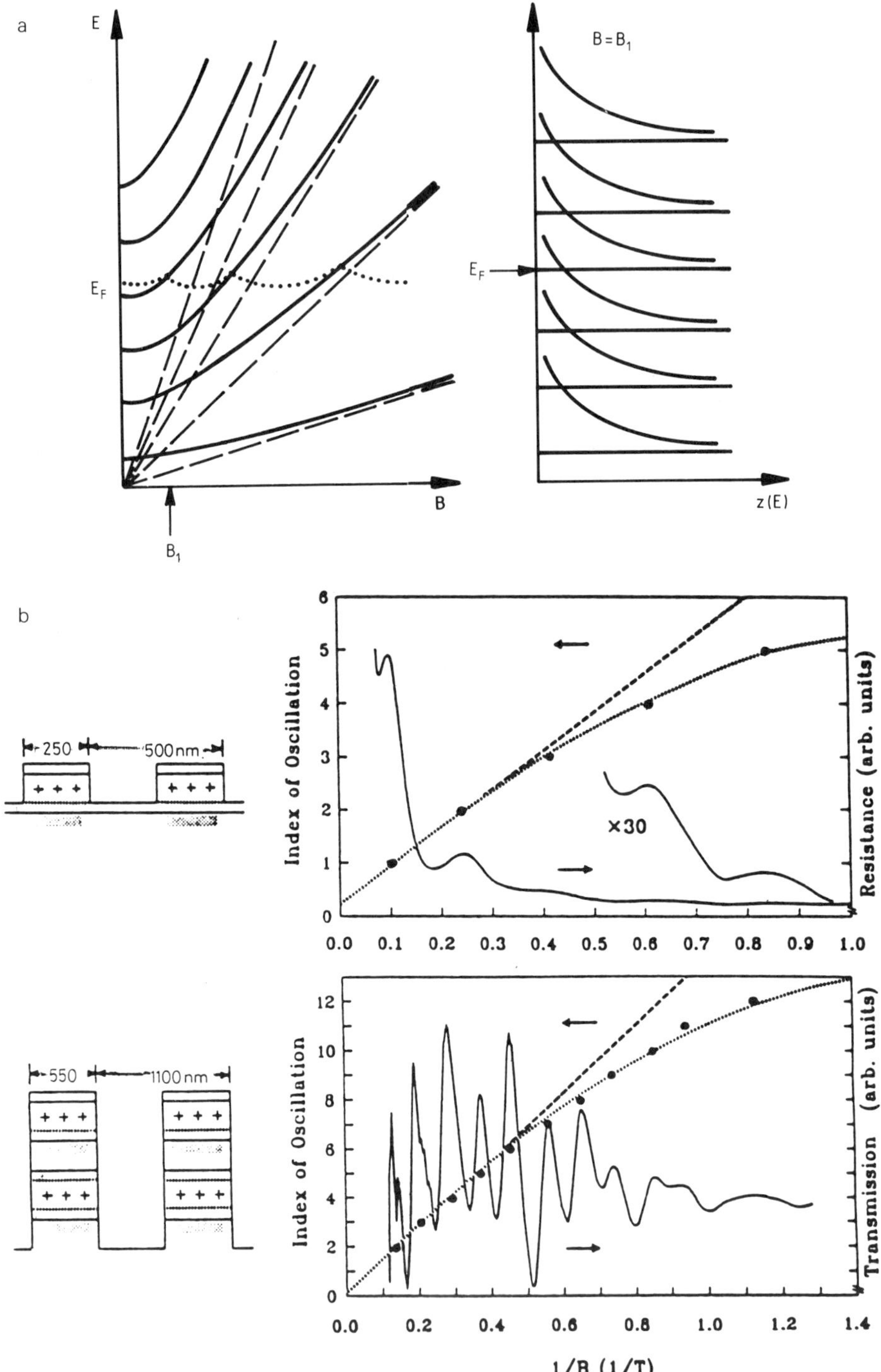

Figure 10. (a) Energies of hybrid oscillator and Fermi energy vs. magnetic field (solid lines). The dashed lines show the Landau levels. Right hand side: density of states for magnetic field B_1. (b) Magnetoconductivity and index of oscillations vs. $1/B$. Deviation from the straight line indicates 1 D behavior. The structures used in the experiments are shown on the left, after [5].

number is limited in the 1 D case. In addition, the period of the SdH oscillations is now proportional to

$$\frac{1}{\omega} = \frac{1}{\omega_c}\left(1 + \frac{\omega_0^2}{\omega_c^2}\right)^{-1/2} \tag{29}$$

which in a plot of the index of oscillation (which corresponds to n) vs. $1/B$ leads to deviations from a straight line for increasing $1/B$. This effect demonstrates the depopulation of Landau levels due to the confinement in x. The strength of the harmonic potential ω_0 can be determined from a fit to the experimental data (Fig. 10 b) [5].

4   Electronic Structure of Quantum Dots

The preparation techniques used for fabrication of quasi-1 D systems have been applied also more recently to realize quasi-0 D systems [20,43,44,45]. Arrays of quantum dots produced by mesa-etching [43] have been designed as double barrier heterostructure to study vertical transport. The depletion effect of a Schottky contact between a metal film (Ni,Cr) and p-type InSb has been used to realize a tunable array of quantum dots (Fig. 11 a) [20]. Before evaporating the metal film the samples were covered with photoresist, which after two exposures (in between which the sample is rotated by 90°) was removed except for a regular array of islands. Under these islands the semiconductor is not depleted and quantum dots are formed. The small effective mass $m^* = 0.014\ m_0$ of electrons in InSb results in a separation of the completely quantized electron states ($L_x \simeq L_y \lesssim 125$ nm) of several meV and allows to reduce the number of electrons per dot to 1. A dual gate device on Si-MOS structures has been used to realize periodic arrays of dots with diameters smaller than 100 nm and about 20-350 electrons [44]. The latter systems[20,44] have been investigated by FIR magnetospectroscopy.

In order to describe the electron states of quantum dots in the presence of a magnetic field in z-direction we assume that all relevant states derive from the lowest electric subband of the inversion layer at the semiconductor-insulator interface. For parabolic confinement in x and y the one-electron Hamiltonian reads

$$H_0 = \frac{1}{2m^*}\left(\vec{p} + \frac{e}{c}\vec{A}\right)^2 + \frac{1}{2}m^*\omega_0^2\left(x^2 + y^2\right) \tag{30}$$

which for $\vec{A} = \frac{1}{2}B(-y, x, 0)$ becomes

$$H_0 = \frac{1}{2m^*}\left(p_x^2 + p_y^2\right) + \frac{1}{2}m^*\left(\omega_0^2 + \omega_L^2\right)\left(x^2 + y^2\right) + \omega_L L_z \tag{31}$$

where $\omega_L = \frac{1}{2}\omega_c$ is the Larmor frequency and L_z the z-component of the angular momentum. This two-dimensional oscillator problem can be formulated in a more elegant way by using oscillator operators $a_\pm$, $a_\pm^\dagger$, where the $\pm$ signs refer to the right and left hand oscillator [46]

$$H_0 = \hbar\omega\left(a_+^\dagger a_+ + a_-^\dagger a_- + 1\right) + \hbar\omega_L\left(a_+^\dagger a_+ - a_-^\dagger a_-\right) \tag{32}$$

with the hybrid frequency $\omega^2 = \omega_0^2 + \omega_L^2$. The eigenvalues of H_0 are [46]

$$E_{nm} = \hbar\omega\left(2n + |m| + 1\right) + m\,\hbar\omega_L \tag{33}$$

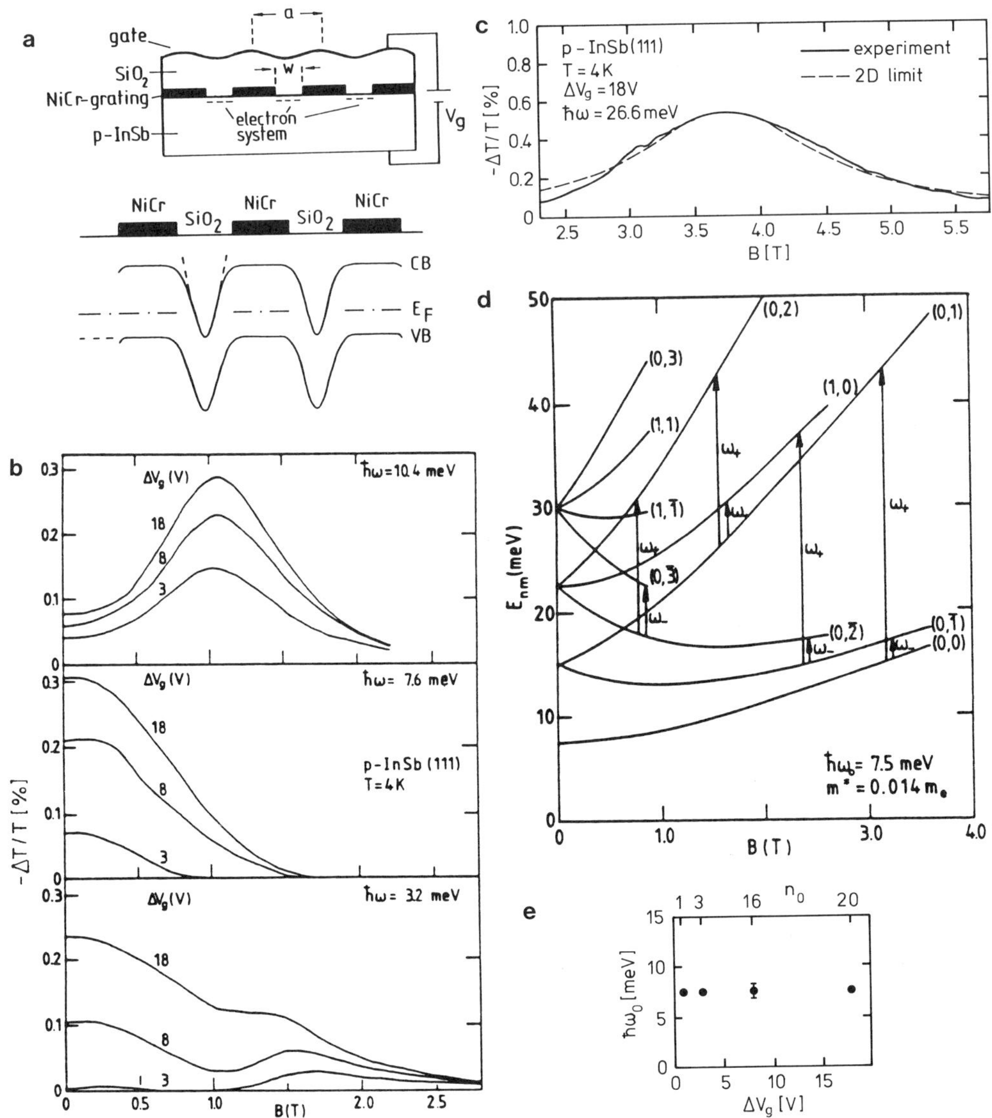

Figure 11. (a) Quantum dots on InSb (upper part) and schematic cross section of the gated MOS system and band structure to show the depletion under the Schottky gate [20]. (b) Relative transmission vs. magnetic field for different gate voltages and excitation energies [20]. (c) Same as (b) but large excitation energy. The dashed curve is calculated for the 2 D limit [20]. (d) Energies of hybrid oscillator states vs. magnetic field for the parameters indicated. For quantum numbers see text. $\omega_{\pm}$ indicate dipole allowed transitions for right and left circular polarized light [20]. (e) Energy of oscillator potential vs. gate voltage or number of electrons per quantum dot [20].

where $n = \mathrm{Min}\,(n_+, n_-) = 0, 1, 2, \ldots$ and $m = n_+ - n_-$. For $B = 0$ the oscillator spectrum of the confinement potential $E_N = N\hbar\omega_0$, $N = 2n + |m| + 1$, is recovered, where N also indicates the degeneracy, which for finite magnetic fields is removed due to the Zeeman splitting described by the last term of (4.4) (see Fig. 11 d). With increasing magnetic field the Zeeman split levels converge towards Landau levels with energies $\hbar\omega_c(n + \frac{1}{2})$ for $m = -|m|$ and $\hbar\omega_c(n + |m| + \frac{1}{2})$ for $m = |m|$, which represent the 2 D limit. This behavior is strongly distorted for a non-parabolic confinement potential [45].

This crossover is demonstrated in FIR magnetotransmission data at fixed photon energy ($\hbar\omega = 10.4$, 7.6 and 3.2 meV) and gate voltage ($\Delta V_G = 18$, 8 and 3 V) (Fig. 11 b,c) [20]. t is again the relative transmission change induced by the gate voltage ΔV_G, which is essentially proportional to the real part of the dynamical conductivity [20,47]

$$\sigma_\pm(\omega) = \frac{e n_0 \tau}{m^*}\left(1 + \left(\frac{\omega_0^2}{\omega} - \omega \pm \omega_c\right)^2 \tau^2\right)^{-1}. \tag{34}$$

If the observed resonance is close to the cyclotron resonance ($\omega_+ \simeq \omega_c$, $\omega_c > \omega_0$) and almost independent of the gate voltage one approaches the 2 D limit (upper spectra in Fig. 11 b), which is reached for about $\hbar\omega = 27$ meV as shown in Fig. 11 c by comparison with the calculated curve for the 2 D case [20]. At intermediate frequencies, $\omega \simeq \omega_0$, close to the characteristic frequency of the confinement potential $\sigma_\pm$ becomes proportional to $(1 + \omega_c^2\tau^2)^{-1}$ in correspondence with the monotonic decrease of t vs. B observed for $\hbar\omega = 7.6$ meV (Fig. 11 b). At lower photon energies ($\omega < \omega_0$) weak resonances are observed which correspond to the transitions ω_- in Fig. 11 d. A quantitative evaluation of the spectra in Fig. 11 b using eq. (34) allows to determine the number n_0 of electrons per dot and the frequency ω_0. As can be seen from Fig. 11 e, the sublevel spacing in the quantum dots does not depend on the number of electrons.

This result is surprising in view of some theoretical studies. The calculated energies of interacting electrons in quantum boxes with a square-well confinement and infinite barriers demonstrated the importance of correlation effects [48]. A coupling of the single-particle excitations to collective modes due to Coulomb interaction between the dots is predicted theoretically for an array of electrically isolated quantum dots [49]. This calculation is performed in the self-consistent field formalism with the assumption of no overlap of the electron wave functions in different dots as for the multiple quantum wire system [39]. This calculation does not consider the background compensating charge which in [20] is connected with the NiCr gate close to the interface and is expected to screen effectively the Coulomb interaction between dots.

On the other hand it has been shown more recently by an extension of Kohn's theorem that magnetooptic absorption between states in a parabolic quantum well takes place at the bare oscillator frequency independent of the number of confined electrons and in spite of Coulomb interaction [51]. These considerations apply also to quantum dots with parabolic confinement as assumed in eq. (30). The extension of eq. (32) to a system with n_0 electrons reads

$$H = \sum_{i=1}^{n_0}\left\{\hbar\omega_+\left(a_{+i}^\dagger a_{+i} + \frac{1}{2}\right) + \hbar\omega_-\left(a_{-i}^\dagger a_{-i} + \frac{1}{2}\right)\right\} + U \tag{35}$$

where $\hbar\omega_\pm = \hbar\omega \pm \hbar\omega_L$, $a_{\pm i}^\dagger$ and $a_{\pm i}$ are the oscillator operators of the individual electrons and U is the electron-electron interaction. Because U commutes with the $a_{\pm i}$, $a_{\pm i}^\dagger$ one finds for $A_\pm = \sum_i a_{\pm i}$, $A_\pm^\dagger = \sum_i a_\pm^\dagger$ that

$$[H, A_\pm] = -\hbar\omega_\pm A_\pm, \qquad \left[H, A_\pm^\dagger\right] = -\hbar\omega_\pm A_\pm^\dagger. \tag{36}$$

As a consequence, if $|n_+, n_- >$ is an eigenstate of the Hamiltonian eq. (35) with energy $E_{n_+ n_-}$, then $A_\pm^\dagger |n_+, n_- >$ is also an eigenstate with shifted energy $E_{n_+ n_-} + \hbar\omega_\pm$. The interaction with circularly polarized light propagating in z-direction is proportional to $A_\pm^\dagger + A_\pm$, where the upper (lower) sign refers to right (left) hand circular polarization and connects eigenstates of H which differ by $\pm\hbar\omega_\pm$. Thus the excitations in a quantum dot with parabolic confinement take place at the single particle energy differences of the bare confinement potential irrespective of the number n_0 of electrons in the dot. Although the experimental results of Fig. 11 can be consistently interpreted by the conjecture of a parabolic confinement potential, this can be valid only for small dots, because experimental results for variable size of the confinement [52] show significant deviations.

5   Concluding Remarks

The electronic structure of small systems realized by lateral microstructures on semiconductor heterostructures and quantum wells, exhibits characteristic features connected with the lateral confinement in 2 or 3 dimensions. At present some of these features, like subband formation, depopulation of Landau levels, crossover from 2 D to 1 D or 0 D, as observed in experiment, are qualitatively understood in terms of simple theoretical models. Future work has to consider more quantitative aspects of e.g. the relation between electronic structure and confinement potential, effects related to the bulk bandstructure, the many particle aspects, and the spatial inhomogeneity of these systems.

References

[1] S. Washburn, R.A. Webb, Adv. Physics **35**, 375 (1986); B.L. Al'tshuler, P.A. Lee, Physics Today, 36 (Dec. 1988); R.A. Webb, S. Washburn, Physics Today, 46 (Dec. 1988)

[2] Physics and Technology of Submicron Structures, H. Heinrich, G. Bauer, F. Kuchar, eds., Springer Series in Sol. St. Sc. **83**, Springer, Berlin (1988)

[3] A.D. Wieck, K. Ploog, Surf. Sci. **229**, 252 (1990)

[4] K. Ismail, D.A. Antoniadis, H.I. Smith, Appl. Phys. Lett. **54**, 1130 (1989)

[5] D. Heitmann, T. Demel, P. Grambow, K. Ploog, Festkörperprobleme/Advances in Sol. St. Phys. **29**, U. Rössler, ed., p. 285, Vieweg, Braunschweig (1989)

[6] Y. Hirayama, H. Okamoto in [2], 45; Y. Hirayama, T. Saku, Y. Horikoshi, Phys. Rev. **B 39**, 5535 (1989)

[7] K. Kash, J.M. Worlock, M.D. Sturge, P. Grabbe, A. Scherer, P.S.D. Lin, Appl. Phys. Lett. **53**, 782 (1988); K. Kash, J.M. Worlock, A.S. Gozdz, B.P. Van der Gaag, J.P. Harbison, P.S.D. Lin, L.T. Florez, Surf. Sci. **229**, 245 (1990)

[8] A.C. Warren, D.A. Antoniadis, H.I. Smith, Phys. Rev. Lett. **56**, 1858 (1986)

[9] W. Hansen, M. Horst, J.P. Kotthaus, U. Merkt, Ch. Sikorski, K. Ploog, Phys. Rev. Lett. **58**, 2586 (1987); J. Alsmeier, Ch. Sikorski, U. Merkt, Phys. Rev. **B 37**, 4314 (1988)

[10] T.P. Smith, H. Arnot, J.M. Hong, C.M. Knoedler, S.E. Laux, H. Schmid, Phys. Rev. Lett. **59**, 2802 (1987); **61**, 585 (1988)

[11] T.J. Thornton, M. Pepper, H. Ahmed, D. Andrews, G.J. Davies, Phys. Rev. Lett. **56**, 1198 (1986)

[12] H.Z. Zheng, H.P. Wei, D.C. Tsui, G. Weimann, Phys. Rev. **B 34**, 5635 (1986)

[13] T. Fukui, H. Saito, Appl. Phys. Lett. **50**, 824 (1987); M. Tsuchiya, J.M. Gaines, R.H. Yan, R.J. Simes, P.O. Holtz, L.A. Coldren, P.M. Petroff, Phys. Rev. Lett. **62**, 466 (1989)

[14] see e.g. N.W. Ashcroft, N.D. Mermin, Solid State Physics, Holt, Rinehart and Winston, New York (1976)

[15] U. Merkt, Ch. Sikorski, Proc. Int. Conf. Narrow Gap Semicond., Gaithersburg 1989, to be published

[16] D. Heitmann, Two-dimensional Systems: Physics and New Devices, G. Bauer, F. Kuchar, H. Heinrich, eds., Springer Series in Sol. St. Sc. **67**, 285, Springer, Berlin (1986)

[17] see e.g. W. Hansen in Festkörperprobleme/Advances in Sol. St. Phys. **28**, U. Rössler, ed., p. 121, Vieweg, Braunschweig (1988)

[18] J.P. Kotthaus, W. Hansen, H. Pohlmann, M. Wassermeier, K. Ploog, Surf. Sci. **196**, 600 (1988)

[19] F. Brinkop, W. Hansen, J.P. Kotthaus, K. Ploog, Phys. Rev. **B 37**, 6547 (1988)

[20] Ch. Sikorski, U. Merkt, Phys. Rev. Lett. **62**, 2164 (1989); Surf. Sci. **229**, 282 (1990)

[21] K.F. Berggren, T.J. Thornton, D.J. Newson, M. Pepper, Phys. Rev. Lett. **57**, 1769 (1986)

[22] H. van Houten, B.J. van Wees, J.E. Mooij, G. Rass, K.F. Berggren, Superlattices & Microstructures **3**, 497 (1987)

[23] J. Cibert, P.M. Petroff, G.J. Dolan, S.J. Pearton, A.C. Gossard, J.H. English, Appl. Phys. Lett. **49**, 1275 (1986)

[24] T. Ando, A.B. Fowler, F. Stern, Rev. Mod. Phys. **54**, 437 (1982)

[25] J. Lee, J. Appl. Phys. **54**, 5482 (1983)

[26] J.S. Weiner, G. Danau, A. Pinczuk, J. Valladares, L.N. Pfeiffer, K. West, Phys. Rev. Lett. **63**, 1641 (1989)

[27] M. Kohl, D. Heitmann, P. Grambow, K. Ploog, Phys. Rev. Lett. **63**, 2124 (1989)

[28] A. Ya. Shik, Sov. Phys. Semicond. **19**, 915 (1985)

[29] W.Y. Lai, S. Das Sarma, Phys. Rev. **B 33**, 8874 (1986); Proc. 18th Int. Conf. Physics Semiconductors, Stockholm 1986, O. Engström, ed., p. 509, World Scientific, Singapore (1987)

[30] S.E. Laux, F. Stern, Appl. Phys. Lett. **49**, 91 (1986)

[31] S.E. Laux, D.J. Frank, F. Stern, Surf. Sci. **196**, 101 (1988)

[32] This effect is used also to realize a quasi-3 D system of free electrons as suggested by B.I. Halperin, Jap. Journ. Appl. Phys. **26** Suppl. 26-3, 1913 (1987)

[33] J.H. Davies, Semicond. Sc. Technol. **3**, 995 (1988)

[34] U. Wulf, Phys. Rev. **B 35**, 9754 (1987); U. Wulf, R.R. Gerhardts, in [2], 162

[35] D.M. Wood, N.W. Ashcroft, Phys. Rev. **B 25**, 6255 (1982); E. Batke, D. Heitmann, Infrared Phys. **24**, 189 (1984)

[36] F. Stern, Phys. Rev. Lett. **18**, 546 (1967)

[37] J. Pillath, E. Batke, G. Weimann, W. Schlapp, Phys. Rev. **B 40**, 5879 (1989)

[38] D.A. Dahl, L.J. Sham, Phys. Rev. **B 16**, 651 (1977)

[39] W. Que, G. Kirczenow, Phys. Rev. **B 37**, 7153 (1988); **B 39**, 5998 (1989)

[40] A.V. Chaplik, Superlattices & Microstructures **6**, 329 (1989)

[41] C. Dahl, Phys. Rev. **B 41** (in print)

[42] D. Childers, D. Pincus, Phys. Rev. **177**, 1036 (1969)

[43] M. Reed, in [2], 64; Festkörperprobleme/Advances in Sol. St. Phys. **29**, U. Rössler, ed., p. 267, Vieweg, Braunschweig (1989)

[44] J. Alsmeier, E. Batke, J.P. Kotthaus, Proc. 4th Int. Conf. on Modulated Semicond. Structures, Ann. Arbor 1989, Surf. Sci. (in print); Surf. Sci.**229**, 287 (1990)

[45] W. Hansen, T.P. Smith III, K.Y. Lee, J.A. Brum, C.M. Knoedler, J.M. Hong, D.P. Kern, Phys. Rev. Lett. **62**, 2168 (1989)

[46] C. Cohen-Tannoudji, Quantum Mechanics Vol. 1, p. 727, Hermann, Paris (1977)

[47] S.J. Allen Jr., H.L. Störmer, J.C.M. Hwang, Phys. Rev. **B 28**, 4875 (1983)

[48] G.W. Bryant, Phys. Rev. Lett. **59**, 1140 (1987)

[49] W. Que, G. Kirczenow, Phys. Rev. **B 38**, 3614 (1988)

[50] W. Kohn, Phys. Rev. **123**, 1242 (1961)

[51] L. Brey, N.F. Johnson, B.I. Halperin, Phys. Rev. **B 40**, 10647 (1989)

[52] J. Alsmeier, E. Batke, J.P. Kotthaus, Phys. Rev. **B 40**, 12574 (1989)

CHAPTER 2

COHERENCE AND DEPHASING

CONDUCTANCE OSCILLATIONS AND PHASE COHERENCE IN SUBMICRON METAL FILMS

Chris Van Haesendonck, Hans Vloeberghs, Yvan Bruynseraede

Laboratorium voor Vaste Stof-Fysika en Magnetisme
Katholieke Universiteit Leuven
B-3030 Leuven, Belgium

1 Electronic Conduction in Disordered Metal Films

Neglecting band structure and correlation effects, the freely moving electrons in a disordered metal film can be described as plane waves $\Psi(\vec{r}) \propto \exp(i\,\vec{k}\cdot\vec{r}\,)$ with $|\vec{k}| = 2\pi/\lambda$ the wave vector. When we restrict ourselves to the low temperature limit, the contribution of the inelastic scattering (at other electrons or phonons) to the electrical conductivity can be neglected. The scattering at lattice defects and impurities will cause an elastic diffusion of the charge carriers. Due to the Pauli principle, only the electrons near the Fermi level contribute to the conductivity which is given by the Einstein relation

$$\sigma_{\mathrm{o}} = e^2 N(E_F)\mathcal{D} \,. \tag{1}$$

$N(E_F)$ represents the electronic density of states near the Fermi level and $\mathcal{D} = \frac{1}{3}v_F\ell_{el}$ is the diffusion constant with v_F the intrinsic electron (Fermi) velocity and ℓ_{el} the elastic mean free path. The diffusion approach will be valid only for the electronic transport in disordered metals and on a length scale L much larger than the mean free path ℓ_{el} (*diffusive regime*). On a length scale $L \ll \ell_{el}$, we enter the *ballistic regime*. The ballistic limit which can be studied in high-quality GaAs/GaAlAs heterostructures, is discussed in detail in other chapters of this volume.

Equation (1) results from a statistical interpretation of the conduction process, where the wave properties of the individual electrons are not taken into account. This approach will be valid for macroscopic samples where the thermodynamic limit (infinite number of electrons) can be applied. For submicron samples with a relatively small number of conduction electrons, one should also take into account interference processes between the diffusively scattered electron waves. The importance of the electron interference becomes obvious for the loop geometry shown in Fig. 1. This experiment is the solid-state analogue of the famous two-slit interference experiment. The four-terminal device is defined in a disordered metal film with thickness t $\sim$ 10 nm. The electrons diffuse elastically along the elastic scattering centers (black dots in Fig. 1) between two large reservoirs which are at chemical potentials μ_L and μ_R.

Quantum Coherence in Mesoscopic Systems
Edited by B. Kramer, Plenum Press, New York, 1991

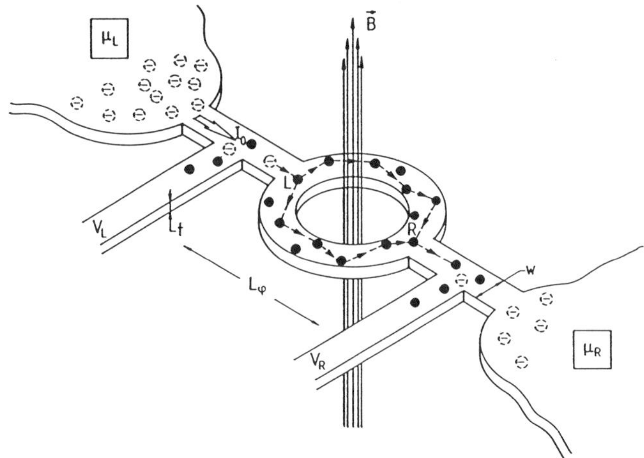

Figure 1. Four-terminal measurement of the solid-state Aharonov-Bohm magnetoconductance oscillations in a disordered metal loop. The interference at point R of the electron waves traveling along both arms of the loop, is tuned by the magnetic flux which is completely confined to the interior of the loop.

At point L, an electron wave splits into two partial waves which interfere again at point R. The quantum-mechanical probability P_d that the electron arrives at point R, is given by the square of the sum of the two scattering amplitudes $\Psi_u(R)$ and $\Psi_\ell(R)$ which describe the electron in the upper and lower part of the loop

$$
\begin{aligned}
P_d &= |\Psi_u + \Psi_\ell|^2 \\
&= |\Psi_u|^2 + |\Psi_\ell|^2 + \Psi_u^*\Psi_\ell + \Psi_\ell^*\Psi_u \\
&= 2|\Psi_u|^2(1 + \cos\theta) .
\end{aligned}
\tag{2}
$$

We have assumed that the probabilities to diffuse along the upper and the lower path are equal, $|\Psi_u|^2 = |\Psi_\ell|^2$. The phase difference θ is determined by the random phase shifts which occur during the elastic diffusion along the upper (lower) path.

The wave character of the conduction electrons has caused extra interference terms whose magnitude is determined by the phase factor θ. For a loop with a width $w \sim \lambda_F$ (λ_F is the Fermi wavelength) the electrons can only move in a single channel and the interference should be observable as giant oscillations of the loop conductance $G = I_o/(V_L - V_R)$. In real experiments, the metal loops will always have a width $w \gg \lambda_F$ and the electrons can move in $w/\lambda_F \gg 1$ channels. Moreover, the elastic scattering will introduce transitions between the different channels. The result eq. (2) has to be averaged over many different θ values and the interference will cause small deviations of the conductance G around the average value G_o which is determined by eq. (1)

$$
G = G_o + \delta G .
\tag{3}
$$

Intuitively, one expects that δG decreases when the number of channels and the amount of impurity scattering increases. Surprisingly, detailed calculations [1, 2] indicate that this stochastic self-averaging does not occur as long as the sample size L remains smaller than the *phase coherence length* L_φ. The diffusion length L_φ is connected to the phase breaking time τ_φ via the classical result $L_\varphi = \sqrt{\mathcal{D}\tau_\varphi}$. While the elastic diffusion does not destroy the phase memory of the electrons, inelastic scattering changes the electron wavelength and will cause a destruction of the interference processes. In section 4, we will show that spin-flip scattering at residual magnetic impurities will cause an additional reduction of the phase coherence. Experiments [3] indicate that due to this spin-flip scattering $L_\varphi \sim 1\,\mu$m below T = 1 K, where the inelastic scattering becomes vanishingly small. In a *mesoscopic sample*, i.e. a sample with size $L < L_\varphi$, the averaging of the interference processes will always be incomplete and δG in eq. (3) will be independent of the sample size L [2].

The exact value and the sign of δG depends on the position of the elastic scattering centers and will vary from sample to sample [1, 4]. When G is measured for many different samples, it is possible to define the root-mean-square (rms) amplitude ΔG for the conductance correction caused by the electron interference

$$(\Delta G)^2 \equiv \langle (\delta G)^2 \rangle \equiv \langle (G - G_{av})^2 \rangle \sim G_{un}{}^2 = (e^2/h)^2 \ . \tag{4}$$

The brackets $\langle \cdots \rangle$ refer to an average over many samples with a different impurity configuration. The universal conductance $G_{un} = e^2/h = 3.9\ 10^{-5}\ \Omega^{-1}$.

The result eq. (4) can be interpreted more generally by stating that the impurity averaging of the direct interference processes is no longer *stochastic* (or statistical) on a mesoscopic scale. On a macroscopic scale $L \gg L_\varphi$, the usual stochastic averaging process is restored. The fluctuations around the average will die out proportional to $N^{-1/2}$ where $N = L/L_\varphi$ is the number of phase-coherent areas within the sample, i.e. the number of independent *impurity ensembles*.

For a *macroscopic* metal film, eq. (2) will have to be averaged over an infinite number of paths with a different θ value. This implies that the ensemble average of the interference terms can be neglected when calculating the electronic conductivity. The *self-averaging* of the interference effects will not occur for the special class of scattering processes in Fig. 1 where the electron returns to its original position. In that case, the electron can e.g. start at point L and, after diffusing completely around the loop, return to the same point L. The partial waves which travel in the clockwise and counter-clockwise direction will always interfere constructively since the interference occurs between *time-reversed scattering sequences* and the phase factor $\theta \equiv 0$ for this *coherent backscattering* [3]. Equation (2) reduces to

$$P_{bs} = 2|\Psi_{cw}|^2 + 2|\Psi_{cc}|^2 = 4|\Psi_{cw}|^2 = 2P_{cl} \ . \tag{5}$$

$|\Psi_{cw}|^2\ (= |\Psi_{cc}|^2)$ is the probability for backscattering in the clockwise (counter-clockwise) direction. The quantum-mechanical probability P_{bs} for backscattering is therefore twice as large as the classical probability P_{cl} for backscattering. The coherent backscattering gives rise to the well-known *weak electron localization* in thin metal films [3]. Due to the rapid decrease of the inelastic scattering rate at low temperatures, more backscattering loops will contribute to the conduction process and cause an anomalous temperature dependence of the conductance G. Since the size of the metallic structures is always considerably larger than λ_F, the electronic structure will be three-dimensional. On the other hand, the effective dimensionality of the interference processes is determined by the sample size with respect to the phase coherence length L_φ. In this contribution, we will consider thin film structures which are either two- or one-dimensional. The two-dimensional (D = 2) limit occurs as soon as the film thickness $t \ll L_\varphi$. When moreover the film width $w \ll L_\varphi$, the sample will behave as one-dimensional (D = 1).

A direct verification of eqs. (3) and (4) would require the fabrication of many identical samples which only differ by their impurity configuration. Experimentally, this is not possible, even when the most advanced lithographic techniques are used. Fortunately, the redistribution of the impurities will occur in a natural way at low, but finite temperatures $(T \sim 1\,K)$. An interstitial atom or a vacancy may migrate through the sample towards the grain boundaries. On the other hand, an impurity may also switch in a reversible way between two equilibrium states which are separated by a tunneling barrier. Experiments have clearly indicated that these so-called two-level systems strongly influence the low temperature properties of disordered materials [5]. When only one scattering center moves at very low temperatures, a typical *telegraph noise* will be observed [6, 7] with conductance jumps δG comparable to $G_{un} = e^2/h$. This confirms that for a mesoscopic line structure, moving a single scattering center over a distance of the order of λ_F ($\sim$ interatomic distance) is sufficient to completely rewrite the interference pattern of the mesoscopic sample! At higher temperatures many defects will start to move and one observes a superposition of many incoherent telegraph traces. This results in a typical $1/f$ noise spectrum [8].

The noise experiments fail, however, to reveal the periodic conductance oscillations which are expected for the loop geometry in Fig. 1 when θ changes continuously (see eq. (2)). Fortunately, a redistribution of the impurity configuration or a change in the path length for the interfering electron waves can also be simulated by applying a perpendicular magnetic field to the loop [2]. As we will discuss in detail in the next section, the Aharonov-Bohm effect [9] will allow to continuously tune the phase difference θ between the interfering electron waves. Moreover, a large enough variation of the magnetic field will also completely scramble the phase differences in a mesoscopic sample and simulate a transition towards a different impurity ensemble. In contrast to the thermally activated motion of lattice defects, the Aharonov-Bohm effect allows to generate a reversible and nicely controllable modification of the impurity configuration.

2   The Solid-State Aharonov-Bohm Effect in Disordered Metal Loops

When a magnetic field $\vec{B} = \vec{\nabla} \times \vec{A}$ is applied, a classical approach indicates that the Lorentz force $\vec{F}_L = -e\vec{v} \times \vec{B}$, will bend the electron path. This bending is most easily observed via the low-field Hall effect. When the magnetic flux is completely confined to the interior of the loop (see Fig. 1), $\vec{F}_L = 0$. Within the quantum-mechanical approach, the magnetic field will still tune the phase φ of the electron wave function via the Aharonov-Bohm (AB) effect [9]

$$\Delta\varphi = \frac{e}{\hbar} \int \vec{A} \cdot d\vec{l}. \tag{6}$$

The phase shift $\Delta\varphi$ is simply given by the line integral of the vector potential $\vec{A}$ along the electron path. The AB effect is a direct consequence of the wave character of the electrons and does not have a classical analogue. Equation (6) has the remarkable implication that the wave function senses the magnetic field through the vector potential $\vec{A}$ even in regions where $\vec{B} = \vec{\nabla} \times \vec{A} = 0$ The AB effect is related to the fact that the phase of the wave function, which is calculated from the Schrödinger equation depends upon the specific choice for the vector potential and is not gauge invariant. On the other hand, the physically observable quantities only depend on the field derived from the potential and are gauge invariant. The AB modulation of the electron phase φ occurs already at fields which are much smaller than the fields required to observe the influence of the quantized Landau levels. Here, we will only concentrate on disordered metal films with an elastic mean free path $\ell_{el} \sim 10\,\mathrm{nm}$, i.e. comparable to the film thickness. In these films, a magnetic field $B > 100\,\mathrm{T}$ would be needed to reach the quantum regime where effects such as the Shubnikov-de Haas-van Alphen magnetoresistance oscillations or the quantized Hall effect may be observed.

When a perpendicular magnetic field pierces the inside of a loop as shown in Fig. 1, the field will cause a periodic modulation of the phase difference between the interfering electron waves. When applying the result eq. (6) between points L and R in Fig. 1, we obtain the extra phase difference $\Delta\varphi$ caused by the magnetic field $\vec{B}$

$$\Delta\varphi(\vec{B}) = \frac{e}{\hbar} \oint \vec{A} \cdot d\vec{l} = \frac{e}{\hbar} \int\int \vec{B} \cdot d\vec{S} = 2\pi \frac{\Phi_B}{\Phi_\circ} , \tag{7}$$

where the flux quantum $\Phi_\circ = h/e$. The phase difference $\Delta\varphi$ varies proportional to the flux Φ_B enclosed by the loop and is as expected a gauge-invariant quantity since it is directly related to the transmission probability of the loop. As already mentioned, the AB effect also occurs at very small magnetic fields, in contrast to the quantum effects which are caused by the formation of the Landau levels. For a mesoscopic loop with a diameter of $1\,\mu\mathrm{m}$, a flux quantum $\Phi_\circ$ corresponds to a field variation of only $0.05\,T$.

From relations eqs. (2) and (7) we calculate that for the direct interference process shown in Fig. 1, the probability P_d to arrive at point R is an oscillating function of the enclosed flux Φ_B

$$P_d(\Phi_B) = 2|\Psi_u|^2[1 + \cos(\Delta\varphi + \theta)] = 2|\Psi_u|^2 \left[1 + \cos\left(2\pi\frac{\Phi_B}{\Phi_\circ} + \theta \right) \right] . \tag{8}$$

We want to stress again that this AB modulation with flux period h/e of the transmission probability should occur even when the magnetic flux is completely confined to the interior of the loop and the electrons do not experience the Lorentz force $\vec{F}_L$ [10].

The tuning of the interference by a magnetic field will also occur for the coherent backscattering around the loop in Fig. 1. Combining relations eq. (5) and eq. (6), we find

$$P_{bs}(\Phi_B) = P_{cl}[1 + \cos(\Delta\varphi)] = P_{cl} \left[1 + \cos\left(2\pi\frac{\Phi_B}{\Phi_\circ/2} \right) \right] . \tag{9}$$

For the backscattering, the flux period $\Phi_\circ/2 = h/2e$ is twice as small as for the direct interference process.

An experimental test of the solid-state analogue of the AB effect, requires the fabrication of a mesoscopic loop with a radius smaller than $L_\varphi \sim 1\,\mu\mathrm{m}$. This implies that the magnetic field can never be confined to the interior of the loop and will also penetrate the arms of the loop. The experimental search for the AB oscillations in metal loops already started a few years before the absence of self-averaging in the mesoscopic regime (see eq. (4)) was proven rigorously. The first experiments were sparked by the theoretical work of Gefen, Imry, and Azbel [11] on the possible existence of a solid-state AB effect. These qualitative theoretical predictions were based on the pioneering work of Landauer [12], who first suggested to treat the disorder as a quantum-mechanical tunneling barrier for the conduction electrons. While isolated, disordered metal loops are able to carry persistent currents [13] similar to a superconducting current, a resistive behavior will appear for the loop in Fig. 1 because the electrons can loose their phase memory in the electrical leads and reservoirs which are connected to the mesoscopic loop. The dissipation and Ohmic heating are caused by the energy exchange between the electrons and the lattice in the leads or reservoirs.

The first experiments [14] failed to show periodic h/e oscillations. Instead, one observed aperiodic (but reproducible) magnetoconductance fluctuations which rapidly grew with decreasing temperature. As first pointed out by Stone [15], the aperiodic oscillations result from the magnetic field penetrating the arms of the loop. Direct interference processes occurring within the arms of the loop will produce sample specific oscillations with different flux periods depending on the enclosed area. For a loop with a poor aspect ratio, i.e. the width w is not much smaller than the loop radius r, the flux Φ_B through the metal lines forming the loop is comparable to the flux enclosed by the partial waves which produce the h/e oscillations. The h/e oscillations will therefore be hidden by the larger aperiodic background, since both effects occur on a comparable characteristic field scale.

In 1984, Webb et al. [16] succeeded in fabricating a Au loop having a very good aspect ratio ($r/w > 10$). Because of the good aspect ratio, the periodic AB h/e oscillations could be easily distinguished from the aperiodic fluctuations which appeared as a slowly varying background. Afterwards, the solid-state AB effect has been confirmed by other experiments in metallic [17, 18] as well as in doped semiconductor [19] structures. Here, we present some typical results which have been obtained for Ag loops at a temperature of 0.3 K [18]. The square loops are prepared by a combination of standard electron-beam lithography and liftoff techniques. Figure 2a shows a compiled electron micrograph for a loop with side $S \simeq 0.95~\mu$m and linewidth $w \simeq 0.075~\mu$m. The low-field magnetoresistance for the loop is presented in Fig. 2b. Near zero field, the dominant flux period for the oscillations is $\Phi_0 = h/2e$. The $h/2e$ oscillations which are caused by the coherent backscattering (see eq. (5)), are rapidly damped at higher fields and are replaced by the AB oscillations with flux period $\Phi_0 = h/e$ (see eq. (2)). In general, the magnetic field dependence of the low-temperature conductance (T < 1 K) can be written as

$$
\begin{aligned}
G = G_0 \; &+ \; A_1(\Phi_B)G_{un} \cos\left[2\pi\frac{\Phi_B}{\Phi_0} + \alpha_1(\Phi_B)\right] \exp\left(\frac{-2S}{L_\varphi}\right) \\
&+ \; A_2(\Phi_B)G_{un} \cos\left[2\pi\frac{\Phi_B}{\Phi_0/2} + \alpha_2(\Phi_B)\right] \exp\left(\frac{-4S}{L_\varphi}\right) \\
&+ \; "higher~harmonics".
\end{aligned}
\tag{10}
$$

The fluctuations around the average conductance G_0 (calculated from the Einstein relation eq. (1)) contain the periodic oscillations with flux period h/e (second term) and flux period $h/2e$ (third term). A_1 and A_2 are slowly varying functions of the magnetic field and have an rms amplitude which is of order unity. α_1 and α_2 are arbitrary phase factors.

It is important to note that the $h/2e$ term contains the coherent backscattering processes as well as the second harmonic AB oscillations which are caused by interference processes that do not obey the time-reversal symmetry. For the latter processes, one can e.g. consider in Fig. 1 a partial wave which travels directly between points L and R,

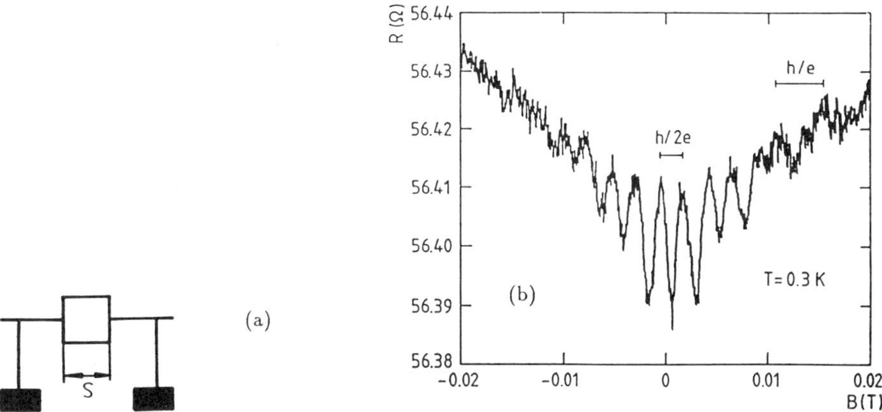

Figure 2. Experimental observation of the solid-state Aharonov-Bohm effect in a mesoscopic Ag loop: (a) shows a compiled electron micrograph of the square Ag loop with linewidth $w \simeq 0.075\ \mu$m and side S $\simeq 0.95\ \mu$m, (b) shows the magnetoresistance trace near zero field.

while the other partial wave travels first from point L to point R, moves back to point L, and finally reaches point R. Experimentally, the time-reversed processes can be easily distinguished by the fact that they do not survive in high magnetic fields (see also section 3). For structures with a finite linewidth w, the magnetic field will cause random phase shifts for paths enclosing a different area. Consequently, the contribution of the backscattering processes (see eq. (5)) will be averaged towards zero at higher magnetic fields. The destruction of the coherent backscattering can be clearly observed in Fig. 2b.

At higher magnetic fields, only the fluctuations caused by processes that do not obey the time-reversal symmetry, will survive. A typical high-field magnetoresistance trace is shown by the full curve in Fig. 3a. As indicated earlier, the h/e oscillations are superimposed upon slowly varying aperiodic fluctuations which are caused by the magnetic field piercing the arms of the loop. The high-field fluctuations can be studied conveniently by classical Fourier analysis. In the Fourier spectrum (see Fig. 3b) of the experimental magnetoresistance, the h/e peak is well separated from the peak near the origin which is caused by the long-wavelength aperiodic fluctuations. The position of the long-wavelength peak is consistent with the fact that the aperiodic fluctuations appear as soon as a magnetic flux comparable to the flux quantum Φ_0 is contained within a phase coherent area wL_φ [2, 15]. For the Ag loop shown in Fig. 2a with $L_\varphi \simeq 2.0\,\mu$m (the methods to determine L_φ are described in detail in section 4), the minimum wavelength should be of the order of 0.1 T, in agreement with the experimental results.

The periodic oscillations given by eq. (10) can be separated from the aperiodic background by digital filtering. When only the region in the vicinity of the h/e peak is retained (dashed curve in Fig. 3b), the inverse Fourier transform produces the beating h/e signal shown by the dashed curve in Fig. 3a. This modulation of the h/e amplitude is caused by the beating between the h/e oscillations and the aperiodic background and corresponds to the magnetic field dependence of the constants A_1 and A_2 in eq. (10). From the dashed curve in Fig. 3a, we find that the h/e oscillations have an rms amplitude $\Delta G \simeq 0.2\ G_{un}$. On the other hand, the higher harmonic $h/2e$ oscillations are much smaller, confirming the exponential damping given by eq. (10).

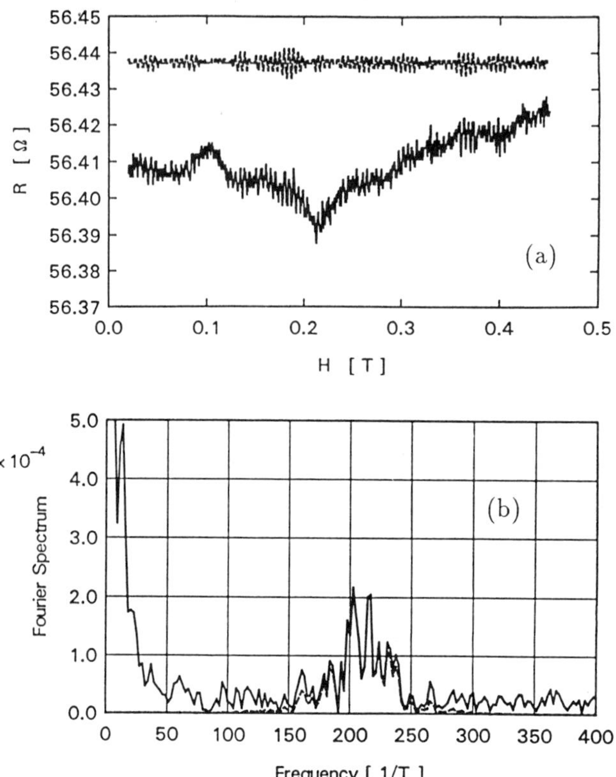

Figure 3. (a) Magnetoresistance trace at higher fields for the Ag loop which is also shown in Fig. 2, before (full curve) and after (dashed curve) removal of the aperiodic background by digital filtering. (b) Fourier transform of the data shown in (a). The dashed curve has been used to calculate the beating h/e signal shown by the dashed curve in (a).

The fact that our experimental result for ΔG is smaller than G_{un} is partially due to the exponential decay which occurs because $2S = 1.9$ μm $\simeq L_\varphi = 2.0\mu$m. On the other hand, the finite temperature (T = 0.3 K) at which the experimental results have been obtained, will cause an additional reduction of the h/e oscillation amplitude. Up to now, our theoretical arguments to describe the solid-state AB effect have been restricted to the limit of very low temperatures, where the thermal energy is much smaller than any characteristic energy scale entering the problem. The destruction of the interference processes will be governed by the energy scale [20]

$$E_c \sim \frac{h}{\tau_\varphi} = \frac{h\mathcal{D}}{L_\varphi^2} \; . \tag{11}$$

E_c is the energy needed to produce uncorrelated interference patterns, i.e the electron energy changes sufficiently to induce a phase shift of the order of 2π for electrons moving around an area with size L_φ. As soon as $k_B T > E_c$, the thermal smearing of the electrons near the Fermi level will introduce an additional *energy averaging* of the interference effects. The finite temperature will cause the mixing of $k_B T/E_c$ uncorrelated interference patterns in the mesoscopic sample. Due to the stochastic averaging of these patterns, the amplitude of the AB effect (constants A_1 and A_2 in eq. (10)) will be reduced by a factor $(k_B T/E_c)^{1/2}$ [2]. For the Ag loop shown in Fig. 2a, E_c corresponds to a temperature of $0.05\,\mathrm{K}$, implying that the interference effects shown in Fig. 2 and Fig. 3 have been reduced approximately by a factor of 3 due to the thermal averaging. More detailed experiments have confirmed that in metal loops the amplitude of both the periodic and the aperiodic AB effect grows $\propto T^{-1/2}$ at low temperatures where L_φ has saturated towards a constant value [21].

We would like to stress that the energy averaging occurs independently of the stochastic self-averaging in samples with a size $L > L_\varphi$ (see section 1). This has been demonstrated experimentally [18] for finite arrays of loops in series. Since the loops can be considered as independent mesoscopic units, the h/e oscillation amplitude decays as expected proportional to $N^{-1/2}$, where N is the number of loops in series.

When the loop geometry is replaced by a simple fine line geometry only the aperiodic fluctuations are present. Experimentally the presence of such aperiodic fluctuations has been confirmed for mesoscopic samples defined in thin metal films [21] as well as in the two-dimensional electron gas of semiconductor devices [22]. Since their rms amplitude is independent of the material structure as well as of the amount of disorder, one usually refers to these magnetoconductance fluctuations as *Universal Conductance Fluctuations* (UCF). The detailed structure of the magnetoconductance depends upon the position of the individual defects [1, 4]. The magnetoconductance traces can therefore be viewed as *magnetofingerprints* of the mesoscopic sample [2]. In section 4, we will discuss in more detail the properties of these aperiodic fluctuations.

Although the loop experiment confirms the importance of the direct interference between partial conduction electron waves in the mesoscopic regime, the presence of a pure AB effect [10] can not be proven unambiguously. Since the magnetic flux is also piercing the arms of the loop, the $\vec{B}$ field interacts with the electrons via the Lorentz force. The clear separation between the h/e signal and the aperiodic background in the Fourier transform does however strongly suggest that the electrons experience the presence of the magnetic flux in the interior of the loop via the AB effect. The bending of the electron paths by the Lorentz force is randomized by the diffuse elastic scattering at the film edges. This randomization prevents the electrons from moving preferentially near the inner edge of the loops, resulting in a constant magnetic field period for the oscillations at higher fields. This argument remains valid as long as the cyclotron frequency ω_c is considerably smaller than the elastic scattering rate τ_{el}^{-1}, i.e. as long as we can neglect the presence of the Landau orbitals. For disordered Au loops with $\ell_{el} \sim 10\,\mathrm{nm}$, periodic h/e oscillations can be observed in magnetic fields up to $15\,\mathrm{T}$ [23].

3 Flux Periodic Oscillations in Cylinders and Weak Electron Localization

As mentioned in section 1, interference processes in the backscattering direction survive even on a macroscopic scale and cause an anomalous temperature dependence of the conductivity. For the loop geometry shown in Fig. 1, these interference processes give rise to oscillations with flux period equal to the superconducting flux quantum $\Phi_o = h/2e$.

In 1981, Altshuler, Aronov, and Spivak (AAS) [24], predicted that the $h/2e$ oscillations also survive in very long (length $H \gg L_\varphi$) thin-walled metal cylinders, provided

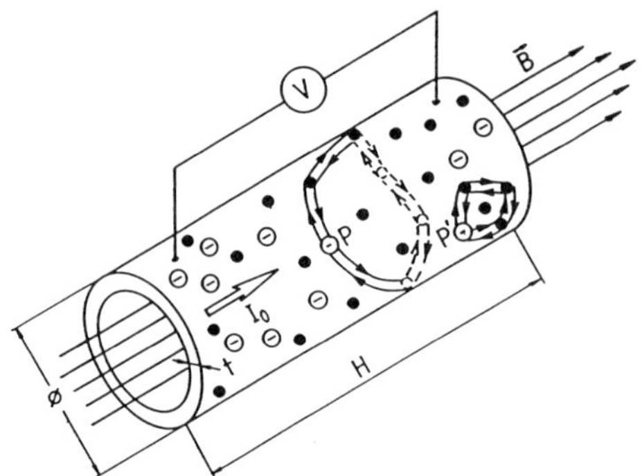

Figure 4. Schematic view of the interference processes causing the $h/2e$ magnetoresistance oscillations in a thin-walled metal cylinder.

the cylinder diameter $\phi \sim L_\varphi$ (see Fig. 4). When such a cylinder is placed in an axial magnetic field, all the electron waves that start e.g. at point P and are scattered elastically around the cylinder, will enclose the same magnetic flux $\Phi_B = \pi\phi^2 B/4$ and contribute to the $h/2e$ oscillations. On the other hand, electron waves starting e.g. at point P' which do not travel completely around the cylinder, enclose zero magnetic flux and do not contribute to the magnetoconductance. When the resistance is measured between the top and the bottom of the cylinder, the amplitude ΔR of the AAS $h/2e$ oscillations is given by [24]

$$\frac{\Delta R_\square}{R_\square^2} \simeq G_{un} K_o \left(\frac{\pi\phi}{L_\varphi}\right) . \tag{12}$$

$R_\square$ is the resistance per square of the metal cylinder wall and $K_o(x)$ is the McDonald function. When the cylinder circumference $\pi\,\phi$ becomes larger than the phase coherence length L_φ, the McDonald function results in an exponential damping of the oscillations. Apparently, the amplitude of the $h/2e$ resistance oscillations can be enhanced considerably by increasing the disorder so that the resistance per square $R_\square$ becomes comparable to G_{un}^{-1}. Unfortunately, the inelastic diffusion length L_{in} varies inversely proportional to $R_\square$ in disordered metal films [25]. The exponential damping caused by the decrease of the phase coherence length L_φ will be much stronger than the enhancement of the oscillation amplitude caused by a larger $R_\square$ value.

Experimentally [26, 27] thin-walled metal cylinders with diameter $\phi \sim 1\,\mu$m can be fabricated by evaporating a 20 nm thick metal film onto a quartz fiber which is stretched over a hole (diameter ~ 1 cm) in a glass substrate. During the metal evaporation, the substrate is rotated in a reduced helium atmosphere to ensure a uniform thickness of the cylinder wall. Figure 5 represents the low-field magnetoresistance trace (field parallel to the fiber axis) for three Mg cylinders with different diameter ϕ at T = 1.5 K. Our experimental results, obtained at the University of Leuven in 1984 [27], clearly confirm the existence of the $h/2e$ oscillations which were observed for the first time by Sharvin and Sharvin in 1981 [26]. The exponential damping of the oscillation amplitude occurring for larger cylinder diameters ($\phi > L_\varphi$) is clearly confirmed. As shown by the full curves in Fig. 5, the $h/2e$ oscillations can nicely be described by the AAS theory when we assume a phase coherence length $L_\varphi \simeq 2.0\,\mu$m.

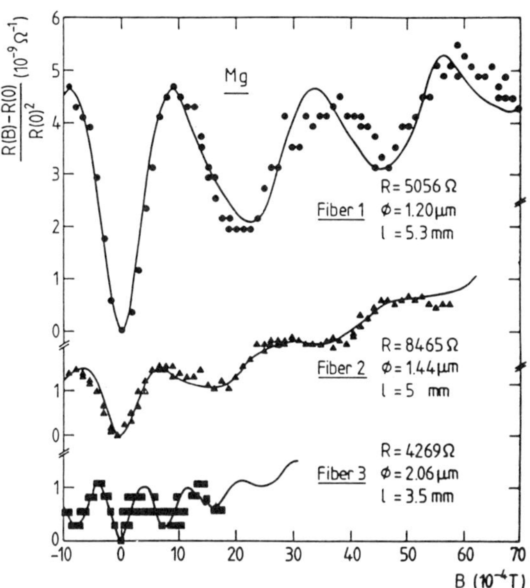

Figure 5. Normalized low-field magnetoresistance at T = 1.5 K for three thin-walled Mg cylinders with a different diameter ϕ. The full curves have been calculated using the weak electron localization theory [27].

The reduction of the $h/2e$ oscillation amplitude at higher magnetic fields is mainly due to the magnetic field penetrating the cylinder walls. The result eq. (12) is valid only for cylinders with wall thickness $t \to 0$. For a finite wall thickness, a high enough magnetic field causes different phase shifts for electron waves diffusing along the inside and the outside of the cylinder wall. The $h/2e$ oscillations will be destroyed as soon as a superconducting flux quantum $h/2e$ fits into the area $\pi \phi t$ (B should not exceed 0.03 T for the cylinders discussed in Fig. 5). The background magnetoresistance appearing at higher fields is largely caused by the fact that the magnetic field cannot be perfectly aligned with the cylinder axis. This implies that the backscattering loop starting at point P' in Fig. 4 will start to contribute to the low-field magnetoresistance.

In a thin metal film where the loop structure is not present, many different areas will be enclosed by the backscattered electron waves. The $h/2e$ oscillations are therefore washed out and are replaced by a uniform Weak Electron Localization (WEL) magnetoresistance [3]. Since the correction caused by the interference processes strongly depends upon the L_φ value, this uniform magnetoresistance rapidly grows at low temperatures. This is illustrated in Fig. 6, where we show the normalized magnetoresistance for a pure Au film with thickness $t = 20.0\,\mathrm{nm}$ and sheet resistance $R_\square = 2.1\ \Omega/\square$. The width of the curves varies inversely proportional to L_φ^2. This can be easily understood by estimating the characteristic field B_φ which is needed to dephase the electron waves traveling in clockwise and counter-clockwise direction by an amount 2π

$$B_\varphi \sim \frac{h/2e}{L_\varphi^2}\;. \tag{13}$$

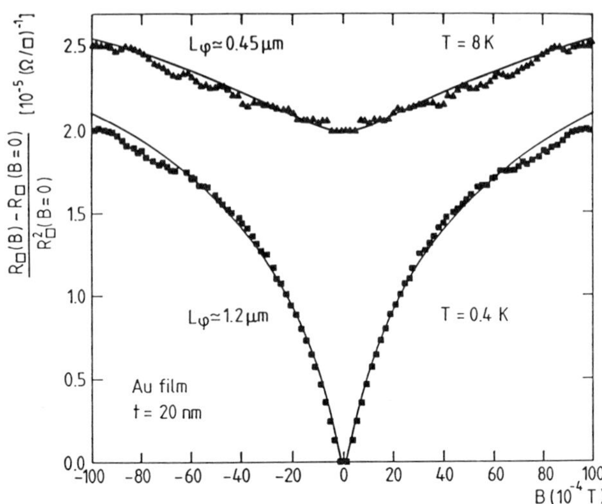

Figure 6. Normalized magnetoresistance near zero field at two different temperatures for a thin Au film. The full curves correspond to the theoretical result [28].

One expects that the coherent backscattering will be influenced by the magnetic field as soon as the field strength exceeds the characteristic field B_φ. More detailed calculations show that for $B \ll B_\varphi$, the magnetoresistance will increase proportional to B^2, independent of the sample dimensionality D. For $B \gg B_\varphi$, the magnetoresistance will vary proportional to $\ln B$ for thin films with a thickness $t \ll L_\varphi$ (D = 2). The full curves in Fig. 6 correspond to the theoretical result [28] with $B_\varphi = 8.0\ 10^{-4}\,\mathrm{T}$ at T = 8.0 K and $B_\varphi = 1.2\ 10^{-4}\,\mathrm{T}$ at T = 0.4 K. This implies that the phase coherence length grows from 0.45 μm to 1.2 μm when the temperature decreases from 8.0 K to 0.4 K.

Although the experimental results in Fig. 6 confirm the B^2 variation at low fields as well as the $\ln B$ behavior at higher fields, the magnetoresistance is positive while one would expect a negative magnetoresistance, since the magnetic field destroys the constructive interference in the backscattering direction (see eq. (5)). When calculating the influence of the electron interference on the electronic conduction, we have, however, not taken into account any effect of the electron spin. Since scattering of the electron spin will give rise to additional phase shifts of the electronic wave function, the spin scattering influences the direct interference (see eqs. (2) and (8)) as well as the coherent backscattering (see eqs. (5) and (9)). As will be discussed in detail in section 4, *spin-flip scattering* at residual magnetic impurities has a similar effect as the inelastic scattering and destroys the interference effects because of a smaller phase coherence length L_φ. On the other hand, the *spin-orbit scattering* occurring during each elastic scattering event, has the remarkable property that, on average, it will change the sign of the interference terms. In the presence of strong spin-orbit scattering, the backscattering probability (see eq. (5)) will be reduced by a factor of 2 with respect to the classical backscattering probability, $P_{bs} = 0.5P_{cl}$ [3]. The coherent backscattering will cause a reduction of the resistance which becomes more pronounced with decreasing temperature (*weak anti-localization*). Although the characteristic field scale for the anomalous magnetoresistance will still be given by eq. (13), the magnetoresistance will be positive and the amplitude will be twice as small as for the case of weak spin-orbit scattering. The relative strength of the spin-orbit scattering is determined by the ratio [29]

$$\frac{\tau_{el}}{\tau_{so}} = (\alpha Z)^4 \, , \tag{14}$$

where $\alpha = 1/137$ is the fine structure constant and Z is the atomic number. The spin-orbit scattering will be much more important in the heavier metals with a high Z value. For Au ($Z = 79$), the spin-orbit scattering rate τ_{so}^{-1} will be much larger than the dephasing rate τ_{φ}^{-1}, implying that we will always observe a positive magnetoresistance.

The spin-orbit scattering will also influence the direct interference processes. A strong spin-orbit scattering will give rise to a reduction of the rms fluctuation amplitude by a factor of 2 [30]. Due to the random phase factor θ (see eq. (2)), the average sign of the interference terms is not a relevant quantity.

Numerous experiments on disordered thin metal films have confirmed that an analysis of the low field WEL magnetoresistance is a unique and extremely powerful tool to study characteristic scattering processes: the inelastic scattering due to the electron-electron or the electron-phonon interaction, the spin-flip scattering in dilute magnetic alloys, the enhancement of the spin-orbit scattering at a disordered film surface, etc. [3].

4 Influence of Magnetic Spin-Flip Scattering on the Phase Coherence Length

The presence of the AB h/e oscillations in mesoscopic loops and the AAS $h/2e$ oscillations in cylinders clearly demonstrate the wave character of the conduction electrons. Since the interference patterns extend over a length L_φ, this inevitably implies the non-local nature of the conductance oscillations [31]. For a *four-terminal measurement*, the interference extends into the voltage and current probes which are also disordered. While the average conductance G_o only depends upon the properties of the material between the voltage probes, the conductance fluctuations δG are influenced by all electron paths within a distance L_φ of the voltage probes. The amplitude of the AB effects will strongly depend upon the distance separating the metal loop from the reservoirs with chemical potential μ_L and μ_R (see Fig. 1). When both the voltage and current probes disappear into the large contact paths, the geometry shown in Fig. 1 corresponds to a *two-terminal measurement*, for which non-local effects can be neglected. As expected, also the WEL corrections which appear at low magnetic fields, are strongly influenced by the details of the probe geometry [32].

Umbach et al. [33] have provided a very graphic and easy to understand demonstration of the non-local electronic conduction in mesoscopic samples. As shown by the compiled electron micrograph in Fig. 7a, two identical mesoscopic Au lines (length $2\,\mu$m and linewidth $0.062\,\mu$m) have been prepared by contamination lithography in a transmission electron microscope. A loop with a diameter of $0.7\,\mu$m has been connected to one of the lines near the junction of a pair of current and voltage leads. Figures 7b and 7c show a typical magnetoconductance trace for the two structures. The single line produces only aperiodic fluctuations which are caused by the penetration of the magnetic field into the line segments. On the other hand, the line with the externally connected loop shows additional oscillations with a flux period corresponding to h/e for the loop.

The four-terminal geometry is also responsible for the *asymmetry* of the magnetoconductance ($G(B) \neq G(-B)$) which is observed for mesoscopic samples (see Fig. 2)

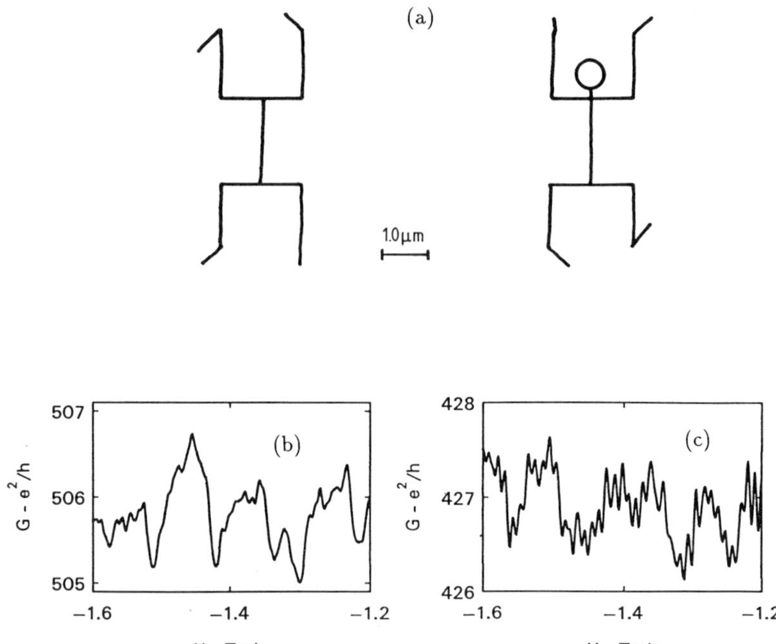

Figure 7. Four-terminal magnetoconductance measurement for a mesoscopic Au line with and without a loop attached to the outside of the contact probes. While the line without the loop only shows aperiodic fluctuations, the line with the loop clearly shows additional periodic oscillations with flux period h/e [33].

[23]. The measured voltage will include an anti-symmetric mesoscopic Hall voltage, even when the voltage probes are not on opposite sides of the fine line [34]. Since both the anti-symmetric Hall fluctuations and the symmetric longitudinal fluctuations have a comparable rms amplitude $\Delta G \sim G_{un}$, a mixture of the two effects will cause an apparent violation of Onsager's principle in mesoscopic samples. The asymmetry is also related to the random phase factors α_1 and α_2 appearing in eq. (10).

In section 3, we have indicated that measuring the WEL magnetoresistance in thin metal films allows a quantitative determination of the phase coherence length L_φ (see discussion of Fig. 5). In Fig. 8, the square symbols represent the experimentally observed temperature dependence of the scattering rate $\tau_\varphi^{-1} = 4e\mathcal{D}B_\varphi/\hbar$ for this pure Au film with thickness $t = 20\,\mathrm{nm}$ and sheet resistance $R_\square = 2.1\ \Omega/\square$. The full curve is given by the empirical formula

$$\tau_\varphi^{-1}(T) = A_\mathrm{o} + A_1 T + A_3 T^3 \ . \tag{15}$$

While the cubic term corresponds to the classical electron-phonon scattering which is independent of the disorder, the linear term describes the disorder enhanced electron-electron scattering for D = 2. As predicted by theory [25], the coefficient A_1 varies proportional to the sheet resistance $R_\square$. The constant rate A_o is probably caused by the spin-flip scattering at residual magnetic impurities (concentration < 1 ppm) in the Au film.

The influence of the magnetic scattering on the phase coherence can be determined quantitatively by doping the Au films with a well defined concentration of magnetic

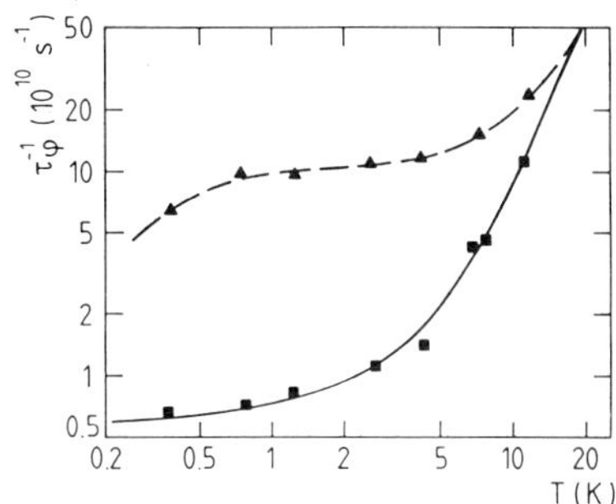

Figure 8. Temperature dependence of the phase breaking rate τ_φ^{-1} for a pure Au film (squares) and a similar Au film doped with 10 ppm of Fe impurities (triangles). The full curve has been calculated using eq. (15). The dashed line is only a guide to the eye.

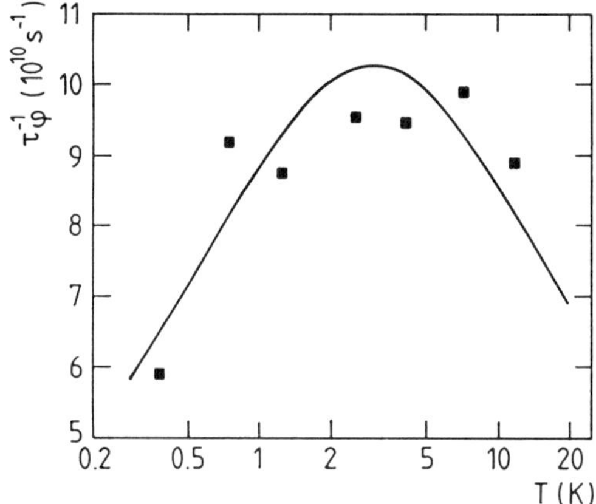

Figure 9. Temperature dependence of the phase breaking rate τ_φ^{-1} caused by the spin-flip scattering at the Fe Kondo impurities in the AuFe film which is also shown in Fig. 8. The full curve corresponds to the Suhl-Nagaoka theoretical result [35].

impurities. We have prepared thin films of a AuFe Kondo alloy by flash evaporation of small pieces of a master alloy from a resistively heated Mo source onto liquid nitrogen cooled substrates. The Fe content of the films is checked by measuring the Kondo anomaly in the low-temperature resistivity and is always comparable to the concentration of the master alloy [35].

As expected, the magnetic doping broadens the low-field magnetoresistance curves (see Fig. 6) because of the larger B_φ values. The triangles in Fig. 8 correspond to the scattering rate τ_φ^{-1} which has been calculated for a AuFe film (thickness $t = 21$ nm and $R_\square = 3.7\ \Omega/\square$) with an Fe concentration of 10 ppm. Subtracting the τ_φ^{-1} values for the pure Au film from the τ_φ^{-1} values for the doped film (taking into account the small difference in $R_\square$ which will influence the coefficient A_1 in eq. (15)), we obtain the dephasing rate caused by the magnetic impurities. The result is shown in Fig. 9. Although the temperature dependence of this dephasing rate is much weaker than the power-law variation of the inelastic scattering rate, the small variations can be reproduced consistently for different AuFe films. The full curve corresponds to the Suhl-Nagaoka formula [35] for the spin-flip scattering rate τ_{sf}^{-1}, assuming a Kondo temperature $T_K \simeq 3.0$ K. This Kondo temperature is somewhat larger than the usual value $T_K \simeq 0.8$ K which is assumed for bulk alloys [36]. The destruction of the coherent backscattering by the spin-flip scattering is very similar to the destruction of Cooper pairs in a superconductor which has been doped with magnetic impurities. In both cases, the violation of the time-reversal symmetry by the magnetic scattering destroys the correlation between electron states with wave vector $\vec{k}$ and $-\vec{k}$.

The results shown in Fig. 9 confirm the validity of the theoretical result for the dephasing of the WEL interference [28]

$$\tau_\varphi^{-1}(T) = \tau_{in}^{-1}(T) + 2\tau_{sf}^{-1}(T) \ . \tag{16}$$

Up to now we have implicitly assumed that the dephasing rate for the WEL interference and the direct interference in mesoscopic samples is the same. More detailed calculations show that the relevant dephasing rate for the AB magnetoconductance oscillations is given by [37]

$$\tau_\varphi^{-1}(T) = \tau_{in}^{-1}(T) + \tau_{sf}^{-1}(T) \ . \tag{17}$$

The discrepancy between the phase coherence length L_φ for the two effects should therefore be smaller than 40%.

Experimentally, it has become clear that even for the purest metal films, the presence of a very small amount of magnetic impurities or a paramagnetic oxide layer at the film surface [38] will limit the value $L_\varphi(T \to 0)$ to a few μm. Since the characteristic length scale for the h/e oscillations is twice as small as for the $h/2e$ oscillations (see eq. (10)), it is much easier to observe the h/e effect in magnetically contaminated metal loops. Moreover, the h/e effects will become stronger at higher fields ($B > 1\,T$), for which the spin-flip scattering process is inhibited by the Zeeman splitting of the electron energy levels. Since the WEL effects are completely destroyed by a high magnetic field, the increase of L_φ at higher fields is not relevant for these effects.

We have also studied the influence of the spin-flip scattering on the AB fluctuations. Therefore, we have prepared submicron samples with pure Au as well as with a AuFe Kondo alloy. The samples are prepared by combining standard electron-beam and liftoff techniques and have a width $w \simeq 0.125\mu$m and a length $L \simeq 4.0\mu$m $> L_\varphi$. Typical magnetoconductance curves, measured at T $= 0.4$ K, are shown in Fig. 10 for a pure Au sample as well as for a AuFe sample with a 10 ppm Fe concentration. These samples have been prepared in the same vacuum run as the two-dimensional samples discussed in Fig. 6, Fig. 8, and Fig. 9, and are expected to have a similar phase coherence length L_φ.

The sharp dip which is observed in Fig. 10 near zero field, is caused by the coherent backscattering and can be described by the WEL theory [32], provided the quasi one-dimensional sample structure is taken into account. As expected, this sharp dip is largely destroyed by the spin-flip scattering in the AuFe. The aperiodic fluctuations have a rms amplitude $\Delta G = 2.0 \ 10^{-6} \ \Omega^{-1} \simeq 0.05 \ G_{un}$ (in the field range $-2T < B < 2T$) for the pure Au sample. On the other hand, for the AuFe sample $\Delta G = 0.75 \ 10^{-6} \ \Omega^{-1} \simeq 0.02 \ G_{un}$. The stronger reduction of the rms amplitude in the AuFe can be explained by the fact that the diffusion constant $\mathcal{D}$ is about 2 times smaller than for the pure Au.

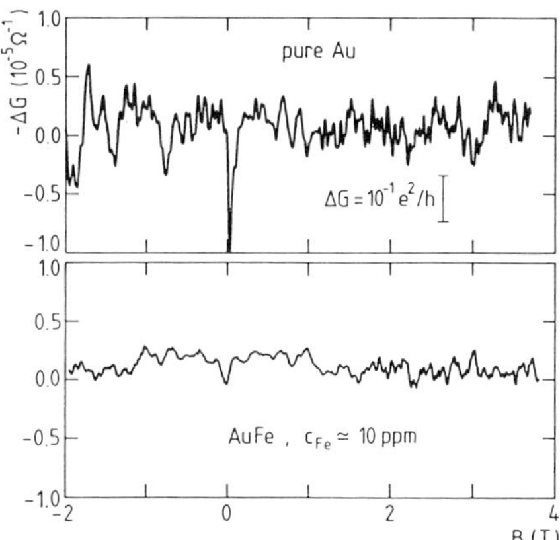

Figure 10. Fluctuations of the magnetoconductance at T = 0.4 K for a pure Au and AuFe (10 ppm of Fe impurities) submicron line (width $w \simeq 0.125\mu$m, length $L \simeq 4.0\mu$m).

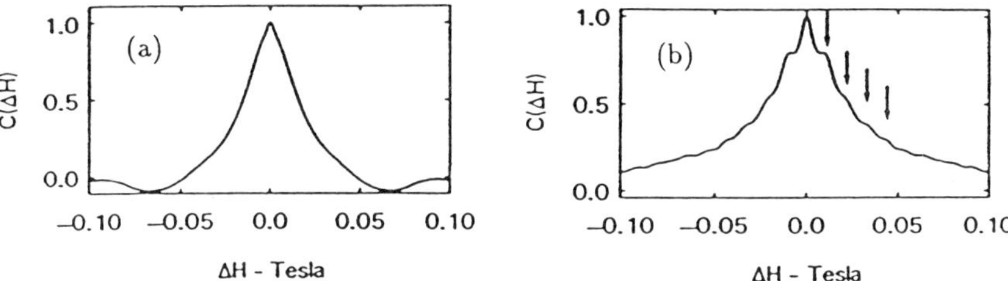

Figure 11. Autocorrelation function for the magnetoconductance traces shown in Fig. 7b (a) and Fig. 7c (b). The arrows indicate the nominal periodicity expected for the h/e oscillations [33].

Since the thermal length $L_T = [(\hbar \mathcal{D})/(k_B T)]^{1/2} < L_\varphi$ for both samples at T = 0.4 K, energy averaging dominates the self-averaging [39]. The influence of the smaller L_φ value for the AuFe [40] should become observable only for T < 0.05 K.

Alternatively the phase coherence length can also be linked to the minimum wavelength for the aperiodic fluctuations via the *critical field* $B_c = (A\Phi_o)/(wL_\varphi)$ with A a constant of order unity [2]. Physically, B_c corresponds to the field variation which is needed to produce uncorrelated interference patterns within a phase-coherent area. One therefore extends the *ergodic hypothesis* by assuming that changing the applied magnetic field by an amount B_c has the same effect as modifying the impurity ensemble. B_c can be obtained directly [41] from the width of the autocorrelation function $C(\Delta B) = \langle G(B)G(B + \Delta B)\rangle$, where $\langle \ldots \rangle$ corresponds to an average over many magnetic fields. Provided the ergodic hypothesis is valid, this average will be identical to an ensemble average. In Fig. 11a and Fig. 11b, we display the autocorrelation functions which have been calculated from the magnetoconductance traces shown in Fig. 7b and Fig. 7c. Neglecting the fine structure caused by the h/e oscillations (see arrows in Fig. 11b), the smooth variation of the autocorrelation function indicates the same nominal value of $B_c \simeq 0.018$ T. This corresponds to a phase coherent area 2.2 times larger than the distance between the voltage probes (see Fig. 7a), confirming the influence of the external loop on the magnetoconductance.

5   Conclusion

In this review, we have presented experimental results which clearly support the idea that interference between conduction electron waves can not be neglected in disordered materials. The interference effects extend over a distance comparable to the phase coherence length L_φ which for pure metals is of the order of a few μm at low temperatures where the inelastic diffusion length becomes very long. The interference between electrons traveling along time-reversed paths in the back-scattering direction, gives rise to the well-known weak electron localization phenomena which are also present in samples much larger than L_φ. The tuning of the coherent backscattering by a magnetic field (Aharonov-Bohm effect) gives rise to an anomalous low-field magnetoresistance from which the characteristic scattering times in thin metal films can be determined. For a cylinder with a diameter comparable to L_φ, the coherent backscattering causes low-field oscillations of the magnetoresistance with flux period $h/2e$.

When all of the sample dimensions are smaller than L_φ, we reach the mesoscopic limit, where the direct interference between splitted electron waves causes sample-specific variations of the conductance G. In the presence of a magnetic field, fluctuations of the magnetoconductance with a universal amplitude $\Delta G \sim e^2/h$ are observed. For a mesoscopic loop, the aperiodic fluctuations are replaced by oscillations with flux period h/e. This effect is the solid-state analogue of the famous two-slit interference experiment with electron waves.

In the mesoscopic regime, many other surprising electronic properties emerge. The results presented in Fig. 7 imply e.g. that a classical four-terminal measurement fails in the mesoscopic regime. Consequently, the classical scaling down approach to modelling small circuits will have to be modified to take into account possible interference effects. On the other hand, when the periodic AB oscillations can be made large enough, it is also conceivable that devices based on the quantum interference will become important in the future.

The experimental results clearly confirm that the elastic scattering at defects does not destroy the phase coherence. This is also the case for the spin-orbit scattering occurring during the elastic scattering events. On the other hand, the spin-flip scattering at magnetic impurities strongly enhances the dephasing caused by the inelastic scattering at phonons or at other electrons. The phase coherence length has a similar magnitude for the direct interference in the mesoscopic regime and the WEL effects in the macroscopic regime.

6   Acknowledgements

The authors would like to thank the Belgian F.G.W.O., I.I.K.W., G.O.A., and I.U.A.P. Research Programs as well as the Research Council of the Katholieke Universiteit Leuven for financial support. C.V.H. is a Research Associate and H.V. is a Research Assistant of the Belgian National Fund for Scientific Research (N.F.W.O.). The authors are also much indebted to M. Büttiker, Y. Gefen, M. Gijs, Y. Imry, R. Laibowitz, R. Landauer, P. Santhanam, C. Umbach, J. Vranken, S. Washburn, and R. Webb for a very fruitful collaboration.

References

[1] B.L. Altshuler, JETP Lett. **41**, 648 (1985); B.L. Altshuler and B.Z. Spivak, JETP Lett. **42**, 447 (1986)

[2] P.A. Lee and A.D. Stone, Phys. Rev. Lett. **55**, 1622 (1985); P.A. Lee, A.D. Stone, and H. Fukuyama, Phys. Rev. **B 35**, 1039 (1987)

[3] G. Bergmann, Phys. Rep. **107**, 1 (1984)

[4] S. Feng, P.A. Lee, and A.D. Stone, Phys. Rev. Lett. **56**, 1960 (1986); **56**, 2772(E) (1986)

[5] N.O. Birge, B. Golding, W.H. Haemmerle, H.S. Chen, and J.M. Parsey, Jr., Phys. Rev. **B 36**, 7685 (1987)

[6] K.S. Ralls, W.J. Skocpol, L.D. Jackel, R.E. Howard, L.A. Fetter, R.W. Epworth, and D.M. Tennant, Phys. Rev. Lett. **52**, 228 (1984)

[7] T.L. Meisenheimer and N. Giordano, Phys. Rev. **B 39**, 9929 (1989)

[8] N.O. Birge, B. Golding, and W.H. Haemmerle, Phys. Rev. Lett. **62**, 195 (1989)

[9] Y. Aharonov and D. Bohm, Phys. Rev. **115**, 485 (1959)

[10] A. Tonomura, N. Osakabe, T. Matsuda, T. Kawasaki, J. Endo, S. Yano and H. Yamada, Phys. Rev. Lett. **56**, 792 (1986)

[11] Y. Gefen, Y. Imry, and M.Ya. Azbel, Phys. Rev. Lett. **52**, 129 (1984)

[12] R. Landauer, IBM J. Res. Dev. **1**, 223 (1957)

[13] M. Büttiker, Y. Imry, and M.Ya. Azbel, Phys. Rev. **A 30**, 1982 (1984)

[14] C.P. Umbach, S. Washburn, R.B. Laibowitz, and R.A. Webb, Phys. Rev. **B 30**, 4048 (1984); G. Blonder, Bull. Am. Phys. Soc. **29**, 535 (1984)

[15] A.D. Stone, Phys. Rev. Lett. **54**, 2692 (1985)

[16] R.A. Webb, S. Washburn, C.P. Umbach, and R.B. Laibowitz, Phys. Rev. Lett. **54**, 2696 (1985)

[17] V. Chandrasekhar, M.J. Rooks, S. Wind, and D.E. Prober, Phys. Rev. Lett. **55**, 1610 (1985)

[18] C.P. Umbach, C. Van Haesendonck, R.B. Laibowitz, S. Washburn, and R.A. Webb, Phys. Rev. Lett. **56**, 386 (1986)

[19] G. Timp, A.M. Chang, J.E. Cunningham, T.Y. Chang, P. Mankiewich, R. Behringer, and R.E. Howard, Phys. Rev. Lett. **58**, 2814 (1987)

[20] A.D. Stone and Y. Imry, Phys. Rev. Lett. **56**, 189 (1986)

[21] S. Washburn, C.P. Umbach, R.B. Laibowitz, and R.A. Webb, Phys. Rev. **B 32**, 4789 (1985)

[22] W.J. Skocpol, P.M. Mankiewich, R.E. Howard, L.D. Jackel, D.M. Tennant, and A.D. Stone, Phys. Rev. Lett. **56**, 2865 (1986)

[23] A.D. Benoit, S. Washburn, C.P. Umbach, R.B. Laibowitz, and R.A. Webb, Phys. Rev. Lett. **57**, 1765 (1986)

[24] B.L. Altshuler, A.G. Aronov, and B.Z. Spivak, JETP Lett. **33**, 94 (1981)

[25] H. Fukuyama, in: Localization, Interaction, and Transport Phenomena, B. Kramer, G. Bergmann, and Y. Bruynseraede, eds., 51, Springer-Verlag, Berlin (1985)

[26] D.Yu. Sharvin and Yu.V. Sharvin, JETP Lett. **34**, 272 (1981)

[27] M. Gijs, C. Van Haesendonck, and Y. Bruynseraede, Phys. Rev. Lett. **52**, 2069 (1984)

[28] S. Hikami, A.I. Larkin, and Y. Nagaoka, Prog. Theor. Phys. **63**, 707 (1980)

[29] R. Meservey and P.M. Tedrow, Phys. Rev. Lett. **41**, 805 (1978)

[30] S. Feng, Phys. Rev. **B 39**, 8722 (1989)

[31] S. Maekawa, Y. Isawa, and H. Ebisawa, J. Phys. Soc. Jpn. **56**, 25 (1987); M. Büttiker, Phys. Rev. **B 35**, 4123 (1987); H.U. Baranger, A.D. Stone, and D.P. Di-Vincenzo, Phys. Rev. **B 37**, 6521 (1988)

[32] B. Douçot and R. Rammal, Phys. Rev. Lett. **55**, 1148 (1985); V. Chandrasekhar, D.E. Prober, and P. Santhanam, Phys. Rev. Lett. **61**, 2253 (1988); P. Santhanam, Phys. Rev. **B 39**, 2541 (1989)

[33] C.P. Umbach, P. Santhanam, C. Van Haesendonck, and R.A. Webb, Appl. Phys. Lett. **50**, 1289 (1987)

[34] M. Büttiker, Phys. Rev. Lett. **57**, 1761 (1986)

[35] C. Van Haesendonck, J. Vranken, and Y. Bruynseraede, Phys. Rev. Lett. **58**, 1968 (1987)

[36] R.P. Peters, G. Bergmann, and R.M. Mueller, Phys. Rev. Lett. **58**, 1964 (1987)

[37] V. Chandrasekhar, *Ph.D. Thesis*, Yale University (1989)

[38] J. Vranken, C. Van Haesendonck, and Y. Bruynseraede, Phys. Rev. **B 37**, 8502 (1988)

[39] J.C. Licini, D.J. Bishop, M.A. Kastner, and J. Melngailis, Phys. Rev. Lett. **55**, 2987 (1985); S.B. Kaplan and A. Hartstein, Phys. Rev. Lett. **56**, 2403 (1986); V.T. Petrashov, P. Reinders, and M. Springford, JETP Lett. **45**, 720 (1987)

[40] A.D. Stone, Phys. Rev. **B 39**, 10736 (1989)

[41] C.W.J. Beenakker and H. van Houten, Phys. Rev. **B 37**, 6544 (1988)

TWO-DIMENSIONAL CONDUCTIVITY OF THIN INHOMOGENEOUS GOLD FILMS

Günter Dumpich and Axel Carl

Experimentelle Tieftemperaturphysik
Universität Duisburg
D-4100 Duisburg, F. R. Germany

1 Introduction

Two-dimensional (2 D) conductivity behavior for thin metallic films has widely been analyzed and discussed in the last ten years [1]. In some cases, e.g. for homogeneous thin films consistency has been achieved determining the relevant electronic scattering length's from the temperature – as well as from the magnetic field dependence of the resistance at low temperatures [2]. However, it has been argued that for inhomogeneous thin films it is difficult to obtain the relevant electron scattering length's, since the diffusion constant is not well defined [3]. Moreover, if the inhomogeneity of the films becomes large, e.g. approaching the percolation threshold, it is proposed that the diffusion constant becomes length dependent [4]. To study these problems we investigated the structure of thin percolating gold films as well as their electrical properties. We find that the length dependence of the diffusion constant can be experienced by the temperature dependence of the phase coherence length.

2 Experimental

Gold films of thicknesses d between 6 nm and 10 nm are prepared by evaporating high purity gold (99.999%) simultaneously onto quartz- and NaCl-crystal substrates at room temperature in an UHV-System. To yield equal condensation conditions during deposition both of the substrates are precoated with carbon prior to the gold deposition. Gold films condensed onto (carbon coated) NaCl are removed and structurally investigated by transmission electron microscopy (TEM). For the corresponding gold films – condensed onto carbon coated quartz – the resistance per square $R_\square$ is measured between 1.5 K and 300 K, in situ as well as in a separate ^{4}He-cryostat where magnetic fields up to $B = 5$ T can be applied. The resistance characteristics are not changed when transferring the samples.

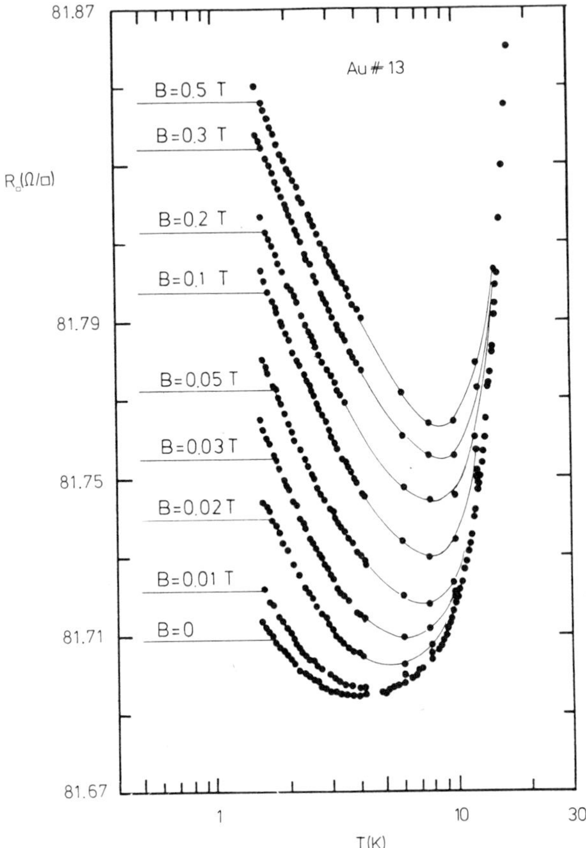

Figure 1. Resistance of a gold film (thickness $d = 7.1\,\text{nm}$, coverage $Z = 76\%$) versus $\log T$ for various magnetic Fields B, applied perpendicular to the film plane.

3   Results

Gold films with typical thicknesses below $10\,\text{nm}$ exhibit pronounced hole – and channel – structures the size of which are increasing when approaching the percolation threshold, which occurs at a coverage of about $Z_c \simeq 68\%$ and a mean thickness of $d_c \sim 6\,\text{nm}$. Complementary to the percolation length being the maximum length of clusters below the percolation threshold we find for the percolation-correlation length of the open channels above the threshold typical values between $300\,\text{nm}$ and $100\,\text{nm}$ for $75\% < Z < 85\%$. The resistance behavior of the films (for $Z > Z_c$) is metallic exhibiting a linear temperature dependence at "high" temperatures.

Figure 1 shows the resistance of a gold film ($d = 7.1\,\text{nm}$; $Z = 76\%$) versus $\log T$ at low temperatures, when various magnetic fields B are applied perpendicular to the film plane. As one can see from Fig. 1, one finds for the resistance per square below $T_{min} = 6\,\text{K}$ a logarithmic temperature dependence. With increasing magnetic field the resistance increases (positive MR) and the logarithmic "slope" saturates. This can be

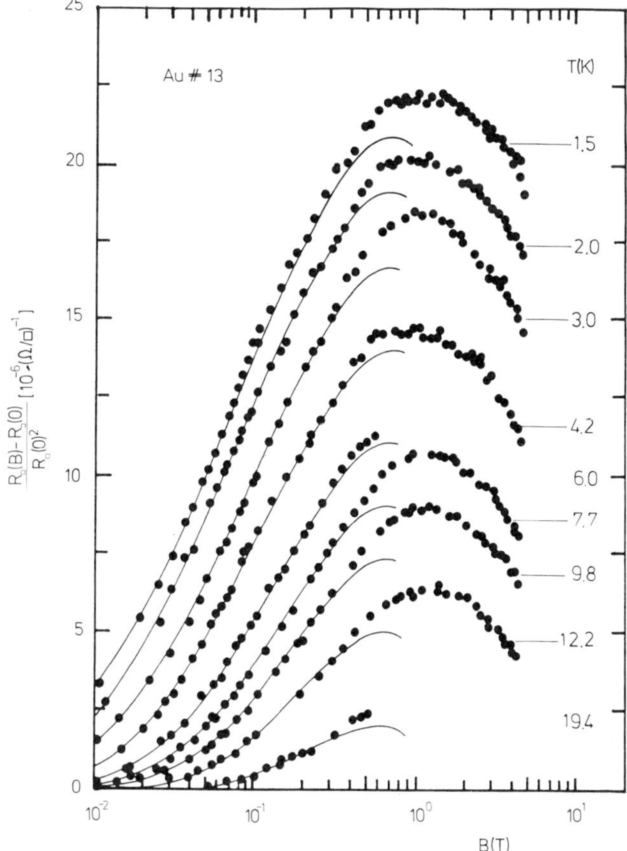

Figure 2. Magnetoresistance $(R_\square(B) - R_\square(0))/R_\square^2(0)$ versus $\log B$ for the gold film (Fig. 1) at various temperatures.

attributed to the fact that weak localization (WL) is (almost) destroyed when imposing "large" magnetic fields, whereas electron-electron interactions (EEI) are still present.

Figure 2 shows the MR-contribution as obtained from Fig. 1 (dots) for various temperatures between 1.5 K and 20 K versus $\log B$. The solid lines are calculated using the expressions given by S. Hikami et al. [5], taking into account the presence of strong spin-orbit scattering.

As one can see from Fig. 2, there is for $B < B_c = 0.5\,T$ good agreement between experimental data and the theoretical curves (lines). For $B > B_c$ the experimental data are systematically larger as compared to those expected from theory [5]. This has already been observed by A. Palevski and G. Deutscher [6] and is attributed to the percolating structure of the gold films. There are also other attempts to explain this behavior [7, 8]. For the present analysis, we emphasize that the observed additional contribution to the MR above $B = B_c$ is temperature *independent*. This is indicated by the dots in Fig. 3, which represent the difference between experimental and calculated MR-data for $B > B_c$ and $1.5\,K < T < 20\,K$. The solid line is only a guide to the eye.

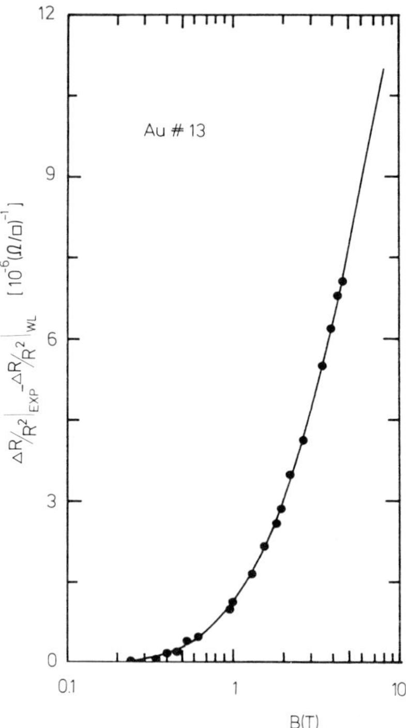

Figure 3. Experimentally obtained MR-data (Fig. 2) minus the calculated MR-contribution arising from WL [5] versus $\log B$ for various temperatures between 1.5 K and 20 K.

4   Discussion

Figure 4 shows the temperature dependence of the phase coherence length L_ϕ^{MR} as obtained by fitting the experimental data in Fig. 2 for $B < B_c$ with theoretical predictions [5]. The experimental data are given in a log-log plot indicating that L_ϕ^{MR} varies with $T^{-0.5}$ for $T < T_\xi = 3.3$ K; for $T > T_\xi$ with $L_\phi^{MR} \sim T^{-0.7}$ and for $T > 10$ K with $L_\phi^{MR} \sim T^{-1.05}$. For homogeneous gold films we find in adequate temperature regions L_ϕ^{MR} varying as $T^{-0.5}$, T^{-1} and $T^{-1.5}$ resp., as indicated by the solid line in Fig. 4. For $T = T_\xi$ a crossover occurs: exhibiting for $T < T_\xi$ a $L_\phi^{MR}(T)$-dependence that corresponds to the MR-behavior of homogeneous gold films, whereas for $T > T_\xi$ the $L_\phi^{MR}(T) = \mathcal{L}_\phi(T)$-behavior is "weaker". Since, we find by TEM-investigations the length's of "open" channels (ξ_p) between 200 nm and 250 nm which yields $L_\phi^{MR} \sim \xi_p$ or even $L_\phi^{MR} < \xi_p$ we attribute the "weakening" of the temperature dependence of $L_\phi^{MR}(T)$ for $T > T_\xi$ to the influence of the percolating structure on the phase coherence of the electrons. Following an approach of A.G. Aronov et al. [9] who introduced a length dependence of the diffusion constant for $L < \xi_p$, the "weaker" temperature dependence of $L_\phi^{MR}(T)$ is given by

$$\mathcal{L}_\phi(T) \sim T^{-P/(2+\tilde{t}-\tilde{\beta})} \qquad \text{for} \qquad \mathcal{L}_\phi < \xi \tag{1}$$

where $\tilde{t} - \tilde{\beta} = 0.87$ as given by the percolation theory for 2 D [10]. This yields with $p = 1, 2, 3$ (in distinct temperature regions, see Fig. 3), for $p = 1$: $L_\phi(T) \sim T^{-0.35}$, for

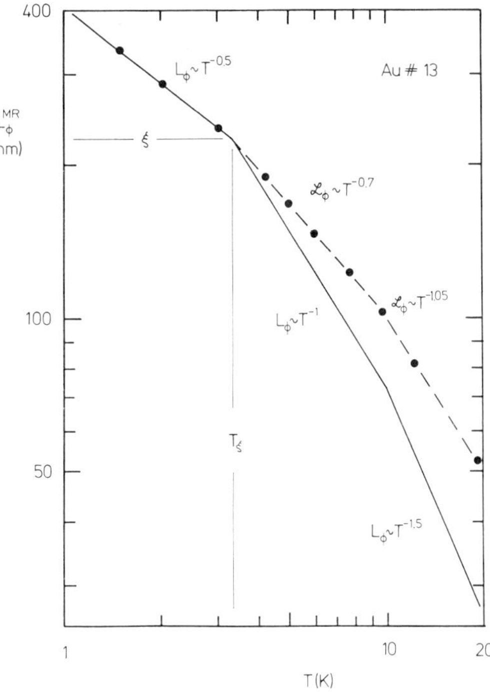

Figure 4. Temperature dependence of the phase coherence length L_ϕ^{MR}, as obtained from Fig. 2 in a log-log presentation (dots). The full lines indicate the $L_\phi(T)$-behavior for a homogeneous thin film.

$p = 2$: $L_\phi(T) \sim T^{-0.7}$ and $L_\phi(T) \sim T^{-1.05}$ for $p = 3$. As one can see from Fig. 4 this is in good agreement to the experimentally obtained temperature dependence of $L_\phi^{MR}(T)$ within $3.3\,\mathrm{K} < T < 10\,\mathrm{K}$ and $10\,\mathrm{K} < T < 20\,\mathrm{K}$, resp. From this we suggest that the weaker temperature dependence of $L_\phi^{MR}(T)$ results from the percolating structure of the gold films, if $L_\phi < \xi_p$. Below $T = 3.3\,\mathrm{K}$ (quasi)-homogeneous behavior with $L_\phi^{MR} \sim T^{-0.5}$ is obtained, since $L > \xi_p$. This yields for $T = T_\xi = 3.3\,\mathrm{K}$ with $L_\phi = \xi_p$ the percolation-correlation length ξ_p as indicated by the horizontal line $\xi_p = 220\,\mathrm{nm} = $ const in Fig. 3 which is in good agreement to $\xi_p \sim 250\,\mathrm{nm}$ as obtained by TEM.

5 Acknowledgements

The authors are grateful to E.F. Wassermann for his steady support of this work.

References

[1] Localization, Interaction and Transport Phenomena, edited by B. Kramer, G. Bergmann and Y. Bruynseraede, Springer Ser. in Sol.-St. Sc. **61** (1985); G. Bergmann, Phys. Rep. **107** (1984); P.A. Lee and T.V. Ramakrishnan, Rev. Mod. Phys. **57**, 287 (1985)

[2] M. Gijs, Ch. Van Haesendonck and Y. Bruynseraede, J. Phys. **F 16**, 1227 (1986)

[3] Ch. Van Haesendonck, M. Gijs and Y. Bruynseraede, in Ref. 1

[4] Y. Imry, Phys. Rev. Lett. **44**, 469 (1980)

[5] S. Hikami, A.L. Larkin and Y. Nagaoka, Prog. Theor. Phys. **63**, 707 (1980)

[6] A. Palevski and G. Deutscher, Phys. Rev. **B 34**, 431 (1986)

[7] T. Kawaguti and Y. Fujimori, J. Phys. Soc. Jpn **52**, 722 (1983)

[8] C.P. Umbach, S. Maekawa, R.B. Laibowitz and S.P. McAlister, Phys. Rev. **B 33**, 4303 (1986)

[9] A.G. Aronov, M.E. Gershenson and Yu. E. Zhuravlev, Sov. Phys. JETP **60**, 554 (1984)

[10] D. Stauffer, Introduction to Percolation Theory, Taylor and Francis, London (1985)

SPIN-ORBIT EFFECTS IN DISORDERED SYSTEMS

Yigal Meir, Yuval Gefen[†], Ora Entin-Wohlmann[‡]

Dept. of Physics, M.I.T., Cambridge, MA 02139, USA
[†]Dept. of Nuclear Physics, Weizmann Institute, Rehovot 76100, Israel
[‡]School of Physics and Astronomy, Tel Aviv University
Tel Aviv 69978, Israel

1 Introduction

When Dirac equation is expanded to second order in v/c, one finds various relativistic corrections to the Hamiltonian. One of these terms, V_{so}, is of the form

$$V_{so} = \frac{\hbar}{2m^2c^2}\sigma \cdot \left(\nabla V \times (\mathbf{p} - e\mathbf{A}/c)\right),\tag{1}$$

where $\hbar\sigma$ is the electrons's spin, $\mathbf{p}$ its momentum, V is the scalar potential and $\mathbf{A}$ the vector potential. The contribution of $e\mathbf{A}/c$ to the momentum can be neglected for most values of magnetic fields. For a spherically symmetric potential, V_{so} takes the form

$$V_{so} = \frac{\hbar}{2m^2c^2}\frac{1}{r}\frac{\partial V}{\partial r}\sigma \cdot \mathbf{L},\tag{2}$$

where $\mathbf{L} = \mathbf{r} \times \mathbf{p}$ is the angular momentum operator. Thus, this term is coined *spin-orbit interaction*.

Although the steps leading to the derivation of V_{so} from Dirac's equation are quite formal, it is easy to see why such a term has to appear in a relativistic theory. Considering a Zeeman interaction, $(e\hbar/2mc)\sigma \cdot \mathbf{H}$, and remembering that $\mathbf{H}$, the magnetic field, transforms the same way as $(\mathbf{v}/c) \times \mathbf{E}$ (where $\mathbf{v}$ and $\mathbf{E}$ are the electron velocity and the electric field, respectively), or, equivalently, as $(1/mc)(\mathbf{p} - e\mathbf{A}/c) \times \nabla V$, one finds the same interaction term as in eq. (1). In fact, even if there is no spin-orbit (SO) interaction in a given system, the Zeeman term will introduce such an interaction in the equivalent Lorentz transformed system.

Two aspects of the SO interaction are worth mentioning. The first is that this interaction is invariant under time-reversal operation. This symmetry strongly restricts the form of the SO scattering matrix, as will be demonstrated in the next section. Secondly, for an electron near a Hydrogen-like atom (i.e. potential of the form Ze^2/r), the interaction increases like Z^4 (since $r \sim 1/Z$). Even when the screening of the inner

Quantum Coherence in Mesoscopic Systems
Edited by B. Kramer, Plenum Press, New York, 1991

electrons is taken into account, it can be shown [1] that the interaction increases like Z^2, which means that it can be quite strong for heavy atoms. Indeed, the effects of SO scattering are usually studied by introducing heavy atoms (such as gold) into a matrix of lighter atoms. These effects have been extensively investigated. Since we will apply our theory mainly to the regime of weak localization, we will mention only experimental and theoretical studies of this effect in this context.

In the weakly localized regime one ignores interference between different paths, since along each path the electron acquires a random phase. Upon averaging the sum of the interference terms between the different paths will be zero. There will be, however, finite contributions when the two paths are exactly identical, or when one of them is the time-reversed path of the other. Since the former is just the classical contribution, the contribution of the time-reversed paths is referred to as the *weak localization* or *quantum* correction, e.g., to the conductance. Thus, in the weakly localized regime one is interested only in sum over independent one-dimensional (1 D) time-reversed paths, or 1 D loops. This is equivalent to a one-loop expansion in the diagrammatic theory. The contribution of the quantum correction to the conductance can be measured by applying a magnetic field, since the classical conductance is not sensitive to small magnetic fields. Moreover, in multiply connected geometries one expects observation of the Aharonov-Bohm (AB) effect. The period of this effect was predicted to be $\phi_0 (\equiv hc/e)$ for a single 1 D ring [2], or $\phi_0/2$ for a cylinder [3]; period-halving results from an effective ensemble averaging [4]. Indeed, all these effects were observed experimentally [5].

The effect of SO scattering on quantum conductance has been studied in the context of weak localization by diagrammatic expansion [6, 7], heuristic arguments [8], semi-classical analysis [9], and random matrix theory [10]. These yielded a sign change of the magnetoconductance due to SO scattering and reduction by a factor of $1/2$, concomitant with a reduction factor of $1/4$ in the conductance fluctuations. The former prediction was confirmed by experiments [11]. No results have been obtained so far concerning specific realizations or disordered AB geometries. Here we report on a systematic, non-perturbative study of SO scattering effects in AB geometries [12].

2 One-Dimensional Ring

We first consider a 1 D ring penetrated by a magnetic flux ϕ. In that geometry one can choose a gauge for which the flux appears only in the boundary condition of the wavefunction, $\psi(L) = \exp(2\pi i \phi/\phi_0)\psi(0)$, where L is the ring circumference. We adopt for the Hamiltonian the tight-binding picture, which in the presence of SO scattering reads [10, 13]

$$H = \sum_{n\sigma} \epsilon_n |n\sigma\rangle\langle n\sigma| + \sum_{n\sigma\sigma'} V_n(S_n)_{\sigma\sigma'} |n\sigma\rangle\langle n+1\sigma'| + h.c. \ . \tag{3}$$

Here S_n is a 2×2 matrix in spin space. Invoking time-reversal symmetry of the Hamiltonian, eq. (3), we may choose V_n to be real and S_n to be a unitary matrix with a unit determinant [10, 13]. Equation (3) with all $S_n \equiv I$ describes the corresponding bare picture (i.e. the same spatial disorder but no SO scattering).

We next derive the transfer matrix T_n for the Hamiltonian eq. (3), defined by

$$\begin{pmatrix} \psi_{n+1} \\ \psi_n \end{pmatrix} = T_n \begin{pmatrix} \psi_1 \\ \psi_0 \end{pmatrix}, \tag{4}$$

where ψ_n is the spinor wavefunction corresponding to the n-th site; T_n is thus a 4×4 matrix. A straightforward inductive procedure yields for T_n the form

$$T_n = \begin{pmatrix} f_n^{(1)} S_n \ldots S_1 & f_n^{(2)} S_n \ldots S_0 \\ f_n^{(3)} S_{n-1} \ldots S_1 & f_n^{(4)} S_{n-1} \ldots S_0 \end{pmatrix}. \tag{5}$$

Here $f_n^{(i)}(\epsilon_0, \epsilon_1 \ldots \epsilon_n; V_0, V_1 \ldots V_n; E)$ describe the bare problem; in the absence of SO interaction they constitute the bare 2×2 transfer matrix $T_n^{(0)}$. The eigenvalue equation for the Hamiltonian eq. (3) is determined by T_N, N being the number of sites in the ring. Assuming for simplicity that $S_0 = S_N = I$ (which should not lead to any loss of generality for large enough N) the eigenvalue equation takes the form

$$e^{2\pi i\phi/\phi_0} \begin{pmatrix} \psi_1 \\ \psi_0 \end{pmatrix} = T_N \begin{pmatrix} \psi_1 \\ \psi_0 \end{pmatrix} = \tilde{S} \otimes T_N^{(0)} \begin{pmatrix} \psi_1 \\ \psi_0 \end{pmatrix}$$
$$\tilde{S} \equiv S_N \ldots S_1, \tag{6}$$

and $\otimes$ denotes a tensorial product. The transfer matrix thus decouples into a SO part and a spatial part. Physically, this stems from the time-reversal invariance of the SO scattering. The spin of an electron scattered back and forth does not change. Hence all trajectories going once around the ring result in the same rotation in spin space, independent of the actual trajectory in real space.

Since $\tilde{S}$ is a unitary matrix of a unit determinant, its eigenvalues are of the form $\exp(\pm i\lambda)$. Writing then the spinors ψ_1 and ψ_0 each as a linear combination of the eigenvectors of $\tilde{S}$, eq. (6) splits into two 2×2 matrix equations

$$e^{i(2\pi\phi/\phi_0 \mp \lambda)} \begin{pmatrix} a_1 \\ a_0 \end{pmatrix}_\pm = T_N^{(0)} \begin{pmatrix} a_1 \\ a_0 \end{pmatrix}_\pm, \tag{7}$$

where $a_+(a_-)$ is the projection of ψ on the eigenstate corresponding to the eigenvalue $e^{i\lambda}(e^{-i\lambda})$. It follows that the eigenvalue equations (7) are the same as those of the bare problem, but with the flux shifted by $\pm\lambda\phi_0/2\pi$. This is the central result of our analysis. It implies that all energies $E_n(\phi)$ and wavefunctions $\psi_n(\phi)$ of the bare problem, spin-degenerate in the absence of SO interaction, split in the presence of SO scattering into the two sets $E_n(\phi \pm \lambda\phi_0/2\pi)$, $\psi_n(\phi \pm \lambda\phi_0/2\pi)$.

This result can be used to deduce the effect of SO scattering on any thermodynamic or transport quantity $Q(\phi)$ that does not depend explicitly upon spin operators (Q may still depend on temperature),

$$Q/\phi) = \frac{1}{2}\left(Q_0(\phi + \delta) + (Q_0(\phi - \delta))\right), \tag{8}$$

where Q_0 is the corresponding bare system quantity and $\lambda = 2\pi\delta/\phi_0$. Although SO scattering is represented as an effective shift in the magnetic flux, time-reversal symmetry is preserved.

The bare (spin-independent) quantity Q_0 is periodic in ϕ and hence can be expanded in Fourier series

$$Q_0(\phi) = \sum_n \left(a_n \cos(2\pi n\phi/\phi_0) + b_n \sin(2\pi n\phi/\phi_0)\right). \tag{9}$$

In the presence of SO scattering, it follows from eq. (8) that

$$Q(\phi) = \sum_n c_n \left(a_n \cos(2\pi n\phi/\phi_0) + b_n \sin(2\pi n\phi/\phi_0) \right), \tag{10}$$

with $c_n = \cos n\lambda$. Thus, each harmonic of Q_0 is multiplied by a universal reduction factor (independent of the quantity at hand), determined solely by the SO interaction. Note that some configurations of SO scattering may lead to $\lambda = \pi/2$ and hence to the vanishing of all odd harmonics in eq. (10). In such cases, the quantity $Q(\phi)$ will exhibit period-halving (from ϕ_0 to $\phi_0/2$) without invoking either energy or ensemble averaging. This probably happens when the SO length is of the order of L, so going once around the ring rotates the spin state into an orthogonal state.

We next consider averaging over the SO scattering (to be distinguished from spatial disorder averaging). The SO parameter λ may be related [8, 14] to the rotation angles α, β and γ in the three-dimensional spin coordinate frame, $\cos\lambda = \cos(\beta/2)\cos((\alpha + \gamma)/2)$. Averaging over strong SO coupling is equivalent [8] to averaging over these angles with a uniform distribution. Thus $\langle \cos^{2n+1}\lambda \rangle = 0$, while $\langle \cos^2\lambda \rangle = 1/4$, $\langle \cos^4\lambda \rangle = 1/8$, etc. It follows that upon averaging over strong SO scattering, all odd harmonics disappear (even if they do not vanish upon spatial disorder averaging). Averaging over the second harmonic, i.e. over $\cos 2\lambda$, yields a factor of $-1/2$, while all higher even harmonics vanish upon averaging. Furthermore, one is also able to consider fluctuations of the harmonics, $\delta Q_n = \langle Q_n^2 \rangle - \langle Q_n \rangle^2$, where Q_n is the n-th harmonic. Considering odd n's, for which $\langle Q_n \rangle$ vanishes, δQ_n is given by $\langle Q_{0n}^2 \rangle \langle \cos^2 n\lambda \rangle$ for any spin-independent quantity (the last reduction factor being $1/4$ for $n = 1$, and $1/2$ for $n \geq 3$). Somewhat more complicated factors can be obtained for the even harmonics.

Notably, our approach provides also information on averages and fluctuations over the whole range of SO scattering strengths. For example, we note that the reduction in $Q_2, R(Q_2)(= c_2)$ is related to the reduction in $Q_1^2, R(Q_1^2)$, by $R(Q_2) = 2R(Q_1^2) - 1$, independent of the SO scattering strength.

3 Application to Higher Dimensions

The result so far pertain to 1 D rings (or double-connected structures for which the effective number of transverse channels is one). Although such systems may be experimentally studied [15], it should be noted that our results may be extended to higher dimensions provided that physical quantities can be calculated as sums over independent 1 D closed paths in real space. As mentioned in the introduction, the conduction and its fluctuations, within weak localization theory, do obey this restriction, and thus our results are applicable to the weakly localized regime in all dimensions.

In the bulk systems, as the weight of closed paths that repeat themselves (e.g. form two overlapping loops) is negligibly small, one expects the main contribution to arise from the lowest harmonics and the reduction factor to be c_1 if the first harmonic does not average out to zero, or c_2 otherwise. Averaging over strong SO scattering, the reduction factor of Q is thus $-1/2$; its fluctuations are reduced by a factor $1/4$. This is in agreement with previous weak localization calculations [6-10] of the reduction factors of the magnetoconductance and the conductance fluctuations, respectively. For arbitrary SO coupling, the reduction factor of the conductance, $R(G)$, is related to that of the conductance fluctuations, $R(G^2)$, by $R(G) = 2R(G^2) - 1$. One may then use the semi-classical result [9] $R(G) = (1/2)\left(3e^{-4\tau_\phi/3\tau_{s0}} - 1\right)$ (τ_ϕ is the phase breaking time and τ_{s0}

the SO scattering time) to predict $R(G^2) = 1/4\left(3e^{-4\tau_\phi/3\tau_{so}} + 1\right)$. We understand that this prediction has been verified recently [16].

Our results are also applicable to studies of the AB effect in cylinders. One expects that as the length of the cylinder increases, the weight of the self-overlapping closed paths decreases. Therefore, as a function of the cylinder length, the reduction factor of the n-th harmonic should cross over from c_n to c_1, if the first harmonic does not average out to zero (either because of the scattering in real space or when SO scattering is strong enough); otherwise, all odd harmonics will disappear as the cylinder length increases, while the reduction factor of the $(2\,n)$-th harmonic will cross over from c_{2n} to c_2.

4 Higher Spins and the "Classical" Limit

So far we have dealt with the physical case of spin-$1/2$ particles. Here we will generalize the results to an arbitrary spin s. Although the physical validity of such a generalization is doubtful (in fact, the Dirac equation, from which the SO interaction emerged, describes only spin-$1/2$ particles), the limit $s \to \infty$ provides one with a natural scheme of investigating the "classical" limit of the AB effect.

In fact, the procedure described in section 2 is readily generalized to any spin s. The transfer matrix T_n in eq. (4) is now a $(4s+2) \times (4s+2)$ matrix, but otherwise the results eqs. (6) and (7) still hold. This means that the $(2s+1)N$ energies and eigenfunctions are given by $E_n(\phi + k\lambda\phi_0/\pi), \psi_n(\phi + k\lambda\phi_0/\pi), k = -s, \ldots, s; n = 1, \ldots, N$. Thus eq. (10) still holds, where c_n is given now (for half-integer s and $t = s + 1/2$) by

$$c_n = \frac{1}{t}\sum_{m=1}^{t}\cos(nm\lambda) = \frac{\cos\left((t+1)n\lambda/2\right)\sin(tn\lambda/2)}{\sin(n\lambda/2)} \tag{11}$$

and a similar expression for integer s. Here again c_n is a universal constant, independent of the physical quantity in hand. For arbitrary irrational λ, c_n (for $n \neq 0$) is an oscillatory function, bounded by the envelope $\pm 1/t$. In the limit $s \to \infty, c_n \to 0$, and $n = 0$ is the only surviving term in eq. (10), $Q(\phi) = a_0$, leading to the vanishing of the AB effect. It should be emphasized that this term still has to be determined quantum-mechanically. The simplest way to see this is to consider 1 D chain. Here the quantum probability to get from the left to the right is independent of any SO scattering along the chain and thus the above limiting procedure will not affect the nature of the wave functions, which, in the presence of disorder, will still be localized, even when the above "classical" limit is taken.

5 Discussion and Conclusions

We have presented a simple theory of spin-orbit effects in disordered mesoscopic systems. We have made predictions concerning the universality of SO reduction factors for a rather large class of physical quantities and their fluctuations. These may be tested by experiments and numerical simulations. In particular our results agree with earlier diagrammatic calculations for weak localization theory. Our theory implies that once the diagrammatic calculation has been made for the system without SO scattering, the latter can be included in a simple manner and one does not have to repeat the diagrammatic expansion including the SO scattering. It should be emphasized that the results of our theory are valid well beyond the weak localization regime, as long as one is interested in quantities like $\langle G \rangle$ or $\langle G^2 \rangle$ and not $\langle \log(G) \rangle$. The latter is the relevant physical quantity in the strongly localized regime, and attempts to generalize out theory to this regime are presently carried out.

We have also presented the conceptual limit of $s \to \infty$, and showed that it corresponds to the "classical" limit as far as the AB effect is concerned, but other quantum phenomena may still persist.

In our theory we have neglected the Zeeman splitting altogether. This is indeed justified when this splitting does not cause the electron to rotate its spin, going once around the ring, by a phase of the order of 2π. Thus the magnetic field should obey the condition $\Delta E \tau \ll h$, where $\Delta E = \mu_e H$ is the Zeemann splitting and τ is the diffusion time around the ring. Since the Thouless energy E_c is defined by h/τ, this condition breaks down when the Zeeman energy becomes of the order of the Thouless energy. This usually happens, for mesoscopic systems, for H of the order of hundreds of Gauss. For bulk systems one should replace E_c by h/τ_ϕ. It should be noted that even for smaller magnetic field the Zeeman term will still be relevant for some high harmonics.

6   Acknowledgements

It is our pleasure to acknowledge discussions with Y. Aharonov, and in particular his comments on the generalization of this work to the $s > 1/2$ case. Y. G. acknowledges the hospitality of the Sackler Institute of Solid State Physics, Tel Aviv University. The research was partially supported by the US-Israel Binational Science Foundation, the Minerva Foundation, Munich, Germany, the fund of basic research administered by the Israel Academy of Sciences and Humanities and by the German-Israeli Foundation.

References

[1] L.D. Landau and E.M. Lifshitz, Quantum Mechanics, p. 266, Pergamon, Oxford (1977)

[2] Y. Gefen, Y. Imry, and M.Ya. Azbel, Phys. Rev. Lett. **52**, 129 (1984)

[3] B.L. Al'tshuler, A.G. Aronov, and B.Z. Spivak, Pis. Zh. Eksp. Teor. Fiz. **33**, 101 (1981) [JETP Lett. **33**, 94 (1981)]; D. Yu. Sharvin and Yu. V. Sharvin, Pis. Zh. Eksp. Teor. Fiz. **34**, 285 (1981) [JETP Lett. 34, 272 (1981)]

[4] Y. Gefen, unpublished (1984); M. Murat, Y. Gefen and Y. Imry, Phys. Rev. **B 34**, 659 (1986); A.D. Stone and Y. Imry, Phys. Rev. Lett. **56**, 189 (1986)

[5] For review see S. Washburn and R.A. Webb, Adv. Phys. **35**, 395 (1986)

[6] S. Hikami, A.I. Larkin, and Y. Nagaoka, Prog. Theor. Phys. **63**, 707 (1980)

[7] B.L. Al'tshuler and B.I. Shklovskii, Zh. Eksp. Teor. Fiz. **91**, 220 (1986) [Sov. Phys. JETP **64**, 127 (1986)]; P.A. Lee, A.D. Stone, and H. Fukuyama, Phys. Rev. **B 35**, 1039 (1987); S. Feng, Phys. Rev. **B 39**, 8722 (1989)

[8] G. Bergmann, Sol. St. Commun. **42**, 815 (1982)

[9] S. Chakravarty and A. Schmid, Phys. Rep. **140**, 195 (1986)

[10] N. Zanon and J.L. Pichard, J. Phys. (Paris) **49**, 907 (1988)

[11] G. Bergmann, Phys. Rev. **107**, 1 (1984)

[12] Y. Meir, Y. Gefen, and O. Entin-Wohlmann , Phys. Rev. Lett. **63**, 798 (1989)

[13] J. Friedel, P. Lenglart, and G. Leman, J. Phys. Chem. Sol. **25**, 781 (1964)

[14] R.P. Feynmann, R.B. Leighton, and M. Sands, The Feynmann Lectures on Physics, **III**, Addison-Wesley Reading (1966)

[15] see e.g. G. Timp, A.M. Chang, P. Markiewich, R. Behringer, J.E. Cunningham, T.Y. Chang, and R.E. Howard, Phys. Rev. Lett. **59**, 732 (1987); M.L. Roukes, A. Scherer, S.J. Allen Jr., H.G. Craighead, R.M. Ruthen, E.D. Beebe, and J.P. Harbison, Phys. Rev. Lett. **59**, 3011 (1987)

[16] O. Millo, S.J. Klepper, M.W. Keller, D.E. Prober, S. Xiong, and A.D. Stone, preprint

LINEAR RESPONSE AND DEPHASING BY COULOMB ELECTRON–ELECTRON INTERACTIONS

Ady Stern[*], Yakir Aharonov[*†], Yoseph Imry[‡]

[*]School of Physics and Astronomy, Tel Aviv University
69978 Tel Aviv, Israel
[†]Dept. of Physics, University of South Carolina
Columbia, SC 29208, USA
[‡]Physics Dept., Weizmann Inst. of Sciences, Rehovot 76100, Israel, and
IBM T.J. Watson Research Center, Yorktown Heights, NY 10598, USA

1   Introduction and Review of Previous Results

Many of the interesting effects in mesoscopic systems are quantum interference phenomena. Such are, e.g., the weak localization corrections to the conductivity, the universal conductance fluctuations, persistent currents and many others. Those effects are known to be affected by the coupling of the interfering particle to its environment, e.g., to a heat bath. The way such a coupling affects quantum phenomena has been studied for a long time, both theoretically [1,2], and experimentally. The effect of the coupling to the environment is characterized by the "phase-breaking" time, τ_ϕ.

In a recent work [3], we have studied the way a coupling of an interfering particle affects a two-wave interference experiment. We have used two descriptions of the way the interaction of an interfering particle with its environment might suppress quantum interference. The first regards the environment as measuring the path of the interfering particle. When the environment has the information on that path, no interference is seen. The second description answers the question naturally raised by the first: how does the interfering particle "know", when the interference is examined, that the environment has identified its path? This question is answered by the observation that the interaction of a partial wave with its environment might induce uncertainty in this wave's phase, thereby turn the interference pattern into a sum of many patterns, shifted relative to one another. We have proven that the two descriptions are equivalent, and applied them to the dephasing by electromagnetic fluctuations in metals, and by photon modes in thermal and coherent states. In the present work, we will review the two descriptions, and examine the dephasing by the electron-electron interaction in metals. Contrary to our previous work, we examine this problem here from the first point of view mentioned

Quantum Coherence in Mesoscopic Systems
Edited by B. Kramer, Plenum Press, New York, 1991

above, rather than the second, that is, we try to find out where the information on the interfering electron path is hidden in the bath of electrons it interacts with.

As a guiding example, we consider an Aharonov-Bohm (AB) interference experiment on a ring. The AB effect has been proven to be a convenient way to observe interference patterns in mesoscopic samples, because it provides an experimentally easy way of shifting the interference pattern. This experiment starts by a construction of two electron wave packets, $l(x)$ and $r(x)$. (l, r stand for left, right), crossing the ring along two opposite sides. We assume that the two wave packets follow well defined classical paths, $x_l(t), x_r(t)$. The interference is examined after each of the two wave packets traverses half of the ring's circumference. Therefore, the initial wavefunction of the electron (whose coordinate is x) and the environment (whose set of coordinates is denoted by η) is

$$\psi(t = 0) = [l(x) + r(x)] \otimes \chi_0(\eta). \tag{1}$$

At time τ_0, when the interference is examined, the wavefunction is, in general,

$$\psi(\tau_0) = l(x, \tau_0) \otimes \chi_l(\eta) + r(x, \tau_0) \otimes \chi_r(\eta) \tag{2}$$

and the interference term is

$$2Re\left[l^*(x, \tau_0)r(x, \tau_0) \int d\eta \chi_l^*(\eta)\chi_r(\eta)\right]. \tag{3}$$

Had there been no environment present in the experiment, the interference term would have been just $2Re\left[l^*(x, \tau_0)r(x, \tau_0)\right]$. So, the effect of the interaction is to multiply the interference term by $\int d\eta \chi_l^*(\eta)\chi_r(\eta)$. The first way to understand this effect is seen directly from this expression, which is the scalar product of the two environment's states coupled to the two partial waves. At $t = 0$ these two states are identical. During the time of the experiment, each partial wave has its own interaction with the environment, and therefore the two states become different. Since the environment is not observed in the interference experiment, its coordinate is integrated upon, i.e. the scalar product of the two states is taken. When the two states do not overlap at all, the final state of the environment identifies the path the electron took. Quantum interference, which is the result of an uncertainty in this path, is then lost. Thus, the phase-breaking time, τ_ϕ, is the time in which the two interfering partial waves shift the environment into states orthogonal to each other, i.e., when the environment has the information on the path the electron takes.

The second explanation for the loss of quantum interference regards it from the point of view of how the environment affects the partial waves, rather than how the waves affect the environment. It is well known that when a static potential $V(x)$ is exerted on one of the partial waves, this wave accumulates a phase (a system of units where $\hbar = 1$ is applied)

$$\phi = -\int V(x(t))dt \tag{4}$$

and the interference term is multiplied by $e^{i\phi}$. "A static potential" here is a potential which is a function of the particle's coordinate and momentum only, and does not involve any other degrees of freedom. For a given particle's path, the value of a static potential is well defined. When V is not static, but created by environment degree(s) of freedom, its value is not well defined any more. The uncertainty in its value results from the quantum uncertainty in the state of the environment. Therefore, ϕ is not definite, too. In fact, ϕ becomes a statistical variable, described by a distribution function $P(\phi)$ (for details of this description see ref. [3]). The effect of the environment on the interference is then to multiply the interference term by the average value of $e^{i\phi}$, i.e.

$$\langle e^{i\phi} \rangle = \int P(\phi) e^{i\phi} d\phi. \tag{5}$$

The averaging is done on the interference "screen" that shows a sum of many interference patterns, corresponding to different environment states. Since $e^{i\phi}$ is periodic in ϕ, $\langle e^{i\phi} \rangle$ tends to zero when $P(\phi)$ is slowly varying over a region much larger than one period, i.e. 2π. When this happens, the interference screen shows a sum of many interference patterns, mutually canceling each other. Hence, the phase-breaking time is also the time in which the uncertainty in the phase becomes of the order of the interference periodicity. In the Feynman-Vernon terminology, $\langle e^{i\phi} \rangle$ is the influence functional of the two paths taken by the two partial waves. This is then the second explanation for the loss of quantum interference.

Our statement of equivalence between the two explanations is then put into an equation

$$\langle e^{i\phi} \rangle = \int d\eta \chi_l^*(\eta) \chi_r(\eta). \tag{6}$$

When the environment measures the path taken by the particle, it induces a phase shift whose uncertainty is of the order of 2π.

2 Dephasing by Electron-Electron Interaction

An interesting application of the general principle is the dephasing of mesoscopic interference effects by electron-electron interaction. In our previous work, we have applied the phase uncertainty approach to dephasing by electron-electron interaction in metals. We have shown that this approach reproduces the results obtained in the pioneering work of Alt'shuler, Aronov and Khmel'nitzkii (AAK). In the following, we consider the dephasing due to electron-electron interactions from the point of view of the changes induced in the state of environment.

The general picture is that of an interfering electron, whose paths are denoted by $x_{r,l}(t)$, interacting with a bath of environment electrons, whose coordinates are y_i. The Coulomb interaction of the interfering electron with the rest of the electrons is, in the interaction picture,

$$\hat{V}_I(x, t) = \int \frac{\hat{\rho}_I(r', t) d^3 r'}{|x - r'|}, \tag{7}$$

where $\hat{\rho}_I(r', t) = \sum_i \delta(r' - \hat{y}_I^i(t))$. For the brevity of the following expressions, we consider only the interaction of the electron bath with the right partial wave of the interfering electron, we omit the corresponding subscript, and we assume that the electron bath is initially in its ground state, $|0\rangle$. Assuming that the left partial wave does not interact with the electron bath, the intensity of the interference pattern is reduced by the probability that the bath's state coupled to the right wave is different from $|0\rangle$. Up to second order in the interaction, this probability is

$$P = \sum_{|n\rangle \neq |0\rangle} \int_0^{\tau_0} dt \int_0^{\tau_0} dt' \langle 0|V_I(x(t),t)|n\rangle \langle n|V_I(x(t'),t')|0\rangle. \tag{8}$$

τ_ϕ is the value of τ_0 for which p = 1. For the ground state $\langle 0|\hat{\rho}|0\rangle = 0$, so that the summation in eq. (8) can be extended to include all states. Assuming that the potential-potential correlation function is translational invariant, eq. (8) can be written in terms of the Fourier-transformed density operator, as

$$P = \sum_{|n\rangle} \int_0^{\tau_0} dt \int_0^{\tau_0} dt' \int d^3q \frac{(4\pi e^2)^2}{q^4} \langle 0|\rho_I^q(t)|n\rangle \langle n|\rho_I^q(t')|0\rangle e^{iq\cdot(x(t)-x(t'))}. \tag{9}$$

Moreover, by transforming into Schrödinger picture operators and inserting a dummy integration variable ω, P can be rewritten in the following form

$$P = \sum_{|n\rangle} \int_0^{\tau_0} dt \int_0^{\tau_0} dt' \int d^3q \int d\omega \frac{(4\pi e^2)^2}{q^4} |\langle 0|\rho_S^q|n\rangle|^2$$
$$\delta(\omega - \omega_{n0}) e^{iq\cdot(x(t)-x(t'))+i\omega(t-t')}. \tag{10}$$

At a first glance, eq. (10) looks useless, due to the practical impossibility of calculating the bath's eigenstates $|n\rangle$. However, the usefulness of eq. (10) stems from its relation to the linear response expression for the imaginary part of the complex dielectric function [5]

$$\mathrm{Im}\left(\frac{1}{\epsilon(q,\omega)}\right) = \frac{4\pi^2 e^2}{q^2} \sum_{|n\rangle} |\langle 0|\rho_S^q|n\rangle|^2 \, \delta(\omega - \omega_{n0}). \tag{11}$$

Thus, eq. (10) becomes

$$P = \int_0^{\tau_0} dt \int_0^{\tau_0} dt' \int d^3q \int d\omega \frac{(4e^2)}{q^2} \mathrm{Im}\left(\frac{1}{\epsilon(q,\omega)}\right) e^{iq\cdot(x(t)-x(t'))+i\omega(t-t')}. \tag{12}$$

Equation (12) is the central result of this section. Before proceeding to a discussion of this result, we comment that the calculation can be generalized to an electron bath initially at a thermal state. The integrant in eq. (12) is then multiplied by $\coth(\omega/2k_BT)$ [6]. We will now draw a few conclusions out of the above calculation:

(1) For good conductors, $\mathrm{Im}\,(1/\epsilon(q,\omega)) = \omega/4\pi\sigma$, and the probability that the state of the electron's bath was changed is

$$P = \int_0^{\tau_0} dt \int_0^{\tau_0} dt' \int d^3q \int d\omega \frac{e^2 \omega}{q^2 \sigma} e^{iq\cdot(x(t)-x(t'))+i\omega(t-t')} \coth \frac{\omega}{2k_BT}. \tag{13}$$

As seen in our previous work [3], this probability is just the uncertainty in the phase accumulated by the right partial wave. We refer the reader to that work for a further discussion of this result, its equivalence to the AAK one, and the extraction of the phase breaking time, τ_ϕ, out of it. The present derivation demonstrates that the origin of this dephasing is in the electrostatic electron-electron interaction, and establishes the connection with the linear response of the bath.

(2) For poor conductors $\mathrm{Im}\epsilon(q,\omega) \ll \mathrm{Re}\epsilon(q,\omega)$. Then, the q,ω integrals in eq. (13) have significant contributions only from those values of q,ω in which $\mathrm{Re}\epsilon = 0$. A typical example is the $\omega = \omega_p$, the plasma frequency.

(3) In both cases mentioned above, the rate of dephasing crucially depends on the imaginary part of dielectric constant. This, in turn, determines the rate at which the electron bath is excited by the interfering electron. It should be emphasized here that the polarization of the electron bath by the interfering electron, reflected in the real part of the dielectric function, does not dephase the interference. This polarization disappears when the electron leaves the polarized region and therefore it does not identify the path taken by the electron. For a general discussion of the relation between dissipation, excitations and dephasing, the reader is referred to ref. [3].

(4) The above discussion of dephasing due to the Coulomb interaction can be easily generalized to any two particle interaction $V(r - r')$. This is done by replacing $e^2/|r - r'|$ in eq. (7) by $V(r - r')$, and following the derivations in eqs. (8) - (13). In particular, it is interesting to consider the case of a short range potential, which can be approximated by $V(r - r') \propto \delta(r - r')$. For such a potential, the probability that the bath's state is changed is proportional to the density-density correlation function of the bath's electrons,

$$P \propto \int_0^{T_0} dt \int_0^{T_0} dt' \langle \rho(x(t), t)\rho(x(t'), t') \rangle. \tag{14}$$

Thus, the intensity of the interference effects provides here information on the density-density correlation function of the bath, and through that, on the dynamical structure factor.

(5) As emphasized above, the interaction of the environment with the interfering partial waves changes the state of the environment, so that the environment acquires information on the path taken by the interfering particle. One might then consider the case of a very slow electron traversing a piece of metal and examine the change it induces in the state of the metal. At a first glance, it looks as if the state of the metal adiabatically follows the motion of the electron, so that when the electron leaves the metal, the metal is back in its initial state. However, since the excitations spectrum of the electron bath is continuous, the adiabatic argument is never applicable. It is true that the electron induces a polarization in the bath, a polarization that follows adiabatically its motion and disappears when it leaves the metal. But, due to the bath's continuous spectrum, this polarization has to involve an excitation of the bath, and this excitation does not disappear when the interaction of the bath with the interfering electron is over.

The situation is different, of course, for insulators. There, due to the gap in the excitations spectrum, a very slow electron can polarize the bath without exciting it, i.e. without identifying its path.

3 Acknowledgements

Research at USC was partially supported by NSF grant no. PHY-8807812. Research at TAU and WIS was partially supported by the fund for basic research administered by the Israel Academy of Sciences and Humanities. Research at TAU was partially supported by the German-Israel foundation for scientific research and development. One of us (A.S.) acknowledges the support of the David Ben-Gurion foundation, administered by the General Labor Federation of Israel. We thank B.L. Alt'shuler and S. Levit for instructive discussions.

References

[1] R.P. Feynman and F.L. Vernon, Ann. Phys. **24**, 118 (1963)

[2] A.O. Caldeira and A.J. Leggett, Ann. Phys. **149**, 374 (1983)

[3] A. Stern, Y. Aharonov, and Y. Imry, Phys. Rev. **A 41**, 3436 (1990)

[4] B.L. Alt'shuler, A.G. Aronov, and D.E. Khmel'nitzkii, J. Phys. **C 15**, 7367 (1982); B.L. Alt'shuler, A.G. Aronov, and D.E. Khmel'nitzkii, Sol. St. Commun. **39**, 619 (1981)

[5] C. Kittel, Quantum theory of solids, J. Wiley and sons (1963)

[6] E.M. Lifshitz and L.P. Pitaevskii, Statiscal Physics **1**, Pergamon Press (1976)

QUANTUM COHERENCE EFFECTS IN ONE-DIMENSIONAL CHAINS WITH INELASTIC SCATTERING

Michael Schreiber, Klaus Maschke[†]

Institut für Physikalische Chemie, Universität Mainz
Jakob-Welder-Weg 11, W-6500 Mainz, F. R. Germany
[†]Institut de Physique Appliquée, Ecole Polytechnique Fédérale
1015 Lausanne, Switzerland

1 Introduction

To describe the ballistic transport in a 1 D chain Landauer [1] has calculated the resistance R of a series of elastic scatterers from their transmission coefficient T

$$R = \frac{h}{e^2} \frac{1 - T}{T} \tag{1}$$

This relation implies complete quantum coherence between incident and all backscattered waves. Dephasing due to irreversible processes has been introduced into this model by Büttiker [2] who added inelastic scatterers coupled to an external heat bath to the chain. In this way it is possible to describe also certain dissipative aspects of electron transport. However, his approach does not allow to study the gradual transition from coherent to incoherent transport with increasing strength of the inelastic coupling.

To overcome these difficulties, we define a generalized scatterer which incorporates elastic and inelastic processes simultaneously. In particular, this ansatz enables us to investigate the dependence of the resistance on the Fermi wave vector q_F and the length N of the chain.

2 Model

The generalized scatterer at the n-th position in the chain is characterized (see Fig. 1) by two channels $2n + 1$ and $2n + 2$ which supply the coupling to the heat bath as well as the two transport channels 1 and 2 which connect the scatterer to its neighbors or to the external contacts at the ends of the chain. Thus for a chain of length N altogether $2N + 2$ channels have to be considered.

Quantum Coherence in Mesoscopic Systems
Edited by B. Kramer, Plenum Press, New York, 1991

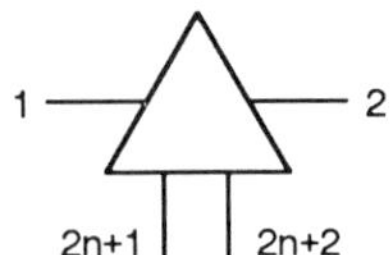

Figure 1. Schematic representation of the n-th scatterer in
a chain with transport channels 1 and 2 and coupling to
the heat bath via channels $2n + 1$ and $2n + 2$.

The scattering matrix corresponding to the n-th scatterer can be composed from
2×2-matrices in the following way

$$
\mathbf{s}_n =
\begin{pmatrix}
\sqrt{1-\varepsilon}\,\sigma & \mathbf{0} & \cdots & \sqrt{\varepsilon}\,\bar{\mathbf{1}} & \cdots & \mathbf{0} \\
\mathbf{0} & \bar{\mathbf{1}} & \cdots & \mathbf{0} & \cdots & \mathbf{0} \\
\vdots & \vdots & \ddots & \vdots & & \vdots \\
\sqrt{\varepsilon}\,\bar{\mathbf{1}} & \mathbf{0} & \cdots & \sqrt{1-\varepsilon}\,\bar{\sigma} & \cdots & \mathbf{0} \\
\vdots & \vdots & & \vdots & \ddots & \vdots \\
\mathbf{0} & \mathbf{0} & \cdots & \mathbf{0} & \cdots & \bar{\mathbf{1}}
\end{pmatrix}
\tag{2}
$$

Here the elastic scattering involving the transport channels is described by the unitary
scattering matrix

$$
\sigma = \begin{pmatrix} r & t \\ t & r \end{pmatrix}
\tag{3}
$$

with the transmission coefficient $t = \sqrt{\delta}$ and the reflection coefficient $r = i\sqrt{1-\delta}$. The
strength of the coupling to the heat bath is determined by the parameter ε: for $\varepsilon = 0$
the scattering is completely elastic, for $\varepsilon = 1$ it is fully inelastic. The matrix $\sqrt{\varepsilon}\,\bar{\mathbf{1}}$ with

$$
\bar{\mathbf{1}} = \begin{pmatrix} 0 & 1 \\ 1 & 0 \end{pmatrix}
\tag{4}
$$

thus reflects the inelastic scattering between the transport channels and the heat bath.
To guarantee the unitary of $\mathbf{s}_n$ one has to define a coupling between the two channels of
the heat bath by

$$
\bar{\sigma} = \begin{pmatrix} r & -t \\ -t & r \end{pmatrix}
\tag{5}
$$

as in eq. (2).

The remaining submatrices $\bar{\mathbf{1}}$ and $\mathbf{0}$ (null matrix) in eq. (2) are chosen in such
a way that the channels by which the generalized scatterers at the other positions are
coupled to the heat bath are not affected by $\mathbf{s}_n$.

3   Calculations

The scattering matrices of the N scatterers can easily be transformed into the respective
transfer matrices which are then multiplied to obtain the transfer matrix of the total
system. For non-vanishing wave vector q the displacements of the scatterers from the
origin have to be taken into account by suitable phase factors or transfer matrices.

Another transformation then yields the scattering matrix of the total system. The squared moduli of its elements determine the scattering probabilities p_{ij} between the channels i and j. The currents in the different channels can be expressed in terms of these probabilities, e.g.

$$I_1 = \frac{e}{h} p_{12} \left(\mu_\ell - \mu_r \right) + \frac{e}{h} \sum_{n=1}^{N} \left(p_{1,1+2n} + p_{1,2+2n} \right) \left(\mu_\ell - \mu_n \right) \tag{6}$$

Here we have introduced the chemical potentials of the left and right contacts (μ_ℓ and μ_r). Since we assume a vanishing temperature, the heat bath at the n-th scatterer is characterized only by the chemical potential μ_n.

Employing the n current conservation conditions for each of the scatterers

$$I_{2n+1} + I_{2n+2} = 0 \tag{7}$$

one can determine the chemical potential μ_n and thus the current in the transport channels in dependence on the external voltage difference $(\mu_\ell - \mu_r)/e$. Adopting the commonly used notation $p_{12} \equiv T$, we obtain from eq. (6) for the resistance R

$$R^{-1} = \frac{e^2}{h} \left[T + \sum_{n=1}^{N} \left(p_{1,1+2n} + p_{1,2+2n} \right) \eta_n \right] \tag{8}$$

with $\eta_n = (\mu_\ell - \mu_n)/(\mu_\ell - \mu_r)$. To end once for all a long-standing dispute about the relevance of the different Landauer expressions for the resistance, namely $R^{-1} \propto T$ or $R^{-1} \propto T/(1 - T)$, we point out that for an ideal lead (i.e. $\delta = 1$ and $\varepsilon = 0$) eq. (8) reduces to $R^{-1} = e^2/h$. This is nothing else but the contact resistance, which is included in our derivation of eq. (8) where we implied that the voltage difference is measured between the contacts. We note that the discrepancy between the two Landauer formulae is just given by this contact resistance. Our numerical results will be presented after subtracting the contact resistance, in correspondence to eq. (1).

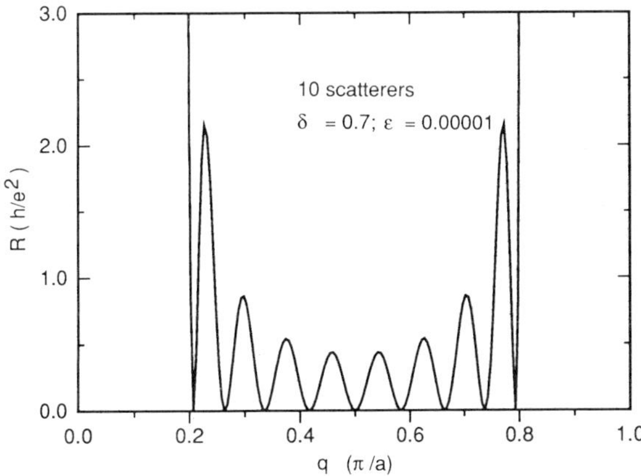

Figure 2. Resistance of a chain of 10 elastic scatterers. The spacing between the scatterers is denoted by a.

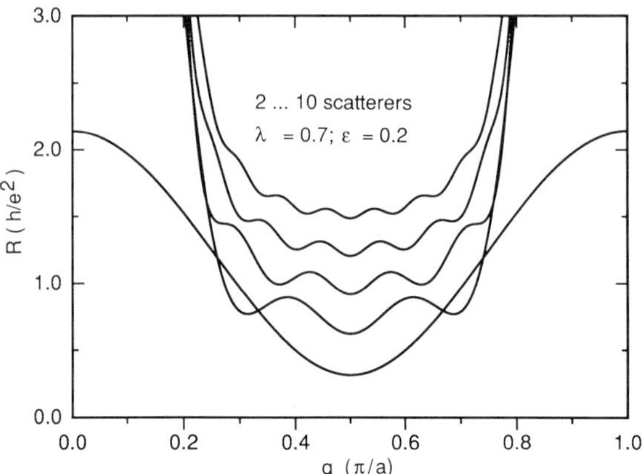

Figure 3. Resistance of 2, 4, 6, 8, and 10 generalized scatterers (from bottom to top).

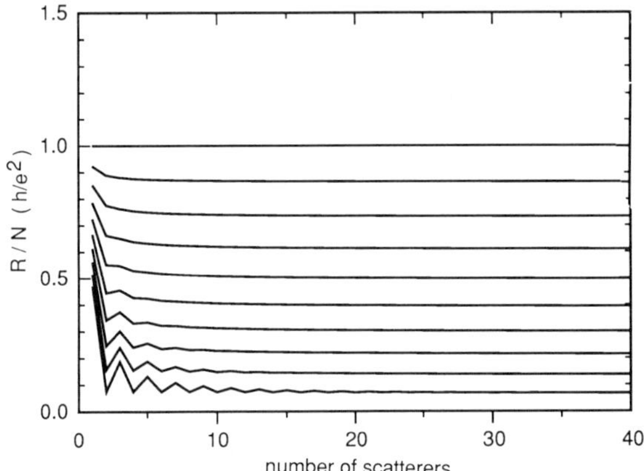

Figure 4. Length dependence of the resistivity for $\varepsilon = 0.1, 0.2, \ldots 1.0$ (from bottom to the top).

4   Results

In Fig. 2 we show the resistance as a function of the Fermi wave vector $q = q_F$ for $N = 10$ scatterers for negligible inelastic and intermediate elastic scattering. Quantum coherence manifests itself in the strong oscillations and the corresponding resonances in the conductance. In the present model in which band structure effects are neglected, there are $N - 1$ resonances.

For small inelastic coupling (see Fig. 3) the oscillations are damped. It can be seen that for a given wave vector the resistance does not increase linearly with the chain length N, and may even decrease. With increasing inelastic scattering the resistance becomes independent of q and finally increases linearly with respect to N, as can be seen from Fig. 4. This linear behaviour is approached asymptotically for comparatively weak inelastic scattering. The characteristic length for this asymptotic decay corresponds to the mean free path in a conventional transport theory. In this regime we have found the expected monotonic increase of the resistance upon increasing the inelastic scattering.

In conclusion, we have generalized the Landauer-Büttiker theory of DC transport to the case of partly inelastic scatterers, coupling the transport channels appropriately to the heat bath. In this new formulation ballistic and dissipative transport are treated in a coherent manner.

References

[1] R. Landauer, Phil. Mag. **21**, 863 (1970)

[2] M. Büttiker, Phys. Rev. **B 33**, 3020 (1986)

CHAPTER 3

QUANTIZATION OF CONDUCTANCE

TRANSPORT IN AN ELECTRON WAVEGUIDE

Gregory Timp, Robert E. Behringer, Eric H. Westerwick,
Jack E. Cunningham

AT&T Bell Laboratories
Holmdel, New Jersey 07733, USA

1 Introduction

An electron waveguide is a wire that is so clean and so small that electron waves can propagate coherently in guided modes, which are characteristic of the geometry, with minimal scattering. An electron waveguide is supposed to be reminiscent of an optical or microwave waveguide, but unlike an electromagnetic wave, an electron wave is sensitive to an applied electric or magnetic field because it possesses a charge. In response to an applied electric field (or to an applied current), an electron waveguide has a resistance which is related to the quantum mechanical transmission through the wire [1]. The transmission probability is affected by both elastic and inelastic scattering. Elastic scattering, such as might occur at an impurity for example, changes the distribution of the electrons between the modes of the guide, but it is phase deterministic; i.e. information associated with the phase of the electronic wave is not ruined by an elastic scattering event. In contrast, inelastic scattering destroys the phase memory in the wave. Coherent electronic transport is possible whenever the wire is smaller than the inelastic scattering length, whether or not there is elastic scattering [2]. Only when a device is smaller than both the inelastic and the elastic scattering lengths and comparable to the wavelength, like it is in an electron waveguide, is the modal distribution of the current important.

It is now possible to make a wire which is reminiscent of an electron waveguide [3]. Using molecular beam epitaxy in conjunction with high resolution electron beam lithography, wires shorter than the mean free path ($\approx 1 - 10\,\mu m$) with a width and thickness comparable to the Fermi wavelength of the electron ($\approx 50\,nm$) have been made by laterally constricting a high mobility, two-dimensional electron gas (2 DEG) in a AlGaAs/GaAs heterostructure to submicron dimensions. At low temperature, the *coherence* and *confinement* of an electron in an electron waveguide give rise to gross deviations from classical charge transport [4,5]. Two spectacular advertisements of the quantum mechanical nature of the resistance of an electron waveguide have been found: the Aharonov-Bohm (AB) effect [6-9], and quantized resistances [10-14]. But the wave nature of the electron can be recognized in other features of the resistance too. In this paper we explore the quantum mechanical aspects of transport in an electron waveguide

Quantum Coherence in Mesoscopic Systems
Edited by B. Kramer, Plenum Press, New York, 1991

using two, three and four terminal resistance measurements. We show many features of the resistance – such as the four terminal resistance of a bend in an electron waveguide, or resonant tunneling through a zero-dimensional (0 D) cavity – which must have a quantum mechanical interpretation to be explained entirely, but we especially emphasize the observations of the AB effect and the quantized resistance as unambiguous manifestations of quantum mechanics. Specifically, we examine how robust the AB effect and the quantization of the resistance are.

To illustrate the role of coherent scattering in the resistance, we examine in the next section the AB effect found in the magnetoresistance of an annulus fabricated from two submicron constrictions in a high mobility 2 DEG. Although the AB effect is resilient with respect to elastic scattering, it is suppressed in intense magnetic fields where the conduction through the annulus occurs via states near the edges of the wires comprising the annulus. In the third section we show that two terminal measurements of variable width, submicron constrictions in a 2 DEG reveal a resistance which is quantized as a function of the width, W, at $h/2e^2N$ where N is an integer and h is Planck's constant [10,11]. The index N counts the number of discrete transverse energy levels occupied in the constriction. We find that the resistance is quantized only in constrictions $< 1\,\mu$m long, however, which may reflect the importance of scattering from disorder in longer constrictions. Scattering from a lead or contact can also be detrimental to the quantization. In the fourth section, measurements of the three terminal resistance, using a lead juxtaposed between two constrictions in series, show that a lead can destroy the quantization associated with the high index N subbands, but the low index subbands can still be accurately quantized. Electrons occupying in the lowest energy levels do not propagate effectively into the voltage lead, but instead are collimated or focussed in the forward direction away from an intervening voltage probe. The distinctive way an electron wave propagates around a bend has important implications for four terminal measurements too. In section five, we consider the four terminal magnetoresistance of a junction between four constrictions and show that the longitudinal resistance can be negative, and that the Hall resistance can be suppressed at low field, and quantized at h/e^2i for integers $i \leq 2N$ where N is the number of subbands occupied in the absence of a field when $He^2W/hc > 1$. In section six, as a final illustration of the effect of coherent scattering and quantization on the resistance, we examine the low temperature transport through a 2 DEG laterally constrained to 0 D. We tentatively interpret peaks found in the two terminal conductance as evidence of resonant tunneling through quasi-bound states of the 0 D cavity.

2   The Aharonov-Bohm Effect

The AB effect occurs because two electron trajectories which encircle a magnetic flux acquire a phase shift proportional to the flux. As the flux changes, the transmission probability oscillates with a periodicity of hc/e, even if there is no classical magnetic field in the path of the electron. A phenomenon related to the AB effect has been recently observed in magnetotransport through an annulus defined lithographically in a high mobility 2 DEG. The procedure used to laterally constrict a 2 DEG in a AlGaAs/GaAs heterostructure to an annulus has already been described elsewhere [8,15].

We have fabricated annuli in δ-doped AlGaAs/GaAs heterostructures. The salient feature of the fabrication is a partial etch of the AlGaAs which laterally constrains the 2 DEG at the AlGaAs/GaAs interface to the region beneath an etch mask defined by

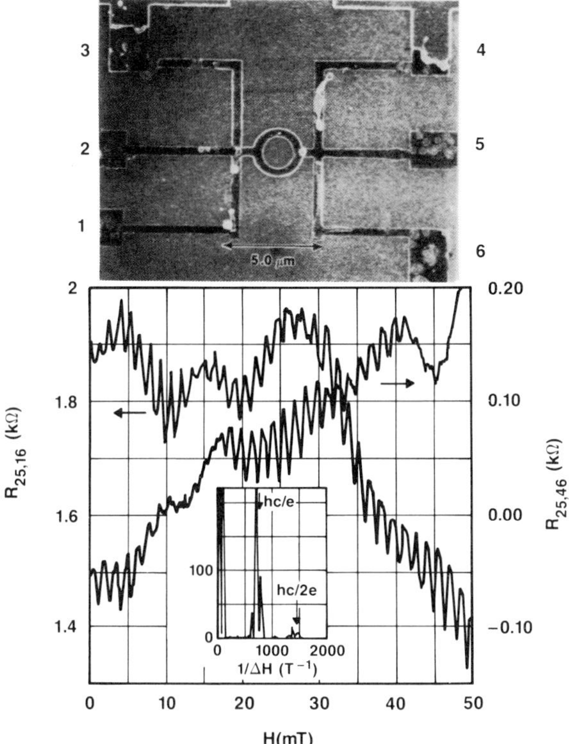

Figure 1. The magnetoresistance, $R_{25,16}$ and $R_{25,46}$ of a nominally 1 μm radius (actually $r = 0.95\,\mu$m) annulus measured at 270 mK. The top portion shows a typical etch mask which is transferred into the heterostructure using reactive ion etching. The convention for numbering the leads is given there. The Fourier power spectrum of $R_{25,16}$ is shown in the inset to the lower portion of the Figure.

electron beam lithography. The annulus pictured in the upper portion of Fig. 1 is an etch mask on top of a AlGaAs/GaAs heterostructure. The wires comprising the mask are about 500 nm wide, but the conducting width of the electron gas beneath the mask is estimated to be considerably narrower (typically $W < 250$ nm) due to depletion from the exposed edges of the device. For economy, we focus here on the results we obtained from devices with a starting two-dimensional (2 D) mobility and carrier density of about $\mu = 50\mathrm{m}^2$ and $n = 4.4 \times 10^{15}\mathrm{m}^{-2}$ respectively. The carrier density in the constrictions deduced from the high field Hall effect is approximately $n = 3.0 \times 10^{15}\mathrm{m}^{-2}$. We measured the four terminal magnetoresistance of nominally 2 μm diameter annuli like that shown in Fig. 1. The current excitation was typically less that 1 nA.

The low temperature magnetoresistance of the annulus oscillates with a period corresponding to a flux of hc/e through the average area of the annulus. Figure 1 shows the oscillatory magnetoresistances, $R_{25,16}$ and $R_{25,46}$, with a period of 1.4 mT found at 270 mK. Under the conditions of the observation, the magnetic field penetrates both the annulus and the constrictions which comprise the annulus. The notation $R_{kl,mn}$ denotes a resistance measurement in which there is a positive current from lead k to l, and a positive voltage is detected between leads m and n. The convention for numbering the leads is given in Fig. 1.

The periodic oscillations observed in the magnetoresistance have been attributed to the AB effect (even though the magnetic field penetrates the wires comprising the annulus). As the magnetic field changes, the flux enclosed by the annulus changes. Consequently, the transmission probability for an electron, and so the resistance, through the annulus oscillates with a periodicity of hc/e. The annulus of Fig. 1 has an average diameter of about $1.9\,\mu$m. In the inset to the lower portion of Fig. 1, our estimate of the frequency corresponding to a hc/e periodicity is indicated along with the Fourier power spectrum of the measurement $R_{25,16}$. Notice that the amplitude of the oscillation in the resistance is a significant fraction, as much as 5-10%, of the resistance. hc/e oscillations with a similar amplitude have been found in the magnetorèsistance of annuli fabricated in heterostructures with a 2 DEG of much lower mobility [16]. The width of the hc/e fundamental in the Fourier spectrum is assumed to be given approximately by $\Delta H^{-1} = 2\pi r W/(hc/e)$ where r is the radius, and W is a lower bound on the width of the constrictions which comprise the annulus. For the device of Fig. 1, we find that $W \approx 150\,$nm. Finally, notice that oscillations with the same period are found in $R_{25,46}$ as well, even though the voltage leads used in the measurement are on the same side of the annulus! Measurements such as $R_{46,46}$ and $R_{46,25}$ show similar effects.

The observation of the AB effect in the magnetoresistance is unequivocal evidence of phase coherent scattering of the electron wave. The phase coherence must be maintained at least on the scale of the circumference of the annulus to observe such an effect. Thus, the phase coherence length, L_ϕ, must be $L_\phi \approx 6\,\mu$m at $270\,$mK. Because of the coherence, the resistance is non-local: i.e. the current at one point in the wire depends not only on the electric field at that point, but on electric fields micrometers away. The non-local nature of the resistance is evident from the measurement $R_{25,46}$ which detects the effect of the annulus even though the voltage probes are on the same side of the annulus.

The penetration of the wires comprising the annulus by the magnetic field has a profound effect especially in an intense magnetic field, and cannot be neglected [7,8,17]. Background fluctuations, which have lower frequency aperiodic and periodic components [18], are found superimposed upon the regular oscillations associated with the annulus. The background fluctuations beat with the periodic oscillations so that for narrow intervals of magnetic field the amplitude can be reduced. The fluctuations are supposed to be due to the flux enclosed by trajectories within the wires comprising the annulus [19].

For magnetic fields $HeW^2/hc \gg 1$, the periodic oscillation amplitude associated with the flux enclosed by the annulus is globally suppressed. The oscillations, with a periodicity near zero field corresponding to a flux of hc/e through the average area of the annulus, decrease exponentially in intensity and shift to a lower frequency as the magnetic field increases [7,8]. Figure 2 shows the magnetoresistance $R_{25,16}$, of a nominally $2\,\mu$m (actually $1.82 \pm 0.05\,\mu$m) diameter annulus found for $70\,$mT intervals starting at magnetic fields of $H = 0.0\,$T, $0.920\,$T and $2.470\,$T respectively. We find that the magnetoresistance oscillates near $0.0\,$T with periodicity corresponding to a flux of hc/e through the area of the annulus defined by the average diameter, but as illustrated by traces (b) and (c), both the amplitude and frequency decrease dramatically as the magnetic field increases. The changes observed in the Fourier power obtained from $R_{25,16}$ as a function of magnetic field are succinctly illustrated in Fig. 3. To avoid complications

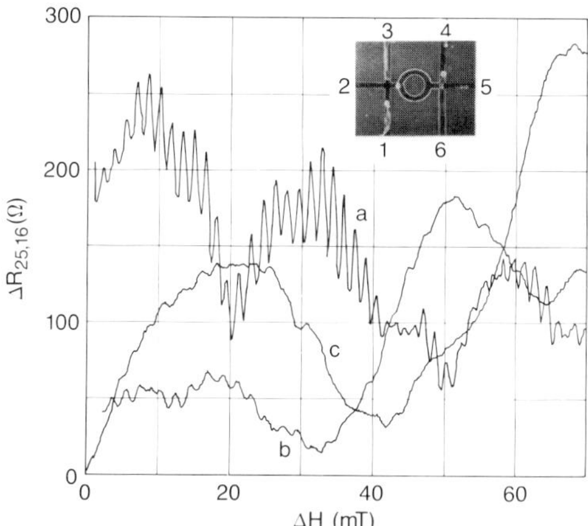

Figure 2. The typical magnetoresistance $R_{25,16}$ of a nominally $1\,\mu$m radius (actually $r = 0.93\,\mu$m) annulus obtained at $T = 280\,$mK for H near (a) 0 T, (b) 0.92 T and (c) 2.47 T. An electron micrograph of a typical device is shown in the inset and the convention for numbering the leads is indicated there.

in the interpretation of the Fourier spectra due to the beating phenomenon mentioned above, the Fourier transforms were taken using a window approximately $220\,$mT wide; much larger than the typical periodicity associated with a beat (approximately $20\,$mT). Therefore, each transform represents an average over many beats. Using the width of the Fourier power spectrum near zero field, we estimate the width of the constrictions comprising the annulus of Fig. 2 to be $W \approx 210\,$nm.

Near zero field, peaks in the Fourier spectra of the magnetoresistance of the annulus are observed at approximately $630\,$T^{-1} and $1250\,$T^{-1}, corresponding to fluxes of hc/e and $hc/2e$ respectively penetrating the area of the annulus defined by the average diameter. Figure 3 shows the shift in frequency and the decrease in Fourier power typically observed. Figure 3(1) shows the magnetoresistances $R_{25,16}$ and $R_{25,46}$ of the same annulus for $0 < H < 5$ T. As the field increases, the centroid of the power spectrum shift monotonically from $630\,$T^{-1} to a lower frequency, eventually reaching approximately $420\,$T^{-1} near 1 T. The Fourier power in the band between $500 - 800\,$T^{-1}, the bandwidth associated with the width of the constrictions comprising the annulus at $H = 0$, is below the noise for $H \geq 1$ T, and no power above the noise is ever observed on the corresponding upper edge of the zero field bandwidth beyond $H = 800\,$mT. The shift in frequency can be readily appreciated by comparing Figs. 3a and 3f, or by examining Fig. 4. In Fig. 4 the spectrum obtained near $H = 1.20$ T (solid line) is juxtaposed with the spectrum obtained near zero field (dashed). It is apparent from Fig. 4 that the shift in the frequency of the hc/e peak corresponds to a shift to the lower frequency edge of the hc/e band obtained near zero magnetic field. According to our interpretation of the spectrum obtained near zero field, the frequency observed at high field corresponds to a flux of hc/e enclosed by the area associated with the inside radius of the annulus.

On comparing Figs. 3a and 3b, we find that the Fourier power in the hc/e band of the spectrum obtained for $H \approx 220\,$mT is below 10% of the peak value observed

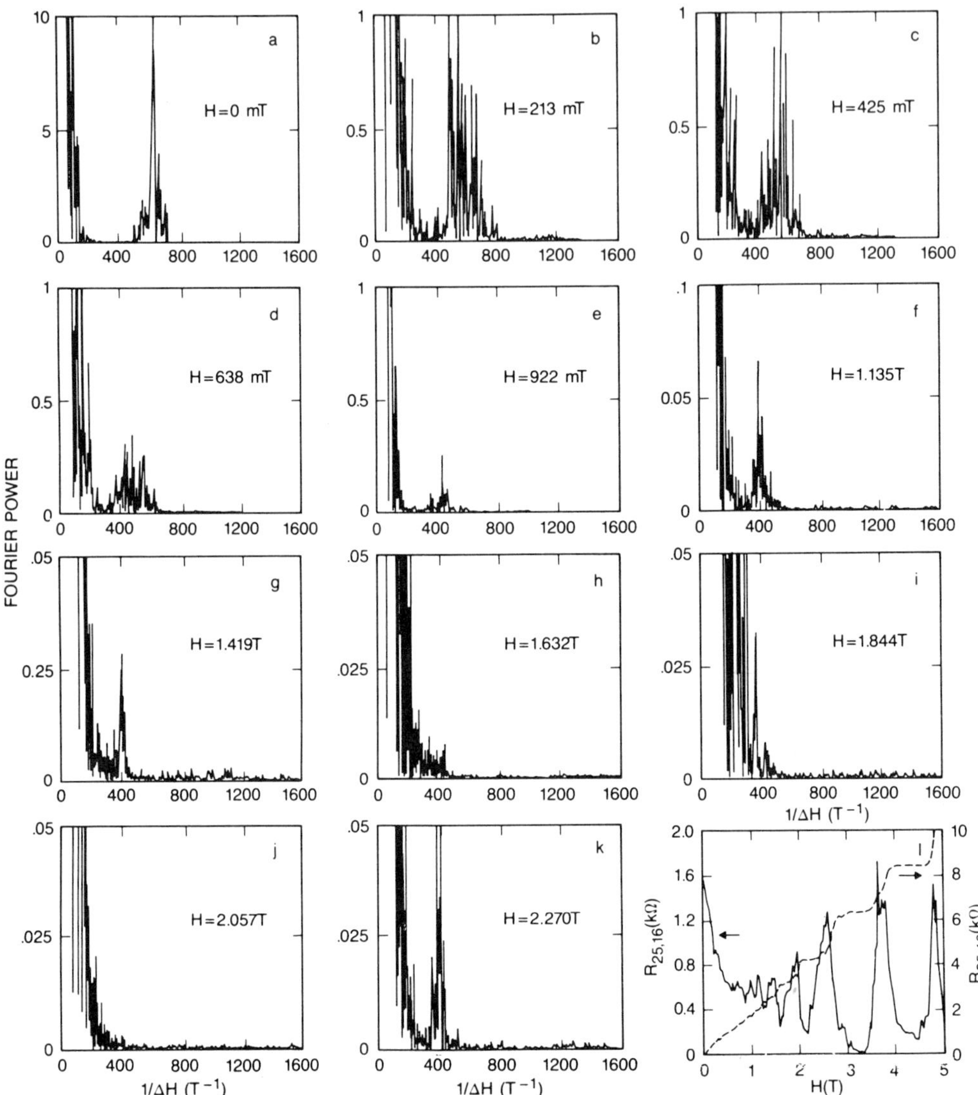

Figure 3. (a) – (k) depict the Fourier power spectra found in $R_{25,16}$ as a function of magnetic field obtained over a 220 mT field range beginning at the field indicated for the device of Fig. 2. The peak in the Fourier spectra corresponds to a single flux quantum hc/e through the area of the annulus. The Fourier power is seen to decrease dramatically and shift to lower frequencies as the magnetic field increases into the quantum Hall regime. (1) shows the magnetoresistances $R_{25,16}$ and $R_{25,46}$ from $H = 0$ to 5 T. Notice that the hc/e peak in the spectra vanishes for $H = 1.632\,\mathrm{T}$ and $H = 2.057\,\mathrm{T}$ corresponding to minima in the resistance $R_{25,16}$ and plateaus in $R_{25,46}$, but reappears beyond the minima.

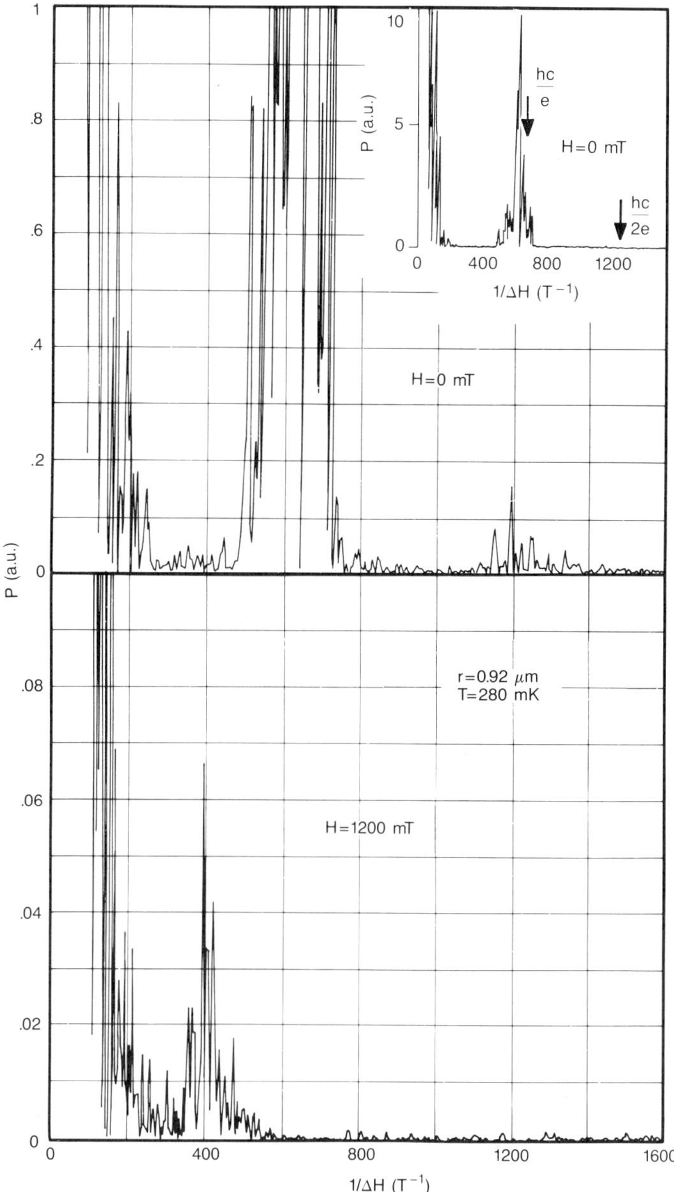

Figure 4. The Fourier power (P) spectra obtained near $H = 0\,\mathrm{T}$ (top) and near $H = 1.20\,\mathrm{T}$ (bottom) with a window $220\,\mathrm{mT}$ wide obtained from the device of Fig. 2 are shown for comparison. The zero field spectra is shown on a scale which exaggerates the power near the baseline. In the inset to the top Figure the spectra obtained near zero field is shown entirely. Notice that the centroid of the spectra for each device obtained near $H = 1.2\,\mathrm{T}$ shifts to the inside edge of the spectra found at $H = 0$ indicative of conduction via the inside edge state.

in the hc/e band near zero field. Generally, we find that $HeW^2/hc > 1$ for magnetic fields where the AB effect is 90% attenuated. Beyond this field range, the Fourier power decreases exponentially with magnetic field. For $200\,\mathrm{mT} < H < 1.6\,\mathrm{T}$, the Fourier power decays with magnetic field according to law: $P = P_0 \exp(-L_0^2/4r_0^2)$ where P is the Fourier power, L_0 is a characteristic length of approximately 100 nm and $r_0^2 = hc/eH$. For example, the fit to the data of Fig. 3 (indicated by the dashed line in the bottom Figure of Fig. 5) yields $L_0 = 95 \pm 7\,\mathrm{nm}$.

The spectral position of the peak does not depend on magnetic field for $1T < H < 2.5\,\mathrm{T}$ for the device of Fig. 2. As the magnetic field increases, the peak in the spectrum near $420\,\mathrm{T}^{-1}$ vanishes near the resistance minima observed in $R_{25,16}$, corresponding to plateaus seen in $R_{25,46}$, but reappears beyond the minima. In Fig. 3(1) minima in the longitudinal resistance, $R_{25,46}$, are observed near $H = 1.63\,\mathrm{T}$ and $H = 2.06\,\mathrm{T}$, and no peak is found in the Fourier power spectra at the same fields as shown by Figs. 3h and 3j. But peaks are found in the spectra of Figs. 3g, 3i and 3k, corresponding to the maxima found in $R_{25,46}$.

The shift in the centroid of the hc/e band in the spectrum and the exponential decrease in the power with magnetic field are summarized in Fig. 5 for three devices. The devices represented in the top and center portions of the Figure correspond to annuli with different radii, $r = 0.47\,\mu\mathrm{m}$ and $0.95\,\mu\mathrm{m}$, fabricated in material with $\mu = 30\,\mathrm{m}^2/Vs$, while the device in the lower portion of the Figure is the device of Fig. 2 which has a $0.93\,\mu\mathrm{m}$ radius with $\mu = 50\,\mathrm{m}^2/Vs$. Generally, the centroid of the Fourier spectrum, represented by the solid triangles in Fig. 5, is shifted monotonically to a lower frequency as the magnetic field increases. The frequency observed at high field, near the quantized Hall regime, corresponds to the edge of the bandwidth (indicated by the error bar in Fig. 5) associated with the hc/e peak observed near $H = 0\,\mathrm{T}$, and a flux of hc/e through the area associated with the inside edge of the annulus.

Thus, we deduce that the frequency of oscillation observed near the quantized Hall regime corresponds fo a flux of hc/e through the area circumscribed by the inside diameter of the annulus, and the exponential decrease in the amplitude is indicative of the exponential decrease in the overlap between the outside and inside edges of the annulus. According to this interpretation, the AB effect is suppressed because of a dramatic change in the current distribution and a reduction in backscattering which develops in an intense magnetic field. As shown schematically in Fig. 6a, for $HeW^2 < 1$ the current is carried in the bulk of the wire and the resistance is due to scattering from either the device geometry, the leads, impurities or other electrons, but for $HeW^2 > 1$, the motion along the transverse and longitudinal directions in a constriction is coupled, the subband energy becomes a function of the position along the width of the wire, and increases abruptly near the edges [20]. Since the Fermi energy crosses the subband energies near the edges of the wire, the net current is carried by states near the edges which have a spatial extent r_0 for $HeW^2 > 1$. The current along the wire is proportional to the slope of the confining potential along the transverse direction [20]. Thus, the driving force provided by the confining potential moves carriers along the edges of the wire with opposite edges carrying oppositely directed currents as illustrated in Fig. 6b.

When the width of the wire becomes much larger than the extent of the wavefunction, a carrier along one edge of the wire cannot backscatter to the opposite edge. The absence of backscattering between the two edges results in dissipationless transport along the wire; i.e. $R_{25,16} \approx 0$ [21]. Consequently, the AB effect vanishes in the quantized Hall regime where $He^2W/hc \gg 1$: (1) because the outer edge states, which are connected to

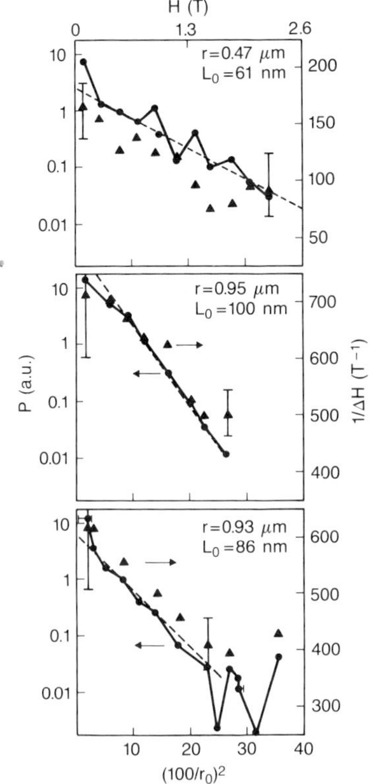

Figure 5. The Figure compares the results obtained for the peak Fourier power, P, associated with the hc/e peak (circle), and the centroid of the hc/e band (triangle) as a function of magnetic field and the reciprocal magnetic length squared for three different annuli. The middle and bottom plots correspond to the devices of Fig. 1 and Fig. 2 respectively. The exponential fit to the data $P = P_0 \exp[-(L_0/r_0)^2]$ used to determine the parameter L_0 is indicated by the dashed line. The changes in the slope of the Fourier power as a function of the r_0^{-2} are indicative of different width wires. The width of the Fourier spectra are indicated by the error bar on the centroid data.

the leads and determine the resistance, do not enclose the flux; (2) because the inner edge states which do enclose a flux are not coupled to the outer edge states; and (3) because the edge states do not backscatter. So we interpret our results to be indicative of the absence of backscattering in the quantized Hall regime [21,17]. For fields just beyond $HeW^2/hc \approx 1$, the AB effect is suppressed because the current carried along the inner edge of the annulus (see Fig. 6) backscatters only weakly to the outlying edge involved in the voltage measurement. The peak amplitude depends upon the backscattered intensity, and we presume that the backscattering depends upon the overlap between edge states. The exponential decay of the Fourier power with $r_0^{-2} \propto H$ between 200 mT and 1.6 T is

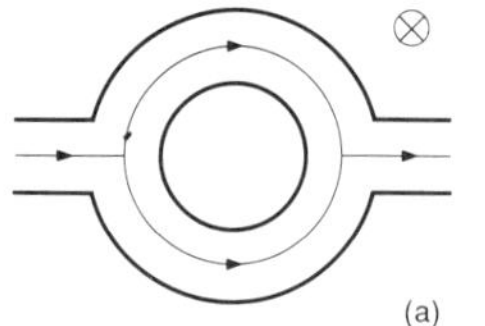
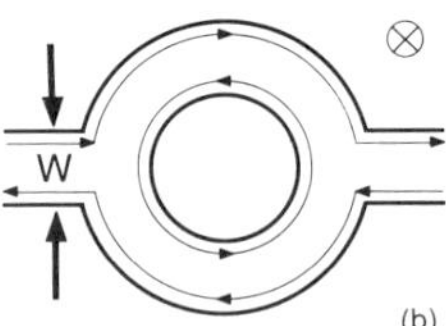

Figure 6. A schematic representation of transport through an annulus near zero magnetic field, $W < 2r_0$ (a) and in the quantum Hall regime $W \gg 2r_0$ (b). The arrows indicate the direction of current. In the quantized Hall regime a net current results from the difference between two oppositely directed currents associated with the edge states.

consistent with a decreasing overlap with increasing magnetic field. The expected form for the overlap in a high magnetic field is proportional to $\exp(-\gamma W^2/4r_0^2)$ [21] where the parameter γ is a constant of order unity and depends upon the shape of the confining potential. Our observation of an exponential decay with magnetic field is consistent with the expected form provided $\gamma \approx 0.25$ since $L_0 = 95\,\text{nm} \approx W/2 = 210/2\,\text{nm}$.

We associate the peak in the Fourier spectrum, found at high field on the lower edge of the hc/e frequency band obtained near zero field, with the area circumscribed by the inner edge states. According to this interpretation, the width of the Fourier transform found at zero field underestimates the actual width of the constriction by a factor of 1.7. For the device of Fig. 2, the mean lithographic radius is $0.93\,\mu\text{m}$. The radius corresponding to the $420\,\text{T}^{-1}$ frequency is $\approx 0.75\,\mu\text{m}$ which is greater than the inside lithographic radius of $0.68\,\mu\text{m}$ by $70\,\text{nm}$, and corresponds to $W \approx 360\,\text{nm}$, where we assume that $W/2$ is given by the difference between the radii associated with the centroid of the hc/e peak in the Fourier spectrum for $H = 0\,\text{mT}$ ($630\,\text{T}^{-1}$, $r \approx 0.91\,\mu\text{m}$) and $1.2\,\text{T}$ ($420\,\text{T}^{-1}$, $r \approx 0.74\,\mu\text{m}$) respectively.

Realistically, the outer edge states are coupled to areas enclosed by localized states throughout the bulk of the constriction as well as states about the inner edge, but the latter states are more prominent in the Fourier spectra presumably because that area, defined by the lithography, does not change appreciably with magnetic field. In the field range beyond the resistance minima and between the quantized Hall plateaus, we assume that the edge states along the inner edge of the annulus cannot tunnel directly to the outer edge, but rather scatter via potential fluctuations within the constriction.

3 Two Terminal Resistance of an Electron Waveguide

3.1 Measurements of the Two Terminal Resistance

Beside phase coherence, another distinctive feature of an electron waveguide is the quantization of the energy and transverse momenta. The quantization of the energy can be observed only if the bandwidth associated with the elastic scattering due to disorder is smaller than the intersubband energy interval. van Wees et al. [10] and Wharam et al. [11] discovered that the two terminal conductance of a short one-dimensional (1 D) constriction in a high mobility 2 DEG is quantized in steps of approximately $2e^2/h$, *in the absence of a magnetic field*, as the width of the constriction is varied. The quantization of the conductance is unequivocal evidence of the discrete energies available in a submicron constriction in a high mobility 2 DEG.

Following Pepper [22], to make a 1 D constriction in a 2 DEG we have used the electrostatic potential provided by a split-gate geometry to laterally constrain the electron gas to the region within the gap between the gates (see Fig. 7). The fabrication procedures are described in detail elsewhere [23]. The device is made on top of a $100\,\mu$m wide Hall bar that had been etched into a heterostructure. On top of the Hall bar, split-gate electrodes are fabricated using electron beam lithography. Electron beam lithography is used to prepare a mask for lift-off A Ti/Au or Ti/AuPd film approximately $7.5/50\,$nm thick is evaporated onto the mask and removed to form the electrodes. Typically, the split-gate electrodes are $200\,$nm wide with an intervening gap of $300\,$nm.

By applying a negative voltage to the split-gates, the 2 DEG gas at the AlGaAs/ GaAs interface immediately beneath the gate electrodes is depleted and so the 2 DEG is laterally constrained within the gap between the electrodes. The electrostatic potential due to split-gates, not accounting for the δ-doped impurity layer or the charge density associated with the 2 DEG or the dielectric constant, is represented in Fig. 7 [24]. The two salient features of the electrostatic confinement potential are represented in Fig. 7. The first is the smooth taper from the 2 DEG to the 1 D constriction. The contact is not abrupt because the constriction is formed by depletion using gate electrodes which are between $60\,$nm – $200\,$nm away from the 2 DEG depending upon the bias voltage. A self-consistent calculation of the potential produces a smooth taper even if there are imperfections in the gate electrodes for this reason [25]. The potential is relatively flat near the center of the constriction, but rises abruptly near the edges. When the gate voltage is large and negative, the constriction gradually widens to the 2 D contact over approximately $250\,$nm which is about $5-10\lambda_F$. A second feature of this solution is the potential barrier between the constriction and the 2 DEG along the z-axis; the constriction is at a higher potential than the 2 D contact. Since we assume that the Fermi energy is constant throughout the device, the higher potential within the constriction means that the carrier density is lower there.

Figure 8 shows the two terminal resistances (conductances), $R_{12,12}$ ($G_{12,12}$), as a function of the applied gate voltage that we found in devices like that shown schematically in Fig. 7. Electron micrographs taken from a top view of actual split-gate electrodes are also shown in top of the Figure. The series resistance found at $V_g = 0V$ was subtracted from the measured resistance obtained as a function of gate voltage to give the data shown. The depletion of the 2 DEG immediately beneath the gate electrodes occurs near -0.325 V, and -0.5 V for the devices represented respectively by the top and bottom of Fig. 8a, and is not shown. As the gate voltage decreases further, the constriction within the gap between the electrodes narrows; the carrier density within the constriction decreases, and plateaus are observed in the resistance. The two terminal conductance obtained by inverting the resistance versus gate voltage is also shown in Fig. 8. The *average* conductance (minus the series resistance) is approximately $2e^2 N/h$, with N an integer ranging from 1 to about 10, and is evidently quantized in steps of $2e^2/h$ with about $1-5\%$ accuracy as an increasingly negative gate voltage makes the constriction narrower. The gate length, L (along the x-axis in Fig. 7), for the device of Fig. 8a is about $200\,$nm. The mobilities of the 2 DEG used to fabricate the devices of Fig. 8 correspond to a low temperature mean free path of at least $L_e = \hbar k_F \mu/e \approx 7\,\mu$m where $\hbar = h/2\pi$. Since the $300\,$nm gap in each device pinches off for large negative gate voltages, the depletion around the gate must be less than $150\,$nm under conditions where

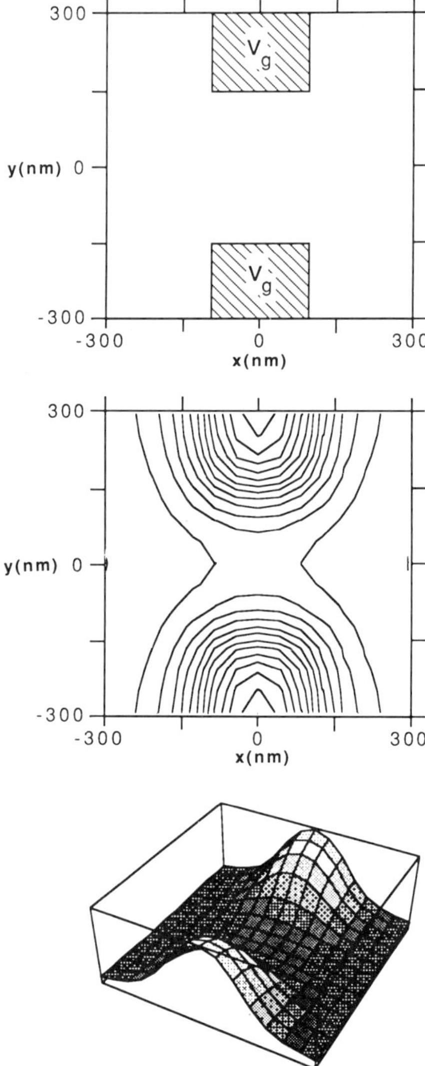

Figure 7. A graphical representation of the potential in the heterostructure near the AlGaAs/GaAs interface defined by the split-gate electrodes on the surface 70 nm above. It is assumed that -1.0 V is applied to each electrode. The top plot represents the disposition of the gate electrodes on the surface of the heterostructure. The split-gate is 200 nm long, with a 300 nm gap between electrodes. In the center is a contour plot of the potential in the plane of the 2 DEG. The equipotential contours are at intervals of 0.066 V. The lower Figure is a 3 D representation of the same potential where the height represents the size of the potential.

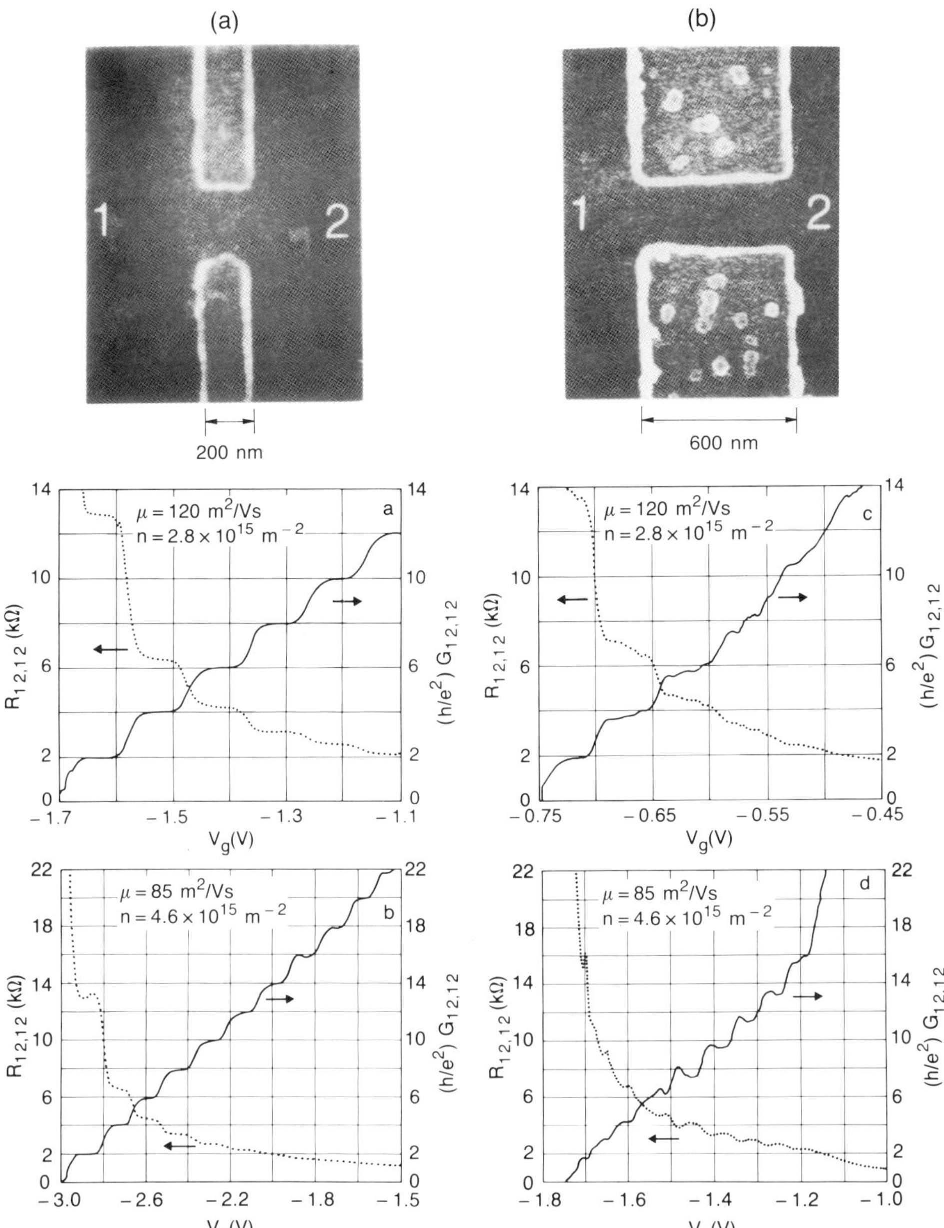

Figure 8. The two terminal resistances (conductance), $R_{12,12}$ ($G_{12,12}$), of a constriction in a 2 DEG as a function of gate voltage, V_g, obtained at 280 mK for two different heterostructures with a mean free paths, estimated from the mobility, of about 7 μm. The data shown in the top of Figs. 8a and 8b were obtained from devices in two different heterostructures. In (a) the split-gate is 200 nm long with a 300 nm gap between the electrodes. (a) shows the conductance, $G_{12,12}$, is quantized in steps of $2e^2/h$ as a function of V_g. In (b) the split gate electrodes are 600 nm long with a 300 nm gap between electrodes. Although the low temperature mean free path in the 2 DEG is much longer than the length of the constriction, (b) shows deterioration of the quantization of the two terminal resistance.

the constriction conducts, and so we estimate the actual length of the constriction to be less than 500 nm in Fig. 8a; much less than the mean free path deduced from the mobility of the 2 DEG.

We find that, as the length of the constriction increases, the quantization deteriorates. Figure 8b shows the resistance and conductance determined as above for a constriction with gate electrodes of length $L = 600$ nm, spaced with a 300 nm gap on the same two heterostructures. For this geometry, the length of the constriction is estimated to be less than 900 nm which is still a factor of eight less than the mean free path estimated from the 2 D mobility and carrier density, yet the quantization has deteriorated dramatically from that shown in Fig. 8a. While the quantization of the resistance is still apparent in the higher mobility device represented in the top of Fig 8b; the accuracy of the quantization is only about 90% for the first three steps with no quantization observed for higher N. The bottom of Fig. 8b shows the complete deterioration of the quantized resistance found in a device 600 nm long in the heterostructure which has lower mobility, but higher carrier density.

3.2   Theoretical Estimate

The measured resistance of a 1 D constriction is due to the redistribution of the current among the electronic states in the wide 2 D contact and the 1 D constriction. The quantization of the two terminal resistance of a 1 D constriction was not anticipated theoretically because it was widely held that the nature of the contact, used to measure the resistance of a ballistic constriction, was not ideal [26]. While Imry [26] observed that, theoretically, the two terminal conductance is a quantized function of the number of occupied 1 D subbands, he concluded that fluctuations of magnitude e^2/h in the measured potential due to scattering at the contacts would obscure the effect. Subsequently, Glazman et al. [27] demonstrated that the conductance of a constriction measured between two semi-infinite contacts could be quantized provided that: (1) the contacts are adiabatically tapered to match to the constriction, and (2) that the constriction is both short enough to be ballistic and long enough to prevent evanescent modes from carrying appreciable current, in qualitative agreement with the observations by van Wees et al. [10].

In a conventional two terminal measurement, two wider conductors contact each end of a constriction. If the contacts are wide enough, they become reservoirs in which the electronic motion approaches thermal equilibrium. Each contact can then be characterized by chemical potentials μ_1 and μ_2. When $\mu_1 > \mu_2$, there is a net current through the constriction. Generally, only a fraction of the flux incident in a particular subband (j) is transmitted through the constriction, i.e. $I_j = e\nu_j\delta n = e\nu_j(dg/dE)_j \sum_{k=1}^{N} t_{jk}\Delta\mu$, where ν_j and $(dg/dE)_j$ are, respectively, the Fermi velocity and the density of states at the Fermi energy for subband j, t_{jk} is the probability intensity for transmission from subband k into subband j, N is the number of subbands in the constriction, and $\Delta\mu$ is the chemical potential difference between the two wide contacts to either side of the constriction, i.e. $\Delta\mu = \mu_1 - \mu_2$. In 1 D, the density of states is inversely proportional to the velocity and so, $I_j = (2e/h) \sum_{k=1}^{N} t_{jk}\Delta\mu$. Summing the contributions from each of the subbands gives the total current, and since the voltage difference between the wide reservoirs is $V = \Delta\mu/e$, the two terminal conductance is

$$G_{12,12} = \left(2e^2/h\right) \sum_{j,k=1}^{N} t_{jk}. \tag{1}$$

This is the Landauer formula for the two terminal conductance. If we assume that (1) the temperature is much smaller than the separation between transverse energy levels, (2) the length of the perfect wire is much longer than the decay length for evanescent modes, and (3) there is no scattering (i.e. $t_{ij} = \delta_{ij}$); then the current injected into subband j is $I_j = (2e/h)\Delta\mu$, independent of the specific 1 D subband under consideration, and so the total current is $I = N(2e/h)\Delta\mu$. The two terminal conductance of constriction with N occupied subbands is then

$$G_{12,12} = \left(2e^2/h\right) N, \tag{2}$$

exactly. Thus, the ideal, two terminal conductance is a quantized function which measures the number of occupied 1 D subbands.

The correspondence between eq. (2) and the data of Fig. 8 is surprising because to obtain eq. (2) it is assumed that (1) $T = 0$, (2) the transmission through the contacts into the constriction and through the constriction is perfect, and (3) the contacts to the constriction are thermal reservoirs. Thus, the observation of a well quantized two terminal conductance implies (1) that the energy separation between subbands is much larger than $280\,\mathrm{mK}$, (2) that the constriction is ballistic for $L = 200\,\mathrm{nm}$ when $L_e \gg L$ and the reflections associated with contacts to the constriction are negligible , and (3) that the finite width of the contacts can be ignored. While phenomenologically the 2 D contacts to the ballistic constriction appear ideal because the resistance is quantized; it is not necessarily so that there is no scattering at the contacts. As Szafer and Stone [28] have shown, the resistance may be well quantized even if there are substantial reflections in the neighborhood of the contacts.

Because of depletion, the 1 D constriction gradually widens to embrace the 2 D contact as shown in Fig. 7. Glazman et al. [27] first appreciated this and developed an adiabatic model for conductance of a constriction which relies upon a gradual taper from the 1 D constriction to the 2 D contact. Following Glazman [27], we imagine an electron waveguide with a width which is a function of the longitudinal position, i.e. $W(x)$, yielding an adiabatically smooth constriction of width W_i tapering from a width W_f. If the variation along x is slow, the Hamiltonian separates into terms in the longitudinal variable, x, and the variable y. The y-dependent part of the Hamiltonian is the hard wall problem with solution $E_N(x) = \hbar^2\pi^2 N^2/2m^*W^2(x)$, where $N = 1, 2, \ldots$ The x-dependent part of the Hamiltonian represents scattering from a barrier. For a given Fermi energy, E_F, only a finite number of modes, N, will be above the barrier. If tunneling below and reflections above the barrier are negligible, the N modes, with $t_{ij} = \delta_{ij}$, carry current while the transmission of all other modes vanishes exponentially and so eq. (2) is recovered from eq. (1).

It is implicit in the model by Glazman that the electron wave is focussed. A wave propagating adiabatically along a variable-width guide conserves the mode number N. As the width of the guide narrows, the transverse energy increases while the longitudinal energy decreases continuously. When the wave abruptly encounters the end of the adiabatic taper, the transverse wavevector is now conserved (at least in the interval $\pm\pi/W_f$) and the wave is focussed to a width $\pi k W_f$. It is unlikely that the actual constrictions

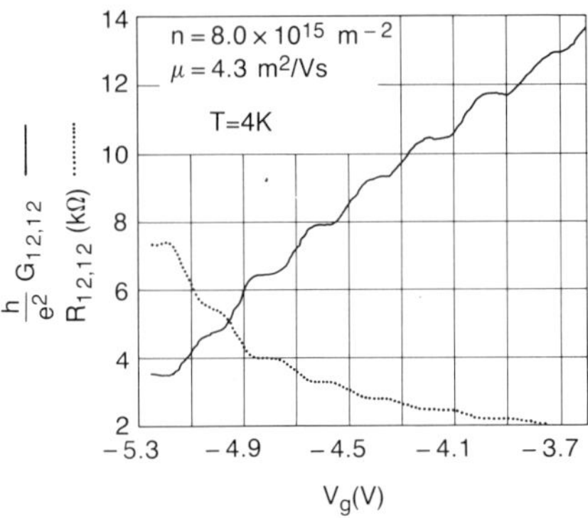

Figure 9. The two terminal resistance (conductance), $R_{12,12}(G_{12,12})$, of a constriction in a 2 DEG as a function of gate voltage, V_g, found at 4 K. The split-gate electrodes used to define the constriction are 200 nm long and 300 nm wide. Even though the low temperature mean free path estimated from the mobility ($L_e \approx 450\,\mathrm{nm}$) is comparable to the length of the constriction, the resistance is still quantized with a precision of 15%.

are entirely adiabatic; Fig. 7 shows that it is not. However, even if the taper is not adiabatic, the reflections associated with the mismatch between the 2 D contact and the constriction may not be large because the taper is *gradual*, occurring on the scale of 250 nm or $5 - 10\lambda_F$. It is sufficient that the adiabaticity holds only up to a certain width beyond which reflections occur or that the reflections are small [29].

We attribute the deterioration of the quantization in the long constrictions to disorder or impurity scattering within the constriction [30]. Depending on the position, impurity scattering may reduce or enhance the conductance. If an impurity is in the constriction, the conductance is reduced from the quantized value. Alternatively, if the impurity is outside the constriction, the conductance may be either enhanced or reduced. In either case, an impurity destroys the quantization of the conductance. While it is widely used as an estimate for the mean free path and a measure of the extent of impurity scattering, L_e, inferred from the mobility of the wide wire is not necessarily an appropriate estimate for the elastic scattering length in the constriction, especially for gate voltages near pinch-off. This hypothesis is dramatically illustrated by Fig. 9 which demonstrates quantization of the two-terminal conductance in steps of $2e^2/h$ with a precision of about 15% at $T = 4$ K in a heterostructure for which $L_e \approx 450$ nm [31]. Parenthetically, the accuracy of the quantization of the conductance of the device of Fig. 9 was not improved by lowering the temperature to 280 mK.

A 2 DEG effectively screens potential fluctuations due to Coulombic impurities, but as the constriction in the 2 DEG becomes narrower with more negative gate voltages and the carrier density is reduced, the Fermi wavevector k_F becomes shorter, and the screening of the impurities in the doped layer by the electron gas becomes less effective [32]. Impurity potentials beneath the gate will be screened by the metal electrodes, but the impurities in the gap between the electrodes are not. We supposed that the elastic scattering length is reduced to approximately $1\,\mu$m or less because the 1 DEG ineffectively screens potential fluctuations.

4 Three Terminal Resistance

4.1 Theoretical Estimates

Two terminal measurements of a 1 D constriction do not directly measure the potential of the constriction because of the contact resistance between the 2 DEG and the constriction[26]. Three and four terminal measurements do not generally measure the local potential of the conductor either, even if the leads are made from the same material and have the same dimensions as the conductor and there is no net current in the voltage leads. Although the net current in the voltage leads vanishes, a carrier can still propagate coherently from the conductor into a lead and back into the conductor. The lead then becomes part of the measured resistance and must be included in the computation of the resistance. As illustrated below by three terminal measurements, the leads or contacts can actually determine the resistance of a 1 D constriction.

Following the developments of the Landauer formula made by Büttiker [33], we can generalize the computation of the resistance to a multi-terminal measurement. The current in lead j is given by

$$I_j = -(e/h) \sum_{k=1}^{N} T_{jk}\mu_k,\tag{3}$$

where T_{jk} is the trace over the subbands, i.e. $T_{jk} = \sum_{mn} t_{tk,mn}$ with $t_{jk,mn}$ representing the probability for a carrier in subband n and lead j to be transmitted to subband m in lead j, and $T_{jj} = \sum_{mn} r_{jj,mn} - N_j$ with N_j the number of subbands in lead j and $r_{jj,mn}$ the probability of a carrier in subband n and lead j to be reflected into subband j and lead m. The potential, V_k, associated with a reservoir of chemical potential μ_k is given by $V_k = \mu_k/e$.

From eq. (3) we can obtain formulae for the three terminal or four terminal resistance by imposing a current, I, between two leads, requiring the net current in the other leads to vanish, and solving for the voltage difference. For the purpose of illustration, we consider the mathematically simpler problem associated with the determination of the three terminal resistance first. If we impose a current between leads 1 and 2 so that a positive current enters through lead 1 and exist through lead 2, then calculate the potential differences $\mu_1 - \mu_3$ and $\mu_3 - \mu_2$, we find the conductances

$$G_{12,13} \equiv eI/(\mu_1 - \mu_3) = \frac{e^2}{h}\frac{D}{T_{32}}\tag{4}$$

$$G_{12,32} \equiv eI/(\mu_3 - \mu_2) = \frac{e^2}{h}\frac{D}{T_{31}}\tag{5}$$

with $D = T_{21}T_{31} + T_{21}T_{32} + T_{31}T_{23}$; whereas the two terminal conductances associated with leads 1 and 2, and 1 and 3 are, respectively

$$G_{12,12} = \frac{e^2}{h}\frac{D}{(T_{31} + T_{32})}\tag{6}$$

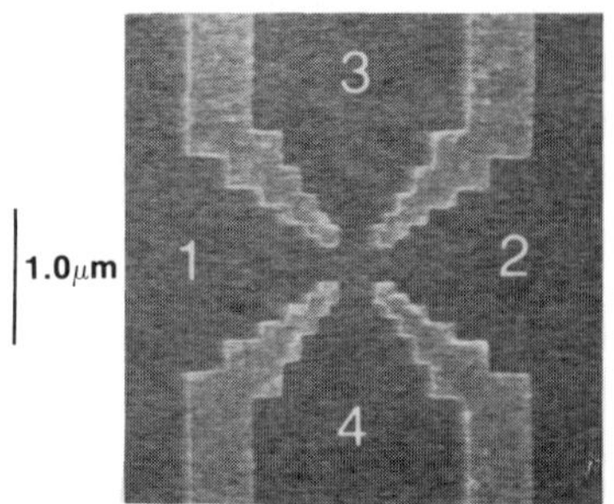

Figure 10. An electron micrograph of two series split-gate electrodes in close proximity fabricated on a high mobility heterostructure. Associated with each port in the device, numbered 1 through 4, there is a contact to the 2 DEG so that multi-terminal resistance measurements can be performed on a junction in which the geometry can be manipulated.

$$G_{13,13} = \frac{e^2}{h} \frac{D}{(T_{21} + T_{23})} \tag{7}$$

The two terminal conductance in the presence of one additional lead generally differs from eq. (1). The additional lead scatters carriers according to the probabilities T_{13} and T_{32}.

4.2 Measurements of the Three Terminal Resistance

To demonstrate the effect of a lead on the conductance, we examined two constrictions in close proximity to one another. Figure 10 shows an electron micrograph of such a device. The device is comprised of two sets of split-gate electrodes in series; each set of electrodes is 200 nm long with a gap of 300 nm between the electrodes. The two sets are separated by 300 nm to make the device symmetric. Associated with each port of the device there are contacts to the 2 DEG which are not shown. The convention for numbering the leads is given in the Figure. Figure 11 illustrates the potential contours calculated for such a device when each of the gate electrodes is biased at $V_g = -1V$. The gate bias is applied between the gate electrode and the source contact. Ideally, with each of the gate electrodes biased similarly, the device forms a symmetric junction comprised of four constrictions in a cross geometry. Each constriction can be used as a current of voltage lead.

Although the disposition of the split-gate electrodes in a typical device is symmetric, each electrode can be biased independently so the symmetry of the device can be manipulated. For example, it is possible to bias the device of Fig. 10 such that only three of the four constriction conduct. Figure 12 shows the conductances $G_{12,12}$, $G_{12,13}$ and $G_{13,13}$ obtained in such a device. The top trace in Fig. 12 shows the dependence of $G_{12,12}$ on the voltage applied to the electrodes which define lead 1, V_{g1}, when the gate voltage on the constriction which defines lead 2, V_{g2}, is zero. Under these circumstances, $G_{12,12} = G_{13,13} = G_{12,13}$ because contacts 2 and 3 are identical. The traces at the bottom of Fig. 12 were obtained by biasing the gate electrodes defining lead 2 in Fig. 10 at

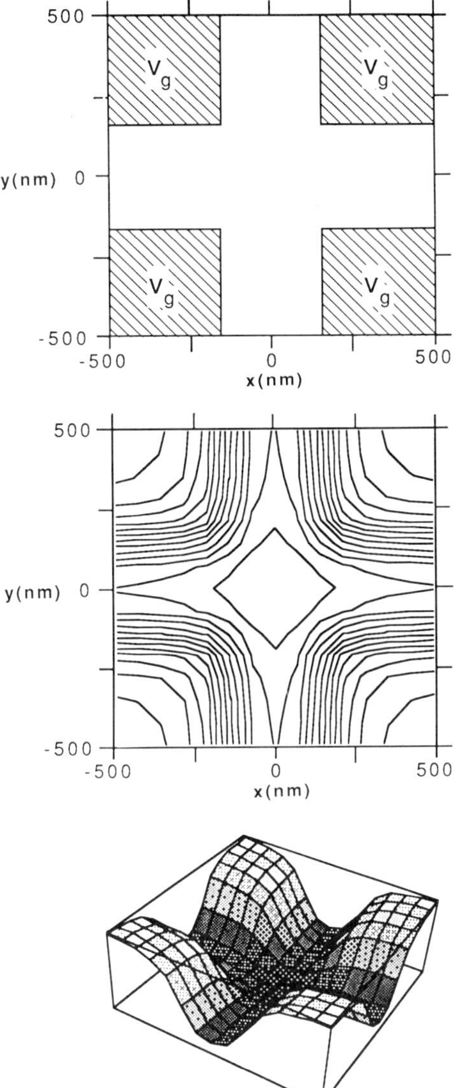

Figure 11. A graphical representation of the potential in the heterostructure near the AlGaAs/GaAs interface defined by two split-gate electrodes in series on the surface 70 nm above. It is assumed that -1.0 V is applied to each electrode. The top plot represents the disposition of the gate electrodes on the surface of the heterostructure. Each split-gate is 200 nm long, with a 300 nm gap between electrodes. The two split-gates are 300 nm apart. In the center is a contour plot of the potential in the plane of the 2 DEG. The equipotential contours are at intervals of 0.066 V. The lower Figure is a 3 D representation of the same potential where the height represents the size of the potential.

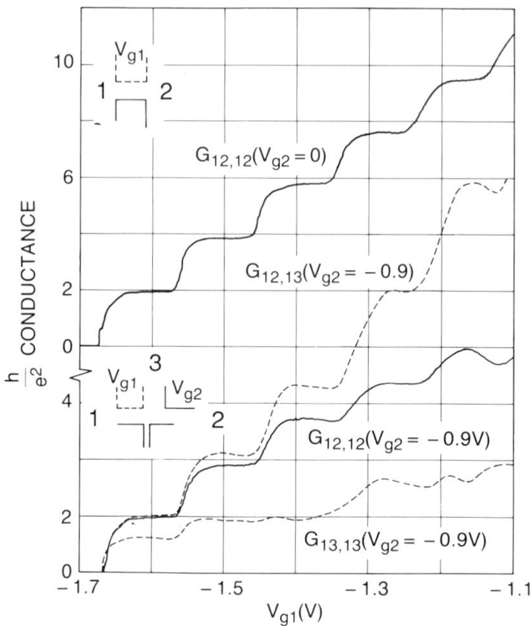

Figure 12. The two terminal resistances, $G_{12,12}$ and $G_{13,13}$, and the three terminal resistance, $G_{12,13}$, as a function of gate voltage for two series constrictions. In the top trace, all three conductances coincide because there is no second constriction. When the second constriction is biased at $V_{g2} = -0.90$ V so that only 6 subbands are occupied, the two and three terminal resistances no longer coincide because of the discrepancy between the transmission coefficients which they measure.

$V_{g2} = -0.9$ V and varying the potential V_{g1} from 0.0 V to -1.7 V. In this particular device, the thresholds for the constrictions comprising the cross were such that lead 4 did not conduct if $V_{g1} < -0.8$ V for $V_{g2} < -0.8$ V. Using eqs. (4)-(5) and (6)-(7) above we can estimate the transmission coefficients directly from the data of Fig. 12. For example, near $V_{g1} = -1.6$ V we find that $T_{13} = 0.00\pm0.02$, $T_{21} = 1.98\pm0.02$ and $T_{23} = 2.94\pm0.04$, while near $V_g = -1.4$ V we find that $T_{13} = 0.6627\pm0.05$, $T_{21} = 5.0\pm0.05$ and $T_{23} = 1.82\pm0.05$. Under these experimental conditions, we deduce that $T_{31} \ll T_{21}$ and $T_{32} \approx 1.5$ for low N although T_{31} increases and becomes comparable with T_{21} as N increases.

In Fig. 13 we compare the dependence of the two and three terminal conductances, $G_{12,12}$ and $G_{12,13}$, as a function of gate voltage applied to the second constriction, V_{g2}. Traces (a), (b), (c) and (d) were obtained by biasing V_{g2} at voltages of: -0.55 V, -0.67 V, -0.9 V, and -1.15 V respectively and varying V_{g1}. In the inset to Fig. 13, the two terminal conductance through the second constriction, $G_{12,12}$, as a function of gate voltage is shown when $V_{g1} = 0$. The voltages corresponding to the traces (a)-(d) are also indicated in the inset. The shift in the threshold for conduction as a function of V_{g1} found in Fig. 13 for different values of V_{g2} is only about 50 mV, but becomes greater as V_{g2} becomes more

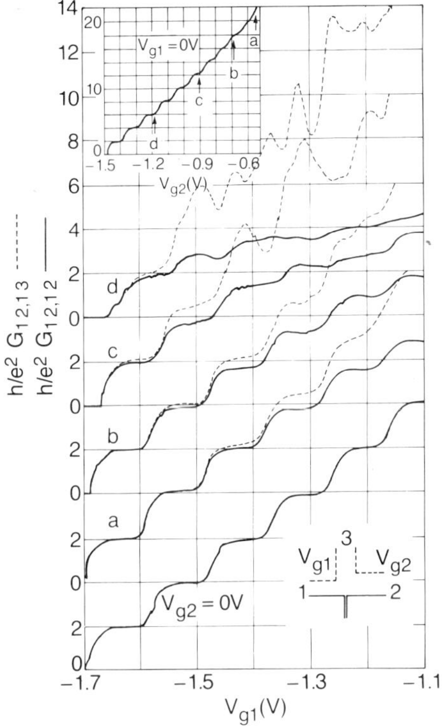

Figure 13. The three terminal conductance, $G_{12,13}$, and the two terminal conductance, $G_{12,12}$, of two series constrictions as a function of gate voltage applied to the forward constriction, V_{g1}. The gate voltage applied to the second constriction, V_{g2} is the parameter varied between traces (a)-(d). For the biases shown, lead ♯4 in Fig. 10 is pinched off and does not conduct. $G_{12,13}$ and $G_{12,12}$ coincide only for low gate voltages $V_{g1} < -1.6$ V where only one subband is occupied in the forward constriction. The inset depicts the quantization of the two terminal conductance, $G_{12,12}$, versus V_{g2} for the second series constriction when there is no forward constriction, i.e. $V_{g1} = 0$ V.

negative. The shift occurs because near threshold the electrostatic potential beneath a particular electrode depends upon the voltage applied to the other electrodes, as well.

The solid lines in traces (a)-(d) of Fig. 13 indicate that the conductance, $G_{12,12}$, is determined by the value of the narrowest constriction [34]. For example, in trace (d) $G_{12,12}$ is suppressed to values below $6e^2/h$ at $V_{g1} = -1.1$ V corresponding to the conductance of the second constriction where only three 1 D subbands are occupied, i.e. $N = 3$. Figures 12 and 13 show that a second constriction in series with and in close proximity (300 nm) to the first can be detrimental to the quantization for $N > 1$. Although regular steps are found in the conductance for $V_{g1} > -1.6$ V, the steps are not quantized precisely at $2e^2/h$. The latter observation differs from a report by Wharam et al. [34] of quantized resistances for constrictions about 1 μm apart. The conductance found for lower index $N \approx 1$ can be well quantized however, even when the second constriction is very narrow (see trace (d) in Fig. 13).

The data of Figs. 12 and 13 demonstrate that a lead can be invasive and so influence the measurement of the resistance of a wire. The dashed lines in Fig. 13 depict the results of the three terminal measurement $G_{12,13}$. We notice that the three terminal conductance is generally larger than $G_{12,12}$ and deviates substantially from the quantized values except for low N in the forward constriction. For low N, $G_{12,13} \approx G_{12,12} \approx 2e^2 N/h$. Under these circumstances, it follows from eqs. (4) and (6) that $T_{31} \ll N$ for low N.

From eq. (5) we infer that the second constriction is practically a perfect conductor when N is small in the first constriction, i.e. $G_{12,32} \gg e^2/h$. The measured resistance of the second constriction was found to be less than the experimental error of $50\,\Omega$. The resistance of the second series constriction vanishes because there is no contact resistance. There is no contact resistance because the first constriction focuses the electron wave on the second series constriction [35]. In addition, it is apparent from Fig. 13 that the resistance of two series constrictions is not additive [34]. For example, when $V_{g1} = -1.6\,\mathrm{V}$ and $V_{g2} = 0\,\mathrm{V}$, $G_{12,12} = 2e^2/h$; when $V_{g1} = 0\,\mathrm{V}$ and $V_{g2} = -1.2\,\mathrm{V}$, $G_{12,12} \approx 6e^2/h$; but when $V_{g1} = -1.6\,\mathrm{V}$ and $V_{g2} = -1.2\,\mathrm{V}$, $G_{12,12} \approx 1.8e^2/h$ which is greater than the value corresponding to the sum of the two resistances.

The collimation of the electron wave in the forward direction, which is a feature of the adiabatic approximation [27] for a tapered contact to the constriction, can produce a vanishing resistance in the second constriction and non-additivity [35]. Because an electron wave in a low index subband in lead 1 does not propagate around the bend into the voltage contact 3, the constriction which defines lead 3 does not scatter or couple to all modes in the same way and a non-equilibrium distribution of the current among the available momenta develops. The widening of the constriction between lead 1 and 3 causes electrons injected into lead 1 from the 2 DEG contact to travel through the constriction to the wide region of the junction while conserving the transverse quantum state. Consequently, there is a transfer of transverse momentum into the longitudinal momentum. Thus, the current injected from the 2 D contact, which was equally distributed between the 1 D modes of the constriction, reaches the wide junction between the three constrictions distributed only among the lowest-energy subbands of those allowed in the wider region of the junction. Ultimately, the junction widens abruptly so that the equipotential lines must turn through 90 degrees. From this point, the transverse momentum is conserved, not the number of occupied subbands, yielding a wave collimated in the forward direction [36].

The reluctance of an electron wave to propagate around a bend, measured by T_{31}, is a manifestation of the collimation of the wave. The narrower the constriction is, the lower the index for the number of occupied subbands, the more tightly focussed the electron wave is.

5   Four Terminal Resistance of an Electron Waveguide

5.1   Theoretical Estimates

The small probability for the low index N modes to propagate around a bend, and the non-equilibrium distribution of the carriers among the subbands in the junction which develops as a consequence of it, has important implications for four terminal resistance measurements of a constriction. Following Büttiker [33], we can express the four terminal resistance in terms of transmission probabilities by using eq. (3). If we impose a current, I, between leads m and n, we can solve eq. (3) for the voltage difference, $V_k - V_l$, between leads k and l. The four terminal resistance, $R_{mn,kl} = (V_k - V_l)/I$, can be expressed as

$$R_{mn,kl} = (h/2e^2)(T_{km}T_{ln} - T_{kn}T_{lm})/D \qquad (8)$$

where D is always positive and depends on the lead geometry but is independent of permutations in the indices mn, kl. If a four terminal junction like that shown in Fig. 10 is approximately symmetric, then we expect that $T_{31}, T_{24} \ll N$ and $T_{21}, T_{34} \approx N$ for low N. So $R_{14,32} = (T_{31}T_{24} - T_{34}T_{21})/D < 0$ and $R_{12,43} = (T_{41}T_{32} - T_{42}T_{31})/D \approx 0$ in the absence of a magnetic field.

Although eq. (8) applies for an arbitrary magnetic field, the transmission coefficients vary with an applied magnetic field because of the Lorentz force [37]. As we showed in the second section, in an intense magnetic field, the net current is carried by edge states and the transmission probabilities between adjacent leads is exactly equal to the number of occupied edge states (times two because of the spin-degeneracy), while the transmission between leads which are not adjacent vanishes. Thus, the Hall resistance measures the transmission probability between adjacent leads. Since $T_{13}, T_{24}, T_{41}, T_{32} = i$ while all other $T_{ij} = 0$, $D = i^3$, we infer from eq. (8) that $R_{14,32} \geq 0$ and $R_{12,43} = h/e^2 i$ where i denotes the number of (spin-polarized) edge states occupied in the 1 D constriction. In contrast with a 2 DEG, a 1 D constriction exhibits the integer quantized Hall effect $R_{12,43} = h/e^2 i$ only for $i \leq 2N$, where N is the number of 1 D subbands occupied in the absence of a magnetic field, because there are only a finite number of edge states to begin with. There is a one-to-one correspondence between the number of spin-degenerate edges states in a magnetic field, and the number of 1 D subbands occupied in the absence of a magnetic field.

5.2   Measurements of the Four Terminal Resistance

If the gate electrodes are biased similarly, then the potential distribution of the junction of Fig. 10 ideally resembles the cross geometry shown in Fig. 11. We can use the four constrictions as the current and voltage leads to measure the resistance. Figures 14a and 14b respectively show the calculated four terminal resistances [38] $R_{12,43}$ (dashed line) and $R_{14,32}$ (solid line) corresponding to an ideal junction with hard walls in the absence of a magnetic field, and the same resistances measured in a device like that shown in Fig. 10. In addition to the four terminal resistances, we also show the two terminal conductances $G_{12,12}$ associated with (1) only one split-gate electrode biased (dot-dashed line), and the two split gate electrodes in series biased (dotted line). The calculated resistance is plotted as a function of the Fermi energy normalized by the ground state energy of the wire, while the measured resistance is plotted versus gate voltage. There is not necessarily a direct correspondence between the two abscissae of Fig. 14.

While the calculated resistance, $R_{12,43}$, is featureless because the junction was assumed to be perfectly symmetric, the calculation of $R_{14,32}$ reveals a negative resistance with sharp minima as a function of energy [39-41]. The sharp minima found in $R_{14,32}$ correspond to the thresholds for occupation of 1 D subbands in the wires comprising the junction [39-41]. The value of $G_{12,12}$ can be used to ascertain the number of occupied subbands since $G_{12,12} = (e^2/h)T_{12} = (2e^2/h)N$. As the energy increases from a value just above threshold for the occupation of a subband to a value below the threshold for occupation of another subband, all of the modes become lower in energy and the

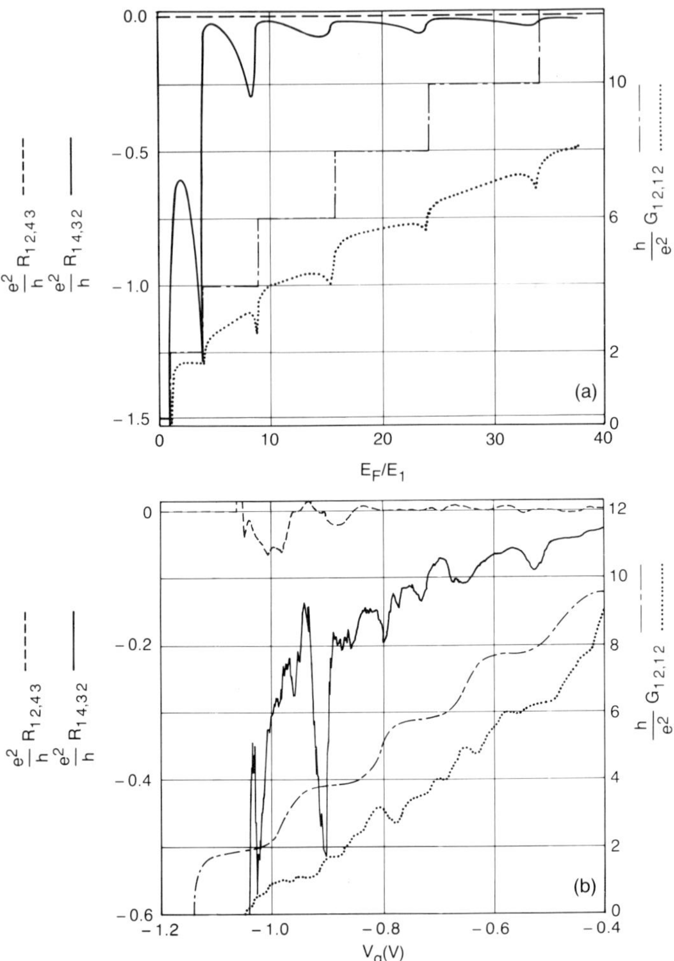

Figure 14. (a) shows the calculated dependence on Fermi energy, E_F, of three characteristic resistance measurements made in a cross geometry with hard walls and sharp corners [48]. The Fermi energy is normalized to the ground state energy of a square well, E_1. The solid line represents the energy dependence of the bend resistance, $R_{14,32}$, the dashed line represents the dependence of the resistance $R_{12,43}$, and the dotted line represents the dependence of the two terminal conductance, $G_{12,12}$. The dashed-dotted line represents the energy dependence of the two terminal conductance of only one constriction without any other constriction; it is used as a measure of the subband occupation. (b) shows the measured dependence of gate voltage, V_g, of the same three characteristic resistance measurements made in the cross geometry between two split-gate electrodes in series. The solid line represents the dependence on the voltage applied to two split-gate electrodes of the bend resistance, $R_{12,32}$, the dashed line represents the dependence of the resistance $R_{12,43}$, and the dotted line represents the dependence of the two terminal conductance, $G_{12,12}$. The dashed-dotted line represents the dependence of the two terminal conductance on the gate voltage applied to only one split-gate electrode with the other grounded.

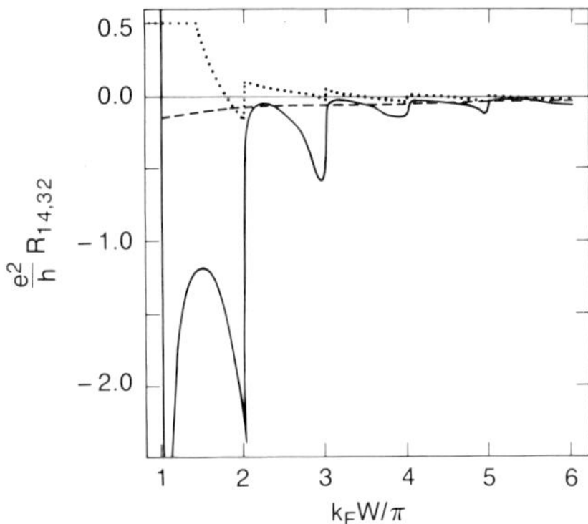

Figure 15. Three calculations of the dependence of the bend resistance, $R_{14,32}$, of a junction with sharp corners and hard walls on the subband occupation measured by $k_f W/\pi$. The dashed, dotted and solid lines depict classical, semi-classical and quantum mechanical calculations of the resistance respectively.

Fermi-wavevector associated with each of the occupied modes becomes more forward-directed. Consequently, T_{21} increases relative to T_{31}, and the resistance $R_{14,32}$ becomes more negative. Globally, $R_{14,32}$ becomes less negative as the energy increases, however. The global change in $R_{14,32}$ occurs because the number of subbands increases.

The sharp minima $R_{14,32}$ are due to the quantum mechanical nature of the transport. Figure 15 compares a classical (dashed line), semi-classical (dotted line) and quantum mechanical (solid line) calculation of the resistance $R_{14,32}$ of a junction with hard walls and sharp corners. The classical calculation of the resistance uses a Boltzmann transport equation assuming that the transport is ballistic without any quantization of the momenta or phase coherent scattering. The injected particles have momenta which is distributed uniformly as a function of the angle. The semi-classical calculation also assumes ballistic transport without phase coherent scattering, but *with* discrete transverse momenta. Consequently, only certain injection angles, of all those which are classically available, are allowed. The quantum mechanical calculation includes both the quantization of the transverse momenta and phase coherent scattering. The importance of quantum mechanical reflections near the threshold for the occupation of a subband which develop from coherence is apparent from the dramatic differences between the classical and semi-classical calculations, and the quantum mechanical calculation for low N. As the number of subbands increases however, the differences between the results of the three calculations diminishes.

Sharp minima are also observed in a measurement [42] of $R_{14,32}$ versus V_g as shown in Fig. 14b. As V_g becomes more negative, globally $R_{14,32}$ becomes more negative, but there are substantial fluctuations about the average at low temperature ($T < 2$ K). The same features with the same magnitude are found in $R_{13,42}$ at the same gate voltages, as well, as shown in Fig. 16a. Similar to the calculation, the measurement of $G_{12,12}$, when only constriction ♯1 is biased, is quantized [10] in steps of $2e^2/h$. The V_g's associated with the threshold for occupation of a subband in a single constriction represents only a lower bound for the threshold voltage when all the constrictions are biased because the potentials of the other gate electrodes add to the potential of the two which define constriction ♯1. The quantization of both the measured and calculated $G_{12,12}$ deteriorates in the junction compared to a single constriction because of scattering by the leads [12]. The minima found in the calculated $G_{12,12}$ in the junction cannot be unambiguously identified in the measured conductance. This lack of correspondence may be related asymmetry, or flaring of the constrictions at the junction. Note that $R_{12,43}$ does not vanish for all V_g which is evidence of some asymmetry in the scattering.

There is a correspondence between (1) the minima observed in $R_{14,32}$ in the absence of a magnetic field, (2) the value of the resistance associated with the first plateau found in the Hall effect with increasing magnetic field, and (3) the number of plateaus found as a function of H, when $N \leq 4$ in the constrictions comprising the junction. The correspondence indicates that the sharp minima found in $R_{14,32}$ versus V_g are associated with a different number of occupied subbands in the constrictions which comprise the junction. Furthermore, the temperature dependence of the minima is consistent with them being near the threshold for occupation of subband [42].

Since $G_{12,12}$, $G_{34,34}$, $G_{13,13}$, and $G_{14,14}$ all vanish beyond $V_g = -1.050\,\text{V}$, we assume that the sharp minima shown in $R_{14,32}$ (and $R_{13,42}$) near $V_g = -1.025\,\text{V}$ in Fig 16a correspond to the threshold for occupation of the $N = 1$ subband in all of the constrictions in the junction. This identification is supported by the Hall resistance shown in Fig. 16b, traces (a), (b), and (c), where $-1.035\,\text{V} \leq V_g \leq -0.98\,\text{V}$. With an applied magnetic field, $R_{12,43}$ changes dramatically from a value near zero at zero magnetic field to a quantized value corresponding to the number of occupied subbands in the constriction. According to Fig. 16b, $R_{12,43}$ reaches a plateau at $(h/e^2)/(2.05 \pm 0.07)$ near $H = 2\,\text{T}$, which we associated with the depopulation of a spin-polarized edge state $i = 2$. There is no evidence of the depopulation of a higher index edge state for $V_g < -1.0\,\text{V}$. Traces (d) and (e) of Fig. 16b show the Hall resistance when the split-gate electrodes which define the junction are each biased near $V_g = -0.90\,\text{V}$, where there is a sharp minima in $R_{14,32}$. We presume that this range of gate voltage corresponds to the threshold for occupation of the second subband in all of the constrictions comprising the junction, i.e. $N = 2$ or $i = 4$ subbands are occupied. This identification is supported by the observation of a plateau in $R_{12,43}$ beyond $H = 0.5\,\text{T}$ at $(h/e^2)/(4.10 \pm 0.2)$ for $-0.936\,\text{V} \leq V_g \leq -0.90\,\text{V}$, which corresponds to the magnetic depopulation of the $i = 4$ edge state. (The shallow minimum found near $V_g = -0.950\,\text{V}$ in $R_{14,32}$ might be associated with the occupation of the $N = 2$ subband in at least one, but not all, of the constrictions.) We associated the third sharp minimum found near $V_g = -0.8\,\text{V}$ with the occupation of $N = 3$. Traces (g) and (h) of Fig. 16b support this interpretation since the first plateau found in the Hall resistance, for $-0.82\,\text{V} < V_g < -0.795\,\text{V}$, is near $h/e^2 6$. While it is possible to identify the minima in $R_{14,32}$ for $V_g = -0.88\,\text{V}$ with the occupation of higher energy subbands, the identification is dubious because of the possibility that the junction may have at least one constriction with a different number of subbands occupied than the others for particular ranges of V_g, and because of the poor accuracy of the quantization of $R_{12,43}$ for $i > 6$.

Figure 17 shows the magnetoresistances found in the device of Fig. 14b for $V_g = -1.025\,\text{V}$ applied to each of the electrodes. Figure 17 shows that for $H < 200\,\text{mT}$ the magnetoresistance, $R_{14,32}$, is asymmetric and negative, and while generally suppressed below the conventional 2 D Hall resistance, $R_{12,43}$ is non-zero at $H = 0$. As the applied magnetic field increases, the resistance changes dramatically. Figure 18 shows the symmetrized magnetoresistance found at $V_g = -1.035\,\text{V}$, $-0.900\,\text{V}$, $-0.770\,\text{V}$, and $-0.560\,\text{V}$

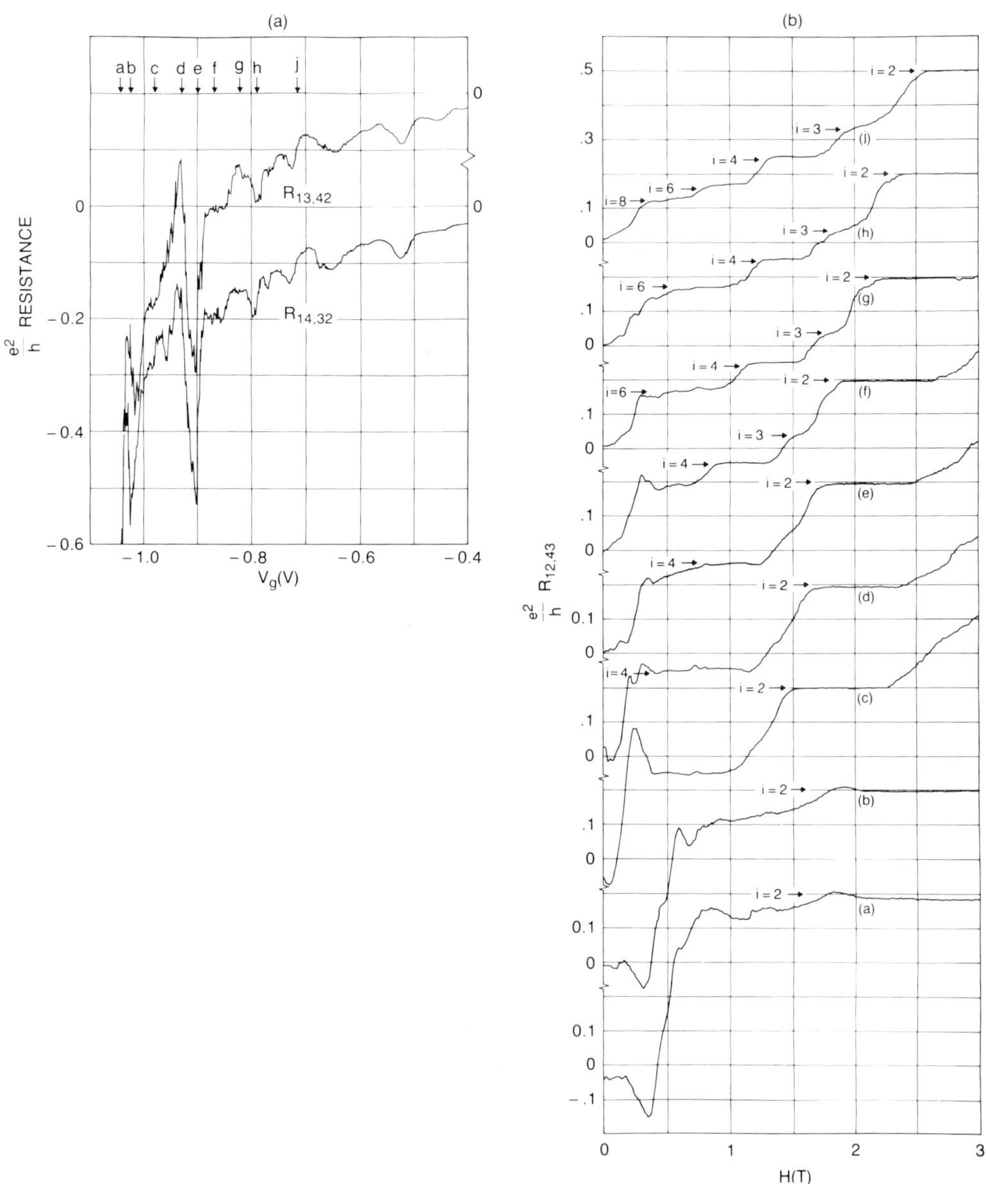

Figure 16. (a) shows the bend resistances, $R_{14,32}$ and $R_{13,42}$, as a function of the gate voltage V_g applied to each of the series split-gate electrodes of a device like that shown in Fig. 10. $R_{13,42}$ is offset by $0.2h/e^2$ from $R_{14,32}$ for clarity. (b) shows the Hall resistance, $R_{12,43}$, versus magnetic field for 9 different gate voltages (a-h, j) applied to the junction of (a). For trace (a-h, j) V_g =-1.035 V, -1.025 V, -0.980 V, -0.936 V, -0.900 V, -0.871 V, -0.817 V, -0.795 V, and -0.712 V respectively. These gate voltages are indicated by the arrows in (a) as well.

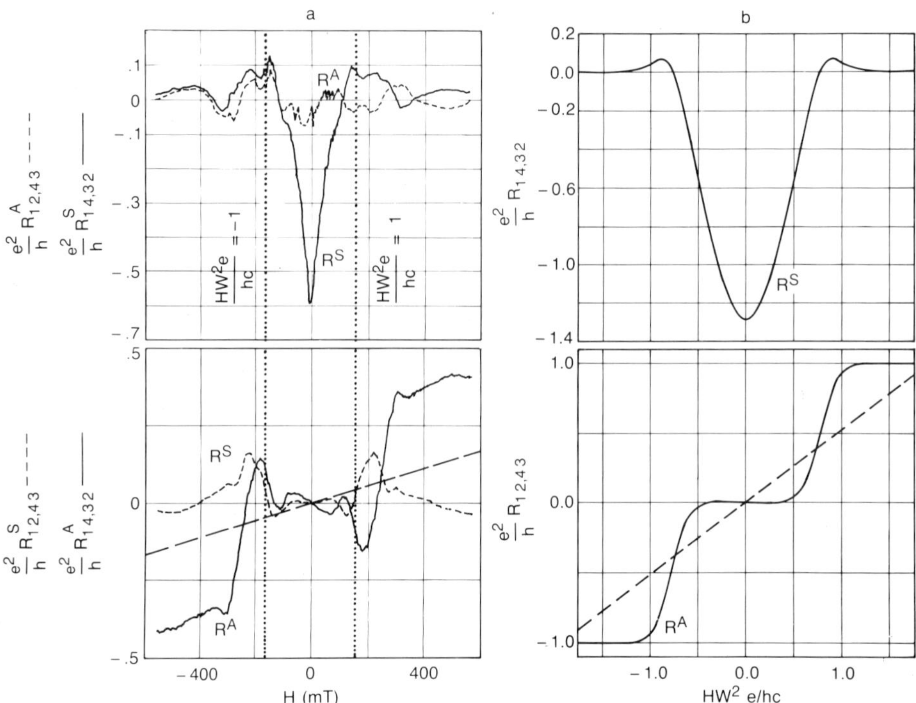

Figure 17. (a) depicts the symmetric, R^S, and anti-symmetric, R^A, components of the magnetoresistances $R_{14,32}$, and $R_{12,43}$ found at 280 mK for the device of Fig. 14b. The decompositions $R^S = 1/2[R_{kl,mn}(H) + R_{mn,kl}(H)]$ and $R^A = 1/2[R_{kl,mn}(H) - R_{mn,kl}]$ are supposed to be symmetric and anti-symmetric respectively as a direct consequence of the symmetry of the transmission coefficients. In a macroscopic junction with the same geometry, 2 D mobility and carrier density, $R_{14,32}$, is a positive number at zero field (see arrow in Fig. 17a top) and $R_{12,43}$ goes to zero linearly as a function of magnetic field (as indicated by the dashed line in the bottom of (a). The dotted vertical line indicates where the condition $HeW^2/hc = \pm 1$ is satisfied assuming a hard wall confining potential. (b) shows the calculated magnetoresistance for a junction made from a hard wall potential with sharp corners [48]. Only one subband is occupied, and the spin-degeneracy is ignored in the calculation. The essential features found in the measurement of the resistance are reproduced by the calculation. Beyond $HeW^2/hc = \pm 1$ both (a) and (b) show that $R_{14,32}$ is no longer negative, and the Hall resistance, $R_{12,43}$, is not suppressed.

respectively in the device of Fig. 14b. Using our estimate for the number of subbands in the constriction and our estimate for the carrier density, and assuming a hard wall confinement potential, we can infer the width of the constrictions comprising the junction. The width of the constrictions is related to the carrier density and the number of 1 D subbands through an integral over the 1 D density of states. We find that for the bias conditions $V_g =$ -1.035 V, -0.900 V, -0.770 V and -0.560 V, which correspond respectively to the conditions shown in Fig. 18a-d, $W = 70 \pm 25$ nm, 80 ± 25 nm, 120 ± 20 nm, and 170 ± 15 nm. The arrows in the top row of Fig. 18 represent the conventional longitudinal resistance expected for a bend in a wide, diffusive junction with the same geometry, 2 D mobility, and 2 D carrier density. The long dashed lines in plots of $R_{12,43}$ represent the conventional 2 D Hall resistance deduced from carrier density which is estimated from the high field Hall resistance. As Figs. 17 and and 18 show, the magnetoresistance $R_{14,32}$ is negative for -200 mT$< H < 200$ mT approximately. The range of field for which $R_{14,32}$ is negative is relatively independent of V_g. Figure 18 also shows that $R_{12,43}$ is suppressed below the conventional 2 D Hall effect, but the range over which $R_{12,43}$ is suppressed changes from 270 mT for Fig. 18a to 0 in Fig. 18d. The discrepancy between the ranges of field for which the longitudinal resistance is negative and the Hall effect is suppresses is not understood.

It has been argued that the intrinsic Hall effect should not be suppressed [43], and Imry has proposed that a measurement of the intrinsic Hall voltage could be accomplished using probes in which only one 1 D subband is occupied, *provided the probes are weakly coupled* [29]. Using elementary perturbation theory to evaluate the effect of a small magnetic field on the wavefunction, it is possible to calculate the change in the electron density across the wire due to the field. This change is proportional to the electrochemical potential, i.e. the intrinsic Hall voltage across the wire. Thus, the polarization of the wavefunction in a 1 D constriction yields a finite Hall effect [44]. Since our measurements show that the Hall effect is still suppressed when $N = 1$ in the constrictions comprising the junction in both the voltage and current probes as shown in Figs. 17 and 18a, we infer that the probes cannot be treated perturbatively.

Kirczenow [45], Ravenhall [46], Baranger and Stone [36] and others [47] have explored the suppression of the Hall resistance numerically, treating the leads as an integral part of the resistance measurement, and have observed that the resistance is not generically suppressed for a junction with hard walls and sharp corners unless $N = 1$ in each of the constrictions comprising the cross. Figure 17 shows the magnetoresistance calculated according to such a model when $N = 1$ in each of the constrictions comprising the junction, beside the magnetoresistance measured when $N = 1 \pm 1$ [36,48]. The field where $HeW^2/hc \approx 1$ is indicated in the experimental data to facilitate a comparison with the calculation. The general form and the magnitude of the measured resistance is reproduced by the calculation. (However, the agreement with the magnitude of the measured resistance $R_{14,32}$ achieved by the calculation may be coincidental.) For $N \geq 1$ in the constrictions comprising the junction, the models of Ravenhall and Kirczenow cannot reproduce all of the features observed experimentally, however. A comparison with Fig. 18 reveals that the suppression of the Hall resistance is much more generic phenomenon [49] applying at least for constrictions where $N \leq 4$.

Baranger and Stone [36] resolved this issue by considering more realistic potential contours for the junctions. In particular, they observed that, for a junction in which the width of the wires are gradually tapered to meet the cross members, the Hall resistance is suppressed generally. According to Baranger and Stone, the gradual widening of the wires comprising the junction causes the electron to travel adiabatically from the narrow to the wide region of the junction conserving the number of subbands. Thus, the current injected into the narrow wire, which is equally distributed between the subbands of the constriction, reaches the wide region distributed only among the low lying modes. A col limation or focussing of the electron wave in the forward direction is a direct consequence of the adiabatic feed into the junction.

While the experimental condition generally necessary for the suppression of the Hall effect are still controversial [50,51], we can deduce two prerequisites from Fig. 18. When the $R_{14,32}$ is negative, the transmission probabilities T_{31}, T_{14}, T_{23}, and T_{42} are all small relative to T_{12}, and T_{34}. We deduce from Fig. 18 and eq. (3) that T_{31}, T_{14}, T_{23}, and

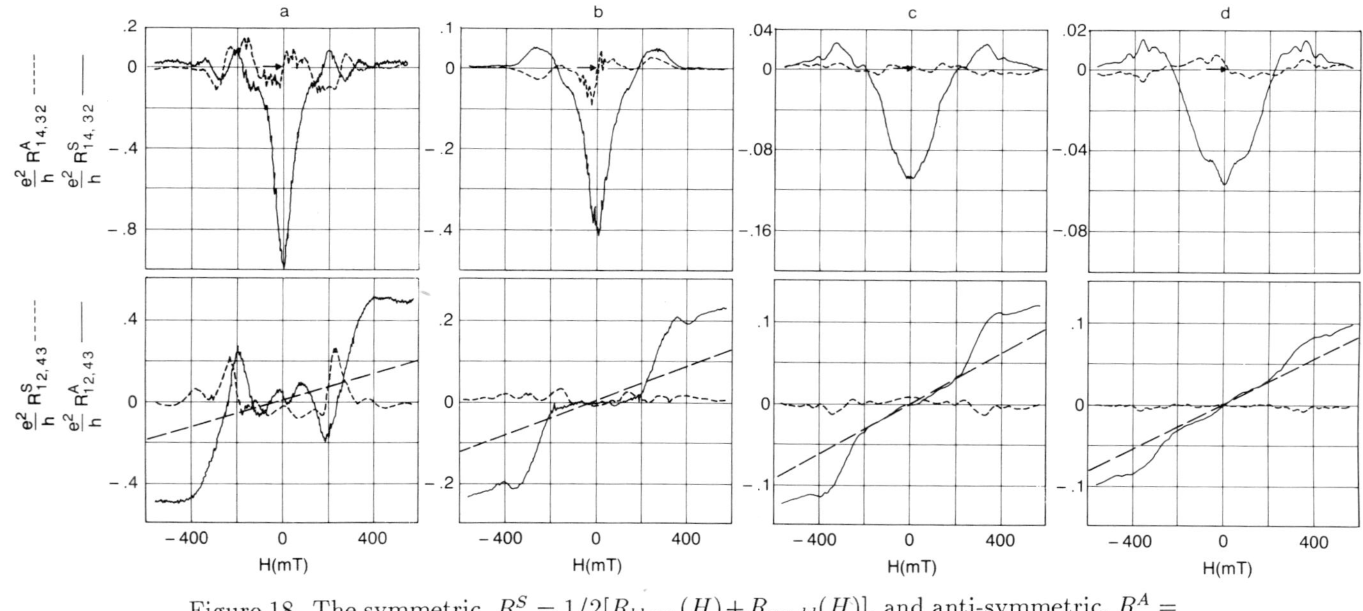

Figure 18. The symmetric, $R^S = 1/2[R_{kl,mn}(H) + R_{mn,kl}(H)]$, and anti-symmetric, $R^A = 1/2[R_{kl,mn}(H) - R_{mn,kl}(H)]$ components to the magnetoresistances $R_{14,32}$ and $R_{12,43}$ when four different gate voltages are applied to a junction comprised of two split-gate electrodes in series. The data is obtained from the device of Fig. 14b. In (a) $V_g = -1.035\,\mathrm{V}$ and $N = 1 \pm 1$ in each of the constriction comprising the junction, in (b) $V_g = -0.90\,\mathrm{V}$ and $N = 2 \pm 1$, in (c) $V_g = -0.770\,\mathrm{V}$ and $N = 4 \pm 1$, and in (d) $V_g = -0.560\,\mathrm{V}$ with $N = 6 \pm 1$.

T_{42} are comparable to T_{12} and T_{34}, for $V_g > -0.9\,\text{V}$ or $N > 2$ since the magnitude of $R_{14,32}$ is small. Under these conditions, Fig. 18d shows that $R^A_{12,43}$ is no longer suppressed below the conventional Hall resistance for magnetic fields beyond $20\,\text{mT}$. Therefore, the Hall resistance is suppressed below the 2 D value when (1) the coefficients T_{31}, T_{14}, T_{23}, and T_{42} are all small, and (2) $T_{31}T_{42} \approx T_{23}T_{14}$. By changing the size of T_{31}, T_{14}, T_{23}, and T_{42} relative to T_{12} and T_{34} through a change in the geometry of the junction, Ford et al. [50] demonstrated this. There is yet a third prerequisite for the suppression of the Hall effect. According to the interpretation by Baranger and Stone [36], and Beenakker and van Houten [5], the suppression of the Hall resistance can only occur if the elastic mean free path is larger than the width of the junction.

Recently, a semi-classical model for the resistance of a ballistic cross with tapered contacts, applicable for N large, has been shown to produce the suppression of the Hall resistance, the negative magnetoresistance found in $R_{14,32}$, and a feature in the Hall resistance resembling a plateau for magnetic fields such that the cyclotron diameter is greater than the width of the wire, i.e. $HeW^2/\hbar k_F < 1$ [5]. Beenakker and van Houten [5] assume in their semi-classical model that the angular distribution of the wavevectors of electrons injected into the tapered constrictions which comprise the cross is cosine-like and that an adiabatic taper focuses the classical electron trajectories. The semi-classical treatment is successful because the classical transmission probabilities are approximately the same as those calculated quantum mechanically, but there are exceptions [48]. For example, Beenakker and van Houten assert that a semi-classical description can produce the average negative resistance found as a function of width, but not the sharp minima shown in Fig. 14. The semi-classical model neglects quantum mechanical reflections which develop at the junction and so underestimates the resistance. In particular, a fit of the semi-classical theory to the data of Fig. 18b can only account for about half of the measured resistance, $R_{14,32}$, found at $V_g = -0.90\,\text{V}$ for $H = 0$, and cannot account for the well quantized Hall effect at $R_{12,43} = h/e^2 4$. (The general form of the magnetoresistance for $HeW^2/hc < 1$ is recovered, however.) The semi-classical Hall resistance calculated by Beenakker and van Houten is not quantized for $HeW^2/hc > 1$; as found experimentally for $N \leq 3$ (e.g. see Fig. 18d). However, the identification of a plateau in the measured resistance with the magnetic depopulation of the highest N subband becomes dubious for $N > 3$. The feature observed experimentally in the Hall resistance may be associated with that found in the semi-classical calculation when $N > 3$.

Figure 19a-c show the magnetoresistances, $R_{14,32}$ and $R_{12,43}$, observed at $100\,\text{mK}$ when each of the constrictions comprising the junction are (a) not biased ($V_g = 0\,\text{V}$), (b) biased so that only the two lowest subbands ($N = 2 \pm 1$) are occupied in each of the constrictions, and (c) biased so that only the lowest subband ($N = 1 \pm 1$) is occupied. N is estimated using (1) the quantization of the two terminal resistances of each of the constrictions versus V_g at $H = 0\,\text{T}$, (2) the peaks found in $R_{14,32}$ vs. V_g at $H = 0\,\text{T}$ associated with the change in the number of occupied subbands, and (3) the quantization of $R_{12,43}$ vs H. Using a hard wall confinement potential, the carrier density deduced from $R_{12,43}$ at $H = 8\,\text{T}$ and our estimate for N, we infer that $W = 105 \pm 25\,\text{nm}$ in Fig. 19b and $W \approx 70 \pm 25\,\text{nm}$ in Fig. 19c. Notice that $R_{14,32} \geq 0$ and $R_{12,43} = H/nec$ for $H < 100\,\text{mT}$ in the 2 D junction of Fig. 19a. In contrast, $R_{14,32}$ is negative and $R_{12,43}$ is suppressed from the conventional 2 D Hall resistance for $H < 100\,\text{mT}$ in Fig. 19b and 19c [13], in correspondence with our expectations for $HeW^2/hc < 1$.

As the magnetic field increases, the four terminal Hall resistance, $R_{12,43}$, of the high mobility 2 DEG of Fig. 19a is precisely quantized in steps of $h/e^2 i$, where i is an integer

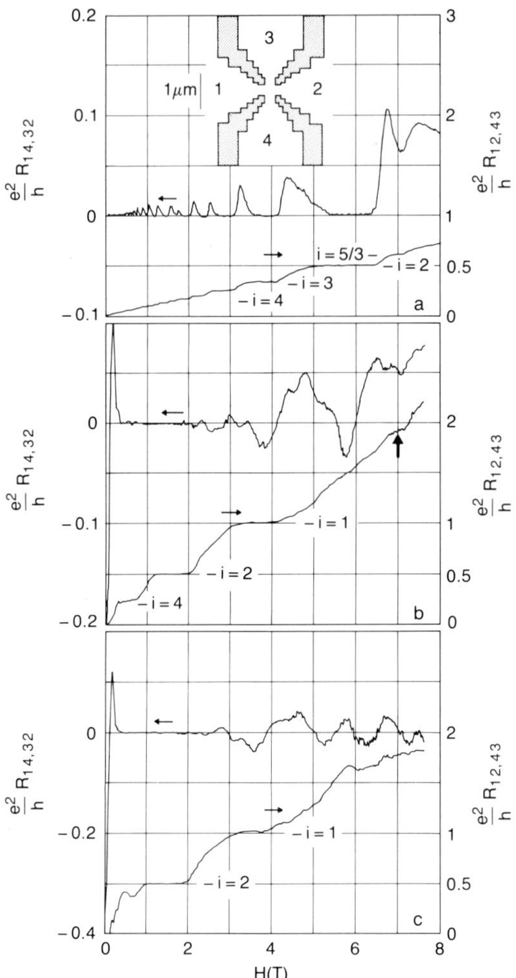

Figure 19. The resistances, $R_{12,43}$ and $R_{14,32}$, versus magnetic field measured at 100 mK. (a) The leads comprising the cross are 100 μm wide ($V_g = 0$ V), while in (b) and (c) the leads are about 105 nm with $N = 2 \pm 1$, and 70 nm wide with $N = 1 \pm 1$ respectively assuming a hard wall confinement potential. The vertical arrow in (b) at $H = 7$ T indicates the position of an anomalous plateau in $R_{12,43}$ near $(h/e^2)/0.53$ (see [21]). The inset to (a) is a schematic representation of the junction; the convention for labeling the leads or the constrictions between the electrodes is given there.

(with 99.9% accuracy here), corresponding to the depopulation of spin-polarized edge states. The four terminal Hall resistance of a cross comprised of four 1 D constrictions in a 2 DEG is also quantized, but as shown in Figs. 19b and 19c there are a finite number of steps corresponding to twice the number of spin-degenerate 1 D subbands in the constriction in the absence of a magnetic field [13], and moreover, the quantization is not precise [52]. For magnetic fields $HeW^2/hc > 1$, $R_{12,43}$ in Figs. 19b and 19c is quantized at approximately $(h/e^2)/i$ for integers $i \leq 2N$. In Fig. 19b $R_{12,43}$ reaches a plateau at approximately $300\,\mathrm{mT}$, i.e. $R_{12,43} = (h/e^2)/(4.10\pm0.15)$, corresponding to the depopulation of the highest spin-polarized 1 D subband, $i = 4$. The plateau associated with $i = 3$ is not well defined. However, with increasing magnetic field other plateaus are observed in the resistance centered at $(h/e^2)/(1.97 \pm 0.03)$ and $(h/e^2)/(0.99 \pm 0.03)$ near $H = 1.6\,\mathrm{T}$ and $4\,\mathrm{T}$ respectively, corresponding to the $i = 2$ and $i = 1$ states. In Fig. 19c, $R_{12,43}$ reaches a plateau at approximately $1\,\mathrm{T}$ and $3.5\,\mathrm{T}$ where $R_{12,43} = (h/e^2)/(2.02 \pm 0.03)$ and $(h/e^2)/(0.95 \pm 0.07)$ respectively corresponding to $i = 2$ and $i = 1$.

Unlike a 2 DEG where the filling factor $i = nhc/eH$, the magnetic field is not a direct measure of the number of occupied subbands in a 1 D constriction. In a 1 D constriction the plateaus in the Hall resistance are not found at regular intervals in H^{-1} because the subbands are defined by both magnetic and electrostatic confinement, and because the carrier density in the constrictions is determined by the Fermi energy in the 2 D contacts which varies as a function of H. For example, for a harmonic confining potential with a classical frequency ω_0, the magnetic field dependence of the filling factor is given by $i = nhc/e(H^2 + H_0^2)^{1/2}$ with $H_0 = \omega_0 m^* c/e$. If $H_0 = 1\,\mathrm{T}$ then $i = 4$ should occur for $H = 0.26\,\mathrm{T}$, $i = 2$ for $H = 1.8\,\mathrm{T}$ and $i = 1$ at $H = 4.0\,\mathrm{T}$ similar to what is found in Fig. 19b.

The value of the plateaus resistance, $R_{12,43}$, is a measure of the number of occupied subbands. However, for $300\,\mathrm{mT} < H < 5\,\mathrm{T}$, $R_{12,43}$ is not exactly quantized, and moreover $R_{14,32}$ can be negative. We attribute these observations to the overlap between edge states due to the narrow width of the constrictions. For $H < 2\,\mathrm{T}$ with $W \approx 70\,\mathrm{nm}$, the overlap between edge states is $\exp[-(W/2r_0)^2] < 0.1$. Due to the overlap, there is $< 10\%$ backscattering between opposite edges, which gives rise to $< 10\%$ deviations in the transmission probability between adjacent leads, and ultimately to $< 0.1 h/e^2$ deviations from exact quantization in $R_{12,43}$ and $-0.1 h/e^2 < R_{14,32} < 0$.

6   Two Terminal Resistance of a Zero-Dimensional Cavity

The effect of coherence and quantization on the resistance of a submicron constriction in a 2 DEG can be exaggerated by eliminating the longitudinal degree of freedom in a wire using a double barrier. Figure 20 is a schematic representation of a configuration of gate electrodes we have used to define a 0 D dot. Figure 21 is an electron micrograph taken from a top view of the device showing the disposition of the gate electrodes on the scale of micrometer. The configuration of electrodes is similar to that in Fig. 10, except that in the center, between the two series constriction, there are four additional electrodes which are used to define and manipulate a scatterer. Although each of the gate electrodes can be biased independently, when these electrodes are biased with the same negative voltage, the resulting potential traps electrons near the center. The potential contours shown in Fig. 20 explicitly represent this situation. As shown in Fig. 20, the 2 DEG is laterally confined in a region near the center of dimensions comparable to the

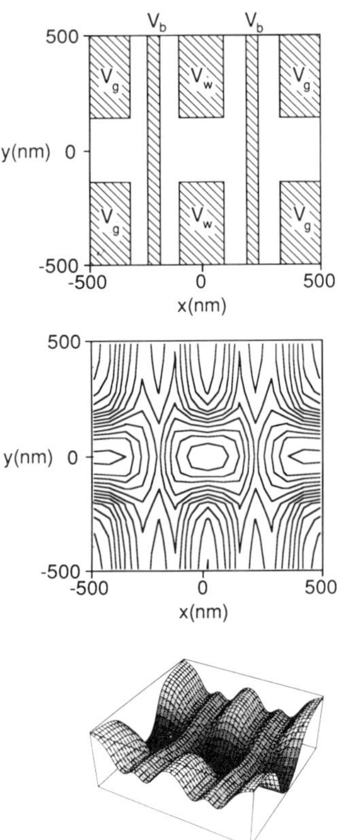

Figure 20. A graphical representation of the potential in the heterostructure near the AlGaAs/GaAs interface defined by two barriers alternating between three split-gate electrodes on the surface 70 nm above. It is assumed that -1.0 V is applied to each gate electrode. The top plot represents the disposition of the gate electrodes on the surface of the heterostructure. Each split gate electrode is 200 nm long with a 300 nm gap between electrodes. The split-gates are 250 nm apart. Centered in the gap between the split-gates there is a long, 40 nm gate used to introduce a barrier. In the center is a contour plot of the potential in the plane of the 2 DEG. The equipotential contours are at intervals of 0.066 V. The lower figure is the 3 D representation of the same potential where the height represents the size of the potential.

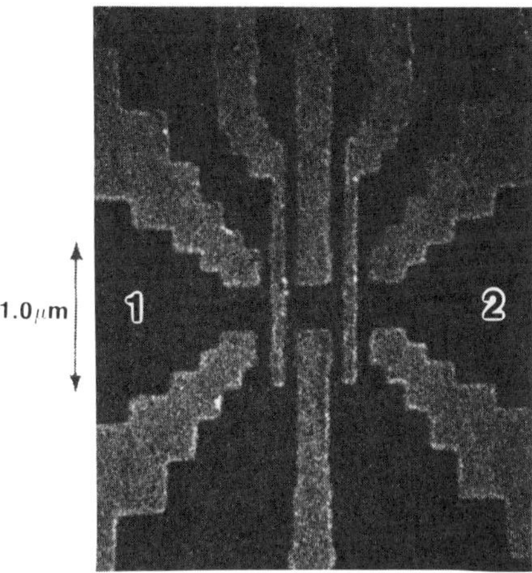

Figure 21. An electron micrograph of three, series, split-gate electrodes in close proximity separated by barriers fabricated on a high mobility heterostructure. The convention for numbering the leads is also given in the micrograph.

electron wavelength ($\approx 50 - 100$ nm). Since all of the lateral dimensions of the dot can be controlled independently with this configuration of electrodes, the number and energy level spacing of the quasi-bound states associated with the dot can be manipulated. Thus, in contrast with earlier work [53], it may be possible to achieve resonant tunneling laterally through the double barrier by manipulating only the height of the barrier using V_b, or the width of the center constriction using V_w without applying a large potential difference across the device.

We have examined the transport through this structure as a function of the barrier height used to define it. An excitation of only $1\,\mu$V was applied between contacts 1 and 2 in Fig. 21. Figure 22 shows one example of the two terminal conductance as a function of the barrier height found at 280 mK. In this example, the gate electrodes which potentially could be used to define contacts 1 and 2 are not biased; i.e. V_g in Fig. 20 is 0 V. When $V_b = 0$ V, $G_{12,12} \approx 6e^2/h$ indicating that only three spin-degenerate 1 D subbands are occupied in the center constriction. As V_b becomes more negative, the conductance and the transmission through the dot decrease non-monotonically. The non-monotonic variation of the conductance near threshold is further evidence of coherent backscattering. The peaks observed in the two terminal conductance near threshold, denoted E_1 and E_2 in the Figure, may be associated with resonant tunneling through quasi-bound states which develop from the complete lateral confinement of the 2 DEG to submicron dimensions [54]. Similar features have been observed by Smith et al. [55] using a single electrode with a hole in the interior and were attributed to resonant tunneling through a 0 DEG associated with the hole. According to this interpretation the features found when $G_{12,12} > 1.5e^2/h$ and $V_b > -0.5$V could be associated with resonant scattering from the barriers when the Fermi energy is larger than the barrier height.

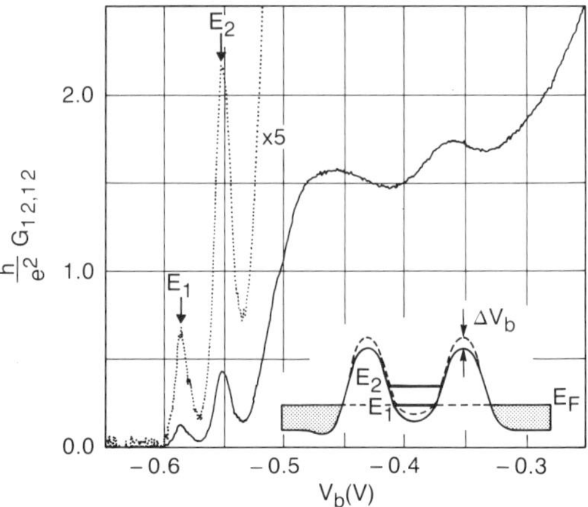

Figure 22. The two terminal conductance, $G_{12,12}$, measured on the device of Fig. 21 as a function of the height of the barriers, V$_b$. During this measurement the center constriction, V_w, was biased to admit 3 1 D subbands while the constrictions to the right and left were grounded; i.e. V$_g$ = 0 V. The conductance is a non-monotonic function of the barrier height.

The spacing of the peaks versus V$_b$ found as a function of V$_b$ may correspond to the intersubband energy, or to the addition of a single electron with a quantized charge, e, to the cavity [56,57]. The ambiguity in the interpretation of the data develops because a naive estimate for the charging energy ($\Delta \equiv e^2/C$) where we estimate the capacitance between the electron gas and the barrier gate electrode to be $C \approx 0.05 fF$ comparable to the intersubband splitting in this device [58]. Finally, the amplitude of the peaks in the conductance are comparable to but less than e^2/h corresponding to perfect transmission through a state near the center of the cavity. Scattering from disorder could reduce the transmission [59]. Moreover, tunneling through quasi-bound states associated with an impurity could give rise to non-monotonic variations in the conductance too.

7 Summary

Our experimental observations of the low temperature resistance demonstrate the predominate roles quantization and coherent scattering have. Our calculation of this resistance corroborate this interpretation. In this paper we have explicitly demonstrated the extent of phase coherence and the role coherent scattering plays through an examination of oscillations in the magnetoresistance of an annulus made from submicron constrictions in a high mobility 2 DEG. We have shown that the two and three terminal resistances of a submicron constriction can be quantized if the constrictions are short enough ($L < 1 \mu$m), and that the four terminal resistance of a constriction can be quantized as a function of magnetic field even though it is only 70 nm wide! The quantization of the resistance is due to the discrete transverse energies which develop from the confinement of the electron on the scale of its wavelength, and the quantization of the electronic charge. The coherent scattering and the quantization of the energy are also manifested in the four terminal resistance of a bend in an electron waveguide and in the two terminal resistance of a 0 D dot which both show non-monotonic variations in the resistance as a function of the size when the dimensions are comparable to the electron wavelength at low temperature.

References

[1] R. Landauer, in: Localization, Interaction, and Transport Phenomena, B. Kramer, G. Bergmann, and Y. Bruynseraede, eds., p. 38, Springer-Verlag, Heidelberg, New York (1985)

[2] S. Washburn and R.A. Webb, Adv. Phys. **35**, 375 (1986)

[3] Proc. Intl. Symp. on Nanostructure Physics and Fabrication, W.P. Kirk and M. Reed, eds., Academic Press, New York (1989)

[4] C.W.J. Beenakker and H. van Houten, 'Electronic Properties of Multilayers and Low-Dimensional Semiconductor Structures', J.M. Chamberlain, L. Eaves, and J.C. Portal, eds., NATO Advanced Study Institute Series, Plenum, London (1990)

[5] C.W.J. Beenakker and H. van Houten, Phys. Rev. Lett. **63**, 1857 (1989)

[6] Y. Aharonov and D. Bohm, Phys. Rev. **115**, 485 (1959)

[7] G. Timp, P.M. Mankiewich, P. de Vegvar, R. Behringer, J.E. Cunningham, R.E. Howard, H.U. Branger, and J.K. Jain, Phys. Rev. **B 33**, 6227 (1989)

[8] C.J.B. Ford, T.J. Thornton, R. Newbury, M. Pepper, H. Ahmed, D.C. Peacock, D.A. Ritchie, J.E.F. Frost, and G.A.C. Jones, Appl. Phys. Lett. **54**, 21 (1989)

[9] B.J. van Wees, L.P. Kouwenhoven, C.J.P.M. Harmans, J.G. Williamson, C.E.T. Timmering, M.E.I. Broekaart, C.T. Foxon, and J.J. Harris, Phys. Rev. Lett. **62**, 2523 (1989)

[10] B.J. van Wees, H. van Houten, C.W.J. Beenakker, J.G. Williamson, L.P. Kouwenhoven, D. van der Marel, and C.T. Foxon, Phys. Rev. Lett. **60**, 848 (1988)

[11] D.A. Wharam, T.J. Thornton, R. Newbury, M. Pepper, H. Ahmed, J.E.F. Frost, D.G. Hasko, D.C. Peacock, D.A. Ritchie, and G.A.C. Jones, J. Phys. **C 21**, L209 (1988)

[12] G. Timp, R.E. Behringer, S. Sampere, J.E. Cunningham, and R. Howard, in: Proc. Intl. Symp. on Nanostructure Physics and Fabrication, W.P. Kirk and M. Reed, eds., p. 331, Academic Press, New York (1989)

[13] M.L. Roukes, A. Scherer, S.J. Allen Jr., H.G. Craighead, R.M. Ruthen, E.D. Beebe, and J.P. Harbison, Phys. Rev. Lett. **59**, 3011 (1987)

[14] G. Timp, A.M. Chang, P. Mankiewich, R. Behringer, J.E. Cunningham, T.Y. Chang, and R.E. Howard, Phys. Rev. Lett. **59**, 731 (1987)

[15] P.M. Mankiewich, R.E. Behringer, R.E. Howard, A.M. Chang, T.Y. Chang, B. Chelluri, J.E. Cunningham, and G. Timp, J. Vac. Sci Technol. **B 6**, 131 (1988)

[16] A.M. Chang, G. Timp, T.Y. Chang, J.E. Cunningham,. B. Chelluri, P.M. Mankiewich, R.E. Behringer, and R.E. Howard, Surf. Sc. **196**, 46 (1988)

[17] M. Büttiker, in: Proc. Intl. Symp. on Nanostructure Physics and Fabrication, W.P. Kirk and M. Reed, eds., p. 319, Academic Press, New York (1989); Phys.Rev. **B 38**, 9375 (1988)

[18] G. Timp, A.M. Chang, P. de Vegvar, R.E. Howard, R. Behringer, J.E. Cunningham, and P. Mankiewich, Surf. Sc. **196**, 68 (1988)

[19] A.D. Stone, Phys. Rev. Lett. **54**, 2692 (1985)

[20] B.I. Halperin, Phys. Rev. **B 25**, 2185 (1980)

[21] J.K. Jain, Phys. Rev. Lett. **60**, 2074 (1988)

[22] K.F. Berggren, T.J. Thornton, D.J. Newson, and M. Pepper, Phys. Rev. Lett. **57**, 1769 (1986)

[23] G. Timp, 'Seminconductors and Semimetals', M.A. Reed, Vol. ed., Academic Press, New York (1990)

[24] This solution is due to J.H. Davies, see [21]

[25] A. Kumar, S.E. Laux, and F. Stern, Appl. Phys. Lett. **54**, 1270 (1989)

[26] Y. Imry, in: Physics of Mesoscopic Systems, Directions in Condensed Matter Physics, G. Grinstein and G. Mazenko, eds., p. 101, World Scientific Press, Singapore (1986)

[27] L.I. Glazman, G.B. Lesovick, D.E. Khmel'nitskii, and R.I. Shekhter, Pis'ma Zh. Eksp. Teor. Fiz. **48**, 218 (1988) [JETP Lett. **48**, 238 (1988)]

[28] A. Szafer and A.D. Stone, Phys. Rev. Lett. **62**, 300 (1989)

[29] Y. Imry, in: Proc. Intl. Symp. on Nanostructure Physics and Fabrication, W.P. Kirk and M. Reed, eds., p. 379, Academic Press, New York (1989)

[30] E.G. Haanappel and D. van der Marel, Phys. Rev. **B 39**, 5484 (1989); D. van der Marel and E.G. Haanappel, Phys. Rev. **B 39**, 77811 (1989); C.S. Chu and R.S. Sorbello, Phys. Rev. **B 40**, 5941 (1989)

[31] Significant parallel conduction was found in the magnetoresistance of this Hall bar

[32] J.H. Davies and J.A. Nixon, Phys. Rev. **B 39**, 3423 (1988)

[33] M. Büttiker, Phys. Rev. Lett. **57**, 1761 (1986); IBM J. Res. Develop. **32**, 317 (1988)

[34] D.A. Wharam, M. Pepper, H. Ahmed, J.E.F. Frost, D.G. Hanko, D.C. Peacock, D.A. Ritchie, and G.A.C. Jones, J. Phys. **C 21**, L887 (1988)

[35] C.W.J. Beenakker and H. van Houten, Phys. Rev. **B 39**, 10445 (1989)

[36] H.U. Baranger and A.D. Stone, Phys. Rev. Lett. **63**, 414 (1989)

[37] H.U. Baranger and A.D. Stone, Phys. Rev. **B 40**, 8169 (1989)

[38] The calculations were performed using a recursive Greens function method as in H.U. Baranger, A.D. Stone, and D.P. DiVincenzo, Phys. Rev. **B 37**, 6521 (1988)

[39] Y. Avishai and Y. Band, Phys. Rev. Lett. **62**, 2527 (1989)

[40] H.U. Baranger and A.D. Stone, in: Science and Engineering of One- and Zero-Dimensional Semiconductors, S.P. Beaumont and C.M. Sotomajor Torres, eds., p. 121, Plenum, New York (1990)

[41] G. Kirczenow, Sol. St. Commun. **71**, 469 (1989)

[42] R. Behringer, G. Timp, H.U. Baranger, and J.E. Cunningham, unpublished

[43] F. Peeters, Phys. Rev. Lett. **61**, 580 (1988)

[44] G. Kirczenow, Phys. Rev. **B 38**, 10958 (1988)

[45] G. Kirczenow, Phys. Rev. Lett. **62**, 2993 (1989)

[46] D.G. Ravenhall, H.W. Wyld, and R.L. Schult, Phys. Rev. Lett. **62**, 1780 (1989)

[47] H. Akera and T. Ando, Phys. Rev. **B 39**, 5508 (1989)

[48] H.U. Baranger, private communication

[49] C.J.B. Ford, T.J. Thornton, R. Newbury, M. Pepper, H. Ahmed, D.C. Peacock, D.A. Ritchie, J.E.F. Frost, and G.A. Jones, Phys. Rev. **B 38**, 8518 (1988)

[50] C.J.B. Ford, S. Washburn, M. Büttiker, C.M. Knoedler, and J.M. Hong, Phys. Rev. Lett. **62**, 2724 (1989)

[51] A.M. Chang, T.Y. Chang, and H.U. Baranger, Phys. Rev. Lett. **63**, 996 (1989)

[52] A.M. Chang, G. Timp, J.E. Cunningham, P.M. Mankiewich, R.E. Behringer, and R.E. Howard, Sol. St. Commun. **76**, 769 (1988)

[53] M.A. Reed, J.N. Randall, R.J. Aggarwal, R.J. Matyi, T.M. Moore, and A.E. Wetsel, Phys. Rev. Lett. **60**, 535 (1988)

[54] A.B. Fowler, G.L. Timp. J.J. Wainer, and R.A. Webb, Phys. Rev. Lett. **57**, 128 (1986)

[55] C.G. Smith, M. Pepper, H. Ahmed, J.E. Frost, D.G. Hasko, D.C. Peacock, D.A. Ritchie, and G.A.C. Jones, J. Phys. **C 21**, L893 (1989)

[56] J.H.F. Scott-Thomas, S.B. Field, M.A. Kastner, H.I. Smith, and D.A. Antoniadis, Phys. Rev. Lett. **62**, 583 (1989)

[57] U. Meirav, M.A. Kastner, M. Heiblum, and S.J. Wind. Phys. Rev. **B 40**, 5871 (1989)

[58] R.H. Silsbee and R.C. Ashoori, comment subm. to Phys. Rev. Lett.

[59] S. Washburn. A.B. Fowler, H. Schmid, and D. Kern, Phys. Rev. **B 38**, 1554 (1988)

MAGNETORESISTANCE AND BOUNDARY SCATTERING IN BALLISTIC WIRES

Trevor J. Thornton, Michael L. Roukes, Axel Scherer,
Bart P. Van der Gaag

Bellcore
331 Newman Springs Road
Red Bank, New Jersey 07701, USA

1 Introduction

Boundary scattering has long been known to influence the magnetoresistance of very thin metal wires and films [1]. In carefully prepared films the electron mean free path at low temperatures can be much larger than the thickness of the film and the electrons are scattered frequently by the surfaces. In a metal, the Fermi wavelength is of the order of the atomic spacing and unless great care is taken to ensure that the surfaces are very smooth the boundary scattering is predominantly diffuse i.e. momentum parallel to the surface is not conserved.

In semiconductors the situation is somewhat different because of the very much longer Fermi wavelength. Although the electrons are now less susceptible to random scattering by atomic imperfections, larger scale fluctuations, such as occur at the Si-SiO_2 interface of a MOSFET [2], can limit the mobility. However, with the precise atomic layer growth of molecular beam epitaxy (MBE) nearly perfect interfaces can be achieved and electron motion in the plane of the 2 DEG of a single heterojunction is virtually unimpeded by the presence of the interface.

This paper is concerned with the additional scattering which arises when a high mobility 2 DEG is further confined by *lateral* boundaries formed during low energy ion exposure [3]. The magnetoresistance of wires formed in this way can be explained in terms of a classical model from which we can estimate the coefficient of specular scattering p. This value is compared to estimates obtained from electron focussing experiments and lastly we consider a few of the implications of diffuse boundary scattering.

2 Classical Description of Boundary Scattering

To accurately describe electrons scattering off the surfaces of a solid we really need a complete microscopic description of all the boundaries involved. Except for the very

Quantum Coherence in Mesoscopic Systems
Edited by B. Kramer, Plenum Press, New York, 1991

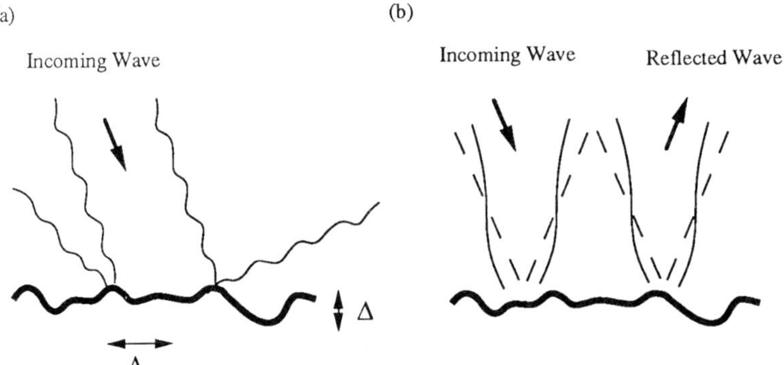

Figure 1. Schematic representation of electron waves scattered by a rough boundary (a) $\lambda_F \ll \Delta$, (b) $\lambda_F \gg \Delta$.

smallest structures this is an almost impossible task and the best we can do is character-ize the surfaces in terms of parameters which describe their average properties. Ziman [4] gives a prescription for a quasi-classical approach by considering the scattering of electrons normal to a rough surface which has asperities of average height Δ and distri-bution Λ (see Fig. 1). The conclusions agree with our intuitive expectations: when the wavelength of electrons incident normal to the surface is much smaller than Δ they are strongly scattered in all directions (Fig. 1a); when $\lambda_F \gg \Delta$ the surface appears nearly smooth to the incident electrons which are scattered almost specularly (Fig. 1b).

A real surface will consist of different crystal facets each with its own terracing, scratches and associated defects. If we were to persist with Ziman's approach the vari-ation of Δ and Λ within different facets would best be described by a distribution over, say, surface normal, from which the additional resistance due to the surface can be cal-culated. Unfortunately we have now lost the appealing simplicity of Ziman's treatment whereby the average properties of the surface are described by just two parameters.

Because of the difficulties involved in estimating the statistical roughness of a sur-face the problem is usually simplified by considering a single parameter p [5]. This coefficient describes the probability of specular reflection and varies between unity for a smooth, mirror-like surface and zero when the scattering is completely diffuse. For a given surface p will depend upon the wavelength of the incident electrons as mentioned above but, for the degenerate systems which we are concerned with, the conductivity will be determined by those electrons at the Fermi surface and the value of p will be constant provided the carrier density, n, is not changed significantly. The value of p will also vary with the angle at which the electrons arrive at the surface electrons at glancing incidence will scatter very differently to those arriving normally. In the measurements described here p must be considered an average value taken over the entire electron distribution.

The resistivity of a ballistic wire with rough edges can be derived using Matthies-sen's rule by replacing the bulk mean free path, l_0, with one that takes account of diffuse scattering at the boundaries. This leads naturally to the concept of a boundary scattering length, l_b [6], which may be defined as the average distance an electron travels along a wire before diffuse scattering events at the edges scatter it into a state with different longitudinal momentum [7]. Typically, such an electron will experience $1/(1-p)$ collisions with the wire edges before scattering diffusely. The same number is given by the value of l_b/W and the boundary scattering length can be written as $l_b = W/(1-p)$. If we then assume that the scattering rate in the bulk of the wire and that at the surface are additive (Matthiessen's rule) the effective mean free path is $1/l_{eff} = [1/l_0 + (1-p)/W]$. To a reasonable approximation the resistivity is therefore

$$\rho_{eff} = \rho_0 l_0 \left[1/l_0 + (1-p)/W \right], \tag{1}$$

where ρ_0 is the resistivity of the bulk (infinite) solid. The quantity $\rho_0 l_0$ is a material parameter independent of the degree of disorder and is given by $\hbar k_F/ne^2$. Equation (1) is of course an over simplification and disorder induced corrections such as weak localization [8] and the electron-electron interaction [9] will give rise to additional terms in the resistivity [10] which are not included in eq. (1). However, for the high mobility wires we are considering, it turns out that provided $l_0 \gg W$ and $p < 1$, the corrections due to disorder are negligible compared to the boundary scattering contribution. Indeed, in the narrowest wires the boundary scattering completely dominates the resistivity.

From the above discussion it is clear that, as long as they are sufficiently rough, the edges of a narrow wire provide an efficient mechanism for random electron scattering. When $l_0 \gg l_b$ boundary scattering dominates the momentum relaxation and therefore limits the mobility. The degree of boundary scattering can be 'tuned' by the application of a magnetic field perpendicular to the wire and this is illustrated very schematically in the diagrams of Fig. 2. At zero or low magnetic field those electrons with a large component of momentum parallel to the boundaries can contribute significantly to the conductivity even for $p \ll 1$ because they interact with the edges relatively infrequently and can travel distances of order l_0 before scattering within the wire. However, as the magnetic field is increased, more and more electrons are forced to interact with the edges and if $p < 1$ random scattering reduces the effective mean free path, l_{eff}, and the resistivity increases. The resulting low field positive magnetoresistance eventually saturates at some maximum when the cyclotron radius is approximately twice the width. Any further increase in the magnetic field will now lead to a drop in the resistivity as the probability of an electron being backscattered by the edges is reduced. Eventually, a cyclotron orbit will fit inside the wire (i.e. $2\hbar k_F/eB = W$) and an electron scattering off one edge can only reach the opposite side if it suffers a collision within the wire. Therefore, electrons scattering off the edges are confined to one particular edge over distances of the order of the bulk mean free path measured at the appropriate magnetic field. If this field is not so large that Landau quantization becomes impor,ant the effective mean free path is approximately equal to l_0 and can be significantly larger than the value at $B = 0$. The magnetic field therefore *quenches* the boundary scattering contribution to the total scattering rate resulting in a resistivity similar to that in a bulk sample with dimensions much larger than l_0. The quenching of the boundary scattering is a manifestation of the suppression of back scattering at high magnetic fields [11]. In the samples we have studied, $p < 1$ and $l \gg W$, and the backscattering arises mainly because of diffuse scattering at the boundaries with only a small contribution due to impurities.

In what follows we discuss our recent results on the boundary scattering in narrow wires defined by low energy ion exposure. The mean free path measured in the bulk is much larger than the wire width and the resistivity is well described in terms of the simple framework discussed above. Surprisingly, although the measured values of p are not much less than unity, boundary scattering dominates the resistivity of the narrowest wires.

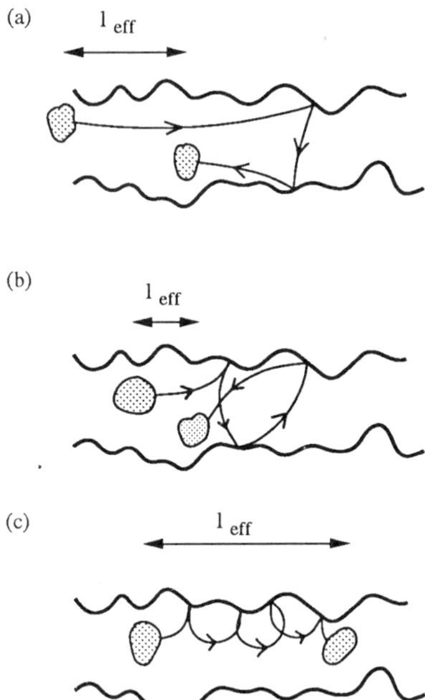

Figure 2. Electron-boundary scattering in narrow wires with rough edges in a perpendicular magnetic field. (a) When $l_0 \gg W$ the effective mean free path, l_{eff}, is dominated by boundary scattering. (b) At higher fields when $W/r_c \sim 0.55$, more electrons are forced to interact with the walls and the resistivity increases. (c) When $2r_c < W$ electrons are confined to either edge and the backscattering due to the boundaries is quenched. r_c is the cyclotron radius. The shaded 'clouds' represent elastic scattering within the wire.

3   Narrow Wires by Ion Exposure

When a modulation doped heterojunction is exposed to a beam of ions the once highly mobile 2 DEG suffers a drastic reduction in the conductivity and can easily be rendered insulating. This has proved to be a useful technique for the lateral confinement of a 2 DEG. There are basically two approaches; (i) a high energy, focussed ion beam is scanned over the surface selectively exposing those regions to be turned into insulators [12]; (ii) alternatively, a mask can be deposited onto the surface and a broad beam of ions is used to transfer the pattern into the 2 DEG [3]. The wires used in this work were made by the latter technique which, when optimized, has a resolution of a few hundred Ångstroms.

Figure 3a shows a cross section through one of the wires. The mask used to protect the underlying 2 DEG is a metal patterned by electron beam lithography and deposited by lift-off. The entire device is then exposed to a beam of 150 eV Neon ions whilst monitoring the channel resistance to ensure that it receives the optimum dose for

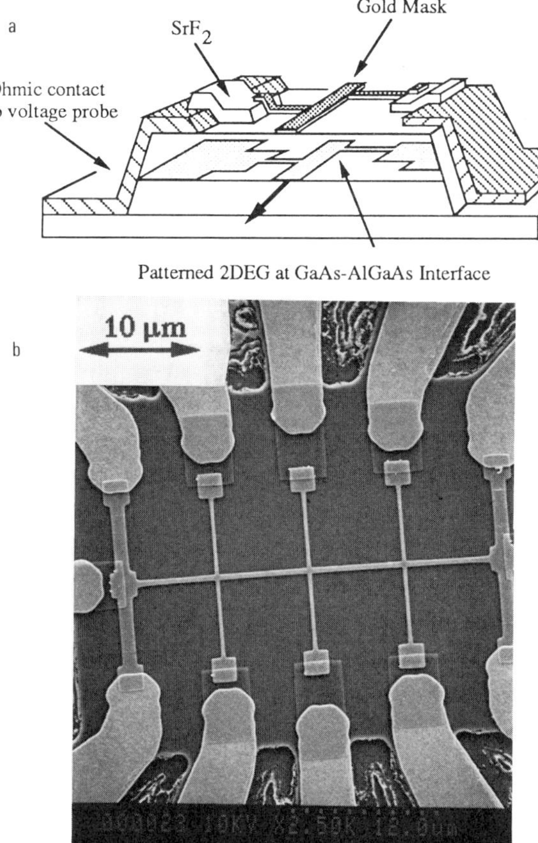

Figure 3. (a) Schematic cross-section through an ion exposed wire with a self aligned gate. The arrow indicates direction of current flow. (b) Photograph of the surface showing the gate and voltage probes.

pattern transfer. Note that the SrF$_2$ pads are essential to maintain connectivity between the voltage probes and the ohmic contacts without shorting the gate to the 2 DEG.

The metal mask acts as a self-aligned gate that can be used to vary the density of the 2 DEG, typically in the range 1 to $5 \times 10^{15} \mathrm{m}^{-2}$. At lower densities the wire becomes very non-uniform eventually approaching a metal-insulator transition, and to induce more than $5 \times 10^{15} \mathrm{m}^{-2}$ requires such a large positive gate voltage that there is a serious risk of achieving a flat band condition resulting in a large current flow between the gate and the 2 DEG.

One advantage of this method of patterning is the fact that the wire width is essentially independent of the carrier density over a large range [13]. This is in contrast to methods which employ split-gates [10,14] or shallow etching [15] where the wire width varies significantly with density. The constant wire width greatly simplifies analysis of the density variations of the magnetoresistance.

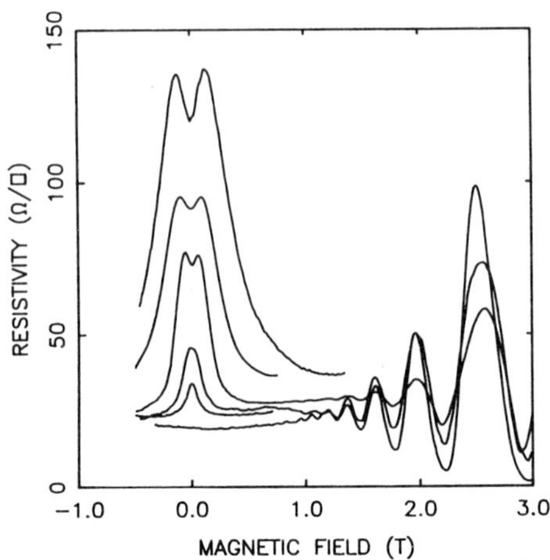

Figure 4. magnetoresistance of ion exposed wires with widths $0.42, 0.74, 1.0, 2.0, 3.6, 6.2\,\mu$m from top to bottom. The carrier density in each case is in the range $4.2 \pm 0.1 \times 10^{15}\mathrm{m}^{-2}$

4   Magnetoresistance

and Boundary Scattering In Fig. 4 we show the magnetoresistance traces for wires of width ranging from $6.2\,\mu$m to $4200\,$Å. The electron density, determined from the periodicity of the Shubnikov-de Haas oscillations, is the same in each case so that we can compare the resistivities. Qualitatively the curves are consistent with the classical theory outlined above. For the widest wires ($l_0 \sim W$) boundary scattering makes a negligible contribution to the resistivity and the magnetoresistance is similar to that in a wide 2 DEG. However, in narrower wires [16], the zero field resistivity first increases somewhat before dropping below the zero field value as B increases [17]. At high fields, when $2r_c < W$ (r_c is the cyclotron radius), the resistivity saturates and is independent of magnetic field before the onset of Shubnikov-de Haas oscillations. The value of the saturated resistance is similar for wires of width $7\,\mu$m $\geq W \geq 1\,\mu$m although there is a small, systematic increase as the width decreases.

Before applying eq. (1) to the data in Fig. 4 we will first consider how the resistance maximum at finite B scales with W as this will give us a fairly accurate method for determining the wire width. With increasing magnetic field the trajectories of those electrons with large parallel momentum are curved into the edges of the wires and if p is less than unity the resistivity of the wire increases. In Fig. 4 this is clearly resolved for wires less than $1\,\mu$m wide. The position of the maximum, call it $B_{\max}$, is determined solely by the ratio of the wire width to the cyclotron radius and calculations by Pippard [1] have shown that $W/r_c(B_{\max}) = 0.55$. This is very similar to a result derived by Forsvoll and Holvech [17] for thin metal films with a magnetic field directed parallel to the plane of the film.

The scaling of r_c and W works remarkably well as shown in Fig. 5 where we have plotted the cyclotron radius at $B_{\max}$ measured for many different wires, in each case taking W as the nominal mask width. The best fit to the data gives $W/r_c = 0.55 \pm 0.1$. As well as looking at different wires we can also change the position of the resistivity

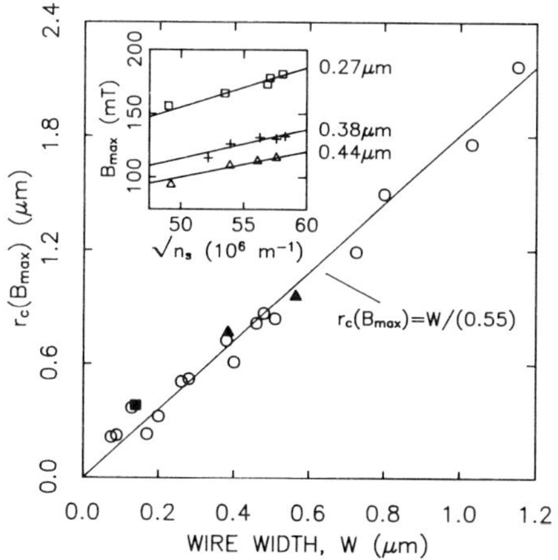

Figure 5. The cyclotron radius at the peak position, $r_c(B_{\mathrm{max}})$, plotted against wire width for many different wires. The variation of B_{max} with carrier density, n, is shown in the inset.

maximum in a given wire by varying the carrier density and therefore the cyclotron length. We expect B_{max} to be proportional to $k_F = (2\pi n)^{1/2}$, provided that the wire width is independent of n, and this is confirmed by the inset to Fig. 5. The straight lines in the inset are derived from the expression $B_{\mathrm{max}} = 0.55h(2\pi n)^{1/2}/eW$. In each case the best fit gives a value for W which is within 10% of the nominal mask width. Whenever possible the wire widths were estimated from the magnetic depopulation of 1D subbands [18]. Again the results were close to the nominal mask widths. When the peak at B_{max} is not resolvable the width of the wire is usually greater than $2\,\mu$m and can be taken as the mask width with no loss of accuracy. Figure 5 demonstrates that for wires with sufficiently diffuse boundary scattering the anomalous resistivity peak at B_{max} provides a useful method for determining the wire width since it depends solely on the geometry with no need to assume a particular form for the confining potential [18].

Having established the wire width we can now determine the resistivity from the measured resistance. The curves in Fig. 5 can be characterized by two resistivity values, that at $B = 0$ and the quenched value at which it eventually saturates ρ_{qu} (see Fig. 6). Both values are strongly dependent upon the carrier density, n, as shown in Fig. 7 for several different wires. For carrier densities less than 1.5×10^{15}m^{-2} both ρ_0 and ρ_{qu} increase dramatically suggesting the onset of strong localization. At higher densities the variation of ρ_{qu} with n (solid symbols) follows an approximate power law. On the other hand the variation of $\rho(0)$ (open symbols) shows no evidence for a single power law behaviour. The trends can best be described by considering how $\rho(0)$ and ρ_{qu} vary in a 1 μm wide wire (open and closed circles in the Figure). At high densities ρ_{qu} is much smaller than $\rho(0)$ but has a significantly steeper gradient and the two curves eventually coalesce when $n \sim 1.5 \times 10^{15}$m^{-2}. The gradient of ρ_0 for the 1 μm wire is very similar to those in 3.6 and 10 μm wide wires (for the latter there is no excess resistivity at $B = 0$ i.e. $\rho(0) = \rho_q$). Indeed, for all the wires we measured the value of the quenched resistivity varies with n as $\rho \sim n^{-\alpha}$ where α is in the range 1.8 to 2.3. Using the Drude value of the conductivity, $\rho = 1/ne\mu$, the measured variation with n suggests

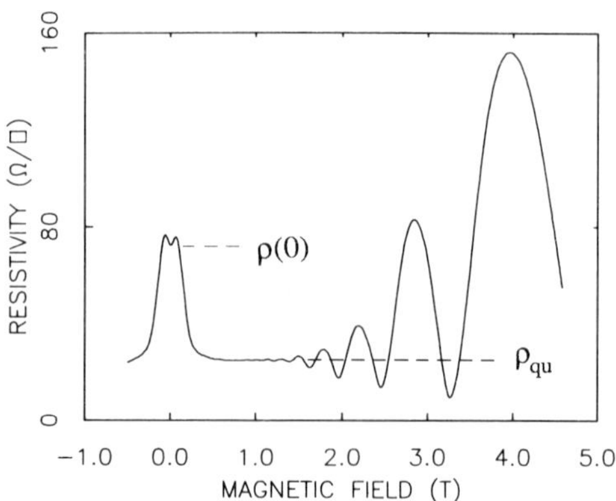

Figure 6. The magnetoresistance of a $1\,\mu$m wide wire demonstrating the complete suppression of the boundary scattering resistance. The values of $\rho(0)$ and ρ_{qu} are used to characterize the data.

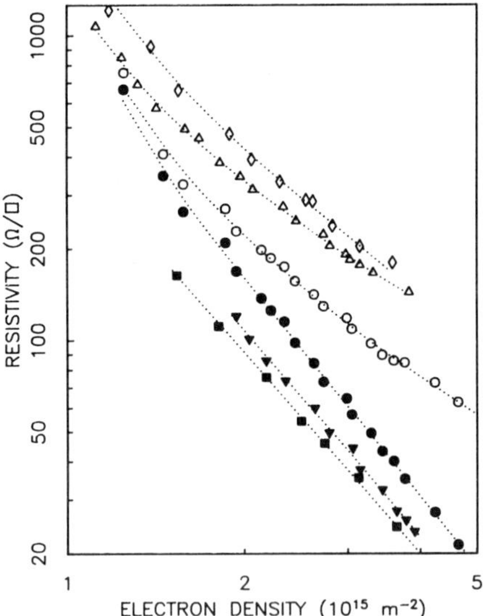

Figure 7. The density dependence of $\rho(0)$ (open symbols) and ρ_{qu} (closed symbols). Wire widths are $10\,\mu$m (squares), $3.6\,\mu$m (solid triangles), $1\,\mu$m (circles), $0.74\,\mu$m (open triangles) and $0.63\,\mu$m (diamonds).

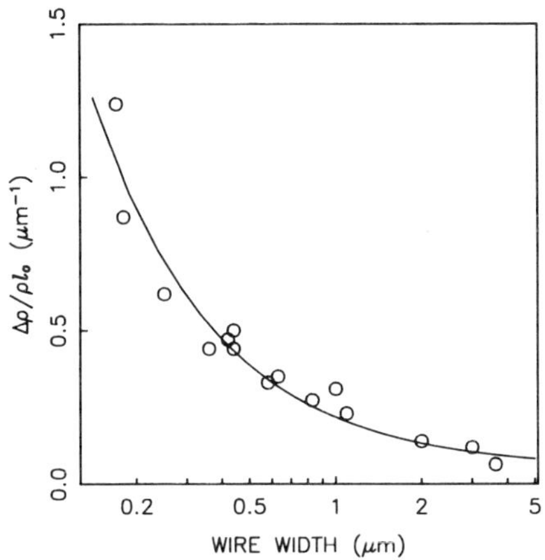

Figure 8. The increase in the zero field resistivity with decreasing wire width. The solid line is the best fit to eq. (1).

that $\mu \sim n^{0.8-1.3}$ which is similar to the density dependence observed in a wide 2 DEG [19]. As well as similar power law behaviour, the magnitude of ρ_{qu} is similar for wires of width $10\,\mu$m $> W > 3\,\mu$m. For narrower wires, however, there is a systematic increase as the width decreases. The gradual increase in ρ_{qu} as the width gets smaller is most likely caused by ions which are scattered sideways under the mask. As the wire width decreases the resulting degradation in mobility becomes more important and might ultimately limit the resolution of this confinement technique.

The fact that ρ_{qu} is similar in both magnitude and n dependence to the resistivity of a wide 2 DEG is consistent with the classical model described above. The boundary scattering is suppressed by the magnetic field so that, according to eq. (1), the effective resistivity approaches the bulk value. In what follows we assume that ρ_{qu} corresponds to the resistivity that would be measured in a wide 2 DEG under similar conditions and provided B is not too large this is approximately equal to ρ_0.

At zero field the additional resistance due to the boundary scattering is given by $\Delta\rho = \rho_{eff} - \rho_0 = (1-p)\rho_0 l_0/\omega$ and an estimate of p can be obtained by measuring $\Delta\rho$ in wires of different width. This of course assumes that the electron-boundary interaction is the same in each wire and in order to ensure that this was the case each was exposed to the same optimum dose of 150 eV neon ions and $\Delta\rho$ was measured for carrier concentrations in the range $3 - 4 \times 10^{15}$m^{-2} i.e. at an approximately constant Fermi wavelength of 350-400 Å. The results are shown in Fig. 8 along with the best fit to $\Delta\rho/\rho_0 l_0 = (1-p)/W$ from which we estimate a value of p of 0.85.

Despite the fact that p is as much as 85%, boundary scattering makes the largest contribution to the resistivity for wires with width less than $1\,\mu$m. On the average an electron suffers $1/(1-p)$ collisions with the wire edges before being scattered into a new state with a different longitudinal momentum - approximately seven times in this case. This means that for a wire of width $1\,\mu$m the boundary scattering length is $7\,\mu$m, already comparable to the bulk mean free path.

The additional resistivity due to the boundary scattering depends on n since $\Delta\rho = \rho_0(1-p)l_0/W$ and $\rho_0 l_0 \approx 1/\sqrt{n}$. Therefore as n decreases the absolute value of $\Delta\rho$

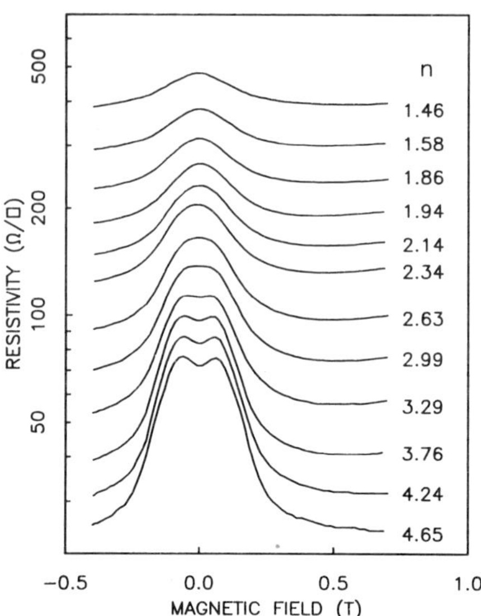

Figure 9. Low field magnetoresistance of a $1\,\mu$m wide wire for different electron densities.

increases but the importance of the boundary scattering relative to internal scattering actually decreases as $\Delta\rho/\rho_0 \approx \sqrt{n}$. This is most clearly seen when the resistivity is plotted on a logarithmic scale as shown in Fig. 9. As n decreases both $\rho_{qu}(\sim \rho_0)$ and $\rho(0)$ increase such that $(\rho_{eff} - \rho_{qu})/\rho_{qu}$ approaches zero.

The resistivity maximum at $B_{\max}$ disappears well before the peak at zero field vanishes. At first sight this appears as something of a contradiction since both effects arise from boundary scattering. However, the increased resistivity at zero field is due to the additional scattering of *all* the electrons whereas the peak at $B_{\max}$ is due to a small subset which are much more sensitive to any reduction in l_0. These electrons are the ones with large parallel momentum and the positive magnetoresistance up to $B_{\max}$ results from a suppression of their effective mean free path. As l_0 decreases the additional current they carry is reduced until eventually, at $l_0 = l_b$, their contribution to the conductivity is no more significant than the average. At this point the positive magnetoresistance vanishes and $\rho_{eff} \approx 2\rho_0$ which is approximately the case for the $n = 2.99$ curve in the figure.

The decay of the peak at $B_{\max}$ is shown in Fig. 10 where the normalized resistance has been plotted. The magnitude of the peak $\delta R(B_{\max})/R(0)$ plotted against the bulk mean free path is shown in the inset. The peak at $B_{\max}$ vanishes at an extrapolated value for l_0 of approximately $1\,\mu$m. If we take this to be the boundary scattering length $l_b = (1 - p)/W$ we obtain a value for p of 0.75. This is close to the previous value but we stress that it is a rather crude estimate.

5   Failure of the Classical Model

The data in Fig. 8 is reasonably well described by eq. (1) over almost the entire range of wire widths. However, for wires of width less than $0.3\,\mu$m the additional resistivity due to the boundary scattering, $\Delta\rho$, seems to be increasing faster than the $1/W$ dependence

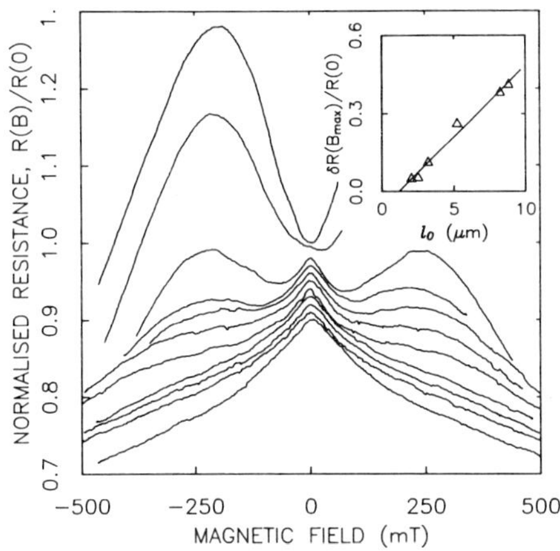

Figure 10. The decay of the peak at B_{max} with decreasing electron density for a wire of width $0.27\,\mu$m. The carrier density varies from $4.5 \times 10^{15}\,\mathrm{m}^{-2}$ (top curve) to $2.2 \times 10^{15}\,\mathrm{m}^{-2}$ (bottom curve). The inset shows the amplitude of the normalized peak plotted against the transport mean free path l_0.

predicted by eq. (1). A similar behaviour has been observed in thin metal films where deviations from eq. (1) occur for films of thickness, t, less than $100\,\text{Å}[20, 21]$.

Equation (1) is unable to describe the resistance of very narrow wires or films because it completely neglects the quantum mechanical nature of the problem. For thin films a fully quantum mechanical treatment is available [22, 23] and the calculations show much better agreement with experiment when $t < 100\,\text{Å}$.

In very thin $\mathrm{Al}_x\mathrm{Ga}_{1-x}\mathrm{As}/\mathrm{GaAs}$ quantum wells, variations in the well thickness, δt, can lead to large fluctuations in the energy of the quantized (2 D) subbands. This randomly varying potential scatters electrons confined in the well and is the dominant scattering mechanism when the well width is less than $100\,\text{Å}$. Sakaki et al. [24] have measured the electron mobility as a function of well width and their results show a very strong dependence, $\mu \propto t^6$. They were able to estimate the height, Δ, and spatial variation, Λ, at the $\mathrm{Al}_x\mathrm{Ga}_{1-x}\mathrm{As}/\mathrm{GaAs}$ interface (see Fig. 1) and obtained values of $\sim 3\,\text{Å}$ and $100\,\text{Å}$ respectively. In principle, a similar approach could be be used to estimate the variation in the width of our wires, δW, which is hard to measure directly. Unfortunately, exact solutions to the boundary scattering problem in a wire do not exist and until such time as they are calculated we have to make do with the phenomenological approach of eq. (1).

6 Electron Focussing

The value of the coefficient p has been estimated for different laterally imposed boundaries by an alternative method known as electron focussing [25]. Electrons emitted from a small hole are forced to skip along the boundary by application of a perpendicular magnetic field. Another hole in the boundary a distance L away acts as a collector for those electrons which satisfy the condition $2ir_c = L$ where $i = 1, 2 \ldots$ etc. As the magnetic field is increased a number of focussing peaks are observed in the collector current with each additional peak corresponding to an extra cyclotron diameter fitting between the emitter and collector. Higher order peaks suffer more collisions with the boundary and the amplitude of successive peaks decrease as $A_{i+1}/A_i = p$. By measuring the attenuation of a number of peaks it is possible to estimate p.

Electron focussing is a particularly elegant method for probing boundary roughness and recent work has used samples defined by gate depletion [26], ion implantation [27] and wet etching [28]. Although the results indicate that ion implanted boundaries are more diffuse ($p > 0.35$) than those defined by gate depletion ($p \sim 1$) mechanisms other than roughness scattering may be contributing to the decay of the focussing peaks. This makes it more difficult to interpret the data unambiguously. For instance, data from a wet etched sample [28] looks qualitatively similar to one defined by ion implantation [27] suggesting that the two have similar values for p even though wet etched boundaries are expected to be highly specular.

A schematic view of an electron focussing sample with boundaries defined by surface depletion is shown in Fig. 11. The structure is similar to those used in refs. [27] and [28]. The voltage on each gate can be varied independently and when reduced below threshold ($V_t \sim -0.5V$) the current can only flow from source to drain via the two point contacts. The magnetoresistance for $V_{g1} = V_{g2} = -0.8\,\text{V}$ is shown in Fig. 12. The structure above 2 Tesla varies little with gate voltage and is due primarily to Shubnikov-de Haas oscillations in the broad regions either side of the point contacts. Below 2 T we see three well defined focussing peaks and a fourth which can be resolved when the background is subtracted. The decay of the focussing peaks gives a value for p of 0.45, similar to the result of Nakamura et al. [27] and Nihey et al. [28] and much smaller than expected for gate defined boundaries. This discrepancy is probably due to the short distance between the emitter and collector. A relatively large magnetic field is needed to focus the electrons ($\Delta B = 2\hbar k_F/L$) and for the $i = 3$, $i = 4$ peaks the cyclotron diameter is comparable to the size of the emitter/collector openings. This will broaden the peaks and reduce the resolution of the measurement. The data of Spector et al. [29] suggests a similar conclusion. Their results for an emitter/collector separation of $4\,\mu\text{m}$ shows a second order peak already significantly smaller than the primary but when $L \geq 16\,\mu\text{m}$ many peaks can be observed before there is significant attenuation.

Clearly, effects other than surface roughness can influence electron focussing spectra. In particular, calculations by Pippard [1] have shown the surprising result that focussing peaks can be observed even when the boundary scattering is completely diffuse. The peaks in this case are due to a correlation in the potential of those electrons which leave the surface almost normally. The correlation persists over a number of orbit diameters giving the impression of multiple specular scattering. In the light of this result and the influence of emitter-collector spacing mentioned above, we should be cautious when interpreting the results of focussing experiments.

7   Implications of Boundary Scattering

Diffuse boundary scattering has some important implications for ballistic transport in small structures. The most obvious effect, an increase in the zero field resistivity, is due to the extra scattering averaged over the entire electron distribution. There are, however, several transport anomalies which can be explained in terms of *specific* electron trajectories [30]. Although these particular trajectories are only a small fraction of the

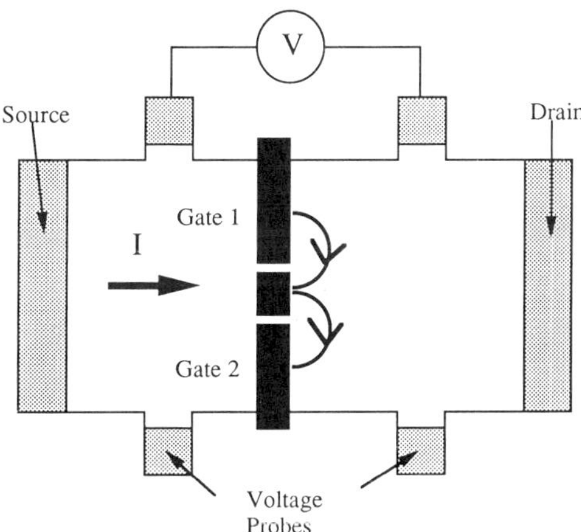

Figure 11. A schematic diagram of the sample used in the electron focussing experiments. The voltage on gates V_{g1} and V_{g2} can be varied independently. The distance between the centers of the point contacts is $\sim 0.88\,\mu$m.

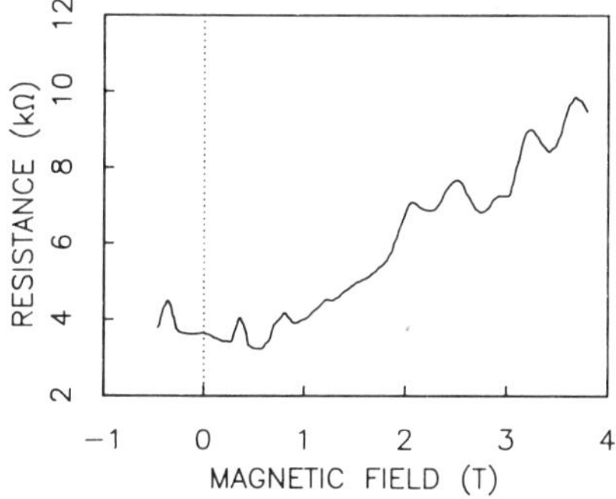

Figure 12. Magnetoresistance of two point contacts in parallel.

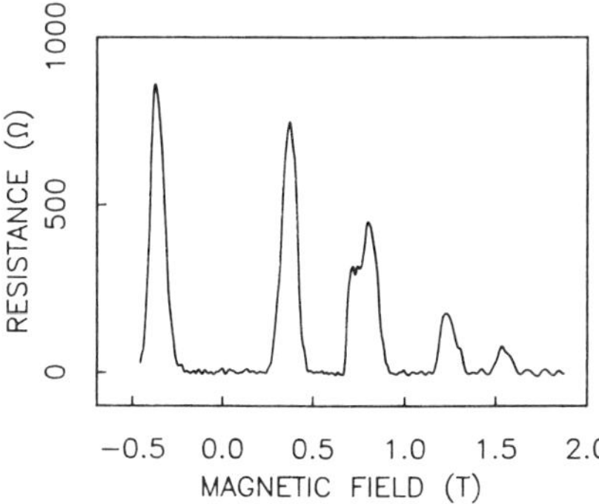

Figure 13. Electron focussing peaks of Fig. 12 after subtraction of the background.

total distribution they can lead to interesting effects in the magnetoresistance. The positive magnetoresistance around zero field is one such example and others include the quenched [16] and negative [31] Hall resistance and the negative bend resistance [32]. All of these phenomena will be influenced by the nature of the boundary scattering [33]. Of particular interest is the collimation of electrons [34] since this has been used to explain the quenching of the Hall effect [35]. Collimation, the focussing of electrons in the forward direction, increases the probability that an electron will be transmitted through a crossed wire junction resulting in a suppression of the Hall resistance. Electron collimation occurs in a flared junction but only after several specular collisions with the boundaries. If p is too small, electrons will scatter diffusely at the edges of the junction and flaring alone will not be sufficient to collimate the beam.

The above discussion would suggest that boundary roughness prevents quenching of the Hall resistance, at least via the collimation mechanism. However, as pointed out by Akera and Ando [36], boundary roughness can itself suppress the low field Hall resistance even in wires with no junction flaring. The reason is that most of the current is carried by the lowest subband (since higher modes are strongly backscattered) and electrons in this subband have a high probability for transmission through the junction.

The distance between diffuse boundary scattering events limits the extent to which electron transport can be considered ballistic. Our estimates of p indicate that this boundary scattering length can be much smaller than the transport mean free path of the original 2 DEG. If long range ballistic transport is to be preserved in narrow wires the microfabrication will have to be optimized to substantially increase the degree of specular scattering – a not insignificant challenge.

References

[1] A.B. Pippard 'magnetoresistance in Metals', Chapter 6, and references therein, Cambridge Univ. Press (1989)

[2] A. Hartstein and A. Fowler, Surf. Sci. **58**, 178 (1976)

[3] A. Scherer and M.L. Roukes, Appl. Phys. Lett. **55**, 377 (1989)

[4] J.M. Ziman, 'Electrons and Phonons', Chapter 11, Oxford Univ. Press UK (1960)

[5] K. Fuchs, Proc. Camb. Phil. Soc **34**, 100 (1938)

[6] T.J. Thornton, M.L. Roukes, A. Scherer, B.P. Van der Gaag, Phys. Rev. Lett **63**, 2128 (1989); M.L. Roukes, T.J. Thornton, A. Scherer, B.P. Van der Gaag in: Electronic Properties of Multilayers and Low Dimensional Solids, J.M. Chamberlain, L. Eaves and J.C. Portal, eds., Plenum, London (1990)

[7] The concept of a boundary scattering length is only valid when $l_0 \gg W$, otherwise most electrons will scatter within the bulk of the wire and only a fraction of order l_0/W will interact with the edges

[8] B.L. Altshuler and A.G. Aronov, JETP Lett **33**, 501 (1981)

[9] B.L. Altshuler and A.G. Aronov, Sol. St. Commun. **46**, 429 (1983)

[10] T.J. Thornton, M. Pepper, H. Ahmed, G.J. Davies, and D.A. Andrews, Phys. Rev. Lett **56**, 1181 (1986)

[11] M Büttiker, Phys. Rev **B 34**, 9375 (1988)

[12] T. Hiramoto, K. Hirakawa, Y. Iye, and T. Ikoma, Appl. Phys. Lett. **51**, 1620 (1987); Y. Hirayama, S. Tarucha, Y. Suzuki, and H. Okamoto, Phys Rev **B 37**, 2774 (1988)

[13] T.J. Thornton, M.L. Roukes, A. Scherer, B.P. Van der Gaag, in: Science and Technology of 1- and 0- Dimensional Semiconductors, S.P. Beaumont and C.M. Sottomayor-Torres, eds., Plenum, London, to be published

[14] C.J.B. Ford, T.J. Thornton, R. Newbury, M. Pepper, H. Ahmed, D. Peacock, D. Ritchie, J. Frost, and G. Jones, Appl. Phys. Lett. **54**, 21 (1989)

[15] A.M. Chang, T.Y. Chang, and H.U. Baranger, Phys. Rev. Lett. **63**, 996 (1989)

[16] M.L. Roukes et al., Phys. Rev. Lett. **57**, 3011 (1987)

[17] K. Forsvoll and L. Holwech, Phil. Mag. **9**, 435 (1964)

[18] K.-F. Berggren, T.J. Thornton, D.J. Newson, and M. Pepper, Phys. Rev. Lett. **57**, 1769 (1986)

[19] K. Hirakawa, H. Sakaki, and J. Yoshino, Phys. Rev. Lett. **54**, 1279 (1985)

[20] G. Fischer and H. Hoffman, Sol. St. Commun. **35**, 793 (1980)

[21] J.C. Hensel, R.T. Tung, J.M. Poate, and F.C. Unterwald, Phys. Rev. Lett. **54**, 1840 (1985)

[22] Z. Tesanovich, M.V. Jaric, and S. Maekawa, Phys. Rev. Lett. **54**, 1840 (1985)

[23] N. Trivedi and N.W. Ashcroft, Phys. Rev **B 38**, 12298 (1988)

[24] H. Sakaki, T. Noda, K. Hirawaka, M. Tanaka, and T. Matsusue, Appl. Phys. Lett. **51**, 1934 (1987)

[25] V.S. Tsoi, JETP Lett. **19**, 70 (1974)

[26] H. van Houten et al., Phys. Rev. **B 39**, 8556 (1989)

[27] K. Nakamura, D.C. Tsui, F. Nihey, H. Toyoshima, and T. Itoh, Appl. Phys. Lett. **56**, 385 (1990)

[28] F. Nihey et al., unpublished

[29] J. Spector, H.L. Stormer, K.W. Baldwin, L.N. Pfeiffer, and K.W. West, in: Proc. 4th Int. Conf. on Modulated Semiconductor Structures (Ann Arbor), to be published in Surf. Sc.

[30] C.W.J. Beenakker and H. van Houten, Phys. Rev. Lett. **63**, 1857 (1989); 'Electronic Properties of Multilayers and Low Dimensional Solids', J.M. Chamberlain, L. Eaves, and J.C. Portal, eds., Plenum, London (1990)

[31] C.J.B. Ford et al., Phys. Rev. Lett. **62**, 2724 (1989)

[32] Y. Takagaki et al., Sol. St. Commun. **68**, 1051 (1988)

[33] see the chapter by M.L. Roukes and O. Allerhand in this volume

[34] C.W.J. Beenakker and H. van Houten in: Electronic Properties of Multilayers and Low-Dimensional Semiconductor Structures, J.M. Chamberlain, L. Eaves, and J.C. Portal, eds., NATO ASI, Plenum, London, to be published

[35] H.U. Baranger and A.D. Stone, Phys. Rev. Lett. **63**, 414 (1989)

[36] H. Akera and T. Ando, Phys. Rev **B**, to be published

ADIABATIC MODE SELECTION AND ACCURACY OF QUANTIZATION OF BALLISTIC POINT CONTACTS

Amir Yacobi, Yoseph Imry

Department of Nuclear Physics
Weizmann Institute of Science
Rehovot 76100, Israel

Recent experiments by van Wees et al. [1] and Wharam et al. [2] revealed that the two-terminal conductance of a narrow constriction is quantized in integer multiples of $e^2/\pi\hbar$ (including spin degeneracy). The experiments were done on a two-dimensional electron gas (2 DEG) in a $Al_xGa_{1-x}AS/GaAs$ heterostructures for which the length of the constriction was smaller than the elastic and inelastic mean free paths. The constriction is regarded as a narrow strip, having its own n modes or "conduction channels", connecting two wide regions that in turn are connected with negligible resistance to electron reservoirs. It has been pointed out [3] that such a quantization follows from the two-terminal Landauer [4-6] formula $G = e^2/\pi\hbar \sum_{ij} T_{ij}$, for full transmission ($\sum_i T_{ij} = 1$) in the conducting channels, i.e. for a ballistic constriction. Glazman et al. [7] have recently shown how an adiabatic (i.e. slowly varying) geometry of the constriction can very easily explain the quantization which is exact in the adiabatic limit. However, in most real samples the variations in the geometry of the confining walls are slow in the region where the constriction is narrowest but are abrupt near its termination. Accurate quantization therefore required good impedance matching between the point where adiabaticity is lost and the wide region that is connected to the constriction. Here it is shown, that in order to get a good impedance match one needs the highest transverse mode which arrives at the non-adiabatic opening to be sufficiently low compared with the maximal one supported at the opening. This mode selection can be achieved by an adiabatic constriction and/or by a potential barrier. These insights can be used to predict which conditions the conductance quantization will become accurate.

Let us begin by describing the quantized conductance in the adiabatic picture introduced by Glazman et al. [7]. Imagine a ballistic (disorder-free) 2 D wire, or electronic waveguide, confining the electrons to the region $|y| \leq d(x)$, yielding a smooth constriction. This is obtained by taking $d(x)$ to be a symmetric smooth function with $\lim_{x\to\pm\infty} = d$, $d(0) = d_0 < d$ and changing slowly from d to d_0 over a scale $L \gg d, d_0, \lambda$ (λ being the electron wavelength). One now makes a Born-Oppenheimer type separation of the "slow" longitudinal variable x and the "fast" transverse one y. The y-problem is

Quantum Coherence in Mesoscopic Systems
Edited by B. Kramer, Plenum Press, New York, 1991

just a square well having energies $E_n(x) = (\hbar^2/2m)(n\pi/2d(x))^2$, and the wave functions $|n_x(y)\rangle$, which satisfy the boundary conditions for every x. As is familiar from the usual separation (valid also for soft wall confinement), $E_n(x)$ play the role of an additional potential for the mode n. This potential depends on n and has the shape of a barrier with a maximum proportional to $(n/d_0)^2$. For a given E_F, a finite number of modes will have their energy above the barrier. This number, $n_{\max}$ is the integral part of $(2k_F d_0)/\pi$. In the adiabatic limit tunneling below the barrier and reflections above it are negligible and the x-problem is well approximated by the WKB method. It follows that $n_{\max}$ modes have $T_{ij} = \delta_{ij}$, and all the others have zero transmission, hence $G = (e^2/\pi\hbar)n_{\max}$. This establishes quantization of G in this limit. An important feature of the adiabatic constriction is that the number of channels occupied by the electrons propagating away from the constriction remains $n_{\max}$ although the region with wider d can support a larger number of channels. This effect is due to the conservation of mode number typical to the adiabatic limit and is referred to as adiabatic mode selection.

In order to discuss the breakdown of adiabaticity [7-9], we use the scaled variable $u \equiv x/L$. Thus for $|u| > 1$, $d(x)$ is constant. Let us introduce the adiabatic basis by considering $|n_x(u)\rangle$ for every n. This basis spans the set of all relevant wave functions at every u. A general wavefunction ψ can be represented as $\sum_{n=1}^{\infty} C_n(u)|n(u)\rangle$ which is written to emphasize the u dependence of the transverse wavefunctions. We insert ψ into the Schrödinger equation, in terms of u, and project on mode m. Then, grouping all the diagonal terms to the LHS and noting that $\langle m|\partial/\partial u|m\rangle = 0$ gives

$$C_m'' + (k_m L)^2 C_m = -\sum_{n\neq m} \left(2C_n'\langle m\left|\frac{\partial}{\partial u}\right|n\rangle + C_n\langle m\left|\frac{\partial^2}{\partial u^2}\right|n\rangle \right), \tag{1}$$

where $q_m = m\pi/2d$, $\langle m|\hat{A}|n\rangle$ means integration over y only and $k_m^2 = 2mE/\hbar^2 - q_m^2 + (\langle m|(\partial^2/\partial u^2)|m\rangle)/(L^2)$. k_m is the local wave number along the wire. We can see that the term q_m^2 and the matrix element serve as a potential barrier that each mode encounters, but this barrier is different for each mode. The asymptotic solutions to eq. (1) $(L \to \infty)$ are the well known WKB solutions in which the different modes decouple and therefore the adiabatic limit is obtained.

The lowest order corrections to the asymptotic solution are obtained by iterating eq. (1) once and give interchannel scattering. The transmission and reflection amplitudes for an electron incident from the right in mode s to the different modes m are

$$t_{ms} = \frac{1}{L} \int_{-1}^{1} \frac{1}{\sqrt{k_m}} \exp\left(iL \int_1^u k_m du\right) \frac{1}{\sqrt{k_s}} \exp\left(-iL \int_1^u k_s du\right) f_m^{(0)}(u)du, \tag{2}$$

$$r_{ms} = \frac{1}{L} \int_{-1}^{1} \frac{1}{\sqrt{k_m}} \exp\left(-iL \int_1^u k_m du\right) \frac{1}{\sqrt{k_s}} \exp\left(-iL \int_1^u k_s du\right) f_m^{(0)}(u)du, \tag{3}$$

where $f_m^{(0)} = (-2iLk_s\langle m|\partial/\partial u|s\rangle + \langle m|\partial^2/\partial u^2|s\rangle)$. In the large L limit, the integrands have quickly varying phases and analyzing them by steepest descent and stationary phase methods, one arrives at the result that *all* the corrections to t_{ms} and r_{ms} are exponentially small in L. The details of the calculation for both types of modes are presented in [8].

We now turn to discuss the conditions for the breakdown of adiabaticity at some $d(x) > d_0$, for gradually increasing $d(x)$. This is relevant for realistic experimental devices. Obviously, for given values for m and s and length L, $(k_m - k_s)L$ will become very small for large enough d, and therefore the phase in eq. (2) will not oscillate rapidly and the previous results will not be valid. We will first calculate the indefinite integrals, denoted by $t_{ms}(u)$ and $r_{ms}(u)$. We then obtain the transmission and reflection amplitudes for a given interval $[u_0, u]$ by $t_{ms}(u) - t_{ms}(u_0)$ and $r_{ms}(u) - r_{ms}(u_0)$ respectively. Assuming that k_m, k_s and $f_m^{(0)}$ are approximately constant over the interval $[u_0, u]$. We find that

$$|t_{ms}(x)| \approx \frac{16}{\pi} \frac{2dk_F d'}{\pi} \frac{ms}{(m^2 - s^2)^2} \quad \text{and} \quad |r_{ms}(x)| \approx \frac{4}{\pi} \frac{\pi d'}{2dk_F} \frac{ms}{(m^2 - s^2)}, \tag{4}$$

where $m \neq s$ and d' is the derivative of d with respect to x, the non-scaled variable. For $dk_s \gg 1$, $|r_{ms}|$ is much smaller than $|t_{ms}|$. The largest value x, x_c, for this approximation to be valid is such that $\max_m |t_{ms}| \leq O(1)$ and it is given by

$$d'(x_c) \leq \frac{\pi}{4} \frac{\pi}{2d(x_c)k_F}. \tag{5}$$

At x_c the perturbative treatment used becomes invalid. This condition can also be obtained by geometrical considerations. At a given $d(x)$ the electron has a determined transverse and longitudinal momentum and therefore can be viewed as moving at an angle $\theta_s(x) = \tan^{-1}(q_s/k_s)$. This angle depends only on $d(x)$ and since in the adiabatic limit the transverse quantum number s is conserved the electron adjusts its angle continuously as it propagates along the waveguide. The motion between the confining walls is ballistic and therefore in order to have a definite angle for every $d(x)$, the distance, Δx, the electron passes without changing its angle (i.e. the distance along the x-direction the electron passes from the point it hits the lower wall to the point it hits the upper wall) must be such that d changes very little over the range. If we now consider deviations from the adiabatic limit and ask what is d' such that $\theta_s(x)$ will correspond to $d(x + \Delta x)$ with quantum number $s + 1$, we obtain eq. (5). Using condition eq. (5) we obtain the total reflection probability from mode s to all other modes at x_c

$$R_s(x_c) = \sum_m |r_{ms}|^2 \approx |r_{s+1,s}|^2 + |r_{s-1,s}|^2 \propto \left(\frac{\pi}{2d(x_c)k_F} \right)^4 s^2 = \left(\frac{d_0}{d(x_c)} \right)^4 \frac{s^2}{n_{\max}^4}, \tag{6}$$

where we have used the relation $k_F \cong (n_{\max}\pi)/(2d_0)$. Note that although the transmission probability is large, the reflections are small even for the maximum mode number $s = n_{\max}$. If the constriction continuous to open up to larger d', the electrons will reach the point $\bar{x}$ where the geometrical-optics rays will be parallel to the confining walls and obviously from the point on the electrons will not be affected by the constriction any more. Therefore we expect that the above constriction will be equivalent to a constriction that is suddenly removed at $\bar{x}$.

Szafer and Stone [10] have calculated the transmission and reflection probabilities in the sudden geometry numerically and showed that the exact numerical results can be very well approximated by a Mean Field Approximation (MFA) which they have introduced. In this approximation it turns out that the total transmission probability from mode s is given in terms of n_a, the number of modes available (not necessarily occupied) at the sudden opening

$$T_s = \sum_i T_{is} \cong 1 - \frac{1}{64} \left(\frac{n_a}{n_a^2 - s^2} \right)^4 = 1 - R_s, \tag{7}$$

where n_a is the integral part of $(2d(\bar{x})k_F)/\pi$. If, for example, the adiabatic part preceding the abrupt termination point selects only the first 10 channels ($n_{\max} = 10$) and $d(\bar{x}) =$

$1.2d_0$ it follows that $n_a = n_{\max} + 2$ and the reflections for $s = n_{\max}$ (which are the largest) are $O(10^{-4})$. Thus, one may say that in order to get accuracy in the quantization of the constriction conductance, one needs that it will open adiabatically to a large enough (but still rather modest) width, so that the sudden opening beyond that width would not cause appreciable reflections. These results are consistent with the previous conclusions of [9].

To achieve a reflectionless transmission, even for a constriction with constant width, d, it is clearly necessary to populate less channels than the maximal allowed number $n_a \cong (k_F 2d)/\pi$. This can obviously be achieved also by a potential barrier along the channel. This barrier should be long enough to be either fully transmitting or fully reflecting and somewhat smooth to avoid resonances. A smooth saddle-point potential, for example, is satisfactory [11]. Due to Coulomb effects, potential variations practically always exist in narrow channels.

The above may be the principal reason for the observability of conductance quantization in the existing experiments. Examining the quantization by mode selection with a potential barrier should therefore have the additional advantage of accurate quantization for adiabatic barriers. An interesting feature of the channel selection is the possibility [12] of focussed emission of electrons from the sudden termination.

Acknowledgements

The authors are grateful to I. Bar Joseph, A.B. Fowler, N. Garcia. Y. Gefen, M. Heiblum, D.E. Khmel'nitskii, R. Landauer, M. Pepper, A. Schwimmer, E. Shimshoni, U. Sivan, U. Smilansky, and Y. Yacobi for instructive discussions. This research was supported by the fund for basic research of the Israeli Academy of Sciences, Jerusalem and by the Minerva Foundation, Munich.

References

[1] B.J. van Wees, H. van Houten, C.W.J. Beenakker, J.G. Williamson, L.P. Kouwenhoven, D. van der Marel, and C.T. Foxton, Phys. Rev. Lett. **60**, 848 (1988)

[2] D.A. Wharam, M. Pepper, H. Ahmed, J.E.F. Frost, D.G. Hasko, D.C. Peacock, D.A. Ritchie, and G.A.C. Jones, J. Phys. **C 21**, L209 (1988)

[3] Y. Imry, "Physics in mesoscopic systems", in: Directions in Condensed Matter Physics, G. Grinstein and G. Mazenko, eds., memorial volume in honor of S.K. Ma, 101, World Scientific, Singapore (1986)

[4] R. Landauer, IBM J. Res. Devel. **1** (1957)

[5] R. Landauer, Phil. Mag. **21**, 863 (1970)

[6] M. Büttiker, Y. Imry, R. Landauer, S. Pinhas, Phys. Rev. **B 31**, 6207 (1985)

[7] L.I. Glazman, G.B. Lesovik, D.E. Khmel'nitskii, and R.I. Shekhter, Pisma Zh. Exsp. Teor. Fiz. **48**, 218 (1988) [Sov. Phys. JETP Lett. **48**, 238 (1988)]

[8] A. Yacobi and Y. Imry, Phys. Rev. **B 41**, 5341 (1990); see also M.C. Payne, J. Phys.: Cond. Matt. **1**, 4939 (1989)

[9] L.I. Glazman and M. Jonson, J. Phys.: Cond. Matt. **1**, 5547 (1989)

[10] A. Szafer and A.D. Stone, Phys. Rev. Lett. **62**, 300 (1989); G. Kirczenow, Sol. St. Commun. **68**, 715 (1988); Phys. Rev. **B 39**, 10452 (1989); I.B. Levinson, Pisma Zh. Exsp. Teor. Fiz. **48**, 273 (1988) [Sov. Phys. JETP Lett. **48**, 301 (1988)]; E. Tekman and S. Ciraci, Phys. Rev. **B 39**, 8772 (1989); L. Escapa and N. Garcia, J. Phys.: Cond. Matt. **1**, 2125 (1989); N. Garcia and L. Escapa, Appl. Phys. Lett., in press; D. van der Marel and E.G. Haanappel, Phys. Rev. **B 39**, 7811 (1989); Y. Avishai and Y.B. Band, to be published

[11] M. Büttiker, Phys. Rev. **B 41**, 7906 (1990)

[12] N. Lang, A. Yacobi, and Y. Imry, Phys. Rev. Lett. **63**, 1499 (1898); C.W.J. Beenakker and H. van Houten, Phys. Rev. **B 39**, 10445 (1989)

CHAPTER 4

QUANTUM HALL EFFECT

ADIABATIC TRANSPORT
IN THE FRACTIONAL
QUANTUM HALL EFFECT REGIME

Carlo W.J. Beenakker

Philips Research Laboratories
P.O. Box 80 000
5600 JA Eindhoven, The Netherlands

1 Introduction

The quantum Hall effect (QHE) is the phenomenon that the Hall conductance G_H is
quantized in units of e^2/h, as expressed by the formula

$$G_H = \frac{p}{q}\frac{e^2}{h} \tag{1}$$

(p and q being mutually prime integers). The *integer* QHE ($q = 1$) was discovered
10 years ago by von Klitzing, Dorda, and Pepper [1] in the two-dimensional electron
gas (2 DEG) confined to a Si inversion layer. The *fractional* QHE ($q > 1$ and odd)
was first observed by Tsui, Störmer, and Gossard [2] in the 2 DEG at the interface
of a $Al_xGa_{1-x}As/GaAs$ heterostructure. Microscopically the two effects are entirely
different. The integer QHE, on the one hand, can be explained satisfactorily in terms
of the states of non-interacting electrons in a magnetic field (the Landau levels). The
fractional QHE, on the other hand, exists only because of electron-electron interactions
[3]. Phenomenologically, however, the integer and fractional QHE are quite similar. In
an *unbounded* 2 DEG this similarity is understood from Laughlin's general argument [4]
that: (1) The Hall conductance shows a plateau as a function of magnetic field (or Fermi
energy) whenever the quasi-particle excitations in the bulk of the 2 DEG are localized by
disorder; and that: (2) The value of G_H on the plateau is precisely an integer multiple p
of ee^*/h, where $e^* = e/q$ is the quasi-particle charge. (The product ee^* appears because
one e is needed to change the unit of conductance from Amperes per electron Volts to
Amperes per Volts). Theory and experiment on the QHE in an unbounded 2 DEG have
been reviewed in the books by Prange and Girvin [5] and by Chakraborty and Pietiläinen
[6] (see also the article by MacDonald in the present volume).

In the past few years, a variety of experiments have uncovered a novel phenomenol-
ogy of the QHE on short length scales. For example, in small sub-micron-size samples

the QHE can occur in the absence of disorder [7,8] and can show deviations from precise
quantization [9]. An anomalous quantization of the Hall conductance has been observed
[10] in samples which are large but which contain a pair of closely spaced current and
voltage contacts: quantization of G_H then occurs at multiples of e^2/h determined by the
properties of the contacts, rather than of the bulk 2 DEG. Indeed, it has been possible
in such an experiment to measure the fractional QHE in a 2 DEG which by conventional
measurements shows the integer effect [11].

These anomalies are not easily understood within the conventional description of
the QHE, which determines the quantized value of G_H from the charge of a quasi-particle
excitation localized in the bulk of the 2 DEG. One needs a description which can be
applied to small samples without disorder and which explicitly includes the properties of
the current and voltage contacts. For the *integer* QHE the Landauer-Büttiker formalism
provides such a description [12]. A central concept in this formulation is the concept of an
edge channel, which is the collection of states at the Fermi energy within a given Landau
level. These states are extended along the edges of the 2 DEG whenever the Fermi level
lies between two Landau levels in the bulk. Many of the anomalies in the integer QHE
can be understood as resulting from the absence of local equilibrium at the edge, which
in turn is a consequence of the reduction of scattering between edge channels in a strong
magnetic field [10,13,14]. On short length scales the electron transport becomes fully
adiabatic, i.e. without inter-edge channel scattering. Edge channels in the integer QHE
are defined in one-to-one correspondence with bulk Landau levels. This single-electron
description is not applicable to the fractional QHE, which is fundamentally a many-body
effect. In this article we review recent work towards a generalization of the concept of
adiabatic transport in edge channels, with the aim of providing a unified description of
anomalies in the integer and fractional QHE.

We first summarize, in section 2, the Landauer-Büttiker formalism for the integer
QHE. Our generalization [15] to the fractional QHE is described in section 3 and applied
to experiments in section 4. Two open problems are addressed in section 5. One is
the question: "What charge does the resistance measure?" The other refers to an alternative generalized Landauer formula proposed by MacDonald [16]. We will argue that
the appearance of both "electron" and "hole" channels in this formula implies a novel
limitation to the accuracy of the fractional QHE. Much of the material in the present
article is based on a review with a wider scope written in collaboration with H. van
Houten [17].

2   Integer Edge Channels

The Landauer-Büttiker formalism [18,19] is a linear response formalism which expresses
the conductance (a non-equilibrium property) in terms of an equilibrium Fermi level
property of the conductor. That property consists of a rational function of transmission
probabilities between current and voltage contacts of propagating modes with the Fermi
energy. In a strong magnetic field in the QHE regime the propagating modes are extended
along the edges of the conductor, because all Fermi level states in the bulk are localized.
The Landauer-Büttiker formalism thus describes the integer QHE in terms of properties
of *edge states*. We review this description in the present section.

We restrict the discussion to the case of a smoothly varying potential energy landscape $V(x,y)$ in the 2 DEG. The smoothness criterion is that V should vary by less
than the Landau level separation $\hbar\omega_c \equiv \hbar eB/m$ over a magnetic length $l_m \equiv (\hbar/eB)^{1/2}$

(which plays the role of the wavelength in a strong magnetic field B). In such a smooth potential the quantized cyclotron motion energy $\left(n - \frac{1}{2}\right)\hbar\omega_c$ (being the energy of the n-th Landau level, $n = 1, 2, \ldots$) is a constant of the motion. The total energy E_F of an electron at the Fermi level is the sum of this Landau level energy and the energy E_G from the electrostatic potential,

$$E_G = E_F - \left(n - \frac{1}{2}\right)\hbar\omega_c. \tag{2}$$

(The spin-splitting of the Landau levels by the Zeeman energy is ignored here, for simplicity.) The constancy of the Landau level index n for smooth V implies that the motion of the electron is along the equipotential $V(x,y) = E_G$. Classically, the center of the cyclotron orbit is guided along equipotentials by the combined effects of the Coulomb and Lorentz forces. Hence the name guiding center energy for E_G. The drift velocity $\mathbf{v}_{drift}$ of the orbit center (known as the guiding center drift) follows by balancing the Coulomb and Lorentz forces,

$$\mathbf{v}_{drift} = \frac{1}{eB^2}\nabla V \times \mathbf{B}. \tag{3}$$

The wavefunctions of states at the Fermi level have an appreciable amplitude within l_m of the equipotentials at E_G. One can distinguish between *extended* states near the sample boundaries, and *localized* states encircling potential maxima and minima in the bulk, as illustrated in Fig. 1. The extended states with the same Landau level index n are referred to collectively as the n-th edge *channel*. The edge channel with the smallest index n is closest to the sample boundary, because it has the largest E_G (eq. (2)). This is seen more clearly in a cross-sectional plot of $V(x,y)$ (Fig. 2). Notice that if the peaks and dips of the potential in the bulk have amplitudes below $\hbar\omega_c/2$, then only states with the highest Landau level index can exist in the bulk at the Fermi level.

The simplicity of the guiding center drift along equipotentials has been originally used in the percolation theory [20-22] of the QHE, soon after its experimental discovery [1]. In this theory the existence of edge states is ignored, and the Hall resistance is expressed in terms of properties of extended states in the bulk of the sample. Since in equilibrium all Fermi level states in the bulk are localized in general, the percolation theory requires for its applicability a threshold electric field to create extended bulk states (it is thus not a linear response theory). A description of the QHE based on extended edge states and localized bulk states, as in Fig. 1, was first put forward by Halperin [23], and further developed by several authors [12,24-27]. With the exception of Büttiker [12], these authors assume local equilibrium at the edge. In the presence of a chemical potential difference $\delta\mu$ between the edges, each edge channel can be shown to carry a current $(e/h)\delta\mu$, and thus to contribute e^2/h to the Hall conductance. The *equipartitioning* of current among the edge channels is characteristic for a local equilibrium. The total number of edge channels N at the Fermi level is equal to the number of bulk Landau levels below the Fermi level (because of the one-to-one correspondence between edge channels and bulk Landau levels). In this case of local equilibrium one thus has the usual integer QHE, $R_H = h/Ne^2$, with $R_H \equiv 1/G_H$ the Hall resistance (we disregard for convenience of notation the two-fold spin degeneracy of each Landau level).

The Hall resistance R_H is a four-terminal resistance, meaning that the voltage contacts are distinct from the contacts through which the current is passed. For the two-terminal resistance R_{2t} the voltage is measured between the current-carrying contacts. In

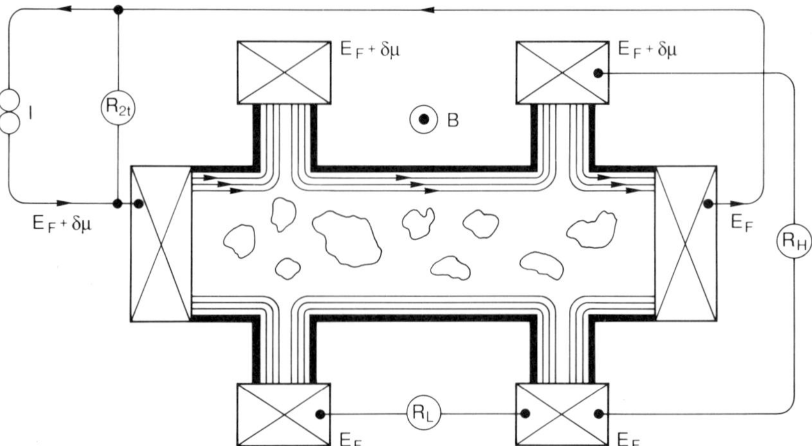

Figure 1. Measurement configuration for the two-terminal resistance R_{2t}, the four-terminal Hall resistance R_H, and the longitudinal resistance R_L. The edge channels at the Fermi level are indicated, arrows point in the direction of motion of edge channels filled by the source contact at chemical potential $E_F + \delta\mu$. The current is equipartitioned among the edge channels at the upper edge, corresponding to the case of local equilibrium (from Ref. [17]).

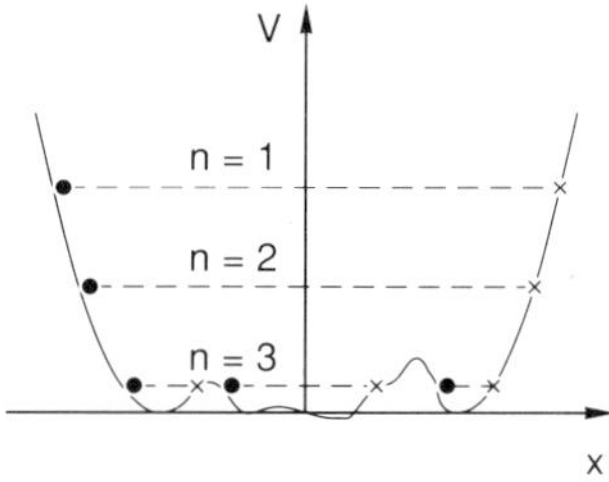

Figure 2. Cross-section of the electrostatic potential $V(x,y)$, along a line perpendicular to the Hall bar in Fig. 1. The location of the states at the Fermi level is indicated by dots and crosses (depending on the direction of motion). The value of E_G for each n is indicated by the dashed line (from Ref. [17]).

the case of local equilibrium these two resistances are the same, $R_H = R_{2t} = h/Ne^2$, see Fig. 1. One can also read off from Fig. 1 that the (four-terminal) longitudinal resistance R_L vanishes, $R_L = 0$. The distinction between a longitudinal and Hall resistance is topological: A four-terminal resistance measurement gives R_H if current and voltage contacts alternate along the boundary of the conductor, and R_L if that is not the case. There is no need to further characterize the contacts in the case of local equilibrium at the edge.

If the edges are not in local equilibrium, the measured resistance depends on the properties of the contacts. Büttiker has developed the formalism [12] to treat anomalies in the integer QHE due to the absence of local equilibrium, when measured with non ideal contacts. To illustrate this formalism we consider, following [17], a situation in which the edge channels at the lower edge are in equilibrium at chemical potential E_F, while the edge channels at the upper edge are not in local equilibrium. The current at the upper edge is then not equipartitioned among the N edge channels. Let f_n be the fraction of the total current I which is carried by states above E_F in the n-th edge channel at the upper edge, $I_n = f_n I$. The voltage contact at the lower edge measures a chemical potential E_F, regardless of its properties. The voltage contact at the upper edge, however, will measure a chemical potential which depends on how it couples to each of the edge channels. The transmission probability T_n is the fraction of I_n which is transmitted through the voltage probe to a reservoir at chemical potential $E_F + \delta\mu$. The incoming current

$$I_{in} = \sum_{n=1}^{N} T_n f_n I, \ \text{ with } \sum_{n=1}^{N} f_n = 1, \tag{4}$$

has to be balanced by an outgoing current

$$I_{out} = \frac{e}{h}\delta\mu \sum_{n=1}^{N} T_n \tag{5}$$

of equal magnitude, so that the voltage probe draws no net current. (In eq. (5) we have applied a sum rule to identify the total transmission probabilities of outgoing and incoming edge channels, see [17].) The requirement $I_{in} = I_{out}$ determines $\delta\mu$ and hence the Hall resistance $R_H = \delta\mu/eI$,

$$R_H = \frac{h}{e^2} \left(\sum_{n=1}^{N} T_n f_n \right) \left(\sum_{n=1}^{N} T_n \right)^{-1}. \tag{6}$$

The Hall resistance has its regular quantized value $R_H = h/Ne^2$ only if *either* $f_n = 1/N$ *or* $T_n = 1$, for $n = 1, 2, \ldots N$. The first case corresponds to local equilibrium (the current is equipartitioned among the edge channels), the second case to an ideal contact (all edge channels are fully transmitted).

A non-equilibrium population of the edge channels is generally the result of *selective backscattering*. Because edge channels at opposite edges of the sample move in opposite directions, backscattering requires scattering from one edge to the other. Selective backscattering of edge channels with $n \geq n_0$ is induced by a potential barrier across the sample, if its height is between the guiding center energies of edge channel n_0 and $n_0 - 1$ (recall that the edge channel with a larger index n has a smaller value of E_G). Selective backscattering can also occur *naturally* in the absence of an imposed potential barrier. The edge channel with the highest index $n = N$ is selectively backscattered when the Fermi level approaches the energy $\left(N - \frac{1}{2}\right)\hbar\omega_c$ of the N-th bulk Landau level. The guiding center energy of the N-th edge channel then approaches zero, and backscattering either by tunneling or by thermally activated processes becomes effective – but for that edge channel only, which remains almost completely decoupled from the other $N - 1$ edge channels over distances as large as [13,28,29] 250 μm (although on that length scale the edge channels with $n \leq N - 1$ have equilibrated to a large extent [14]).

We conclude this section by emphasizing that the edge channel formulation of the QHE by no means implies that the current flows within a few magnetic lengths from the edge. This assumption would be untenable experimentally, see [30]. The flow lines in Fig. 1 only show the location of the extended states at the equilibrium Fermi level. A determination of the spatial current distribution, rather than just the total current, requires consideration of all the states below the Fermi level, which acquire a net drift velocity because of the Hall field. Within the range of validity of a linear response theory, however, knowledge of the current distribution is not necessary to know the resistance (see [17] for a further discussion of this point).

3   Fractional Edge Channels

In this section we show, following [15], how the concept of an edge channel can be generalized to the fractional QHE, in the case of a smoothly varying electrostatic potential. This is the case of relevance for experiments on adiabatic transport in the fractional QHE, see section 4. Our result is phrased in terms of a generalized Landauer formula, in which the edge channels contribute with a fractional weight. Hence the name "fractional" edge channels. The different generalization of the Landauer formula proposed by MacDonald [16] is discussed in section 5.

Consider first the equilibrium state of the system. If the electrostatic potential energy $V(x,y)$ varies slowly in the 2 DEG, then the equilibrium density distribution $n(x,y)$ follows by requiring that the local electrochemical potential $V(\mathbf{r}) + du/dn$ has the same value μ at each point $\mathbf{r}$ in the 2 DEG. Here du/dn is the chemical potential of the *uniform* 2 DEG with density $n(\mathbf{r})$. It is a remarkable fact [3,5,6] that the internal energy density $u(n)$ of a uniform interacting 2 DEG in a strong magnetic field has downward cusps at densities $n = \nu_p Be/h$ corresponding to certain fractional filling factors ν_p. The chemical potential du/dn thus has a discontinuity (an energy gap) at $\nu = \nu_p$, with du_p^+/dn and du_p^-/dn the two limiting values as $\nu \to \nu_p$. The size of the gap is the cyclotron energy $\hbar\omega_c$ when ν_p is an integer, and of the order of the Coulomb energy $e^2/\varepsilon l_m$ when ν_p is a fraction (ε is the dielectric constant). An order of magnitude for the energy gap is 10 meV at $B = 6$ T. As noted by Halperin [31], when $\mu - V$ lies in the energy gap the filling factor is pinned at the value ν_p:

$$n = \nu_p Be/h, \quad \text{if} \quad du_p^-/dn < \mu - V < du_p^+/dn,$$

$$\frac{du}{dn} + V(\mathbf{r}) = \mu, \quad \text{otherwise.} \tag{7}$$

Note that $V(\mathbf{r})$ itself depends on $n(\mathbf{r})$, and thus has to be determined selfconsistently from eq. (7) taking the electrostatic screening in the 2 DEG into account. We do not need to explicitly solve for $n(\mathbf{r})$, but can identify the edge channels from the following general considerations [15].

At the edge of the 2 DEG the electron density decreases from its bulk value to zero. Equation (7) implies that this decrease is stepwise, as illustrated in Fig. 3. The requirement on the smoothness of V for the appearance of a well-defined region at the edge in which ν is pinned at the fractional value ν_p, is that the change in V within the magnetic length l_m is small compared to the energy gap du_p^+/dn and du_p^-/dn . This ensures that the width of this region is large compared to l_m, which is a necessary (and presumably sufficient) condition for the formation of the fractional QHE state. Depending on the smoothness of V, one thus obtains a series of steps at $\nu = \nu_p (p = 1, 2, \ldots P)$, as one moves from the edge towards the bulk. The series terminates in the filling factor $\nu_P = \nu_{bulk}$ of the bulk, assuming that in the bulk the chemical potential $\mu - V$ lies in an energy gap. The regions of constant ν at the edge form bands extending

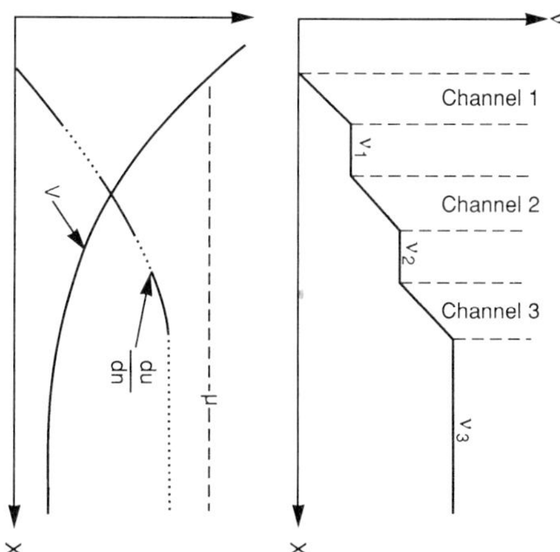

Figure 3. Schematic drawing of the variation in filling factor ν, electrostatic potential V, and chemical potential du/dn, at a smooth boundary in a 2 DEG. The dashed line in the bottom panel denotes the constant electrochemical potential $\mu = V + du/dn$. The dotted intervals indicate a discontinuity (energy gap) in du/dn, and correspond in the top panel to regions of constant fractional filling factor ν_p which spatially separate the edge channels. The width of the edge channel regions shrinks to zero in the integer QHE, since the compressibility χ of these regions is infinitely large in that case (from Ref. [15]).

along the conductor. These *incompressible bands* (in which the compressibility $\chi \equiv (n^2 d^2u/dn^2)^{-1} = 0$) alternate with bands in which $\mu - V$ does not lie in an energy gap. The latter compressible bands (in which $\chi > 0$) may be identified as the *edge channels* of the transport problem, as will be discussed below. To resolve a misunderstanding [32], we note that the particular potential and density profile illustrated in Fig. 3 (in which the edge channels have a non-zero width) assumes that the compressibility of the edge channels is not infinitely large – but that the analysis given below is independent of this assumption.

The conductance is calculated by bringing one end of the conductor in contact with a reservoir at a slightly higher electrochemical potential $\mu + \Delta\mu$. We are concerned with the linear response current, so that the electrostatic potential landscape $V(\mathbf{r})$ is kept at its equilibrium form. The resulting change Δn in electron density is

$$\Delta n = \left.\frac{\delta n}{\delta \mu}\right|_V \Delta\mu = -\left.\frac{\delta n}{\delta V}\right|_\mu \Delta\mu, \tag{8}$$

where δ denotes a functional derivative. In the second equality in eq. (8) it has been used that n is a functional of $\mu - V$, by virtue of eq. (7). In a strong magnetic field, this excess density moves along equipotentials with the guiding-center-drift velocity given by eq. (3). The component v_{drift} of the drift velocity in the y-direction (along the conductor) is

$$v_{drift} = \hat{\mathbf{y}} \cdot \left(\nabla V \times \mathbf{B}/eB^2\right) = -\frac{1}{eB}\frac{\partial V}{\partial x}. \tag{9}$$

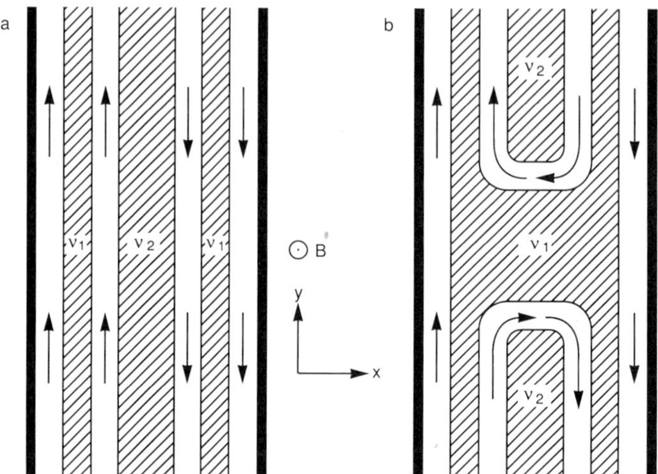

Figure 4. Schematic drawing of the incompressible bands (hatched) of fractional filling factor ν_p, alternating with the edge channels (arrows indicate the direction of electron motion in each channel). (a) a uniform conductor; (b) a conductor containing a barrier of reduced filling factor (from Ref. [15]).

The current density $j = -e\Delta n v_{drift}$ becomes simply

$$j = -\frac{e}{h}\Delta\mu\frac{\partial\nu}{\partial x}.$$ (10)

It follows from eq. (10) that the incompressible bands of constant $\nu = \nu_p$ do not contribute to j. The reservoir injects the current into the compressible bands at one edge of the conductor only (for which the sign of $\partial\nu/\partial x$ is such that j moves away from the reservoir). The edge channel with index $p = 1, 2, \ldots P$ is defined as that compressible band which is flanked by incompressible bands at filling factors ν_p and ν_{p-1}. The outermost band from the center of the conductor, which is the $p = 1$ edge channel, is included by defining formally $\nu_0 \equiv 0$. The arrangement of alternating edge channels and compressible bands is illustrated in Fig. 4a. Note that different edges may have a different series of edge channels at the same magnetic field value, depending on the smoothness of the potential V at the edge (which, as discussed above, determines the incompressible bands that exist at the edge). This is in contrast to the situation in the integer QHE, where a one-to-one correspondence exists between edge channels and bulk Landau levels (section 2). In the fractional QHE an infinite hierarchy of energy gaps exists, in principle, corresponding to an infinite number of possible edge channels – of which only a small number (corresponding to the largest energy gaps) will be realized in practice.

The current $I_p = (e/h)\Delta\mu\,(\nu_p - \nu_{p-1})$ injected into edge channel p by the reservoir follows directly from eq. (10), on integration over x. The total current I through the conductor is $I = \sum_{p=1}^{P} I_p T_p$, if a fraction T_p of the injected current I_p is transmitted to the reservoir at the other end of the conductor (the remainder returning via the opposite edge). For the conductance $G \equiv eI/\Delta\mu$ one thus obtains the generalized Landauer formula for a two-terminal conductor [15]

$$G = \frac{e^2}{h} \sum_{p=1}^{P} T_p \Delta\nu_p, \tag{11}$$

which differs from the usual two-terminal Landauer formula [18] by the presence of the fractional weight factors $\Delta\nu_p \equiv \nu_p - \nu_{p-1}$. In the integer QHE, $\Delta\nu_p = 1$ for all p, so that eq. (11) reduces to the Landauer formula with unit weight factors.

A multi-terminal generalization of eq. (11) for a two-terminal conductor is easily constructed, following Büttiker [19]:

$$I_\alpha = \frac{e}{h}\nu_\alpha\mu_\alpha - \frac{e}{h} \sum_{\beta} T_{\alpha\beta}\mu_\beta, \tag{12}$$

$$T_{\alpha\beta} = \sum_{p=1}^{P_\beta} T_{p,\alpha\beta}\Delta\nu_p. \tag{13}$$

Here I_α is the current in lead α, connected to a reservoir at electrochemical potential μ_α, and with fractional filling factor ν_α. Equation (13) defines the transmission probability $T_{\alpha\beta}$ from reservoir β to reservoir α (or the reflection probability, for $\alpha = \beta$), in terms of a sum over the generalized edge channels in lead β. The contribution from each edge channel $p = 1, 2, \ldots P_\beta$ contains the weight factor $\Delta\nu_p \equiv \nu_p - \nu_{p-1}$, and the fraction $T_{p,\alpha\beta}$ of the current injected by reservoir β into the p-th edge channel of lead β which reaches reservoir α. Apart from the fractional weight factors, the structure of eqs. (12) and (13) is the same as that of the usual Büttiker formula [19].

Applying the generalized Landauer formula eq. (11) to the ideal conductor in Fig. 4a, where $T_p = 1$ for all p, one finds the quantized two-terminal conductance

$$G = \frac{e^2}{h} \sum_{p=1}^{P} \Delta\nu_p = \frac{e^2}{h}\nu_P. \tag{14}$$

The four-terminal Hall conductance G_H has the same value, because each edge is in local equilibrium. In the presence of disorder this edge channel formulation of the fractional QHE is generalized in an analogous way as in the integer QHE, by including localized states in the bulk. In a smoothly varying disorder potential these localized states take the form of circulating edge channels, as in Fig. 1. In this way the filling factor of the bulk can locally deviate from ν_P without a change in the Hall conductance, leading to the formation of a plateau in the magnetic field dependence of G_H. In a narrow channel, localized states are not required for a finite plateau width, because the edge channels make it possible for the chemical potential to lie in an energy gap for a finite magnetic field interval. The Hall conductance then remains quantized at $\nu_P(e^2/h)$ as long as $\mu - V$ in the bulk lies between du_P^+/dn and du_P^-/dn.

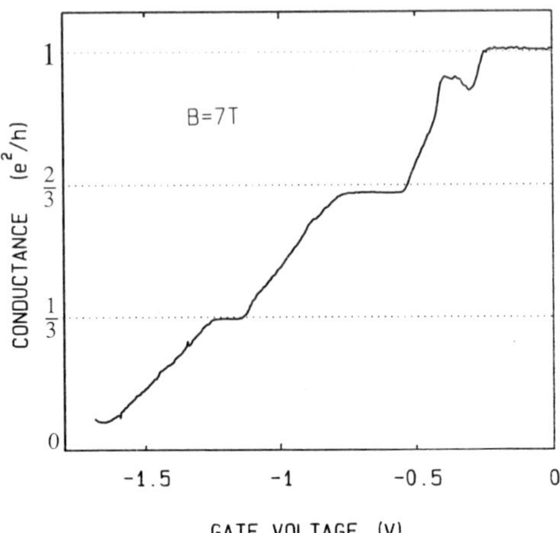

Figure 5. Two-terminal conductance of a constriction containing a potential barrier, as a function of the voltage on the split gate defining the constriction, at a fixed magnetic field of 7 T. The conductance is quantized according to eq. (15) (from Ref. [33]).

4   Experiments

We now apply the generalized Landauer formula eq. (11) to some recent experiments on adiabatic transport in the fractional QHE regime. Consider first a conductor containing a potential barrier. The potential barrier corresponds to a region of reduced filling factor $\nu_{P_{\min}} \equiv \nu_{\min}$ separating two regions of filling factor $\nu_{P_{\max}} \equiv \nu_{\max}$. The arrangement of edge channels and incompressible bands is illustrated in Fig. 4b. We assume that the potential barrier is sufficiently smooth that scattering between the edge channels at opposite edges can be neglected. All transmission probabilities are then either zero or one: $T_p = 1$ for $1 \leq p \leq P_{\min}$ and $T_p = 0$ for $P_{\min} < p \leq P_{\max}$. Equation (11) then tells us that the two-terminal conductance is

$$G = \frac{e^2}{h}\nu_{\min}. \tag{15}$$

In Fig. 5 we have reproduced experimental data by Kouwenhoven et al. [33] on the fractionally quantized two-terminal conductance of a constriction containing a potential barrier. The constriction (or point contact [29]) is defined by a split gate on top of a GaAs–AlGaAs heterostructure. The conductance in Fig. 5 is shown for a fixed magnetic field of 7 T as a function of the gate voltage. Increasing the negative gate voltage increases the barrier height, thereby reducing G below the Hall conductance corresponding to $\nu_{\max} = 1$ in the wide 2DEG. The curve in Fig. 5 shows plateaus corresponding to $\nu_{\min} = 1, 2/3$ and $1/3$ in eq. (15). The 2/3 plateau is not exactly quantized, but is too low by a few percent. The constriction width on this plateau is estimated [33] at $W = 500$ nm, which is a factor of 50 larger than the magnetic length at $B = 7$ T. It would seem that scattering between fractional edge channels at opposite edges (necessary to reduce the conductance below its quantized value) can only occur via states in the bulk for this large ratio of W/l_m.

Timp et al. [34] have measured the four-terminal Hall conductance in a narrow cross geometry ($W = 90\,\text{nm}$). They find, in addition to quantized plateaus near $1/3$, $2/5$, and $2/3 \times e^2/h$, also a plateau-like feature around $1/2 \times e^2/h$. (This even-denominator fraction is special because it is not observed as a Hall plateau in a bulk 2 DEG). Notice, however, that the 500 nm wide constriction of Fig. 5 has a conductance which is featureless at $e^2/2h$. A narrower constriction ($W = 150\,\text{nm}$) studied by Kouwenhoven et al. [33] shows more fluctuations on the plateaus at $1/3$ and $2/3 \times e^2/h$, but no plateau-like feature at $1/2 \times e^2/h$. The origin of the difference between these two experiments[33,34] remains to be understood.

A four-terminal measurement of the fractional QHE in a conductor containing a potential barrier can be analyzed by means of eqs. (12) and (13). The longitudinal resistance R_L of the barrier (measured by two adjacent voltage probes, one at each side of the barrier) is given by

$$R_{\mathrm{L}} = \frac{h}{e^2} \left(\frac{1}{\nu_{\min}} - \frac{1}{\nu_{\max}} \right). \tag{16}$$

This result follows from eqs. (12) and (13) provided *either* the edge channels transmitted across the barrier have equilibrated with the extra edge channels available outside the barrier region; *or* the voltage contacts are ideal, i.e. they have unit transmission probability for all fractional edge channels. In the case of the integer QHE, eq. (16) (with ν integer) was derived some time ago by Van Houten et al. [35] and (independently) by Büttiker [12], and was found to be in agreement with experiments [35-37]. Chang and Cunningham [38] have measured R_L in the fractional QHE, using a 1.5μm wide 2 DEG channel with a gate across a segment of the channel. Contacts to the gated and ungated regions allowed $\nu_{\min}$ and $\nu_{\max}$ to be determined independently. Equation (16) was found to hold to within 0.5% accuracy.

adiabatic transport in the fractional QHE has been demonstrated [11] by the selective population and detection of fractional edge channels, achieved by means of barriers in two closely separated current and voltage contacts. The geometry is illustrated in Fig. 6a. It is essentially the same as the geometry employed by Van Wees et al. [10] for the selective population and detection of Landau levels in the integer QHE. Fig. 6b illustrates the arrangement of fractional edge channels and incompressible bands for the case that the chemical potential lies in an energy gap for the bulk 2 DEG (at $\nu = \nu_{\mathrm{bulk}}$), as well as for the two barriers (at ν_I and ν_V for the barrier in the current and voltage lead, respectively). Adiabatic transport is assumed over the barrier, as well as from barrier I to barrier V (for the magnetic field direction indicated in Fig. 6). Equation (12) for this case reduces to

$$I = \frac{e}{h}\nu_I \mu_I,$$
$$0 = \frac{e}{h}\nu_V \mu_V - \frac{e}{h}\min(\nu_I,\nu_V)\mu_I, \tag{17}$$

so that the Hall conductance $G_{\mathrm{H}} = eI/\mu_{\mathrm{V}}$ becomes

$$G_H = \frac{e^2}{h}\max(\nu_I,\nu_V) \leq \frac{e^2}{h}\nu_{\mathrm{bulk}}. \tag{18}$$

The quantized Hall plateaus are determined by the fractional filling factors of the current and voltage leads, not of the bulk 2 DEG.

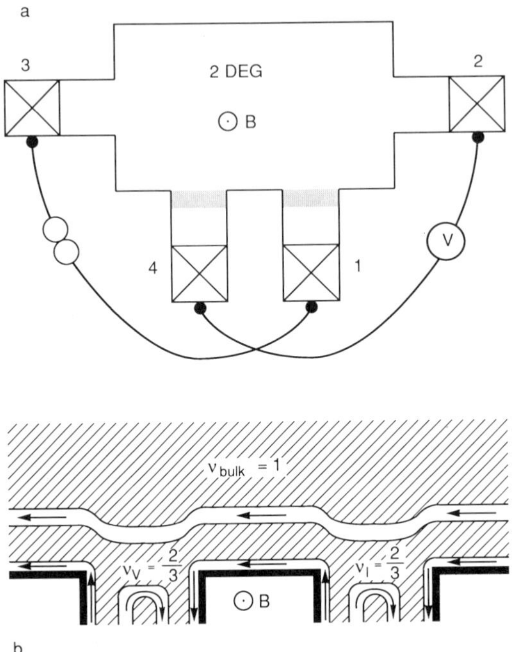

Figure 6. (a) Schematic drawing of the experimental geometry of Kouwenhoven et al. [11]. The crossed squares are contacts to the 2 DEG. One current lead and one voltage lead contain a barrier (shaded), of which the height can be adjusted by means of a gate (not drawn). The current I flows between contacts 1 and 3, the voltage V is measured between contacts 2 and 4. (b) Arrangement of incompressible bands (hatched) and edge channels near the two barriers. In the absence of scattering between the two fractional edge channels one would measure a Hall conductance $G_H \equiv I/V$ which is fractionally quantized at $\frac{2}{3} \times e^2/h$, although the bulk has unit filling factor (from Ref. [15]).

Kouwenhoven et al. [11] have demonstrated the selective population and detection of fractional edge channels in a device with a $2\,\mu$m separation of the gates in the current and voltage leads. The gates extended over a length of $40\,\mu$m along the 2 DEG boundary. In Fig. 7 we reproduce one of their experimental traces. The Hall conductance is shown for a fixed magnetic field of $7.8\,$T as a function of the gate voltage (all gates being at the same voltage). As the barrier heights in the two leads are increased, the Hall conductance decreases from the bulk value $1 \times e^2/h$ to the value $\frac{2}{3} \times e^2/h$ determined by the leads – in accord with eq. (18). A more general formula for G_H valid also in between the quantized plateaus is shown in [11] to be in quantitative agreement with the experiment.

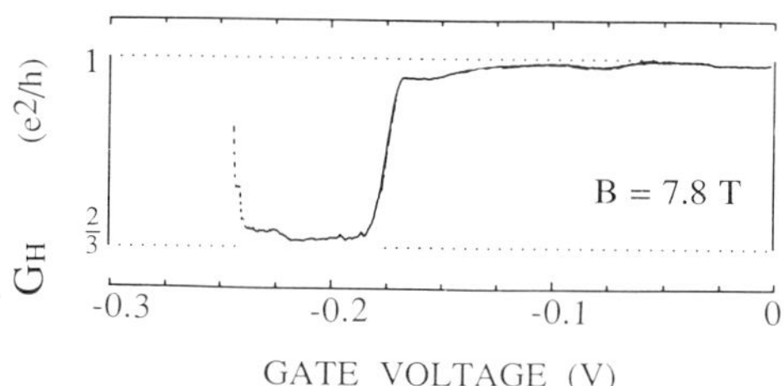

Figure 7. Anomalously quantized Hall conductance in the geometry of Fig. 6, in accord with eq. (18) ($\nu_{\text{bulk}} = 1, \nu_I = \nu_V$ decreases from 1 to 2/3 as the negative gate voltage is increased). The temperature is 20 mK. The rapidly rising part (dotted) is an artifact due to barrier pinch-off (from Ref. [11]).

5 Open Problems

5.1 What Charge does the Resistance Measure?

The fractional quantization of the conductance in the experiments discussed above is understood as a consequence of the fractional weight factors in the generalized Landauer formula eq. (11). These weight factors $\Delta\nu_p = \nu_p - \nu_{p-1}$ are *not* in general equal to e^*/e, with e^* the fractional charge of the quasi-particle excitations of Laughlin's incompressible state. The reason for the absence of a one-to-one correspondence between $\Delta\nu_p$ and e^* is that the edge channels themselves are not incompressible [15]. The transmission probabilities in eq. (11) refer to charged "gapless" excitations of the edge channels, which are not identical to the charge e^* excitations above the energy gap in the incompressible bands (the latter charge might be obtained from thermal activation measurements, see [39]).

It is an interesting and (to date) unsolved problem to determine the charge of the edge channel excitations. Kivelson and Pokrovsky [40] have suggested performing tunneling experiments in the fractional QHE regime for such a purpose, by using the charge dependence of the magnetic length $(\hbar/eB)^{1/2}$ (which determines the penetration of the wave function in a tunnel barrier, and hence the transmission probability through the barrier). Alternatively, one could use the h/e periodicity of the Aharonov-Bohm magnetoresistance oscillations as a measure of the edge channel charge. Simmons et al. [41] find that the characteristic field scale of quasi-periodic resistance fluctuations in a 2μm wide Hall bar increases from 0.016 T $\pm$ 30% near $\nu = 1,2,3,4$ to 0.05 T $\pm$ 30% near $\nu = \frac{1}{3}$. This is suggestive of a reduction in charge from e to $e/3$, but not conclusive since the area for the Aharonov-Bohm effect is not well-defined in a Hall bar.

5.2 Electron and Hole Channels

MacDonald has, independently of [15], proposed a different generalized Landauer formula for the fractional QHE [16] in a smooth electrostatic potential. The difference with eq. (11) is that the weight factors in MacDonald's formula can take on both positive *and negative* values – corresponding to electron and hole channels, respectively. In the case of local equilibrium at the edge, the sum of weight factors is such that the two formulations give identical results. The results differ in the absence of local equilibrium, if fractional edge channels are selectively populated and detected. For example, MacDonald predicts a *negative* longitudinal resistance in a conductor at filling factor $\nu = 2/3$ containing a segment at $\nu = 1$. Another implication of [16], as we understand it, is that the two-terminal conductance G of a conductor at $\nu_{\max} = 1$ containing a potential barrier at filling factor $\nu_{\min}$ is reduced to $\frac{1}{3} \times e^2/h$ if $\nu_{\min} = 1/3$ (in accord with eq. (15)), but remains at $1 \times e^2/h$ if $\nu_{\min} = 2/3$. That this is not observed experimentally (see Fig. 5) could be due to inter-edge channel scattering, as argued by MacDonald. The experiment by Kouwenhoven et al. [11] (Fig. 7), however, is apparently in the adiabatic regime, and was interpreted in Fig. 6 in terms of an edge channel of weight 1/3 at the edge of a conductor at $\nu = 1$. In MacDonald's formulation, the conductor at $\nu = 1$ has only a single edge channel of weight 1. This would have to be reconciled with the experimental observation of quantization of the Hall conductance at $2/3 \times e^2/h$. What is needed is a theory which allows one to introduce edge channels not only for the case of a smooth potential at the edge (considered in [15] and [16]), but also for an abrupt confinement. Such a theory exists for the integer QHE [23] but not yet for the fractional effect.

The presence of both positive and negative weights in a generalized Landauer formula has an interesting implication for the accuracy of the fractional QHE. As we discussed in section 2 for the integer QHE, accurate quantization of the Hall resistance requires either a local equilibrium at the edge, or ideal contacts (i.e. contacts which fully transmit all available edge channels [12]). The reduction of inter-edge channel scattering in strong magnetic fields leads to deviations from local equilibrium (i.e. the current is not equipartitioned among the edge channels). Ideal contacts then become necessary for accurate quantization. A contact is essentially a region with a high electron density connected to the low-density electron gas, see Fig. 8a. An ideal contact is realized by a smooth increase in density in the contact region, so that the edge channels in the 2 DEG are transmitted *adiabatically* into the contact. The contact then *induces* a local equilibrium by redistributing the current among the edge channels. This is illustrated in Fig. 8b by means of arrows, which indicate the current-carrying edge channels: One incoming edge channel carries the current, whereas both outgoing edge channels are populated by the contact. These considerations for the integer QHE carry over completely to the fractional edge channels described in section 3. However, if both electron and hole channels are present in the 2 DEG, then the situation is different. A hole channel is reflected on approaching a region with a smoothly increasing electron density [16]. In other words, a contact can not be "ideal" (i.e. fully transmitting) for both electron and hole channels. As shown in Fig. 8c, the contact is then not able to redistribute the current among the edge channels. The accuracy of the fractional QHE would thus be limited by the extent to which inelastic scattering is effective in establishing a local equilibrium between electron and hole channels – regardless of the ideality of the contacts. This conclusion is of importance not only for adiabatic transport in the fractional QHE, but for other situations as well in which coexisting electron and hole channels are believed to occur. One example is the integer QHE in a periodic potential discussed by MacDonald in this volume. Another is the integer QHE in parallely conducting electron and hole gases, present in certain semiconductor heterostructures.

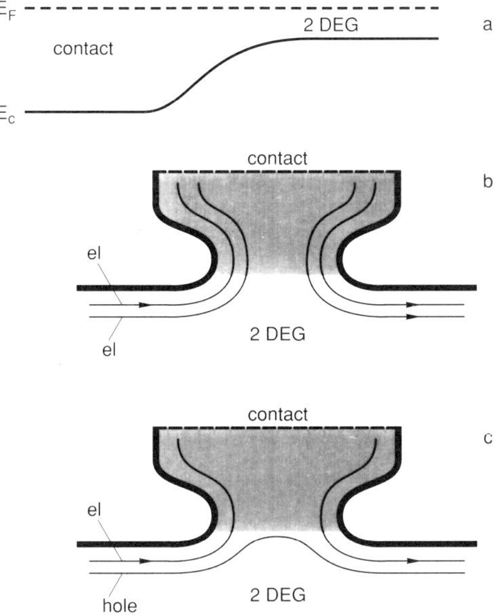

Figure 8. (a) Schematic drawing of the bottom of the conduction band E_c and the Fermi energy E_F at the transition from a low-density to a high-density region in a 2 DEG. (b,c) Top view of a 2DEG near a contact, modeled by a high-density region (shaded) as in (a). The contact is ideal (i.e. fully transmitting) for electron channels (b), but not for hole channels (c). The arrows indicate the current-carrying edge channels. This figure illustrates why a contact is effective in establishing local equilibrium among electron channels, but not among electron and hole channels. In case (c) one would measure anomalies in the Hall conductance, due to the absence of local equilibrium.

References

[1] K. von Klitzing, G. Dorda, and M. Pepper, Phys. Rev. Lett. **45**, 494 (1980)

[2] D.C. Tsui, H.L. Störmer, and A.C. Gossard, Phys. Rev. Lett. **48**, 1559 (1982)

[3] R.B. Laughlin, Phys. Rev. Lett. **50**, 1395 (1983)

[4] R.B. Laughlin, Phys. Rev.**B 27**, 3383 (1983)

[5] The Quantum Hall Effect, R.E. Prange and S.M. Girvin, eds., Springer, New York (1987)

[6] T. Chakraborty and P. Pietiläinen, The Fractional Quantum Hall Effect, Springer, Berlin (1988)

[7] D.A. Wharam, T.J. Thornton, R. Newbury, M. Pepper, H. Ahmed, J.E.F. Frost, D.G. Hasko, D.C. Peacock, D.A. Ritchie, and G.A.C. Jones, J. Phys. **C 21**, L209 (1988)

[8] B.J. van Wees, L.P. Kouwenhoven, H. van Houten, C.W.J. Beenakker, J.E. Mooij, C.T. Foxon, and J.J. Harris, Phys. Rev. **B 38**, 3625 (1988)

[9] A.M. Chang, G. Timp, J.E. Cunningham. P.M. Mankiewich, R.E. Behringer, and R.E. Howard, Sol. St. Commun. **76**, 769 (1988)

[10] B.J. van Wees, E.M.M. Willems, C.J.P.M. Harmans, C.W.J. Beenakker, H. van Houten, J.G. Williamson, C.T. Foxon, and J.J. Harris, Phys. Rev. Lett. **62**, 1181 (1989)

[11] L.P. Kouwenhoven, B.J. van Wees, N.C. van der Vaart, C.J.P.M. Harmans, C.E. Timmering, and C.T. Foxon, Phys. Rev. Lett. **64**, 685 (1990)

[12] M. Büttiker, Phys. Rev. **B 38**, 9375 (1988)

[13] S. Komiyama, H. Hirai, S. Sasa, and S. Hiyamizu, Phys. Rev. **B 40**, 12566 (1989)

[14] B.W. Alphenaar, P.L. McEuen, R.G. Wheeler, and R.N. Sacks, Phys. Rev. Lett. **64**, 677 (1990)

[15] C.W.J. Beenakker, Phys. Rev. Lett. **64**, 216 (1990)

[16] A.H. MacDonald, Phys. Rev. Lett. **64**, 220 (1990)

[17] C.W.J. Beenakker and H. van Houten, 'Quantum Transport in Semiconductor Nanostructures', to appear in: Solid State Physics, H. Ehrenreich and D. Turnbull, eds., Academic Press, New York (1991)

[18] R. Landauer, IBM J. Res. Dev. **1** , 223 (1957); ibidem, **32**, 306 (1988)

[19] M. Büttiker, Phys. Rev. Lett. **57**, 1761 (1986); IBM J. Res. Dev. **32**, 317 (1988)

[20] R.F. Kazarinov and S. Luryi, Phys. Rev. **B 25**, 7626 (1982); S. Luryi and R.F. Kazarinov, Phys. Rev. **B 27**, 1386 (1983); S. Luryi, in: High Magnetic Fields in Semiconductor Physics, G. Landwehr, ed., Springer, Berlin (1987)

[21] S.V. Iordansky, Sol. St. Commun. **43**, 1 (1982)

[22] S.A. Trugman, Phys. Rev. **B 27**, 7539 (1983)

[23] B.I. Halperin, Phys. Rev. **B 25**, 2185 (1982)

[24] A.H. MacDonald and P. Streda, Phys. Rev. **B 29**, 1616 (1984)

[25] S.M. Apenko and Yu.E. Lozovik, J. Phys. **C 18**, 1197 (1985)

[26] P. Streda, J. Kucera, and A.H. MacDonald, Phys. Rev. Lett. **59**, 1973 (1987)

[27] J.K. Jain and S.A. Kivelson, Phys. Rev. **B 37**, 4276 (1988)

[28] B.J. van Wees, E.M.M. Willems, L.P. Kouwenhoven, C.J.P.M. Harmans, J.G. Williamson, C.T. Foxon, and J.J. Harris, Phys. Rev. **B 39**, 8066 (1989)

[29] H. van Houten, C.W.J. Beenakker, and B.J. van Wees, 'Quantum Point Contacts', to appear in: Semiconductors and Semimetals, M.A. Reed, volume ed., Academic Press, New York (1991)

[30] P.F. Fontein, J.A. Kleinen, P. Hendriks, F.A.P. Blom, J.H. Wolter, H.G.M. Lochs, F.A.J.M. Driessen, L.J. Giling, and C.W.J. Beenakker, submitted to Phys. Rev. B

[31] B.I. Halperin, Helv. Phys. Acta **56**, 75 (1983)

[32] A.M. Chang, Sol. St. Commun. **74**, 871 (1990)

[33] L.P. Kouwenhoven, B.J. van Wees, N.C. van der Vaart, C.J.P.M. Harmans, C.E. Timmering, and C.T. Foxon, unpublished

[34] G. Timp, R.E. Behringer, J.E. Cunningham, and R.E. Howard, Phys. Rev. Lett. **63**, 2268 (1989)

[35] H. van Houten, C.W.J. Beenakker, P.H.M. van Loosdrecht, T.J. Thornton, H. Ahmed, M. Pepper, C.T. Foxon, and J.J. Harris, Phys. Rev. **B 37**, 8534 (1988)

[36] R.J. Haug, A.H. MacDonald, P. Streda, and K. von Klitzing, Phys. Rev. Lett. **61**, 2797 (1988)

[37] S. Washburn, A.B. Fowler, H. Schmid, and D. Kern, Phys. Rev. Lett. **61**, 2801 (1988)

[38] A.M. Chang and J.E. Cunningham, Sol. St. Commun. **72**, 651 (1989)

[39] R.G. Clark, J.R. Mallett, S.R. Haynes, J.J. Harris, and C.T. Foxon, Phys. Rev. Lett. **60**, 1747 (1988)

[40] S.A. Kivelson and V.L. Pokrovsky, Phys. Rev. **B 40**, 1373 (1989)

[41] J.A. Simmons, H.P. Wei, L.W. Engel, D.C. Tsui, and M. Shayegan, Phys. Rev. Lett. **63**, 1731 (1989)

THE QUANTUM HALL EFFECTS

Allan H. MacDonald

Department of Physics
Indiana University
Bloomington, IN 47405, USA

1   Introduction

The quantum Hall effect [1] occurs in two-dimensional electronic systems (2 DES) [2] in strong perpendicular magnetic fields. It differs from many of the phenomena which are being discussed at this ASI in that it does not depend directly on quantum coherence for its occurrence. Quantum coherence plays an important supporting role rather that a starring role in *this* drama. At low temperatures and strong magnetic fields the magnetoresistance of two-dimensional (2 D) metals becomes extremely small and the Hall resistance nearly constant over certain ranges of carrier density and magnetic field strength. Although this effect seemed quite mysterious at the time of its discovery, the principle elements of the physics which is responsible are now understood. In these notes I hope to explain these elements and to point out some aspects of our current theoretical picture of the phenomenon which seem to be incomplete.

For orientational purposes it is useful to begin with a simple semi-classical description of transport in a 2 DES. In the presence of an electric field, an electron is accelerated for an average time τ_0, the relaxation time, before being scattered to a state which has, on average, zero velocity. In this picture the electrons have an average drift velocity

$$\vec{v}_D = -e\vec{E}\tau_0/m \tag{1}$$

where $\vec{E}$ is the electric field and m is the electron effective mass. The current density in an electric field is then given by

$$\mathbf{j} = -en\vec{v}_D = \sigma_0\vec{E} \tag{2}$$

where

$$\sigma_0 = ne^2\tau_0/m \tag{3}$$

and n is the areal electron density. In the presence of a steady perpendicular magnetic field the Lorentz force must be added to the force from the electric field so that

$$\vec{v}_D = -e(\vec{E} + \vec{v}_D \times \hat{z}B/c)\tau_0/m \tag{4}$$

Quantum Coherence in Mesoscopic Systems
Edited by B. Kramer, Plenum Press, New York, 1991

and the current density is no longer parallel to the magnetic field. In eq. (4) $\hat{z}B$ is the magnetic field. Using eq. (4) for the drift velocity we find that the resistivity tensor $(\mathbf{j} \equiv \sigma\mathbf{E}, \mathbf{E} \equiv \rho\mathbf{j}, \rho = \sigma^{-1})$ is given by,

$$\rho_{xx} = \rho_{yy} = \sigma_0^{-1}, \tag{5}$$

$$\rho_{xy} = -\rho_{yx} = B/nec. \tag{6}$$

Note that the diagonal components of the resistivity tensor are unchanged by the magnetic field; i.e., there is no magnetoresistance. The effect of the magnetic field is to introduce an electric field, the Hall field, in the direction perpendicular to the current flow. The force on the electrons from the Hall field exactly cancels the Lorentz force which the electrons experience as they drift.

This simple theory does a good job of describing the effect of a weak magnetic field on transport. At strong fields, however, there are important corrections which are due to the magnetic quantization of the kinetic energy of 2 D electrons. To go any further toward understanding magnetotransport in strong magnetic fields we must first consider the quantum treatment of an electron in a magnetic field. The Hamiltonian for a free electron in a magnetic field is

$$t = \boldsymbol{\pi}^2/2m, \tag{7}$$

where

$$\boldsymbol{\pi} = -i\hbar\nabla + e\mathbf{A}/c \tag{8}$$

and $\mathbf{A}$ is the vector potential, related to the magnetic field by

$$\nabla \times \mathbf{A} = \mathbf{B}. \tag{9}$$

It follows from eq. (9) that

$$[\pi_x, \pi_y] = -i\hbar^2\ell^{-2} \tag{10}$$

and hence that

$$t = \hbar\omega_c(a^\dagger a + 1/2), \tag{11}$$

where $\omega_c = eB/mc$,

$$a^\dagger = \frac{\ell}{\sqrt{2}\hbar}(\pi_x + i\pi_y), \tag{12}$$

and $[a^\dagger, a] = 1$. Here $\ell = (\hbar c/eB)^{1/2}$ is the classical cyclotron orbit radius of an orbit with kinetic energy $\hbar\omega_c/2$.

The ladder operator commutation relations imply that the allowed kinetic energies are restricted to the discrete set of values $\hbar\omega_c(n+1/2)$ where n is an integer. Classically, however, it is clear that a cyclotron orbit is not specified completely by its energy. The solution to the equation of motion for a classical electron in 2 D moving in a magnetic field is

$$z(t) = C - iv(t)/\omega_c, \tag{13}$$

where $z(t) = x + iy$ is the electron position, $v(t) = v_x(t) + iv_y(t)$ is the electron velocity and $C = C_x + iC_y$ is the cyclotron orbit center. For a given orbit velocity, and hence a given orbit energy, the center is arbitrary. Quantum mechanically we might also expect that there be many possible states for one of the allowed kinetic energies, corresponding to different orbit centers. Expressing the orbit center as a quantum mechanical operator it is easy to show that

$$[C, C^\dagger] = 2\ell^2, \tag{14}$$

where we have substituted $\mathbf{v} = \pi/m$ in eq. (13). For the quantum case the orbit center cannot be precisely determined. From eq. (14) it is clear that we can define a set of ladder operators based on the orbit center operator just as we defined a set of ladder operator based on the orbit velocities. Let

$$b = C/\sqrt{2}\ell = \frac{2}{\sqrt{2}} \left((z + iA/B)/\ell + 2\ell\frac{\partial}{\partial z^*} \right) \tag{15}$$

so that $[b, b^\dagger] = 1$. (Here $A = A_x + iA_y$). Then $[a, b] = [a, b^\dagger] = 0$ and a complete orthonormal set of eigenfunctions of t is given by

$$\phi_{n,m} = \frac{(a^\dagger)^n (b^\dagger)^m \phi_{0,0}}{\sqrt{n!m!}} \tag{16}$$

where $a\phi_{0,0} = b\phi_{0,0} = 0$. The set of single-particle states with a given value for the kinetic energy is called a Landau level. The ladder operators $a, a^\dagger$ change the kinetic energy and are therefore inter-Landau level ladder operators. while the ladder operators $b, b^\dagger$ are closed in a subspace of any given kinetic energy and are called intra-Landau level ladder operators. The ability to separate operators into a factor associated with the cyclotron orbit velocity and a factor associated with the cyclotron orbit center will be very valuable when it comes to discussing the fractional quantum Hall effect.

Everything we have discussed to this point is independent of the choice of gauge. In the remaining portion of this section we carry the discussion a bit further for the gauge choice which has been most valuable for fractional quantum Hall effect studies, the symmetric gauge,

$$\mathbf{A} = (-By/2, Bx/2, 0). \tag{17}$$

For this choice

$$b^\dagger = \left(z^*/2\ell - 2\ell\frac{\partial}{\partial z} \right) / \sqrt{2} \tag{18}$$

and

$$a^\dagger = i \left(z/2\ell - 2\ell\frac{\partial}{\partial z^*} \right) / \sqrt{2}. \tag{19}$$

Note that $(\partial/\partial z)^\dagger = -(\partial/\partial z^*)$. Using the algebra of the ladder operators it is easy to show that

$$\phi_{n,m} = \frac{(-i)^n G^{m,n}(iz^*/\ell) \exp(-|z|^2/4\ell^2)}{\sqrt{2\pi}\ell}, \tag{20}$$

where

$$G^{m,n}(\alpha) = \sqrt{n!/m!} \left(-i\alpha/\sqrt{2}\right)^{m-n} L_n^{m-n}\left(|\alpha|^2/2\right) \tag{21}$$

and $L_n^k(x)$ is a generalized Laguerre polynominal. The areal density which can be accommodated by each Landau level is

$$n_1 = \sum_m |\phi_{n,m}|^2 = (2\pi\ell^2)^{-1} = B/\Phi_0, \tag{22}$$

where $\Phi_0 = hc/e$ is the electron's magnetic flux quantum. Equation (22) follows from eqs. (20), (21) and the frequently useful identity,

$$\sum_k G^{m,k}(\alpha_1)G^{k,m}(\alpha_2) = \exp(-\alpha_1^*\alpha_2/2)G^{n,m}(\alpha_1 + \alpha_2). \tag{23}$$

The density measured in units of n_1 is referred to as the Landau level filling factor, $\nu \equiv 2\pi\ell^2 n$. This quantity plays an essential role in both the integer and fractional quantum Hall effects.

2   Incompressibility Implies Quantization

In Fig. 1 we show the dependence of the chemical potential of a 2 D gas of non-interacting spinless electrons on density at fixed magnetic field and zero temperature. The chemical potential is defined by

$$\mu \equiv \frac{\partial F}{\partial N} \tag{24}$$

where F is the free energy and N is the number of electrons. At zero temperature μ is the change in the ground state energy when one electron is added to the system. For free-electrons it follows that μ jumps by $\hbar\omega_c$ whenever a Landau level is filled, i.e. whenever ν is an integer. The densities at which the chemical potential jumps occur are proportional to magnetic field since n_1 is proportional to magnetic field. We show below that the quantum Hall effect will occur whenever the chemical potential jumps at a magnetic-field-dependent density. At densities where the chemical potential jumps the electron system is incompressible, since the compressibility is inversely proportional to

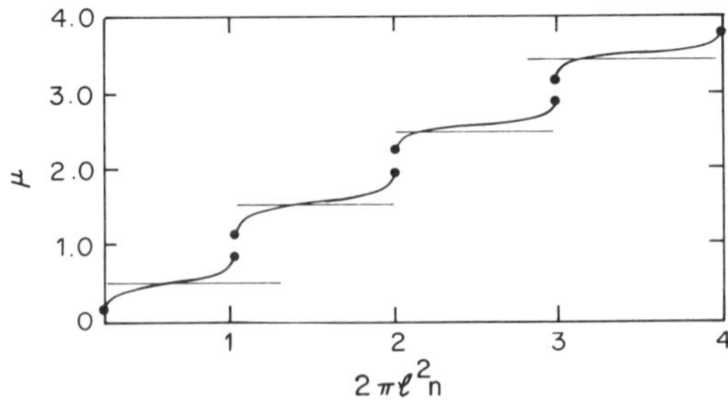

Figure 1. Density dependence of the chemical potential of a non-interacting 2 DES with and without disorder broadening of the Landau levels.

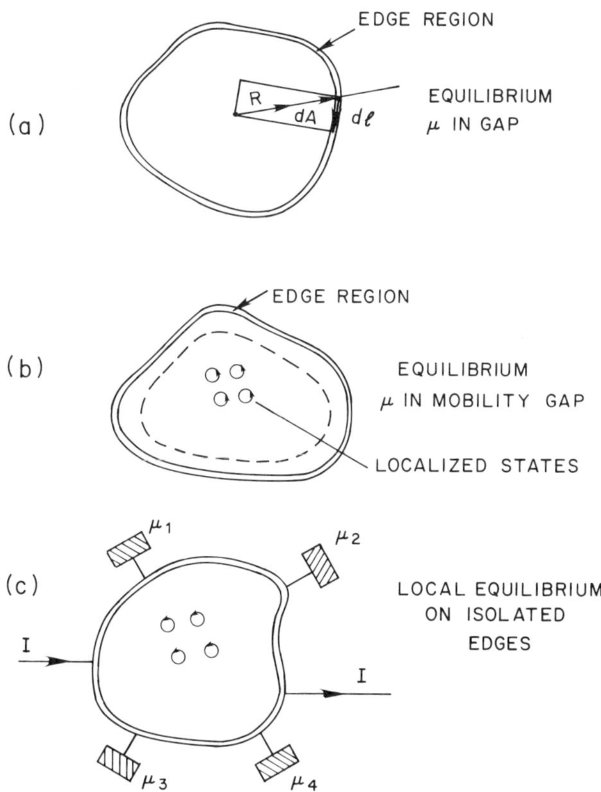

Figure 2. Derivation of the Streda formula. When the chemical potential has a gap in the thermodynamic limit the local properties of the system change with chemical potential only at the edge. The rate of change of edge current with chemical potential is related to the rate of change of electron density with magnetic field because the current operator is related to the derivative of the Hamiltonian with respect to magnetic field.

the density dependence of μ. It is useful to separate an explanation of the quantum Hall effects into: (1) an explanation of the fact that *incompressibility implies quantization* and (2) an explanation of the physical origin of the incompressibilities. For the integer ($\nu = i$) quantum Hall effect the chemical potential jumps are clearly due to the quantization of the kinetic energy of 2 D electrons by a magnetic field as discussed above. For interacting electrons and for electrons in an external periodic potential the electronic system can become incompressible at non-integral values of ν. In this section we will discuss the transport consequences of chemical jumps without reference to their origin. In the following sections we will turn our attention to aspects of the physics which are particular to the various possible origins of incompressibilities.

Consider a finite 2 D electron system with a confining potential of arbitrary shape as illustrated in Fig. 2. Assume that the chemical potential has a value which would be in a gap for a bulk system in the thermodynamic limit. We want to evaluate the change in the magnetic moment of the system when the chemical potential of the system changes at constant magnetic field. The operator for the magnetic moment in the $\hat{z}$ direction is given by

$$\hat{M} = -\frac{\partial H}{\partial B} \tag{25}$$

where H is the Hamiltonian for electrons in an arbitrary external potential and with arbitrary electron-electron interactions.

$$H = \sum_i \frac{\pi_i^2}{2m} + V_{ext}(\mathbf{x}_i) + \sum_{i<j} V_{ee}(\mathbf{x}_i - \mathbf{x}_j) \tag{26}$$

Only the kinetic term in H depends on the magnetic field. Using eq. (8) in eqs. (26) and (25) we find that [3]

$$\hat{M} = \frac{1}{2c} \int d^2\mathbf{x}\, \mathbf{x} \times \hat{\mathbf{j}}(\mathbf{x}), \tag{27}$$

where $\hat{\mathbf{j}}(\mathbf{x})$ is the current density operator,

$$\hat{\mathbf{j}}(\mathbf{x}) = \frac{-e}{2m} \sum_i \left(\delta(\mathbf{x} - \mathbf{x}_i)\boldsymbol{\pi}_i + \boldsymbol{\pi}_i \delta(\mathbf{x} - \mathbf{x}_i) \right). \tag{28}$$

Because the chemical potential has been chosen so that it would lie in the gap for an infinite bulk system, when it changes the expectation value of the current density operator can change only at the system's edge. The change in the expectation value of the magnetic moment, δM, with a small change in chemical potential is

$$\delta\langle\hat{M}\rangle = \frac{1}{2c} \int d^2\mathbf{x}\, \mathbf{x} \times \delta\langle\hat{\mathbf{j}}(\mathbf{x})\rangle. \tag{29}$$

The integrand in eq. (29) is non-zero only in the edge region. Taking the position, $\mathbf{x}$, outside the integral when integrating across the edge (see Fig. 2), and using current conservation when integrating around the edge, eq. (29) simplifies to

$$\delta\langle\hat{M}\rangle = \frac{\delta I}{2c} \oint \mathbf{R} \times d\mathbf{l} = A\delta I_e/c, \tag{30}$$

where A is the area of the finite 2 D system and δI_e is the change in the current circulating around the edge. In the intermediate form of eq. (30) $\mathbf{R}$ is a position on the edge and $d\mathbf{l}$ points in the local direction of the change in edge current. Equation (30) provides us with a very general expression for the ratio of the change in the edge current to the change in the chemical potential.

$$\frac{\delta I_e}{\delta\mu} = \frac{c}{A} \left.\frac{\partial M}{\partial\mu}\right|_B = \frac{c}{A} \left.\frac{\partial N}{\partial B}\right|_\mu. \tag{31}$$

The second form for the right hand side of eq. (31) follows from the thermodynamic identity.

The range of filling factors for which only edge states are at the chemical potential vanishes in the thermodynamic limit since the number of edge states scale with the system perimeter while the Landau level degeneracy scales with the system area. An essential part of von Klitzing's discovery was that the Hall conductance is precisely constant and the magnetoresistance extremely small over a finite range of filling factors. To explain the occurrence of these plateaus, which allow the quantum Hall effect to be visible in large samples, we must invoke the phenomenon of localization. The states in the tails of the disorder broadened Landau levels, will be localized near minima or maxima in the random potential. For our finite system we will say that the chemical potential lies in a mobility gap whenever the bulk states are sufficiently localized to allow an unambiguous separation between bulk and edge contributions to the change in the local current density in eq. (29) (see Fig. 2) . In this case the bulk contribution to the magnetization, M_B satisfies

$$\left.\frac{\partial M_B}{\partial \mu}\right|_B = \left.\frac{\partial N_B}{\partial B}\right|_\mu , \tag{32}$$

where N_B is the contribution to the density from the bulk region. (In Fig. 2 the dashed line encloses the bulk region.) It follows that in the mobility gap

$$\partial^2 I_e \,\partial\mu^2 = \frac{c}{A}\frac{\partial}{\partial B}\left(\left.\frac{\partial N}{\partial \mu}\right|_B - \left.\frac{\partial N_B}{\partial \mu}\right|_B\right) = 0. \tag{33}$$

The last equality in eq. (33) is true in the thermodynamic limit since the contribution to the rate of change in particle number with chemical potential from the edge region will become negligible. Equation (33) says that the rate of change of edge current with chemical potential is constant within a mobility gap.

The above discussion was for a finite 2 D system in equilibrium. When current leads are attached to the system, the edge is divided into two branches (see Fig. 2). Electrons injected into the system from the leads cannot communicate with the localized states in the bulk. The net current carried through the system, I, can be accommodated by establishing separate local equilibria on the two edges. The magnetoresistance, determined by the voltage difference between two points along the same edge, vanishes. The Hall conductance, G_H, is the ratio of the net current (i.e. the difference in edge currents) to the electrochemical potential difference between the two edges and it can be evaluated using eq. (31). (Since we incorporate any electric fields which may be present in the Hamiltonian the electrochemical potential difference is just e times the chemical potential difference.) If the chemical potential lies within a mobility gap associated with an incompressibility pinned to filling factor ν_0, the derivative appearing in eq. (31) is known and the Hall conductance is given by

$$G_H = \frac{\delta I_e}{\delta \mu} = e\nu/h. \tag{34}$$

From the discussion above we see that whenever the Fermi level lies in a gap the Hall conductance will be given by

$$G_H = ec\,\left.\frac{\partial n(\varepsilon F)}{\partial B}\right|_\mu . \tag{35}$$

As far as we are aware, this formula was first derived for non-interacting electrons, by Streda [4] using the Kubo formula. The argument given above which leads to this formula has a more general validity and explains its origin in a more physical way. It is closely related to other [5,6] simple arguments for the quantization of the Hall conductance. As was emphasized early in the development of the quantum Hall effect theory by Prange [7] the mysterious aspect of the integer quantum Hall effect is the fact that the free gas formula for G_H is exactly obeyed between Landau levels, despite the fact that localization reduces the number of current carrying states in a Landau level. The Streda formula shows that the Hall conductance in a mobility gap depends only on the magnetic field dependence of the carrier density when the Fermi level lies within the corresponding gap. In the case of the integer quantum Hall effect, for example, the Hall conductance depends only on the density which can be accommodated by each Landau level, something which is clearly preserved as long as the disorder is not so strong that different disorder broadened Landau levels overlap. (In fact it is believed that the integer quantum Hall effect can occur even when Landau levels overlap.)

3   The Integer Quantum Hall Effect

The integer quantum Hall effect is the simplest of the three quantum Hall effects which we will discuss. In this case it is possible to provide a concrete example which is useful in understanding the general discussion of the preceding section. The most elementary model which exhibits all the essential features of the quantum Hall effect is that of a strip of non-interacting electrons in an external potential which depends only on the x coordinate. The single-particle Hamiltonian for this system is given by

$$h = t + V(x), \tag{36}$$

where t is the kinetic energy operator discussed in section 1 and $V(x)$ is the external potential. We imagine $V(x)$ goes to infinity in some unspecified way at large and small x so that the 2 DES is confined to a strip of finite width, but is otherwise arbitrary. In discussing this problem it is convenient to choose a gauge, the Landau gauge, in which the vector potential is independent of the y coordinate,

$$\mathbf{A} = (0, Bx, o). \tag{37}$$

This allows us to choose eigenstates of h which have a plane-wave dependence on the y coordinate,

$$\psi(x,y) = \phi/x)\exp(ik_y y)/\sqrt{(L_y)}. \tag{38}$$

Here L_y is the length of the system in the $\hat{y}$ direction and, adopting periodic boundary conditions in this direction, the allowed values of k_y are separated by $2\pi/L_y$. For reasons which will become apparent we choose to label states by $X \equiv -\ell^2 k_y$ rather than by k_y. Substituting eq. (38) into eq. (36), we see that

$$h_x(X)\phi_{n,X}(x) = \varepsilon_{n,X}\phi_{n,X}(x), \tag{39}$$

where

$$h_x(X) = p_x^2/2m + \frac{1}{2}m\omega_c^2(x - X)^2 + V(x). \tag{40}$$

$\phi_{n,X}(x)$ provides the quantum description of a cyclotron orbit which is centered near $x = X$ and is free to drift along the $\hat{y}$ direction in response to the external potential. To calculate the Hall conductance in this circumstance we need only note that the total current flow is given by

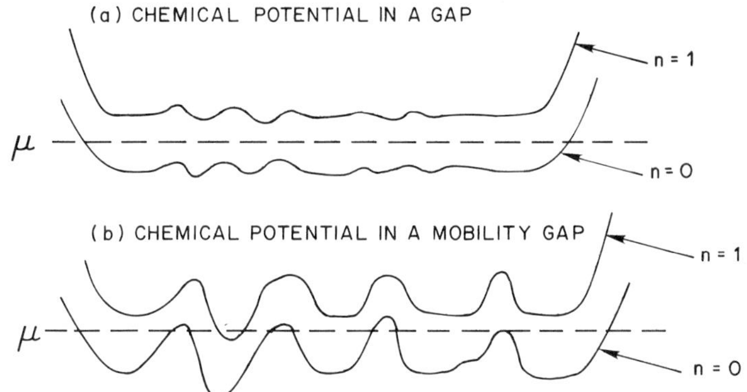

Figure 3. Derivation of the Streda formula for the simple case of non-interacting electrons in a disorder potential which depends on only one coordinate.

$$I_y = -e \sum_{n,X} \langle n, X | \pi_y | n, X \rangle / (m L_y) = -e \sum_{n,X} \langle n, X \left| \frac{\partial h_x}{\partial k_y} \right| n, X \rangle / (\hbar L_y). \tag{41}$$

The second form of eq. (41) is obtained from the first by comparing with eq. (40). Since adjacent values of X allowed by periodic boundary conditions are separated by $2\pi\ell^2/L_y$. Equation (41) leads to

$$I_y = \frac{e}{h} \sum_n \int dX \frac{\partial \varepsilon_n(X)}{\partial X}, \tag{42}$$

where we have invoked the Hellman-Feynman theorem to evaluate the expectation value. In eqs. (41) and (42) the sums and integrals are over occupied states. When the chemical potential lies in a gap between Landau levels the lower and upper limits of the integration over X in eq. (42) are at the left and right edge of the sample and in evaluating this integral we are led to introduce the edges explicitly, although we shall see that nothing depends on the specific model used to describe the edges (see Fig. 3). When an integer number, i, of Landau levels are full we may assume that, except at the edges, all states associated with that Landau level index are occupied. At either edge, we can assume that there is a well defined local chemical potential (equal to the energy of the highest occupied level at that edge). This chemical potential is what is probed when a voltage measurement is performed on that edge. Thus eq. (42) implies that

$$I_y = eN(\mu_R - \mu_L)/h = -ie^2 V/h, \tag{43}$$

where V is the voltage difference from the left hand side to the right hand side of the sample. When the chemical potential lies in a mobility gap (see Fig. 3), it is clear from eq. (42) that the rate of change of I_e with changes in the local chemical potential at the edge is e/h for each Landau level whose edge states cross the Fermi level, even if the Landau level is not completely full or completely empty in the bulk. The quantized Hall current is carried by establishing local equilibria on the edges as discussed in the preceding section.

As implicitly assumed in the above discussion, the confining potential which defines the edges of the system causes $\varepsilon_n(X)$ to increase near the edges of the sample. The way in which this occurs depends on the confinement potential: two simple models for which exact results can be obtained are the hard wall [8] case ($V(x) = \infty$ outside the sample) and the case of a parabolic confinement potential. From eq. (42) we see that this leads to large current densities flowing around the edge of a 2 D system. For a system with edges the increase in Landau level degeneracy with magnetic field is accomplished by having edge states adiabatically move toward the bulk with increasing magnetic field. The edge states energies decrease rapidly with magnetic field giving a paramagnetic contribution to the orbital magnetic moment [9,10] of the system which is comparable in size to the diamagnetic contribution from the bulk region even in the thermodynamic limit. The rate of change of the number of states in a Landau level with magnetic field is thus determined by the rate at which edge state energies change with magnetic field and, by eq. (41), to the current carried by the edge states within a given range of energies. This is a particular example of the more general argument given in the preceding section. The change in edge current for a given change in the local chemical potential is fixed by the rate of change of the number of occupied states at fixed chemical potential because the current operator is proportional to the derivative of the Hamiltonian with respect to the vector potential, and hence to the derivative of the Hamiltonian with respect to the magnetic field. The number of states below a gap cannot be altered by weak disorder.

For the model discussed above where the disorder potential is independent of the y coordinate, k_y is a good quantum number. There is a close analogy between the passage from eq. (41) to eq. (42) and the arguments used to justify Landauer type conductance formulas [11,12]. Indeed, Tthe Büttiker [12] formula for multi-probe conductance measurements is readily applied to the strong magnetic field regime [12,14]. This description of strong field transport allows the integer quantum Hall effect to be seen as a particularly simple case of the tremendous variety of quantum transport phenomena which can be understood in terms of the Büttiker formula [12,13]. The passage from weak fields to strong fields is a continuous one; the channel number evolves continuously from being the number of occupied transverse momenta to being the number of occupied Landau levels as magnetic depopulation of subbands occurs in a wire of finite width. From this point of view the unique aspect of the strong field regime is the spatial separation of left-going and right-going states and it is this which leads to the remarkable accurate quantization of transport coefficients. The Büttiker [12] conductance formula can also be used to explain experiments in which non-equilibrium distributions are created at the edges [15]. The understanding of these experiments depends on having a microscopic description of the electronic structure at the edge of the system. For the quantum Hall effect, on the other hand, a local equilibrium always exists at the edge and as we have emphasized the quantization follows solely from the magnetic-field dependence of the density at which an incompressibility occurs.

From our discussion of the quantum Hall effect we see that the Hall conductance should be independent of the Landau level filling factor as long as the chemical potential lies in a region where the localization length, ξ, is small compared with the system size. Since localization is dependent on the maintenance of phase-coherence this condition should be replaced at finite temperatures by the requirement that the Hall conductance will be quantized when ξ is smaller than the phase coherence length. The fact that there must be some extended states within each Landau level is required by the fact that the Landau level degeneracy depends on magnetic field.) The phase coherence length is the distance an electron diffuses between inelastic scattering events. In the strong magnetic field limit we are interested in the diffusive motion of the cyclotron orbit center since the kinetic energy of each cyclotron orbit is quantized. The diffusion constant may be obtained from the Einstein relationship using the fact [16] that, in the absence of localization effects, the conductivity is $\sim e^2/h$. The result is given below and may be interpreted as follows. Classically a cyclotron-orbit center moves ballistically along equipotential contours with velocity

$$v = cE_{\mathrm{loc}}/B \tag{44}$$

in a direction perpendicular to the local electric field, E_{loc}. The typical time between scattering events, τ, can be related to the Landau level width Γ by

$$\tau \sim \hbar/\Gamma. \tag{45}$$

Thus the typical distance moved between scattering events is

$$\sim cE_{\mathrm{loc}}\hbar/(B\Gamma) \sim \ell, \tag{46}$$

where we have noted that typical values of E_{loc} are about $\Gamma/(e\ell)$. Assuming a random distribution of local electric field directions the cyclotron orbit center moves diffusively with a diffusion constant given by

$$D \sim \ell^2/\tau \sim \Gamma\ell^2/\hbar, \tag{47}$$

and the phase coherence length is

$$L_\phi \sim \ell(\tau_{\mathrm{inel}}/\tau)^{1/2}. \tag{48}$$

(Note that, in contrast to the zero field case, L_ϕ decreases with τ.) As the temperature goes to zero, the inelastic scattering rate vanishes as T^p and L_ϕ diverges as $T^{-p/2}$. Early experiments [17] on the integer quantum Hall effect showed that the width of the Hall plateaus in disorder systems approached a filling factor change of one at low temperatures, suggesting [18,19] that extended states exist at only one energy, E_c, within each disorder broadened Landau level. Recent numerical calculations [20] find that for E near E_c

$$\xi \sim (E - E_c)^{-\nu}, \tag{49}$$

where the critical exponent $\nu = 2.34 \pm 0.04$. This result is in agreement with the analytic result obtained by Mil'nikov and Sokolov [21], who included the effects of quantum mechanical tunneling between orbits along percolating equipotentials [22], and also in agreement with experiment [23,24] if the inelastic scattering rate is assumed to vanish as $T^2(p = 2)$ at low temperatures.

4   The Quantum Hall Effect in a Periodic Potential

The integer quantum Hall effect occurs because of gaps which occur at integer Landau level filling factors, rather than at constant density. (The reader should realize that there is something unusual about a gap which occurs at a magnetic-field dependent density.) In this section we discuss a qualitatively different example of the quantum Hall effect in which the gap occurs neither at constant density nor at constant Landau level filling factor. The gaps in this case are those which occur when the 2 DEG is placed in both a constant magnetic field and a periodic potential. The effect of the periodic potential is both to broaden a Landau level and to open up an intricate array of gaps within a each Landau level.

The main features of the quantum Hall effect in a periodic potential can be understood on the basis of qualitative arguments. For a 2 D system in a strong magnetic field and a periodic potential, there are two length scales. One is the magnetic length, ℓ, introduced earlier which was important in the integer quantum Hall effect. The second length scale is the period of the external potential. As we learned above, we can understand the quantum Hall effect if we understand the magnetic-field dependence of the densities at which gaps occur in the spectrum. In the case of a free gas, gaps occur at integer filling factors. The effect of the periodic potential is to allow other gaps to occur.

It is clear from a simple scaling argument that the filling factor at which a gap occurs can depend only on the ratio of the two length scales in the problem. This ratio is conveniently parameterized in terms of the quantity

$$\alpha = 2\pi\ell^2/A_0, \tag{50}$$

where A_0 is the unit cell area of the periodic potential. To determine the spectrum of the system we use translational periodicity, applying periodic boundary conditions to a finite system whose volume can ultimately be taken to infinity. It will not be necessary for our purposes to describe in detail how this is done in the presence of a magnetic field. We merely note that it is possible to apply periodic boundary conditions to the system, only if the number of units cells of the periodic potential in the system is an integer and if $A/(2\pi\ell^2) \equiv N_L$, is an integer. (Here A is the area of the system.) The first condition is obvious and is required at zero magnetic field as well. The second condition is due to the magnetic field , and is required even in the absence of a periodic potential; in that case it requires N_L, the number of states per Landau level, to be an integer in a finite system. The main features of the intricate gap structure can be understand from these observation and the scaling argument.

It follows from the above that in the thermodynamic limit,

$$\sigma \equiv \frac{\partial n(\varepsilon_F)}{\partial B}/(2\pi\ell^2/B) \tag{51}$$

and

$$s \equiv -A_0^2 \frac{\partial n(\varepsilon_F)}{\partial A_0} \tag{52}$$

are integers when ε_F lies in a gap. (ε_F is the Fermi energy which is the zero temperature limit of the chemical potential.) σ is the change in the number of states below the gap when the magnetic field changes by an amount sufficient to change N_L by one, while s is the change in the number of states below the gap when A_0 is changed by an amount sufficient to increase the number of unit cells in the system by one. Note that the Streda formula implies that

$$G_H = (e^2/h)\sigma. \tag{53}$$

The change in the filling factor at which a gap occurs when A_0 and the magnetic field change c an be expressed in terms of σ and s. Using eqs. (50) and (51), in $\nu = 2\pi\ell^2 n$ gives,

$$\delta\nu - [\upsilon - \nu]\frac{\delta B}{B} - \alpha s\frac{\delta A_0}{A_0}. \tag{54}$$

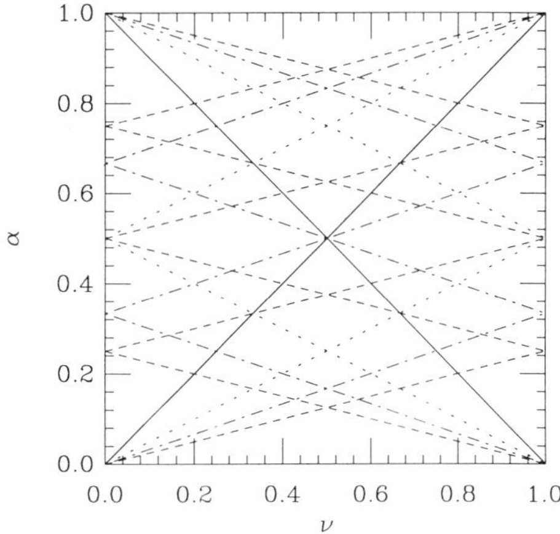

Figure 4. Dependence of the filling factor at which a gap occurs on the number of magnetic flux quanta per unit cell of an external periodic potential. All gaps but one will go to zero at the points of intersection. The Hall conductivity when the chemical potential lies in a gap is e^2/h times the value of ν at which the line intersects $\alpha = 0$.

Similarly the change in α for a change in B and A_0 is

$$\delta\alpha = -\alpha(\delta B/B + \delta A_0/A_0). \tag{55}$$

It follows that the change in the filling factor at which a gap occurs at constant α is given by

$$\delta\nu = [\sigma - \nu + \alpha s]\delta B/B. \tag{56}$$

However, it follows from the scaling argument that the filling factor at which a gap occurs cannot change if α does not change and hence that the filling factor at which a gap occurs is related to σ and s by

$$\nu = \sigma + \alpha s. \tag{57}$$

Equation (57) is a deceptively simple equation from which many conclusions can be drawn. Consider the situation when α has an arbitrary rational value, $\alpha = q/p$. Then any ν at which a gap occurs, using the fact that σ and s in eq. (57) are integers, must also occur at a rational filling factor with the same denominator, $\nu = t/p$. Thus each Landau level is split into p subbands when $\alpha = q/p$. Each gap in the spectrum characterized by two integer quantum numbers σ and s and follows the straight line in ν, α space dictated by eq. (57). A countably infinite number of such lines intersect at any rational value of ν and α, characterized by the integers t, q and p, and their values of σ and s are given by

$$(s, \sigma) = (s_0 + kp, \sigma_0 - kq), \tag{58}$$

where k is an integer (see Fig. 4). The allowed values of s are separated by p and the allowed values of σ are separated by q.

All the above conclusions can be drawn without specifying the periodic potential, which determines which of the possible values of σ is selected at a given value of α and ν. The case of a weak periodic potential of the form

$$V(x) = V_0 \left(\cos(2\pi x/a) + \cos(2\pi y/a)\right) \qquad (59)$$

is equivalent to the model studied numerically ba Hofstader [25] and the complicated recursive structure he discovered in the spectrum, *The Hofstader Butterfly*, is equivalent to the statement [26,27,28] that the correct value of s at any gap for this model is the smallest possible one [29], i.e. the one obeying $|s| \leq p$. This is not a general result ; in particular the gap structure is quite different in the case of a hexagonal external potential [30,31]. From our discussion it is clear that the quantum Hall effect should provide a sensitive probe of the exotic gap structure in a periodic potential and a magnetic field. The prospects of observing exotic transport effects in artificially structured [32,33] 2 D system in the future are promising.

It is easy to verify that in the case of an external periodic potential the values of the integral quantum number related to the Hall conductance, σ are very different from ν and, in particular can have large and even negative values within any Landau level. Thus when a periodic potential is present the quantum Hall effect can occur at fractional filling factors but the filling factor dependence of the gap is always such that the Hall conductance is an integral multiple of e^2/h. In the next section we discuss gaps which originate from electron-electron interactions and occur at fixed fractional Landau level fillings at any field, and hence yield a fractional Hall conductance.

We have followed Streda [34] in discussing the quantum Hall effect periodic potential in terms of the Streda formula. Thouless et al. [26] obtained equivalent results starting directly from the Kubo formula. These authors found that the Hall conductivity is related to the change of the phase of the Bloch wavefunction on moving around the perimeter of the Brillouin zone, which must be an integer (σ) times 2π. The Hall conductance quantum number is thus related to a topological invariant of a mapping of the Brillouin zone by the Bloch wavefunction. It has subsequently been realized that the Hall conductance may be regarded as a topological invariant in more general circumstances [35]. The fact that the Hall conductance in a periodic potential turns out to always be an integral multiple of e^2/h despite gaps at fractional filling is then seen to be a very general feature of the quantum Hall effect. Fractional values of quantized Hall conductance can only be produced by electron-electron interactions and, as we will see, they are accompanied by a fractionalization of the electron charge.

5   The Fractional Quantum Hall Effect

We have learned that the quantum Hall effect will occur for a 2 D system in a magnetic field whenever there is a gap, i.e. whenever the chemical potential has a discontinuity as a function of density at fixed magnetic field. The integer quantum Hall effect is caused by the quantization of the allowed kinetic energy values for an electron moving in a plane perpendicular to a steady magnetic field. In the case of an electron system moving in a periodic potential we learned in the last section that gaps open up within Landau levels. Even in this case, however the Hall conductance is an integer multiple of e^2/h. The fractional quantum Hall effect [28,36,37,38] occurs near a certain set of fractional values of ν and has plateaus in the Hall conductance equal to $\nu e^2/h$. We saw in the last section that a fractionally quantized Hall conductance is not possible for non-interacting electrons, even in an external potential which produces gaps within each Landau level. Our object here will be to explain how electron-electron interactions can give rise to chemical potential discontinuities which are pinned to *fractional* Landau level filling factors.

We will restrict our discussion to the extreme quantum limit in which the Landau level degeneracy is large enough that all electrons can be accommodated within the lowest Landau level and the Landau level separation is large enough that quantum mechanical Landau level mixing by disorder or by interactions can be ignored. Since each electron present has the same kinetic energy, only electron-electron interactions can produce the gap which we require to explain the quantum Hall effect. We will see that the kinetic energy quantization indirectly can produce gaps at fractional filling factors, because the restriction to a single Landau level puts powerful filling-factor-dependent restrictions on the properties of the many-electron wavefunctions. According to eq. (20), the single-particle wavefunctions in the lowest Landau level in the symmetric gauge are

$$\phi_m(\mathbf{x}) = \frac{z^{*m} \exp(-|z|^2/4)}{\sqrt{2^{m+1} m! \pi}} \tag{60}$$

where $z = x + iy$, we have dropped the Landau level index on the wavefunction and we have adopted the magnetic length, ℓ, as the unit of length for the discussion of the fractional quantum Hall effect. (In what follows we will redefine $z \equiv x - iy$.) Note that these wavefunctions describe electrons located within one magnetic length of a circle centered on the origin and enclosing an area

$$\pi \langle m||z|^2|m\rangle = 2\pi \ell^2 (m + 1) \ . \tag{61}$$

Any many-electron wavefunction formed entirely within the lowest Landau level must be a sum of products of one-electron orbitals for each coordinate which are of the form given by eq. (61). It follows that the many electron wavefunction must take the form

$$\Psi[z] = \left(\prod_{k=1}^{N} \exp\left(-|z_k|^2/4\right) \right) P[z] \tag{62}$$

where $P[z]$ is a polynomial, and in the $N \to \infty$ limit an analytic function, in each of the z_k's [39]. N is the number of electrons. At zero temperature the chemical potential equals the change in the ground state energy when one electron is added to the system. Since any state of the form of eq. (62) has the same kinetic energy, $T = N\hbar\omega_c/2$, the discontinuity in the chemical potential responsible for the fractional quantum Hall effect must come from electron-electron interactions. We know that the electron-electron interaction is increasingly repulsive at short distances, so the ground state will be determined, qualitatively, by minimizing the probability of electrons being close together. For an isotropic system of identical particles the pair correlation function, which measures the probability of two electrons being at a certain separation compared to the same probability in an uncorrelated system of the same density, is given by

$$g(|\mathbf{x}_1 - \mathbf{x}_2|) = n^{-2} N(N - 1) \prod_{k=3}^{N} \int d^2 z_k |\Psi[z]|^2 \ . \tag{63}$$

Substituting eq. (62) into eq. (63) gives for the case of interest,

$$g(|z_1 - z_2|) \sim \exp\left(-(|z_1|^2 + |z_2|^2)/2\right) \prod_{k=3}^{N} \int d^2 z_k \exp\left(-|z_k|^2\right) P^*[z]P[z] \,. \tag{64}$$

It will prove convenient to replace the coordinates z_1 and z_2 by a relative coordinate, $\zeta \equiv z_2 - z_1$, and a mean coordinate, $\bar{z} \equiv (z_2 + z_1)/2$. As noted in eqs. (63) and (64) the dependence of g on the mean coordinate is expected to be removed by the integrations over $d^2 z_k$ for an isotropic fluid state, such as is generally expected in an electron gas [40]. Since $P[z]$ is an analytic function of z_1 and z_2, it will be an analytic function of ζ, and we can expand

$$P[z] = \sum_p \zeta^{2p+1} F_p(\bar{z}, z_3, \ldots, z_N) \,. \tag{65}$$

Note that because of the anti-symmetry requirement on the many-electron wavefunction for electrons only odd powers of ζ appear in eq. (65). Substituting eq. (65) into eq. (64) gives

$$g(|\zeta|) = \exp\left(-|\zeta|^2/4\right) \sum_{p,p'} |\zeta|^{*(2p'+1)} \zeta^{2p+1} f_{p',p} \tag{66}$$

where $f_{p',p}$ is given by an integral over the other coordinates. For an isotropic system g can depend only on the magnitude and not on the orientation of the separation ζ. It follows that $f_{p',p}$ must be zero when p is not equal to p' and hence that

$$g(|\zeta|) = \exp(-|\zeta|^2/4) \sum_p |\zeta|^{2(2p+1)} f_{p,p} \,. \tag{67}$$

Equation (67) tells us that for any isotropic state formed in the lowest Landau level of a 2 DEG, the pair correlation function must vanish as an odd power of $|\zeta|^2$ at small $|\zeta|$. The fractional quantum Hall effect is due to a connection between the small-separation behavior of g and the Landau level filling factor, which we now establish. Assume that $g(|\zeta|)$ varies as $|\zeta|^{2m}$ at small $|\zeta|$. It follows from the argument leading to eq. (67) that $P[z]$ has $(z_1 - z_2)^m$ as a factor and hence, since all particles are identical, that $P[z]$ has

$$P_m[z] \equiv \prod_{i<j} (z_i - z_j)^m \tag{68}$$

as a factor. For a wavefunction representing a large but finite number of electrons, N, the maximum power to which z_1 (or any other coordinate) appears in $P[z]$ is therefore

$$M_1 \geq m(N-1) \tag{69}$$

and hence the area occupied by the wavefunction, according to eq. (61), is

$$A \geq 2\pi\ell^2 m(N-1). \tag{70}$$

It follows from eq. (70) that in the thermodynamic limit $A/2\pi\ell^2 N = \nu^{-1} \geq m$. Thus as the electron density is increased at constant magnetic field so that the filling factor crosses $\nu = 1/m$ we go from a regime where it is possible to form states with $g(|\zeta|) \sim |\zeta|^{2m}$ to a regime where $g(|\zeta|)$ vanishes only a $|\zeta|^{2(m-1)}$. This qualitative change in the ability of electrons to avoid each other causes a jump in the chemical potential when the filling

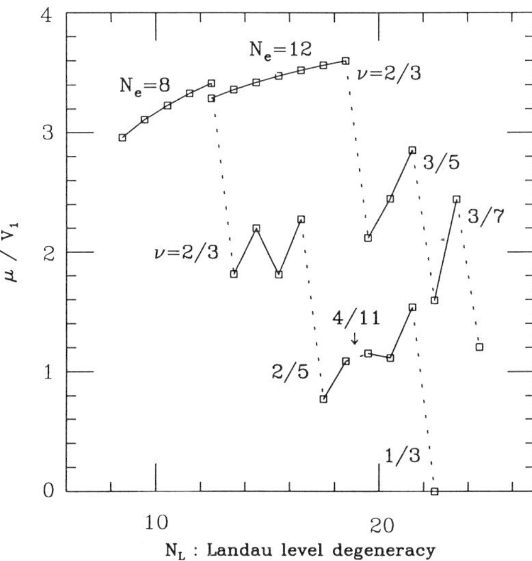

Figure 5. Finite-size exact-diagonalization estimate of the chemical potential's dependence on the number of states in a Landau level for eight and twelve electrons on the surface of a sphere.

factor crosses $1/m$ and, invoking the Streda formula, also causes the Hall conductance to be quantized at $e^2/(mh)$ at filling factor $1/m$.

At $\nu = 1/m$ the ground state wavefunction corresponds to the analytic function $P_m[z]$ (eq. (68)). It was first introduced by Laughlin [41]. When the electrons are in this state we see by inspection that any pair of electrons are always in a state of relative angular momentum at least m. As first emphasized by Haldane [42], the electron-electron interaction within the lowest Landau level is completely specified by a set of pseudopotential parameters, $\{V_m\}$, which give the interaction strength in each relative angular momentum channel. (Only odd values of m are relevant if the electrons are completely spin polarized.) An extremely attractive picture of the fractional quantum Hall effect can be given in terms of these pseudopotential parameters. For filling factors less than $1/m$, where we have shown that it is possible to form states in which no pair of electrons has a relative angular momentum less than m, there are zero-energy eigenstates in models which have repulsion only in channels of relative angular momentum less than m. At filling factor $1/m$ our discussion shows that there is only one state, the Laughlin state, which has zero energy for such a model. Thus associated with the chemical potential jump which occurs as the filling factor crosses $1/m$, there is an excitation gap which occurs when the filling factor equals $1/m$. In Fig. 5 we show the chemical potential as a function of Landau level filling factor for a system in which electrons repel each other only when they have relative angular momentum one as estimated from finite-system exact diagonalization studies [43].

The pseudopotential parameters for realistic electron-electron interaction decrease monotonically with m. Exact diagonalization studies for small systems, pioneered by Yoshioka et al. [44], have provided convincing evidence that the realistic interactions are sufficiently close to the simple pseudopotential models described above, that the ground states maintain a nearly quantitative similarity.

To explain the plateaus which occur in connection with the fractional quantum Hall effect we must also consider states which occur at filling factors close to $1/m$. The expectation is that some type of localization behavior must occur which is analogous to the localization of solutions to the one-body Hamiltonian responsible for the plateaus in the integer quantum Hall effect. For example, consider the case in which the area of the system is larger than that at filling factor $1/m$ by $2\pi\ell^2$ and the disorder potential has a repulsive peak at some point, which we take to be the origin. Laughlin [41] suggested that the ground state in this circumstance could be obtained from the ground state at filling factor $1/m$ by increasing the single-particle angular momentum labels of all the occupied states by one. This would give a state in which the $m = 0$ single-particle state is never occupied. It is easy to show that away from the origin this state has the same correlations as the Laughlin fluid state, while near the origin there is a decrease in the charge density with a total deficiency of charge equal to e/m. One can form similar states in which there is an excess of charge equal to e/m at some point in space. The picture of the situation on the plateaus near $\nu = 1/m$, is therefore as follows. The ground state may be thought of as consisting of a Laughlin fluid, plus some number of quasi-holes or quasi-particles localized [45] near maxima or minima of the disorder potential. The novel feature in the fractional case is that these quasi-particles have *fractional* charge. The fractional charge also leads to the possibility that the quasi particles may be most conveniently described as having neither Fermi or Bose statistics, but rather an intermediate *fractional* statistics [46].

To avoid giving the reader the incorrect impression that everything is now explained we must emphasize that the fractional quantum Hall effect also occurs at filling factors not equal to $1/m$ (see Fig. 5). Some of these fractions can be explained by invoking the particle-hole symmetry which exists in the strong magnetic field limit. For example, a fractional quantum Hall effect occurs at filling factor 2/3, due to the formation of the 1/3 Laughlin state in the holes of a full Landau level. This notion can be generalized so that, for example the 2/5 fractional effect can be explained as being due to the formation of a Laughlin fluid in the *fractionally charged* quasi-particles of the $\nu = 1/3$ Laughlin fluid. It should be recognized, however, that the understanding of these so-called hierarchy states is not yet nearly as complete as our understanding of the Laughlin states.

For a non-interacting metal in the absence of a magnetic field, the excited states which are coupled to the ground state by the one-particle operators relevant to most experiments are particle-hole excitations in which an electron in one of the single particle states inside the Fermi surface is promoted to a single-particle state outside the Fermi surface. For an interacting metal the particle-hole excitations still exist and their excitation energies are, for the most part, only slightly changed. However the excitation energy of one particular linear combination of particle-hole excitations is changed qualitatively, especially at long wavelengths. This linear combination is formed by operating on the ground state with the density operator,

$$\hat{\rho}(\mathbf{k}) = \sum_i \exp(-i\mathbf{k} \cdot \mathbf{x}_i) \, . \tag{71}$$

This state is the plasmon state in which the electrons are collectively oscillating with wavevector $\mathbf{k}$. For a 2 D electron system in a magnetic field,

$$\hat{\rho}(\mathbf{k}) = \sum_i A_i(k) B_i(k) \tag{72}$$

where $k = k_x + ik_y$,

$$A_i(k) = \exp\left(-k^* a_i^\dagger / \sqrt{2}\right) \exp\left(k a_i / \sqrt{2}\right), \tag{73}$$

and

$$B_i(k) = \exp\left(-ik^* b_i / \sqrt{2}\right) \exp\left(-ik b_i^\dagger / \sqrt{2}\right). \tag{74}$$

In eq. (72) we have separated the density operator into a factor involving inter-Landau-level transitions ($A_i(k)$) and a part involving intra-Landau-level transitions ($B_i(k)$). Equation (72) is obtained from eq. (71) by using eqs. (18), (19) and the fact that the inter-Landau-level and intra-Landau-level ladder operators commute. We see that in a magnetic field, the density operators creates particle-hole excitations which involve a change of Landau level index as well as particle-hole excitations within a Landau level. In a strong magnetic field the Landau level separation, $\hbar\omega_c$, is a large energy and the electron-electron interaction will be important in coupling only particle-hole excitations with the same change of Landau level index. Of particular importance to the fractional quantum Hall effect is the coupling of intra-Landau-level excitations since they all have zero excitation energy, in the absence of interactions. We may expect that a collective mode is formed, at least at long wave lengths, from the intra-Landau level particle-hole excitations. The excitation energy for collective modes may be estimated by a generalization of the theory used by Feynman [47] to estimate the collective phonon-roton excitation spectrum in superfluid ^{4}He. We assume that the collective mode state is given approximately by the projection of $\hat{\rho}(k)|\Psi_0\rangle$ onto the lowest Landau level, i.e. by the part of the density wave excitation which does not produce an increase in kinetic energy;

$$|\Psi(k)\rangle = \frac{\bar{\rho}(k)|\Psi_0\rangle}{\sqrt{\langle\Psi_0||\Psi_0\rangle}} \tag{75}$$

where the projected density wave operator $\bar{\rho}(k)$ is given by

$$\bar{\rho}(k) = \sum_i B_i(k). \tag{76}$$

and $|\Psi_0\rangle$ is the ground state wavefunction. The excitation energies of these magnetoroton [48,49] states are given by

$$E_{MR}(k) = \langle\Psi(k)|\bar{V}|\Psi(k)\rangle - E_0 \tag{77}$$

where E_0 is the ground state energy and $\bar{V}$ is the projection of the electron-electron interaction onto the lowest Landau level,

$$\bar{V} = \int \frac{d^2\mathbf{q}}{(2\pi)^2} V(q) \bar{\rho}(-q) \bar{\rho}(q). \tag{78}$$

In eq. (78), $V(q)$ is the Fourier transform of the electron-electron interaction. $E_{MR}(k)$ is readily evaluated by expressing the energy difference in eq. (77) in terms of commutators, expressing $\bar{V}$ in terms of projected density operators using eq. (78), and using the following expression for the commutator of projected density operators:

$$[\bar{\rho}(k_1), \bar{\rho}(k_2)] = (\exp(k_1^* k_2/2) - \exp(k_1 k_2^*/2))\, \bar{\rho}(k_1 + k_2) \tag{79}$$

Equation (79) follows from the commutation relations of the intra-Landau-level ladder operators. The result for the excitation energies is

$$E_{MR}(k) = \int \frac{d^2\mathbf{q}}{(2\pi)^2} V(q) \exp(-|q|^2/2)$$

$$(\exp((q^*k - k^*q)/2) - 1)\, (\tilde{s}(q) - \tilde{s}(k+q))\, /\tilde{s}(k) \tag{80}$$

where $\tilde{s}(k) = \exp(|k|^2/2)\bar{s}(k)$ and

$$\bar{s}(k) = s(k) - \left(1 - \exp(|k|^2/2)\right) , \tag{81}$$

the projected static structure factor, is the difference between the static structure factor of the partly filled state and the structure factor when the Landau level is filled. Small system calculations have verified that collective density modes do indeed exist within a partly filled Landau level *when the ground state is incompressible* and that their dispersion is given extremely accurately by eq. (80). As expected $\lim_{k\to 0} E_{MR}(k)$ is finite and $E_{MR}(k)$ has a minimum where $\bar{s}(k)$ has a maximum, analogous to the roton minimum in Helium.

In two dimensions it is possible for particles to have statistics intermediate between those of Fermi and Bose particles [50]. In fact the Fermi electrons in a two dimensional electron gas can be *equivalently* considered to be bosons provided that a contribution is added to the Hamiltonian which corresponds to an odd number of magnetic flux quanta piercing the system at the position of each particle. Girvin and MacDonald [51] showed that when the particles at $\nu = 1/m$ are considered to be bosons they show quasi-long range order at zero temperature if and only if the ground state is incompressible. This property has been verified by numerical calculations [52] and has been exploited to develop Landau-Ginzburg theories [53] for the fractional quantum Hall states. The incompressible states of the fractional quantum Hall effect thus exhibit a kind of Bose condensation in which the objects condensing consist of an electron and its associated flux quanta. Very recently there has been a great deal of interest in the possibility that the ground state of high-temperature superconductors may be related to fractional Hall incompressible states [54]. The superconducting property of these states would then emerge as a consequence of the long-range order discussed above. It may turn out that it is this aspect of the surprising incompressible states which are responsible for the fractional Hall effect which has the greatest impact on other sub-fields of physics.

6   Conclusions

In these notes we have discussed only the simplest version of the fractional quantum Hall effect where all the electrons are in the lowest orbital Landau level of the conduction band of a semiconductor. The theory has been generalized to higher orbital Landau levels [55] and to electrons in the valence band [56,57]. The most intricate and interesting complications, however, come when more than one Landau level has its energy close enough to the chemical potential to be relevant. This happens, for example, when the Zeeman energy is small enough that the spin degree of freedom becomes important. Such a *two component* system can also be created by fabricating a system in which two 2 D electron layers are in close proximity. The richness of the fractional quantum Hall effect in multi-component systems has been anticipated theoretically [58,59]. Continuing technical advances in material fabrication techniques have recently made it possible to study such systems experimentally [60,61]. This area of study is likely to bring new surprises in the future.

References

[1] K. von Klitzing, G. Dorda, and M. Pepper, Phys. Rev. Lett. **45**, 454 (1980)

[2] The properties of two dimensional electronic systems are reviewed by T. Ando, A. B. Fowler and F. Stern Rev. Mod. Phys. **54**, 437 (1982)

[3] The integrand in the expression for the magnetization operator depends on a choice of gauge and we have chosen the symmetric gauge. It is easy to show, however, using the continuity equation and the fact that the current density vanishes outside the sample that the total magnetization is gauge independent

[4] P. Streda, J. Phys. **C 15**, L717 (1982)

[5] R.B. Laughlin, Phys. Rev. **B 23**, 5632 (1981)

[6] B.I. Halperin, Phys. Rev. **B 25**, 2185 (1982)

[7] R.E. Prange, Phys. Rev. **B 23**, 4802 (1981)

[8] A.H. MacDonald and P. Streda, Phys. Rev. **B 29**, 1616 (1984)

[9] E. Teller, Z. f. Phys. **67**, 311 (1931)

[10] L. Landau, Z. f. Phys. **64**, 629 (1930)

[11] R. Landauer, Philos. Mag. **21**, 863 (1970)

[12] M. Büttiker, Phys. Rev. Lett. **57**, 1761 (1986)

[13] See the article by C.W.J. Bennakker in this volume.

[14] P. Streda, J. Kucera, and A.H. MacDonald, Phys. Rev. Lett. **59**, 1973 (1987); J.K. Jain and S.A. Kivelson, Phys. Rev. **B 37**, 4276 (1988); M. Büttiker, Phys. Rev. **B 38**, 9375 (1988)

[15] B.J. van Wees, E.M.M. Willems, C.J.P.M. Harmans, C.W.J. Bennakker, I.L. van Houton, J.G. Williamson, C.T. Foxon, and J.J. Harris, Phys. Rev. Lett. **62** (1989); S. Komiyama, H. Hirai, S. Sasa, and S. Hiyamizu, Phys. Rev. B **40**, 12566 (1989)

[16] Y. Murayama and T. Ando, Phys. Rev. **B 35**, 2252 (1987) and work cited therein

[17] M.A. Paalanen, D.C. Tsui, and A.C. Gossard, Phys. Rev. Lett. **25**, 5566 (1982)

[18] Y. Ono, J. Phys. Soc. Jpn. **51**, 2055 (1982)

[19] A.A.M. Pruisken, Nuclear Physics b **235**, 277 (1984)

[20] Bodo Huckestein and Bernhard Kramer, Phys. Rev. Lett. **64**, 1437 (1990), and work cited therin.

[21] G.V. Mil'nikov and I.M. Sokolov, Pis'ma Zh. Eksp. Teor. Fiz. **48**, 494 (1988) [JETP Lett. **48**, 536 (1988)]

[22] S.A. Trugman, Phys. Rev. **27**, 7539 (1983)

[23] H.P. Wei D.C. Tsui M.A. Paalanen, and A.M.M. Pruisken Phys. Rev. Lett. **61**, 1294 (1988)

[24] H.P. Wei, D.C. Tsui, and A.M.M. Pruisken, Phys. Rev. **B 33**, 1488 (1986)

[25] D. Hofstader, Phys. Rev. **B 14**, 2239 (1976)

[26] David Thouless, Phys. Rep. **110**, 279 (1984)

[27] D.J. Thouless, M. Kohomoto, M.P. Nightingale, and M. den Nijs Rev. Lett. **49**, 405 (1982)

[28] D.C. Tsui, H.L. Störmer, and A.C. Gossard, Phys. Rev. Lett. **48**, 1559 (1982)

[29] A.H. MacDonald Phys. Rev. B **28**, 6713 (1983)

[30] F. Claro and G.H. Wannier Phys. Rev. **B 19**, 6068 (1979)

[31] A.H. MacDonald Phys. Rev. B **29**, 3057 (1984)

[32] R.R. Gerhardts, D. Weiss, and K. von Klitzing, Phys. Rev. Lett. **62**, 1173 (1989)

[33] R.W. Winkler J.P. Kotthaus, and K. Ploog, Phys. Rev. Lett. **62**, 1177 (1989)

[34] P. Streda, J. Phys. **C 15**, L1299 (1982)

[35] B. Simon, Phys. Rev. Lett. **51**, 2167 (1983); Q. Niu, D. Thouless, and Wu Y-S Phys. Rev. **B 31**, 3372 (1985); D.J. Thouless, Phys. Rev. **B 40**, 12034, 1989 and work quoted therein

[36] R.G. Clark, R.J. Nicholas, A. Usher, C.T. Foxon, and J.J. Harris, Surf. Sci. (Netherlands) **170**, 141 (1986)

[37] G.S. Boebinger, A.M. Chang, H.L. Stormer, and D.C. Tsui, Phys. Rev. Lett. **55**, 1606 (1985)

[38] J. Wakabayashi, S. Sudou, S. Kawaji, K. Hirakasw, and H. Sakak Phys. Soc. Jpn. **56**, 3005 (1987)

[39] S.M. Girvin and T. Jach, Phys. Rev. **B 29**, 5617 (1984)

[40] S.M. Girvin, Phys. Rev. **B 30**, 558 (1984); D. Levesque and J.J. Weis, Phys. Rev. **B 30**, 1056 (1984)

[41] R.B. Laughlin, Phys. Rev. Lett. **50**, 1395 (1983)

[42] F.D.M. Haldane, Phys. Rev. Lett. **51**, 605 (1983)

[43] C. Gros and A.H. MacDonald, submitted to Phys. Rev. **B** (1990)

[44] D. Yoshioka, B.I. Halperin, and P.A. Lee, Phys. Rev. Lett. **50**, 1219 (1983)

[45] J.K. Jain, S.A. Kivelson, and N. Trivedi, Phys. Rev. Lett. **64**, 1297 1990

[46] B.I. Halperin, Phys. Rev. Lett. **52**, 1583 (1984)

[47] see R.P. Feynman, Statistical Mechanics, Benjamin, Read Mass (1972), and references therein

[48] S.M. Girvin, A.H. MacDonald, and P.M. Platzman, Phys. Rev. Lett. **54**, 581 (1985)

[49] S.M. Girvin, A.H. Macdonald, and P.M. Platzman , Phys. Rev. Lett. **33**, 2481 (1986)

[50] F. Wilczek, Phys. Rev. **B 49**, 957 (1982); D.P. Arovas, J.R. Schrieffer, F. Wilczek, and A. Zee, Nucl. Phys. **251**, 117 (1985)

[51] S.M. Girvin and A.H. MacDonald, Phys. Rev. Lett. **58**, 1252 (1987)

[52] F.D.M. Haldane, Phys. Rev. Lett. **61**, 1985 (1988)

[53] S.M. Girvin , in : The Quantum Hall Effect, R.E. Prange and S.M. Girvin, eds., Chapter 10, Springer-Verlag, New York (1986); N. Read, Phys. Rev. Lett. **62**, 86 (1989); S.C. Zhang, T.H. Hansson, and S. Kivelson, Phys. Rev. Lett. **62**, 82 (1989)

[54] V. Kalmeyer and R.B. Laughlin, Phys. Rev. Lett. **59**, 2095 (1987); R.B. Laughlin, Phys. Rev. Lett. **60**, 2677 (1988)

[55] F.D.M. Haldane, in: The Quantum Hall Effect, R.E. Prange and S.M. Girvin, eds., Chapter 8, Springer-Verlag, New York (1986), and references therein

[56] A.H. MacDonald and U.Ekenberg, Phys. Rev. **B 39**, 595 (1989)

[57] S.-R. Eric Yang, A.H. MacDonald, and D. Yoshioka, Phys. Rev. **B 41**, 1290 (1990)

[58] B.I. Halperin, Helv. Phys. Acta **56**, 75 (1983)

[59] B.I. Halperin, Helv. Phys. Acta **56**, 75 (1983); T. Chakraborty and F.C. Zhang, Phys. Rev. **B 29**, 703 (1984) 703 (1984)

[60] J.P. Eisenstein, H.L. Störmer, L. Pfeiffer, and K.W. West, Phys. Rev. Lett. **62**, 1540 (1989); R.G. Clark, S.R. Haynes, A.M. Suckling, J.R. Mallett, P.A. Wright, J.J. Harris, and C.T. Foxon, Phys. Rev. Lett. **62**, 1536 (1989)

[61] G.S. Boebinger, H.W. Jiang, L.N. Pfeiffer, and K.W. West , Phys. Rev. Lett. **64**, 1793 (1990)

CHAPTER 5

PERSISTENT CURRENTS

PERSISTENT CURRENTS IN MESOSCOPIC NORMAL METAL RINGS

Yoseph Imry

Department of Nuclear Physics
Weizmann Institute of Science
Rehovot 76100, Israel

1   Introduction, Generalities

It is well known that persistent currents which "never decay" can flow in superconducting systems. These can be of fundamentally two different varieties: equilibrium currents, such as those shielding the magnetic field in a bulk superconductor or quantizing the magnetic flux in a ring, and extremely long-lived metastable currents (for example those induced in a ring by a time-dependent magnetic flux) which may survive for astronomical times after the flux is appropriately reduced. Basic theoretical considerations for such currents in a ring (we shall use this term for a general doubly-connected geometry, including cylinders etc.) were presented by Byers and Yang [1] and Bloch [2, 3]. There is no reason why the former (equilibrium currents induced by magnetic fields) should not exist in principle in normal (i.e. non-superconducting) systems. In fact, such currents do obviously exist in atoms and molecules when an equilibrium diamagnetic moment is induced in them by a magnetic field. Do such currents exist in finite macroscopic [4] systems? The answer (while still surprising to many) is of course positive. Diamagnetic currents do flow in equilibrium in metals. In fact, whenever the appropriate thermodynamic potential, J, (e.g. E at zero temperatures and F at finite temperatures for canonical systems, $F - \mu N$ for grand canonical ones) depends on the magnetic field, H, the system has an equilibrium magnetization, M, given by $M = -\frac{\partial J}{\partial H}$. This magnetization can be regarded as due to some non-zero circulating currents, which for homogeneous systems are easily shown to flow on the surface of the specimen. Let us now consider a general ring geometry with an Aharonov-Bohm (AB) [5] flux Φ through its hole. The above considerations show that if the thermodynamic potential, J, depends on Φ, then an equilibrium current will circulate around the hole, given by

$$I = -\frac{1}{c}\frac{\partial J}{\partial \Phi} \ . \tag{1}$$

I may often be zero (including the case where the material of the rim of the ring is diamagnetic – it is then possible that reverse circulating currents will flow on the inside and outside surfaces of the rim, and will cancel each other). However, two important facts will be established below.

(1) I is exactly a periodic [1] function of Φ with a period

$$\Phi_0 = \frac{hc}{e}\,. \tag{2}$$

This obviously includes the case $I = 0$. But if $I \neq 0$ at some point it must change periodically with Φ. The relationship of this theorem with Josephson effect was demonstrated by [2, 3] and with the quantized Hall effect [6] by Laughlin [7]; see also [8]. In the latter case, dissipationless currents do flow in a normal conductor.

(2) The condition that I be non-zero in general is that the wavefunctions of the charge carriers should not be localized in a small part of the ring and that they should stay coherent along its circumference L. Denoting the length over which phase memory is retained by L_ϕ and the localization length, if any, by ξ, this means that

$$L \lesssim L_\phi \quad , \quad L \lesssim \xi\,. \tag{3}$$

The second inequality is self-evident and the first one will be discussed below [9]; see also [10] (more discussion on dephasing in sections 4 and 5).

Let us prove the first statement. Consider a general doubly-connected system with an AB flux Φ through its opening, as schematically depicted in Fig. 1 for a regular ring. The theorem due to [1, 2, 3] states that all physical properties of this "ring" are periodic in Φ with a period Φ_0. The proof proceeds by eliminating Φ with the gauge transformation

$$\Psi' = \Psi e^{(ie/\hbar c)\sum_j \chi(r_j)}\,, \tag{4}$$

where r_j are the electron coordinates and χ is defined by $\vec{A}_\Phi = \vec{\nabla}\chi$ where $\vec{A}_\Phi$ is the vector potential whose curl is the AB magnetic field (i.e. curl $\vec{A} = 0$ in the material and $\oint \vec{A}_\Phi \cdot d\vec{l}$ on any path circulating the ring's opening is equal to Φ). The gauge-transformed many electron Schrödinger equation has $\vec{A}_\Phi = 0$. The price for this is, of course that the transformed wave function Ψ', does not in general satisfy periodic boundary conditions around the ring, in fact, the phase of Ψ' changes by

$$\alpha \equiv \Delta\phi = 2\pi\Phi/\Phi_0, \tag{5}$$

when one electronic coordinate is rotated once around the ring. Thus the fluxes Φ and $\Phi + n\Phi_0$ are *indistinguishable*. In addition to establishing the exact claimed periodicity of any physical property (energy levels, matrix elements, etc. ...), eq. (5) also tells us that a non-integer flux is mathematically equivalent to a change in boundary conditions. This concept will prove extremely useful to us and we shall discuss an even more elementary way to establish it later. In any system obeying the *classical* laws, the AB flux Φ is clearly irrelevant. All physical properties do *not* depend on it. This is, of course, trivially consistent with the above theorem (a constant is also periodic, but not very interesting). The real issue is whether there is a sizeable sensitivity of, say, the energy levels or of the transition probabilities to Φ (periodicity is always guaranteed!). Here, one can make connection [9] with an important idea due to Thouless [11] and Edwards and Thouless

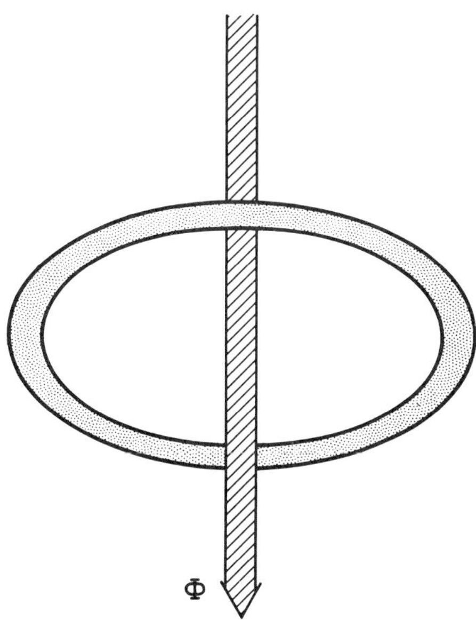

Figure 1. A ring threaded by an AB flux Φ (schematic).

[12]. According to this the conductance of a system is on the order of $e^2/\hbar$ times the (dimensionless) ratio between the sensitivity, V_T, of its energy levels to changes in the boundary conditions, and the separation, w, between levels (both at the Fermi energy, E_F). Thus, knowledge of the conductance of the system will enable us to estimate its flux sensitivity. The Thouless criterion will be reviewed in detail in section 3.

What about statement 2? Let us start with *free* electrons. It has first been noticed by Dingle [13] that equilibrium properties such as the average energy or magnetization of such a small system with a simple, *ideal*, geometry, e.g., a perfect disc or ring, are sensitive to a magnetic field. Oscillatory behavior is obtained as a function of the field, where the scale is set by the magnetic flux through the system being on the order of a flux quantum, $\Phi_0 = hc/e$. In the particular case of a simple ring with an AB flux Φ through its opening, the thermodynamic functions are periodic in Φ with a period Φ_0.

Dingle's type of results have been obtained later by several researchers in different contexts [14, 15, 16], see also [17]. However, an important difficulty in the background has been that electron scattering was almost universally expected [15, 18] to eliminate these effects in any realistic system.

It is very difficult to expect a real system to be not only impurity and defect free but also to have perfect surfaces. Some surface roughening will practically always exist. Thus, the elastic mean free path, l, will be usually limited, in metals, at best by the ring arms' width and thickness. So that l will be typically much smaller than, say, the ring's circumference, L – which is the distance over which the electron's wavefunctions experience interference. One's first intuition would say that the many scatterings the electron has to experience when travelling around the ring would completely eliminate any interference pattern. This is, in fact, in agreement with common notions on electron-beam diffraction experiments where care has to be taken to perform the experiment in a high enough vacuum to reduce the electron random scattering.

We shall see, however, that the above expectation is badly wrong and the analogy with beam experiments misleading. The point is that there exists an important distinction between *elastic* scattering, due to some *static* potential – in which wavefunctions with well-defined phases exist, and *inelastic* scattering. In the latter case, the electron may for example excite a phonon or alter the state of a "dust" particle, etc. The electron will as a result not have a definite phase, and thus will not be able to sense the change of boundary conditions unless eq. (3) holds. Models for the destruction of the persistent currents by coupling to a reservoir were discussed by Büttiker [19, 20] and by Landauer and Büttiker [21]. The general connection between loosing phase memory and flipping the environment to an orthogonal state has been recently established by Stern et al. [22]. The important distinction between the effects of elastic and inelastic scattering has become clear through the recent understanding [11, 23, 24] of conduction in disordered systems via localization theory [25, 26]. Prior to that, Landauer [27] has informally expressed similar insights (based on [28]) and Gunther and Imry [14] predicted persistent diamagnetic currents in a system with a finite resistance.

The persistent currents are another example of sample-specific, non-ensemble-averaged, "mesoscopic" fluctuation effects, such as the h/e oscillations in the transport in normal metal rings [29, 30, 31, 32, 33]. In that problem, averaging over the impurity ensemble [34] destroys the h/e oscillation but the $h/2e$ harmonic [18, 35, 36] survives (see e.g. [10]). This survival is essential in our case if one would like to enhance the magnetic signal of the persistent currents by using a large number of rings, as successfully done by Levy et al. [37]. It turns out that the h/e-periodic persistent current indeed practically vanishes upon ensemble-averaging, so does the $h/2e$ component [38, 39] in the grand canonical ensemble. However, if an ensemble of systems each with fixed N, but N possibly varying from system to system is used – the $h/2e$-periodic component survives! This was found by Cheung et al. [40] for 1 D electrons, including disorder. It was Bouchiat and Montambaux [41] who realized the generality of this effect, understood it in the presence of weak impurity scattering and obtained it from simulations with stronger disorder. The various ensemble-averaging procedures will be briefly reviewed in section 4 where a new proof of the surviving $h/2e$ periodicity, which clarifies its physics, will also be presented. Further issues and the relationship with experiment will be discussed in section 5.

2   One-Dimensional Ring with Disorder, Free-Electron Multi-Channel Rings

Büttiker et al. were the first [9] to understand the effect of elastic scattering using the simple model of a 1 D ring with disorder. They noted that the boundary condition eq. (5) is similar to that satisfied by the Bloch function ψ_k in a periodic potential across the unit cell of size L. Thus, identifying $\Delta\phi$ with kL establishes a one-to-one correspondence between the two problems. In fact, the condition eq. (5) can then be understood since the electron experiences the same potential by moving again and again around the ring, i.e., an effectively periodic situation, where the whole circumference of the ring plays the role of the unit cell. The electronic energy levels of the ring as functions of Φ are like those of 1 D Bloch electrons. This is applicable for an arbitrary random potential along the ring, since it can be shown [42] that in 1 D the only extreme of $E(k)$ are at $k = 0, \pm\pi/a$. For nearly free electrons one gets the usual wide bands and narrow gaps , i.e. $V \sim w$, while for a strong potential (small transmission along the ring) the opposite tight-binding situation (narrow bands, large gaps), i.e. $V \ll w$, is obtained. The latter case corresponds to strong localization at E_F. It is possible to estimate the flux-dependence of the total energy, E, at low temperatures ($k_B T \lesssim w$) since E is the sum of all occupied levels. Due to the alternating signs of $\partial E_j / \partial \Phi$ for consecutive levels, one has a strong cancellation and the sum is of the order of the last term around E_F. The circulating current is given [1, 3] by $-\frac{1}{c}\frac{\partial E}{\partial \Phi}$, thus, for N electrons, assuming $V_T \sim w$

$$I = \frac{-1}{c}\frac{dE}{d\Psi} \sim \frac{eE_F}{Nh} \sim \frac{ev_F}{L} \; . \tag{6}$$

For a strictly 1 D ring made from metal with $E_F \sim 2\,\mathrm{eV}$ with a circumference of a micron, $I \sim 10^{-8}$ amp. Since $w \sim 10\,\mathrm{K}$, having $k_B T < w$ is quite feasible. A further condition necessary to observe the oscillation is a long enough inelastic time i.e.

$$\hbar/\tau_{in} \ll w, V_T \; , \tag{7}$$

namely, that the level width is much smaller than the level separation or the bandwidth, which ever is smaller. The condition $\hbar/\tau_{in} \ll V_T$ is seen to be equivalent to $\sqrt{D\tau_{in}} \ll L$. An unfavorable case is the limit $V \ll w$, where the effects are very small – localized states are not sensitive to boundary conditions.

Note that in the presence of a finite non-integral Φ, the small diamagnetic-type currents flowing around the ring are *persistent* once the temperature is low enough so that eq. (7) is satisfied. The currents do *not* decay if the inelastic scattering is weak enough – the latter just establishes, within a few τ_{in}, an equilibrium population among the states and the current is given by the appropriate average, but there is no way for the persistent currents to decay. This result was quite surprising to many but it is correct. The decrease of this equilibrium current amplitude with decreasing τ_{in} and increasing dissipation has been discussed by Landauer and Büttiker [21]. These currents yield an orbital magnetic moment, M (sometimes referred to as "diamagnetic", although M may be parallel or antiparallel to H) and the magnetic susceptibility χ oscillating as functions of Φ, with a period Φ_0.

So far we discussed the case where all the magnetic fields were pure AB type. In the case where there are also some non-zero magnetic fields inside the metal itself leading to a flux Φ_M, their (paramagnetic as well as orbital) effects have to be added. There will be no periodicity in the total magnetic flux, only in the dependence on the AB part Φ, but that should be on top of the effects due to the magnetic fields inside the material. If the ratio of the area of the hole to that of the material is large enough ("good aspect ratio") one may expect the fast periodic dependence on Φ to be still visible on top of the slower variation due to Φ_M. Thus aperiodic effects analogous to conductance fluctuations should occur. An interesting case where the effective ring is provided by edge states was recently considered by Sivan and Imry [43].

Many interesting questions occur when the flux Φ changes with time to yield an e.m.f. $V = -\frac{1}{c}\frac{d\Phi}{dt}$. For the case where V is pure *dc*, the resulting current will oscillate [2, 3] with a Josephson frequency (e being the charge of the effective carrier)

$$\omega = eV/\hbar \; . \tag{8}$$

When the change of Φ is not slow enough (for *ac* voltages or for a finite *dc* one), Zener-type transitions may occur among the bands. This necessitates a dynamical treatment which we shall not touch upon here.

Until now, we discussed only the pure 1 D case. However, in most conceivable experiments the wires making the ring have a finite cross-section, A. Thus, the number of transverse states (across the wire) below E_F (equal to the number of "conducting channels") is on the order of

$$N_\perp \sim k_F^2 A \tag{9}$$

and the total number of electrons is

$$N \sim k_F L N_\perp \ . \tag{10}$$

Here, the levels as functions of Φ display a much more complicated structure with many maxima and minima, than in the 1 D case. It is not straightforward to estimate even the order of magnitude of these currents. This will be done in the next section. Here, we briefly review the free-electron case. Before that we repeat that persistent currents do exist in diamagnetic samples, they cancel in the bulk, for homogeneous samples, but a non-zero surface contribution remains. An interesting case is that of a non-uniform system – e.g. a mixture of a metal and an insulator – where non-zero diamagnetic currents may exist in the bulk, e.g. along the metal-insulator interfaces. In the ring geometry, these currents flow along the inner and outer surfaces, and the novel property is the periodicity of the total circulating current as a function of Φ.

Since elastic scattering by itself is not detrimental to the persistent currents, the very simple case of free electrons on a multi-channel ring is instructive. (The case of interest is $N_\perp \gg 1$. For simplicity we consider only a narrow, short height ring; the long cylinder can also be straightforwardly treated [14, 40]). The interesting result is that both the total persistent current and its temperature dependence are determined in this case by the energy level separation for motion around the ring

$$E_1 \equiv \frac{\hbar v_F}{L} \sim N_\perp w \ . \tag{11}$$

Since for most physical properties the channel number $N_\perp$ in the free regime is replaced in the diffusive one by the effective channel number, which is of the order of g [44], one is led to expect that the appropriate energy scale for the diffusive regime should be $gw \sim V_T$. It turns out that this is exactly what happens, as we shall see.

3 The Sensitivity of Energy Levels to Boundary Conditions, Estimates and Results for the Persistent Currents in the Diffusive Many-Channel Case

The spectrum (i.e. positions of energy levels) of a quantum-mechanical system is determined by the Hamiltonian *and* by the boundary conditions imposed on the wavefunctions. As explained in the last section, the mathematical condition of a phase change α imposed on the wavefunction between the two "sides" of a singly-connected system is equivalent to closing it on itself into a ring and employing an AB flux $\Phi = \alpha \Phi_0 / 2\pi$. For a general many-body system this is valid when the coordinate of any particle is taken once around the ring (the charge appearing in Φ is then the charge of that particle [1]). We now restrict ourselves to non-interacting quasi-particles and the above condition applies then for the single-particle wavefunction $\Psi_n(x)$. The AB flux in a ring is thus a convenient device for probing the dependence of ϵ_n (the single-particle levels) on the boundary conditions. It makes sense that the more delocalized the wave function is (i.e. the more easily the electrons move around the ring), the more sensitive would ϵ_n would be to α. One would also expect that, thus, the mobility (and therefore the conductivity) would increase with the sensitivity to α [45].

In 1972, Edwards and Thouless [12] proved, using second-order perturbation theory in α, the Kubo formula and some assumptions on the spectra, that the conductivity of the system is proportional to the typical sensitivity of the energy levels at E_F to α (for a discussion of linear response in a related context, see [46]). The latter can be parameterized by what we shall refer to as the Thouless energy V_T, defined by

$$V_T \equiv \left. \overline{\frac{\partial^2 E_n}{\partial \alpha^2}} \right|_{\alpha=0} , \tag{12}$$

where the bar denotes a typical absolute value or appropriate rms average for E_n near E_F. Thouless [11] showed that the above relationship can be cast into

$$g \equiv \frac{G\hbar}{2e^2} = \frac{V_T}{w} , \tag{13}$$

for the dimensionless conductance (conductance, G in units of $2e^2/\hbar$). Thouless [11] used this fundamental relationship to develop a theory for localization in a narrow wire which started the modern scaling theory of localization and laid (with the work of Wegner [47]) the basis for it.

The above picture is so relevant for our purpose that we shall present here another physical argument for, as well as an independent derivation of it. It is also possible to obtain this result by assuming the spectrum to obey Wigner-Dyson type [48, 49] correlations rather than the Cauchy distribution assumed by Edwards and Thouless [50]. The physical significance of V_T is best appreciated using the following quasi-classical argument. For a typical classical path diffusing along the system and experiencing the boundary conditions across its length L, the diffusion time is

$$t_L \sim L^2/D . \tag{14}$$

D being the diffusion coefficient. From eq. (13) the Einstein relation and the geometrical relation between G and σ one finds [11] that

$$V_T \sim \hbar/t_L . \tag{15}$$

The length of the path is $v_F t_L \sim L^2/l$ (where l is the elastic mean-free-path) and the phase acquired along it is $\sim E_F t_L/\hbar$. Thus V_T is immediately seen as the energy scale for which this phase change, χ, along the typical classical path increases by π

$$\frac{d\chi}{dE_F} \sim \frac{\pi}{V_T} . \tag{16}$$

This will be seen to be a useful relationship for the characteristic phase (or flux!) - dependence of the energy.

For a more formal new derivation of the Thouless relation eq. (13) we employ the exact eigenstates method used in a related context by McMillan [51] and Kaveh and Mott [52]. It employs [53, 8] the quasi-classical approximation for certain matrix elements for a diffusing particle between exact eigenstates whose energies E_n and E_m differ by $\hbar\omega$,

$$|\langle m| e^{i q \cdot r} |n\rangle|^2 = \frac{1}{\pi n_0} \frac{Dq^2}{(Dq^2)^2 + \omega^2} . \tag{17}$$

Using the small q limit of this expression and the relation between the matrix elements of x and p we find for the matrix element of p in an arbitrary direction

$$|\langle n|p|m\rangle|^2 = \frac{D}{\pi n_0 \hbar} \ . \tag{18}$$

This is equivalent to the result obtained by Thouless and Edwards from the Kubo formula. Thus the second-order perturbation theory result for the second derivative of the level E_n with respect to the AB flux is

$$V_T = \frac{De^2}{\hbar \pi n_0 c^2} \left(\frac{\Phi_0}{4\pi L}\right)^2 \sum_{m \neq n} \frac{1}{(E_n - E_m)} \ . \tag{19}$$

Approximating the sum by $1/w$ yields eq. (14). V_T can be interpreted [11] as the uncertainty in energy associated with the dwell-time t_L. The relation with the transmission between two adjacent blocks of the system was discussed by Imry [54]. The Thouless energy gives immediately the typical current response to the AB flux α at small α. As long as $E_n(\alpha)$ can be approximated by its quadratic $\sim V_T \alpha^2$ dependence we find for the typical first derivative

$$|I_{n,typ}| \sim V_T \alpha / \phi_0 \qquad (\alpha \ll 1) \ . \tag{20}$$

What is the typical single-level current in the whole interval $\alpha \leq \pi$? Cheung et al. calculated the typical current in an energy interval of $0(w)$, which should be of the same order of magnitude as the typical single-level current. Their result would thus be

$$|I_{n,typ}| \sim \frac{ev_F}{L} \frac{g^{1/2}}{N_\perp} \sim \frac{w}{\Phi_0} g^{1/2} \qquad (\text{all } \alpha) \ , \tag{21}$$

$N_\perp$ being the channel number. This result is best interpreted by saying that eq. (20) applies as long as the variation of the single level energy is of order w. This means that eq. (20) is valid up to $\alpha^2 \sim w/V_T \sim 1/g$, which yields a typical current in the whole interval of order $V_T/\sqrt{g}$, in agreement with eq. (21). To get the total current at $T = 0$, one has to sum all the single-level currents up to the Fermi-level. As in the 1 D case, there is a tremendous cancellation in the sum. The result is dominated by the last levels around E_F. It is not qualitatively clear, however, how many of these contribute and what are the precise correlations among them [55]. The Bouchiat-Montambaux paper [41] demonstrates how to do this in the extremely weak disorder case. There the currents of the levels cancel in pairs and the total current, I_{tot} is thus of the order of eq. (11). The typical value of I_{tot} has been calculated perturbatively by Cheung et al. [39]. Their result being

$$I_{tot,typ} \sim V_T/\Phi_0 \ , \tag{22}$$

so that $I_{tot,typ} \sim \sqrt{g} I_{n,typ}$. The result eq. (22) has been obtained also by Altshuler and Spivak [56] on the related model of an SNS (superconducting-normal-superconducting) junction. (A related ring with an AB flux inducing a phase shift has been treated by Büttiker and Klapwijk [57] in a study of the h/e vs the $h/2e$ periodicity. These are due to electron tunneling across the superconducting part and to Cooper pair motion across the normal part, respectively). The similarity of the SNS model with our wholly normal AB ring stems from the following observation: an electron hitting the NS boundary from the normal side is Andreev-reflected [58] as a hole into its time-reversed path acquiring an additional phase of χ_1 (χ_1 is the phase of the superconducting order parameter in the reflecting S part). The hole in turn gets to the other NS boundary and is Andreev-reflected there as an electron with an additional phase of $-\chi_2$ (χ_2 is the superconducting order parameter phase in the second S region). Thus, the net result of the two Andreev reflections is that the electron comes back to the same path it started in, with an additional phase of $\chi_1 - \chi_2$. This is equivalent to our AB ring with $\alpha = \chi_1 - \chi_2$.

4 A Path-Integral Formulation for Flux-Sensitivity, and the Various Ensemble-Averaging Procedures

Let us start from the Green's function $G(\vec{r}, \vec{r'}, E)$. The density of states (DOS) is given by

$$n(E) = \pi \int d^d r \, Im G(\vec{r}, \vec{r}, E) \,. \tag{23}$$

In the following, we shall leave the space integration understood and use the approximation for $G(r, r, E)$ as a sum over all (classical) paths of given energy which start at and come back to $\vec{r}$ [59]. Let us denote by A_n the contribution to $G(E)$ from paths winding n times in the clockwise direction around the hole of ring (see, in this connection [60, 61]). We [62] denote at zero magnetic flux $A_n = A_n e^{i\varphi_n}$ and for systems with time-reversal symmetry $A_n = A_{-n}$. The presence of the AB flux, ϕ, will multiply A_n by $e^{in\alpha}(\alpha = 2\pi\Phi/\Phi_0$. Thus, in the presence of the flux the modified DOS at zero temperature is given by

$$
\begin{aligned}
n_\Phi(E) &= n_0(E) + |A_1| \sin \varphi_1 \cos \alpha + |A_2| \sin \varphi_2 \cos 2\alpha + \ldots \\
&\equiv n_0(E) + \delta n_1 \cos \alpha + \delta n_2 \cos 2\alpha + \ldots \equiv n_0(E) + \delta n_\Phi(E)
\end{aligned}
\tag{24}
$$

$n_0(E)$ being the (usually dominant) flux-independent part and $\delta n_\Phi(E)$ is the flux-dependent correction. [1] A_n and φ_n are functions of E. From eq. (16) and the discussion around it we expect φ_1 to generally increase with E at a rate of $\sim \pi/V_T$, and this rate for φ_n will increase strongly with n. This tendency for a periodic dependence of $n_\phi(E)$ on E is in qualitative agreement with numerical results by [40, 41, 63]. Equation (24) is a very convenient presentation of the flux-dependence of $n_\Phi(E)$ as a series of harmonics. Several important and non-trivial insights can be gained almost by inspection from this presentation.

(1) Given a dephasing time [22] τ_φ, all paths such that the time they require is much longer than τ_φ will be exponentially cut-off. Thus, all details of the DOS on scales finer than $\hbar/\tau_\varphi$ will be smeared out. Concerning the A_1 contribution: it requires typically a time t_L (eq. (14)). Thus, the condition for survival of the first harmonic of the flux-dependence is

$$\tau_\varphi \gtrsim t_L \Longleftrightarrow L_\varphi \gtrsim L \,. \tag{25}$$

and higher harmonics will be cut-off by dephasing progressively faster. The energy scale w is irrelevant for this issue [64].

---

[1] To think of $n(E)$ as a continuous function, one employs a level broadening, $\hbar/\tau_\varphi \gtrsim w$; in some of the estimates below we shall even take $\hbar/\tau_\varphi \sim V_T$.

(2) In addition to dephasing, the flux-dependent harmonics will be progressively damped with increasing temperatures by energy-averaging. Here, too, the relevant energy scale to be compared with kT (disregarding the relatively unimportant energy dependence of A_i) is V_T for $n = 1$ and smaller and smaller scales for increasing n. It can be shown that this averaging also cuts off the flux-dependence in an exponential manner.

(3) When a quantity such as $\sin \varphi_n$ is averaged over the usual impurity-ensemble, i.e. an ensemble of systems where the macroscopic parameters are identical, but the positions of the impurities are arbitrary, the phases are understood to fluctuate so widely that the ensemble-average $\langle \sin \varphi_n \rangle_{av}$ vanishes. It is this impurity ensemble-averaging which destroys the h/e periodic AB oscillation and the UCF in rings and wires, respectively, as first pointed out by Gefen [34] (see also [65, 66, 67, 10]). This is expected to happen once the system is large enough compared with the impurity scattering scales. It was indeed found [38, 39] that under appropriate conditions, ensemble-averaged persistent currents vanish exponentially with L/l.

(4) The presentation of δn by sums over closed paths (see in this connection [68]) is related to the weak-localization corrections to the conductivity, σ [61, 23, 69]. However, these quantities are determined respectively by ImG and $|G|^2$. To estimate the h/e-periodic correction one notes that the corresponding correction to the propagator in the time-domain reaches its long-time limit of the inverse volume of the specimen within times of the order of L^2/D. Doing the Fourier transform to the energy domain we find

$$\delta n_1 \sim 1/V_T \tag{26}$$

It is interesting that the flux-dependent contributions to the DOS are related to those of the conductivity! It turns out [55] that the estimate eq. (26) is valid when the first harmonic, δn_1, mainly exists, i.e. for $\hbar/\tau_\varphi \sim V_T$.

A powerful result that follows from the observations (2) and (3) below is that the *ensemble-averaged DOS is flux-independent for $L \gg l$*. The same statement is valid for a given sample with $k_B T \gg V_T$. This is the basis for the belief commonly expressed in the literature (see e.g. [70]) that equilibrium properties should *not* show flux-dependence after ensemble-averaging. This result follows immediately from the observation that the partition function is an integral (Laplace transform) over the DOS, so that its flux-sensitivity is destroyed by ensemble-averaging!

The above is indeed valid for the single-particle partition function. The situation is different, however, for a many-particle system, even without the effect of interactions. For a grand-canonical system – i.e. one where the chemical potential rather than the particle number is given, all equilibrium properties are again given by integrals of $n(E)$ times the Fermi function and/or its derivatives, etc. Thus, in this case too, ensemble-averages of equilibrium properties should have *no* flux dependence for a large system. This is in agreement with the result of Entin-Wohlmann and Gefen [38] and Cheung et al. [39].

To get flux-dependence in ensemble-averaged equilibrium properties for non-interacting quasi-particles, one clearly needs a non-linear dependence on $\delta n_\varphi(E)$. It turns out that this is exactly what happens in an ensemble which is actually more appropriate for many experimental configurations (probably including that of Levy et al. [37]). Consider an ensemble of systems where, again, their macroscopic parameters are identical but they differ in the specific arrangement of the impurities in each of them. Suppose that each of these rings has a fixed number of electrons which *can not change when Φ is varied*. This should happen, for example, when the metallic rings are deposited on a neutral insulating substrate. We shall refer to this as a canonical ensemble, although the number of electrons, N, can vary among members of the ensemble. The crucial point is that N is *constant in each member*. The variability of N among members was emphasized by Bouchiat and Montambaux [41], but we believe it is not crucial for the diffusive regime. This variation of N from system to system is, of course possible – but it does not change the result. Moreover, the systematic dependence of the results on the parity of N is lost in the diffusive regime. For clarity, we present this part of the treatment at zero temperature, where the N lowest levels are filled, up to some E_F, and since N is constant, E_F may vary with the flux α. Writing

$$E_F(\alpha) = E_F^0 + \Delta(\alpha) , \tag{27}$$

we find from the constancy of N that to second order in the flux-dependent terms δn_ϕ

$$\Delta(\alpha) + \delta N_\phi(E_F)/n_0 = -\frac{1}{2}\frac{n_0'}{n_0}\Delta^2 . \tag{28}$$

Here n_0 stands for $n_0(E_F^0)$, n_0' for the derivative at E_F^0 and $\delta N_\phi(E)$ is defined by

$$\delta N_\phi(E) \equiv \int_0^E \delta n_\phi(\epsilon)d\epsilon . \tag{29}$$

We next evaluate the canonical average of the energy for a specific system, subtracting the value, E_0, of the energy at $\phi = 0$

$$E - E_0 = \int_{E_F^0}^{E_F^0+\Delta} n_0(\epsilon)\epsilon d\epsilon + \int_0^{E_F} \delta n_\phi(\epsilon)\epsilon d\epsilon . \tag{30}$$

We expand the first term in powers of Δ (or δn) and evaluate the second one by parts. The first-order term in the former cancels and the total first-order-contribution to $E - E_0$ becomes

$$\Delta E^{(1)}(\alpha) = - \int_0^{E_F^0} \delta N_\phi(\epsilon)d\epsilon . \tag{31}$$

This is a sample-specific contribution which varies randomly from sample to sample but retains the symmetries – periodicity in ϕ, and symmetry in ϕ for the systems obeying time-reversal symmetry we consider here. It is immediately seen that a sample-specific h/e-periodic term similar to eq. (31) exists in the grand-canonical case too.

Using the estimates of eq. (26), eq. (31) can be used to make estimates for the size of $\Delta E^{(1)}(\alpha)$, and hence of the persistent currents. To do that, one needs more information on the delicate cancellations in integrals such as eqs. (29) and (31). All we can do, at this stage, is make reasonable but *ad hoc* and *tentative* assumptions on this issue. However, these estimates are still problematic as far as small powers of g are concerned. A better understanding will have to await a more complete treatment of the appropriate spectral correlations. Let us thus *tentatively* assume that the oscillating integrands in eqs. (29) and (30) cancel except the last shell of width V_T (since by eq. (16)

this is the average "period" of $\delta n(E)$). The number of levels in this interval is of order g. *Assuming* that the contributions of that interval add coherently *and* that the integral eq. (31) is similarly determined by the "last shell", one obtains

$$\delta E_{typ} \sim V_T ,\tag{32}$$

in agreement with refs. [39, 56].

Upon ensemble-averaging, the contribution $\Delta E^{(1)}(\alpha)$ will, of course, vanish, as will all other direct higher order terms. However, in the second order (in δN and Δ), the terms of eq. (30) are

$$\Delta E^{(2)}(\alpha) = \left[\frac{1}{2}\frac{\partial}{\partial\epsilon}(n_0\epsilon)\right]_{E_F^0}\Delta^2 + \Delta\delta N_\phi(E_F^0)$$
$$- \frac{1}{2}E_F^0\frac{\partial n_0}{\partial\epsilon}\bigg|_{E_F^0}\Delta^2 - \Delta\delta N_\phi(E_F^0) = \frac{n_0}{2}\Delta^2 ,\tag{33}$$

where we have used eq. (28). Upon ensemble-averaging we find, using eq. (24) and the first-order part of eq. (28)

$$< \Delta E^{(2)}(\alpha) >_{av} = const \times \cos^2(\alpha),\tag{34}$$

where the general expression of eq. (33) for the *const* agrees very well with the analytical results of [41] obtained for the case of strong disorder and with their numerical results in general. In the comparison one uses the observation that the variations of Δ are similar to those of a typical level. At small α, we obtain

$$\Delta E^{(2)}(\alpha) = const - const' \times \alpha^2 .\tag{35}$$

The small-α persistent current is linear and odd in α and "paramagnetic" in sign!

The ensemble-averaged persistent current is in general odd in α, $h/2e$-periodic, was found numerically to be of the order of $(wV_T)^{1/2}/\Phi_0$ [71], and is "paramagnetic" in the simplest case. The survival of the $h/2e$ harmonic is essentially (and very simply) related to the constancy of N as function of ϕ. The "periodic halving" is related but not identical to the analogous effect in σ. The latter is obtained from $|G|^2$ using terms such as A_1^2 in which φ_1 has totally cancelled out. According to recent calculations [55] based on eq. (33) and the perturbative results of [56], the lowest harmonic of the ensemble-averaged persistent current is only of order w for $\hbar/\tau_\varphi \sim V_T$, which also agrees with estimates similar to eqs. (26) and (32). The numerical result of [71] may still apply for $\hbar/\tau_\varphi \sim w$.

5   Concluding Remarks

We reviewed in this paper the persistent currents that flow in normal-metal rings in periodic response to AB fluxes for independent Fermions. The magnetic flux through the conductor itself should lead ([43, 72]) to an aperiodic flux-dependence – similar to the UCF. In both grand-canonical and canonical situations there exist sample-specific, h/e-periodic contributions which are generally on the order of V_T/ϕ_0 . When ensemble-averaging on the impurities is done, as appropriate to multi-ring systems, all contributions vanish for large systems in the grand-canonical case. However, the constancy of $N(\phi)$ in canonical systems leads to a surviving $h/2e$-periodic contribution, in agreement with the Bouchiat-Montambaux results and small disorder argument. However, for

$\hbar/\tau_\varphi \sim V_T$, the theoretical results [55, 73] are smaller by at least one and a half orders magnitude than the estimates of [37].

The sign of this contribution is "paramagnetic" at small fluxes. In the recent impressive experiment [37], such an ensemble-averaged $h/2e$ periodic variation of the persistent (in their case – on the time scale of the experiment, $\sim 10^{-1}$ sec) currents was indeed observed. Its sign, however, was *tentatively* identified as "diamagnetic" at small fluxes. [2] Levy et al. suggested that the possible change of sign may be due to a strong enough spin-orbit coupling, according to the general results of Meir et al. [74]. It is indeed easy to check that the results of [74], when applied to the $h/2e$-periodic canonical-ensemble-averaged oscillation do in fact lead to a sign change in the 1 D case, but the sign does not [55] change for a finite cross section, "many channel" case.

The situation is not unique: also because the electron-electron interaction calculation in the related SNS model by Altshuler et al. [75] (see also [76]) leads to an $h/2e$-periodic ensemble-averaged persistent current whose sign depends on the sign of the coupling constant. Using the appropriate coupling constant for copper the magnitude of this term is also smaller by an order of magnitude from [37]. In this case, the canonical averaging is not needed and that fact together with a careful examination of the size of the effect and its sign might be the way to experimentally distinguish between the two pictures. [3]

It is of interest to understand the physical origin of the effect of Altshuler et al. [75]. It is due to correlations of electron pairs. For attractive interaction one can think about Cooper pair correlations above the superconducting transition temperature. In fact, flux-quantizing persistent currents were theoretically predicted and discussed by Gunther et al. [14] and Imry [77] for systems that do not have superconducting long range order and do show a finite resistance. What happens with repulsive interactions in less intuitive: basically, just the sign of the currents is reversed with respect to the attractive case. The physics of this fascinating effect still needs clarification.

We conclude this review by emphasizing the difference between the damping of these effects by energy averaging and by real dephasing. In this case both are exponential (one involving $k_B T/V_T$ and the other $k_B T \tau_\phi/\hbar$) so it is non-trivial to distinguish between them. However, in principle they are totally different. Dephasing is equivalent to leaving a trace in the environment [22] and is usually irreversible. Energy-averaging is a simple superposition of many interference patterns. It might, in principle, be eliminated, for example, by injecting a "monoenergetic" electron distribution, as recently done in the ballistic regime [78].

Persistent currents in normal rings is a subject that has been for great theoretical interest for decades. Now that the technology for experiments exists, this promises to be an exciting subject for studying fundamental issues on a controllable quantum system with a non-trivial, but variable, coupling to the external world.

6   Acknowledgements

The author is grateful to B. Altshuler, O. Entin-Wohlmann, Y. Gefen, S. Levit, U. Smilansky and A. Stern for instructive discussions. This research was supported by the

---

[2]The author is informed by Levy et al. that this sign is to be regarded at present as open.

[3]Note that the interaction picture will, for example, yield a paramagnetic contribution for electron-electron interactions that are repulsive. The above sign can *not* be reversed with spin-orbit coupling. This exemplifies the relevance of this sign determination.

Israeli Academy of Sciences and Humanities, Jerusalem, and by the Minerva Foundation, Munich.

References

[1] N. Byers and C.N. Yang, Phys. Rev. Lett. **7**, 46 (1961)

[2] F. Bloch, Phys. Rep. **137**, A787 (1965)

[3] F. Bloch, Phys. Rep. **B 2**, 109 (1970)

[4] Y. Imry, Ann. Phys. NY **51**, 1 (1969) (in this paper some of the subtleties with taking the "thermodynamic limit" too early have been discussed)

[5] Y. Aharonov and D. Bohm, Phys. Rev. **115**, 485 (1959)

[6] K. von Klitzing, G. Dorda, M. Pepper, Phys. Rev. Lett. **45**, 494 (1980)

[7] R.B. Laughlin, Phys. Rev. **B 23**, 5632 (1981)

[8] Y. Imry, J. Phys. **C 16**, 3501 (1983)

[9] M. Büttiker, Y. Imry, and R. Landauer, Phys. Lett. **96 A**, 365 (1983)

[10] Y. Imry, in: directions in condensed matter physics, G. Grinstein and G. Mazenko, eds., World Scientific, Singapore (1986)

[11] D.J. Thouless, Phys. Rev. Lett. **39**, 1167 (1977)

[12] J.T. Edwards and D.J. Thouless, J. Phys. **C 5**, 807 (1972)

[13] R.B. Dingle, Proc. Phys. Soc. **A 212**, 47 (1952)

[14] L. Gunther and Y. Imry, Sol. St. Commun. **7**, 1391 (1969); unpublished results (1969-1970)

[15] I.O. Kulik, JETP Lett. **11**, 275 (1970)

[16] N.B. Brandt, E.N. Bogachek, D.V. Gitsu, G.A. Gogadze, I.O. Kulik, A.A. Nikolaeva, and Ya.G. Ponomarev, JETP Lett. **24**, 273 (1976)

[17] M. Schick, Phys. Rev. **166**, 401 (1968)

[18] B.L. Altshuler, A.G. Aronov, and B.Z. Spivak, JETP Lett. **33**, 94 (1981)

[19] M. Büttiker, Phys. Rev. **B 32**, 1846 (1985)

[20] M. Büttiker, Ann. NY Acad. Sci. **480**, 194 (1986)

[21] R. Landauer and M. Büttiker, Phys. Rev. Lett. **54**, 2049 (1985)

[22] A. Stern, Y. Aharonov, and Y. Imry, Phys. Rev. **A 41**, 3936, (1990)

[23] G. Bergmann, Phys. Rep. **107**, 1 (1984)

[24] P.A. Lee and R.V. Ramakrishnan, Revs. Mod. Phys. **57**, 287 (1985)

[25] P.W. Anderson, D.J. Thouless, E. Abrahams, and D.S. Fisher, Phys. Rev. **B 22**, 3519 (1980)

[26] E. Abrahams, P.W. Anderson, D.C. Licciardello, T.V. Ramakrishnan, Phys. Rev. Lett. **A 42**, 673 (1979)

[27] R. Landauer, unpublished IBM proposal (1966)

[28] R. Landauer, IBM J. Res. Dev. **1**, 223 (1957)

[29] Y. Gefen, Y. Imry, and M.Ya. Azbel, Phys. Rev. Lett. **52**, 129 (1984)

[30] M. Büttiker, Y. Imry, and M.Ya. Azbel Phys. Rev. **A 30**, 1982 (1984)

[31] R.A. Webb, S. Washburn, C.P. Umbach, and R.B. Laibowitz, Phys. Rev. Lett. **54**, 2696 (1985)

[32] V. Chandrasekhar, M.J. Rooks, S. Wind, and D.E. Prober, Phys. Rev. Lett. **15**, 1610 (1988)

[33] S. Datta, M. Melloch, S. Bandyopadhyay, R. Noren, M. Vaziri, M. Miller, and R. Reifenberger, Phys. Rev. Lett. **55**, 2344 (1986)

[34] Y. Gefen, private communication (March, 1984)

[35] D.V. Sharvin and V.V. Sharvin, JETP Lett. **32**, 272 (1981)

[36] B.L. Altshuler, A.G. Aronov, D.E. Khmel'nitskii, A.I. Larkin, in: Quantum Theory of Solids, I.M. Lifschitz, ed., p. 146, Mir Publishers, Moscow (1982)

[37] L.P. Levy, G. Dolan, J. Dunsmuir, and H. Bouchiat, Phys. Rev. Lett. **64**, 2074 (1990)

[38] O. Entin-Wohlmann and Y. Gefen, Europhys. Lett. **8**, 477 (1989)

[39] H.F. Cheung, Y. Gefen, and E.K. Riedel, Phys. Rev. Lett. **62**, 587 (1989)

[40] H.F. Cheung, Y. Gefen, E.K. Riedel, and W.H. Shih, Phys. Rev. **B 37**, 6050 (1988)

[41] H. Bouchiat and G. Montambaux, J. Phys. (Paris) **59**, 2695 (1989)

[42] R. Peierls, Quantum Theory of Solids, p. 29, Oxford (1955) (comment attributed to W. Shockley)

[43] U. Sivan and Y. Imry, Phys. Rev. Lett. **61**, 1001 (1988)

[44] Y. Imry, Europhys. Lett. **1**, 249 (1986)

[45] W. Kohn, Phys. Rev. **133**, A161 (1964)

[46] N. Trivedi and D. Brown, Phys. Rev. **B 38**, 9581 (1988)

[47] F. Wegner, Z. Phys. **25**, 327 (1976)

[48] F.J. Dyson, J. Math. Phys. **3**, 140, 157, 166 (1962)

[49] M.L. Mehta, Random Matrices, Academic Press, New York, London (1967)

[50] G. Lopez, P.A. Mello, and T.H. Seligman, Z. Phys. **A 392**, 351 (1981)

[51] W.L. McMillan, Phys. Rev. **B 24**, 2739 (1981)

[52] M. Kaveh and N.F. Mott, J. Phys. **C 14**, 2179 (1981)

[53] M.Ya. Azbel, private communication (1980)

[54] Y. Imry, Phys. Rev. **B 21**, 2042 (1980)

[55] B.L. Altshuler, Y. Gefen, and Y. Imry, unpublished results (1990)

[56] B.L. Altshuler and B.Z. Spivak, Sov. Phys. JETP **65**, 343 (1987)

[57] M. Büttiker and T.M. Klapwijk, Phys. Rev. bf 33, 5114 (1986)

[58] A.F. Andreev, JETP **46**, 1823 (1964)

[59] M.C. Gutzwiller, J. Math. Phys. **12**, 343 (1971)

[60] A.I. Larkin and D.E. Khmel'nitskii, Sov. Phys. Uspekhi. **25**, 185 (1982)

[61] D.E. Khmel'nitskii, Physica **126 B**, 235 (1984)

[62] Y. Gefen and Y. Imry, unpublished results (1990)

[63] D. Divincenzo and M.P.A. Fisher, unpublished results (1988)

[64] A. Stern and Y. Imry, unpublished results (1990)

[65] M. Murat, Y. Gefen, and Y. Imry, Phys. Rev. **B 34**, 659 (1986)

[66] A.D. Stone and Y. Imry, Phys. Rev. Lett. **56**, 189 (1985)

[67] Y. Imry and N. Shiren, Phys. Rev. **B 33**, 7992 (1986)

[68] M. Berry, Proc. Roy. Soc. Lond. **A 400**, 229 (1985)

[69] S. Chakravarty and A. Schmid, Phys. Rep. **140**, 195 (1986)

[70] B.L. Altshuler, A.G. Aronov, B.Z. Spivak, D.Yu.Sharvin, and Yu.V. Sharvin, JETP Lett. **35**, 5898 (1982)

[71] G. Montambaux, H. Bouchiat, D. Szigeti, and R. Friesner, preprint (1990)

[72] H. Fukuyama, J. Phys. Soc. Jpn. Lett. **58**, 47 (1989)

[73] A. Schmid, unpublished results (1990)

[74] Y. Meir, Y. Gefen, and O. Entin-Wohlmann, Phys. Rev. Lett. **63**, 798 (1989)

[75] B.L. Altshuler, D.E. Khmel'nitskii, and B.Z. Spivak, Sol. St. Commun. **48**, 10 (1983)

[76] V. Ambegaokar and U. Eckern, Phys. Rev. Lett. **65**, 381 (1990)

[77] Y. Imry, in: Proc. of the 1969 Stanford Superconductivity Conf., F. Chilton ed., 344, North Holland (1971)

[78] M. Heiblum, M.I. Nathan, D.E. Thomas, and C.M. Knoedler, Phys. Rev. Lett. **55**, 2200 (1985)

SEARCH FOR PERSISTENT CURRENTS

Alain Benoit, Dominique Mailly*,
Mohamed El-Khatib, Pierre Perrier

CNRS CRTBT, 25 av. des martyrs
38042 Grenoble, France
*)L.M.M., 196 av. H. Raverra
92220 Bagneux, France

1 Theoretical Predictions

The existence of persistent currents in a mesoscopic ring in presence of disorder was suggested by Büttiker, Imry and Landauer [1]. However a complete theoretical approach, easy in the case of one-dimensional (1 D) system [2], is very difficult to extend to a real three-dimensional (3 D) system with disorder and at finite temperature [3]. The direct observation of such currents would give very important informations for the understanding of such mesoscopic systems.

1.1 Perfect One-Dimensional Ring

For a perfect ring in the absence of disorder, the magnetic flux ϕ inside the loop affects the periodic boundary conditions of the wave functions $\Psi(x)$ of the electrons

$$\Psi(x + L) = \Psi(x)e^{2i\pi\phi/\phi_0} \tag{1}$$

where L is the circumference of the loop and $\phi_0 = h/e$ the flux quantum. These conditions imply that the energy levels and any physical property of the system are a periodic function of the flux with period ϕ_0. For a given energy level n, the current flowing in the ring is given by the derivative of the energy E_n and is therefore proportional to the velocity of the electrons

$$I_n(\phi) = dE_n/d\phi = -ev_n/L. \tag{2}$$

This current vanishes for zero flux and the sign associated to each level is random. The total current at zero temperature is given by the sum over all occupied energy levels. In the case of free electrons, we can observe that each successive contribution has opposite sign and therefore the total current has the same order of magnitude than the current created by one electron at the Fermi level

$$I_0 = -ev_f/L. \tag{3}$$

In this model, the exact description of the system is important since the dependence of the current versus flux has a different behaviour if the chemical potential is constant (fixed Fermi energy) or if the number of electrons is fixed to an odd or even number.

1.2   Multi-Channel Ring

In the absence of disorder, we can compute the effect of the transverse dimension of the ring by taking into account the number of channels M defined by the expression $2L_y = M\lambda_f$ where L_y is the transverse dimension and λ_f the Fermi wavelength. The total current is the sum of currents in different channels. The result depends on the correlations between channels. In the absence of disorder there is no correlation on the average and the total current is $\sqrt{M}$ times the one channel current. However this result is only correct in the ballistic regime. In the diffusive regime this multiplicative factor disappears, due to the compensation between currents in differents channels.

1.3   Effect of Finite Temperature

The first effect of temperature is to smear the Fermi surface and then average the current over levels in an energy interval $E = k_b T$. The characteristic temperature T^* where the persistent currents begin to decrease is determined by the level spacing. However, in the multi-channel system, the derivatives of the energy levels are correlated and the characteristic temperature is given by the level spacing of the equivalent 1 D ring

$$K_b T^* = \Delta = h v_f / L. \tag{4}$$

This characteristic temperature is the same that the one used to describe the decrease of conductance fluctuation or Aharonov-Bohm (AB) oscillation in a ring. The decrease of persistent currents is clearly exponential for a 1 D ring but in the 3 D case, either a power law or an exponential law are expected.

1.4   Effect of Inelastic Scattering

This effect can be studied theoretically considering a perfect ring coupled with multiple reservoirs with total phase randomization. The result is simply an exponential decreasing when the length L of the sample exceeds the coherence length L_ϕ

$$I = I_0 \exp(-L/2L_\phi). \tag{5}$$

1.5   Effect of Disorder

The disorder in the ring can be characterized by the elastic mean free path l and the localization length which is related to the mean free path by $\xi = Ml$. In the case of strong disorder where the localization length ξ is smaller than the sample dimension, the wave functions of the electrons are connected only by exponentially small tails and the persistent currents decrease very rapidly

$$I = I_0 \exp(-L/\xi). \tag{6}$$

However, in the diffusive regime characterized by $l < L < \xi$, the effect of disorder is to open a gap at each crossing point of the energy levels $E(\phi)$. This effect induces a small decrease of current due to the slope reduction of the curves $E(\phi)$. The result in this case is simply a linear decrease with l

$$I = I_0(l/L). \tag{7}$$

This result is easy to understand, since considering the time for one electron to make one turn around the ring and replacing in the expression of I_0 the time $t = L/v_f$ by the diffusive time $t = L^2/D$, we obtain the same expression. For the same reason, the disorder reduces the characteristic temperature to the Thouless energy

$$K_b T^* = (h v_f/L)(l/L) = h D/L^2. \tag{8}$$

1.6   Ensemble Averaging

We consider here the current resulting from averaging over a number N of rings. The current of a single ring has a random sign depending on the microscopic realization of the ring. One would expect that the total current increase as $\sqrt{N}$. However, in 1D system, the $I(\phi)$ curve is asymmetric and if we decompose this periodic function in Fourier series, the even harmonics of current are mostly positive and give a contribution to the persistent current proportional to the number of rings ($I = N I_0$). This result is not obtained if we average over different configurations of a multi-channel ring with fixed chemical potential but H. Bouchiat and G. Montambaux [4] have shown that this even harmonic contribution persists if we use the condition of a fixed number of electrons. Recent experiments [5] confirm this property.

1.7   Effect of Magnetic Field Inside the Wire

All the previous theoretical predictions are calculated in the absence of magnetic field applied to the electrons. However, in real experiments, it is difficult to put a flux inside the loop without applying some magnetic field on the sample. Let us divide the effect of the magnetic field in two contributions: (1) the direct force applied on the electrons, responsible of changes in the current flow and of the quantum Hall effect; (2) the change of the phase of wave functions due to the vector potential. The first effect becomes important only when $\omega_c \tau \geq 1$. It can be neglected since this condition is not fulfilled in this type of experiment. The second effect will change the persistents currents only if the flux created inside the wire approaches one flux quantum ϕ_0 and changes significantly the phases of the electron wavefunctions. In conclusion, we can estimate that, as for the AB oscillations, the effect of a homogeneous applied field on a ring is similar to the effect of a flux inside the loop, as long as the aspect ratio of the ring is large.

2   Persistent Currents Experiment

We present here the experiment we have build to observe the persistent currents in a single ring. We do not discuss the experiment on multiple rings [5] which are directly connected to the ensemble averaging problem discussed by H. Bouchiat and G. Montambaux in this volume.

2.1   Choice of the Ring

In order to observe this current we have to choose a material with long phase coherence length. We can use pure metals as Gold or Copper for example, that provide a coherence length of a few microns. Another possibility is to use a semiconductor with a smaller number of electrons (smaller velocity at the Fermi level) but with a higher mobility (smaller disorder). Such properties are obtained in a GaAs-GaAlAs heterojunction, which gives a two-dimensional electron gas with elastic mean free path higher than one micron. The table below shows the values of persistent current expected in a gold or in a semiconducting ring, using the parameters measured on real experimental rings. From this table, we can observe that the GaAs ring is the best choice to measure persistent currents.

Table 1

	metallic ring	GaAs ring
diameter	$1.5\,\mu$m	$2.5\,\mu$m
L_ϕ	$2\,\mu$m	$8\,\mu$m
k_f	$1.2 \cdot 10^{10} m^{-1}$	$0.9 \cdot 10^{8} m^{-1}$
v_f	$10^6\,$m/s	$1.5 \cdot 10^5\,$m/s
I_0	$30\,$nA	$6\,$nA
l	$40\,$nm	$5\,\mu$m
M	10^5	15
T^*	$4\,$mK	$40\,$mK
I	**0.8 nA**	**5 nA**

2.2   General Configuration

In order to observe the current created by a single ring, we need a sensitive flux detector very well coupled to the ring. It is clear that a good coupling with a two micron diameter ring can only be done by lithography. The best sensitivity can be obtain by placing a DC-SQUID close to the sample to eliminate the losses due to the self inductance of the connection line between ring and detector. To satisfy this condition, we designed a simple DC-SQUID which can be built directly by lithography on the substrat supporting the ring.

The second experimental problem is to vary the flux through the ring to probe the ϕ_0 periodicity of persistent currents without direct effect on the detector. In standard susceptibility experiments, the detection coil is compensated near the sample and the detector is screened by a superconducting shield to avoid any flux generated by the polarization field into the SQUID itself. This arrangement is difficult, due to the short distance between ring and detector. To avoid these problems, we have realized the compensation by applying a non-uniform magnetic field on the dispositif.

The geometry adopted is shown in Fig. 1. The same polarization current circulates in the left and right loop but in opposite direction and the resulting magnetic field is zero on the SQUID junctions. Also there is no total flux created in the SQUID and in the flux transformer. With this geometry, a current of $2\,$mA in the polarization coil is sufficient for applying one quantum flux into the ring. In real sample, the flux transformer is much wider than the ring and deposed on the top of the ring, screening all magnetic field inside the GaAs wire. A test loop placed in a position symmetric to the GaAs ring is used for direct calibration of the dispositif sensitivity.

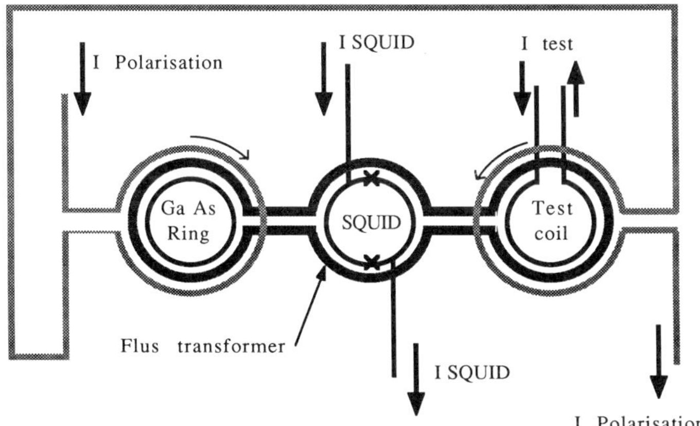

Figure 1. Schematic drawing of the dispositif used for persistent currents measurement.

2.3   Realization of the SQUID

In order to simplify the realization of the SQUID, we replace the classical tunnel Josephson junction by microbridge junctions. The increase of the resolution in nanolithography allow us to obtain very good Josephson junctions (typical dimensions $\ll$ superconducting coherence length ξ_s) with reproducible critical currents [7]. We use Aluminium as superconducting material because the desired experimental temperature is of order of $10\,\mathrm{mK}$. Aluminium is easy to evaporate, very stable, and has a long coherence length.

With a SQUID inductance similar to the inductance of the ring of the order of $4\,\mathrm{pH}$, the optimal value of critical current for the SQUID is $200\,\mu\mathrm{A}$, corresponding to a microbridge $60\,\mathrm{nm}$ wide (Fig. 2). Such a SQUID can be done in one operation using the lift-off technique. To obtain the best resolution, each Josephson junction has to be shunted with resistors in order to suppress the hysteresis of the junctions. In a first time, we did not use such shunt resistors for simplicity and we have developed a special electronic (similar to the one used for microscopic quantum tunneling) to measure directly a SQUID with hysteresis. With this technique, the measure of the critical current is converted in time measurement with a recycling frequency of typically $10\,\mathrm{kHz}$. Figure 3 shows a typical histogram of critical current with a width of $5\,\mu_s$ corresponding to a resolution in critical current of $100\,\mathrm{nA}$ and a flux change of $10^{-3}\phi_0$. This corresponds to a typical noise

$$\Delta\phi = 10^{-5}\phi_0/\sqrt{\mathrm{Hz}}. \tag{9}$$

[tb] In order to eliminate all slow variations and $1/f$ noise, a triangular signal was applied to the polarization coil and the SQUID measurements were accumulated with a computer. With this method, no flux was applied to the ring by the measure, since we did not modulate the flux in the SQUID. Therefore we obtained directly the complete response curve giving the susceptibility of the ring as a function of the polarization flux.

2.4   Experimental Setup

The first operation consists in growing by MBE the GaAs-GaAlAs heterojunction on a GaAs substrate. We then characterize the 2 D electron gas: electron density $n =$

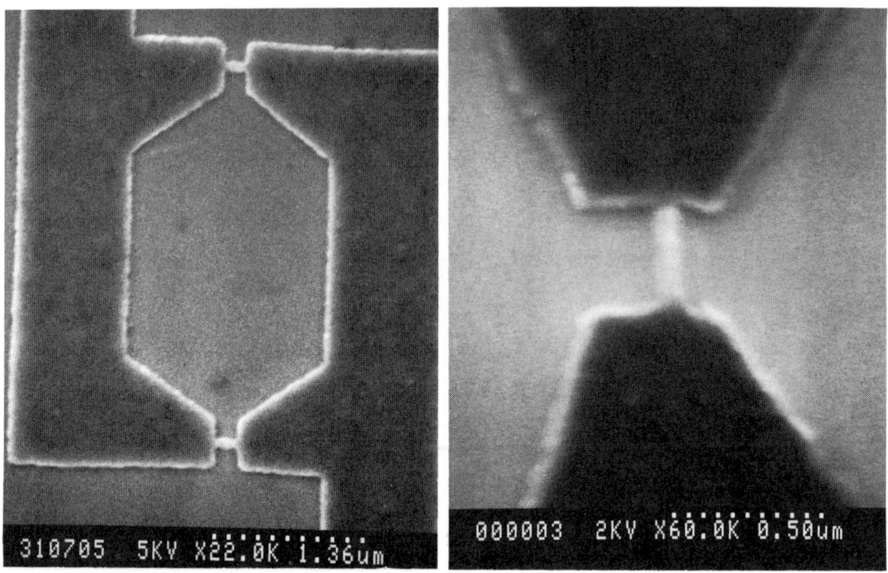

Figure 2. Aluminium DC-SQUID with two (left) microbridge junctions (right).

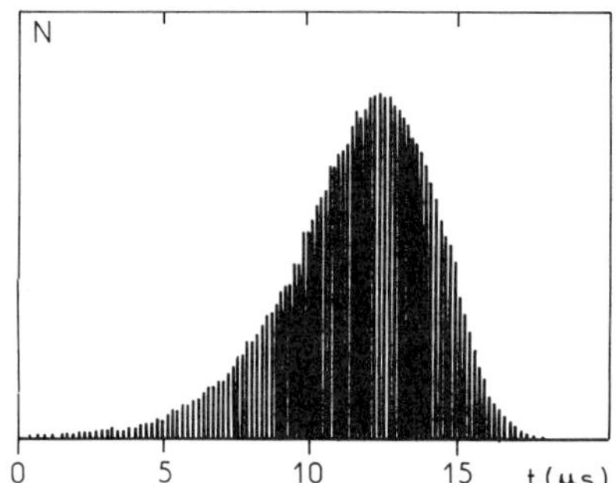

Figure 3. Histogram of critical currents for an Aluminium DC-SQUID at a temperature of 100 mK. The measured time on the X axis corresponds to a variation of the SQUID critical current of $20\,\mathrm{nA}/\mu s$ or a flux of $2 \cdot 10^{-4}\phi_0/\mu s$.

$4.5 \cdot 10^{11}$ electrons/cm^2 and mobility $\mu = 1.1 \cdot 10^6\,\mathrm{cm^2/Vsec}$. All the lithography made in this fabrication process uses the lift-off technique on PMMA with a JEOL 5 DIIU electron beam writer. The first step is to pattern the metal mask used to define the ring by ion beam etching. After removal of the mask, the SQUID and the calibration coil are made with a 50 nm thick Aluminium layer. The chip is then covered by 100 nm SiC as insulator layer. The next step is to pattern the flux transformer (50 nm Aluminium). Then, after a second insulated layer (100 nm SiC) the polarization coil is realized using 100 nm Aluminium. All different layers in the process are aligned using alignment marks near the device. The sample is placed into a dilution cryostat and cooled down to 10 mK.

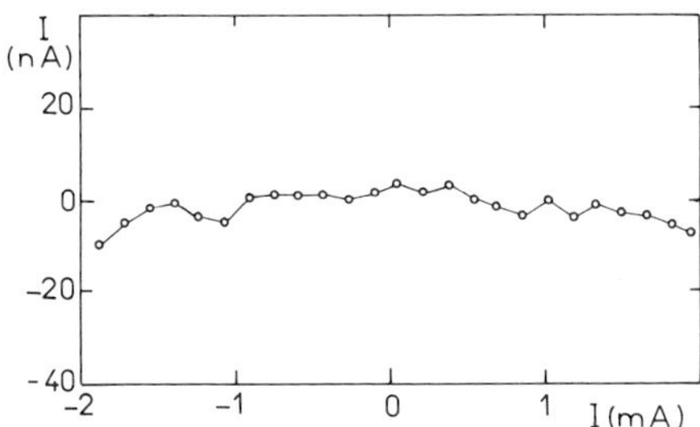

Figure 4. Measured signal as a function of current in the polarization coil, after subtraction of linear and parabolic contributions. The signal on the Y axis is converted into the equivalent current in the test coil producing the same effect.

3 Results

The calibration of the sensitivity using the test loop gives a value of $3\,\mathrm{mA}$ in the loop for a quantum flux ϕ_0' in the SQUID ($\phi_0' = h/2e$), in good agreement with calculation. From this result, a resolution of $10^{-6}\phi_0'$ obtained in $100\,\mathrm{sec}$, is sufficient to detect a persistent current of $3\,\mathrm{nA}$. However, a residual signal appears in the SQUID when we change the polarization of the ring, due to imperfection in the symmetry of the device. This contribution of the order of 10^{-4} to $10^{-3}\phi_0'$ (depending on the sample) can be eliminated by numerical subtraction or by applying a small current in opposite direction into the test coil. After this subtraction, a residual quadratic signal corresponding to about $10^{-5}\phi_0'$ persist in all samples, probably due to a modification of the geometry caused by field penetration in the flux transformer. The curve on Fig. 4 presents the signal obtained after subtraction of these terms. On the Y axis, the signal measured by the SQUID is converted into the current in the test coil producing the same effect. We observe that the experimental noise is in good agreement with the value deduced from the width of the histogram. However, we do not observe any periodic variation which would indicate the existence of persistent currents.

4 Discussion

Several reasons can be invoked to explain the absence of signal. The first point is the limitation of the flux we can induce into the ring due to a maximum current of $3\,\mathrm{mA}$ in the polarization coil. This flux has been checked with a SQUID in place of the ring and corresponds only to $0.5\phi_0$ in this experiment. We already design and realize a new device with better coupling between polarization coil and ring which can induce up to $3\phi_0$ into the ring.

The second important point is the thermalization of the ring. The temperature of the sample holder is very well controlled but the thermal coupling between electrons in the ring and sample holder is unknown. To control directly the temperature of the electrons in the ring, we design a sample with four wires connected to the ring to measure the AB oscillations in the same conditions. A gate on the top of the GaAs allows us to connect or disconnect the ring easily.

The third problem is the sensitivity of the detector. If the persistent currents are smaller than the predicted value, it will be necessary to increase the sensitivity of the SQUID. We think that using a shunt resistor on each Josephson junction will increase the sensitivity by one order of magnitude.

5 Possibilities of the Method

The first advantage of this experimental method is that we obtain directly the complete response curve $I(\phi)$. This allow us to check directly the periodicity of the response.

The temperature dependence of persistent currents can also be measured up to $400\,\mathrm{mK}$. This limitation is due to the temperature domain where the SQUID can be used.

In the studied ring, the disorder is weak and by reducing the mobility, it will be easy to measure the change in persistent currents, going from a ballistic regime to a diffusive one.

The effect of the number of channels can also be checked by changing the width of the ring.

One interesting possibility to separate the persistent currents from all other parasital signals consists to change the number of electrons in the ring by applying light on the heterojunction. The effect of light is to trap some electrons changing the total number of electrons in the ring. A previous experiment on AB oscillations in the same GaAs sample has shown a complete change of the sample after each illumination at low temperature. By subtraction of a persistent current measurement before illuminating the sample and after, we can eliminate all external contributions.

This also allow us to average the signal over an ensemble of different samples and check the averaging of the $\phi_0/2$ periodic oscillations.

Finally, by applying a voltage bias on the gate, we can switch between the canonical ensemble (fixed chemical potential) to the grand canonical ensemble (N fixed) and measure the averaging process in the two statistics ensembles.

6 Acknowledgements

We wish to thank R.Webb who gave us the first idea of this experiment. We also thank B. Etienne and F.R. Ladan for their participation in the elaboration of the samples.

References

[1] M. Büttiker, Y. Imry, and R. Landauer, Phys. Rev. **A 96**, 365 (1983)

[2] R. Landauer and M. Büttiker, Phys. Rev. Lett. **54**, 2049 (1985); H.F. Cheung, Y. Gefen, E.K. Riedel, and W.H. Shih, Phys. Rev. **B 37**, 6050 (1988)

[3] H.F. Cheung, Y. Gefen,and E.K. Riedel, Phys. Rev. Lett. **62**, 587 (1989); N. Trievedi and D.A. Browne, Phys. Rev. **B 38**, 9581 (1988)

[4] H. Bouchiat and G. Montambaux, J. Phys. **50**, 2695 (1989)

[5] L.P. Levy, G. Dolan, J. Dunsmuir and H. Bouchiat, Phys. Rev. Lett. **64**, 2074 (1990)

[6] M. El-Khatib, Thesis (Grenoble, France)

PERSISTENT CURRENTS IN MESOSCOPIC RINGS: ENSEMBLE AVERAGE AND HALF–FLUX–QUANTUM PERIODICITY

Hèléne Bouchiat*, Gilles Montambaux*,
L.P. Levy[†], G. Dolan[†], J. Dunsmuir[†]

*Laboratoire de Physique des Solides Associé au CNRS, Bât. 510
Université Paris Sud, 91405 Orsay, France
[†]AT&T Bell Laboratories, Murray Hill, NJ 07974, USA

1 Introduction

The transport properties of micron size metallic samples at low temperature have been shown to exhibit features characteristic of the quantum coherence of the electronic wave function along the whole sample. One of the most striking has been the detection of Aharonov-Bohm oscillations in the resistance of a loop pierced by a magnetic field perpendicular to its plane [1]. In such resistivity measurements, the electric probes induce a coupling of the sample with reservoirs of electrons in which dissipation occurs. Isolated metallic rings are predicted to present an even more spectacular behavior: Büttiker, Imry and Landauer [2] suggested the existence of persistent currents in the presence of a magnetic field. These currents are a consequence of the sensitivity of the eigenstates to the boundary conditions along the ring. In the presence of a magnetic field, the periodic boundary conditions are indeed modified into

$$\Psi(x + L) = \Psi(x)e^{i\varphi},\tag{1}$$

where $\varphi = 2\pi\phi/\phi_0$. ϕ is the magnetic field through the loop. ϕ_0 is the flux quantum h/e. The persistent current is related to the flux derivative of the eigenenergies by

$$I(\phi) = \sum_n \partial E_n/\partial\phi,\tag{2}$$

where n labels the N filled energy levels. The current has thus the period ϕ_0. There has been recently a growing experimental interest in the detection of this persistent current. On the theoretical side, it is of course of essential interest to estimate the amplitude of this current in a model as close as possible to the experimental situation. Some important questions are of interest: what is the order of magnitude of this current, how does it depend on disorder, on the size of the ring, on temperature, etc....

Quantum Coherence in Mesoscopic Systems
Edited by B. Kramer, Plenum Press, New York, 1991

Some of these questions have been addressed in recent papers [2-11]. One of the most important goals was to go beyond the simple description of a one-dimensional (1 D) ring and to study this current in a multi-channel situation. In the multi-channel case, three regimes can be distinguished, depending of the strength of disorder W. In the ballistic regime, the perimeter L of the ring is smaller than the elastic mean free path $l_e(W)$. In the localized regime, the length L is larger than the localization length ξ . In between, there is the metallic or diffusive regime when $l_e < L < \xi$. The localization length has been found to be of order of $M l_e$, M being the number of channels (at least in the 3 D case) [12]. Recently, Cheung et al. estimated the typical current in these three regimes [6,7]. Let us summarize their results. In the ballistic case, the typical current $I_{\text{typ}} = \sqrt{\langle I^2 \rangle}$ is of the order of $\sqrt{M} I_0$. $\langle ... \rangle$ is the ensemble average on different realizations of disorder or number of particles. I_0 is the 1 D current. It is given by $I_0 = e v_F / L$, v_F being the Fermi velocity. In a metal with micronic size, it is of the order of 10^{-7} A. In the localized case, the typical current decreases exponentially with the length L and inversely with the number of channels $I_{\text{typ}} = (I_0/M) \exp(-L/\xi)$. In the metallic regime it is proportional to the elastic mean free path and *independent of the number of channels*. $I_{\text{typ}} = e v_D / L$ where $v_D = v_F l_e / L$ is the diffusion velocity along the ring. The magnetic moment for a typical mesoscopic metallic ring is thus expected to be at most 10^2 to 10^3 Bohr magnetons.

Experimentally, two kinds of investigations are possible. One is the study of a single loop. It requires an extremely sensitive detector. The other is the detection of the magnetic moment of a large number $\mathcal{N}$ of independent loops. This requires a less sensitive detector but rises the problem of ensemble averaging. Stimulated by the second type of experiment, we have recently studied the ensemble average of the persistent current [10]. The current of a single ring has a priori a random sign which depends on the microscopic configuration of the disorder. As the result, one would expect naively the magnetic response of $\mathcal{N}$ rings to be zero (or of the order of $\sqrt{\mathcal{N}}$). We show that it is not the case . *The ensemble average is finite (of order $\mathcal{N}$) and its periodicity is $\phi_0/2$ instead of ϕ_0.* This result is based on both analytical arguments and numerical simulations of the Anderson model. Other calculations performed at fixed chemical potential do not yield this period halving but on the other hand a zero ensemble average. We emphasize the importance of fixing the number of electrons or the chemical potential μ in the calculation of the persistent current. Our result are obtained with a fixed number of electrons in each ring (this number being different from one ring to another) .

We finally briefly present the results of the recent experiment [13] done with $\mathcal{N} \sim 10^7$ rings which shows the periodicity $\phi_0/2$ as predicted from this theory and a reasonably good quantitative agreement.

2   The Model

We have used the Anderson model [14] to simulate the disorder in a ring of length L and finite width. The transfer term is taken as a constant t between first neighbors. The field effect is simply to change the boundary condition along the ring so that the transfer term gets a phase factor $\exp(2i\pi\phi/\phi_0)$ after one loop around the ring. Open boundary conditions are taken in the two other directions. The disorder is given by a random choice of the on-site energy between $-W/2$ and $W/2$. We are interested in the flux dependence of the total current $I(\phi)$ as a function of L, the number of channels M and the number of electrons N The current carried by each band is $i_n = \partial E_n / \partial \phi$. At

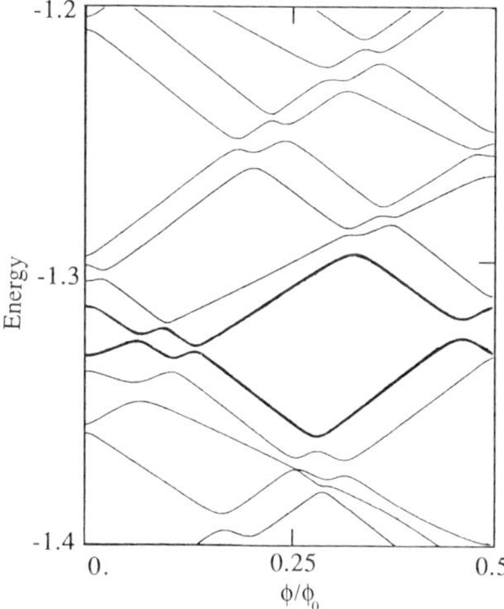

Figure 1. The effect of a weak disorder is to open small gaps at each crossing point of two energy levels $E(k_x(\phi), k_y, k_z)$ and $E(k'_x(\phi), k_y, k_z)$. $L = 50, M = 10, W = 0.2t$ in this figure. The currents carried by the levels 130 and 131 (shown in thick curves) are depicted in Fig. 2.

$T = 0\,\mathrm{K}$, the total current in a ring with N electrons is found from summation eq. (2). In the following we only discuss the currents at zero temperature. In order to investigate the diffusive regime $l_e \ll L \ll \xi$ and study scaling in this regime, we must have the localization length ξ as large as possible compared to the elastic mean free path l_e, since $\xi = M l_e$. This requires the study of systems with as many channels as possible. We have studied systems of size $L^* \sqrt{M}^* \sqrt{M}$ with L up to 128 and M up to 100. For such large systems we have used a program operating on a Cray-2 computer based on a Lanczos algorithm well adapted to the diagonalization of sparse matrices [15].

3   Weak Disorder

We first discuss the situation of very weak disorder. In this case, the only averaging possibility is on the number of electrons which may vary in different rings. We want to show that, in this case, the even harmonics survive to averaging as in the 1 D case. The energy levels can be deduced from the zero disorder levels $E(k_x(\phi), k_y, k_z)$ where $k_x(\phi), k_y$ and k_z are determined by the above boundary conditions. The effect of a weak disorder is to open small gaps at each crossing point of two energy levels $E[k_x(\phi), k_y, k_z]$ and $E[k'_x(\phi), k'_y, k'_z]$ (Fig. 1). The state n, defined as the n-th highest energy level, is thus related to different values of k_y and k_z when the flux varies. The current contribution of such a state $i_n(\phi) = -\partial E_n(\phi)/\partial\varphi$ presents a number of discontinuities $\Delta_i = \Delta(\phi_i)$ at the positions ϕ_i where level crossings occur, as shown on Fig. 2a. The maximum number of discontinuities is $2M$ in one period $[-\phi_0/2, \phi_0/2]$. One now considers the sum of the currents i_{n-1} and i_n carried by two neighboring bands. It is straightforward to notice that all the *negative* discontinuities in $i_n(\phi)$ are exactly compensated by all the *positive* discontinuities in $i_{n-1}(\phi)$ (Fig. 2a). When repeating this argument to each couple of

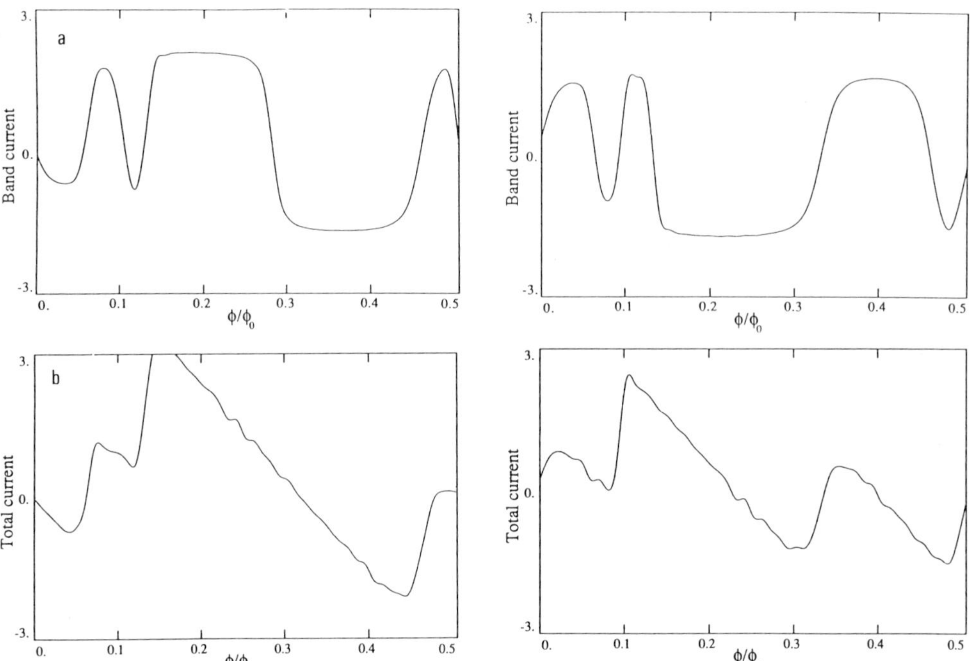

Figure 2. (a) Currents carried by the levels $n = 130$ and $n = 131$ shown in thick curves on Fig. 1. Negative discontinuities in $i_n(\phi)$ are equal to positive discontinuities in $i_{n-1}(\phi)$; (b) total currents $I_n(\phi)$ and $I_{n+1}(\phi)$ for $n = 130$. $I_N(\phi)$ strongly fluctuates with N but has only positive discontinuities.

energy levels, starting from the lowest energy level, it is possible to justify the shape of $I_n(\phi)$, the total current. There is no negative discontinuity left in $I_n(\phi)$. The amplitude and position of the discontinuities in $I_n(\phi)$ are identical to those of the positive ones in $i_n(\phi)$ (Fig. 2b). We have computed the average of $I_N(\phi)$ on a range ΔN of the number of electrons

$$I(\phi) = \frac{1}{\Delta N} \sum_{N-\Delta N/2}^{N+\Delta N/2} I_N(\phi), \tag{3}$$

with $1 \ll \Delta N < N$. This quantity is shown on Fig. 3. This is a periodic function of ϕ *with a period $\phi_0/2$ instead of ϕ_0* (and whose shape and amplitude are independent of M).

In this limit of very weak disorder, it is easy to prove this result analytically using a free electrons picture. In this case, the current of a state $i_n(\phi)$ in linear in ϕ between the discontinuities. The total current $I_N(\phi)$ has only positive discontinuities because of the compensation phenomenon depicted above. It has the following form in $[-\phi_0/2, \phi_0/2]$

$$I_N(\phi) = -\frac{Ne^2}{mL^2}\phi + \sum_i \frac{\Delta_i}{2}\mathrm{sgn}(\phi - \phi_i), \tag{4}$$

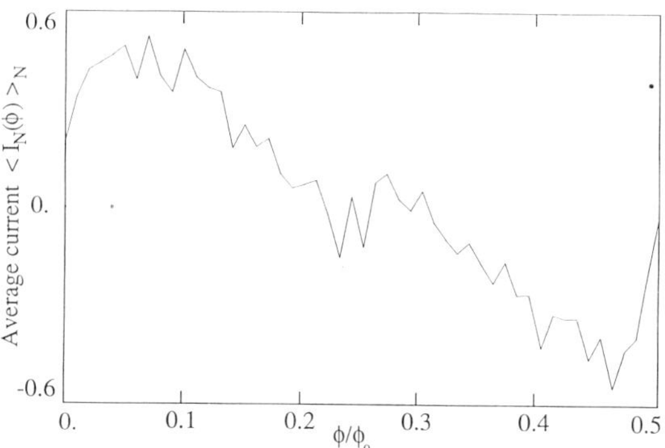

Figure 3. This curve is an average of $I_N(\phi)$ over $150 \leq N \leq 250$, in the range $[0, \phi_0/2]$. It has the period $\phi_0/2$. $L = 50, M = 10, W = 0.2t$.

where ϕ_i, Δ_i label now only *positive* discontinuities. This current can be expanded in a Fourier series whose coefficient of harmonics l yield

$$x_l = \frac{2}{l\pi} \sum_i \Delta_i \cos(2l\pi\phi_i/\phi_0). \tag{5}$$

The ϕ_i's depends on N, M and are randomly distributed excepted $\phi_i = 0$ and $\phi_i = \pm\phi_0/2$, because there is always a crossing point between levels of the same channel (same k_y and k_z), either at the center of zone or at the zone edge. This simply reflects the periodicity of the spectrum $E(\phi + \phi_0) = E(\phi)$. The ensemble average of the x_l, when N varies on a range $\Delta N \gg 1$, can be deduced

$$\langle x_l \rangle = \frac{2}{l\pi} \langle \Delta(0) + (-1)^l \Delta(\phi_0/2) \rangle. \tag{6}$$

In average, $\langle \Delta(0) \rangle = \langle \Delta(\phi_0/2) \rangle$ so that the ensemble average of $I_N(\phi)$ is a periodic function of ϕ with period $\phi_0/2$. It is written as

$$\begin{aligned} \langle I_N(\phi) \rangle &= \langle \Delta(0) \rangle \sum_{p=1}^{\infty} \frac{2}{\pi p} \sin(4p\pi\phi/\phi_0) \\ &= \langle \Delta(0) \rangle (\mathrm{sgn}(\phi) - 4\phi/\phi_0) \end{aligned} \tag{7}$$

in $(-\phi_0/2, \phi_0/2)$. Note that this result is only true for the *total* current, the single level has a zero ensemble average in contrast with the one channel case. The average discontinuity $\langle \Delta(0) \rangle$ is proportional to the average velocity along the x axis at the Fermi level, $\langle \Delta(0) \rangle = ev_F/L$.

As a result, the ensemble average of the current response in a multi-channel ring is just identical to the response of a single channel ring (with the same average Fermi energy). It can be also shown that the average value of odd harmonics when taken on rings with odd (even) number of electrons is also finite with a negative (positive) sign. This result is exactly equivalent to the one concerning the positive sign of the average second harmonic since the second harmonic of the current of a ring of length L with N electrons is identical to the first harmonic of the current of a ring of size $2L$ with $2N$ electrons.

At this stage it is important to note that our results are obtained when averaging over a great number of rings with a *fixed* number of electrons in each ring, this number being possibly different from one ring to another. Other authors [6,7] have assumed instead a fixed value μ of the chemical potential. The quantity they compute $I_\mu(\phi)$ is thus very different from $I_N(\phi)$ we have discussed. It is very easy to figure out that the average of $I_\mu(\phi)$ over the chemical potential is zero. Indeed, the level crossings occurring at $\phi = 0$ or $\phi = \pm\phi_0/2$ (see Fig. 2a) do not contribute to $I_\mu(\phi)$. This contribution is

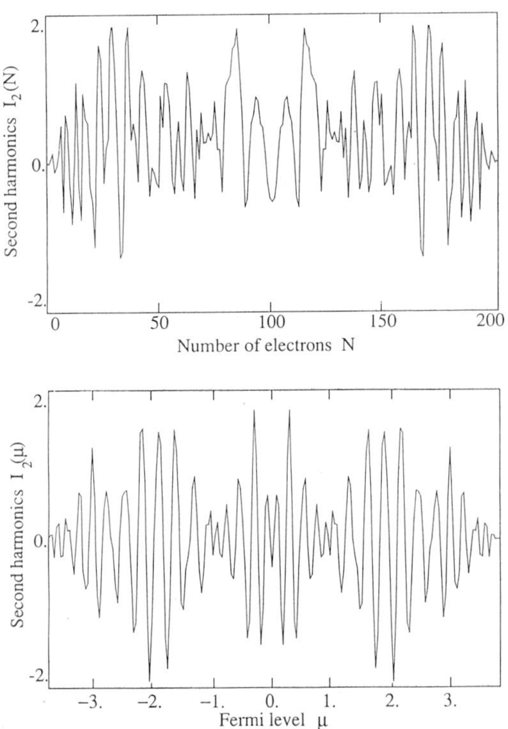

Figure 4. Second harmonics of the current $I_\mu(\phi)$ versus μ and $I_N(\phi)$ versus N for one configuration of disorder. $L = 20, M = 10, W = 2t$.

precisely at the origin of the non-zero average of the even harmonics of $I_N(\phi)$. Note that its amplitude corresponds to the current induced by one single electron at the Fermi energy. It is precisely of the order of magnitude of the difference between $\langle I_\mu(\phi)\rangle$ and $\langle I_N(\phi)\rangle$. Figure 4 shows the second harmonics versus chemical potential. It is seen that it averages to zero in contrast with the second harmonics of $I_N(\phi)$. This difference between the quantities $\langle I_N(\phi)\rangle_N$ and $\langle I_N(\phi)\rangle_\mu$ was already pointed out in ref. [4], for the analysis of ensemble averages in 1 D rings.

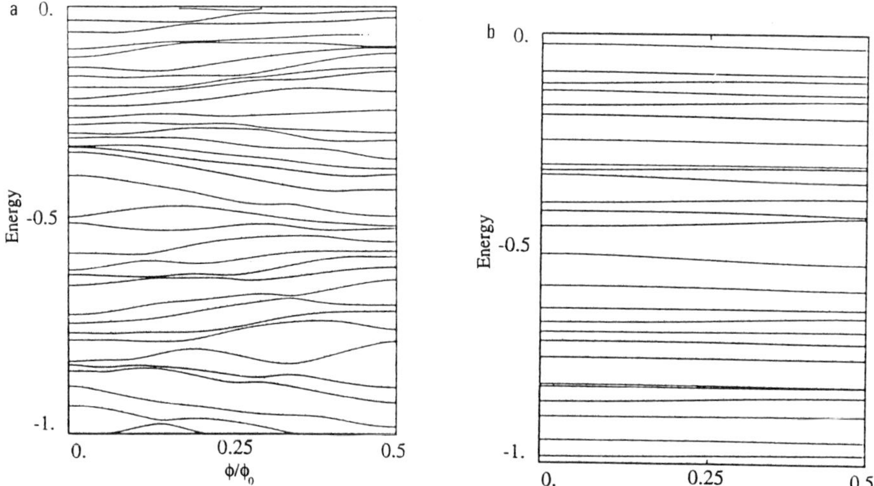

Figure 5. Evolution of the spectrum with increasing disorder. $L = 20, M = 10$, (a) $W = t$; (b) $W = 4t$.

4 Increasing Disorder: Diffusive Regime

It is easy to extend the result obtained above (in the limit of zero disorder), in the presence of a disordered potential whose variance W is smaller than the level spacing. When the disorder can be treated in first order of perturbation, the energy levels are only modified in the very vicinity of free electron level crossings in a symmetric way which preserves the compensation effect between current contributions of adjacent bands. The overall effect of the disorder is just to give a finite width to the current discontinuities discussed above and the conclusions concerning the ensemble average of the even harmonics are not modified. When increasing the disorder, it is not possible to extend the above arguments in a straightforward way and one has to rely on the results of numerical simulations. One can distinguish three different regimes: the ballistic, the localized and, between them, the diffusive regime. The energy spectra depicted in Fig. 5 illustrate the difference already pointed out in [16], between the diffusive regime, where the sensitivity of the energy levels to the boundary conditions E_c is larger than the level spacing η, and the localized regime, where E_c is smaller than η. When W increases, the levels repel each other, become flatter, and their current contributions decrease. The highest Fourier components vanish faster than the lower ones. In the following, we focus our discussion on the two first harmonics of the currents. For this purpose we have computed the quantities $i_n^{(1)} = \pi[E_n(0) - E_n(1/2)]/\phi_0$ and $i_n^{(2)} = \pi[E_n(0) - E_n(1/2) - 2E_n(1/4)]/\phi_0$ where $E_n(\phi/\phi_0)$ is the energy of a single electron. Since the typical values of the $\lambda_n^{(p)}$ decay faster than $1/p$, these expressions yield reasonable approximations of the two first harmonics of the currents. The total currents $I_N^{(1)}$ and $I_N^{(2)}$ are then obtained from summation on the number of electrons. Figure 6 shows the variation with N of the quantities $I_N^{(1)}$ and $I_N^{(2)}$ in the diffusive and localized regimes. We note that E_c, which is also of the order of the characteristic energy range between two successive maxima of the function $I_N^{(1)}$, strongly decreases with W. It has indeed been shown by Thouless that E_c is proportional to the conductivity of the system [16,17].

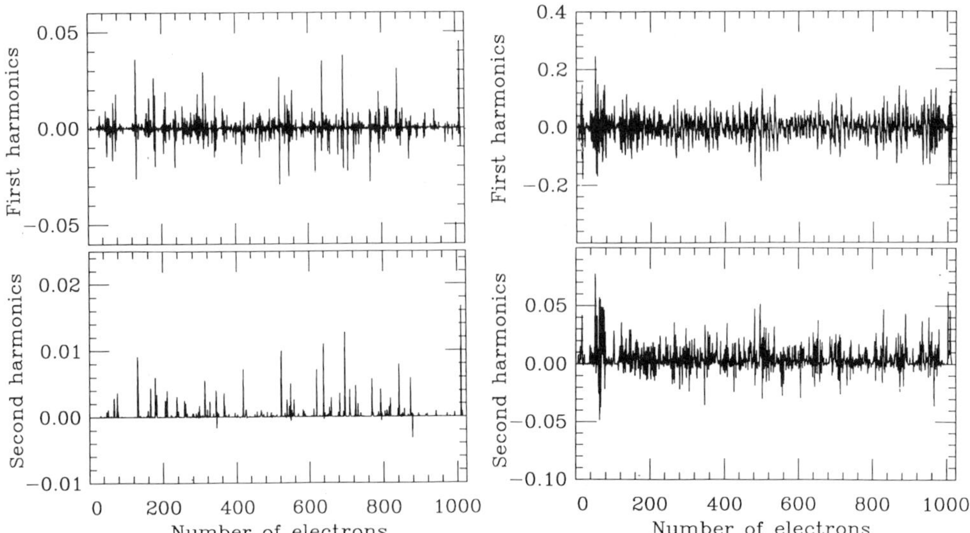

Figure 6. Variation with N, number of electrons in the band, of the first – $I_N^{(1)}$ – and second – $I_N^{(2)}$ – harmonics of the current, in the diffusive ($W/t = 3, left$), and in the localized regime ($W/t = 6, right$), for a size $64 \times 4 \times 4$.

It is seen that $\langle I_N^{(1)} \rangle_N$ is zero and $\langle I_N^{(2)} \rangle_N$ is finite and positive. $I_N^{(2)}$ takes less and less negative values when the disorder increases. Neglecting other harmonics, the average current $\langle I \rangle$ (which is periodic in $\phi_0/2$) is simply equal to $\langle I_N^{(2)} \rangle_N$.

We also discuss the typical values of the first and second harmonics defined as $I_{\text{typ}}^{(1)} = \sqrt{\langle (I_N^{(1)})^2 \rangle}$ and $I_{\text{typ}}^{(2)} = \sqrt{\langle (I_N^{(2)})^2 \rangle}$. Note that this definition is only meaningful in the diffusive regime where we find a normal distribution for the currents. On the other hand, in the localized regime we find a log-normal distribution which should be rather characterized by the variance of the logarithm of the current. In the following, we discuss the dependence of the quantities I_{typ} and $\langle I \rangle$ with the different parameters of the ring. If the averages presented in this work are made on the number of electrons, we have checked that averaging at the same time on the disorder configurations (D) and on the number of particles (N) lead to the same result $\langle \ldots \rangle_N = \langle \ldots \rangle_{N,D}$.

Figure 7 shows the typical first and second harmonics of the current versus disorder W for various lengths and number of channels. The three regimes are well visible in this picture. In the diffusive regime, both typical harmonics do not depend on the number of channels and vary as $1/W^2$ and thus as l_e (according to elementary scattering theory, $l_e \propto 1/W^2$) in agreement with the results of [7]. More precisely, using the relation $l_e = At^2/W^2$ and the 1 D estimation of the coefficient $A \approx 105$ [18], we obtain the following results: $I_{\text{typ}}^{(i)} = K^{(i)} I_0 l_e/L$, where $K^{(1)} = 0.25 \pm 0.02$ and $K^{(1)}/K^{(2)} = 3.5 \pm 0.3$. The ratio of these typical values is a universal number: independent of the dimensions of the ring. The onset of the localized regime shows up as a departure from this power law to an exponential decay. This cross-over occurs at a value of $1/W^2$ which increases linearly as L/M. We find that the cross-over occurs just when $L \approx \xi$ and $L \approx 2\xi$ respectively for the first and second harmonics.

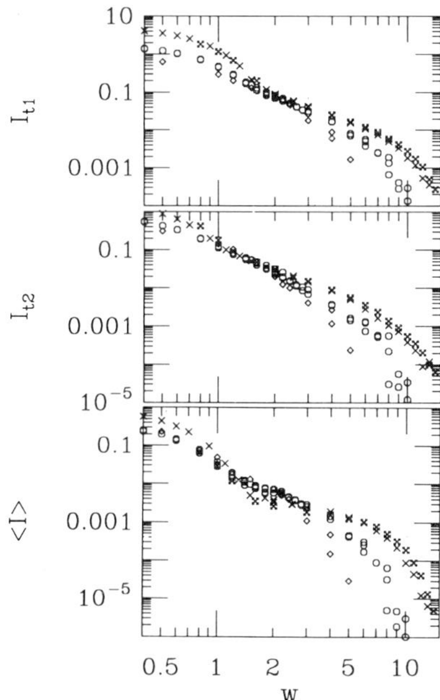

Figure 7. Typical first I_{t1} and second I_{t2} harmonics of the current and average current $\langle I \rangle$ versus disorder for various sizes of the ring $64 \times \sqrt{M} \times \sqrt{M}$. ($\Diamond$) $\sqrt{M} = 2$, ($\circ$) $\sqrt{M} = 4$, ($\times$) $\sqrt{M} = 8$, ($\times$) $\sqrt{M} = 10$. Note that in the diffusive regime, the typical currents decay as $1/W^2$ and that the average current decays as $1/W$.

The most striking and original result concerns the behavior of the average current $\langle I \rangle$. In the diffusive regime, we have found that it varies as $1/W^x$ where $x = 1 \pm 1$. It is thus also proportional to $\sqrt{l_e}$. Contrary to the typical current which is independent of M, it also decreases with the number of channels like $\sqrt{M}$. More generally Fig. 8 shows that the average current varies like $I_0 \sqrt{l_e/ML}$. This result is different from those of refs. [7,8] who performed averages on disorder at fixed chemical potential and found an exponential decay of $\langle I \rangle$ versus L/l_e. We find that the typical and average current can be related in the diffusive regime through

$$\langle I \rangle = C I_{\mathrm{typ}}/\sqrt{M_{\mathrm{eff}}}.$$

The dimensionless coefficient C is 0.2 ± 0.05. M_{eff} is the effective number of conducting channels which is reduced compared to the number of channels: $M_{\mathrm{eff}} = M l_e/L$ [7] .

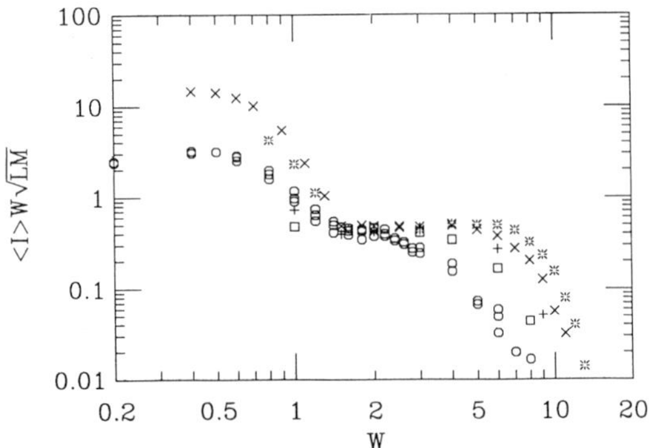

Figure 8. Plot of the quantity $\langle I \rangle^* \sqrt{ML^*W}$ versus disorder W (in t units) for various sizes of the ring. (o) $64 \times 4 \times 4$, ($\times$) $64 \times 8 \times 8$, ($\square$) $128 \times 8 \times 8$, ($*$) $64 \times 10 \times 10$, ($+$) $128 \times 10 \times 10$. Note that $\langle I \rangle^* \sqrt{ML^*W}$ is independent of W only in the diffusive regime whose extension in W corresponds to $l_e < L < Ml_e$.

5   Localized Regime: Perturbation Expansion

In the highly disordered regime, the eigenenergies can be developed in successive powers of t using the Brillouin-Wigner expansion. The perturbed energy of a state is given by

$$E_p = E_p^0 + \sum' t_{pq} t_{qr} \ldots t_{zp} / \left(E_p - E_q^0 \right) \ldots \left(E_p - E_z^0 \right), \tag{8}$$

where the summation Σ' is running over all the paths starting from p and returning to p, (without visiting p). We limit the following discussion to the case where $t < \min \left| E_j^0 - E_{j+1}^0 \right|$. The products $t_{pq\ldots zp}$ depend on the magnetic field through the phase factor $e^{2i\pi\phi/\phi_0}$ if the path $pq \ldots zp$ encloses the flux f. Otherwise they do not have any flux dependence. As a result, the first harmonics of $E_p(\phi) = \lambda_p^{(0)} + \lambda_p^{(1)} \cos \left(2\pi\phi/\phi_0 \right) + \lambda_p^{(2)} \cos \left(4\pi\phi/\phi_0 \right) + \ldots$ is of order t^L. We first consider a 1 D system for which the first harmonics can be simply written as

$$\lambda_p^{(1)} = \frac{1}{\prod_{q \neq p} \left(E_p^0 - E_q^0 \right)}. \tag{9}$$

The second harmonics $\lambda_p^{(2)}$ is obtained by expanding E_p in the denominator of eq. (8)

$$\lambda_p^{(2)} = - \left(\lambda_p^{(1)} \right)^2 \cdot \left(\sum_q \frac{1}{E_p^0 - E_q^0} \right), \tag{10}$$

$\lambda_p^{(2)}$ is proportional to the square of $\lambda_p^{(1)}$. From expression eq. (9) it is clear that the contribution to the persistent current is essentially related to pairs of levels p and $p+1$ which are close in energy. Assuming for all q distinct from p and $p+1$, $|E_q - E_p| \gg |E_p - E_{p+1}|$ and $|E_q - E_{p+1}| \gg |E_p - E_{p+1}|$, in that case $\lambda_{p+1}^{(1)} \approx -\lambda_p^{(1)}$ and

$$\lambda_{p+1}^{(2)} = -\lambda_p^{(2)} = - \left(\lambda_p^{(1)} \right)^2 / \left(E_{p+1} - E_p \right) < 0. \tag{11}$$

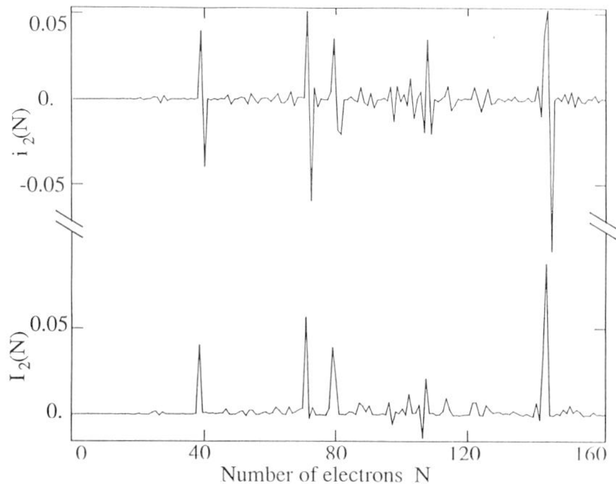

Figure 9. Second harmonics $i_2(N)$ and $I_2(N)$ of the band current and of the total current, respectively, for $L = 50, M = 10, W = 4t$.

The current contributions of such a pair of levels nearly cancel one another. The sign of their first harmonics depends on p. On the other hand, the sign of the second harmonics is always positive for the lowest energy level. As a result, the second harmonics of the total current I_N is positive for $N = p$ and nearly equal to zero for $N = p + 1$. This behaviour of pairs of energy levels close in energy is illustrated by the results of numerical simulations in Fig. 9. This result can be easily extended to a multi-channel ring. In that case, there are two kind of contributions in $\lambda_p^{(2)}$. The first one is related to paths enclosing two times the flux ϕ, the second one is obtained when expanding the energy denominators in the graphs C_{1p} enclosing only one time the flux ϕ. The first term has a random sign whereas the second one contains the sum of the squares of the diagrams contributing to $\lambda_p^{(1)}$. This sum gives rise to a positive contribution in the second harmonics of the total current I_2 for the same reasons than the one dimensional case. The averaged second harmonics (with respect to N) is thus positive but presents more fluctuations due to the multiplicity of diagrams with a random sign which also contribute to $\lambda^{(2)}$. The order of magnitude of $\langle I_2 \rangle_N$ is

$$\langle I_2 \rangle_N \sim \langle i_2 \rangle_N \sim \langle \lambda_p^{(2)} \rangle / \phi_0 \sim \langle (\lambda_p^{(1)})^2 \rangle / \phi_0 \eta, \tag{12}$$

where η is the level spacing. As shown in Fig. 10a, our numerical simulations agree well with this result. Since $I_0 \sim \eta M / \phi_0$, then

$$\langle I_2 \rangle_N \sim \frac{M \langle i_1^2 \rangle_N}{I_0}. \tag{13}$$

This last result has to be taken with a pinch of salt since in the localized regime as we already mentioned the distribution of the currents is log-normal. We find that eq. (12) is also valid in the whole diffusive regime. This result is in agreement with a recent derivation by Imry of the canonical average of the persistent current [19] (see Fig. 10b).

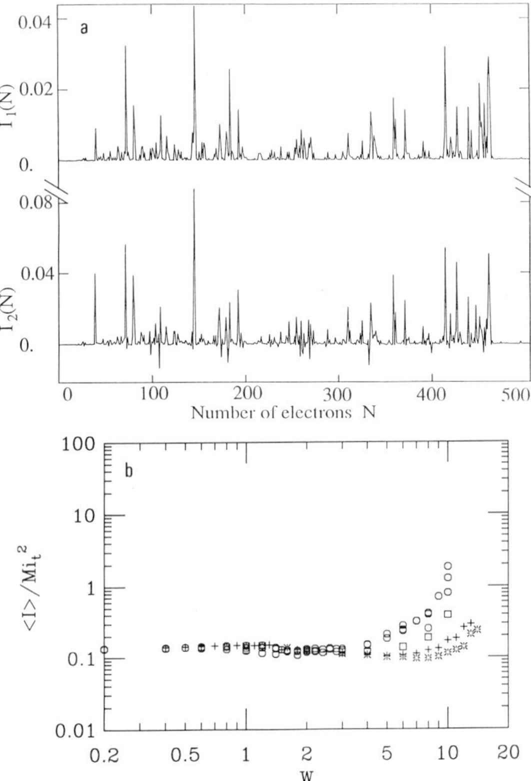

Figure 10. Second harmonic of the total current compared to the square of the first harmonic, (a) $L = 50$, $M = 10$ and $W = 4t$; (b) ratio $\langle I \rangle / M i_t^2$ versus the disorder for different sizes of the ring. (o) $64 \times 4 \times 4$, (+) $64 \times 8 \times 8$, (□) $128 \times 8 \times 8$, (∗) $64 \times 10 \times 10$.

6   Comparison with Experimental Results

We briefly describe the experimental observation by Levy et al. [17] of the persistent currents in an array of mesoscopic rings. The sample consists of 10^7 disconnected squares (0.55μ long) deposited by high resolution electron beam lithography on an insulating sapphire substrate. The magnetic moment induced by the persistent currents in the rings is measured by an ultra sensitive SQUID magnetometer. The magnetic flux applied to the rings is the sum of a DC part and an AC one oscillating at low frequency, $\Omega \sim 0.3\,\text{Hz}$. Since spurious contributions to the magnetic signal essentially give rise to a linear response, it has appeared very efficient to use the expected strong non-linear flux dependence of the persistent currents and to study the second and third harmonics response of the existing frequency Ω.

Indeed if we assume that the magnetic response of one ring is a periodic function of the total applied flux of period ϕ_p the magnetic response of the rings is expected to be $\mu(t) = \mu_a \sin 2\pi(\phi_c + \delta_{dc} \sin \Omega t)$ whose Fourier expansion is

$$\mu(t) = \sum_n \mu_{2n} \cos 2n\Omega t + \mu_{2n+1} \cos(2n + 1)\Omega t. \tag{14}$$

The Fourier amplitudes depend on the DC and AC amplitudes of the magnetic flux through $\mu_{2n} = 2\mu_a J_{2n}(\delta)\sin\theta$ and $\mu_{2n+1} = 2\mu_a J_{2n+1}(\delta)\cos\theta$ where

$$\delta = 2\pi\phi_{\mathrm{AC}}/\phi_p \quad \text{and} \quad \theta = 2\pi\phi_c/\phi_p. \tag{15}$$

The measured temperature and DC flux dependences of μ_2 and μ_3 are depicted in Fig. 11 and Fig. 12. One can see that according to eq. (15) both quantities are periodic functions of ϕ_{DC} which can be respectively fitted by a cosinus and a sinus whose period corresponds to half a flux quantum, $\phi_p = h/2e$, in a ring. This result is thus consistent with the existence of persistent currents in the rings, whose ensemble average has a $\phi_0/2$ periodicity.

The amplitude of the measured magnetic moment corresponds to $\mu_a = \mathcal{N}s\,j_m$ where s is the surface of one ring and $\mathcal{N} = 10^7$ is the total number of rings. The average current j_m expressed in $I_0 = ev_F/L$ units is equal to $3 \cdot 10^{-3}$.

Our numerical estimate yield on the other hand (see Fig. 8)

$$j_m = I_0\sqrt{\frac{l_e}{AML}} \approx 10^{-1}I_0\sqrt{\frac{l_e}{ML}},$$

where the estimated number of channels is $M = s/\lambda_F^2 \approx 7000$ (where s is the transverse section of the ring and λ_F the Fermi wavelength, the factor 2 compared to Fig. 8 comes from spin degeneracy). The number obtained, $j_m \approx 1.5 \cdot 10^{-4}I_0$, is an order of magnitude smaller than the experimental value.

However this numerical estimate strongly depends on the coefficient A in the relation between l_e and W only valid in principle for a 1 D ring. This may explain the roughness of the agreement.

We also note that, as recently suggested [20], some other physical mechanisms like electron-electron interactions [21] could also contribute to a $\phi_0/2$ periodic ensemble average of the persistent currents even in a *grand canonical system*. Experiments performed on a connected array of rings should enable us to distinguish between the different physical mechanisms.

7 Conclusion

We have studied the ensemble average of persistent currents in multi-channel rings at zero temperature and shown that it is periodic in $\phi_0/2$, i.e. it can be expanded in terms of the successive harmonics of $\sin(\pi\phi/\phi_0)$, all of them being positive. The current $\langle I_N(\phi)\rangle$ has been computed by fixing the number of electrons N in each sample and averaging with respect to N. Our results do not describe the same physical situation than the study of other authors [6-9] who have computed the current $\langle I_\mu(\phi)\rangle$ within the assumption of a fixed value of the chemical potential instead of maintaining constant the number of electrons. In the limit of weak disorder $\langle I_N(\phi)\rangle_N$ has been computed exactly and is found to depend only on the Fermi velocity and to be equal to the value in the 1 D ring independently of the number of channels. When increasing the disorder $\langle I_N(\phi)\rangle_N$ is

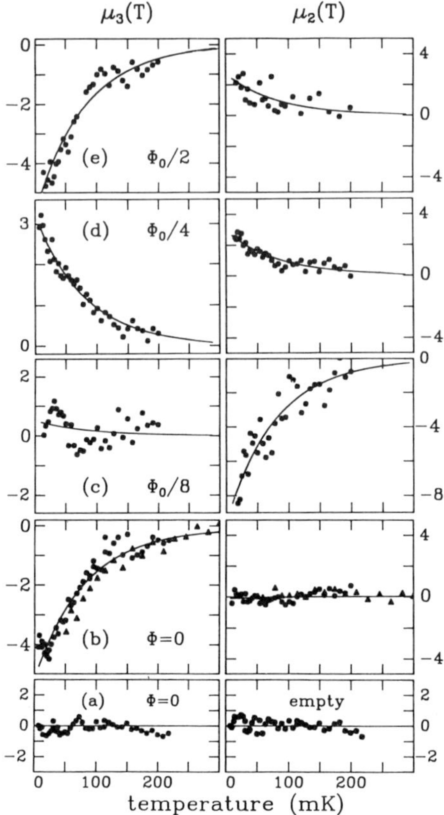

Figure 11. Temperature dependence of the second (μ_2, right column) and third (μ_3, left column) harmonics (plotted in arbitrary units) of the magnetization response using a 15G AC drive, (a) empty magnetometer, (b) sample with no DC field. The solid lines are one parameter fits to $\mu_{2,3}(0)\exp(-kT/E_c)$ adjusting only $\mu_{2,3}(0)$, where E_c is the computed Thouless energy $h v_F l_e/L^2$.

still independent of the disorder realization and $\langle I_N(\phi)\rangle_N \sim \langle I_N(\phi)\rangle_{N,D}$. In the metallic regime the average current varies as a power law of a mean free path

$$\langle I \rangle \sim I_0/\sqrt{l_e/LM}$$

and not exponentially as proposed by other authors. Contrary to the typical current it varies with the number of channels, as $\langle I \rangle \approx I_{\mathrm{typ}}/\sqrt{M_{\mathrm{eff}}}$ where M_{eff} is the number of conducting channels. In the localized regime $\langle I \rangle$ decays exponentially with L/ξ where ξ is the localization length.

The experimental discovery of the persistent currents in an assembly of mesoscopic copper rings has been recently achieved, with a $\phi_0/2$ periodicity as predicted in our work. This is a remarkable confirmation of the results obtained in this work. The amplitude of the measured average is in reasonably good agreement with our predictions in the diffusive regime.

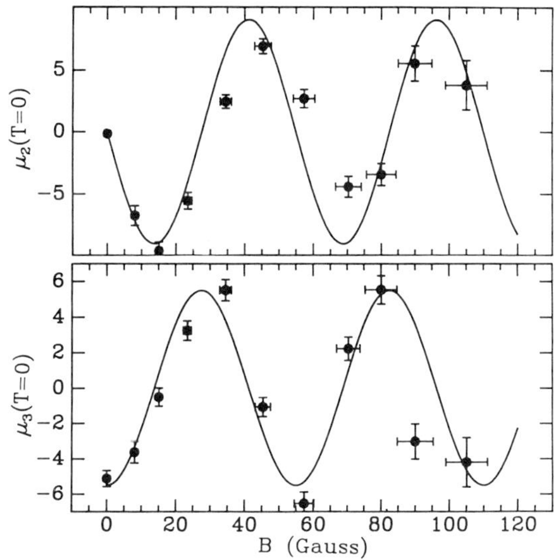

Figure 12. Dependence of the fitted second and third harmonics at $T = 0$ with DC field as detected with a 15 G amplitude AC field. 130 G correspond to one ϕ_0. The horizontal error bars arise from the uncertainties in the field compression induced by the superconducting shield. Vertical error bars are statistical.

8 Acknowledgements

The numerical simulations have been performed at the CRAY2 of the CCVR of the Ecole Polytechnique at Palaiseau. We acknowledge very fruitful and stimulating discussions with Y. Imry and E. Riedel.

References

[1] S. Washburn and R.A. Webb, Adv. in Phys. **35**, 395 (1986)

[2] M. Büttiker, Y. Imry, and R. Landauer, Phys. Rev. **A 96** , 365 (1983)

[3] R. Landauer and M. Büttiker, Phys. Rev. Lett. **54**, 2049 (1985); M. Büttiker, Phys. Rev. **B 32**, 1846 (1985)

[4] H.F. Cheung, Y. Gefen, E.K. Riedel, and W.H. Shih, Phys. Rev. **B 37**, 6050 (1988)

[5] N. Trivedi and D.A. Browne, Phys. Rev. **B 38**, 9581 (1988)

[6] H.F. Cheung, Y. Gefen, and E K. Riedel, IBM J. Res. Develop. **32**, 359 (1988)

[7] E.K. Riedel, H.F. Cheung, and Y. Gefen, Physica Scripta **25**, 357 (1989); H.F. Cheung, E.K. Riedel, and Y. Gefen, Phys. Rev. Lett. **62**, 587 (1989)

[8] D. DiVincenzo and M.P.A. Fisher, preprint

[9] O. Entin-Wohlmann and Y. Gefen, Europhys. Lett. **5**, 447 (1989)

[10] H. Bouchiat and G. Montambaux, J. Physique **50**, 2695 (1989)

[11] G. Montambaux, H. Bouchiat, D. Sigeti and R. Friesner (preprint)

[12] J.-L. Pichard and G.J. Sarma, J. Phys. **C 14**, L127, L617 (1981)

[13] L. Levy, G. Dolan, J. Dunsmuir and H. Bouchiat, Phys. Rev. Lett. **64**, 2074 (1990)

[14] P.W. Anderson, Phys. Rev.**109**, 1492 (1958)

[15] M. Friedrichs and R. Friesner, Phys. Rev.**B 57**, 303 (1988)

[16] J.T. Edwards and D.J. Thouless, J. Phys. **C 5**, 807 (1972)

[17] D.J. Thouless, Phys. Rep. **132**, 93 (1974); Phys. Rev. Lett. **18**, 1167 (1977)

[18] M. Kappus and M.J. Wegner, Z. Phys. **B 45**, 15 (1981); P.D. Kirkman and J.B. Pendry, J. Phys. **C 17**, 4327 (1984)

[19] Y. Imry, these volume

[20] V. Ambegaokar and U. Eckern, Phys. Rev. Lett. **65**, 381 (1990)

[21] B.L. Alt'shuler, D.E. Khmel'nitzkii and B.Z. Spivak, Sol. St. Commun. **48**, 841 (1983)

PERSISTENT CURRENTS IN MESOSCOPIC RINGS

Eberhard K. Riedel

Department of Physics
University of Washington
Seattle, WA 98195, USA

1   Introduction

The status of the field is excellent, much is in flux. For the first time we have experimental evidence [1] for persistent currents in small normal metal rings threaded by a magnetic flux. And there are new theoretical developments that indicate that in addition to the single-electron (or quasi-particle) contribution [2, 3] to the persistent current there are also collective ones [4].

I will place the experiment by Lévy, Dolan, Dunsmuir, and Bouchiat [1] at the center of my discussion and, proceeding more or less historically, ask three questions: What were the theoretical predictions (before the experiment)? What are the experimental results? What are the new questions?

In section 2, I summarize recent developments in theory and experiment of the persistent current problem [1-5]. In section 3, I develop the physical picture for the single-electron and collective contributions to the persistent current, including several new results, and discuss averaging questions. Then I will return to the experiment and its interpretation and end with proposing an experiment to sort out the different mechanisms for persistent currents. In the remainder of section 1, I present some basic facts about the effect.

Persistent currents in mesoscopic rings are an electronic quantum coherence effect. They are reminiscent of dia- and paramagnetism in aromatic molecules, but now in normal metal rings (e.g., Cu or Au) of micrometer circumference that are threaded by a magnetic flux. The key is the possibility of large coherence lengths at low temperatures.

That such persistent currents are real was first proposed for one-dimensional (1 D) rings by Büttiker, Imry, and Landauer [6] in 1983, and elaborated by others [7-9]. Discussion followed whether the effect would survive in multi-channel rings with disorder and finite temperature [10-12]. Formulating a Green's function approach, Cheung, Riedel, and Gefen [2] proved that temperature and disorder are much less disruptive than one might have thought. The amplitude and temperature sensitivity in the metallic region are governed by a Thouless correlation energy E_c rather than the level spacing. Lévy,

Quantum Coherence in Mesoscopic Systems
Edited by B. Kramer, Plenum Press, New York, 1991

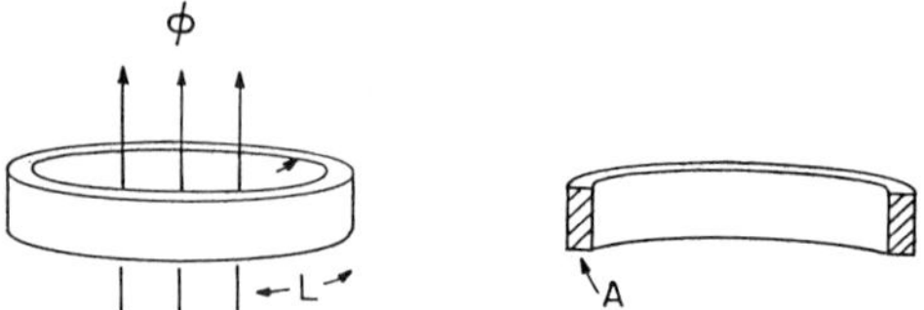

Figure 1. Thin-walled ring of circumference L and cross-sectional area A threaded by a magnetic flux ϕ.

Dolan, Dunsmuir, and Bouchiat [1] performed the first persistent current experiment on an ensemble of many rings [5]. Recently Ambegaokar and Eckern [4] pointed out that electron-electron interaction leads to an additional contribution to the persistent current. A mathematically similar problem occurs in the study of small SNS-junctions [13, 14], but these papers do not mention rings.

Consider a single thin-walled normal metal ring (e.g., Cu or Au) threaded by a magnetic flux ϕ, as shown in Fig. 1 [2]. The material is resistive and contains many impurities. I will concentrate on the diffusive region, which is defined by $\ell_{el} < L < \xi$ in terms of the ring circumference L, elastic mean free path ℓ_{el}, and localization length ξ. I will assume that the *inelastic* scattering length $\ell_{\text{inel}}(T)$, which sets the phase coherence length ℓ_φ (neglecting spin scattering), is of the order of or larger than the ring circumference, $\ell_{\text{inel}}(T) \sim \ell_\varphi > L$ (mesoscopic region). Temperature corrections due to Fermi statistics become important when the thermal diffusion length $\ell_T = (2\pi\hbar D/k_B T)^{1/2}$ becomes smaller than the ring circumference L. A ring with a finite cross-sectional area A contains many transverse channels (in momentum space), namely $M = Ak_F^2/4\pi$. I denote the level spacing at the Fermi surface by $\Delta_M \propto \Delta_1/M$, where $\Delta_1 = 2\pi\hbar v_F/L$ is the one-channel level spacing. A useful unit for the current is $I_0 = ev_F/L$.

Now suppose the ring encloses a magnetic flux ϕ. Using a gauge transformation one can eliminate the vector potential from the Schrödinger equation and incorporate it into the boundary conditions, which become flux periodic around the ring, $\psi_n(x + L) = \exp(2\pi i\phi/\phi_0)\,\psi_n(x)$ [15, 16]. Then the energy levels $E_n(\phi)$ are periodic functions of ϕ with period $\phi_0 = hc/e$. This is illustrated in Fig. 2 for different degrees of disorder in a ring with 4 channels. For the perfect ring one can distinguish the levels within each channel. Note the anti-correlations between the slopes of successive levels within the same channel and the correlations between slopes of levels from different channels. The details of the correlations depend also on the geometry of the rings [12].

Associated with each level is a single-level current given by $I_n = -c\partial E_n/\partial\phi$. A flux $\phi \neq$ integer $\times \phi_0/2$ breaks time reversal symmetry and makes possible a non-zero total current. The total current is the sum over all single-level currents weighted with a level occupation probability f,

$$I = \sum_n f(E_n)I_n. \tag{1}$$

More generally, it can be obtained from the thermodynamic potential Ω of the system by the relation [15, 16]

$$I = -c\frac{\partial\Omega}{\partial\phi} \tag{2}$$

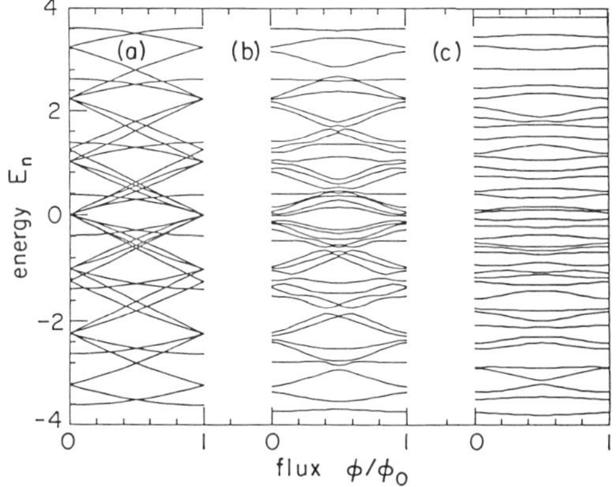

Figure 2. Energy levels as function of flux from the tight-binding model for a cylinder of size $(L, M) = (10.4)$ for (a) zero disorder and (b) $U_R = 0.04$, (c) $U_R = 0.1$ weak disorder, where $U_R = W(L/12M)^{1/2}/4\pi$ [3, 34].

This equation shows clearly that the quantum persistent current is not a transport current but an *equilibrium* property of rings threaded by a magnetic flux.

Though most intriguing the effect is rather small. For Cu or Au rings of dimensions $L = 1\,\mu\mathrm{m}$, $A = (0.02\,\mu\mathrm{m})^2$, $M \approx 5000$ channels, $\ell_{el} \approx 0.02\,\mu\mathrm{m}$, and $T < 50\,\mathrm{mK}$, we predicted [2] a typical persistent current amplitude of order $I^{(\mathrm{typ})} \approx 1.2 \times 10^{-2} I_0 \approx 3\,\mathrm{nA}$. Semiconductor rings can be fabricated with a few channels and elastic mean free paths exceeding the ring circumference [17]. Experimental arrangements can consist of a single ring [18, 19] or an ensemble of a large number of disconnected rings; the latter was used in the successful experiment by Lévy et al. [1].

2 Status of the Field

The status of the field is excellent. The new excitement is due to the experiment by Lévy et al. [1] and recent theoretical developments [2-5].

2.1 Single Particle Current

For multi-channel rings in the diffusive region, Cheung, Riedel, and Gefen [2] proposed that the quantum persistent current is due to quasi-particles diffusing phase coherently around the ring. The characteristic energy is

$$\hbar/\tau_D \propto E_c = \pi^2 \hbar D / L^2, \tag{3}$$

where $\tau_D = L^2/D$ is the time for diffusion around the ring ($D = v_F \ell_{el}/3$). The persistent current I in a ring is sample specific. We predicted the typical amplitude of that current via the rms average of I over all microscopic configurations of disorder,

$$I^{\mathrm{typ}} = \langle I^2 \rangle_D^{1/2} \propto \pm \frac{e}{\tau_D} \sin\left(\frac{2\pi\phi}{\phi_o}\right) + \text{higher harmonics, when } k_B T \ll E_c. \tag{4}$$

There are different ways of writing the amplitude,

$$\frac{e}{\tau_D} \propto \frac{E_c}{\Delta_1} I_0 \propto \frac{\ell_{el}}{L} I_0. \tag{5}$$

[1] The temperature corrections (due to Fermi statistics) are best expressed in terms of the thermal diffusion length ℓ_T. When $\ell_T < L$, or $k_B T > E_c$, the pth harmonic of the current amplitude is reduced by $\exp(-\pi p L/\ell_T)$, which amounts to a square root dependence on temperature in the exponent, $L/\ell_T \propto (k_B T/E_c)^{1/2}$. The range of correlations in the current-current correlation function as function of energy is E_c [2, 3].

For comparison, in the ballistic regime ($\ell_{el} > L$), the magnitude of the typical current (rms over the M single-channel currents) is $(e/\tau_{\text{ball}})M^{1/2} = I_0 M^{1/2}$, with $\tau_{\text{ball}} \approx L/v_F$, and, when $k_B T > \Delta_1$, the temperature reduction factor is $\exp[-2\pi^2 p(k_B T/\Delta_1)]$, with a linear temperature dependence in the exponent and the characteristic energy $\Delta_1/2\pi^2$ [12]. In the strongly localized regime ($L > \xi \propto M\ell_{el}$), the magnitude of the typical current is $(I_0/M)\exp(-L/2\xi)$ with different temperature reduction factors depending on the relative size of ℓ_T, ξ, and L [2, 3].

The result for the current amplitude eq. (4) in the diffusive regime shows several important features. It is independent of the number of channels. It is independent of filling (i.e., number of electrons N or chemical potential μ). It exhibits power law dependence ℓ_{el}/L on disorder, and a temperature dependence governed by the characteristic energy E_c (rather than Δ_M). More insight is gained by noting that

$$E_c \propto \Delta_M M_{\text{eff}}, \quad \text{with} \quad M_{\text{eff}} = M\frac{\ell_{el}}{L}, \tag{6}$$

where M_{eff} is referred to as the "active number" of channels [20]. Hence over the diffusive regime, $\ell_{el} < L < \xi \propto M\ell_{el}$, the correlation energy E_c decreases from Δ_1 to Δ_M. Though the expression eq. (4) looks like a single-level current it is not. It is the rms superposition of M_{eff} single-level currents, cf., eq. (10). In the diffusive regime the single-level currents depend on M and disorder. The temperature dependence may be explained by noting that, at $k_B T > E_c$, electrons are excited to higher levels that encompass ones with slopes of opposite sign, so that additional cancellations in the sum of single-level currents occur. In the strongly disordered regime the current is exponentially small, which reflects that there the electron wavefunctions overlap around the ring only via exponentially small tails, governed by the localization length ξ. Hence the sensitivity to changes in the flux is small.

Computing the typical current requires as input the disorder-averaged one- and two-particle Green's functions, $\langle G(k, k, E)\rangle_D$ and $\langle G(k, k, E)G(k', k', E')\rangle_D$ [2]. The diffusion and Cooperon poles, or the ladder and maximally crossed diagrams, determine $\langle I^2\rangle_D$ to leading order. Via those poles the correlation energy E_c and other energy and length scales enter. A discussion of the crossover from diffusive to strongly-localized behavior involves higher order diagrams that yield diffusion constant renormalization [3].

2.2 Sensitivity to Filling

Two other important questions are the harmonic content of the persistent current, $I = \sum_p I_p \sin(2\pi p\phi/\phi_0)$, and its sensitivity to the number of electrons in the ring, N. Interestingly the questions are related.

---

[1] The coefficient in eq. (4) is $4\sqrt{2}/\pi \approx 1.8$, including a factor 2 for spin.

Already for aromatic molecules one finds that whether the response is diamagnetic or paramagnetic depends on the number of participating electrons [21]. For the quantum persistent current in 1 D disordered rings Cheung and I found the following theorem [22]. The current of any single particle level is sufficient to cancel the sum of the currents of all previous levels. This means that the sign and amplitude of the total current (at zero temperature) are determined by that of the highest occupied level and the sign alternates with the number of electrons N in the ring. The reason is the anti-correlation in the slopes of successive energy levels.

This sensitivity to filling determines also the harmonic content. Start for simplicity with the current for a perfect 1 D loop, [8]

$$I = \sum_{p=1}^{\infty} \frac{2I_0}{p\pi} \cos(p\pi N) \sin\left(2\pi p \frac{\phi}{\phi_0}\right). \tag{7}$$

The amplitudes with odd p have different signs for even and odd N, whereas those with even p are always positive. Figure 3 shows the same feature; while the first harmonics have different signs (corresponding to paramagnetic and diamagnetic behaviors), the amplitudes of the second harmonics are positive in both cases. At finite disorder gaps open in the energy spectrum at zero and integer multiples of $\phi_0/2$ so that the sharp corners in the current-flux characteristics become rounded. For large disorder, defined by $\xi_1 < L$, where ξ_1 is the 1 D localization length, the current amplitudes are reduced by factors of $\exp[-p(L/2\xi_1)]$. Note also that $|I_p(L, N)| = |I_{p=1}(pL, pN)|$.

From the above results we drew the conclusion [8] that for an ensemble of many disconnected 1 D rings, where the number of electrons in each ring is fixed but varies randomly from ring to ring, the average current per ring contains only even harmonics. This is referred to as *period halving*. Furthermore, since the average over N of the second harmonic is positive, $\langle I_{p=2}\rangle_N = I_0 > 0$, the sum of the single-ring currents scales with the number of rings N_r, rather than $N_r^{1/2}$. (Note that $I_{p=2}$ as function of chemical potential μ has zero average, $\langle I_{p=2}\rangle_\mu = 0$ [8]).

From computer simulations Bouchiat and Montambaux inferred an analogous result for multi-channel rings [5]. Their work showed that it is a general feature that the second harmonics of the persistent current $I_{p=2}$ as function of filling N have non-zero averages

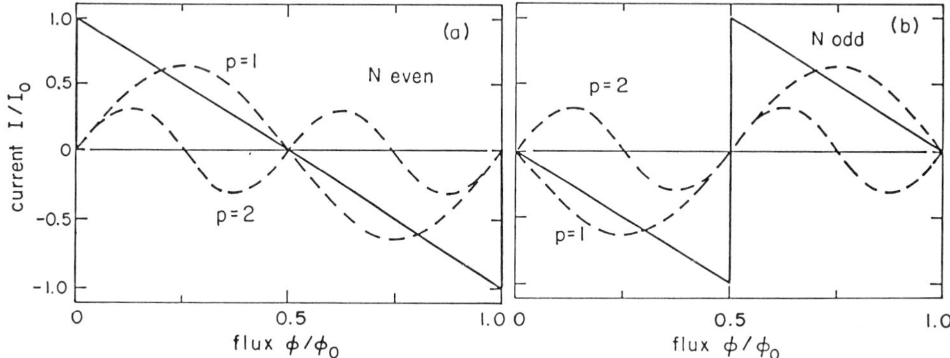

Figure 3. I-ϕ characteristic of a one-channel ring at zero temperature for (a) even and (b) odd number of electrons. Also shown are the first and second harmonics (dashed lines).

(over N or over N and disorder). The reason is that gaps from level crossings in the same band occur at $\phi = 0$ or $\pm\phi_0/2$ etc., regardless of the microscopic configuration of disorder, while gaps from other level crossings shift slightly from one configuration to another. As a consequence, for an ensemble of many multi-channel rings the sum of the (single-electron) currents is not proportional to $N_r^{1/2}\langle I^2\rangle_D^{1/2}$ but $N_r\langle I_{p=2}\rangle_{D,N}\sin(4\pi\phi/\phi_0)$ to leading order [23]. However, questions remained as to the average amplitude of the second harmonic in the different regimes, especially the diffusive one. For weak disorder Bouchiat and Montambaux [5] concluded $\langle I_{p=2}\rangle_N = I_0$ independent of M. That means also that for $M > 1$, $I_{p=2}$ is positive only on average and not for all N, as for $M = 1$. I will return to these questions in section 3.2.

2.3   Experimental Evidence

Lévy, Dolan, Dunsmuir, and Bouchiat presented the first experimental evidence for persistent currents in normal metal rings [1]. They studied a system of $N_r = 10^7$ isolated Cu rings, each of circumference $L \approx 2.2\,\mu m$, transverse dimension $A = 0.06 \times 0.03\,\mu m^2$, elastic mean free path $\ell_{el} \approx 0.02\,\mu m$, which corresponds to $2e/\tau_D \approx 7 \times 10^{-3} I_0 \approx 0.8\,nA$, and phase coherence length $\ell_\varphi > 2\,\mu m$ at $1.5\,K$. The experiment was performed by applying a flux $\phi_{\text{ext}} = \phi + \phi_{ac}\sin\omega t$, with both a *dc* and an *ac* component, and measuring the second and third harmonics $\mu_{2,3}(\phi)$ (with respect to the time dependence) of the total magnetic moment $\mu(t)$ of the ensemble of rings.

The results are as follows. (1) The $\mu_{2,3}(\phi)$ oscillate with period $\phi_0/2$, rather than ϕ_0 (Fig. 3 of [1]). (2) In terms of the moments of a single ring, the $\mu_{2,3}(\phi)$ scale proportional to the number of rings N_r, rather than $N_r^{1/2}$. (3) The temperature dependence is described by $\exp(-k_B T/\bar{E}_c)$ with $\bar{E}_c \approx 80\,mK$, rather than $\exp[-\text{const}(k_B T/E_c)^{1/2}]$. (Fig. 2 of [1].) (4) The tentative assignment of the sign of the effect is diamagnetic at low fields, $\mu_{2,3}(\phi_0/8) < 0$, rather than paramagnetic.

Qualitatively the results (1) and (2) are consistent with those expected for an ensemble of rings. However, as discussed below there are both single-particle and collective contributions to the persistent current and the latter may be larger in the present case. The exponential [2] $\exp[-\text{const}(k_B T/E_c)^{1/2}]$ also fits the temperature dependence of the data (footnote 12 of [1]). The tentative assignment of the sign of the current is opposite to that expected [5]. Spin-orbit scattering [24] has been offered as a possible explanation, [1] since the measured $\ell_{so} \approx 0.3\,\mu m \ll L$. This question is not understood yet. A detailed comparison between theory and experiment is presented in section 3.4.

2.4   Effects of Interaction

For electrons in a ring threaded by a magnetic flux electron-electron interaction leads to additional contributions to the persistent current with period $\phi_0/2$ [4]. I refer to those as collective or Cooperon contributions in contrast to the single-electron contributions discussed above. A mathematically analogous problem occurred in the study of SNS-junctions [14, 25].

Consider the standard Hartree and Fock diagrams to the thermodynamic potential Ω and redraw them in ring form, as shown in Fig. 4. Repeated elastic scattering in the particle-hole channel along time reversed paths leads to Cooperon correlations. A flux through the center of the ring then results in a contribution to Ω with periodicity $\phi_0/2$. The corresponding contribution to the persistent current $\langle I^{(c)}\rangle_n$ was calculated by

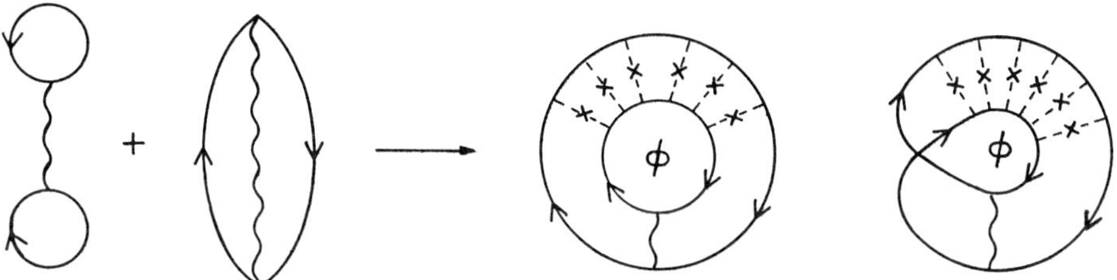

Figure 4. Hartree and Fock contributions to the grand potential Ω of electrons in a ring threaded by a flux ϕ. The wavy lines denote screened Coulomb interactions V while the dashed lines denote impurity scattering.

Ambegaokar and Eckern [4] via eq. (2), and is discussed in section 3.3. The sign of the effect can be positive (paramagnetic) or negative (diamagnetic) depending on the sign of the effective electron-electron interaction. Strong spin-orbit scattering neither changes the magnitude nor the sign of the current amplitude [26].

These effects of electron-electron interaction on the current are real, but it would be incorrect to ignore now the single-particle contributions, as in [4]. The question of the relative size of the various contributions to the persistent current needs to be answered. For an ensemble of rings the following contributions are periodic in $\phi_0/2$ and scale with the number of rings N_r.

(i) Single-electron contribution with an average amplitude $\langle I_{p=2}\rangle_{D,N} \propto (e/\tau_D)M_{\mathrm{eff}}^{-1/2}$, cf., eq. (13). The contribution is paramagnetic. The possibility of a sign change due to spin-orbit scattering has been mentioned but not yet proved.

(ii) Collective or Cooperon contribution due to electron-electron interactions with an amplitude $\langle I_{p=2}\rangle_D \propto (e/\tau_D)N(0)\bar{V}$, cf., eq. (14). This fluctuation contribution can be positive or negative depending on whether the effective electron-electron interaction is repulsive or (phonon induced) attractive.

Finally, for a single ring, the persistent current has a single-electron contribution containing ϕ_0 and $\phi_0/2$ periodic terms of amplitude $I_{p=1,2}^{\mathrm{typ}} \propto (e/\tau_D)$,cf., eq. (4), plus the collective contributions (ii).

3   Persistent Current. Theory and Experiment

Single and multi-ring experiments measure quantities that involve different averaging procedures. They are contrasted in sections 3.1 and 3.2 below. Effects of interactions are presented in section 3.3. The model is an Anderson model with random site disorder and an Aharonov-Bohm (AB) flux. That means, the electrons are assumed to move in a field-free space and a small self-inductance term is neglected. In section 3.4, theory and experiment are compared.

3.1   Typical Current Amplitude

First we consider a single isolated ring. The quantum persistent current is a sample specific property and, at fixed flux, depends on the number of electrons N in the ring

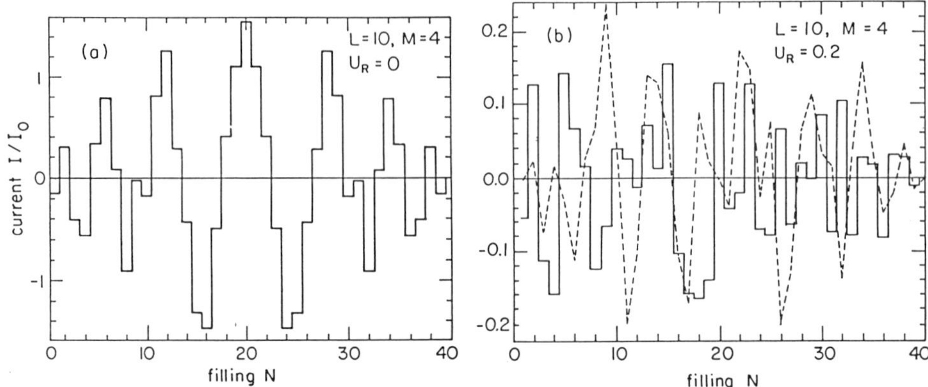

Figure 5. Persistent current I as function of filling N for (a) zero disorder $U_R = 0$ and (b) two microscopic configurations of disorder with $U_R = 0.2$, from the tight-binding model [3, 34]. The (average) period with which sign changes occur is $\propto M$ at zero disorder and $\propto M_{\text{eff}}$ in the diffusive regime.

(filling), the microscopic configuration of disorder, the number of channels,   etry, etc. [2, 3]. This sample specificity results from the *interference* of diffusing p.   es so that one sees scattering from particular impurity configurations rather than   mble averaged behavior.

Figure 5 shows the total current I as function of filling N [3]. For a p M-channel ring the current oscillates with period M, as shown in Fig. 5a. When one adds of the order of M electrons the sign of the current changes. This can be understood in terms of the correlations in the energy spectrum, see, e.g., Fig. 2a.

Figure 5b shows the analogous graph of current I versus filling N for a ring with disorder, for two microscopic configurations of impurities. Now the current I changes sign with an average period M_{eff} with fluctuations in the period of order of its size [2]. The trace of I versus filling is sample specific, it changes from configuration to configuration. Also the current amplitude is smaller for the disordered ring.

Figure 6 shows the average of I^2 over many such sample specific traces. It indicates that the rms average of I is independent of filling. It does not matter whether one calculates I^2 at constant N or μ and then averages. In either case, the typical amplitude of the sample specific (single-electron) current in an isolated ring is $\langle I^2 \rangle_D^{1/2}$, with the result shown in Fig. 6 and eqs. (4) and (5). Indeed the analytical work of [2] was performed at constant μ rather than N. [2] This insensitivity to the averaging process is lost when computing the average current, $\langle I(N) \rangle_D \neq \langle I(\mu) \rangle_D$ [8, 5]. Specifically, $\langle I(N) \rangle_D$ is given by eq. (13) and $\langle I(\mu) \rangle_D \propto I_0 \exp(-L/2\ell_{el})$ is exponentially small [2, 27]. The latter is *not* a physical quantity in these experiments [28].

Let us now interpret the results for the typical current in the "channel picture."

(1) For a one-channel ring, with or without disorder, the highest occupied level determines the magnitude and sign of the current (compare section 2.2). For weak disorder, $\xi_1 > L$, the typical current amplitude is $I^{\text{typ}} \propto I_0$, where $I_0 = ev_F/L = e/\tau_{\text{ball}}$. The temperature sensitivity is governed by Δ_1.

---

[2] The total current I as function of μ exhibits sign changes with an average period $E_c \propto M_{\text{eff}}\Delta_M$, cf., eq. (6).

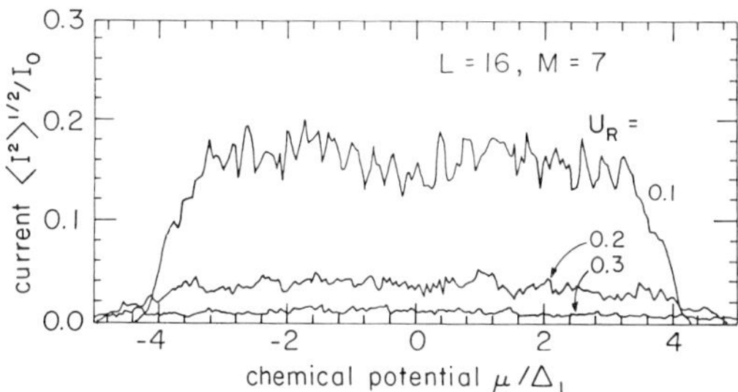

Figure 6. Typical current $\langle I^2 \rangle_D$ as function of filling μ/Δ_1 for three values of disorder $U_R = 0.1,\ 0.2,\ \text{and}\ 0.3$ [3, 34]. The root mean square average of I is independent of filling.

(2) For a perfect ring with M channels the last M levels determine the amplitude of the total current (see section 2.1),

$$I^{\text{typ}} \propto I_0 M^{1/2} = (e/\tau_{\text{ball}}) M^{1/2}. \tag{8}$$

Again the temperature sensitivity is governed by $\Delta_1 \propto M \Delta_M$. The reason is that for rings with height much smaller than ring circumference the currents of different channels are virtually uncorrelated [12]. Only the longitudinal or winding quantum number depends on flux. That is the physics behind what I call the channel picture.

(3) For an M-channel ring in the diffusive regime, a relation analogous to eq. (8) exists between the amplitudes of the typical total and typical single-level currents at constant flux, but now with $M \to M_{\text{eff}}$,

$$\langle I^2 \rangle_D^{1/2} \propto \langle I_n^2 \rangle_D^{1/2} M_{\text{eff}}^{1/2}. \tag{9}$$

The temperature sensitivity is now determined by $E_c \propto M_{\text{eff}} \Delta_M$. This is a remnant of the channel picture, however with an important difference. Since impurity scattering correlates the longitudinal and transverse degrees of freedom, the typical single-level current near the Fermi surface is now not only smaller than I_0 but also M dependent, [2, 3]

$$\langle I_n^2 \rangle_D^{1/2} \propto \pm I_0 \frac{1}{M^{1/2}} \left(\frac{\ell_{el}}{L} \right)^{1/2}, \tag{10}$$

at constant flux (e.g., $\phi/\phi_0 = 0.25$). This is how one obtains that $I^{\text{typ}} \propto e/\tau_D$ is both the superposition of M_{eff} single-level currents and independent of M. (The result can also be expressed [2] in terms of the dimensionless conductance g, since $M_{\text{eff}} \propto g$.)

Relation eq. (9) can be easily tested. It predicts that the ratio of the typical total and single-level currents is proportional to the square root of M_{eff} in the diffusive regime. As function of disorder, $\ell_{el} \propto 1/W^2$ in the diffusive regime in terms of the disorder parameter W in the Anderson model. Figure 7 shows numerical results for a small cylinder [3]. The current ratio is constant in the ballistic regime, then shows the

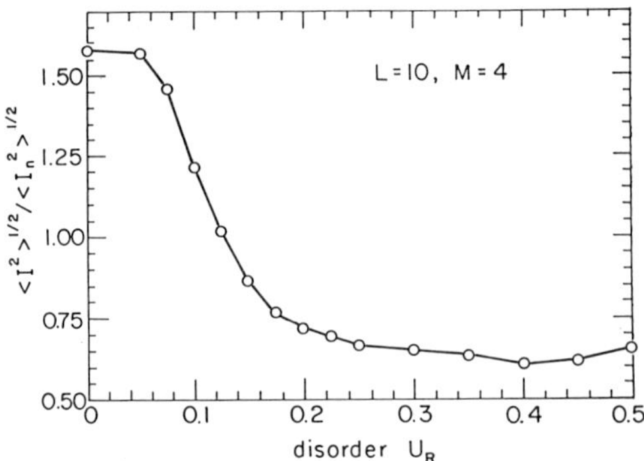

Figure 7. Ratio of the typical total and single-level currents $\langle I^2 \rangle_D^{1/2}/\langle I_n^2 \rangle_D^{1/2}$ as function of disorder U_R at constant flux $\phi/\phi_0 = 0.25$ for a small cylinder of dimensions $(L, M) = (10.4)$ [3, 34]. In the diffusive region the ratio falls off proportional to $M_{\text{eff}}^{1/2}$, cf., eq. (9).

$1/W$ drop-off in the diffusive regime, and approaches a non-zero constant value (≈ 0.6) for large disorder.

The discussion so far in this section concerned current amplitudes at constant flux. Strong disorder ($\xi < L$) and higher temperatures ($\ell_T < L$) both lead to an exponentially fast suppression of the higher Fourier coefficients in I^{typ}. However, in the diffusive regime at $T = 0$ many Fourier coefficients contribute, $I^{\text{typ}} \propto (e/\tau_D)[\sum_p p^{-3} \sin^2(2\pi p\phi/\phi_0)]^{1/2}$. One may also view this as a remnant of the channel picture.

3.2   Ensembles of Many Rings

To describe experiments on ensembles of many rings one needs to deal with the sample specificity in a different way [8, 5]. See also section 2.2.

Denote by $I^{(i)}$ the sample specific current of the i-th ring with fixed number of electrons N, disorder, number of channels, etc. Decompose it into Fourier components, $I^{(i)} = \sum_p I_p^{(i)} \sin(2\pi p\phi/\phi_0)$. Suppose the quantity measured is proportional to the sum of the currents of all N_r rings. How do they add up? Let us define the average current J per ring by

$$J = \sum_p J_p \sin\left(2\pi p\frac{\phi}{\phi_0}\right) = \sum_p \frac{1}{N_r}\sum_{i=1}^{N_r} I_p^{(i)} \sin\left(2\pi p\frac{\phi}{\phi_0}\right). \tag{11}$$

The sum $I_p^{(i)}$ over i presents the theorist with a complicated averaging problem, since both the filling and the configuration of impurities (disorder) differ from ring to ring.

The results of the previous section suggest that the average may be estimated as follows. Figure 8 exhibits schematically the coefficients of the first and second harmonic of the current in a single ring as function of N for three cases: (a) $M = 1$, no disorder; (b) M-channel, no or weak disorder; and (c) M-channel, diffusive regime. First, the I_p change sign as function of N (at fixed disorder and flux) on the scale of the number of channels or effective channels. Second, the second harmonics have a non-zero average over N, $\langle I_{p=2} \rangle_N > 0$, while the averages of the first harmonics vanish, $\langle I_{p=1} \rangle_N - 0$.

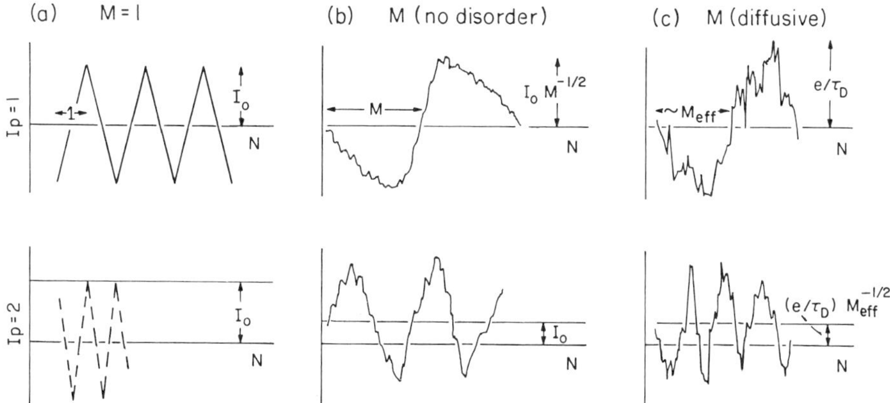

Figure 8. Comparison (schematically) of the first and second Fourier coefficients of the persistent current, $I_{p=1}$ and $I_{p=2}$, versus filling N for (a) $M = 1$, no disorder, (b) M-channel, no disorder, and (c) M-channel, diffusive regime. The typical amplitudes of I_p are proportional to I_0, $I_0 M^{1/2}$, and $(e/\tau_D) \propto I_0(\ell_{el}/L)$, respectively, while the averages are $\langle I_{p=1}\rangle_N = 0$, and $\langle I_{p=2}\rangle_N \propto I_0$, I_0, and $(e/\tau_D)M_{\text{eff}}^{-1/2}$, as indicated.

The typical amplitudes do not depend on N. Changing the microscopic configuration of disorder produces different curves, but with the same typical amplitudes, cf. Fig. 7. Therefore, in an experiment on an ensemble of very many rings, where the number of electrons N is fixed in each ring (with $N \gg M$) but varies randomly from ring to ring (within $\Delta N \gg M_{\text{eff}}$), one may approximate J of eq. (11) to leading order by

$$J \approx \langle I_{p=2}\rangle_{N,D} \sin\left(4\pi\frac{\phi}{\phi_0}\right) + ..., \quad \text{when} \quad \ell_T \gg L, \tag{12}$$

i.e., express it in terms of averages $\langle I_p\rangle_{N,D}$ for a *single* ring [5, 8]. Since in the diffusive regime the *average* period M_{eff} of sign change has large fluctuations, [2, 3] one may moreover replace $\langle I_p\rangle_{N,D}$ by $\langle I_p\rangle_N$ for a single configuration. Equation (12) gives the leading single-electron contribution to the average current per ring J in an ensemble of rings.

The magnitude of $\langle I_{p=2}\rangle_N$ in the diffusive regime was an open question [5]. However, for weak disorder Bouchiat and Montambaux [5] found $\langle I_{p=2}\rangle_N \propto I_0$, which I write as $(I_0 M^{1/2})M^{-1/2}$. Employing the correspondence relations suggested in section 2.1 and eqs. (9) and (10), i.e., $I_0 M^{1/2} \to e/\tau_D$ and $M \to M_{\text{eff}}$, I then conjecture that in the diffusive regime $\langle I_{p=2}\rangle_{N,D} \propto (e/\tau_D)M_{\text{eff}}^{-1/2} > 0$, or using eq. (5),

$$\langle I_{p=2}\rangle_{N,D} = C I_0 \frac{1}{M^{1/2}} \left(\frac{\ell_{el}}{L}\right)^{1/2}. \tag{13}$$

This result was obtained recently also by Montambaux et al., [29] who used large scale simulations. From their work one estimates $C = 0.1 \pm 0.02$, if one assumes $\ell_{el} \approx 105/W^2$ also for multi-channel rings and includes a factor 2 for spin.

This discussion and eq. (13) imply that, for a single ring at constant N, the ratios of the average current $\langle I(N)\rangle_D$ and the root mean square average current $\langle I^2(N)\rangle_D^{1/2}$ decrease like $1/M^{1/2}$, in the ballistic regime, and like $1/M_{\text{eff}}^{1/2}$, in the diffusive regime.

That the average is smaller than the fluctuation is a mesoscopic effect. It is interesting that the average current is proportional to the typical single-level current eq. (10) also in the diffusive region.

3.3   Interaction Effects

Finally, the collective or Cooperon contributions to the persistent current need to be considered. The study of the effects of electron interactions on weak and strong localization has a long history [30]. For a ring geometry something new happens, as mentioned in section 2.4.

The diagrams in Fig. 4 describe Cooperon fluctuations produced by electron-electron interactions. It is significant that this process involves interference between electrons on time reversed paths, as in the theory of weak localization. Therefore, the disorder average $\langle I^{(c)}\rangle_D$ is insensitive to the averaging issues discussed above and can be performed at constant μ [4]. The periodicity of these Cooperon contributions is $\phi_0/2$, in contrast to the AB period of ϕ_0 of single electrons circulating once around the ring.

The result by Ambegaokar and Eckern [4] for the average current in a single ring due to this collective or Cooperon effect, in the zero temperature limit ($\ell_T \gg L$), is

$$\langle I^{(c)}\rangle_D = 2N(0)\bar{V}\frac{4}{3\pi}\frac{\ell_{el}}{L}I_0 \sin\left(4\pi\frac{\phi}{\phi_0}\right) + \text{higher even harmonics.} \tag{14}$$

Here the screened Coulomb interaction was approximated by its angular average over the Fermi surface, which yields

$$2N(0)\bar{V} \approx a\ln[(1+a)/a], \text{ with } a = (q_{TF}/2k_F)^2, \tag{15}$$

in terms of the Thomas-Fermi wave number $q_{TF} = 8\pi e^2 N(0)$ and the density of states at the Fermi surface $N(0)$ (note that $q_{TF}^{-1} \ll \ell_{el}$). Perturbation theory to higher order in the electron-electron interaction leads to an expression similar to eq. (14) with an effective coupling λ [14]. The sign of the collective contribution eq. (14) to the persistent current is determined by the sign of the effective interaction. For a multi-ring experiment the contribution scales proportional to N_r. The temperature dependence is $\exp[-\text{const}\,(k_BT/E_c)^{1/2}]$, when $k_BT \gg E_c$, as for the single-electron current [2, 3]. Also here a linear temperature dependence in the exponential, $\exp(-k_BT/\bar{E}_c)$, is found to yield a good interpolation over a wide range of temperatures. [4].

3.4   Experiments

Now I will compare theory and experiment. I trust the orders of magnitude of the numbers but regard all figures as preliminary. I note that I do *not* discuss or account for the possibility that the measured persistent current effect may be diamagnetic (tentative assignment of sign in [1]) [31].

Lévy et al. [1] investigated an ensemble of 10^7 disconnected rings of circumference $L \approx 2.2\,\mu\text{m}$ and transverse area $A \approx 1.8 \times 10^{-5}\,\mu\text{m}^2$, with disorder characterized by $\ell_{el} \approx 0.02\,\mu\text{m}$, and at temperatures ranging from 7 to 300 mK.

(1) Consider a single Cu ring of the kind studied experimentally. With the above parameters, the theory by Cheung, Riedel, and Gefen [2] predicts a typical (single-electron) current amplitude of $I_{p=1}^{\text{typ}} \approx 0.6\,(\ell_{el}/L)I_0 \approx 6 \times 10^{-3}I_0 \approx 0.7\,\text{nA}$, where

$I_0 = ev_F/L \approx 0.12\,\mu\text{A}$, cf., eq. (4). The first and second harmonics are of the same order and differ only by a factor 0.35. Strong spin-orbit scattering would reduce the amplitudes by $1/2$ [26]. The correlation energy E_c of eq. (3) is $E_c/k_B \approx 16\,\text{mK}$ and the condition $\pi(L/\ell_T) \leq 1$ yields a characteristic temperature of about $10\,\text{mK}$.

(2) Lévy et al. [1] quote an experimentally observed magnitude for the ensemble averaged current of $J^{\text{exp}}_{p=2} \approx 3 \times 10^{-3} I_0 \approx 0.4\,\text{nA}$ per ring (within a factor of 2) and use the correlation energy, $\bar{E}_c/k_B = hD/(2L)^2 \approx 80\,\text{mK}$, to fit the temperature dependence of their data. (These authors use $d = 1$ in the definition of the diffusion constant D, so that their formulas for $\bar{\tau}_D$ and $\bar{E}_c$ differ from mine [32].) Although the measured current is roughly of the same magnitude as the above typical current it must not be identified with it. Let us assume that $J_{p=2}$ contains two contributions, a single-electron and a collective one, $J_{p=2} = J^{(s)}_{p=2} + J^{(c)}_{p=2}$. I identify the single-electron contribution with eqs. (12) and (13), i.e., $J^{(s)}_{p=2} = \langle I_{p=2}\rangle_{N,D}$, and the collective one with eq. (14), i.e., $J^{(c)}_{p=2} = \langle I^{(c)}_{p=2}\rangle_D$ [33].

(3) The average single-electron current $\langle I_{p=2}\rangle_{N,D}$ is roughly by a factor of $0.17/M_{\text{eff}}^{1/2}$ smaller than that of the typical current $I^{\text{typ}}_{p=1}$, where I take the proportionality constant from Montambaux et al. [29]. This yields $J^{(s)}_{p=2} \approx 6 \times 10^{-5} I_0 \approx 0.008\,\text{nA}$, which is by a factor 50 smaller than the experimental value of $0.4\,\text{nA}$ for the current per ring. I note that my estimate for $M_{\text{eff}} = M\ell_{el}/L \approx 250$ (using $M = Ak_F^2/4\pi \approx 25000$) is larger by π than that of [29]. Even if experimental effects would lead to a much reduced value, e.g., $M_{\text{eff}} \approx 80$, $J^{(s)}_{p=2}$ would still be only of order $0.01\,\text{nA}$, since $\langle I_{p=2}\rangle_{N,D}$ in eq. (13) depends on the square roots of ℓ_{el} and M.

(4) The magnitude of the collective contribution to the persistent current equals $J^{(c)}_{p=2} \approx 0.7\,[2N(0)\bar{V}]I^{\text{typ}}_{p=1}$, in terms of the typical (single-electron) current, cf., eqs. (14) and (15) derived by Ambegaokar and Eckern [4]. Using Thomas-Fermi screening, one estimates $2N(0)\bar{V} \approx 0.5$. This would yield $J^{(c)}_{p=2} \approx 2 \times 10^{-3} I_0 \approx 0.25\,\text{nA}$, which is within a factor of 2 of the experimental value $0.4\,\text{nA}$. However, higher order diagrams will reduce the theoretical number, and a very rough estimate is a factor of about $1/3$ (cf. Note added in proof). The largest experimental uncertainties are in ℓ_{el} and $N(0)\bar{V}$, which enter linearly into the theoretical expression for the current. The characteristic temperature $T^{(c)} = 3D\hbar/L^2 \approx 50\,\text{mK}$ is of the same order as that used by Lévy et al. [4].

From the above discussion I draw tentatively the *conclusion* that for experiments on ensembles of rings, as performed by Lévy et al., [1] the collective contribution to the persistent current wins, as long as the number of active channels in the rings is large. In contrast, experiments on isolated single rings will show both ϕ_0-periodic and $(\phi_0/2)$-periodic persistent currents of comparable magnitude (I^{typ}), with the collective contribution to the $(\phi_0/2)$-periodic term being *smaller* than the single-particle one by about $0.25\,[2N(0)\bar{V}]$, already to lowest order perturbation theory. I note that this discussion applies to thin-walled short cylinders, for which the height is much smaller than the circumference.

The single-electron and collective persistent current effects can occur in the same system, but these two effects arise from different interference mechanisms. Can they be separated experimentally? Here one can employ to advantage the fact that at constant μ

all Fourier coefficients of $\langle I(\mu)\rangle_D$ are exponentially small [2, 27]. Hence one can suppress the single-particle persistent current by working either with an ensemble of many rings that are connected and held at constant chemical potential or with a tall thin-walled cylinder of height much larger than L, and at temperatures such that $\ell_\varphi(T)$ is greater than L but much smaller than the height.

Finally, it would be interesting to test the sign of the collective or Cooperon contribution by replacing Cu with a material that exhibits phonon induced *attractive* interactions, i.e., 'para' superconducting fluctuations. In that case one expects a diamagnetic collective contribution to the persistent current.

Note added in proof. An analytical calculation of the coefficient C in eq. (13) yields $C = 4d^{1/2}/9\pi^{3/2} \approx 0.08$ (with $d = 1$) [26], including a factor 2 for spin, which is slightly smaller than the numerical one. Spin-orbit scattering does not change the sign of this term but only reduces its amplitude; it is not included in the estimates below. The effect of higher order diagrams on the collective contribution to the persistent current $J_{p=2}^{(c)}$ is more drastic than mentioned in the text [35]; it yields a reduction factor of $\approx 1/10$ rather than $1/3$. Since in the experiment by Lévy et al [1] ℓ_{el} is comparable to the transverse dimensions of the rings, I prefer to treat the system as exhibiting quasi-1 D diffusion, i,e. use $d = 1$ in the diffusion constant D. Also, the transverse dimensions are slightly smaller than originally stated [29], which leads to a lower estimate for the effective number of channels, $M_{\text{eff}} \approx 170$. Incorporating these changes, I propose the revised theoretical estimates $I_{p=1}^{\text{typ}} \approx 1 \times 10^{-2} I_0 \approx 1.2\,\text{nA}$, $J_{p=2}^{(s)} \approx 6 \times 10^{-5} I_0 \approx 0.007\,\text{nA}$, and $J_{p=2}^{(c)} \approx 7 \times 10^{-4} I_0 \approx 0.09\,\text{nA}$ (with $2N(0)\lambda \approx 0.06$) versus the quoted experimental result of $J_{p=2}^{\text{exp}} = J_{p=2}^{(s)} + J_{p=2}^{(c)} \approx (-)0.4\,\text{nA}$. The characteristic temperature is $\hbar/\tau_D = \hbar D/L^2 \approx 50\,\text{mK}$, with $d = 1$. The conclusions concerning the relative size of the single-electron and collective contributions to the current remain unchanged. However, the theoretical estimate for $J_{p=2}^{(c)}$ agrees with the experimental result only to within a factor of 4 or 5.

4   Acknowledgements

I enjoyed stimulating contacts with many people. I am particularly grateful to my collaborators H.F. Cheung, and Y. Gefen. I thank L.P. Lévy and colleagues, V. Ambegaokar and U. Eckern, and G. Montambaux and H. Bouchiat for sending me copies of their papers prior to publication. I acknowledge with thanks stimulating conversations concerning aspects of these recent works with V. Ambegaokar, H. Bouchiat, Y. Gefen, L.P. Lévy, P.A. Mello, P. Montambaux, B.Z. Spivak, D.J. Thouless, F. von Oppen, and others. I thank S. Fain and F. von Oppen for comments on the manuscript. I am most grateful to B. Kramer for organizing and inviting me to this most stimulating Study Institute. This work was supported in part by the U.S. National Science Foundation under Grant No. DMR-88-13083.

References

[1] L.P. Lévy, G. Dolan, J. Dunsmuir, and H. Bouchiat, Phys. Rev. Lett. **64**, 2074 (1990)

[2] H.F. Cheung, E.K. Riedel, and Y. Gefen, Phys. Rev. Lett. **62**, 587 (1989); E.K. Riedel, H.F. Cheung, and Y. Gefen, Physica Scripta **T25**, 357 (1989)

[3] H.F. Cheung and E.K. Riedel, unpublished

[4] V. Ambegaokar and U. Eckern, Phys. Rev. Lett. **65**, 381 (1990)

[5] H. Bouchiat and G. Montambaux, J. Phys. (France) **50**, 2695 (1989)

[6] M. Büttiker, Y. Imry, and R. Landauer, Phys. Lett. **A 96**, 365 (1983)

[7] M. Büttiker, Phys. Rev. **B 32**, 1846 (1985); R. Landauer and M. Büttiker, Phys. Rev. Lett. **54**, 2049 (1985)

[8] H.F. Cheung, Y. Gefen, E.K. Riedel, and W.H. Shih, Phys. Rev. **B 37**, 6050 (1988)

[9] N. Trivedi and D.A. Browne, Phys. Rev. **B 38**, 9581 (1988)

[10] S. Washburn and R.A. Webb, Adv. Phys. **35**, 375 (1986)

[11] Y. Imry, in: Directions in Condensed Matter Physics, G. Grinstein and G. Mazenko, eds., p. 101, World Scientific, Singapore, (1986)

[12] H.F. Cheung, Y. Gefen, and E.K. Riedel, IBM J. Res. Develop. **32**, 359 (1988)

[13] B.L. Altshuler and B.Z. Spivak, Zh. Eksp. Teor. Fiz. **92**, 609 (1987) [Sov. Phys. JETP **65**, 343 (1987)]

[14] B.L. Altshuler, D.E. Khmel'nitskii, and B.Z. Spivak, Sol. St. Commun. **48**, 841 (1983)

[15] N. Byers and C.N. Yang, Phys. Rev. Lett. **7**, 46 (1961)

[16] F. Bloch, Phys. Rev. **B 2**, 109 (1970)

[17] Such rings would also allow the investigation of persistent currents carried by edge states [X.C. Xie and E.K. Riedel, Bull. Am. Phys. Soc. **34**, 547 (1989)]

[18] R.A. Webb, private communication

[19] A. Benoit, this volume

[20] Y. Imry, Europhys. Lett. **1**, 249 (1986)

[21] G. Kirczenow, Phys. Rev. **B 32**, 7952 (1985)

[22] H.F. Cheung and E.K. Riedel, Phys. Rev. **B 40**, 9498 (1989)

[23] I retract a remark about ensembles of rings [in [2], p. 590] that said otherwise

[24] See, e.g., Y. Meir, Y. Gefen, and O. Entin-Wohlman, Phys. Rev. Lett. **63**, 798 (1989)

[25] I learned about this interaction effect in rings first from B.Z. Spivak [private communication (February 1990)], who referred me to [14]

[26] F. von Oppen and E.K. Riedel, unpublished

[27] O. Entin-Wohlman and Y.Gefen, Europhys. Lett. **8**, 477 (1989)

[28] There is some confusion on this point, e.g. [5]

[29] G. Montambaux, H. Bouchiat, D. Signetti, and R. Friesner, preprint (April 1990); and this volume

[30] P.A. Lee and T.V. Ramakrishnan, Rev. Mod. Phys. **57**, 287 (1985), and references therein.

[31] The effect of spin-orbit scattering on $\langle I_{p=2}\rangle_{N,D}$ in eq. (13) is not known, though a factor $(-1/2)$ has been mentioned in [1]. There is no sign change in the collective contribution (14) due to spin-orbit scattering [26]. However, a diamagnetic collective current contribution occurs when the effective electron-electron interaction is attractive

[32] In deriving eqs. (4) and (14) one can replace the Green's functions by their standard bulk forms (except for discrete momentum variables) and consequently I use $d = 3$ in the diffusion constant D. The overall d-dependence is $1/d^{1/2}$ in eq. (4) and $1/d$ in eq. (14). The coefficients are model dependent, while the powers of ℓ_{el}/L, L/ℓ_T, etc. are expected to be universal

[33] I assume that all 10^7 rings contribute uniformly to the total measured signal

[34] Disorder is characterized by $U_R = W(L/12M)^{1/2}/4\pi$, where W is the disorder parameter in the Anderson model with random site disorder. The hopping matrix element is set equal to 1

[35] I thank B.L. Altshuler for a discussion of this point

CHAPTER 6

QUANTUM TRANSPORT THEORY

RIGOROUS PERTURBATION RESULTS IN THE THEORY OF MESOSCOPIC FLUCTUATIONS: DISTRIBUTION FUNCTIONS AND TIME-DEPENDENT PHENOMENA

Igor V. Lerner

Institute of Spectroscopy
USSR Academy of Sciences
142092 Troitsk, Moscow Region, USSR

1 Basic Results of a Simple Perturbation Theory

During five years after the theoretical prediction [1-3] and the experimental discovery [4-9] of conductance fluctuations in small ('mesoscopic') samples, the 'mesoscopics' became quite a wide field where a lot of experimental results were obtained and a lot of theoretical approaches were developed. Many of them shall be presented in a wide spectrum of contributions to our meeting. In my lectures I will try to review the theoretical results obtained in the framework of rigorous perturbative calculations. By perturbative ones, I mean not only calculations performed to the first few orders but also those including summation of the leading infinite series.

The mesoscopic phenomena which are associated, first of all, with 'magnetofingerprints' and other 'ID's' of specific samples provide, side by side with weak localization effects, another example of revealing quantum interference macroscopically. I would like to analyze these two types of interference effects and concentrate on situations of their simultaneous revealing.

In this first section the basic results obtained in the lowest orders of perturbation theory [2, 3, 10] will be described. The next section concerns the distribution of conductance [11, 12] and current density [13, 14] fluctuations. Their description requires summation of the infinite series which has been performed via renormalization group analysis of the effective field theory [11], [15–17]. The last section treats the time-dependent phenomena, namely, relaxation processes in disordered systems [18, 19] which are tightly connected to mesoscopic effects. The basic model for a perturbational treatment is the model of free electrons in a random potential which is equivalent for large distances to the Anderson tight-binding model with either diagonal or off-diagonal

Quantum Coherence in Mesoscopic Systems
Edited by B. Kramer, Plenum Press, New York, 1991

disorder. The Hamiltonian is given by[1]

$$\hat{H} = \frac{\hat{p}^2}{2} - \epsilon_F + V(\mathbf{r}). \tag{1}$$

Here the random potential $V(\mathbf{r})$ is chosen to be Gaussian with zero average and short-range correlator

$$\langle V(\mathbf{r})V(\mathbf{r}')\rangle = \frac{1}{\pi\nu_0\tau}\delta(\mathbf{r} - \mathbf{r}'),$$

where ν_0 is the mean density of states, $\tau = l/v_F$, v_F is the Fermi velocity, and l is the length of the mean free path between elastic scatterings. The small parameter of the perturbation theory is the parameter of weak disorder

$$(\epsilon_F\tau)^{-1} \ll 1. \tag{2}$$

So, under consideration is the metallic side of the Anderson transition in $d > 2$ dimensions, or the region of weak localization in $d = 2$.

In this first section I would like to demonstrate what is possible to obtain based on a simple perturbational analysis in lowest order. A point to start is to recall the basic

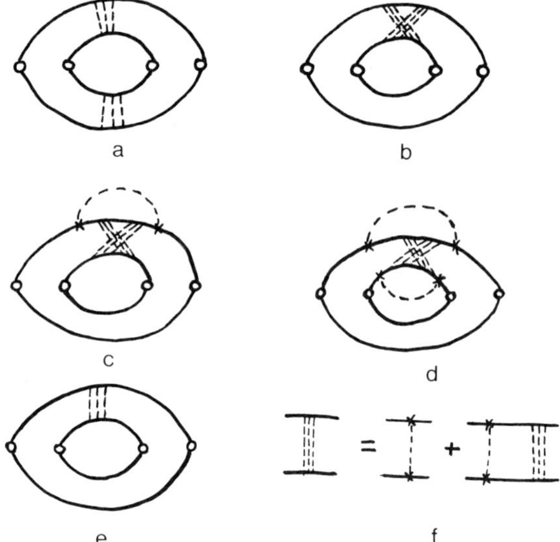

Figure 1. Diagrams for the variance of conductance fluctuations. Solid lines stand for the Green functions, dashed lines for the correlator of the impurity potential giving a factor $(2\pi\nu_0\tau)^{-1}$. Circles are for external vertices (spatial derivatives) entering the definition of conductance in terms of the Green functions, triple lines for diffusons or Cooperons are defined in (f), (a–d) give the expression eq. (3) for the variance, while (e) gives a small correction to it.

---

[1]Here and elsewhere in intermediate expressions the atomic system $\hbar = e = m$ is used.

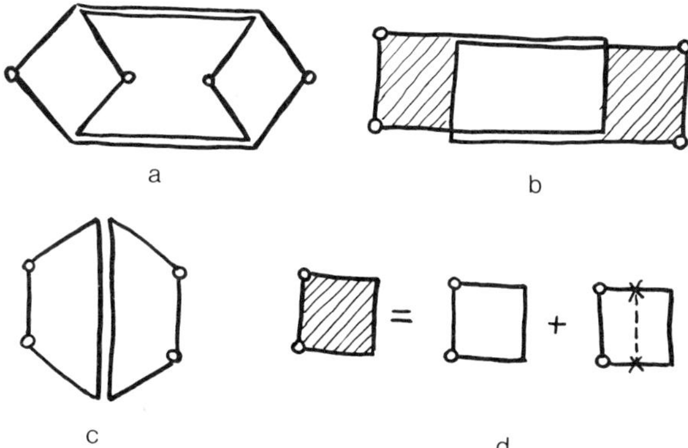

Figure 2. Diagrams for the variance in the Hikami representation. Double lines correspond to diffusons or Cooperons defined in Fig. 1f. Diagram (a) corresponds to that in Fig. 1a, (b) corresponds to the sum of the diagrams in Fig. 1b-d, (c) to Fig. 1e.

results obtained in this way by Altshuler [2] and Lee and Stone [3]. The variance of conductance sample-to-sample fluctuations $\delta G \equiv G - \langle G \rangle$ has been obtained as follows

$$\langle\!\langle G \rangle\!\rangle \equiv \left\langle \left(\delta G\right)^2 \right\rangle = A\left(e^2/h\right)^2, \tag{3}$$

where $\langle\!\langle \ldots \rangle\!\rangle$ stand for irreducible averages (cumulants). It turned out to be a universal quantity which gave rise for a commonly used term for this phenomenon, namely *universal conductance fluctuations* (UCF). The numerical factor A depends slightly on the geometrical shape of the sample. Here it is useful to explain the origin of this dependence which gives also an opportunity to discuss the fundamentals of the perturbative approach in the framework of the standard Green functions technique.

All the diagrams which contribute [3, 10] to eq. (3) are given in Fig. 1. These diagrams are just irreducible (connected) averages of two current loops, each of the loops representing the local conductivity $\sigma(\mathbf{r}, \mathbf{r}')$ which gives the conductance G after integrating over both $\mathbf{r}$ and $\mathbf{r}'$

$$G_{\alpha,\beta}(\omega) = \frac{1}{L^2} \int \sigma_{\alpha\beta}(\mathbf{r}, \mathbf{r}'; \omega) d^d r d^d r'. \tag{4}$$

The averages in the diagrams are taken to the lowest order in the disorder parameter eq. (2). Contributions of higher orders in this parameter may be obtained by adding more 'diffuson' or 'Cooperon' ladders.

For our analysis it is more convenient to redraw (Fig. 2) all of the diagrams into the Hikami representation [20] where diffusons and Cooperons are separated from the polygons made of the electron Green functions (see also [21]). A diffuson (or Cooperon) is given in frequency-momentum space by

$$D(q, \omega) = \frac{1}{2\pi\nu_0\tau^2} \frac{1}{Dq^2 - i\omega}, \tag{5}$$

where the bare value of the diffuson coefficient is $D = v_F l/d$. The diffuson eq. (5) depends on small momentum $q \lesssim l^{-1}$ and frequency $\omega \lesssim \tau^{-1}$. Such a dependence gives only minor corrections to the 'polygons' which are contributed by momentum $p \sim p_F \gg l^{-1}$ and energy $\varepsilon = \varepsilon_F \pm \omega/2 \approx \varepsilon_F \gg \tau^{-1}$. Due to this, a calculation of polygons in typical diagrams of any order (in particular, in those in Fig. 2) is independent of calculating the diffuson (and/or Cooperon) contribution. Each rhombus or square in Fig. 2 gives a factor proportional to $\pi \nu_0 D \tau^2$ so that these diagrams yield the following expression for the variance of the dimensionless conductance $g \sim hG/e^2$ (the tensor indices structure is omitted here until later)

$$\langle (\delta g)^2 \rangle \sim \frac{(\pi \nu_0 D \tau^2)^2}{L^{4-d}} \int D^2(q, 0) d^d q \sim \frac{1}{L^{4-d}} \int \frac{d^d q}{q^4}. \tag{6}$$

Here the factor L^{d-4} results from the definition eq. (4), L^d being given by integrating over $\mathbf{r} + \mathbf{r}'$, and frequency ω set equal to zero. The cut-off for infrared divergence $q_0 \sim L^{-1}$ is due to the finite size of a sample, so that one arrives at the universal expression eq. (3). Certainly, the divergent integral in eq. (6) depends numerically on the geometrical shape of sample. This dependence is hidden in the only non-universal factor in eq. (3), the numerical factor A.

The evident consequence of eq. (3) is an absence of self-averaging of the conductance in $d < 4$ dimensions. For $d = 2$ the relative value of variance does not depend on sample size while for $d = 3$ it decreases proportional to L^{-2}, instead of the classical thermodynamic result L^{-d}. Certainly, self-averaging is restored at finite temperatures. First, a cut-off factor $DL_\phi^{-2} = \tau_\phi^{-1}$ arises in the diffuson denominator eq. (5) due to inelastic interactions (L_ϕ and τ_ϕ are phase-breaking length and time, respectively). In addition, the relevant energies in calculating Green function 'polygons' are spread over a band of a width T near ε_F which adds an energy difference $\varepsilon \sim T$ in denominators of diffusons thus resulting in another cut-off length $L_T = \sqrt{D/T}$. As a result, one finds the relative value of the variance at $T \neq 0$ as follows:

$$\frac{\langle (\delta G)^2 \rangle}{\langle G \rangle^2} \sim (\varepsilon_F \tau)^{-2} \times \begin{cases} (l/L)^{2d-4}, & L < \mathcal{L}_0 \\ (l/\mathcal{L}_0)^{2d-4}(\mathcal{L}_0/L)^d, & L > \mathcal{L}_0 \end{cases}, \tag{7}$$

where

$$\mathcal{L}_0 = \min\left\{ L_T, L_\phi \right\}. \tag{8}$$

Therefore, at $T \neq 0$ the relative value of the variance decreases thermodynamically ($\propto L^{-d}$) for sample sizes $L > \mathcal{L}_0$. Here is the origin of the term "mesoscopics" for the field: self-averaging is absent in samples with sizes obeying the conditions

$$l \ll L \leq \mathcal{L}_0,$$

so that a typical mesoscopic length is large as compared to the microscopic length l (elastic mean free path) but does not reach a macroscopic limit $L \gg \mathcal{L}_0$ where self-averaging occurs.

The absence of self-averaging is a generic feature of an ensemble of macroscopically identical samples of mesoscopic sizes. It means that the conductance (as well as other physical properties) fluctuates from sample to sample. Experimentally, 'mesoscopics' deals, e.g., with such characteristics of a particular sample as reproducible conductance aperiodic oscillations with changing magnetic fields ("magnetofingerprints") [4, 22] or similar oscillations with changing Fermi energy (due to changing a gate voltage in heterostructures) [23]. In order to relate the theoretical ensemble fluctuations to experimentally observable properties of a particular sample, an 'ergodic' hypothesis has been suggested [3], namely the configurational averaging over realizations is equivalent to the averaging over some range of magnetic field or Fermi energy. This suggestion has been proved [24] in case of averaging over ε_F. I want to demonstrate what is necessary for such a prove.

Let us introduce formally the averaging over the interval of (Fermi) energies in a particular sample as follows

$$\overline{\sigma(\varepsilon)_{\Delta\varepsilon}} \equiv \frac{1}{(\Delta\varepsilon)} \int_{\varepsilon-\Delta\varepsilon/2}^{\varepsilon+\Delta\varepsilon/2} \sigma(\varepsilon')d\varepsilon'. \tag{9}$$

Certainly, there are no means to calculate this average directly. Our goal, however, is to compare it with the configurational averaging. For comparison, it is sufficient to calculate the following quantity

$$\left\langle \left[\overline{\sigma(\varepsilon)_{\Delta\varepsilon}} - \langle\sigma(\varepsilon)\rangle \right]^2 \right\rangle = \frac{1}{(\Delta\varepsilon)^2} \int_{\varepsilon-\Delta\varepsilon/2}^{\varepsilon+\Delta\varepsilon/2} \left\langle\!\!\left\langle \sigma(\varepsilon_1)\sigma(\varepsilon_2) \right\rangle\!\!\right\rangle d\varepsilon_1 \, d\varepsilon_2. \tag{10}$$

This quantity is readily obtained even within the standard perturbational approach. The proof of the ergodicity is in a fact that the value eq. (10), as well as values of the appropriate higher order cumulants [24], decreases rapidly for $\Delta\varepsilon > 1/\tau$ so that the difference between the configurational averaging and the averaging over ε_F is vanishing. The ergodicity provides a connection of all the theoretical results based on the configurational averaging to experimentally observable phenomena.

In a similar way one may establish the ergodicity with respect to averaging over magnetic fields [3]. For what follows it is useful to discuss the fluctuations in a less formal quasi-classical way which reveal their origin in quantum interference effects. In Fig. 3a two paths making contributions to conductance and its moments are drawn schematically. Let A_j be the probability amplitude of the j-th path

$$A_j = \sqrt{W_j}\exp(i\phi_j),$$

where W_j is the probability for the j-th path and ϕ_j is a random phase. Then, the two paths make the following contribution to the total probability

$$W = |A_1 + A_2|^2 = W_1 + W_2 + 2\sqrt{W_1 W_2}\cos(\phi_1 - \phi_2), \tag{11}$$

and the appropriate contribution to the averaged value of any physical quantity is, due to the randomness of ϕ, proportional to

$$\langle W \rangle = \langle W_1 \rangle + \langle W_2 \rangle.$$

The contribution to the variance of the same quantity, due to $\langle\cos^2\Delta\phi\rangle = 1/2$, is proportional to

$$\left\langle W^2 \right\rangle - \left\langle W \right\rangle^2 = 2\left\langle W_1 W_2 \right\rangle \neq 0,$$

whenever coherence occurs in the region of the paths 1 and 2.

When imposing a magnetic field B, the phase factor in eq. (11) is shifted as

$$\delta\phi \to \delta\phi + 2\pi\Phi/\Phi_0,$$

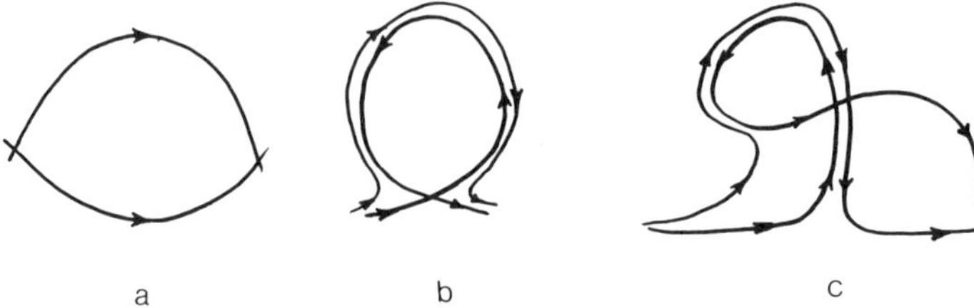

Figure 3. Two types of the interference phenomena. (a) Two typical paths which interfere in contributing to the variance but make independent contributions to the average conductance, (b) time-inverted paths making interference contribution to the averaged conductance, (c) time-inverted paths are shown as a part of two paths contributing to the variance. This coexistence of both types of the interference phenomena illustrates the origin of the weak localization corrections to the variance.

where $\Phi = BS$ is a magnetic flux through an appropriate area S, and $\Phi_0 = hc/e$ is the magnetic flux quantum. Then it is easy to evaluate that the correlation function of conductance in the magnetic field $\langle\langle G(B)G(0)\rangle\rangle$ decreases rapidly when $\Phi \gtrsim \Phi_0$. It means that after changing appropriately the magnetic field, the conductance acquires a 'random' shift $\sim e^2/h$. In other words, the conductance of a coherent 'mesoscopic' sample undergoes Aharonov-Bohm (AB) oscillations with a quasi-period equal to hc/e and an amplitude of the order of e^2/h [22]. This phenomenon is discussed widely in other lectures in this book. I want to mention also another type of conductance oscillations in a magnetic field in metal rings or cylinders of sizes greater than the usual micron or submicron mesoscopic scale. These AB oscillations with a period that corresponds to the change of the magnetic flux inside a cylinder by a half of the flux quantum $\Phi_0/2 = hc/2e$ [25] reveal another type of quantum interference as has been predicted in [26]. To discuss it, let us come back to Fig. 3.

Typical paths make independent contribution to averaged quantities but interfere in contributing to the variances. But there are mutually time-inverted paths (Fig. 3b) which differ only by direction. In this case the phase factors are cancelled in calculating $\langle W \rangle$ so that in eq. (11) the term proportional to $\sqrt{W_1 W_2}$ survives the averaging. This means that such paths make interference contributions to the averaged quantities, e.g. to the averaged conductance. This term is well-known to be responsible for the effects of weak localization, another example of revealing macroscopically quantum interference phenomena. Since the coherence effects can be suppressed by a magnetic field, it leads to the well-known explanations of both the negative magnetoresistance and $hc/2e$ AB oscillations (for a review, see e.g. [27]).

A usual weak-localization correction to conductance may be calculated directly in the quasi-classical way. However, I will describe here standard perturbational calculations similar to those used above in calculating the expression for the variance eq. (3). The lowest order perturbation diagram for conductance corresponding to the interference picture of Fig. 3b is the well-known maximally crossed diagram given in Hikami notation in Fig. 4. It gives the following result [21] for the average dimensionless conductance $\langle g \rangle$

$$\langle g \rangle = g_{cl}(1 - \xi_0). \tag{12}$$

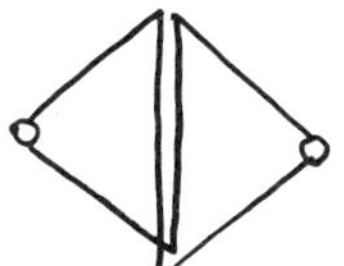

Figure 4. This diagram in the Hikami representation corresponds to the maximally-crossed diagram for conductance
in the standard impurity diagram technique.

Here the classical (Drude) value g_{cl} for the dimensionless conductance is given in dimensionality $d = 2 + \epsilon$ by $g_{cl} = (\pi^2 \nu_0 v_F l / d)(\mathcal{L}/l)^\epsilon$, and

$$\xi_0 \sim g_{cl}^{-1} \times \begin{cases} \ln(\mathcal{L}/l), & d = 2 \\ [1 - (l/\mathcal{L})^\epsilon]/\epsilon, & d = 2 + \epsilon, \end{cases} \tag{13}$$

is the usual perturbation parameter of the weak localization theory where $\mathcal{L}$ is the
minimum of the sample size L and the phase-breaking length L_ϕ. Note that, in contrast
to the case of the fluctuations, the thermal length L_T that enters the definition eq. (8)
of $\mathcal{L}_0$ does not enter the value of average conductance.

All the leading singular corrections to eq. (12) proportional to the higher powers
of ξ_0 are cancelled. To the first non-trivial perturbation order, this cancellation has
been proved by Gor'kov et al. [21]. This provided the possibility to formulate the
renormalization group (RG) equation for the conductance. In $d = 2 + \epsilon$ dimensions, such
RG equation may be written down as follows

$$\frac{dg}{d \ln \lambda^{-1}} = \epsilon g - 1 + o\left(g^{-2}\right). \tag{14}$$

Here λ is a scaling factor. Later this equation has been proved in the whole region $g \gtrsim 1$ in
a more rigorous way which I discuss in the next section. But what is most important now
is that this equation is just the metallic-side ($g > 1$) part of the famous one-parameter
scaling equation stated by Abrahams et.al. [28]. The one-parameter scaling is believed
to describe the Anderson transition from metal to insulator in $d > 2$ dimensions and the
crossover from weak to strong localization region in $d = 2$, both caused by the quantum
interference in elastic scattering by impurities.

Therefore, one type of the interference phenomena is responsible for the localization
effects (and $hc/2e$ oscillations) and the other is for the mesoscopic fluctuations (and hc/e
oscillations). While the one-parameter scaling is the most proper way of describing the
first type of such phenomena, it is not _a priori_ evident that it is also a way of describing
the other type, i.e. the mesoscopic effects. Moreover, in the mesoscopics these two types
of the interference phenomena may coexist, as illustrated in Fig. 3c. The paths given
in this figure describe quasi-classically the weak localization corrections to the variance
of the mesoscopic fluctuations. The problems that I want to discuss in the rest of my
paper are related to this coexistence of the two types of the interference.

The question of such a coexistence arises naturally with increasing effective disorder
$\mathcal{L}/l$ when, according to eq. (14), the average conductance $\langle g \rangle$ decreases. It is hard
to believe that the variance of the fluctuations would be still of the order of unity
when $\langle g \rangle \ll 1$. Even if this was so, the distribution of the fluctuations would be very
non-trivial and definitely non-Gaussian. In any case, with decreasing $\langle g \rangle$ one faces a

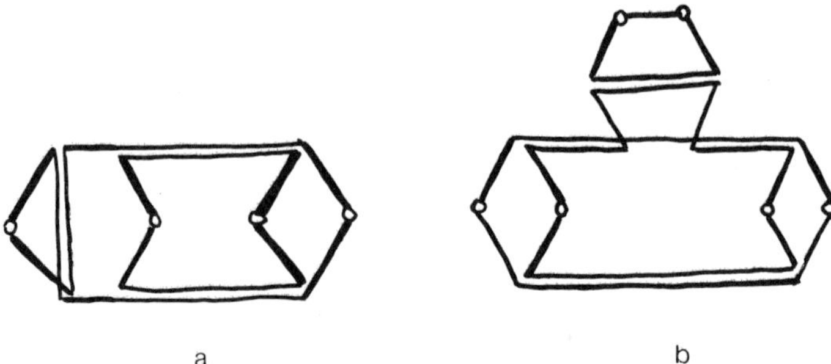

a   b

Figure 5. Examples of the singular corrections to conductance cumulants. (a) A correction to the variance corresponding to quasi-classical picture in Fig. 3c, (b) a strongly diverging correction to the cumulant of the third order. In calculations, both diagrams are cancelled with others which are not drawn here.

situation where it is necessary to calculate corrections to the variance of the fluctuations and find the higher moments of the fluctuations. However it is possible to say that possible perturbation corrections to the variance eq. (3) result just from allowing for the coexistence of different types of interference effects, as in Fig. 3c, i.e. from the same weak localization effects which are responsible for decreasing the conductance eq. (12). In the same way, i.e. allowing for the weak localization effects, one should calculate the higher order moments of the fluctuations.

Certainly, there is no way of calculating these corrections directly in the quasi-classical approach. Moreover, even the standard perturbation technique is of little help in deriving the final results. This technique may help, however, to estimate a possible 'danger' in such corrections. For an illustration, some examples of diagrams making singular corrections to the variance and the higher order cumulants are given in Fig. 5. These diagrams represent corrections due to the coexistence of interference processes, like in Fig. 3c. A possible presence of singular corrections to the variance makes the statement on its universality questionable. Corrections to the higher order cumulants seem to be even more dangerous because of the higher power of divergence, like for the diagram in Fig. 5b which diverges $\propto L^{6-d}$. So, calculations were required to include all of such corrections. These calculations have shown [11, 45] that all the singular corrections to the variance, as well as all the corrections to the higher order cumulants with infrared divergence more dangerous than in eq. (6), cancel.

An way of calculating the moments of conductance fluctuations will be outlined in the next section. Here some results are presented in advance. With allowance for the leading singular corrections in the parameter ξ_0, the conductance cumulants of low[2] orders have been obtained [24] in the whole region $\langle g \rangle \gtrsim 1$ as follows

$$\left\langle\!\left\langle (G)^n \right\rangle\!\right\rangle \propto \langle g \rangle^{2-n} (e^2/h)^n. \tag{15}$$

So even with all the singular corrections taken into account, the variance $\left\langle\!\left\langle G^2 \right\rangle\!\right\rangle$ remains universal in this whole region. By using eq. (12) for $\langle g \rangle$, one gets the cumulants depending on the effective disorder parameter $\mathcal{L}/l$, as is shown for the two-dimensional (2 D) case in Fig. 6.

---

[2] An explanation of what is "low" and what is "high" order cumulants will be given later.

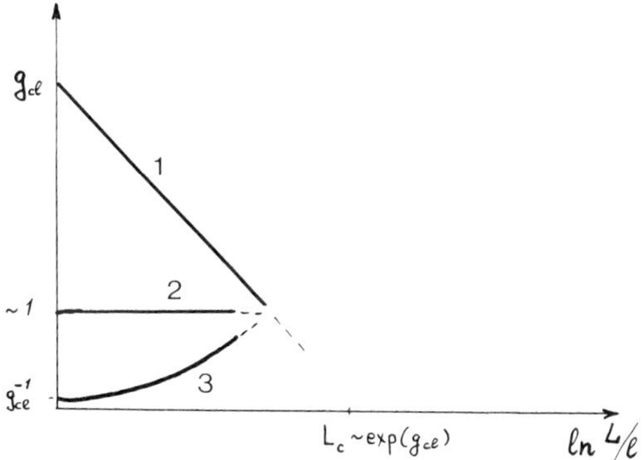

Figure 6. The cumulants of the conductance fluctuations vs the effective disorder parameter $\ln(L/l)$. 1 – the average conductance $\langle g \rangle$, 2 – the variance $\langle\langle g^2 \rangle\rangle^{1/2}$, 3 – the higher order cumulants $\langle\langle g^n \rangle\rangle^{1/n}$ for $n < g_{cl}$.

In the metallic region where $\langle g \rangle \approx g_{cl} \gg 1$, the higher order $(n > 2)$ cumulants are small in comparison with the variance. This is just a generic feature of the close-to-Gaussian distribution, because in the Gaussian case all the higher order cumulants vanish. Note that even in this region the high order cumulants with $n \gtrsim g_{cl}$ are not given by eq. (15) and the distribution is not so trivial, as will be discussed later. Anyway, in this region the average value is well defined, and the fluctuations which may be described only by the variance, remain small.

But with $\langle g \rangle$ decreasing, the variance remains approximately universal while all the higher order cumulants increase so that we finally get a region where all the cumulants, including the variance and the average value, are of the same order in magnitude. In this region, the distribution becomes wide and completely non-Gaussian. Then neither the average value, nor the variance do describe in a proper way the conductance of a particular sample. It means that a complete description of the Anderson transition for mesoscopic samples (or for any disordered conductor in the formal limit $T = 0$) in terms of only average values is impossible. From this viewpoint, some features of $\langle g \rangle$ in the transition, like the presence or the absence of a jump, become less important. Instead, a description is required in terms of the whole distribution function: how is its shape changing with increasing disorder parameter. Such a description will be the subject of the next section.

Now let me recall that despite of the absence of self-averaging, the UCF are small in the whole metal region $\langle g \rangle \gtrsim 1$ because here the variance $\langle\langle g^2 \rangle\rangle \sim 1$. Then it is interesting to ask whether or not there are quantities which fluctuate relatively stronger than the conductance? The answer is positive, and I want to present here two examples of such quantities. The first example is more or less trivial. There are a lot of quantities with average values being equal to zero due to symmetry requirements. One of them is the Hall conductance g_{xy} in the absence of a magnetic field. But the variance of this quantity does not vanish. On restoring the tensor index structure of the expression eq. (3) given by the diagrams of Fig. 1, one finds [3, 10]

$$\langle \delta g_{\alpha\beta}\, \delta g_{\gamma\delta} \rangle = A \left(\delta_{\alpha\beta}\delta_{\gamma\delta} + \delta_{\alpha\gamma}\delta_{\beta\delta} + \delta_{\alpha\delta}\delta_{\beta\gamma} \right).$$

So, the value of $\langle\langle g_{xy}^2 \rangle\rangle$ is of the order of unity, and although it is small in comparison with $\langle g \rangle$, it is infinitely large in comparison with $\langle g_{xy} \rangle$.

The second example is more interesting. It concerns the current density $\mathbf{j}(\mathbf{r})$ which is given by the non-local conductivity $\sigma(\mathbf{r}, \mathbf{r}')$,

$$j_\alpha(\mathbf{r}) = \int \sigma_{\alpha\beta}(\mathbf{r}, \mathbf{r}')E_\beta \, d\mathbf{r}', \tag{16}$$

where $\mathbf{E}$ is a homogeneous classical electric field. It fluctuates from sample to sample much stronger than the conductance. Its variance was found [29] to be

$$(\Delta_j)^2 \equiv \left\langle\!\left\langle j_\alpha(\mathbf{r})\, j_\beta(\mathbf{r}) \right\rangle\!\right\rangle \propto \gamma \mathbf{j}_0^2 \,\delta_{\alpha\beta}, \qquad \gamma \equiv g^{-1} L^2 \Big/ l^2. \tag{17}$$

Here $\mathbf{j}_0 \equiv \langle \mathbf{j}(r) \rangle = \sigma\mathbf{E}$ is the averaged current density. For temperatures $T \neq 0$ one should substitute into this expression $\min(L, \mathcal{L}_0)$ for L where the thermal length $\mathcal{L}_0$ is given by eq. (8). For typical mesoscopic samples, the value of γ may be much greater than 1 so that the variance eq. (17) may be much greater than the average value of j. Note that the fluctuations of the current density are not only of theoretical interest. They are manifest in the fluctuations of a measurable potential difference $\delta V_{a,b}$ between two close contacts a and b

$$\delta V_{a,b} \propto \int_a^b \delta j(\mathbf{r}) d\mathbf{r}$$

in experiments with multi-terminal devices (see, e.g., [22]). Such voltage (or, equivalently, resistance) fluctuations have also been studied by using other approaches [30–35]. For the present purposes, these fluctuations will be treated just because their relative magnitude is high, so that some features that are typical for mesoscopic fluctuations of different quantities are more clearly exhibited.

2   Distribution Functions of Conductance and Current Density Fluctuations

In this section I consider the change of the shapes of distribution functions of both conductance and current density fluctuations with increasing the effective disorder parameter L/l. Again, I would like to derive the rigorous results obtained in the region $\langle g \rangle \gtrsim 1$ with using the small parameter eq. (2). These results required the summation of the infinite series in powers of the singular parameter ξ_0 defined by eq. (13). Finally, some hypothesis concerning the non-perturbative region $\langle g \rangle \ll 1$ will be also discussed.

The direct perturbational analysis of all possible corrections to the variance is quite cumbersome even to the second order. Although in this way it is possible to prove [11] the cancellation of all the corrections proportional to ξ_0^2 where ξ_0 is the weak localization parameter eq. (13), it is hardly possible to consider the diagrams of the next order. Definitely, this is not a suitable way either to prove the cancellation of all the higher order singular corrections to the variance, or to treat the higher moments of the fluctuations when approaching the strong localization region in $d = 2$ dimensions or the region of Anderson transition in $d > 2$.

In the weak localization theory the usual way to sum the leading perturbation corrections to the averaged conductance is based on the renormalization group analysis of the density auto-correlation function which is represented in terms of the non-linear

σ model. This approach has been introduced by Wegner [36] and was further developed by many authors [37–43], [20].

In treating mesoscopic fluctuations, this approach is not very convenient. Instead, it is useful to apply the field-theoretical (σ model) representation directly for the average conductance [44] and its cumulants [11, 15]. In this way, results for all the cumulants of the fluctuations with allowance for weak localization corrections (i.e. with allowance for both types of the interference effects) are obtained via the RG analysis of the appropriate field theory. Certainly, both the derivation of this theory and its RG analysis are the most heavy and cumbersome parts of the whole story. However, I will not jump from here directly to the results but try to indicate the principal steps of the derivation. Details are given in our recent review paper [45].

The first essential step is a field-theoretical representation of the Kubo formula for conductivity with the usual model of free electrons in a random potential eq. (1) chosen as a starting point (conductance is related to conductivity by eq. (4))

$$\sigma_{\alpha,\beta}(\mathbf{r},\mathbf{r}';\omega) = \frac{4}{\pi N^2}\Big\langle \mathrm{Tr}\Big[J_\alpha^+(\mathbf{r})J_\beta^-(\mathbf{r}')\Big]\Big\rangle_\psi, \tag{18}$$

where the brackets denote the functional averaging

$$\langle\ldots\rangle_\psi \equiv \frac{\int \mathcal{D}\overline{\psi}\mathcal{D}\psi \,(\ldots)\exp(iS)}{\int \mathcal{D}\overline{\psi}\mathcal{D}\psi \,\exp(iS).} \tag{19}$$

The effective action S is given by

$$S = \int \overline{\psi}(\mathbf{r})\Big[\hat{H} - (\frac{\omega}{2}+i\eta)\Lambda\Big]\psi(\mathbf{r})d^d r, \quad \eta \to +0. \tag{20}$$

Here $\hat{H}$ is the Hamiltonian of eq. (1), the symmetry-breaking matrix Λ reflects the presence of retarded and advanced Green functions in the initial quantum mechanical representation of the Kubo formula, and $\overline{\psi}$ and ψ are "spinor" N-replicated Grassmann fields [38]. The current matrices $\mathbf{J}^\pm(\mathbf{r})$ are defined in terms of these fields as follows

$$\begin{aligned}
\mathbf{J}^\pm(\mathbf{r}) &\equiv C^\pm\mathbf{J}(\mathbf{r}) \\
J_\alpha(\mathbf{r}) &= \frac{i}{2}\Big\{\psi(\mathbf{r})\otimes\partial_\alpha\overline{\psi}(\mathbf{r}) - \partial_\alpha\psi(\mathbf{r})\otimes\overline{\psi}(\mathbf{r})\Big\},
\end{aligned}$$

where $C^\pm$ are some projection operators [45], their explicit expressions being unimportant here.

The second step is to exponentiate eq. (18) by introducing a matrix source $\mathbf{A}^{ext}$ of some definite structure. As a result we get

$$\sigma_{\alpha,\beta}(\mathbf{r},\mathbf{r}';\omega) = \frac{1}{16\pi N^2}\mathrm{Tr}\left\{\frac{\delta^2 Z[\mathbf{A}^{ext}]}{\delta A_\alpha^{ext}(\mathbf{r})\,\delta A_\beta^{ext}(\mathbf{r}')}\right\}\bigg|_{A^{ext}=0}, \tag{21}$$

where

$$Z[\mathbf{A}^{ext}] = \Big\langle\exp\Big[\int \mathrm{Tr}(\mathbf{A}^{ext}\mathbf{J})d^d r\Big]\Big\rangle_\psi. \tag{22}$$

Then it is straightforward to generalize eq. (18) for the n-th irreducible moment (cumulant) of the local conductivity

$$\prod_{i=1}^{n}\sigma_{\alpha_i\beta_i}(\mathbf{r}_i,\mathbf{r}_i';\omega) = \left(\frac{1}{16\pi N^2}\right)^n\left\{\prod_{i=1}^{n}\mathrm{Tr}\left(\frac{\delta^2}{\delta A_{\alpha_i}^{ext}(\mathbf{r})\,\delta A_{\beta_i}^{ext}(\mathbf{r}')}\right)\right\}Z[\mathbf{A}^{ext}]\bigg|_{A^{ext}=0}. \tag{23}$$

To obtain now the expression for the n-th cumulant of conductance, one has to integrate over all the pairs $\mathbf{r}_i, \mathbf{r}'_i$ and multiply the result by L^{-2n}.

The third step of the derivation is the averaging over all the realizations of the impurity potential $V(\mathbf{r})$ which is performed either by using the standard replica trick [37, 38, 45] or by introducing supersymmetric variables [41]. The configurational averaging is performed by taking a Gaussian integral over V, and one of these tricks is required to remove a denominator in the functional averaging in eq. (19). For a perturbational analysis both approaches are equivalent, and here the replica trick is used. After averaging, eqs. (21)–(23) remain formally unchanged but the functional averaging eq. (19) should be performed with the action eq. (20) which includes the following biquadratic term

$$S_{imp} = -\frac{i}{2\pi \nu_0 \tau} \int \mathrm{Tr}\, q^2 \, d^d r, \qquad q(\mathbf{r}) \equiv \psi(\mathbf{r}) \otimes \overline{\psi}(\mathbf{r}). \tag{24}$$

This term appeared upon configurational averaging instead of the random part of the action eq. (20) which was proportional to $\overline{\psi} V \psi$ where V is the random potential in the Hamiltonian eq. (1). Eventually, all expressions are subject to the replica condition $N = 0$ which should be applied in the final results.

The fourth step is to remove the biquadratic term by means of the Hubbard-Stratonovich transformation of the effective action. The physical meaning of this transformation is changing the variables from "fast" ones $\psi(\mathbf{r})$ and $\overline{\psi}(\mathbf{r})$ with a range of the order of electron wavelength, to "slow" ones $Q(\mathbf{r})$ which describe diffusive motion [36, 37, 38, 11, 45]. To delete the "fast" variables one can multiply both the numerator and denominator of the generating functional eq. (22) by

$$\int \mathcal{D}Q \, \exp\left(-\frac{\pi \nu_0}{8\tau} \int \mathrm{Tr}\, Q^2 \, d^d r\right).$$

Here the Hermitian matrix field $Q(\mathbf{r})$ has the same symmetry as $q(\mathbf{r})$ in eq. (24). Then the shift of variables

$$Q \to Q - \frac{2}{\pi \nu_0} q$$

removes the biquadratic term eq. (24) reducing the integral to quadratic in the "fast" variables. The integration over these variables leads to the following expression

$$Z[\mathbf{A}^{ext}] = \frac{\int \mathcal{D}Q \, \exp\left\{-F[Q; \mathbf{A}^{ext}]\right\}}{\int \mathcal{D}Q \, \exp\left\{-F[Q; 0]\right\}} \tag{25}$$

with the "free energy" functional given by

$$F[Q; \mathbf{A}^{ext}] = -\mathrm{Tr}\, \ln\left(\hat{\xi} - \left(\frac{\omega}{2} + i\eta\right)\Lambda + \frac{1}{2}\{\partial_\alpha, A_\alpha^{ext}\} - \frac{i}{2\tau}Q\right)$$
$$+ \frac{\pi \nu_0}{8\tau} \int \mathrm{Tr}\, Q^2 \, d^d r \tag{26}$$

where $\{\,,\,\}$ stands for anti-commutator.

The last step of the derivation is to get the explicit expression for the functional eq. (26) from the symbolic trace–log expression. It is possible to start from a saddle-point approximation which is given in a class of spatially homogeneous fields Q for $\mathbf{A}^{ext} = \omega = 0$ by the following condition [37, 38]

$$Q^2 = \mathrm{I}, \qquad \mathrm{Tr}\,Q = 0. \tag{27}$$

Then, the effective functional is obtained by expanding the trace–log in eq. (26) in powers of gradients of Q and in powers of ω and $\mathbf{A}^{ext}$ in the class of the fields Q obeying the saddle-point condition eq. (27). Expanding in gradients within the saddle-point manifold means to neglect the transverse modes given by the second term in eq. (26). From the perturbation theory viewpoint, this is justified by the weak-disorder condition eq. (2) and corresponds to neglecting corrections in powers of $(\epsilon_F\tau)^{-1}$ but keeping (in principle) all the singular corrections proportional (in $d = 2$ dimensions) to arbitrary powers of $(\epsilon_F\tau)^{-m}\ln(L/l)$.

The first non-vanishing term of the gradient expansion (for $\omega = \mathbf{A}^{ext} = 0$) gives the standard σ model functional

$$F[Q] \equiv \frac{1}{t}\int \mathrm{Tr}\,(\partial Q)^2\, d^d r. \tag{28}$$

Here ∂ stands for a usual gradient, and the coupling constant t is related to dimensionless conductance g as $t = 16\pi g^{-1}L^{d-2}$, the bare value of g being equal to $g_0 = \pi^2\nu_0 v_F l^{d-1}$.

The expansion in ω will be discussed in the last part of these lectures. Before expanding in powers of $\mathbf{A}^{ext}$, let us note that neglecting the corrections in powers of $(\epsilon_F\tau)^{-1}$, the initial expression eq. (26) is invariant under the following gauge transformation [19, 15]

$$Q \to U^+QU, \qquad A^{ext}_\alpha \to U^+A^{ext}_\alpha U + U^+\partial_\alpha U,$$

where U are unitary matrices $(U^+U = 1)$ which represent the matrices Q eq. (27) as $Q = U^+\Lambda U$. The gauge invariance enables us to obtain the dependence on $\mathbf{A}^{ext}$ just by substituting the covariant derivative ∇Q for the usual gradient ∂Q [16]

$$\partial_\alpha Q \to \nabla_\alpha Q \equiv \partial_\alpha Q - [A^{ext}_\alpha, Q]. \tag{29}$$

This functional gives the conductance directly via eqs. (21), and (4) was first obtained [44, 11] by direct expansion of the trace-log in eq. (26) instead of applying gauge-invariance considerations. Its symmetric form was explained by using a current conservation law. This idea was then applied [46] also to investigate the high gradient terms of the expansion.

However, the generalized σ model obtained after substituting eq. (29) into the functional eq. (28) does not describe completely all relevant perturbative contributions to the cumulants of the conductance fluctuations. To see this, let us look at the Hikami diagrams for the variance (Fig. 2) and for the higher order cumulants (Fig. 7) from the field-theoretical viewpoint. Double lines (i.e. Cooperons or diffusons) represent now Goldstone modes of the σ model functional eq. (28). Each polygon corresponds to $\int \mathrm{Tr}\,(A^{ext})^k\, d\mathbf{r}$ (where k is a number of external vertices in the polygon) which should appear after expanding in powers of t and performing the functional integration in eq. (25). But if such an expansion involved only the functional eq. (28), then only powers of $\int \mathrm{Tr}\,(A^{ext})^2\, d\mathbf{r}$ could appear. In this way it is possible to recover diagrams like that in Fig. 7a which

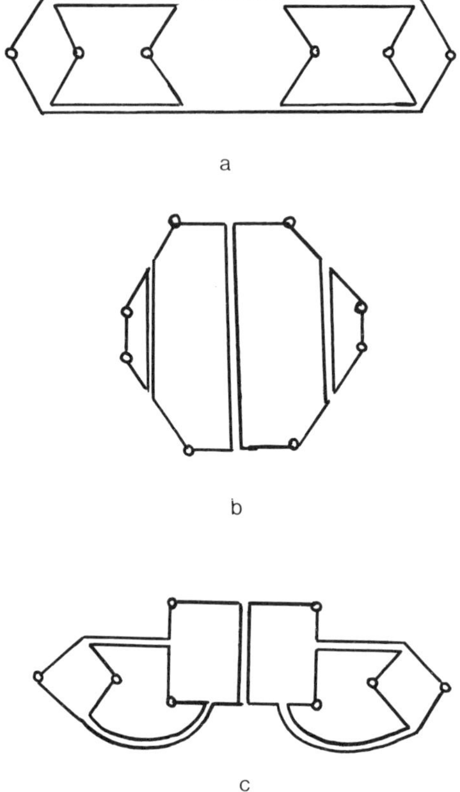

Figure 7. Examples of diagrams for the higher order cumulants. (a) One of the diagrams representing the leading contribution to the low order conductance cumulants, (b) one of the diagrams representing the leading contribution to the high order conductance cumulants, (c) one of the diagrams representing the leading contribution to the current density cumulants.

represent the dominating contribution to the low order cumulants of the conductance but not all the other diagrams in the figure. Even an additional contribution to the variance (Fig. 2c) can not be obtained within the standard σ model.

To derive the complete functional recovering all the diagrammatical contributions, further expansion of eq. (26) in powers of ∂Q is required. This expansion results in the following functional of the extended non-linear σ model [15, 45]

$$F\left[Q; \mathbf{A}^{ext}\right] = \sum_{n=1}^{\infty} F_n\left[Q; \mathbf{A}^{ext}\right]. \tag{30}$$

Here the substitution eq. (29) based on the gauge invariance [19, 15] (or, equivalently, on the current conservation law [11, 46]) was also used. In eq. (30), the first vertex $F_1\left[Q; \mathbf{A}^{ext}\right]$ is the functional of the generalized standard σ model which is obtained from eq. (28) by substituting eq. (29). The vertices with the higher order gradient operators are given by the following expressions

$$F_n\left[Q; \mathbf{A}^{ext}\right] \equiv z_n s_{\alpha_1,\ldots,\alpha_{2n}} \int \mathrm{Tr}\left(\prod_{i=1}^{2n} \nabla_{\alpha_i} Q\right) d^d r. \tag{31}$$

The bare values of the charges z_n are inversely proportional to the bare value of the coupling constant t. The tensor $s_{\alpha_1,\ldots,\alpha_{2n}}$ in this expression arose from angular integration of a product $n_{\alpha_1}\ldots n_{\alpha_{2n}}$ so that it is proportional to the sum of all the possible products of the Kronecker symbols. The terms with repeated spatial derivatives of Q which also appear when expanding eq. (26) are neglected here because they do not contribute essentially to the cumulants [15].

The functional eq. (30) allows to include all the leading singular corrections to the cumulants eq. (23). Firstly, it is necessary to calculate the required quantities to the lowest non-vanishing order of the perturbation, and then to substitute the renormalized values of the charges of the functional eq. (30) for the bare ones. I will not describe here the procedure of renormalization but just use the results [15].

Before doing so, it is important to separate the results obtained with the help of the standard σ model functional alone from those which required the extended functional eq. (30). The former ones depend on the single parameter of the standard σ model dimensionless conductance $g \propto t^{-1}$. The reason is that cumulants of low orders turned out to be governed by one-parameter scaling when calculated with the help of the functional expressions eq. (15) for the conductance. Technically, this means the following. The RG procedure is reduced in the one-parameter case to substituting the bare value of conductance g_0 by the physical one, $\langle g \rangle$, which is given by the solution eq. (12) to the Gell-Mann–Low equation eq. (14). Such a substitution would be equivalent to a summation of all the singular corrections in power of ξ_0, if the one-parameter scaling were applicable.

From this viewpoint, the perturbative result for the variance eq. (3) is specific in the sense that it does not depend on g_0, in contrast to the higher order cumulants eq. (15). Since this result for the variance is reproduced with the help of only the one-parameter functional eq. (28), all the singular corrections to it are cancelled so that the variance remains universal in the region $\langle g \rangle \gg 1$. There are, however, even within the one-parameter description some non-singular contributions to the variance [11] proportional to the higher powers of g_0^{-1}. Again, the RG substitution of $\langle g \rangle$ for g_0 makes these contributions formally relevant in the region of the strong localization $\langle g \rangle \lesssim 1$. The RG approach gives no possibility to establish whether or not the variance remains universal in this region. This question is of little importance, though, because in this region the distribution function of the fluctuations becomes completely non-Gaussian and is not governed by the variance, as I will show in the rest of this section.

In contrast to the functional eq. (28) the complete functional of the extended non-linear σ model eq. (30) depends on the whole set of renormalizable parameters z_n. Therefore, the complete contribution to conductance cumulants appears to be many-parameter. It is not *a priori* evident, though, that the additional corrections are of importance. According to simple perturbation estimations these corrections appear to be small in powers of the parameter l/L. But the RG analysis of the functional eq. (30) showed the charges z_n to be increasing extremely rapidly with the number n of gradients in the appropriate vertex

$$z_n \propto z_n(0) \exp\left[u(n^2 - n)\right]. \tag{32}$$

It is just this rapid increase of the charges which makes them relevant for high n despite their negative naïve dimensionality $d - 2n$, and makes appropriate "additional" contributions to the high order cumulants of conductance dominating over the one-parameter contributions eq. (15). These additional corrections are obtained [11, 15] by using eqs. (21), (25), (30) and (31) and integrating over all pairs $\mathbf{r}, \mathbf{r}'$ as follows

$$\left\langle\!\!\left\langle (\delta g)^n \right\rangle\!\!\right\rangle^{\text{add}} \propto g_0 (l/L)^{2n-d} \exp\left[u(n^2 - n)\right]. \tag{33}$$

Certainly, such a contribution to the low order cumulants (in particular, to the variance) are negligible as compared to eq. (15).

In eqs. (32), (33) the parameter u is defined by

$$u = \ln \frac{\sigma_0}{\sigma}. \tag{34}$$

It is important that this parameter is non-universal in a sense that it depends on both bare, σ_0, and renormalized, i.e. physical, σ, values of conductivity. It means that all the results depending on this parameter are also non-universal, and depend on such characteristics as electron density, mean free time, etc., absorbed by the classical Drude expression for the bare conductivity.

A direct comparison of eqs. (15) and (33) shows the additional contribution to dominate for the cumulants of the n-th order with n obeying the following inequality

$$n \gtrsim n_0 \equiv u^{-1} \ln(L/l). \tag{35}$$

The n-th cumulants eq. (33) with n obeying inequality eq. (35) are large as compared to the appropriate power of the variance eq. (3). Therefore, it turns out that the distribution function of conductance fluctuations is characterized by long tails, these tails being governed by the high order cumulants eq. (33).

The restoration of the distribution function of conductance fluctuations has been described in detail elsewhere [11, 45]. Here it is only worth noting that the $\exp(n^2)$-dependence of the n-th moment is a characteristic feature of the logarithmically normal (LN) distribution as one can see from the following identity

$$\exp(un^2) = \frac{1}{(4\pi u)^{1/2}} \int_0^\infty w^n \exp\left[-\frac{1}{4u}\ln^2 w\right] \frac{dw}{w}. \tag{36}$$

For the problem at hand, the low cumulants eq. (15) are small in comparison with the universal variance eq. (3) so that the bulk of the distribution function proves to be Gaussian with a dispersion of the order of unity (the exact value of the dispersion depends on the geometrical factor A in eq. (3)). The LN asymptotics of the distribution function governing the high order cumulants

$$f(\delta g) \propto \frac{1}{\delta g} \exp\left[-\frac{1}{4u}\ln^2\left(\frac{\delta g}{g}\frac{1}{\tau\Delta}\right)\right] \tag{37}$$

characterizes comparatively large fluctuations

$$\delta g \gtrsim (g_0)^{1/2}. \tag{38}$$

In eq. (37), $\Delta = (L^d \nu_0)^{-1}$ is the spacing between energy levels. Note that a similar long tail characterizes large mesoscopic fluctuations of the density of states [45]. A sketch of the complete distribution function is given in Fig. 8.

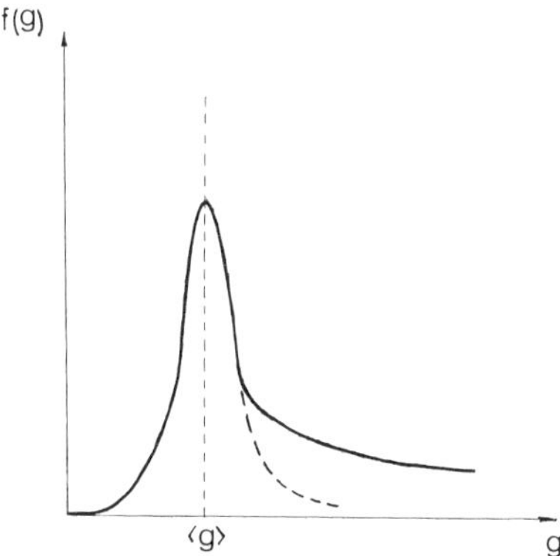

Figure 8. The distribution function of the conductance. The dashed line corresponds to normal Gaussian tails while the solid line shows the considerably slower LN decay of the probability of large (compared to the variance) fluctuations. The distribution function of density of states has qualitatively the same behavior.

This LN probability of large conductance fluctuations exceeds strongly the Gaussian one. However, the tails eq. (37) contain only a small part of the entire distribution function. This part becomes more important when approaching the region of strong localization (where $\langle g \rangle \lesssim 1$). As it follows from inequality eq. (35), in the region $\langle g \rangle \sim 1$ the fluctuations with a probability given by LN asymptotics obey the condition $\delta g \gtrsim (g_0)^{1/2} / \ln g_0$ instead of that in eq. (38). Unfortunately, our technique is limited by condition $\langle g \rangle \gtrsim 1$ so that it is impossible to investigate directly how the LN part of the distribution is changing for $\langle g \rangle < 1$.

The situation is drastically different for the fluctuations of local quantities, such as current density. It turns out [13, 14] that the LN part of the current density distribution function becomes dominating even in the region $\langle g \rangle > 1$ where the one-loop RG approach can be applied.

Formally, the expressions for the current density cumulants are derived from the same eqs. (23),(25),(30) and (31), as the conductance cumulants. However, in this latter case one should integrate the expressions eq. (23) over all $\mathbf{r}'$ while in case of conductance fluctuations one has to integrate over all the pairs $(\mathbf{r}, \mathbf{r}')$. This is just the reason for quite a different selection of the leading diagrams for higher ($n > 2$) cumulants of the conductance and current density fluctuations (Fig. 7).

In the region of weak localization, the leading diagrams for the current density cumulants (Fig. 7c) give immediately (i.e. with bare values of all the charges in the functional of the extended non-linear σ model eq. (30)) the expressions for the low cumulants of the current density fluctuations. For the high order cumulants, the appropriate expressions were obtained after the RG substitution of the renormalized values of the charges z_n in eq. (32) for the bare ones which is equivalent to summarizing the leading singular (logarithmic at $d = 2$) contributions. However, in contrast to the case of the conductance fluctuations, a definition of "high" and "low" cumulants depends strongly on the value of $\langle g \rangle$ so that all the cumulants become "high" with $\langle g \rangle$ decreasing. The appropriate simplified (with dependence on vector indices being omitted[3]) expressions for the even cumulants are given by

$$\left\langle\!\!\left\langle \prod_{i=1}^{2n} j_{\alpha_i}(\mathbf{r}) \right\rangle\!\!\right\rangle \sim \left\{ \begin{array}{ll} \xi_0^{n-1}(\Delta_j)^{2n} & n \lesssim u^{-1/2} \quad (a) \\ \exp\left[u(n^2 - n)\right](\Delta_j)^{2n} & n \gtrsim u^{-1} \quad (b). \end{array} \right. \tag{39}$$

Here ξ_0 is the standard perturbational parameter of the weak localization theory eq. (13). These expressions contain no information about cumulants of intermediate orders $u^{-1/2} < n < u^{-1}$ that are of no importance here.

Therefore, the definition of "high" and "low" cumulants depends directly on the value of the parameter u. As it follows from the eqs. (34) and (12), in the region of weak localization where $\xi_0 \ll 1$ this parameter is small, $u \approx \xi_0 \ll 1$. But when quantum corrections become essential (i.e. when $\xi_0 \sim 1$) with disorder and/or system size increasing, we come to a region where $\langle g \rangle$, although being a large quantity, may be substantially different from g_0 (e.g., $\langle g \rangle = g_0/2 \gg 1$). In this region $u \sim 1$, so that there is no room for the first line in eq. (39), and all the cumulants become "high".

As a result, the shape of the current density distribution function is quite different in these two regions, in contrast to the conductance distribution function which for $u \sim 1$ remains qualitatively the same as in the weak localization region. In the region of weak localization, $\xi_0 \ll 1$, all the cumulants except those of higher order, $n > u^{-1/2} \gg 1$, are given by the perturbative expression eq. (39a) and, therefore, are negligible in comparison with the variance eq. (17). It means that in this region the distribution of the current density fluctuations is close to Gaussian but has LN tails, similar as in the case of the conductance. However, when u increases with disorder, the LN tails in distribution increase very rapidly. Finally, in the region where $u \sim 1$, and all the cumulants eq. (39b) are large in comparison with the variance Δ_j, eq. (17), the distribution function becomes completely LN [13, 14]

$$\phi(\mathbf{j}) = \frac{1}{2\sqrt{\pi u}\,|\mathbf{j} - \mathbf{j}_0|} \exp\left[-\frac{1}{u} \ln^2\left(\frac{|\mathbf{j} - \mathbf{j}_0|}{\widetilde{\Delta}}\right)\right], \tag{40}$$

where $\widetilde{\Delta} \equiv \Delta_j e^{-u}$. Note that the distribution of the local densities of states also becomes LN still in the region $\langle g \rangle \gtrsim 1$ [13], in contrast to that of the global densities of states [11].

The function eq. (40) (Fig. 9) differs significantly in shape from a Gaussian one. As can be seen from eq. (40), the typical currents $j \sim \Delta$ exceed considerably the average current j_0. Moreover, due to the slow decrease of $\phi(\mathbf{j})$, the probability of large currents turns out not to be small. Another interesting feature of the LN distribution is that the variance $\widetilde{\Delta}$ determines not the width but the maximum probability of the distribution, while the probability of a current taking the average value turns out to be zero. The latter statement is only approximate, though. Allowing for corrections to the current density cumulants, eq. (39), makes this probability small but finite.

Note that typical fluctuation currents are large compared with the average value just because they are not correlated in the direction of the external field $\mathbf{E}$. The largest fluctuation currents are determined by the additional contributions to the cumulants eq. (39) which are determined by the (renormalized) diagrams like those in Fig. 7b. Their probability remains to be LN but with the exponent $1/4u$ instead of $1/u$ [14]. These large current density fluctuations are fully correlated in the direction of $\mathbf{E}$.

---

[3]The expressions for the even cumulants where all the vector indices were taken into account, as well as those for the odd cumulants, have been given elsewhere [13].

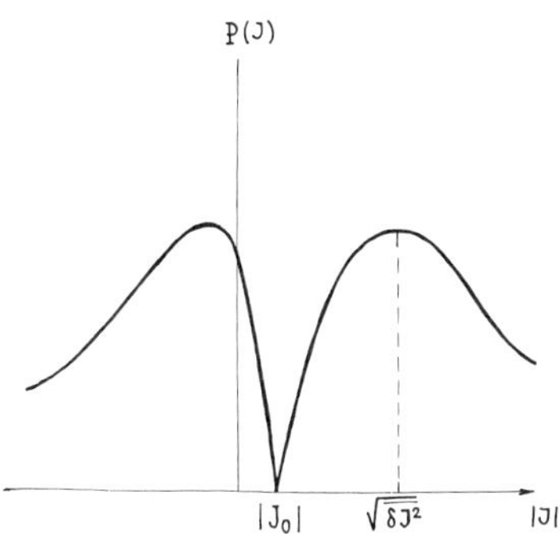

Figure 9. The LN distribution function of the local fluctuations of the current density. This distribution is valid in the region $u \gtrsim 1$ where (for $d = 2$) $g_{cl} - \langle g \rangle \sim g_{cl}$ but still $\langle g \rangle \gg 1$. In the region of weak localization where $g_{cl} - \langle g \rangle \ll g_{cl}$ so that $u \ll 1$, the current density distribution function remains Gaussian in the bulk but with LN tails like in the case of the conductance fluctuations in Fig. 8.

Therefore, there is a crossover in the shape of the current density distribution function from a close-to-Gaussian distribution with LN tails to a completely LN one. The crossover takes place with decreasing $\langle g \rangle$ (i.e. disorder and/or sample size increasing), however still in the metallic region, $\langle g \rangle \gtrsim 1$. In the case of the conductance fluctuations, the distribution function remains close-to-Gaussian with LN tails in the whole region of applicability of the perturbative RG analysis. However, the importance of the tails eq. (37) increases with decreasing $\langle g \rangle$. So, the hypothesis [12] seems quite plausible that a similar crossover exists also in the shape of the distribution function of the conductance fluctuations, although for stronger disorder.

This hypothesis is confirmed indirectly by the results of various calculations for one-dimensional (1 D) systems. For such systems which are always in the insulating state, the distribution functions of the resistance [47, 48, 49] and the global and local density of states [50, 51] are LN. It is interesting to note that (unjustified by any parameter) analytical continuation to $d = 1$ of our $d + \epsilon$ results obtained in a 'metal' phase reproduces exactly, including the numerical factors in the exponents of LN laws, the rigorous 1 D results. Finally, the existence of the limiting LN distribution in the insulating phase has been predicted by Shapiro [52] by using calculations within the Migdal–Kadanoff scheme. Although such calculations are not rigorous, the coincidence of results of different approaches appears to be a good argument in favour of such a behaviour.

A relation of the 2 D case at hand to the 1 D case can be found [45] based on the multi-channel approach [53] which is related to the Landauer formalism [54]. According to this approach, a conductance is contributed by $\langle g \rangle$ independent channels, each contribution being of the order of unity. In a metallic region, the number of independent channels is large so that the central limit theorem leads to the normal bulk distribution. When approaching the region of the strong localization, $\langle g \rangle$ decreases corresponding to a decrease in the number of the channels resulting in the wide non-Gaussian distribution described above. Finally, in the dielectric region $\langle g \rangle \ll 1$ one comes to a situation where the transparency of each channel is exponentially small so that conductance is governed by the most transparent channel. Then it is natural to assume that the conductance distribution is defined by the LN distribution in a single dielectric channel.

In the next section, after presenting time-dependent phenomena, I will discuss the question of universality of all of these distributions.

3 Mesoscopic Phenomena in Relaxation Rrocesses

In this section I will turn to a subject which seems to be quite different from that treated above: relaxation processes in disordered quantum conductors [18, 19]. Nevertheless, this subject is closely connected to the problem of mesoscopic fluctuations. The point is that the relaxation times are mesoscopically dispersed in an ensemble of disordered samples as I hope to show here. But first of all I would like to recall some very simple, basic properties of relaxation processes in disordered conductors.

One of the processes governed by relaxation is a transient characteristic, i.e. the time dependence of an electric current when stepwise changing a voltage. This current $\mathbf{j}(t)$ is determined by the frequency-dependent conductivity $\sigma(\omega)$

$$\mathbf{j}(t) = \int_0^\infty \mathbf{E}(t - t')\sigma(t')\,dt', \tag{41}$$

where $\sigma(t)$ is the Fourier transform of $\sigma(\omega)$

$$\sigma(t) = \int_{-\infty}^\infty \sigma(\omega)e^{-i\omega t}\,\frac{d\omega}{2\pi}. \tag{42}$$

The 'response function' $\sigma(t)$ describes the response to a $\delta(t)$-shaped pulse of the electric field. For any shape of the external electric field $\mathbf{E}(t)$, the transient current is given by the convolution eq. (41) of the external field and the response function. This quantity is also interesting by itself. Due to a direct analogy between free electrons transfer in quantum conductors and light scattering in opaque media, it gives a shape of short light pulse after propagating through a thick slab of non-absorbing material containing random scatterers [55].

The classical Drude formula $\sigma(\omega) = \sigma_0(1 - i\omega\tau)^{-1}$ gives the exponential response function

$$\sigma_0(t) = \frac{\sigma_0}{\tau}\,\exp\left(-\frac{t}{\tau}\right). \tag{43}$$

However, this classical answer is only valid for the time scale corresponding to the elastic mean free path $t \sim \tau$. For $t \gg \tau$, the response function is determined by quantum corrections to $\sigma(\omega)$. At the diffusion time scale $t \sim t_D = L^2/D$ the response function is completely determined [18] by the standard Cooperon correction [21]. This correction which led to the expression eq. (12) for the static conductance, may be written as follows

$$\sigma_1(\omega) = -\frac{4\,e^2}{h}\int_{q>L^{-1}} \frac{1}{q^2 - i\omega/D + L_\phi^{-2}}\,\frac{d^d q}{(2\pi)^d}.$$

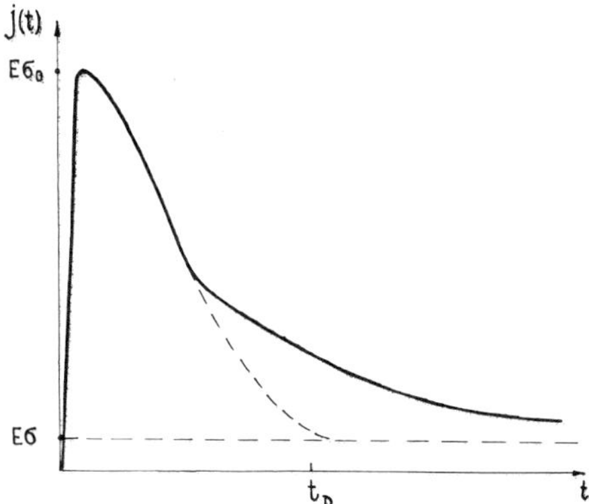

Figure 10. The relaxation current after stepwise switching
on the electric field. The dashed line shows an exponential
decay while the solid line shows the LN long-time-tail which
characterizes the relaxation properties. At large values of
u (i.e. for $g_{cl}/\langle g \rangle \gg 1$) the relaxation is strongly delayed.

Naturally, the integral can be explicitly calculated in any dimensionality also for $\omega \neq 0$. However, it is more convenient to use this form for calculating the quantum contribution to the response function eq. (42)

$$\sigma_1(t) = -\frac{e^2}{\pi h}\frac{1}{t^{d/2}}(4\pi D)^{1-d/2}e^{-t/t_0}, \tag{44}$$

where $t_0 = \min(t_D, \tau_\phi)$, $t_D = L^2/D$ is a characteristic diffusive time, and τ_ϕ is a phase-breaking time due to inelastic interactions. The shape of the entire response for a stepwise inclusion of the external electric field (a transient characteristics) is illustrated in Fig. 10. It is interesting to note that the quantum corrections are separated from the classical result on the time scale. So it is possible, in principle, to study these corrections independently of classical Drude conductivity.

However, the main subject of this section is to prove that the response function is not so simple. The reason is that the quantum corrections, eq. (44), do not describe properly the response function at very long times. The purpose of the following consideration is to show that an inclusion of the higher orders quantum corrections to $\sigma(\omega)$ changes the asymptotics of the response function to the LN one

$$\sigma_2(t) \propto \frac{\sigma}{\tau}\exp\left[-\frac{1}{4u}\ln^2\left(\frac{t}{\tau}\right)\right]. \tag{45}$$

This law becomes valid for a time obeying the following condition

$$t \gtrsim \frac{t_D}{4u}\ln\frac{t_D}{\tau}. \tag{46}$$

Here the parameter u is given by the same expression eq. (34) as in case of the tails of the mesoscopic fluctuations eq. (37). I will discuss later the reasons of this similarity of the LN laws in both cases.

Thus, by allowing for the long-time tails eq. (45), we find eventually a shape of the transient characteristics as shown in Fig. 10. After turning on the external electric field $\mathbf{E}$, the current rises in a short time $t \sim \tau$ up to its classical value $\mathbf{j}_0 = \sigma_0 \mathbf{E}$. After that it falls off to a steady-state value $\mathbf{j} = \sigma \mathbf{E}$, initially (at $t \lesssim t_D$) by a power (at $d > 2$) or logarithmic (at $d = 2$) law in accordance with eqs. (41), (44), then exponentially, eq. (44), and, finally, for times obeying the inequality eq. (46) with the LN law, eq. (45).

Now I want to outline a way of obtaining the long-time tails of the response function eq. (45). Naturally, a long-time behaviour of $\sigma(t)$, eq. (42), is defined by the low-frequency behaviour of $\sigma(\omega)$. It turns out that the higher order quantum corrections to $\sigma(\omega)$ are quite different from the corrections to the static conductivity. I have mentioned in the first section that all the leading singular corrections to $\sigma(0)$ cancel each other. As has been discussed above, this cancellation which results directly from the one-parameter scaling hypothesis was demonstrated diagrammatically [21] and then proved rigorously by RG analysis of the standard σ model [38, 20]. But no such cancellation takes place when taking into account the corrections [4] in the (small) parameter $\omega\tau$. Note that the absence of this cancellation was an evidence of breaking the one-parameter scaling in the case of frequency-dependent quantities [43].

To proceed, it is convenient to represent $\sigma(\omega)$ as a power series

$$\sigma(\omega) = \sigma_0 \sum_{n=0}^{\infty} C_n (i\omega\tau)^n + \frac{e^2}{h} \sum_{n=0}^{\infty} S_n (i\omega t_D)^n. \tag{47}$$

The coefficients C_n and S_n have to be calculated taking into account the quantum corrections. Note that without these corrections $C_n = 1$ and $S_n = 0$, so that eq. (47) leads to the Drude formula eq. (43). In the framework of the impurity diagram technique, the corrections to the coefficients C_n result from expanding the electron Green's function in ω, while the corrections to the coefficients S_n result from expanding the Cooperons or diffusons. The expansion includes, of course, the crossover terms proportional to $\tau^{\tilde{n}} t_D^m$ which were omitted here because they are irrelevant for the further analysis.

An absence of cancellation of the leading quantum corrections to $\sigma(\omega)$ indicates an insufficiency of the standard σ model in treating this quantity. Indeed, this model results directly in the one-parameter scaling description which, in turn, requires such a cancellation. Therefore, similar to the situation with the mesoscopic fluctuations, it is necessary to use the extended non-linear σ model. The appropriate additional vertices are derived by expanding eq. (26) also in powers of ω. The simplest vertices of such a type have the form

$$F[\omega] = \sum_{m=1}^{\infty} F_m[\omega] \equiv \frac{i\pi\nu_0}{4} \sum_{m=1}^{\infty} \left\{ (i\omega\tau)^m z_{0m} \int \mathrm{Tr}\,(\Lambda Q)^m \, d^d r \right\},$$

where the bare values of the charges z_{0m} are dimensionless numbers which only slowly depend on m. Mixed vertices containing both ω-terms and covariant derivatives eq. (29) are also of importance

$$F_{l,m} = z_{l,m} \int \mathrm{Tr}\left[(\nabla Q)^{2l} (\Lambda Q)^m \right] d^d r.$$

In particular, just the vertices $F_{2,m}$ with the renormalized charges $z_{2,m}$ contribute directly to the coefficients C_m of the expansion eq. (47).

---

[4] Here and in what follows I describe the situation of finite size or finite temperature systems when $L, \mathcal{L}_0 < L_\omega \equiv (D/\omega)^{1/2}$. Then it is either the parameter $\mathcal{L}_0$ defined by eq. (8) or the sample size L which cuts off logarithmically divergent (at $d = 2$) integrals while a dependence on ω may be taken into account in a power expansion.

I will not describe here the mechanism of such contributions: it is similar to that described in connection with the mesoscopic fluctuations. The main result is that the coefficients increase with n similar to the cumulants in eq. (33)

$$C_n \propto \exp\left[u(n^2 + 2n)\right]. \tag{48}$$

This rapid increase is the reason for the first sum in eq. (47) to dominate the calculation of the asymptotics of the response function while the second sum proves to be completely negligible, in spite of the inequality $\tau \ll t_D$.

Now, by substituting eq. (48) into the series eq. (47) for $\sigma(\omega)$, and using the definition eq. (42) together with the identities similar to eq. (36), we obtain the response function as follows

$$\sigma(t) \propto -\frac{\sigma_0}{\tau} \int_0^\infty e^{-t/t_\phi} \exp\left[-\frac{1}{4u} \ln^2 \frac{t_\phi}{\tau}\right] \frac{dt_\phi}{t_\phi}. \tag{49}$$

Note that the direct comparison of this law with that of eq. (44) yields the estimate eq. (46) for the time of the beginning of the non-exponential decay eq. (49). After integrating over t_ϕ, eq. (49) results in the LN asymptotic expression of the response function eq. (45). However, this expression is also of independent interest because it gives a convenient interpretation of the obtained results.

Namely, one can treat the long-time non-exponential decay of the response function eq. (45) as arising from averaging exponential contributions $\exp(-t/t_\phi)$ with different relaxation times t_ϕ. The asymptotics of the distribution function of these relaxation times turns out to be LN, eq. (49). If we assume that this distribution function has also a comparatively sharp maximum at $t_\phi \sim t_D$, we find that it describes both the asymptotics eq. (45) and the ordinary exponential decay eq. (44). Therefore, we obtain a distribution function with the same form as that of the conductance or density of states mesoscopic fluctuations (see Fig. 8).

One may interpret this distribution of relaxation times in two different ways. It may characterize either the entire ensemble of samples or the properties of a particular sample. In the first case, it would appear that LN tails result only from averaging which includes contributions of samples with untypically long relaxation times so that the LN decay is not an intrinsic property of any particular sample. To make the situation clear, one should investigate not only the average response function $\langle \sigma(t) \rangle$ but also its mesoscopic fluctuations.

For estimations, it is sufficient to calculate the variance of $\sigma(t)$ in the statistical ensemble. This variance has been obtained as follows [19]

$$\frac{\langle \delta\sigma^2(t) \rangle}{\langle \sigma(t) \rangle^2} = \left(\frac{t}{t_D}\right)^{d/2} = \frac{(Dt)^{d/2}}{L^d}. \tag{50}$$

The expression for $\langle \delta\sigma(t_1)\,\delta\sigma(t_2) \rangle$ at $t_1 \neq t_2$ is a little bit more complicated but contains no essential new information. Therefore, the relative value of the dispersion in the response function $\sigma(t)$, while being small for $t < t_D$, becomes substantial and increasing for $t > t_D$. If t obeys the inequality eq. (46), the relative value becomes even larger: it is proportional to $\exp(+u^{-1}\ln^2 t)$ [19].

To conclude, we can state the following. Due to the large dispersion of the response functions from sample to sample, eq. (50), the long-time tail in the average response function, eq. (45), and even the exponential decay, eq. (44), has nothing to do with relaxation processes in a particular sample. However, *this conclusion is valid only in case of zero temperature.*

An insight may be the following. Any dispersion in the response function should disappear in the thermodynamic limit $L \to \infty$. Indeed, the variance eq. (50) is proportional to L^{-d}. However, at $T = 0$ relaxation processes occur only in the leads at the boundary of the sample so that the average relaxation time $t_D = L^2/D$ also tends to infinity with L. For finite temperatures relaxation times τ_ϕ which are due to inelastic processes become shorter than t_D and independent of L. Then, the fluctuations in the response function vanish in the thermodynamic limit, so that this function becomes self-averaging. Moreover, even for $\tau_\phi > t_D$ the temperature effects prove to suppress the fluctuations of $\sigma(t)$.

In the case when $L_\phi > L > L_T \equiv \sqrt{D\hbar/T}$, phase relaxation caused by inelastic interactions in the sample is negligible. In the formal limit $L_\phi \to \infty$, a finite temperature does not change the average response function. But even in this limit the variance of the response function is decreasing in comparison with eq. (50) due to the smearing of the Fermi distribution $n(\varepsilon)$. The calculations yield [19]

$$K_T(t_1, t_2) = K_{T=0}(t_1 + t_2)\, R^2(t_1 - t_2), \tag{51}$$

where K_T is the variance of the response function at temperature T, and $R(t)$ is a damping factor

$$R(t) = \int_{-\infty}^{\infty} d\varepsilon \frac{\partial n(\varepsilon)}{\partial \varepsilon}\, \cos(\varepsilon t) = \frac{\pi T t/\hbar}{\sinh(\pi T t/\hbar)} \tag{52}$$

that decays exponentially for $t > \hbar/T$.

The exponential decay of the correlation function eq. (51) for $|t_1 - t_2| > \hbar/T$ evidently means that the response function is jagged in a specific sample. In other words, there are aperiodic oscillations in this function with an average period of the order of $\hbar/T$ which are reproducible for a given sample. Thus, temperature effects result in "mesoscopic grass" in the response function. It leads to a specific "ergodicity" of the time response: averaging of $\sigma(t)$ over a time interval Δt obeying $\hbar/T \ll \Delta t \ll t$ is equivalent to averaging over an ensemble of samples. This ergodicity has been proved [19] with the help of considerations similar to those used above in eqs. (9), (10) to prove the ergodicity of the conductance fluctuations with respect to averaging over Fermi energies. Since for relatively "high" temperatures $T \sim 1$K a typical oscillation time $\tau \sim 10^{-11}s$, it is more easy to measure just an averaged response. Certainly, a temporal "mesoscopic grass" would be easier to be observed in the milli-Kelvin temperature region.

In a response function averaged over Δt, it is possible to observe the LN long tails in a specific sample. However, the requirements for such an observation are such that it is formally possible only in $d > 2$ in the vicinity of the Anderson transition [45]. However, the situation becomes completely different when we take into account inelastic processes in the bulk.

Firstly, it is worth to note that the relative magnitude of the variance of the response function eq. (50) remains unchanged at finite temperatures: no inelastic processes affect the averaging of two loops representing different realizations. Then, in a time interval $\tau_\phi \ll t \ll t_D$ when relaxation processes are substantial but the variance remains relatively small, the response function is self-averaging. To calculate the response function, one can calculate directly the distribution of relaxation times which turns out to be applicable to a particular sample at $L \gg L_\phi$.

The idea is the following. The inverse of the phase relaxation time τ_ϕ^{-1} that maybe due to electron-phonon interaction, for instance, is determined by a one-loop diagram which is formally the same as that for the conductance. Similar to the density of states, the averaged τ_ϕ^{-1} is not influenced by a weak disorder. Nevertheless, the disorder influences its cumulants, including the variance, causing this quantity to fluctuate "mesoscopically".

There is a direct analogy in calculating these cumulants and the cumulants of the conductance: the diagrams which yield the most significant contributions to the cumulants $\langle\!\langle (\tau_\phi^{-1})^k \rangle\!\rangle$ for large values of k look exactly like the diagrams which describe the mesoscopic fluctuations of the static conductance (Fig. 2). The only difference is that in case of the relaxation time, a vertex in each loop corresponds to a traceless tensor $q_\alpha q_\beta - q^2 \delta_{\alpha\beta} d^{-1}$ instead of a vector q_α in the case of the conductance. This distinction is irrelevant for evaluating the diagrams, though. The appropriate calculations are based on including into the extended non-linear σ model some new additional vertices [19] which I will not specify here. The RG analysis yields for large k

$$\left\langle\!\!\left\langle (\tau_\phi^{-1})^k \right\rangle\!\!\right\rangle \propto (l/L)^{2(k-1)} \exp(uk^2)\left\langle\!\!\left\langle (\tau_\phi^{-1}) \right\rangle\!\!\right\rangle^k .$$

For small values of k, the higher order cumulants are small as compared to the variance which again resembles the case of the conductance fluctuations. The distribution function $f(\tau_\phi)$ is reconstructed from its cumulants in the same way as in the previous case. As the result we find again a distribution function with a comparatively sharp Gaussian peak at $\tau_\phi \approx \tau_{\phi 0}$ and a LN asymptotic behaviour at large τ_ϕ.

The presented brief analysis of the influence of inelastic processes has been carried out for samples of mesoscopic dimensions. In this case one may ignore the higher orders of the electron-phonon interaction and define τ_ϕ^{-1} in a self-consistent way. It is physically clear, however, that in the limit $L \to \infty$ the subject of interest are the fluctuations in τ_ϕ^{-1} not throughout the volume but within domains with dimensions of the order of $L_{\phi 0} = (D\tau_{\phi 0})^{1/2}$. In this case, the LN asymptotic behaviour of the relaxation current in a specific sample is found to result from averaging the contributions from all such domains. This is the reason why there exists a fairly large region of applicability of the asymptotics eq. (45) to a specific sample. Although even in this case the fluctuations of $\sigma(t)$ become large at $t \to \infty$ and L fixed, the time after which they occur can be arbitrarily long. (This time is finite because of the finiteness of the set of relaxation times in any particular sample.) This means that the long-time-tails in relaxation processes may be observed in real disordered conductors. The most interesting region is in the vicinity of the Anderson transition ($d > 2$) or when approaching the regime of strong localization ($d = 2$) when the parameter u governing the asymptotics eq. (45) may become arbitrarily large.

Summarizing, it is worth to discuss some features of the LN asymptotics in the distributions of mesoscopic fluctuations and relaxation processes. There are the two closely related questions about universality and one-parameter scaling. The parameter u depends on both the renormalized and the bare values of the conductivity σ (eq. (34)). The dependence on the bare value means that the system retains some knowledge about its microscopic characteristics, so that a non-universality manifests itself in this way. It means also that the one-parameter scaling is formally invalid. Indeed, samples with the same average conductance $\langle g \rangle$ may be characterized by different distribution functions of the fluctuations and/or by different rates of relaxation processes. However, in the metallic region the LN tails contain only the small part of probability and LN relaxation occurs only at very long time. What is very important is that this part does not vanish with increasing the disorder parameter L/l but, on the contrary, becomes more essential. Finally, for $u \sim 1$ in the case of current density, and for $u \gg 1$ in the case of conductance, the distribution functions become completely LN. Yet, there is a possibility that this distribution is described by one-parameter scaling in the limit $g \ll 1$, i.e. deeply in the insulating phase, as was obtained with the Migdal-Kadanoff approximation [52]. There is no contradiction of this hypothesis with the results of rigorous considerations performed for the region of metallic conductivity. To make an extrapolation to the insulating phase, it is possible to substitute into the parameter u the expression for the conductivity in the Anderson insulating phase $\sigma = \sigma_0 \exp(-L/R_{\mathrm{loc}})$ where this phase is implied to be characterized with the single localization radius R_{loc}. In this way we obtain the "one-parameter" LN distribution. Just by such an extrapolation of our $(2+\epsilon)$-results to the dimensionality $d = 1$ (and substituting for this case $R_{\mathrm{loc}} = 2l$) we obtained the coincidence mentioned above of our results with the appropriate exact 1 D ones. However, another possibility is not excluded, namely that the disordered systems of higher (than 1) dimensionality are characterized in the phase of the Anderson insulator by a set of mesoscopically fluctuating localization radii. Then, no one-parameter description of the distributions of mesoscopic fluctuations would be possible. This question remains one of the most interesting open questions in connection with the description of the Anderson transition in terms of the distribution functions.

References

[1] A.D. Stone, Phys. Rev. Lett. **54**, 2692 (1985)

[2] B.L. Altshuler, Sov. Phys. JETP Lett. **41**, 648 (1985)

[3] P.A. Lee and A.D. Stone, Phys. Rev. Lett. **54**, 1622 (1985)

[4] R.A. Webb, S. Washburn, C.P. Umbach, and R.B. Laibowitz, Phys. Rev. Lett. **54**, 2696 (1985)

[5] S. Washburn, C.P. Umbach, R.B. Laibowitz, and R.A. Webb, Phys. Rev. **B 32**, 4789 (1985)

[6] C.P. Umbach, C. Van Haesendonck, R.B. Laibowitz, S. Washburn, and R.A. Webb, Phys. Rev. Lett. **56**, 386 (1985)

[7] V. Chandrasekhar, M.J. Rooks, S. Wind, and D.E. Prober, Phys. Rev. Lett. **55**, 1610 (1985)

[8] S. Datta, M. Melloch, S. Bandyopadhyay, R. Noren, M. Vaziri, M. Millier, and R. Reifenberger, Phys. Rev. Lett. **55**, 2344 (1985)

[9] D.C. Licini, D.J. Bishop, M.A. Kastner, and J. Melngailis, Phys. Rev. Lett. **55**, 2987 (1985)

[10] B.L. Altshuler and D.E. Khmelnitskii, Sov. Phys. JETP Lett. **42**, 359 (1985)

[11] B.L. Altshuler, V.E. Kravtsov, and I.V. Lerner, Sov. Phys. JETP **64**, 1352 (1986)

[12] B.L. Altshuler, V.E. Kravtsov, and I.V. Lerner, Phys. Lett. **134 A**, 488, (1989)

[13] I.V. Lerner, Phys. Lett. **133 A**, 253 (1988)

[14] I.V. Lerner, Sov. Phys. JETP **68**, 143 (1989)

[15] V.E. Kravtsov, I.V. Lerner, and V.I. Yudson, Sov. Phys. JETP **94**, 259 (1988)

[16] V.E. Kravtsov, I.V. Lerner, and V.I. Yudson, Phys. Lett. **134 A**, 245 (1989)

[17] I.V. Lerner and F. Wegner, Z. Phys. B, in print (1990)

[18] B.L. Altshuler, V.E. Kravtsov, and I.V. Lerner, Sov. Phys. JETP Lett. **45**, 199 (1987)

[19] B.L. Altshuler, V.E. Kravtsov, and I.V. Lerner, Sov. Phys. JETP **67**, 795 (1988)

[20] S. Hikami, Phys. Rev. **B 24**, 2671 (1981)

[21] L.P. Gor'kov, A.I. Larkin, and D.E. Khmel'nitskii, Sov. Phys. JETP Lett. **30**, 228 (1979)

[22] S. Washburn and R.A. Webb, Adv. Phys. **35**, 375 (1986)

[23] S.B. Kaplan and A. Hartstein, Phys. Rev. Lett. **56**, 2403 (1986)

[24] B.L. Altshuler, V.E. Kravtsov, and I.V. Lerner, Sov. Phys. JETP Lett. **43**, 441 (1986)

[25] D.Yu. Sharvin and Yu.V. Sharvin, Sov. Phys. JETP Lett. **34**, 272 (1981)

[26] B.L. Altshuler, A.G. Aronov, and B.Z. Spivak, Sov. Phys. JETP Lett. **33**, 94 (1981)

[27] P.A. Lee and T.V. Ramakrishnan, Rev. Mod. Phys. **57**, 287 (1985)

[28] E. Abrahams, P.W. Anderson, D.C. Licciardello, and T.V. Ramakrishnan, Phys. Rev. Lett. **42**, 673 (1979)

[29] A.G. Aronov, A.I. Zyuzin, and B.Z. Spivak, Pis'ma v ZhETP **43**, 431 (1986)

[30] A.I. Zyuzin and B.Z. Spivak, Zh. Eksp. Teor. Phys. **93**, 994 (1987)

[31] S. Maekawa, Y. Isawa, and H. Ebisawa, J. Phys. Soc. Jpn. **56**, 25 (1987)

[32] M. Büttiker, Phys. Rev. **B 35**, 4123 (1987)

[33] C.L. Kane, R.A. Serota, and P.A. Lee, Phys. Rev. **B 37**, 6701 (1988)

[34] H.U. Baranger, A.D. Stone, and D.P. DiVincenzo, Phys. Rev. **B 37**, 6521 (1988)

[35] S. Hershfield and V. Ambegaokar, Phys. Rev. **B 38**, 7909 (1988)

[36] F. Wegner, Z. Phys. **B 35**, 207 (1979)

[37] L. Schäfer and F.Wegner, Z. Phys. **B 38**, 113 (1980)

[38] K.B. Efetov, A.I. Larkin, and D.E. Khmel'nitskii, Sov. Phys. JETP **52**, 568 (1980)

[39] A. Houghton, A. Jevicky, R.D. Kenway, and A.M.M. Pruisken, Phys. Rev. Lett. **45**, 394 (1980)

[40] A.J. McKane and M.Stone, Ann. Phys. (N.Y.) **131**, 36 (1981)

[41] K.B. Efetov, Sov. Phys. JETP **82**, 872 (1982)

[42] A.M.M. Pruisken and L. Schäfer, Nucl. Phys. **B 200**, 20 (1982)

[43] V.E. Kravtsov and I.V. Lerner, Sov. Phys. JETP **61**, 758 (1985)

[44] V.E. Kravtsov and I.V. Lerner, Sov. Phys. Sol. St. **29**, 259 (1987)

[45] B.L. Altshuler, V.E. Kravtsov, and I.V. Lerner, NORDITA Preprint **89/56 S**, 1-93 (1989)

[46] R.A. Serota, F.P. Esposito, and M. Ma, Phys. Rev. **B 39**, 2952 (1989)

[47] V.I. Mel'nikov, Sov. Phys. Sol. St. **23**, 444 (1981)

[48] A.A. Abrikosov, Sol. St. Commun. **37**, 997 (1981)

[49] P.W. Anderson, D.C. Thouless, E. Abrahams, and D.E. Fisher, Phys. Rev. **B 22**, 3519 (1980)

[50] B.L. Altshuler and V.N. Prigodin, Sov. Phys. JETP Lett. **45**, 687 (1987)

[51] B.L. Altshuler and V.N. Prigodin, Sov. Phys. JETP **67**, 135 (1988)

[52] B. Shapiro, Phil. Mag. **B 56**, 1031 (1987)

[53] Y. Imry, Europhys. Lett. **1**, 249 (1986)

[54] R. Landauer, IBM J. Res. Dev. **1**, 223 (1957)

[55] B.L. Altshuler, V.E. Kravtsov, and I.V. Lerner, in: Laser Optics of Condensed Matter, 217-222, Plenum Publishing Corp., N.Y. (1988)

HIGH–FREQUENCY QUANTUM TRANSPORT

Friedemar Kuchar*, Josef Lutz*, Kim Y. Lim*,
Ronald Meisels*, Günter Weimann°, Wilfried Schlapp⋄,
Alfred Forchel†, Arnd Menschig†, Detlef Grützmacher‡

*Inst. für Festkörperphysik, Universität, 1090 Wien, Austria, and
Ludwig Boltzmann Inst. für Festkörperphysik, Kopernikusgasse 15
1060 Wien, Austria
°Walter Schottky Institut, W-8000 München, F. R. Germany
⋄Forschungsinstitut der Deutschen Bundespost
W-6100 Darmstadt, F. R. Germany
†4. Physikalisches Inst., Universität, W-7000 Stuttgart, F. R. Germany
‡Inst. für Halbleitertechnik, RWTH, W-5100 Aachen, F. R. Germany

1 Introduction

The Quantum Hall effect and the quantum interference phenomena in mesoscopic systems are two prominent examples of quantum transport in semiconductors. In the past, they have mostly been studied using DC or low-frequency measuring techniques [1,2]. The use of high-frequency electric fields, however, is of particular interest if the product $\omega\tau_\phi \simeq 1$ [3-6] where $1/\tau_\phi$ is the phase breaking rate which determines the phase coherence of the wave function. High-frequency experiments regarding the integer Quantum Hall effect (IQHE) are summarized in [7]. In the microwave range they were performed by two groups [8,9]. Recently, more detailed data about the Hall plateau width and the disappearance of the plateaus at high frequencies were published [10,11]. In quantum wires high-frequency effects were studied using narrow silicon MOSFETs and GaAs epitaxial layers [6] and metal films [5,12]. In the first experiments on heterostructure quantum wires [13] the influence of a microwave field on the weak electron localization (WEL) and the universal conductance fluctuations (UCF) was studied.

In this paper we present new results regarding the microwave IQHE and the localization in the Landau levels as well as the effect of microwave radiation on the DC conductivity of quantum wires. For both, the IQHE and the transport in quantum wires, the theoretical description of the high-frequency effect is based on the introduction of a frequency-equivalent length L_ν which replaces the phase coherence length L_ϕ at high frequencies. L_ϕ is the parameter giving the size of a mesoscopic system and also determines the effective sample size in the DC IQHE [14].

2 Experimental Techniques

The effect of high-frequency radiation can be investigated in principle in two ways:
(a) The "photoconductivity" technique records the change of the DC or low-frequency
resistance due to the irradiation. (b) A "transmission"-type technique where the sample
sees the high-frequency field only. We have applied technique (a) to the experiments
on quantum wires made from $In_xGa_{1-x}As/InP$. They were mounted at the end of a
rectangular waveguide with the high-frequency field parallel to the DC field. Lock-
in detection allowed to vary the modulation frequency of the DC current and/or the
microwave radiation.

A transmission technique (b) was applied for the study of the IQHE [7,8] where
the Hall conductivity σ_{xy} can be measured. In the Ka-band (26.5 - 40 GHz) a waveguide
bridge as described in [17] was used, in the D-band (110-170 GHz) a single waveguide
arrangement. In both cases the essential part of the experimental set-up consists of
crossed rectangular waveguides which act as polarizer and analyzer. In this way the
component of the wave transmitted through the 2 DEG with a polarization perpendicular
to the incident wave can be selected. Its electric-field amplitude is proportional to σ_{xy}/N
with $N = (1+\sigma_{xx}Z/2)^2 + (\sigma_{xy}Z/2)^2$. Z is the impedance of the waveguide. N approaches
1 at high magnetic fields, i.e. far above the cyclotron resonance [17,18]. In that case the
microwave power transmitted through the sample is proportional to $\sigma_{xy}{}^2$.

The IQHE samples, $Al_xGa_{1-x}As/GaAs$ single heterostructures, were grown by
MBE (Table 1). For the quantum wires MOVPE grown high-quality modulation-doped
$In_xGa_{1-x}As/InP$ single quantum well structures were used as the starting material. Long
wires (width $\geq$ 120 nm) were produced by electron beam lithography and a combination of dry and wet etching ($HCl:HNO_3:H_2O=1:1:2$). The processing allowed a variation
of the electron mobility from 170000 in the starting material to 15000 cm^2/Vs in the
narrowest wire.

For both kind of experiments (a) and (b) room temperature radiation was filtered
out by a black polyethylene foil mounted in the waveguide in front of the sample.

3 DC Transport in $In_xGa_{1-x}As/InP$ Wires under Microwave Irradiation

Several effects can cause a photoconductivity signal due to the microwave irradiation
(26.5-40 GHz was the frequency range used in our experiments): (a) "Dephasing": The
high-frequency field destroys the constructive interference between electron waves backscattered along time-reversed paths [3]. This implies that the negative WEL magne-

Table 1. Parameters of the $Al_xGa_{1-x}As/GaAs$ samples.
From DC measurements performed on Hall bars from the
same wafer as the microwave samples.

Sample no.	1	2	3	4
$n_s[10^{11}\,cm^{-2}]$	2.9	5.5	2.5	1.4
$\mu[10^5\,cm^2/Vs]$	1.2	1.4	1.7	6.5

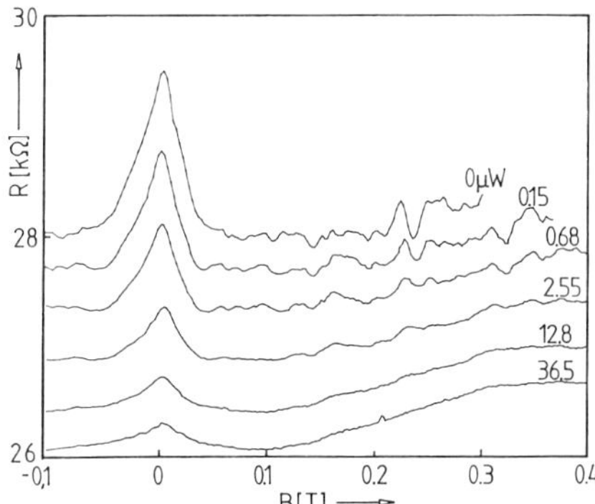

Figure 1. Magnetoresistance of a 260 nm wide $In_xGa_{1-x}As/$ InP wire with the incident microwave power (measured in the waveguide) as the parameter, $\mu = 56000\,cm^2/Vs$, $\nu = 35\,GHz$, $T = 1.5\,K$.

toresistance of the wire should disappear. The UCF are not expected to be destroyed by this dephasing since there are no contributions of time-reversed paths at fields above the negative WEL magnetoresistance and a phase change should only change the UCF pattern but not the amplitude of the fluctuations. (b) "Heating": Electron as well as lattice heating can destroy the WEL and the UCF as soon as the change of the energy of occupied conduction band levels is larger than the correlation energy E_c [15]. We note that dephasing in high-frequency fields, process (a), can occur even if the increase in electron energy is much less than E_c. (c) "Photovoltaic Effects" can be of different origin. Rectification of microwaves can be caused by mesoscopic junctions [16] or by the contacts (if non-ohmic). Hot electron effects can cause a non-linear current-voltage characteristic which can lead to a rectification at non-zero bias. (d) "Thermoelectric Effects" due to asymmetric heating of the contacts.

The dephasing and the electronic heating were experimentally distinguished from the other effects by a proper choice of the modulation of the DC current and the microwave radiation. The microwave field strength E_ω in the wire can be only roughly estimated from the incident microwave power since E_ω very sensitively depends on the coupling into the sample at the open end of the waveguide. This yields $E_\omega \approx 1V/cm$ for the data point at the highest field strength applied which certainly is an overestimate because of inefficient coupling and because of the large metallic contact areas attached to the wires. Figure 1 shows experimental results concerning the WEL and the UCF in a 260 nm wide wire of $In_xGa_{1-x}As/InP$ (length $L = 50\mu m$, $n = 8 \times 10^{11}\,cm^{-2}$, $\mu = 5.6 \times 10^4\,cm^2/Vs$) at $T = 1.5\,K$.

The UCF's are weaker in wires with lower mobility. In these wires, we observe the WEL peak at $B = 0$ as well as a magnetoresistance peak followed by a very strong negative magnetoresistance at fields of the order $1\,T$ (Fig. 2). This kind of magnetoresistance was also observed in $Al_xGa_{1-x}As/GaAs$ and attributed to a geometrical effect [19,20]. At the magnetoresistance peak, the cyclotron radius fits into the wire width causing a maximum of the interaction with the side walls. This interaction is drastically reduced when shrinking the cyclotron radius by increasing the magnetic field. When increasing the wire width the "geometrical" peak in the magnetoresistance shifts towards zero field and

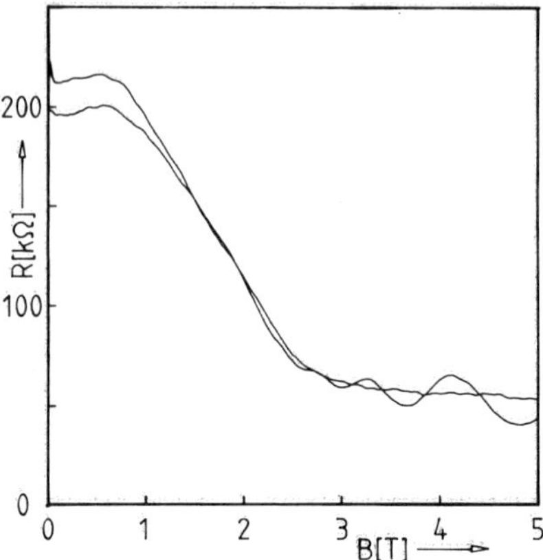

Figure 2. DC magnetoresistance of the 120 nm wire with $\mu = 15000\,\mathrm{cm^2/Vs}$. The curve without the SdH oscillations was recorded under microwave irradiation (power comparable to the lowest trace in Fig. 1), $T = 1.5\,\mathrm{K}$.

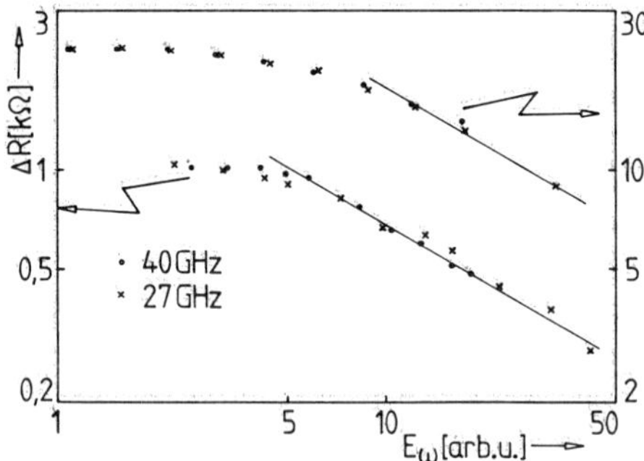

Figure 3. Weak localization contribution ΔR (see text) as a function of the microwave field strength for two $In_x Ga_{1-x} As/InP$ wires. The maximum applied field is approximately $1V/\,\mathrm{cm}$. The horizontal scales for the different sets of data are adjusted so that the steep sections can be directly compared. Lower set of data: wire with $w = 260\,\mathrm{nm}$, $\mu = 56000\,\mathrm{cm^2/Vs}$; upper set: 120 nm, $15000\,\mathrm{cm^2/Vs}$. $T = 1.5\,\mathrm{K}$.

eventually overlaps with the weak localization peak as observed in our samples. Strong microwave fields totally suppress the WEL peak and the SdH oscillations but leave the geometrical effect essentially unchanged. This shows that the latter phenomenon is not sensitive to electron heating (as long as $\omega_c \tau > 1$) and that it actually is of geometrical origin.

The dependence of the WEL contribution to the resistance at $B = 0$, ΔR, on the microwave field is plotted in Fig. 3. ΔR is defined as $R(B = 0) - R_{\min}(B)$. In the upper range of microwave fields, the data can be described by a dependence on field as $\Delta R \propto E_\omega^s$, with $s = -0.55$ (straight lines in Fig. 3). A complete comparison with theoretical results [3,4] is difficult since they are only given in the form of a complicated integral. Approximations for (a) low field amplitudes E_ω and high frequencies and (b) high field amplitudes and low frequencies are given in [3]. The two cases are distinguished by the value of a high-frequency parameter $\alpha = 2e^2 DE_\omega^2/\hbar^2\omega^3$ which contains the high-frequency field amplitude E_ω and the frequency ω. D is the diffusion constant. Case (a) corresponds to $\alpha \ll 1$, case (b) to $\alpha \gg 1$. The results are applicable to the frequency range where the effective inelastic scattering rate $1/t_0$ is larger than $1/\tau_\phi$. $1/t_0$ is connected with the absorption of high-frequency photons. t_0 is given by:

$$t_0 = \frac{1}{\omega\alpha} \qquad\qquad \alpha \ll 1 \tag{1}$$

$$t_0 = \frac{1}{\omega}\left(\frac{45}{2\alpha}\right)^{0.2} \qquad\qquad \alpha \gg 1 \tag{2}$$

The frequency dependent contribution to the conductivity due to two-dimensional WEL is

$$\Delta\sigma = \frac{e^2}{\pi h}\ln(t_0/\tau) \tag{3}$$

where τ is the elastic scattering time.

For one-dimensional systems $\Delta\sigma$ depends on frequency and field strength as follows:
(a) for $\alpha \ll 1$

$$\Delta\sigma \propto \omega E_\omega^{-1}\ln(\tau_\phi/t_0) \tag{4}$$

(b) for $\alpha \gg 1$

$$\Delta\sigma \propto (\omega E_\omega)^{-0.2}. \tag{5}$$

From our estimate of the electric field amplitude E_ω we calculate that $\alpha < 1$ in the whole field range. This indicates that $t_0 \approx \tau_\phi$ at the field strength where ΔR becomes strongly dependent on E_ω (Fig. 3). ΔR is independent of E_ω if $t_0 \gg \tau_\phi$. The value of τ_ϕ obtained from the fit to the DC WEL magnetoresistance is 18 ps. Setting this value equal to t_0 and using eq. (2) we obtain $\alpha = 0.1$ ($E_\omega \approx 0.6\,\text{V/cm}$) justifying our assumption that $\alpha \ll 1$ over the whole experimental field range. The value of 0.6 V/cm is in order of magnitude agreement with the estimate given above. However, from a comparison with ΔR data as a function of lattice temperature we conclude that electron heating dominates the behavior of ΔR under microwave irradiation [23]. The pure high-frequency dephasing process is not expected to dominate unless τ_ϕ is unrealistically large ($\sim 10^{-9}\,s$).

4   Localization and the Integer Quantum Hall Effect

The width of $i = 2$ and $i = 4$ Hall plateaus in σ_{xy} were measured on four $\text{Al}_x\text{Ga}_{1-x}\text{As}/$ GaAs samples as a function of temperature in the interval 1.5 - 4.2 K on a large homogeneous samples ($3 \times 3\,\text{mm}^2$) at Ka-band frequencies. The data shown in Fig. 4 were taken at a microwave power of about $1\,\text{mW}$. Reducing the power by a factor of 100 increases the noise but leaves the shape of the σ_{xy} curve unchanged. In all the samples the plateaus narrow above 2.4 K, below 2.4 K their widths remain constant. In DC experiments on a sample from the same wafer as that of Fig. 4, the width is observed

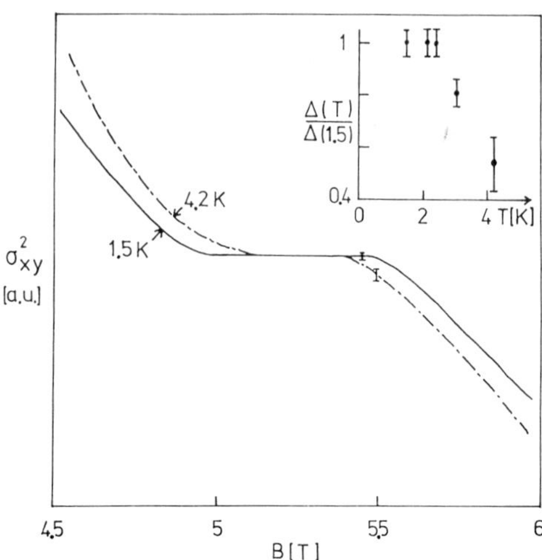

Figure 4. $\sigma_{xy}{}^2$ at the $i = 4$ plateau, sample 2, 35 GHz.
Insert: Temperature dependence of the plateau width Δ.

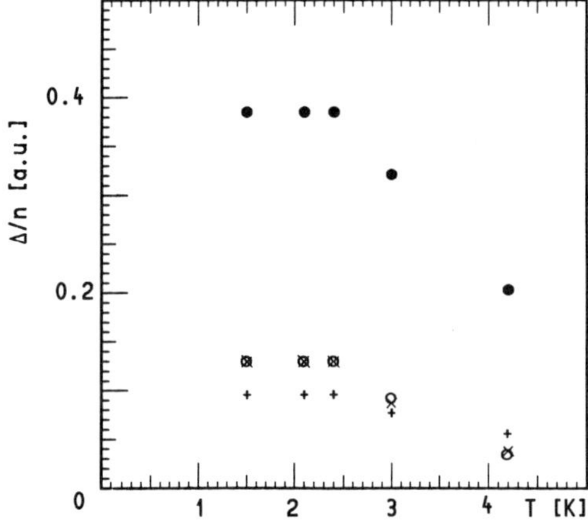

Figure 5. Temperature dependence of the plateau width Δ
at 35 GHz for four samples: 1 ... ●, 2 ... +, 3 ... o, 4
... ×. Sample 1, 3, 4: $i = 2$ plateau, 2: $i = 4$. Since the
magnetic field positions of the plateaus scale with n_s, Δ is
divided by n_s. Relative errors as in the insert of Fig. 4.

to increase steadily with decreasing temperature down to 0.4 K. The width plotted in
Fig. 5 is defined by a 0.5% deviation from the average plateau value. At a frequency
of 160 GHz, again the typical $\sigma_{xy}{}^2$ behavior (c.f. [8,17]) is observed, i.e. a peak at the
cyclotron resonance (at $B \simeq 0.5T$) and a $1/B^2$ decay at high magnetic fields. However,
just weak structure is present in the positions of the plateaus [11]. Similar observations
were made by Galchenkov et al. [10] at 30 and 60 GHz but at $T = 4.2$ K and on
lower-mobility samples where the Landau levels presumably overlap strongly.

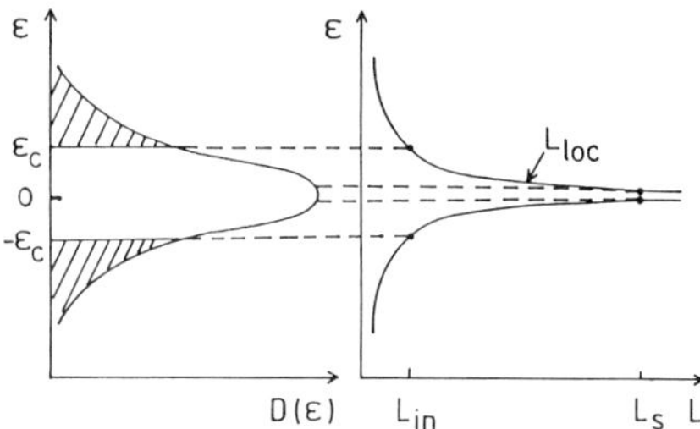

Figure 6. Schematic diagram of the density-of-states ($D(\epsilon)$, left) and characteristic lengths (L_s, L_{in}, L_{loc}, see text) in a Landau level. Shaded areas represent localized states. $\pm\varepsilon_c$ are the energies of the mobility edges relative to the center of the Landau level.

For the interpretation of the observed width of the IQHE plateaus we suggest a unified phenomenological model [11] for zero (low) and high frequencies. The discussion is based on the model shown in Fig. 6. The generally accepted phenomenological description of the IQHE assumes delocalized (extended) states near the center and localized states in the tails of the Landau levels. The plateaus in σ_{xy} or ρ_{xy} and the minima in σ_{xx} and ρ_{xx} occur whenever the Fermi level lies within the range of the localized states. At finite temperatures, the transition region between plateaus is finite and a measure for the number of delocalized states [14]. Wei et al. [14] found for the magnetic field regions between the plateaus that the maximum slope in the ρ_{xy} curve varies with temperature as $T^{-\kappa}$ and the half-width in the ρ_{xx} curve as T^κ with $\kappa = 0.42\pm0.04$. The characteristic lengths are the localization length L_{loc}, the sample length L_s, and the inelastic scattering length L_{in} ($\simeq L_\phi$). L_{loc} diverges at the center of the Landau level and decreases in the tails [21]. If the condition for the existence of localized states in the broadened Landau level were $L_{loc} < L_s$ only an extremely small number of delocalized states would exist and consequently the regions between the Hall plateaus would be extremely narrow. This is in contradiction to the general experimental DC results at not too low temperatures. At high enough temperatures in the presence of inelastic scattering, however, the localization of those states will be destroyed where $L_{in} < L_{loc}$. In other words, localization is only effective if $L_{loc} < L_{in}$ and localized states can be visualized as closed phase-coherent loops (closed equipotential lines in the percolation picture) as long as their size is smaller than L_{in}. L_{in} is a well established parameter for the description of mesoscopic systems and is given by $\sqrt{D\tau_{in}}$. The diffusion constant $D = v_F^2\tau/2$ is $5.6 \times 10^3\,\mathrm{cm^2/s}$ (τ from the mobility) for the sample of Fig. 4. $1/\tau_{in}$ is the inelastic scattering rate. For high frequencies we define a frequency-equivalent length scale $L_\nu = \sqrt{D\tau_\nu}$. As a first estimate τ_ν is set equal to the period of the high-frequency electric field since the results for t_0 of section 3 are applicable for the case $B = 0$ only. L_ν takes over the role of L_{in} as soon as it becomes smaller than L_{in}. L_ν is only very weakly temperature dependent (via D) below $T = 4.2$ K. Therefore one can expect a constant number of localized states and consequently a constant Hall plateau width whenever $L_\nu < L_{in}$. The data shown in Fig. 5 imply that this condition is fulfilled at $T \leq 2.4$ K. This also yields an estimate of $L_{in} \simeq 4\mu$m at the edges of the $i = 4$ Hall plateau. At high enough frequencies (but

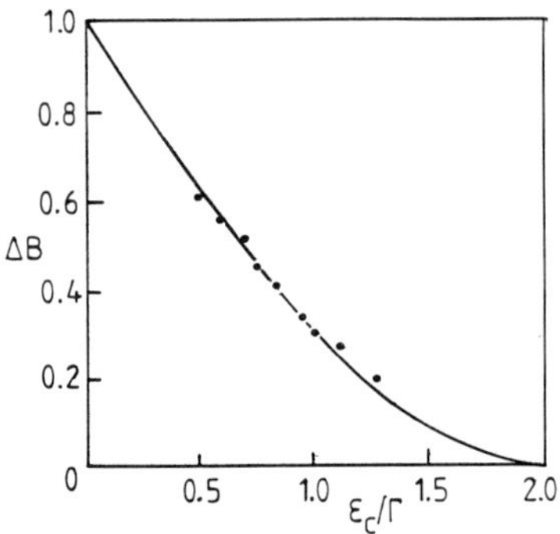

Figure 7. Normalized d. c. plateau width ΔB of sample 2. $\Delta B = 1 - 2F(T)$, see eq. (4). $\epsilon_c \propto T^\kappa$. Γ is the broadening parameter of the Landau levels. Full curve: theoretical results [22].

still far below the cyclotron resonance frequency) the plateaus should disappear. This obviously occurs already below 160 GHz ($L_\nu = 1.9\mu$m) in the sample of Fig. 4.

In this discussion the zero-field diffusion constant D was used. However, it is not clear whether this can be done in the case of high magnetic fields particularly near the mobility edge. Therefore, as regards the quantitative results, it might be better to use just the primary parameters appearing in the expressions for L_{in} and L_ν, viz. the scattering rate $1/\tau_{in}$ and the radiation frequency ν. Then, delocalization due to the high frequency occurs where $1/\nu < \tau_{in}$. This does not change the general conclusion that a physically macroscopic sample under IQHE conditions effectively behaves as a small system whose size is determined by inelastic scattering and frequency-induced delocalization, respectively. The delocalization can be only dominated by the high-frequency effect under the additional condition $h\nu > kT$. For 40 GHz this is fulfilled at $T \leq 2\,$K. This is also the range of temperatures where the plateau width is constant.

Recently, the results of a theoretical scaling study of the localization problem in a strong magnetic field were presented [22]. Based on the ideas mentioned above, quasi-mobility edges ε_c in the Landau levels were introduced which occur at those energies where $L_{loc} = L_{in}$. With $L_{in} \propto T^{-p/2}$ this allowed to calculate the temperature dependence of ε_c and of the Hall plateau width Δ. Δ is defined by the number of localized states between the upper and lower mobility edges of two successive Landau levels.

$$\Delta(T) \approx \frac{n_s h}{e} \frac{1}{(N + 1/2)(N + 3/2)}(1 - 2F(T)) \tag{6}$$

$$F(T) = 0.37(\varepsilon_c - 0.08\varepsilon_c^3). \tag{7}$$

n_s is the two-dimensional electron density. $N = i - 1$ ($i \ldots$ number of the Hall plateau), $\epsilon_c \propto T^\kappa$. A fit of the theoretical results to our DC data of sample 2 between $T = 0.4$ and 4.2 K (Fig. 7) yields the broadening parameter $\Gamma \approx 1\,meV$ of the Landau levels involved in the $i = 4$ plateau [22]. At $T = 0.4\,$K ε_c is found to be $\Gamma/4$ and $p \approx 2$ using $\kappa = 0.42$. We consider these results as the first quantitative evaluation of the Hall plateau width of the IQHE and of the position of the mobility edges in the Landau levels.

5 Acknowledgements

This work was supported by the "Fonds zur Förderung der wissenschaftlichen Forschung", Austria, project no. 6437 and by "Stiftung Volkswagenwerk", F.R.G.

References

[1] R.E. Prange, S.M. Girvin, The Quantum Hall Effect, Springer, Heidelberg (1987)

[2] Physics and Technology of Submicron Structures, H. Heinrich, G. Bauer, F. Kuchar, eds., Springer Series in Solid State Sciences **83** (1988)

[3] B L. Altshuler, A G. Aronov, D.E. Khmel'nitskii, Sol. St. Commun. **39**, 619 (1981)

[4] D E. Khmelnitskii, Physica **126 B**, 235 (1984)

[5] J. Liu, N. Giordano, Springer Proc. in Physics **28**, 159 (1988)

[6] A.A. Bykov, G.M. Gusev, Z.D. Kvon, D.I. Lybyshev, and V.P. Migal, JETP Lett. **49**, 13 (1989)

[7] F. Kuchar, R. Meisels, K.Y. Lim, P. Pichler, G. Weimann, and W. Schlapp, Physica Scripta **19**, 79 (1987)

[8] F. Kuchar, R. Meisels, G. Weimann, W. Schlapp, Phys. Rev. **B 33**, 2965 (1986)

[9] V.A. Volkov, D.V. Galchenkov, L.A. Galchenkov, I.M. Grodnenskii, O.R. Matov, S.A. Mikhailov, A.P. Senichkin, and K.V. Starostin, JETP Lett. **43**, 328 (1986)

[10] L.A. Galchenkov, I.M. Grodnenskii, M.V. Kostovetskii, O.R. Matov, JETP Lett. **46**, 542 (1988)

[11] K.Y. Lim, I. Auer, F. Kuchar, G. Weimann, W. Schlapp, A. Forchel, A. Menschig, Surf. Sc., in print

[12] P.E. Lindelof, S. Wang, Phys. Rev. Lett. **59**,1156 (1987)

[13] F. Kuchar, J. Lutz, K.Y. Lim, G. Weimann, W. Schlapp, A. Forchel, A. Menschig, Extended Abstracts, Symposium "New Phenomena in Mesoscopic Structures", p. 200, Hawaii (1989)

[14] H.P. Wei, D.C. Tsui, M.A. Paalanen, A.M.M. Pruisken, Phys. Rev. Lett. **61**, 1294 (1988)

[15] S. Washburn, R.A. Webb, Adv. in Phys. **35**, 375 (1986)

[16] V. Fal'ko, Europhys. Lett. **8**, 785 (1989)

[17] F. Kuchar, Adv. Sol. St. Phys. **28**, 45 (1988)

[18] R. Meisels, F. Kuchar, Z. Phys. **67**, 443 (1987)

[19] T.J. Thornton, M.L. Roukes, A. Scherer, B.P. van de Gaag, Phys. Rev. Lett. **63**, 2128 (1989)

[20] A. Menschig, A. Forchel, B. Roos, R. Germann, K. Pressel, W. Heuring, D. Grützmacher, submitted to Appl. Phys. Lett.

[21] B. Kramer, Y. Ono, T. Ohtsuki, Springer Ser. Sol. St. Sci. **87**, 24 (1989)

[22] B. Huckestein, W. Apel, B. Kramer, Springer Ser. in Sol. St. Sc., in print

[23] J. Lutz, F. Kuchar, A. Menschig, A. Forchel, D. Grützmacher, P. Benton, S.P. Beaumont, and C.D.W. Wilckinson, Proc. 20th ICPS, Thessaloniki, in print (1990)

FREQUENCY DEPENDENT TRANSPORT IN QUANTUM COHERENT SYSTEMS

Bernhard Kramer and Jan Mašek[†]

Physikalisch-Technische Bundesanstalt
Bundesallee 100, W-3300 Braunschweig, F. R. Germany
[†]Institute of Physics, Czechoslovak Academy of Sciences
Na Slovance 2, 180 40 Prague, Czechoslovakia

1 Introduction

At low enough temperatures the electrical transport properties of small conductors exhibit a rich variety of quantum phenomena due to the suppression of inelastic, phase randomizing scattering. They are related to finite size *quantization*, and to the *coherence* of the electron states throughout the whole sample. For the theoretical description of these *mesoscopic effects* one needs

1. information about the single particle quantum states in finite, but not atomic(!), systems in the presence of disorder,

2. a suitable transport theory that includes the effect of contacts and leads, and

3. information about the influence of interactions on the states, and on the transport processes.

The theory of the single particle states is comparatively simple. It requires essentially the solution of a random Schrödinger equation under certain boundary conditions. By means of modern computational facilities this is possible for surprisingly large system sizes [1].

More complicated is the development of a suitable transport theory. Two approaches have been considered. Most important was the discovery that in the quantum coherent regime the transport properties of a sample must be characterized by size and shape dependent conductances or resistances, instead of using the geometry independent components of a local, size and shape-independent conductivity or resistivity tensor. It was Landauer [2], who first pointed out in the context of one-dimensional (1 D) disordered conductors that in the DC limit coherent transport can be related to quantum mechanical transmission. In his approach, phase randomization is excluded completely from the "sample" which is considered as only a part of a larger system containing in

Quantum Coherence in Mesoscopic Systems
Edited by B. Kramer, Plenum Press, New York, 1991

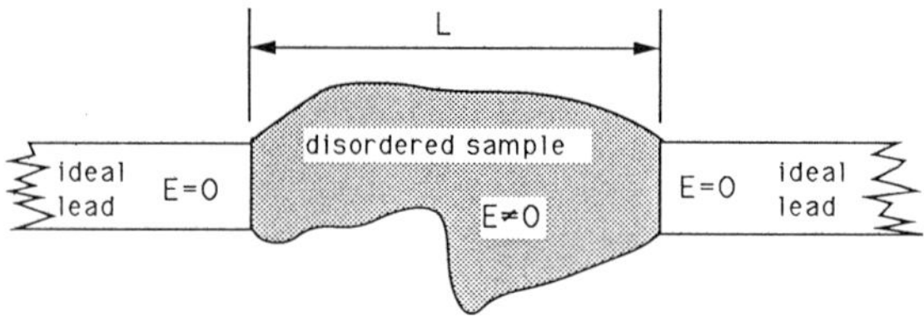

Figure 1. Disordered sample of arbitrary shape connected to ideal, quasi-1 D leads. The sample may be represented by a potential that includes not only the "internal" disorder due to impurities, for instance, but also the geometrical shape. The leads are assumed to be ideal quasi-1 D far away from the sample, not containing any randomness, and to have some finite, asymptotically constant cross-section. They serve to define the "transport channels". These correspond to the asymptotically free, geometrically quantized quasi-1 D electron states. In the Landauer theory it is assumed that the chemical potentials differ in the reservoirs (not shown in the Figure) attached on the left and on the right hand sides of the sample via the leads. In the linear response theory it is assumed that the electric field is non-zero only within the sample.

addition "reservoirs", and "ideal leads" that connect the "reservoirs" with the "sample" (Fig. 1). The former are characterized by their chemical potentials and simulate the "world" behind the contacts. The driving forces for the currents are the differences in the chemical potentials. All inelastic and phase randomizing processes necessary for energy and phase relaxation are incorporated into the "reservoirs" . The (random) potential that represents the "sample" scatters the electrons only elastically. Landauer's approach has not only been conceptually extremely useful. In addition, it provided an excellent scheme for the explanation of the transport features of 2 D ballistic point contacts, and certain aspects of the quantized Hall effect [3].

On the other hand, quantum mechanical linear response theory relates the local current density $\vec{j}$, and the electric field $\vec{E}$ via the non-local conductivity tensor σ.

$$\vec{j}(\vec{r},t) = \int d^3r' dt' \sigma(\vec{r},t;\vec{r}',t')\vec{E}(\vec{r}',t') \tag{1}$$

Equation (1) should be able to provide the general microscopic framework for the quantum transport theory of coherent systems. Thus, the Landauer theory must follow from linear response for the proper geometry as a special limiting case. In the DC limit the conditions for the equivalence was discussed by several authors [4, 5, 6, 7, 8, 9, 10]. We will not repeat here the various arguments. We only mention a few facts that strongly motivate the reformulation, and the application, of linear response theory for mesoscopic systems [11, 12].

1. Since it starts from time-dependent currents and fields it can be used for systematic studies of frequency and time dependencies.

2. As a microscopic theory it allows, at least in principle, to discuss the influence of all of the scattering processes (inelastic and elastic) on an equal (microscopic) footing,

3. it provides the framework for the identification of the "transport channels" in the Landauer theory. The quantum mechanical transmission probability amplitudes of the channels can be calculated microscopically, and

4. it can straightforwardly be generalized to discuss non-linear mesoscopic effects [13, 14].

Most difficult is the inclusion of the phase and energy randomizing processes. Although they can be treated microscopically, in principle, for instance by diagrammatic techniques [15], it turned out to be extremely complicated in practice. For most purposes it suffices to consider these processes to be incorporated into an *inelastic scattering time* τ_{in} that depends on the temperature[15, 16, 17]. If the temperature is low enough such that this "mean time between two successive inelastic events" is of the order of or larger than the characteristic travelling time of an electron through the sample, quantum transport phenomena will become experimentally accessible.

In this paper, we want to reformulate the linear response theory for the time-dependent transport in quantum mechanically coherent systems. We consider the zero temperature limit, and neglect phase and energy randomization completely. Results for a quasi-1 D system are reported that, on the one hand, provide insight into the microscopic nature of the "transport channels" discussed in the context of the Landauer approach, and, on the other hand, allow for some predictions concerning the frequency dependence of the conductance for systems without and with a perturbing (random) potential.

2   States in Quantum Wires

As the startpoint, we consider independent electrons in an ideal "quantum wire" of the length $L(\to \infty)$ in the x-direction (periodic boundary conditions), and of finite width M in the perpendicular directions. The "sample" can then be defined by adding to the Hamiltonian an additional potential that describes the influence of the impurities, and of the geometrical shape . Without potential we consider the portion of the quantum wire containing the applied electric field as the "sample". The eigenvalues and eigenstates of the Schrödinger equation for the ideal quantum wire are

$$E_\mu(k) = E_\mu + \frac{\hbar^2 k^2}{2m}, \tag{2}$$

$$\Psi_{\mu k} = \phi_\mu e^{ikx}. \tag{3}$$

Due to the geometrical constriction of the motion of the electrons perpendicular to the x-direction the spectrum consists of well separated quasi-1 D energy bands. In the presence of a magnetic field the motions in the x- and in the perpendicular directions do no longer decouple. The shape of the energy bands and the corresponding wave functions are more complicated, although the changes remain small as long as the cyclotron energy is small compared with the distance between two successive energy bands (Fig. 2). A moderate perturbing potential (disorder, boundaries) introduces strong mixing of the bands apparent near the onsets E_μ due to the associated van Hove singularities in the

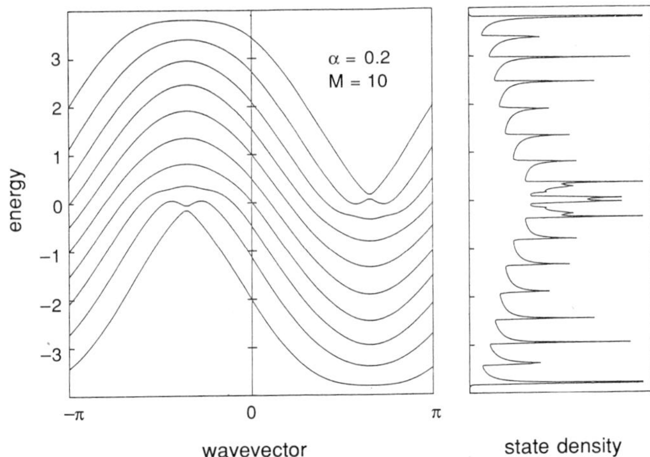

Figure 2. Quasi-1 D energy bands (in units of V, the nearest neighbor matrix element of the Hamiltonian) of a tight binding system of the width M in the presence of a perpendicular magnetic field of the strength $B = \alpha\phi_0$ (ϕ_0 magnetic flux quantum).

density of states [18, 19, 20]. As we shall see in the next section, the quasi-1 D subbands define the "transport channels" occurring in the Landauer approach: at zero temperature transport takes place only in states with wave-vectors that correspond to intersections of the band structure $E_\mu(k)$ with the Fermi energy E_F. Thus, a given 1 D subband may correspond to several channels depending on its shape (Fig. 2).

3   Linear Response Approach to AC Conductance

The linear response expression for the conductance is obtained straightforwardly in the usual way. Startpoint is the Hamiltonian

$$\mathbf{H} = \mathbf{H_0} - \int dx' A(x',t)\mathbf{j}(x'). \tag{4}$$

The influence of the (inhomogeneous) electric field is included linearly via the vector potential, $E(x,t) = \partial A/\partial t$. For simplicity, it is assumed that only its x-component is non-zero, in view of the quasi-1 D nature of the system considered that is described by H_0, the unperturbed Hamiltonian. The current density operator is defined as $\mathbf{j} = (e/2m)[\mathbf{p}, \delta(x - \mathbf{x})]_+$, $\mathbf{p} = -i\hbar\partial_x$. For a monochromatic field the vector potential is

$$A(x,t) = \frac{i}{\omega + i\eta}E(x)e^{-i(\omega+i\eta)t} \tag{5}$$

The small imaginary part of the frequency, $i\eta$, guarantees the vanishing of the electric field for $t \to -\infty$. The dissipative conductance $\Gamma(\omega)$, in units of e^2/h, is defined by the absorbed power

$$P(\omega) = \frac{1}{2}\int dx' j(x')E(x') = \frac{e^2}{2h}\Gamma(\omega)U^2 \tag{6}$$

where $U = \int dx E(x)$ is the applied voltage. For plane waves in the x-direction (ideal system without perturbing potential) the non-local conductivity (eq. (1)) may be evaluated

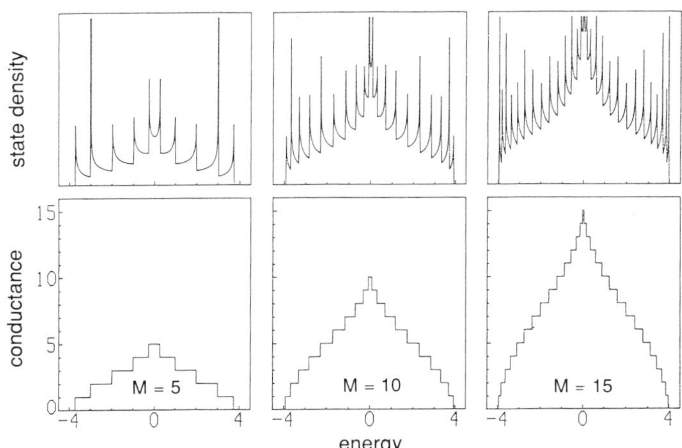

Figure 3. Density of states (in arbitrary units, top), and the DC-conductance (in units of e^2/h) of a tight binding system of the width M, as a function of the Fermi energy (in units of V).

explicitly [21]. In this case the frequency dependent conductance is given by the Fourier transform of the auto-correlation function of the field, $L(q) = \int dx\, dx'\, E(x) E(x+x') e^{iqx'}$.

$$\Gamma(\omega) = \sum_{\mu=1}^{\mu_{occ}} \frac{L(\omega/v_\mu)}{L(0)} \tag{7}$$

$v_\mu = [2(E_F - E_\mu)/m]^{1/2}$ is the Fermi velocity in the μth subband. μ_{occ} is the total number of the occupied channels.

The linear response expression for the conductance of a coherent quasi-1 D quantum system shows several interesting general features that are very important for the understanding of the experiments [18, 19, 20, 21, 22, 23].

1. For an ideal system (without perturbing potential) the conductance may be written as a sum of the contributions of independent transport channels. This remains true even in the presence of a homogeneous magnetic field.

2. The single contributions to the conductance of the individual subbands scale as $(\sin(x)/x)^2$, $x = \omega L/2v_\mu$, if the electric field is assumed to be homogeneous within an interval of the length L and zero outside.

3. In general the conductivity tensor is non-local in space and time. The conductance depends on the geometry in a non-trivial way.

4. The AC conductance depends on the spatial shape of the electric field. In order to predict the result of an experiment one would have to know the spatial dependence of the field in the sample, i. e. one would have to consider interaction effects and screening.

5. In the DC limit only the applied voltage is of importance. For the ideal quasi-1 D system the DC conductance is quantized and given by $(e^2/h)\mu_{occ}$ independent of the length of the system, and of the spatial behavior of the electric field (Fig. 3).

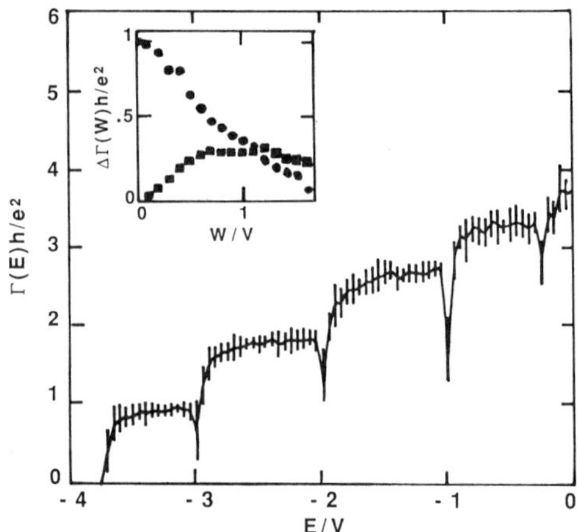

Figure 4. The average of the DC-conductance (in units of e^2/h) of a quasi-1 D disordered tight binding model of the width $M = 15$ that contains disorder of the strength $W = 0.2V$ as a function of the Fermi energy E (in units of V). The length of the sample is $L = 200$. The insert shows the average height of the conductance steps ($\bullet$), and the square root of the variance of the conductance fluctuations ($\square$) as a function of the disorder W for $L = 100$.

4 Influence of a Perturbing Potential on DC Transport

Perturbing potentials may have various physical origins. Impurities, defects, dislocations, and surface roughness lead unavoidably to a certain degree of disorder in any real sample. Also the man-made shape of a sample may be represented by some additional potential energy. In order to compare with experiments it is necessary to study the influence of these perturbations on the conductance. Here we consider only the case of a random potential.

Figure 4 shows the average of the DC conductance as a function of the Fermi energy of a quasi-1 D tight binding system subject to a random potential energy that is distributed according to a box distribution of a width W [19]. The average conductance is still quantized, although the unit of quantization (heights of the steps) is smaller than e^2/h. Due to the disorder, the conductance fluctuates within the statistical ensemble. The square root of the variance (rms) of the conductance fluctuations, $\Delta\Gamma$, are shown as vertical bars in the Figure. As can be seen from the insert, approximate quantization persists as long as $\Delta\Gamma$ does not exceed the average step height which happens here at $W = V$ (V is the off-diagonal matrix element of the Hamiltonian between nearest neighbor lattice sites). At the onset of each of the plateaus one observes a sharp anti-resonance like quenching of the conductance which can be attributed to an enhancement of the impurity scattering due to the van Hove singularities in the density of states at the onsets of the 1 D subbands. The width of the anti-resonances increases with increasing disorder and length of the sample. Instead of quantized plateaus the conductance shows a non-monotonous increase with regular oscillations. The effect predicted in [19, 20] has been observed in recent experiments on narrow constrictions prepared on materials with

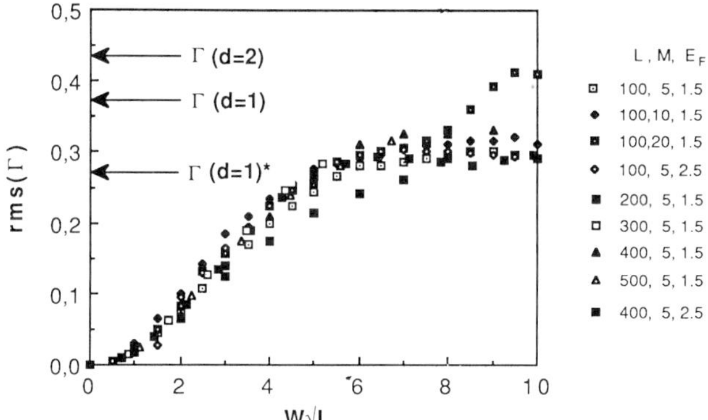

Figure 5. The root mean square deviation of the conductance, $rms(\Gamma)$, of a quasi-1 D disordered tight binding system (disorder W) of the length L and the width M at a Fermi energy E_F as a function of $W\sqrt{L}$. The arrows indicate the values obtained for the universal conductance fluctuations in 1 D and 2 D [24]. The value $\Gamma(d=1)^*$ was obtained by using results of random matrix theory [25].

lower mobility (85 m^2/Vs, see Fig. 8 of the paper by Timp et al. in this volume).

The behavior of the fluctuations of the conductance is shown in more detail in Fig. 5 where we have plotted $rms(\Gamma)$ of a tight binding system with randomness W, of the width M, and of the length L for different Fermi energies as a function of $W\sqrt{L}$ [23]. First of all, we observe that $rms(\Gamma)$ scales as a function of $W\sqrt{L}$. Secondly, for $W\sqrt{L} \leq 2$, $rms(\Gamma) \propto W^2 L$ whereas for $W\sqrt{L} \geq 6$, $rms(\Gamma)$ saturates at a value of about 0.3. This value is in reasonable agreement with the results of the theories of the universal conductance fluctuations (UCF) [24, 25] which occur as soon as $L \geq \ell$ where $\ell \propto W^{-2}$ is the mean free path due to the disorder. For smaller systems the conductance fluctuations are not universal. They depend on the disorder, and on the geometry of the system but still show the scaling behavior mentioned above.

5   AC Conductance and Disorder

As a function of the frequency ω of the applied electric field the conductance of a random system fluctuates around its average value, as shown in Fig. 6 for the case of a single channel (1 D system) [22]. If the disorder is not too large the average conductance still shows oscillatory behavior, reminiscent of the oscillations for $W = 0$.

The magnitude of the fluctuations is of the same order as the average conductance. It depends in a characteristic manner on ω and on the length L of the sample. This can be seen in Fig. 7 where the configurational average of the logarithm of the conductance, and the corresponding rms, Δ, are shown for several frequencies as a function of L [22]. As long as the localization length λ which is determined by the disorder and the energy is much larger than $\lambda(\omega) \equiv 2\pi v_F/\omega$ the oscillations $\langle \log \Gamma \rangle$ (and correspondingly in Δ) as a function of L can clearly be detected (Fig. 7, $W = 0.2V$). For $\lambda < \lambda(\omega)$ the behavior of $\langle \log \Gamma \rangle$, and Δ is monotonically decreasing and increasing , respectively, with increasing

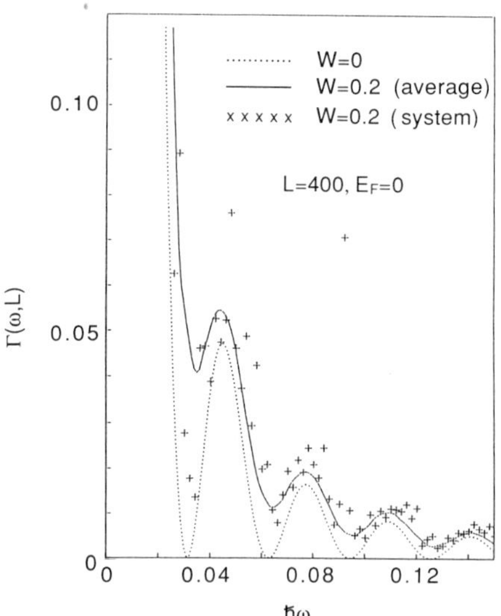

Figure 6. The conductance Γ of a 1 D disordered tight binding system of the length $L = 400$ for disorder $W = 0$ (dotted curve) and $W = 0.2V$ (configurational average: full curve; single configuration: crosses) at the centre of the band, $E_F = 0$.

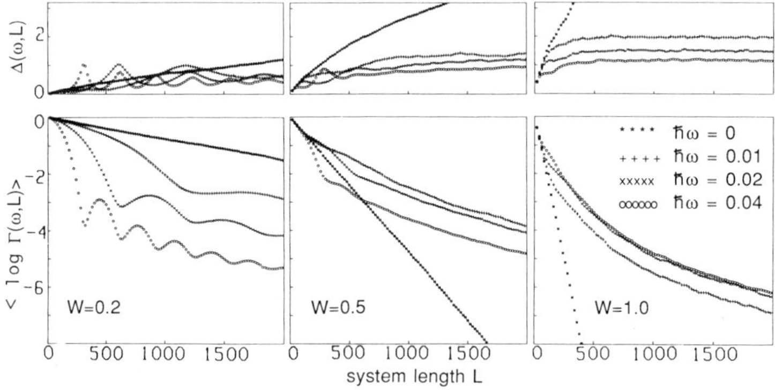

Figure 7. The configurational average of the logarithm of the conductance, $\langle \log \Gamma \rangle$, and the corresponding rms, Δ, as a function of the system length L for different disorder W, and different frequencies ω as indicated.

L (Fig. 7, $W = 1.0V$). Analysis of the data shows that for $W \neq 0$ the relative rms

$$rrms(\omega, L) \equiv -\frac{\Delta(\omega, L)}{\langle \log \Gamma(\omega, L) \rangle} = \frac{const(\omega)}{\omega^{1/2} \log L} \tag{8}$$

behaves qualitatively different from that in the DC limit

$$rrms(0, L) = \frac{const}{L^{1/2}}. \tag{9}$$

In the latter case $\langle \log \Gamma \rangle$ is self-averaging in the conventional sense, i.e. the relative fluctuations die out as the inverse of the square root of the size of the system. However, for $\omega \neq 0$ $rrms(\omega, L)$ decreases only logarithmically with L. Thus, $\log \Gamma$ *is, in the strict sense, not self-averaging with respect to the system size* in contrast to the DC limit. It is then obvious that also $\Gamma(\omega)$ cannot be self-averaging.

On the other hand, the fluctuations decrease with increasing frequency as $\omega^{-1/2}$. This suggests that the contributions to the conductance of the states that lie energetically within the interval $\hbar\omega$ around the Fermi energy are statistically independent, and contrasts the behavior in the metallic limit $L \ll \lambda$ were the existence of the UCF in the DC limit indicate strong correlations [26].

6 Conclusion

We have pointed out in the preceding sections a number of unusual and surprising mesoscopic effects in the frequency dependent linear transport properties of quantum coherent systems. Whereas in the DC limit experiments are comparatively simple and show clearly quantum interference effects, and quantization phenomena similar to those discussed above (cf. the papers by Timp et al., and Thornton and Roukes in this volume) experiments at finite, and high enough frequencies are difficult to perform, due to the smallness of the samples. It remains to be seen whether or not frequency or time dependent mesoscopic effects are experimentally accessible. However, independent of experimental accessibility the theoretical treatment of frequency dependent quantum transport is an interesting and important task in itself since it provides new insights into the *quantum* and especially also the *statistical* behavior of transport in macroscopic but coherent systems.

References

[1] M. Schreiber, Physica A **167**, 188 (1990)

[2] R. Landauer, in: Localization, Interaction, and Transport Phenomena, B. Kramer, G. Bergmann, Y. Bruynseraede, eds., Springer Series in Solid-State Sciences **61**, 38, Springer Verlag, Berlin (1985)

[3] M. Büttiker, in: Festkörperprobleme/Advances in Solid State Physics, U. Rössler, ed., vol. **30**, 42, Vieweg Verlag, Braunschweig (1990)

[4] D. S. Fisher and P. A. Lee, Phys. Rev. **B 23**, 6851 (1981)

[5] D. C. Langreth and E. Abrahams, Phys. Rev. **B 24**, 2978 (1981)

[6] E. N. Economou and C. M. Soukoulis, Phys. Rev. Lett. **46**, 618 (1981)

[7] H. L. Engquist and P. W. Anderson, Phys . Rev. **B 24**, 1151 (1981)

[8] M. Ya. Azbel, J. Phys. **C 14**, L225 (1981)

[9] D. J. Thouless, Phys. Rev. Lett., **47**, 972 (1981)

[10] M. Büttiker, Y. Imry, R. Landauer, and S. Pinhas, Phys. Rev. **B 31**, 6207 (1985)

[11] A. D. Stone and A. Szafer, IBM J. Res. Develop. **32**, 384 (1988)

[12] H. U. Baranger and A. D. Stone, Phys. Rev. **B 40**, 8169 (1989-II)

[13] V. Fal'ko and D. E. Khmel'nitskii, ZETF **95**, 349 (1989)

[14] V. Fal'ko, Europhys. Lett., **8**, 785 (1989)

[15] B. L. Altshuler and A. G. Aronov, in: Electron-Electron Interactions in Disordered Systems, A. L. Efros and M. Pollak, eds., 1, North Holland (1985)

[16] D. J. Thouless, Phys. Rev. Lett. **39**, 1167 (1977); Sol. St. Commun. **34**, 683 (1980)

[17] P. W. Anderson, E. Abrahams, and T. V. Ramakrishnan, Phys. Rev. Lett. **43**, 718 (1979)

[18] B. Kramer and J. Mašek, J. Phys. C: Sol. State Phys. **21**, L1147 (1988)

[19] J. Mašek and B. Kramer, Z. Phys. **B 75**, 37 (1989)

[20] J. Mašek, P. Lipavský, and B. Kramer, J. Phys.: Cond. Matter **1**, 6395 (1989)

[21] B. Velický, J. Mašek, and B. Kramer, Phys. Lett. **A 140**, 447 (1989)

[22] J. Mašek and B. Kramer, Sol St. Commun. **68**, 611 (1988)

[23] B. Kramer, J. Mašek, V. Špička, and B. Velický, Surf. Sci. **229**, 316 (1990)

[24] P. A. Lee, A. D. Stone, and H. Fukuyama, Phys. Rev. **B 35**, 1039 (1987)

[25] Y. Imry, Europhys. Lett. **1**, 249 (1986)

[26] A.D. Stone, K.A. Muttalib, J.-L. Pichard, Springer Procs. in Physics **28**, 315 (1987)

HIGH-FREQUENCY PROPERTIES OF MESOSCOPIC JUNCTIONS

Vladimir I. Fal'ko

Institute of Solid State Physics
Chernogolovka, Moscow District, 142432, USSR

1 Introduction

The statistics of electron levels in an open disordered quantum system is one of the general problems solved in the theory of mesoscopics. As the random potential of elastic scatterers in disordered conductors leads to repulsion of electron levels, the latter are strongly correlated near the Fermi energy. In an open system, such as a mesoscopic junction of the characteristic size L joining bulky electrodes, this correlation exists only within intervals of the width $\Delta\epsilon \approx h/\tau_f$ [1], where $\tau_f \approx L^2/D$ is the time of diffusion through the junction. The electron spectrum can be conveniently divided into uncorrelated intervals with almost independent fluctuations of the number of transport levels. The temperature dependence of the mesoscopic resistance [2], and the non-linear conductance fluctuations of small metallic particles [3] were shown to be sensitive to this correlation. Here it is argued that the structure of the electron spectrum predicted by Altshuler and Shklovskii [1] manifests itself in random frequency dependences of AC-kinetic coefficients of microjunctions. Thus, all of them can serve as spectral "fingerprints" of a sample in the same way as the magnetoresistance serves as its "magneto-fingerprint" [4].

2 General Results

This suggestion is based on detailed calculations of the correlation functions of mesoscopic fluctuations of linear and non-linear terms in currents induced in microjunctions by microwave electric fields of various frequencies [5, 6]. The results of these calculations can be symbolically presented in the form of a series

$$\delta I = e\sqrt{\omega/\tau_f}\left\{a_1(eLE_\omega/\hbar\omega) + a_2(eLE_\omega/\hbar\omega)^2 + a_3(eLE_\omega/\hbar\omega)^3 + \ldots\right\} \tag{1}$$

where E_ω is the amplitude of the applied field. The expansion parameters in eq. (1) gives the probability of a testing electron interaction with the alternating field during the time τ_f.

Quantum Coherence in Mesoscopic Systems
Edited by B. Kramer, Plenum Press, New York, 1991

3 AC Conductance Fluctuations

The first term in the series eq. (1) describes the mesoscopic fluctuations of an AC conductance with the mean square value

$$\langle |\delta G_\omega|^2 \rangle = \left(\frac{e^2}{h} \right)^2 \frac{V_d}{L^d} \frac{\langle |a_1(\omega)|^2 \rangle}{\omega \tau_f},$$

(2)

where

$$\langle |a_1(\omega)|^2 \rangle \sim \begin{cases} 1 & d=1 \\ \ln(\omega \tau_f) & d=2 \\ \sqrt{\omega \tau_f} & d=3 \end{cases}$$

which are suppressed in comparison with the (e^2/h) DC-conductance fluctuations. (A very close result was obtained in [7] in numerical simulations of 1 D chain conductance.) Consider excitation of a current by a high-frequency field inducing quasi-particle transitions between states differing in energy by a field quantum $\hbar\omega$. The suppression mentioned above is caused by a partial self-averaging of random contributions from $n = \omega \tau_f \gg 1$ independent spectral intervals. On the other hand, the increase in frequency by an amount $\Delta\omega > \pi^2/\tau_f$ is followed by an involvement of new uncorrelated spectral intervals in the current formation, and therefore δG_ω can change both its magnitude and phase. The correlation function $\langle \delta G_\omega \delta G_{\omega'}^\star \rangle$ of two mesoscopic conductances at different frequencies ω and ω', calculated by use of the diagram technique, shows that at $\Delta\omega = \omega - \omega' > \pi^2/\tau_f$ correlations between AC conductance fluctuations disappear and

$$\langle a_1(\omega) a_1^\star(\omega') \rangle \approx \langle |a_1(\omega)|^2 \rangle \begin{cases} (\Delta\omega \tau_f)^{-1/2} & d=1 \\ 1 - (\Delta\omega/\omega) & d=2,3. \end{cases}$$

(3)

Thus, the coefficient $a_1(\omega)$ in the series eq. (1) is a random function of the frequency specific for each structure and varying over a scale π^2/τ_f.

A random frequency dependence of δG_ω interplays with weak localization corrections having a drastic frequency dependence in the same range of frequencies. In order to separate them one can apply a magnetic field to suppress localization effects or irradiate with a field polarized perpendicular to the sample axis. The macroscopic part of the current in the latter case is absent, while the mesoscopic part is given by eq. (2) due to the anisotropy of the random potential.

4 Mesoscopic Photovoltaic Effect

The second term in the series eq. (1) gives the magnitudes $I_{0,2}$ of the photovoltaic (PV) effect [5], and the second harmonic generation [6] caused by inversion symmetry breaking due to randomity of distribution of impurities. An alternating field excites in the junction a number of non-equilibrium carriers which amounts to the number $N \sim VE^2\sigma/\hbar\omega$ of field quanta absorbed per unit time. In the diffusion time τ_f these excited carriers leave the junction and reach the bulky electrodes. As the random potential in the junction has no inversion center, different numbers of carriers diffuse to the left and the right banks of the junction. From the point of view of the theory of the electron spectrum [1] the fluctuation states added in each spectral interval h/τ_f to the background density of states ν_d are responsible for this transport. The part of non-equilibrium carriers that fill these asymmetrical states and contribute to the rectified current is

$$\delta N \sim N\tau_f/\nu_d. \tag{4}$$

All of these states belong to $n = \omega\tau_f$ independent spectral intervals, thus a partial mutual cancellation of their contributions having random signs gives

$$I_0 \sim e\delta N/\sqrt{n} \sim e\sqrt{\omega/\tau_f}\,[eE_\omega L/\hbar\omega]^2 . \tag{5}$$

A random frequency dependence of a rectified current is also obvious [5], and therefore the coefficient $a_2(\omega)$ in eq. (1) is also a random function of the frequency,

$$\langle a_2(\omega)a_2^\star(\omega')\rangle \sim \left(\pi^2/\Delta\omega\tau_f\right)^{2-d/2} , \qquad |a_2(\omega)| \sim 1 \tag{6}$$

with a characteristic scale $\omega \sim \pi^2/\tau_f$.

As to the polarization properties , the mesoscopic PV effect is anisotropic and linear, and the polarization correlation function

$$\left\langle I_0\left(\vec{E}_\omega\right) I_0\left(\vec{E}_{\omega'}\right)\right\rangle \sim \left|\vec{E}_\omega \vec{E}_{\omega'}^\star\right|^2 \tag{7}$$

shows that two currents excited by fields with perpendicular linear polarizations are almost independent. In systems with a strong spin-orbital coupling an admixture of a circular PV effect occurs,

$$\left\langle I_0^C\left(\vec{E}_\omega\right) I_0^C\left(\vec{E}_{\omega'}\right)\right\rangle \sim \left|\left[\vec{E}_\omega \vec{E}_{\omega'}^\star\right]\right|^2 , \tag{8}$$

but generally it is comparatively small.

5   Higher Harmonics in the Current

Other higher terms in the series eq. (1) describe higher even and odd harmonics in the current with the correlation properties almost analogous to those discussed above,

$$\langle|I_N(\omega)|^2\rangle \sim 8\pi^2\frac{\omega}{\tau_f}\left[2^{3/2}eEL/\hbar\omega\right]^{2N} , \tag{9}$$

$$\langle a_N(\omega)a_N^\star(\omega')\rangle \sim \left(\pi^2/\Delta\omega\tau_f\right)^{N-d/2} , \qquad |a_N(\omega)| \sim 1. \tag{10}$$

The correlations between even and odd harmonics are absent because even harmonics arise from inversion symmetry breaking by the random potential, while the odd ones originate from electron-hole asymmetry near the Fermi level. Moreover, the correlations between different even (or odd) current components that are maximal at low frequencies gradually vanish at $\omega \sim \pi^2/\tau_f$,

$$\langle a_{2N}(\omega)a_{2N'}^\star(\omega)\rangle \sim \left(\pi^2/\omega\tau_f\right)^{N-N'}/(N-N')!, \tag{11}$$

since different modes are related to electron transitions between states within spectral intervals of different widths $N\hbar\omega$ and $N'\hbar\omega$. Therefore, all functions $a_i(\omega)$ are practically independent, and their complete set contains the full information about th electron spectrum of a specific sample.

6 Mesoscopic Microwave Photoeffect

Microwave irradiation also affects the DC-properties of microjunctions, for instance its conductance. Dynamical suppression of weak localization [8] in the mesoscopic regime gives rise to an increase in the DC-conductance measured in fractions of the conductance quantum. In a specific conductor this effect interplays with the alternating field induced change $\delta G(E_\omega)$ in the mesoscopic conductance. The latter originates from electron-hole asymmetry near the Fermi level [6], and its correlation properties can be described by the correlation function $K_G(\omega, \Delta\omega)$ of two photoconductances induced by fields $\vec{E}_\omega$ and $\vec{E}_{\omega'}$. An applied magnetic field H effectively suppresses the localization contribution to the photoeffect extracting its mesoscopic part. At $H > \Phi_0/L^2$ and $eE_\omega L/\hbar\omega \ll 1$ the rms photoconductance is

$$\left\langle \left[\delta G(\vec{E}_\omega)\right]^2 \right\rangle = K_G(\omega, 0) = \left(\frac{e^2}{h}\right)^2 \{F_1 + F_2\} \xi_d \frac{V_d}{L^d}. \tag{12}$$

where $F_1 = (9/8)|eE_\omega L/\hbar\omega|^2|eE_{\omega'}L/\hbar\omega'|^2$, $F_2 = |e^2 L^2 \vec{E}_\omega \vec{E}_{\omega'}/\hbar^2\omega\omega'|^2$ and $\xi_1 = 0.21$, $\xi_2 = 0.1, \xi_3 = 0.055$. Some reduction of correlations at $\Delta\omega > \pi^2/\tau_f$,

$$K_G(\omega, \Delta\omega) = \left(\frac{e^2}{h}\right)^2 \left\{F_1 + \left(\frac{\pi^2}{\Delta\omega\tau_f}\right)^2 F_2\right\} \xi_d \frac{V_d}{L^d}, \tag{13}$$

means that $\Delta G(\vec{E}_\omega)$ comprises two parts with monotonic and random frequency dependences with the ratio $\kappa_d = K_G(\Delta\omega \gg \pi^2/\tau_f)/K_G(\Delta\omega = 0) = 9/17$ being a universal value determined only by the sample geometry. In summary, the photoconductance of a small disordered conductor can also serve as its spectral "fingerprint".

Note that the correlation function eq. (13), and the parameter κ_d can be compared with experimentally derived autocorrelation functions of photoconductance fluctuations in a magnetic field. In the same way another universal value can be used as an E_ω-independent experimental test. The calculations show that the correlation function F of $\delta G(E_\omega, H)$ and the magnetoresistance fluctuation $\delta G(H)$ normalized to their mean square values

$$F = \frac{\langle \delta G(E_\omega, H)\delta G(H)\rangle}{\langle \delta G(E_\omega)^2\rangle^{1/2}\langle \delta G(H)^2\rangle^{1/2}} = \left(\frac{6\zeta^2(7-d)}{17\zeta(5-d)\zeta(9-d)}\right)^{1/2} \Theta_d \tag{14}$$

(with $\Theta_{1,3} = 1$ and $\Theta_2 = 0.92$) is a constant at $eE_\omega L/\hbar\omega < 1$ and increases up to unity for $eE_\omega L/\hbar\omega \gg 1$.

Finally, as preliminary observations are promising [9, 10, 11], a further experimental investigation of the predicted microwave effects in mesoscopic conductors will give us a powerful tool to verify our understanding of the statistical properties of open disordered systems in a wide range of energies.

References

[1] B.L. Al'tshuler and B.I. Shklovskii, Sov. Phys. JETP **64**, 127 (1986)

[2] B.L. Al'tshuler, Sov. Phys. JETP **41**, 648 (1985); P.A. Lee and A.D. Stone, Phys. Rev. Lett. **55**, 1622 (1985)

[3] D.E. Khmel'nitskii and A.I. Larkin, Sov. Phys. JETP **64**, 1075 (1986)

[4] S. Washburn and R.A. Webb, Adv. Phys. **35**, 375 (1986)

[5] V.I. Fal'ko and D.E. Khmel'nitskii, ZETF **95**, 349 (1989)

[6] V.I. Fal'ko, Europhys. Lett. **8**, 785 (1989)

[7] J. Măsek, B. Kramer, Sol. St. Commun. **68**, 611 (1988)

[8] B.L. Al'tshuler, A.G. Aronov, and D. E. Khmel'nitskii, Sol. St. Commun. **39**, 619 (1981)

[9] A.A. Bykov, G.M. Gusev, Z. D. Kvon et al., Sov. Phys. JETP Lett. **49**, 13 (1989); A.A. Bykov, G.M. Gusev, Z. D. Kvon et al., Sov. Phys. JETP **70**, 140 (1990)

[10] N. Giordano and J. Liu, Physica **B 165**, 279 (1990)

[11] F. Kuchar, private communication

RESPONSE TIME IN HIGH-FREQUENCY QUANTUM TRANSPORT

Yaotian Fu

Department of Physics
Washington University
Box 1105, St. Louis, Missouri 63130, USA

1 Introduction

The theory of quantum transport is the theoretical basis for nanostructure devices. In the last few years, the theory of quantum transport has been extensively discussed, developed, and tested [1]. Much of this discussions, however, has been centered on DC transport. The question naturally arises as to whether, and if so to what extent, we can generalize the results of DC Quantum transport theory and apply them in the high frequency regime. In this paper we study this question. Specifically, we consider the question of the intrinsic time scale of AC operation. Just like the drift diffusion process in a conventional semiconductor device sets one such time scale, so too do various physical processes in a quantum device. We concentrate on two such processes: the RC response and the time scale related to tunneling (see below). It will become clear in the following that this problem is a complicated one; while various formulations in the literature capture *some* aspects of the problem, a complete description does not appear possible, and any oversimplification can only lead to erroneous results.

Throughout our discussion we will emphasize the basic physics of the problem. No attempt has yet been made to relate our theoretical discussion to experimental results. Our primary goal here has been the development of techniques and the qualitative discussion of the problem. We intend to turn to the quantitative analysis in the near future.

2 Quantum Response Time

It is not always easy to define precisely what we mean by an intrinsic time scale. If we confine ourselves to the specific problem of a resonant tunneling diode, we can find four such scales in the literature. They are: the tunneling time as originally proposed by Wigner and by Smith [2] and applied by Coon and Liu [3] to the resonant tunneling diode, the traversal time as proposed by Büttiker and Landauer [4], the RC time constant of the equivalent circuit, and the inverse of the level width as a measure of the time an

Quantum Coherence in Mesoscopic Systems
Edited by B. Kramer, Plenum Press, New York, 1991

electron spends inside the device [5]. The first two concepts and their applications are discussed extensively in ref. [6].

While all these time scales are important in some restricted applications, none of them provides a complete and convenient description. Consider, first of all, the scattering time. It is given by the formula [2]

$$\tau_{WS}(E) = \hbar \frac{\partial}{\partial E} \delta(E),$$

(1)

where $\delta(E)$ is the phase shift of scattering. This formula is derived by considering a wavepacket transmitted through a potential region, and the center of the wavepacket is delayed by an amount given by eq. (1). One then assumes that the same time scale governs the high frequency operation of the device. This is suggestive but not persuasive. It is not clear that the peak of a wavepacket should correspond to the flow of electrons in a *steady* state AC transport problem. While the concept of scattering time may be useful as a bound on the causal response rate of a system, it is not useful when applied to a steady state problem. The utility of eq. (1) is also restricted by the fact that the time constant thus obtained is not always positive definite. In addition, since eq. (1) is derived using a wavepacket of infinitesimal energy spread and gives an energy-specific time scale, whereas in an AC transport problem electrons of very different and in general time-dependent energies are involved, it cannot be used directly. It is not known whether some kind of average over energy may be carried to make eq. (1) more useful in AC analysis.

The traversal time was discovered by Büttiker and Landauer [4] and is given by

$$\tau_{BL}(E) = \int_{x_1(E)}^{x_2(E)} \frac{dx}{\sqrt{2m(V(x) - E)}},$$

(2)

where m is the mass of the electron, E its energy, $V(x)$ the potential, and $x_1(E)$ and $x_2(E)$ are the energy-dependent turning points defined by $V(x_1) = V(x_2) = E$ and define the classically forbidden region in a tunneling problem. This result can be interpreted in a crude sense as the time that a tunneling electron spends under the barrier. This result can be derived, as Büttiker and Landauer have done, by considering the transmission through a *time-dependent* barrier. It is a simple, elegant, and useful concept and has been used to analyze, for example, the motion of an electron tunneling in a transverse magnetic field through a thick insulating barrier, where the Lorentz force deflects the electron while it is under the barrier and shifts the tunneling path by an amount determined by the traversal time [6]. On the other hand, it is not immediately obvious that the AC response time of a quantum device is directly controlled by $\tau_{BL}(E)$, for tunneling may not be involved at all. The traversal time is also an energy-dependent quantity, and as such cannot be used directly in an AC problem (except possibly in the small-signal limit) where, as we have discussed earlier, the energy of the electrons is generally not constant.

The RC time constant is familiar from conventional electronics, and there is ample experimental evidence that charging and discharging of the device govern its response time. In a lucid discussion, Luryi [8] showed how the RC delay sets one frequency limit for the resonant tunneling diode. Close inspection shows, however, that is not the only frequency limit, and the application of the concept of RC response to quantum transport is not straightforward. This is because in the extremely quantum mechanical limit, the current operator and the voltage operator do not commute with each other (see below), and a quantum mechanical treatment of the RC problem is needed. In

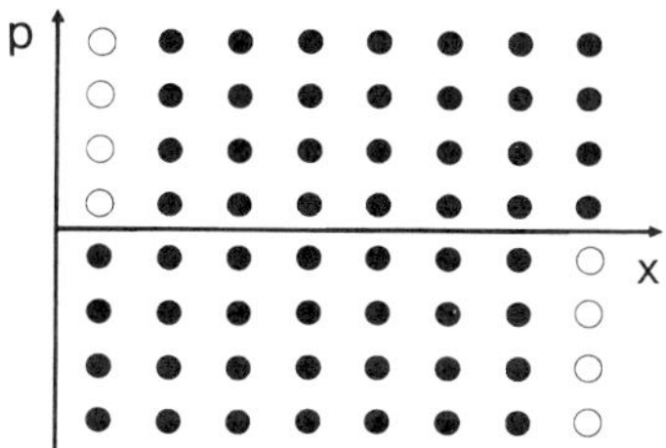

Figure 1. Discretized phase space. Open circles: boundary points.

section 3 we discuss a simple model for that. An additional complication is that, quantum mechanically, the relaxation of an interacting electron liquid may be anomalously slow. Under suitable conditions, this slow quantum relaxation may become bottleneck. This is discussed in details elsewhere [9] and summarized below.

3 Numerical Solution

In view of the experimental and technological importance of this problem [10-14], a quantitative analysis is of great interest. To this end, we have performed numerical simulations using the well-known technique of Wigner function [15-18]. Previous workers have emphasized the transient response of the device. It is our desire to study the AC steady state response in order to settle the question of the intrinsic time constant. We have therefore carried out the simulation in frequency-domain as opposed to the time-domain as it has been done previously. We avoided the use of linear theory which relates the time-domain behavior and the frequency-domain behavior. Beyond the linear regime no such relation exists, and a simulation of the transient response does not by itself answer the question of the steady state AC response and the intrinsic frequency limit.

Since our purpose is to test the direct applicability of the two times scales τ_{WS} and τ_{BL}, we did not impose Coulomb self-consistency. We have carried out simulations on a one-dimensional system as discussed by Frensley, and by Ferry and coworkers. The phase space is discretized and is shown schematically in Fig. 1. Open circles represent points where boundary conditions may be applied to connect the device smoothly to the metal leads [15,18]. As emphasized by Ringhofer et al. [18]. the boundary condition of this problem is non-trivial for it must satisfy two requirements: the electrons entering the device must be of the appropriate thermal distribution, and the electrons leaving the device must not be spuriously reflected back. We have implemented their boundary condition, suitably modified for our frequency-domain simulation [19]. The device parameters used are the same as those of Frensley [15].

The equation of motion in the frequency-domain has the form

$$-in\omega W_n(R,p) = -\frac{p}{m}\partial_R W_n(R,p) + \frac{1}{2}F\partial_p\left[W_{n+1}(R,p) + W_{n-1}(R,p)\right]$$

$$+ \int \frac{dq}{2\pi}V(R,p-q)W_n(R,q). \tag{3}$$

where $W_n(R,p)$ is the n-th Fourier coefficient of the Wigner function,

$$W(R,p,t) = \sum_{-\infty}^{\infty} W_n(R,p)e^{-in\omega t}, \tag{4}$$

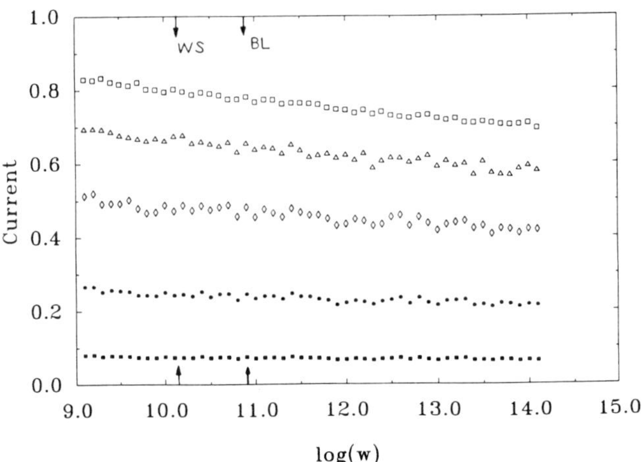

Figure 2. AC response of the model device. The two markers BL and WS refer to the time scale predicted by Büttiker and Landauer and that by Wigner and Smith, respectively.

$F = eE_0$, E_0 is the magnitude of the AC field. V is the same as that given in Frensley [15].

As it stands, eq. (3) is an infinite set of linear differential-integral equations. Once the phase space is discretized, there are still an infinite number of equations labeled by n. One can show that for large n, W_n generally decreases as n^{-2} under our boundary conditions [20]. We therefore adapted the following procedure to reduce the equations to a manageable size. We terminate the equations at a certain order, typically $n = 10$, and solve the resulting closed set of equations. We then repeat the process with a slight larger n. If the difference between the two solutions is sufficiently small for the first five harmonics we regard the n chosen as an acceptable size of truncation. Once the Wigner function is known, the current may be calculated using

$$j_n(R) = \frac{e}{2} \int \frac{dp}{2\pi} \frac{p}{m} \left[W_n(R,p) + W_{-n}(R,p) \right]. \tag{5}$$

Our final results are summarized in Fig. 2 [19]. The electric field strength is fixed and corresponds to a voltage $V = 0.1\,V$. We plot the current (in arbitrary units) as a function of the frequency over the range of $\omega = 2\pi\nu = 10^9\,$Hz to $10^{14}\,$Hz. It is clear that while there is a gradual decrease in current as the frequency is increased, there is no sharp cutoff at either the Wigner-Smith time of the Büttiker-Landauer time. Our results are not expected to agree with experiments, because an important aspect of the problem [8], that of RC transient, is not included in our analysis.

4   RC Constant: a Quantum Mechanical Model

Luryi [8] has given a quantum mechanical treatment of RC time constant. In this section we generalize his analysis and consider several new effects. An RC problem is necessarily an interacting problem, and the single electron theory is no longer useful. Consider first of all a single barrier problem. The Hamiltonian for the system can be written as

$$H \;=\; \sum_i \epsilon_i a_i^\dagger a_i + \sum_j \epsilon_j b_j^\dagger b_j + \sum_{ij} t_{ij} a_i^\dagger b_j$$

$$+\text{H.C.} + \frac{e^2}{2C}\left(N_1 - N_2\right)^2 + \frac{eV(t)}{2}\left(N_1 - N_2\right), \tag{6}$$

where $a_i, a_i^\dagger$ are the electron operators for the electrons to the left of the barrier and the b's are for the electrons to the right. The tunneling between left and right is described by the tunneling matrix element t_{ij}. C is the effective capacitance of the tunneling junction. $N_1 = \sum_i a_i^\dagger a_i$ is the number of electrons on the left; N_2 the number of electrons on the right, is similarly defined. H.C. stands for Hermitian conjugation. $V(t)$ is the voltage. We note that the current is proportional to $d(N1 - N2)/dt$. The states on the two sides of the barrier can be labeled by the wavevector of the electrons. If we assume that tunneling can only take place between states of the same wavevector (magnitude), the number of electrons in a given wavevector state k, left plus right, is conserved. We can then introduce a set of fictitious spin operators to describe the system. We define: $S_k^z = a_k^\dagger a_k - b_k^\dagger b_k, S_k^+ = a_k^\dagger b_k, S_k^- = b_k^\dagger a_k$. The Hamiltonian can be rewritten as

$$H = \frac{e^2}{2C}\left(\sum_k S_k^z\right)^2 + \frac{eV(t)}{2}\sum_k S_k^z + \sum_k t_k S_k^x, \tag{7}$$

and the current is given by $j = -e\sum_k t_k S_k^y$. Formally, then, the problem of AC response is transformed to one of magnetic resonance, in which a large number of "spins" are labeled by k and undergo resonance in an "RF field" $eV(t)/2$ in the presence of an external field t_k which varies from site to site and an electric quadrupole interaction given by the first term of eq. (7). It is clear that the current operator and the voltage operator correspond to two non-commuting quantum mechanical operators, and the classical description is incomplete. Indeed, the Hamiltonian eq. (7) has quantum resonances which have no classical counterpart. In a real sample the values of t_k are distributed and no sharp resonances are expected to occur except under special conditions.

A more systematic treatment of the quantum RC problem has been given elsewhere [9]. The central result can be summarized as follows. A peculiar feature of the interacting Fermions is that their relaxation following a sudden local change in potential (such as a tunneling diode in a switching process) is often anomalously slow. This is because the Pauli principle and the many body nature of the system prohibit a direct relaxation to screen out the potential after the change and many small steps are needed to accomplish that. The most famous example of this effect is the so-called X-ray edge problem [21], in which the transient dynamics of an electron fluid following the sudden knock-out of one core electron by the X-ray exhibits non-exponential decay. We have performed a similar analysis [9] and found that in the quantum regime the RC transient does not decay exponentially but only by a power law. This suggests that the quantum slow relaxation may be the main bottleneck at high speed. The frequency limit posed by the quantum relaxation can be the dominating factor.

5 Acknowledgements

This work was supported in part by ONR and by NSF National Center for Computational Electronics and by the National Center for Supercomputer Application, Champaign, Illinois.

References

[1] For a general review see, e.g. the articles in: The Physics of Nonlinear Transport in Semiconductors, D.K. Ferry, J.R. Baker, C. Jacoboni, eds., Plenum, New York (1980); The Physics of Submicron Structures, H.L. Grubin, K. Hess, G.J. Iafrate, D.K. Ferry, eds., Plenum, New York (1982); IBM J. of Res. & Dev., **32**, issues 1 and 3 (1988)

[2] E.P. Wigner, Phys. Rev. **98**, 145 (1955); F.J. Smith, Phys. Rev. **118**, 349 (1960)

[3] D.D. Coon and H.C. Liu, Appl. Phys. Lett. **49**, 94 (1986)

[4] M. Büttiker and R. Landauer, Phys. Rev. Lett. **49**, 1739 (1982)

[5] S. Collins, D. Lowe, and J.R. Barker, J. Phys. **C 20**, 6233 (1987)

[6] E.H. Hauge and J.A. Støvneng, Rev. Mod. Phys. **61**, 917 (1989)

[7] P. Gueret, A. Baratoff, and E. Marclay, Verhandlungen der Deutschen Physikalischen Gesellschaft, Reihe 6, Band **21**, 1446 (1986)

[8] S. Luryi, Appl. Phys. Lett. **47**, 490 (1985)

[9] Y. Fu, 'Switching speed of a tunneling junction and the X-ray edge problem', to be published

[10] T.C.L.G. Sollner, W.D. Goodhue, P.E. Tannenwald, C.D. Parker, and D.D. Peck, Appl. Phys. Lett.**43**, 588 (1983)

[11] T.C.L.G. Sollner, P.E. Tannenwald, D.D. Peck, and W.D. Goodhue, Appl. Phys. Lett. **45**, 1319 (1984)

[12] F.J. Whitaker, G.A. Mourou, T.C.L.G. Sollner, and W.D. Goodhue, Appl. Phys. Lett. **53**, 385 (1988)

[13] E.R. Brown, T.C.L.G. Sollner, C.D. Parker, W.D. Goodhue, and C.L. Chen, Appl. Phys. Lett. **55**, 1777 (1989)

[14] M. Tsuchiya, T. Matsusue, and H. Sakaki, Phys. Rev. Lett. **59**, 2356 (1987)

[15] W.R. Frensley, Phys. Rev. **B 36**, 1570 (1987)

[16] N.C. Kluksdahl, A.M. Kriman, D.K. Ferry, and C. Ringhofer, IEEE Electron Device Lett. **9**, 457 (1988)

[17] N.C. Kluksdahl, A.M. Kriman, D.K. Ferry, and C. Ringhofer, Phys. Rev. **B 39**, 7720 (1989)

[18] C. Ringhofer, D.K. Ferry, and N.C. Kluksdahl, Transport Theory and Statistical Physics **18**, 331 (1989)

[19] Y. Fu, 'AC I-V characteristics of a resonant tunneling diode: a steady state numerical study', to be published

[20] C.M. Bender, private communication

[21] G.D. Mahan, Phys. Rev. **163**, 612 (1967); P. Nozières and C.T. De Dominicis, Phys. Rev. **178**, 1097

CHAPTER 7

UNIVERSAL CONDUCTANCE FLUCTUATIONS

RESISTANCE FLUCTUATIONS IN SMALL SAMPLES: BE CAREFUL WHEN PLAYING WITH OHM's LAW

Sean Washburn

IBM Thomas J. Watson Research Center
P.O. Box 218
Yorktown Heights, NY 10598, USA

1 Introduction

The discovery of random fluctuations [1,2] in the resistance of metal samples has wrought a small revolution in the understanding of electronic transport in samples with weak disorder (commonly known as metals). The fluctuations arise from quantum mechanical interference among the various trajectories that the electrons "follow" as they migrate from one end of the sample to the other. Obviously then the electrons must retain the coherence of their wavefunctions from one end to the other. By carefully summing up all the possible trajectories and interferences among them, Al'tshuler [3] and Lee et al. [4] have calculated the ensemble averaged variance of the conductance of wires of length L_φ to be $\langle(\Delta G)^2\rangle \equiv \langle G^2\rangle - \langle G\rangle^2 \simeq (e^2/h)^2$. From an ensemble of classically identical samples (the theorists assume that the samples are carved from some mythical film where, on any scale larger than the mean free path length, the resistivity is utterly homogeneous, and the angular brackets refer to the average over this ensemble), one would find that the conductance fluctuates among the members by an amount characterized by a distribution of width $\simeq e^2/h$. Owing to a random Aharonov-Bohm (AB) effect, the conductance of a particular sample fluctuates by the same amount as a function of magnetic field [5], and the result of the sensitivity of the conductance to the impurity potential, it also fluctuates by $\simeq e^2/h$ as a function of the Fermi energy [4] or when minute rearrangements occur in the impurity configuration [6]. In principle, this "universal" feature of the conductance of metal wires can be used as a measure of the phase coherence of the carriers and as a exquisitely sensitive probe of changes in the impurity potential [6].

In order to calculate the conductance of a disordered wire, one usually adopts a model in which the disordered potential is suspended between two particle reservoirs. If the disorder is weak (as in a metallic sample), then any difference between the two chemical potentials of the reservoirs μ_1 and μ_2 will allow a net current to flow from the higher potential to the lower [7]. The amount of current that flows is governed by the transmission capability of the impurity potential. For the measurement of the two probe conductance $G_{12} = I_{2\to1}/(V_1 - V_2)$, the current is proportional to the total probability T of transmission through the disorder for electrons with energies between μ_1 and μ_2. Assuming that the response is linear in the applied voltage difference, one obtains a conductance proportional to the transmission probability of the disordered potential [8]

$$G = \frac{e^2}{h} \sum_{mn} |t_{mn}|^2 , \tag{1}$$

where the spin degeneracy of the electrons levels has been ignored. (Non-linear response can be computed if the energy dependence of the transmission coefficients is properly accounted for [9].) Equation (1) may be written more compactly as a dimensionless conductance $g \equiv hG/e^2$

$$g = T. \tag{2}$$

To calculate the transmission amplitude t_{mn} one sums over all the allowed paths (such trajectories are illustrated in Fig. 1a) of the particles from channel m on one side of the sample to channel n on the other [8]. The amplitude to traverse each path $Q_j = q_j \exp\{i\varphi_j\}$ is a complex number, and the transmission coefficient is $t_{mn} = |\sum_j Q_j|^2$, where the summation on j runs over all paths that connect the two channels. In many practical cases (such as large samples at room temperature) the series may be handily approximated by $\sum_j |Q_j|^2$, which just gives the classical conductance (and corresponds to real number probability amplitudes). Ignoring the phase coherence (the complex factors) is usually safe, because any inelastic scattering destroys memory of it, and thermal smearing of the Fermi surface adds dispersion, but *in principle* it is a dangerous practice. To describe measurements at low temperatures (where there is not much smearing of the Fermi surface) and at short distances (where the sample length L not much greater than the phase memory length L_φ), one must keep account of the phases and the cross-terms in the summation.

The transmission matrix T can be explicitly calculated for some models for the propagation and disorder. In metals, it is an excellent approximation to assign the appropriate mass to the electrons and treat them as non-interacting particles in the usual manner [10]. Since the impurities are sparse, the electron wavefunctions are nearly plane waves of number $\simeq k_F$ that scatter from the impurities after "travelling" freely a distance $\sim \ell \gg 1/k_F$ as depicted in Fig. 1b. The standard method for calculating T involves perturbation expansions (in $1/k_F\ell$) of the propagation amplitudes. In a metal the sum encompasses many paths that span the sample from one reservoir to the other. The dimensionless conductance is therefore much larger than ℓ, even though the typical transmission coefficient might be rather small.

Given a certain concentration of impurities with a certain potential and cross-section, the theory predicts the classical conductance if phase coherence is ignored, and it predicts "weak localization" if phase coherence is retained in large samples [11]. For a given local configuration of impurities, the electrons have some amplitude to scatter back to their starting point. In fact, circling around any such closed loop, the partial waves going clockwise accumulate the same phase increment as those going counter-clockwise so that the interference is always constructive at the origin. Even if one sums over the myriad different configurations of impurities that would be found in a sample if size $L \gg L_\varphi$, the coherence effects do *not* average away, because the constructive interference

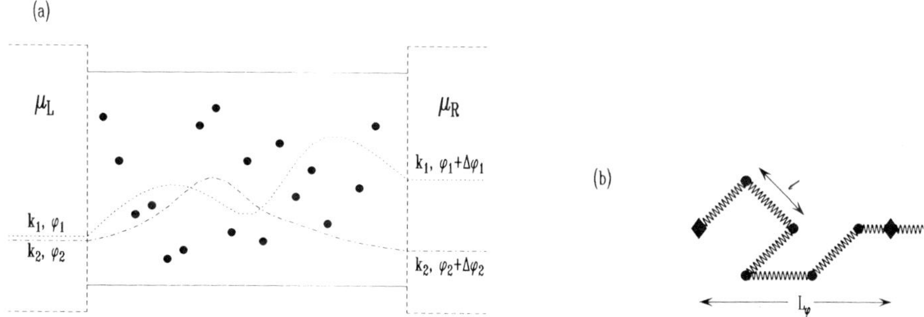

Figure 1. (a) The standard theoretical model of a disordered conductor consists of a random array of impurities connected to particle reservoirs by clean wires (without any impurities). The clean wires are artificial, and serve only to count the channels coming out of the reservoirs. The conductance is calculated through a sum over all of the paths connecting the left hand reservoir to the right hand reservoir for a given potential bias between the two. (b) In the semi-classical picture for the propagation of electrons through the weak disorder in metals, the electron of wavelength λ_f executes a random walk of step size $\sim \ell \gg \lambda_F$ among the impurities, and after some distance L_φ loses memory of the phase of the wavefunction.

is independent of the precise impurity configuration. As the temperature decreases and phase relaxation mechanisms become weaker, the electrons retain phase coherence over larger distances. The phase coherence length $L_\varphi = \sqrt{(D\tau_\varphi)}$, where $D = \ell^2/d\tau$ is the diffusion coefficient in d dimensions (almost always $d = 3$ in the metal experiments and 2 in MOSFET and heterostructure experiments) and τ_φ is the mean time between collisions that relax the phase of the wavefunction. The typical diffusion coefficient in Au is 0.01 m^2/s, and the typical value of L_φ at $T = 1\,\mathrm{K}$ is around 1 μm. As T decreases, $\tau_\varphi \propto T^{-p}$ (where $(2/3) < p < 3$, depending on the relaxation mechanism) increases, and ever larger loops contribute to the "coherent back-scattering": the resistance increases.

2   Conductance Fluctuations in Coherent Conductors

If one somehow contrives to measure the response at length scale $L = L_\varphi$, then not only the time-reversed paths that generate the weak localization, but all paths must be included in the interference contributions to the conductance. Since the uncorrelated paths have no clamp that requires constructive interference, their contribution is random from one pair of paths to the next. At first it would seem that summing over many such paths, one would obtain an average contribution that is negligible. For instance the number of paths through the sample is $N \simeq k_F^2 wt$, where wt is the area in cross-section to the net current flow, which for a very small gold wire 50 nm × 50 nm is $\gtrsim 10^5$ – a rather large number. There are N^2 individual elements t_{mn} in the matrix of transmission amplitudes, if the t_{mn} are not correlated one with another, then the relative fluctuation $\Delta G/G$ in the conductance should be $\simeq 1/N$ [8]. This naive guess that the channels are uncorrelated is quite wrong. In fact there is a rather strong correlation which results from repulsion among the electron levels in the conduction band. The electron traverses the sample on a random walk trajectory in a time L^2/D, and so the individual energy levels have a width $E_C \simeq \hbar D/L^2$ [12] (for the generic $1\,\mu$m Au wire, $E_C/k_B \simeq 0.07$ K). Within this span of energies, the levels repel to have equal spacings just as the levels in a nucleus do [13]. The equal spacing leads to an effective number of channels (to which the conductance is proportional) that just balances the change of the conductance as length increases; it leads to universal fluctuations in the conductance of size $\Delta G = Ce^2/h$ [14,15]. The parameter C is slightly dependent on sample shape [16], but it is independent of $G, L, \ell \dots$

The measurement of the coherent effects need not be performed at scale L_φ, because the fluctuations do not decay strongly when the sample length exceeds L_φ. For $L \gg L_\varphi$, the sample effectively comprises $N = (L/L_\varphi)^d$ coherent clocks stacked together. Since the conductance of the blocks are all independent random variables, the total resistance fluctuation amplitude ΔR (for the case where $L_\varphi > w$ so that $d = 1$) increases by $\sqrt{N}$ because of the addition of N uncorrelated random patterns. The relative fluctuation amplitude $\Delta R/R \simeq \Delta G/G \propto 1/\sqrt{N}$. The absolute conductance fluctuation amplitude $\Delta G = G^2 \Delta R$ decreases by factor $1/\sqrt{N^d}$ [14,16]. The decay is a power law in L/L_φ not an exponential as might be expected from analogy with Shubnikov-de Haas oscillations, which means that the fluctuations still appear in rather large samples $L > 10000 \times L_\varphi$ [17].

The interference condition is critically dependent on the precise pattern of trajectories that interfere. Moving so much as *one* of the impurities in one of the paths by more than λ_F significantly alters the interference amplitude for that pair of paths [6]. This seems a trivial concern until one notices that for one- and two-dimensional (1 D, 2 D) samples, each electron trajectory visits essentially *all* of the impurities, which implies that moving one impurity in the whole sample can change the conductance by e^2/h. This sensitivity is an astonishing result that implies that conductance fluctuations can be used to probe minute changes in the impurity configuration such as are expected to occur in spin-glasses [6,18]. It also implies that the amplitude of flicker noise can be much enhanced beyond what one naively guesses based on the assumption that the fractional change in conductance would be about equal to the fraction of impurities that move [19,20]. The conductance fluctuations serve to amplify the motion, so that if one impurity moves in a volume of size L_φ it appears (from the naive viewpoint) that all of them have. Whether or not this actually has anything to do with observed room temperature $1/f$ noise has been debated, but at temperatures below 100 K, there is a good deal of evidence that the amplification contributes significantly [21]. In a series of beautiful experiments, it was proved that the noise decreases by a predictable fraction when a small magnetic field is applied [22-24]. The reduction occurs because the field dephases the time-reversed paths that contribute to interference (see below) and hence to the conductance fluctuations [25].

So far the argument has been made that changes in impurity configurations from sample to sample result in fluctuations of a universal amplitude among the conductances of members of an ensemble of classically similar samples. The fluctuations also appear if electrons trajectories are altered by any other method. For instance if the Fermi energy moves to a different place in the conduction band, the electrons prefer different trajectories, and the conductance changes randomly by $\Delta G = Ce^2/h$. Not surprisingly, one must move E_F by an amount that is greater than E_C (the range over which the electron paths are correlated), and sweeping E_F across many E_C generates a random pattern of fluctuations of the universal amplitude. The half-width of the auto-correlation

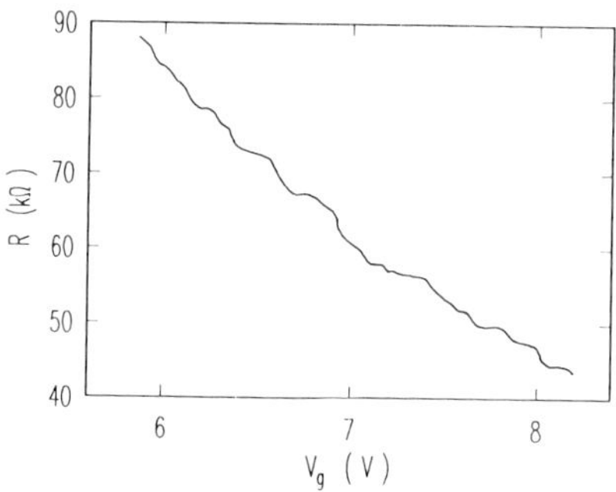

Figure 2. The variation of $R(V_g)$ for a $7\,\mu$m long $0.25\,\mu$m wide MOSFET (operating in accumulation mode) measured at $E = 0.13$ K. Both random conductance fluctuations and the usual smooth increase of conductance with increasing gate voltage are apparent [26].

function of such data provides a nice measurement of E_C, which is independent of the estimate based on the diffusion coefficient [4]. An example of such random fluctuations is displayed in Fig. 2 where one sees both a random fluctuation [26] and a smooth monotonic change (because conductance is proportional to electron density) as a function of the gate voltage on a MOSFET.

The AB effect [27] can also alter the interference among the paths. For any pair of paths encircling a magnetic flux Φ, the interference condition changes

$$\sum Q \to \sum Q \exp\left\{\frac{ie}{\hbar}\int \vec{A}\cdot d\vec{x}\right\}, \tag{3}$$

where $d\vec{x}$ is the element of the path and $\vec{A}$ is the vector potential which accompanies the magnetic induction $\vec{B} \equiv \nabla \times \vec{A}$. Since the wavefunction at the terminus of the paths is single valued [28], the interference causes an oscillation with period $\Delta\Phi = h/e$. Quite obviously any smearing of the wavefunction phase by collisions or thermal smearing will wipe out the oscillations. The oscillations have been observed in vacuum where the electrons in the two paths suffer no collisions at all [29], and in super-conducting cylinders [30] where the rigidity of the Cooper pair wavefunctions enforces quantization of the flux threaded through the cylinder [31]. (Since the Cooper pair comprises two electrons, the flux quantum is $h/2e$.) Similar oscillations appear in thin single crystals of Bi [32], where ℓ is long enough to allow the electron to traverse the circumference without scattering: once again there are no collisions. In this case the oscillations are Shubnikov-de Haas oscillations resulting from the Landau energy quantization. They are periodic in B (rather than the usual $1/B$) because the cyclotron orbit size is clamped by the diameter of the crystal. Larger diameter crystals reduce the oscillation amplitude exponentially because of collision broadening of the Landau levels.

Much more interesting in the present discussion are experiments on long, thin cylinders made from disordered metals and placed in magnetic field directed along the cylinder axis [11,33]. The magnetoresistance of a thin cylinder (diameter $\simeq 1\,\mu$m) contains periodic oscillations in the region near $B = 0$ [34]. A 2 D array of loops (all having the same diameter) also exhibits similar oscillations [11]. The oscillations arise from the same phase coherent paths that cause the weak localization of the electrons at low temperatures. Here, since the area enclosed by the paths is restricted by the walls of the annulus (the cross-section of the cylinder perpendicular to the field) AB oscillations appear in the magnetoresistance. They have a flux period $h/2e$ because the coherent back-scattering involves two complete circuits of the annulus. The experiments have occasionally been mislabeled as "flux quantization", but there is no quantization of the field, it is homogeneous as it is in any other metal object. This homogeneity means that there is some flux in the annulus. The flux there is randomly enclosed by the tangle of trajectories, and it causes a smooth decay of the oscillation amplitude as different pairs of paths get out of step with one another. One may understand this schematically by observing that paths along the outside of the annulus enclose more of the uniform field than paths along the inside. (Of course the random walk trajectories cover the annulus, so that the correct argument contains a numerical factor that accounts for this averaging.)

For measurements on samples of size L_φ, the random uncorrelated pairs of paths again enclose random amounts of flux Φ [5], so sweeping the magnetic field generates a random fluctuation in conductance just as sweeping the Fermi energy does [35]. For a 1D wire, the correlation range is $B_C = zh/eLw$, which corresponds to placing an extra unit of flux zh/e in the sample area [4]. The prefactor z is another number, which like C above, depends weakly on the shape of the sample and on whether thermal smearing or inelastic scattering dominates the phase relaxation averaging [4]. Many experiments have confirmed the existence of random magnetoresistance fluctuations in metals, semi-metals, MOSFETs and heterostructures [35], and usually the amplitudes of the fluctuations have been consistent with $\Delta G \simeq e^2/h$ once reasonable considerations have been brought to bear about the phase coherence lengths. An early example appears in Fig. 3, which contains the magnetoresistance of an Au wire. In contrast to the coherent back-scattering effects, the fluctuations do not die out as the field strength increases, and measurements in heterostructures have shown that they persist even after strong fields (in the classical sense) have been reached where $\omega_C\tau \gg 1$ [36]. This persistence at high fields is natural because the field is scrambling the interference conditions for a particular set of paths; scrambling them again only makes the conductance fluctuate again by about the same amount [4,37]. When strong fields are reached, the paths themselves are altered, but with no great consequence (the fluctuation amplitude results from the number of paths – not their precise configuration) except under peculiar circumstances when there is no scattering at all (such as in the quantized Hall effect [38]).

The complete experiment which corresponds directly to the theoretical ΔG, i.e. measuring the conductance of each member of a large ensemble of samples, is difficult [39]. The magnetic field and Fermi energy dependence allow an experimentalist to characterize the distribution of conductances using only one sample rather than having to measure dozens in order to gather the same statistics. Of course, they also allow him direct access to E_C and B_C, which contain fundamental information about the transport mechanisms (namely correlations among the paths in energy and space). The Fermi energy dependence leads to a rather spectacular non-linear response, which will be discussed in detail below, and the magnetic field dependence also has a number of beautiful consequences.

The sum rule that causes the random conductance fluctuations to have universal amplitude is the same that governs the amplitude of any periodic AB oscillations that might occur in a ring shaped structure [4,40]. The complete resistance formula for such

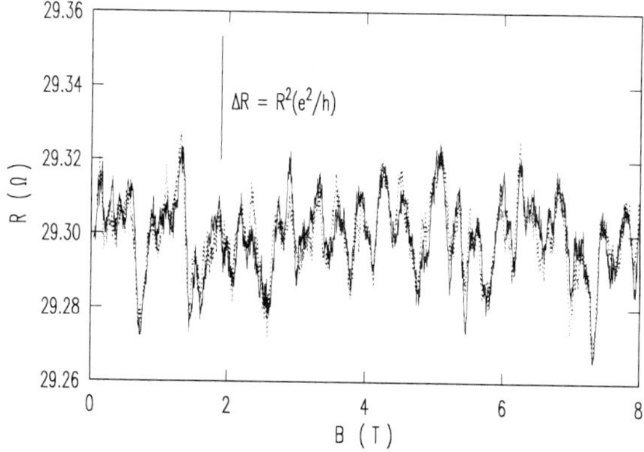

Figure 3. The magnetoresistance of a 350 nm long Au wire of cross-section $(37\,\text{nm})^2$ at $T = 0.01$ K contains reproducible fluctuations that result from the random AB effects among the complicated electron trajectories through the wire. The proof that this is not simply temporal noise is that a second recording of the magnetoresistance (dotted line) contains the same random wiggles as the first trace (solid line).

a sample will be [41]

$$g = g_{cl} + \sum_{m=0} g_m(B) \cos \left\{ \frac{2\pi m \Phi}{h/e} + \alpha_m(B) \right\}, \tag{4}$$

where g_{cl} is the classical conductance, Φ is the average flux enclosed by the loop, and $g_m(B)$ and $\alpha_m(B)$ are random functions of the magnetic field. The amplitude g_m is the universal conductance fluctuation amplitude if all of the inelastic scattering occurs in the reservoirs at the ends of the sample. This is not a very realistic model in most cases, because inelastic scattering occurs randomly throughout the sample (but one could certainly contrive a sample where it would apply by artificially enforcing inelastic scattering [42]). For homogeneous phase relaxation, the g_m are successively smaller because the paths that contribute to them are longer. The decrease is an exponential function of m because in order to generate oscillations of period h/me, the electrons must complete m circuits of the loop retaining coherence the whole way. The comparatively benign algebraic averaging mentioned above relied on the ability to relax the length scale for the coherent region, but here the scale is fixed by the circumference of the loop. The correlation fields $B_C^{(m)}$ that describe the fluctuations in g_m and α_m also decrease, because the paths for each term cover more sample area as m increases. That is, paths contributing to g_0 cover only one arm of the ring, but paths for g_1 cover both arms, paths for $h/2e$ circle twice and therefore cover twice as much area as for g_1, and so on.

All of these assertions have been confirmed in experiments on loops [43]. The dimensionless conductance of an 820 nm diameter Au ring (pictured in Fig. 4a) is displayed in Fig. 4b. The resistance clearly exhibits periodic oscillations across the whole sweep. To ascertain this one simply computes the Fourier spectrum which is displayed in Fig. 4c. Near zero "frequency", there is a peak that results from the "long wavelength" random

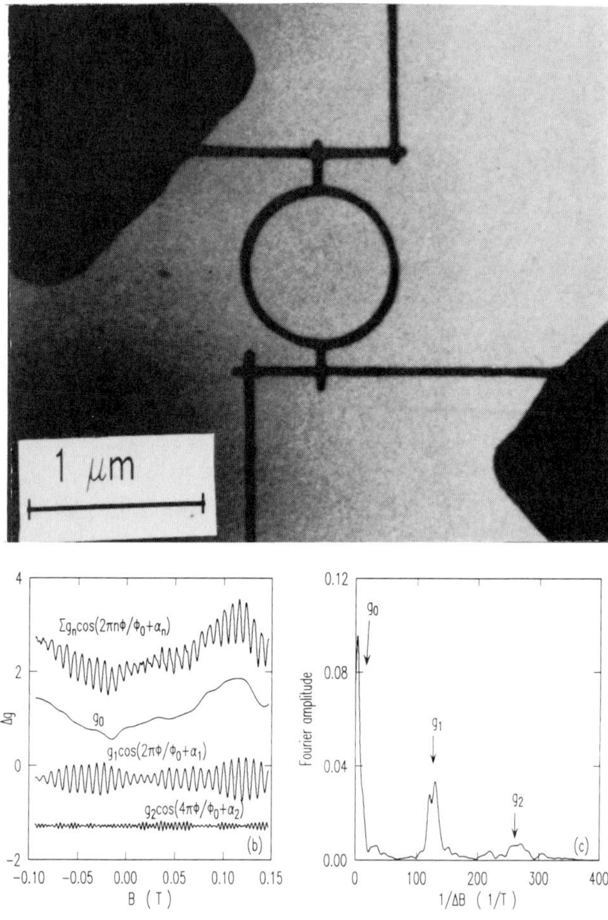

Figure 4. (a) An electron microscope photograph of a loop of gold that was 820 nm in diameter formed of 40 nm wide, 37 nm thick wires. (b) The dimensionless conductance $g(H)$ from the loop contains obvious periodic oscillations. The oscillation period is precisely that expected for adding a unit of flux h/e to the area encircled by the loop. (c) The Fourier spectrum (the square-root of the power spectrum) contains 3 peaks corresponding to a slow random fluctuation $(1/\Delta H \lesssim 30/T)$ and h/e oscillation and a fainter $h/2e$ oscillation. By selecting a region of the Fourier spectrum (of appropriate width, say $\simeq 50/T$ for g_0) and transforming back, one may easily view the respective components of the conductance individually.

magnetoresistance fluctuations (g_0); near $1/\Delta H = 130/T$ is a peak corresponding to adding flux units h/e to the average area enclosed inside the ring; and a much weaker peak occurs at twice that frequency. The structure in the peaks reflects the random dependence of the amplitudes g_1 and g_2 and their associated offsets α_1 and α_2 on the field strength. The smaller peak results because a fraction of the electrons complete two trips around the loop (in opposite senses) and therefore effectively enclose twice as much flux. If one wishes to view directly any term in eq. (4), one can simply select that peak in the Fourier spectrum and invert the transform. This has been performed for the first three terms in the series and the results are displayed in Fig. 4b. As expected, the root-mean-square conductance fluctuation $\sqrt{(\langle g^2 \rangle - \langle g \rangle^2)}$ decreases as m increases and so does B_C. Of course this kind of statistical information can be obtained directly from the Fourier spectrum; it is apparent that the areas beneath the peaks decrease as m increases while the widths of peaks increase.

A simple experiment was performed to study the loss of the h/e period in the magnetoresistance of cylinders and arrays [44]. Series arrays containing different numbers (N = 1,3,10 and 30) of similar Ag loops were examined. The loops in the arrays were strung together at a separation of about L_φ, so that each loop should have contributed its own set of AB oscillations that were uncorrelated with the AB spectrum from neighboring rings. The coherent back-scattering $h/2e$ oscillations were observed near $H = 0$ in all samples, and at higher fields, h/e oscillations appeared. The coherent back-scattering oscillations were relatively stronger in the samples with more loops. If the $h/2e$ oscillations near zero field were subtracted from the magnetoresistance, there remained h/e oscillations, that were *randomly maximum or minimum* at $H = 0$. Since the h/e oscillations arise from uncorrelated paths, there is no preference for one polarity over the other [45], but the coherent back-scattering always generates a particular polarity at $H = 0$ (maximum resistance for no spin-orbit scattering and minimum for strong spin-orbit scattering). It is not too difficult to surmise that upon summing a large number of random polarities (as is done in the cylinder and array experiments), one finds that the h/e amplitude is negligible. That the ensemble average is the correct scheme was proved by comparing the N dependence of the $h/2e$ oscillation amplitude

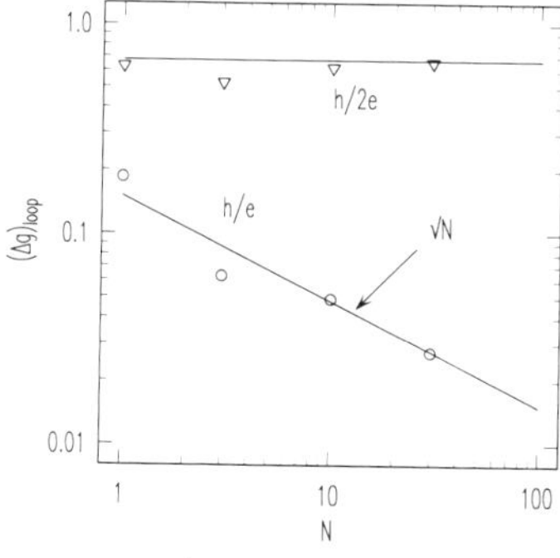

Figure 5. The ensemble averaging of the h/e oscillations (circles) results in a decay that is $\propto 1/\sqrt{N}$, but the coherent back-scattering $h/2e$ oscillations (triangles) are not destroyed by the average. The data are from experiments on series arrays of Ag loops at $T = 0.4$ K [44].

to that for h/e oscillations. The oscillation amplitude *per loop* is plotted in Fig. 5 for both periods. This rendering of the data effectively treats the sample as a cylinder of height NL_φ. The coherent back-scattering amplitude is independent of N; it does not decay when the "cylinder" grows very long. On the other hand, the amplitude for the h/e oscillations decays as $\sqrt{N}$ – just as one expects for averaging N random variables.

Ensemble averaging also occurs if the energies of the electrons are broadly distributed. Temperatures larger than E_C/k smear the Fermi surface so that electrons at different energies contribute uncorrelated patterns to the magnetoresistance fluctuations. The number of patterns is $N \simeq kT/E_C$, so that as T increases, one finds the fluctuations decaying [5] as $\sqrt{T}$. This decay has been observed in magnetoresistance experiments [46] on Au rings where the inelastic scattering length is independent of temperature below about 1 K. Representative temperature dependence data for the loop of gold pictured in Fig. 4a are displayed in Fig. 6 along with their Fourier spectra. The temperature dependence of any individual feature in the magnetoresistance of the Fourier spectrum is nearly useless information. Its interpretation is very complicated because of the random mixing of fluctuation patterns. For instance a peak in the Fourier spectrum (marked by the dotted line) first shrinks and then grows as the temperature increases. The more useful quantities are the statistical averages obtained by summing the Fourier weights in the appropriate "frequency" ranges. The amplitudes g_0, g_1 and g_2 from two different rings were obtained in this way, and the results followed the $\sqrt{T}$ dependence above a characteristic temperature. The characteristic temperatures E_C/k were in good agreement with the values of E_C inferred from the diffusion coefficients from the two loops. This thermal average leads to a second dephasing length $L_T = \sqrt{(hD/kT)}$, the distance electrons with energies differing by kT diffuse before their wavefunctions are significantly out of phase. A similar decay occurs if the voltage bias $V = (\mu_1 - \mu_2)/e$ across the disordered potential exceeds E_C/e. In this case the electrons are injected into several uncorrelated regions of the conduction band and so generate different conductance fluctuation patterns, the number of patterns being $N \simeq eV/E_C$. As V increases beyond E_C/e, the fluctuation amplitude decays according to $\Delta g \propto 1/\sqrt{N} \simeq \sqrt{(E_C/eV)}$, and for smaller bias, the fluctuation amplitude is constant (see Fig. 6).

One other common cause of dephasing is spin-exchange interactions between the electrons and paramagnetic impurities. Since the electron trajectories pervade the sample, the presence of even a very few spins can completely obliterate the conductance fluctuations. This has been observed in experiments where the gold was lightly doped ($\sim 10\,\mathrm{ppm}$ of about 100 spins interspersed among the 10^7 Au atoms in the loop) with Mn by ion implanting [47]. The Mn atoms afix randomly to the Au lattice creating paramagnetic moments. Even with this tiny doping fraction, at low fields, the h/e oscillations are absent – they have vanished completely. At higher fields, however, they recover to the full amplitude g_1 expected for the sample without the extra dephasing from spin-spin scattering. The reason for the recovery, is that the field aligns the impurity spins and the electron spins, and the exchange no longer destroys the wavefunction. Even placing a small number of impurities on the surface of the loop will accomplish this quenching of the conductance fluctuations [36]. Spin-orbit scattering, on the other hand, involves no exchange of energy and preserves phase coherence, but it changes the sum rules so that the conductance fluctuation amplitude is reduced by a fraction 1/4 [4].

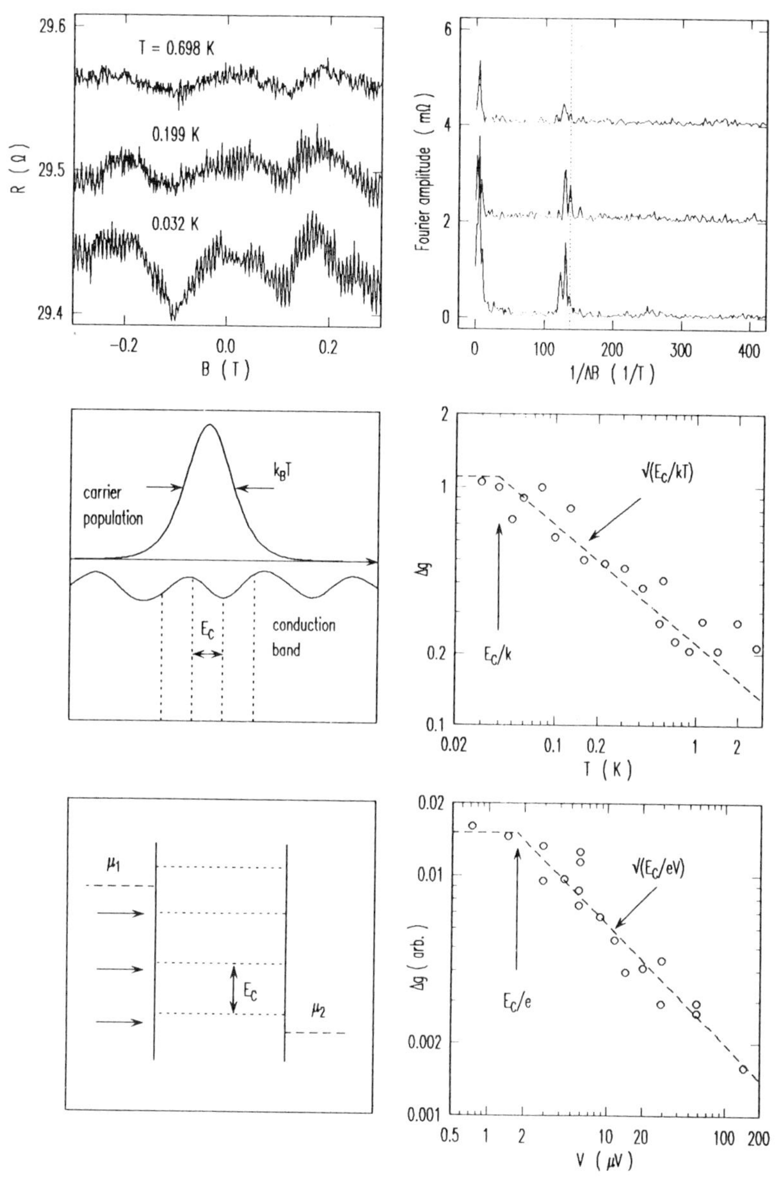

Figure 6. (top panel) An example of the temperature dependence of the magnetoresistance, and its Fourier transform from a Au loop (the same one pictured in Fig. 4); (middle panel) the average amplitude of both the random fluctuations and the periodic oscillations decay as $1/\sqrt{T}$ at high temperatures, but are constant at temperatures below E_C/k; (bottom panel) an equivalent average occurs when the voltage bias exceeds E_C/e at temperature of 0.01 K as indicated by these data from the same ring.

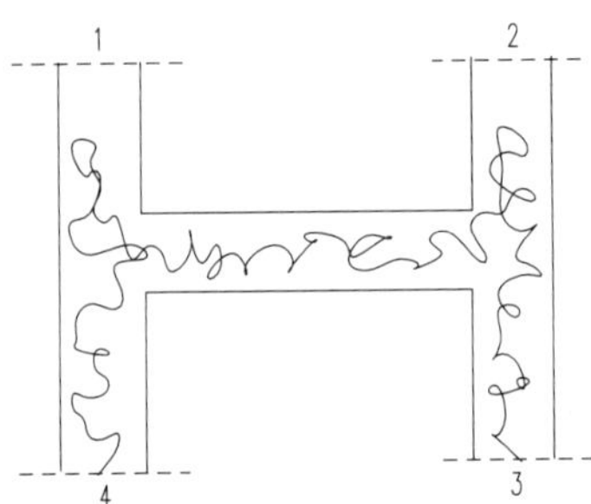

Figure 7. A picture of a four-probe sample where the electron trajectories wander throughout the sample including into and out of the voltage probes. The particle reservoirs at the ends of the leads are numbered and discussed in the text.

3   Universal Scaling of Resistance Fluctuations

The measurement of the sample conductance is never managed in exact analogy to the theoretical models. There is no box labelled "phase randomizing reservoir" that one can attach to a wire to mimic the theorists' particle baths. In fact the probes that connect the tiny samples to large area pads (where wires connected to volt-meters and current sources are attached) are generally of some resistive material and frequently the same material as the sample. Typically, there are no physical barriers that delimit "the sample" from the rest of the probes and pads. The electron trajectories, therefore, respect no imaginary boundaries around "the sample" but wander blindly into the current and voltage probes for distances of order L_φ. An example of a non-local path is shown in Fig. 7; interferences among pairs of such trajectories contribute differently to the resistance depending on which wires are used for current leads and which are used for voltage leads. Hence, the phrase "the conductance of the sample" is rather vague and is wisely avoided. To specify a conductance, one must specify the geometry of the sample *and the measurement*. Certain probes are chosen to inject and sink current and certain probes are chosen to measure the voltages. In the standard four-probe method, a resistance (not a conductance) is measured by voltage probes connected to a meter with very high impedance when a current is driven through the sample in other probes. The meter sinks essentially no current – no *net* current that is; the electron trajectories still freely wander into and out of the probes. Such a four-probe measurement is defined by

$$R_{ij,mn} = \frac{V_i - V_j}{I_{m \to n}}. \tag{5}$$

Büttiker has proved that the four-probe resistance can be described by the same method of transmission coefficients used in the two-probe model [48]. Here of course, the formulae are more complicated because they involve mixtures of two-probe conductances among the various current and voltage leads. Although cumbersome, the complicated nomenclature is necessary. In the phase coherent case, the classical notions of longitudinal (ρ_{xx}) and transverse (ρ_{xy}) resistance are inherently fallacious because different measurements of "$(\rho_{\alpha\beta})$" yield completely uncorrelated results. For instance (referring to Fig. 7) one might try both $R_{12,43}$ or $R_{43,12}$ for the longitudinal resistance and arrive at disparate answers. The answer depends on which voltage probes are used because the resistance fluctuations arise from interference patterns in the probes, as well as along what one thinks of as "the sample".

Because the electron trajectories wander throughout the leads, the resistance has a "non-local" character [49-54]. Instead of adopting the conventional assumption that the conductivity (or resistivity) can be defined at point of the sample (or a region no larger than ℓ^d), one must recognize that the conductivity is a long-range function [52] that cannot be defined in regions smaller than L_φ^d. One consequence of the long-range phase coherence is that the resistances have a complicated symmetry with respect to zero magnetic field [55]. There is a component that is strictly anti-symmetric in field $[R(B) = -R(-B)]$, which arises solely from regions of size L_φ around the probe junctions, and another symmetric component that accumulates all along the region between the probes as well as in the regions around the probes. The resistances here are analogous to measurements on inhomogeneous classical conductors. Any given measurement will be neither symmetric nor anti-symmetric, but the decomposition to symmetric and anti-symmetric components and the relations among these components for various measurements are quite the same as for the inhomogeneous classical conductors [48].

Moreover since the phase coherent regions extend into the leads, the placement of the leads does not determine the "length of the sample" [56,57]. For large separation L between the voltage probes, this is not particularly important because the main contribution to the conductance comes from the $(L/L_\varphi) - 2$ coherent chunks between the voltage probes (all of the segments expect the two where the voltage probes attach), which are effectively two-probe measurements. Also in this large L situation the measured resistance will be nearly symmetric around $B = 0$, because these two-probe measurements are symmetric in field. At small separations, however, L_φ controls the length of the sample, with the arresting consequence that "non-local" measurements (such as $R_{14,23}$ on Fig. 7) exhibit fluctuations similar to those observed in the "local" measurements [56-58]. An example of the "non-local" fluctuations appears in Fig. 8 along side of a local measurement employing the same sample probes. Generally, there are several differences between the local and the non-local measurements. Most obvious is that there is no classical component in the non-local case, i.e. the average resistance is zero. (There is in principle an average classical resistance – a "spreading resistance", but this component is an exponentially decreasing function of L/w, where w is the width of the wire. We know that the fluctuations do not arise from an inhomogeneous spreading resistance because if they did, then they would be no larger than the average resistance.) More subtle differences are also apparent. First, the non-local fluctuations are smaller, because the amplitude is proportional to the fraction of carriers retaining phase coherence when they reach the voltage probes. This fraction is $\simeq \exp\{-L/L_\varphi\}$ (which follows from the definition of L_φ) and strongly decreases at large L. The local fluctuations, on the other hand, tend to grow with L because they comprise incoherent sums of fluctuations along L. Second, the field scale is smaller for the non-local fluctuations, and their auto-correlation (see Fig. 8) generally has a different shape. The differences here can be explained qualitatively if we consider the different kinds of paths that contribute to the two resistances. Generic paths for the two kinds of measurement are contrasted in Fig. 9. The local measurement is dominated by the numerous short loops in the region between the voltage probes and since the field scale for the fluctuations is set by the addition of a flux unit h/e to the interference area, the field scales for these short loops can be rather long. The auto-correlation function therefore has a long tail. The short loops have no effect on the non-local measurements since they don't reach the voltage

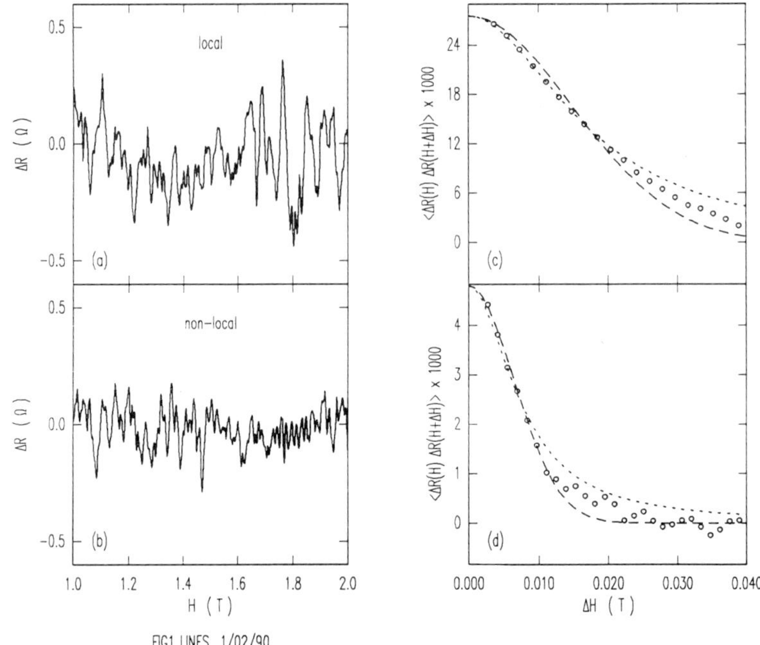

Figure 8. Examples of local (a) and non-local (b) resistance fluctuations measured in a $0.66\,\mu m$ long by $0.125\,\mu m$ wide Sb wire which had a classical resistance of $\langle R \rangle = 49\,\Omega$. Both measurements were performed at $T = 0.048\,K$ with a $20\,nA$ drive current. Comparing auto-correlation functions for the two sets of data, one sees that the non-local resistance fluctuations (d) are smaller and have a smaller field scale that the local fluctuations (c) [57].

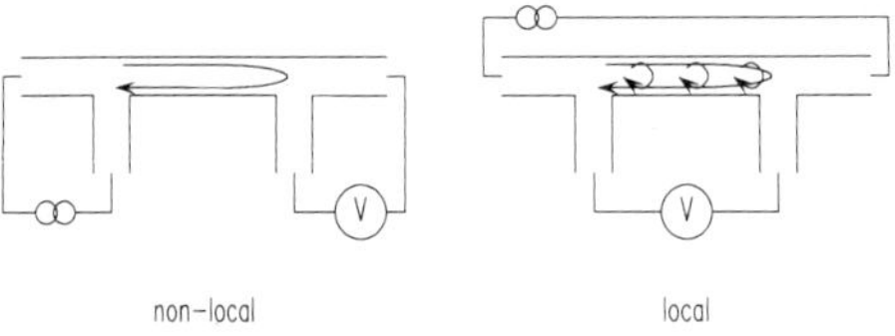

Figure 9: Illustration of the kind of trajectories that contribute to fluctuations in the non-local measurements (a) and the local measurements (b).

probes; only the long paths (loop size $\gtrsim L$) affect the measured resistance. Since the distance L places a minimum bound on the size of the loop, there is a maximum bound on the correlation scale which is always smaller that the average loop size in one of the corresponding local measurements.

It was mentioned above that, in wires longer than L_φ, the total resistance fluctuation amplitude would be $\propto \sqrt{(L/L_\varphi)}$ and concomitantly the conductance fluctuations scale as $g_0 \propto 1/\sqrt{(L/L_\varphi)^3}$ [16]. This scaling law arises in the two-probe model from addition of many uncorrelated coherent segments as one would sum the resistances in a classical network. Quite obviously it must break down for small L – else the conductance fluctuation amplitude approaches infinity as $L \to 0$, which would contradict the assertion that, for two probe samples, g_0 reaches its maximum value in a completely coherent sample, i.e. at length scale L_φ. (Moreover, the condition $L < L_\varphi$ has no meaning in the two-probe conductor terminated at either end by reservoirs.) It also breaks down in the four-probe measurements, because the area covered by the coherent electrons cannot be altered by moving the probes closer together [56]. Once the probes are separated by less than L_φ, the "length of the sample" is governed by L_φ. Measurements of the length dependence of the resistance fluctuations confirm this point. The change of the resistance fluctuation amplitude with $\sqrt{(L/L_\varphi)}$ ceases when $L \lesssim L_\varphi$, and the amplitude becomes essentially length independent.

The cross-over in length scaling has been studied in experiments on several materials. All of the data can be reduced to a single scaling plot where $\Delta R/\Delta R_\varphi$ is graphed as a function of the reduced probe separation L/L_φ. ΔR_φ is the resistance fluctuation amplitude expected at length scale L_φ, namely $(Ce^2/h)R_\varphi^2$, where R_φ is the resistance of the L_φ long segment. Measurements on several different Sb wires (all contained in

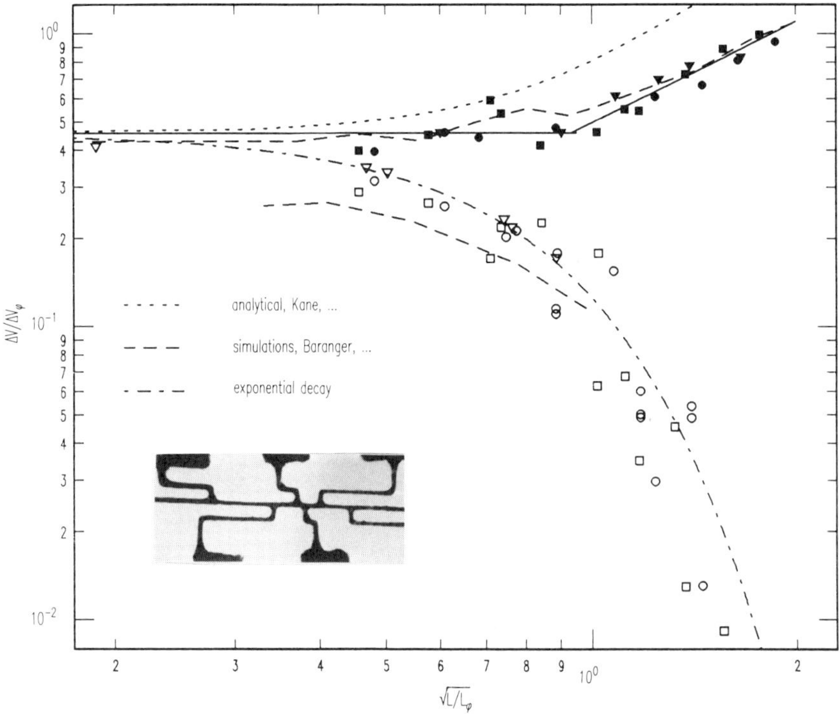

Figure 10. The length dependence of the local resistance fluctuations (solid symbols) $\Delta R/\Delta R_\varphi$ on L/L_φ. The solid line is the result from the numerical simulations [63], and the dotted line is the analytical theory [53]. The solid lines indicate constant amplitude and the classical addition of fluctuations ($\propto \sqrt{L}$). The non-local fluctuations from the same (open symbols) wires. The dashed lines are the results from the numerical simulations [63], and the dash-dot line is the exponential decay described in the text [57]. The inset shows a photograph of the sample used. It was made by C.P. Umbach (c.f. plates at the beginning of the volume).

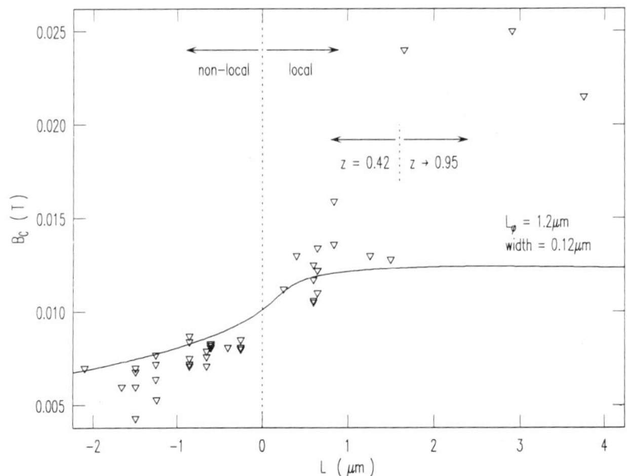

Figure 11. Correlation field scale from various line segments. The results from local measurements are plotted with positive L and the non-local results with negative $L(L \rightarrow -L)$. The solid line illustrates the L dependence expected [61] when $L_\varphi = 1.2\,\mu$m, $w = 0.12\,\mu$m and $L_T \rightarrow \infty$ [57].

one "spider sample": see inset) with differing L are displayed in Fig. 10, and similar measurements from Au wires and Silicon MOSFETs [56,57] are consistent with these results. In the figure, both the length independence at small L and the classical addition at large L are apparent (represented by solid lines). Also plotted for comparison are the results of non-local measurements, which decay exponentially with increasing L as expected from the discussion above.

In order to make a quantitative statement about the correct scaling law for the fluctuations, one must answer a couple of difficult questions. Foremost is: what is L_φ? One can measure L_φ in a number of ways. The magnetoresistance associated with weak localization is a standard way of determining the inelastic scattering length, but as it happened, only one of the Sb wires studied exhibited the characteristic magnetoresistance cusp. By applying the appropriate four-probe model, one can extract an accurate value of L_{in} – at least in principle [59]. The correlation scale B_C also measures L_φ, but the prefactor z depends on geometry and on the kind of phase relaxation – whether L_{in} or L_T is shorter [4]. If $L_{in} \gg L_T$ then $z \simeq 0.95$ in a two-probe sample; in the opposite limit, $z = 0.042$ [60]. The theoretical estimate of B_C for a four-probe sample [61] has been made only at $T = 0$ ($L_T \gg L_{in}$) and only for constant width samples with constant width voltage probes. The situations in the experiments share none of these characteristics. L_{in} and L_T are not very different, and because of dispersion in the lithography, the sample geometries are not rectilinear – the corners where the probes attach are always rounded so that the effective width varies. The same considerations about the non-ideal geometry create uncertainty in the assessment of L_{in} in the weak localization experiment and, therefore, about ΔR_φ in the fluctuation experiments [57].

To obtain L_φ, we have calculated B_C for the magnetoresistance fluctuations as a function of L, and the results are plotted in Fig. 11. For convenience, the non-local mea-

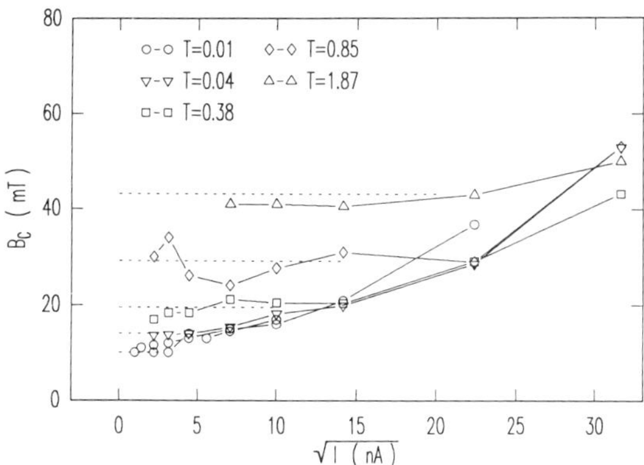

Figure 12. The dependence of B_C on drive current from a series of local measurements on the $0.66\,\mu$m Sb wire. Data obtained at several different temperatures are displayed, and for each temperature the zero current asymptote (dashed lines) agrees with a scaling $B_C(T, I = 0) \simeq \sqrt{T}$ as expected from a simple voltage averaging [57].

surements are plotted with $L \to -L$, which method illuminates the smooth cross-over from non-local to local measurements. They are compared with a theoretical calculation based on the assumptions of $L_\varphi = 1.2\,\mu$m, $L_T \to \infty$, and that the samples have constant width and rectilinear geometry. Considering the naivety of the 2nd and 3rd assumptions, the agreement is respectable. There is a convincing trend at B_C decreases as the "length" decreases toward $-\infty$, and the fractional change between positive and negative L is about right. Part of the remaining discrepancy can be easily removed by noticing that, in the Sb sample, the widths were larger for smaller $|L|$.

The sudden rise in B_C for local fluctuations in wires with $L > 1.6\,\mu$m probably results from the prefactor changing from 0.42 to 0.95 when the voltage across the distance L exceeds the correlation energy. At this point an effective $L_T \propto 1/\sqrt{T_{eff}} \simeq \sqrt{eV/k}$ is shorter than L_{in} and so dominates L_φ. (This is not proper heating but is instead energy averaging; one does not expect inelastic collision rates to change.) The point where the transition occurs is $eV \simeq kT$ as expected. Measurements at lower currents would have reduced B_C in the longer wires. Measurements at successively higher temperatures (displayed in Fig. 12) confirm this model of the effective temperature for energy averaging being the minimum of T and eV/k. The saturation values of B_C as $I \to 0$ are handily described by $1/\sqrt{T}$. With this in mind, one expects that the trend for $I \to 0$ would be given by the solid line in Fig. 11 if the prefactor $z = 0.42$ were maintained for all lengths. From all of the analysis brought to bear on the problem, one concludes that the value $L_\varphi = 1.1 \pm 0.3\,\mu$m is the average phase breaking length for the whole sample.

The decay of the non-local fluctuations then is described by the exponential law $\Delta R/\Delta R_\varphi = \exp\{-aL/L_\varphi\}$ with the prefactor $a = 1.2 \pm 0.3$ (dash-dot line). The prefactor is completely consistent with theoretical modelling (again based on a particular choice of constant width geometry) which obtains $a = 1.1$ for $L \simeq L_\varphi$; $a \equiv 1$ only in the case that L approaches infinity [62]. Numerical simulations [63], where the inelastic scattering was invoked only in the leads, provide a very good *quantitative* description of the data if this value of L_φ is used (these results are dashed lines in Fig. 10). The agreement is easily within the statistical errors in the experiment and modelling for both the non-local and the local fluctuations.

Several analytical calculations [49,53,54,61] of the local fluctuation amplitudes (ΔR vs D/L_φ) have been made, and they are all in agreement with one another except for different asymptotes at $L \to 0$. They are, however, in conflict with the experiments as demonstrated by the dotted line in Fig. 10. It is astounding that the same formalism gives a rather exact description of the non-local fluctuations [62]. One can enforce agreement only by trickery. For instance, one might arbitrarily stretch the abscissa for the theory by a factor of 2.5, but this would be an absurd tack because one should then stretch the axis by the same amount for the non-local fluctuations. It also forces the reduction in the $L \to 0$ asymptote since $\Delta R_\varphi \propto L_\varphi^2$. Eliminating certain terms from the perturbation series in the theory also allows agreement with the experiment [61], but here again the choice is capricious and not supported by any physics. We are left with the inescapable conclusion that something is askew in the theory.

The resolution of this conflict between the local fluctuation measurements and the analytical calculations has not been resolved. It seems clear that at least part of the problem lies in the theoretical method used to kill the phase of the propagating electrons [54]. This is the major difference between the numerical simulations (which provide an excellent description of both local and the non-local data) and the analytical work (which can describe only the scaling of the non-local fluctuations). In the former, all of the inelastic scattering occurs in the reservoirs so that L_φ is fixed by the placement *and the length* of the voltage leads. The simulations, therefore, rely on an artificial change in the voltage lead length to maintain constant L_φ when L changes. In contrast, L_φ enters as a homogeneous (occurring uniformly throughout the sample) relaxation of the propagators in the analytical theory. It would seem that the latter method has by far the better chance to get the correct answer. The question of just how uniformly the phase relaxation occurs in the sample is subtle, and it is at least conceivable that relaxation occurs primarily at fixed grain boundaries or other static potential barriers. It seems certain, however, that there is no direct mimic of the shortening of the voltage leads as L increases. Another possible resolution may be that while the theory calculates strictly in linear response ($I \to 0$) and $T = 0$, there is always finite voltage and temperature in the measurements. The finite temperature is probably not at fault since the non-local fluctuations are adequately described by all theoretical methods, and substituting L_T for L_φ in the local measurements does not recoup the agreement with the experiment [54]. There is no finite voltage drop across the segment being measured in the non-local measurements; the average voltage along the segment is zero, where as there is certainly a voltage dropped in the local case. Even in these measurements, however, the *amount* of voltage dropped is quite small – smaller than any other relevant energy scale (kT or E_C). For this reason, it is doubtful that the voltage drop can significantly alter the picture. Experiments at lower currents certainly do not yield a different scaling plot. This leaves a mystery: why does the analytical calculation fail for the local fluctuations after describing the non-local fluctuations?

4   The I–V Curve of a Metal Wire

Having stumbled upon a severe discrepancy between theory and experiment that may be related to the subtle absence of linear response, one is wise to study more closely the response of such a sample to finite voltage bias. The picture of a disordered potential pinned at either end by reservoirs with fixed Fermi energies makes it obvious that the ham-handed application of classical laws is doomed [7]. The average response to a bias between the reservoirs will still be described by Ohm's law, but "on top of" the average will be random fluctuations. The fluctuations do not result from AB interference; instead, they are due to random changes in the transmission coefficients as the potential tilts. Furthermore, since there is no inversion symmetry for a random potential, the current-voltage (I-V) characteristic will generally not be anti-symmetric when $V \rightarrow -V$ [16]. Fluctuations of $\Delta G(V)$ are not correlated with $\Delta G(-V)$. The random I-V curve also serves to generate large response at harmonics of an alternating bias at some fundamental frequency [16].

Non-linear response is familiar from classical electronics from diodes, the I-Vcurves of which are typically described by $I \propto 1 - \exp\{V/V_0\}$, V_0 being some characteristic energy defined by temperatures and materials. The I-V curve of metal is quite different [64]. The smooth monotonic exponential dependence is replaced by a random fluctuation. The fluctuations occur on energy scale E_C in analogy with shifting the Fermi level in the MOSFET experiment [26] except that here the Fermi level is changed for one reservoir only. Not surprisingly the amplitude of the fluctuations is $\Delta G \simeq e2/h$ for voltage bias $\sim E_C/e$. Examples of the conductance fluctuations from one Sb and one Au wire are displayed in Fig. 13 [65]. In both cases, the fluctuations are large amplitude with short range near $I = 0$, but the amplitude shrinks and the range increases as I increases beyond a few E_C/eR_φ or a few kT/eR_φ. As expected, the fluctuations have no discernible symmetry when current is reversed. The increasing range and decreasing amplitude result from voltage averaging and, possibly, from extra inelastic scattering at the large currents. These two traces are typical of the response at low temperatures. The auto-correlation functions of the two traces are also displayed. Except for the widths of the auto-correlation functions, they are similar: a narrow component dominated by the response near $I = 0$ and a broad component that arises from the voltage averaged part of the I-V curve. From the narrow component, one can ascertain the correlation energy E_C. From these two traces one obtains $E_C = 0.08\,\mathrm{K}$ for the Sb wire and $0.03\,\mathrm{K}$ for the Au. (The Au data in the figure are somewhat broadened by temperature at $T = 0.045\,\mathrm{K}$.) Both numbers are in reasonable agreement with the estimates from the diffusion coefficients.

At higher temperatures ($kT \gg E_C$), the fluctuation amplitude is much reduced near zero bias, but then it grows back and the low temperature envelope $\langle(\Delta G(V))^2\rangle$ re-appears (see Fig. 13). The trace exhibits a "quiet" region of width $2\,kT$ around zero bias which is predicted by the theory [64]. It apparently results from thermal averaging over the kT/E_C fluctuation patterns that are proximate to the Fermi surface. The existence of the quiet region in the I-V characteristic appears to indicate that linear response has been recovered. The theoretical description of the I-V curve [64] provides a solid qualitative description of the fluctuation amplitude as a function of bias [65]. In contrast to the diode case, the I-V curve is quite random. The smooth monotonic variation of $G(V)$ familiar from classical electronics is replaced here with a disturbingly complicated function (just how disturbing will discussed below). The contrast is illustrated in Fig. 14, which contains schematic representations of the non-linear I-V curves for the diode, a metal wire, and the curve that would result from simple heating in a sample where the resistance is temperature dependent. Aside from the qualitatively different dependences of conductance on bias, there will also be very different response to alternating bias. Any non-linear device will respond at harmonics of the drive frequency ω even if the drive is

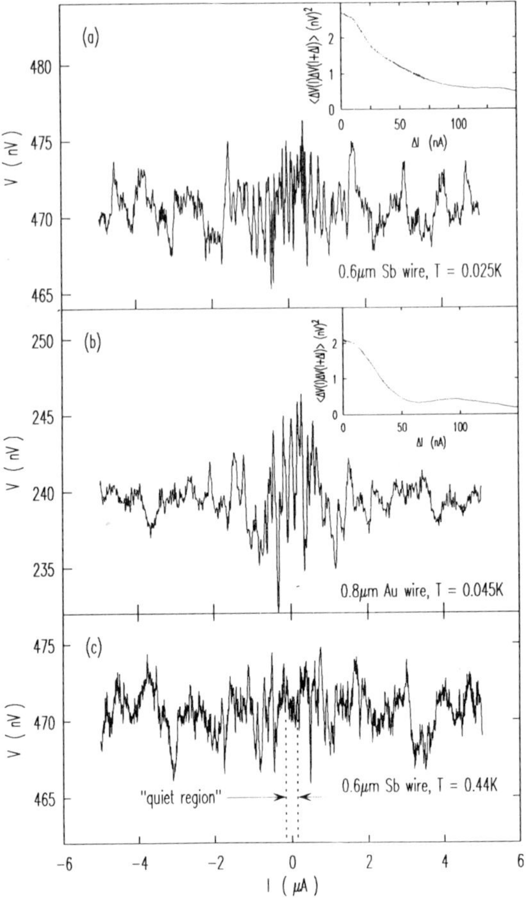

Figure 13. The fluctuations in the voltage (with an AC sampling current of 10 nA) as a function the DC current bias in a $0.6\,\mu$m Sb (a) wire at $T = 0.010$ K, and a $0.8\,\mu$m Au wire (b) at $T = 0.045$ K. The auto-correlations of the data are inset. Similar data for the Sb wire at $T = 0.45$ K (c) contain a quiet region near $|I| = 0$ [65].

a pure tone devoid of harmonic content. For instance, when the diode is driven with the bias $I\cos\{\omega t\}$, there will be signal at harmonics $n\omega$ (where n is a positive integer) that will have amplitudes that are reduced exponentially as n increases. The heating model, in contrast, will generate harmonics only for odd n (again decreasing exponentially as n increases) because the I-V curve is anti-symmetric when the bias is reversed. The anti-symmetry occurs, of course, because $G(V)$ is dependent only on the power dissipated, not on the absolute direction of the bias. The harmonic response from the metal has not been determined theoretically but it has been measured in experiments. Depending on the conditions, it might approximate the response of the diode [66] or it can be utterly unusual [65].

As an illustration one can obtain the harmonic response from the diode by expand-

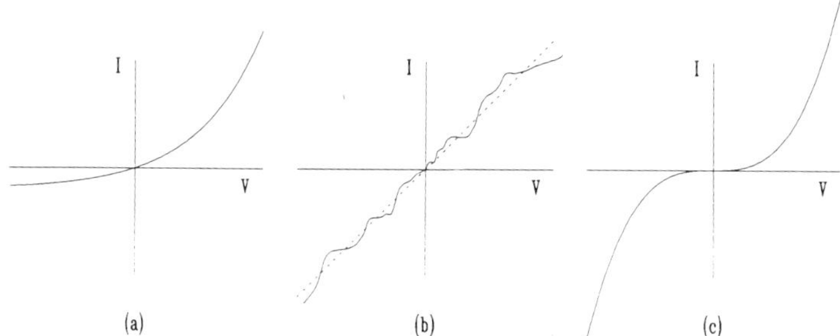

Figure 14. Schematic I-V curves for a diode (a), disordered metal (b) and simple heating by the drive current (c).

ing the I-V curve in a Taylor series. The conductance will be

$$G(V) = \sum_{p=0} G_p V^p \propto \frac{1}{V_0} \sum_{p=0} \left(\frac{V}{p! V_0} \right)^p , \tag{6}$$

where V_0 is the voltage scale to turn on the diode. Since the I-V curve is a smooth monotonic function on the scale of the bias V, the power series is an excellent approximation. Assuming an alternating drive and Fourier transforming each term separately one easily obtains the response at $n\omega$

$$I_{n\omega}(V) \propto \left(\frac{V}{(n-1)! V_0} \right)^n . \tag{7}$$

A metal sample with temperature dependent resistance will yield the same result except that the even harmonics are suppressed by the symmetry of the I-V curve. Tests of the non-linear response were performed in a GaAs-Al$_x$Ga$_{1-x}$As heterostructure, and the results were in partial agreement with eq. (7) for $n = 2$ and 3 (the only harmonics studied) [66]. The response from the magnetoresistance fluctuations were measured over a range of field encompassing many correlation scales so that a reasonable "ensemble" average was obtained for the fluctuations amplitude [65]. The harmonic responses $\langle \Delta V_{2\omega}(I) \rangle$ and $\langle \Delta V_{3\omega}(I) \rangle$ were found to be of the same order of magnitude (in disagreement with the diode model) but proportional to I^2 and I^3, respectively (see Fig. 15). In these experiments, the voltage drop across the sample V was smaller than kT/e and E_C/e, and the transport was nearly ballistic, $\ell \simeq L$.

In rather stark contrast are the experiments on metals [65] displayed in Fig. 16a, where the ratio of energy scales is either essentially the same, $eV \lesssim kT < E_C$. The harmonic response at low temperatures is quite unusual. $\langle \Delta V_{n\omega}(I) \rangle \propto I$ *regardless of the harmonic number n!* This true for $1 \leq n \leq 6$ even down to the smallest biases measured in which the voltage drop across the sample was less than $0.3\mu V$. This voltage drop corresponds to a temperature of $4\,\mathrm{mK}$ which is less than the measurement temperature $T \simeq 10\,\mathrm{mK}$ and considerably less than the correlation energy $100\,\mathrm{mK}$ for the Sb wire being studied. Even allowing for the non-local character of the measurement and doubling the effective length of the sample (L_φ was about $2\,L$ in this experiment), the voltage bias is still smaller than either of the important energy scales. From this inequality, one would like to draw the conclusion that the data are in the linear response regime. (The average voltage is certainly proportional to the current throughout the

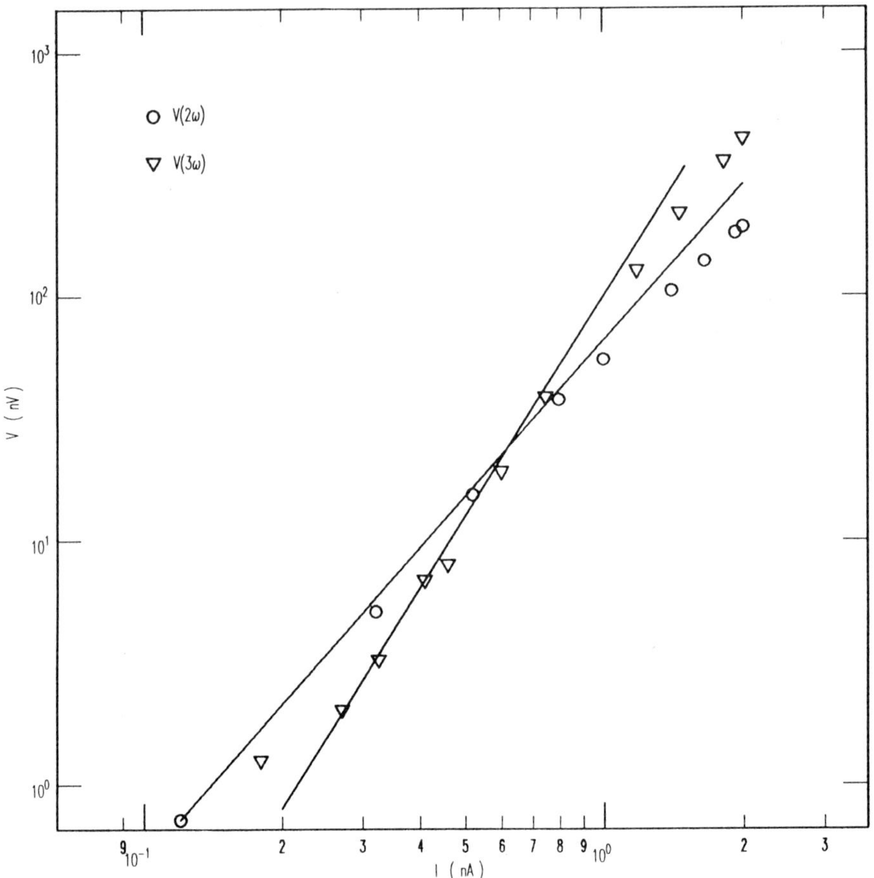

Figure 15. The harmonic response at 2ω (circles) and 3ω (triangles) from a ballistic wire of GaAs-Al$_x$Ga$_{1-x}$As driven by a cosine input signal at frequency ω [66].

range.) Unfortunately, the data refute this possibility; near linear-response one expects small deviations and a "diode I-V curve". Quite obviously, the data have not reached this regime. At higher current biases, the response crosses over to be proportional to $\sqrt{I}$ for all n. The square root appears here as a signature of voltage averaging when eV exceeds kT.

Even if the temperature is increased so that $kT \gg E_C \gtrsim eV$, and the "quiet region" appears in the I-V curve, the harmonic response is still peculiar [65]. As verified in Fig. 16b, $\langle \Delta V_{n\omega}(I)\rangle \propto I$ for $n = 1$, and $\propto I^{3/2}$ for other n (only 2 and 4 were studied in the experiment). In spite of the apparently linear region in the I-V curve, the harmonic response indicates that the response is still far from linear.

It can be shown that a fluctuating $G(V)$ generates the harmonic behavior found in the experiments [67], and curiously, the same behavior is obtained in tunnelling conductance (where the states are narrow and widely separated in space as well as energy), when more than one impurity governs the tunnelling resistance [68]. When only one impurity figures in the tunnelling event, then the "diode response" recovers with the exception that the even harmonics are suppressed in relation to the odd. This suppres-

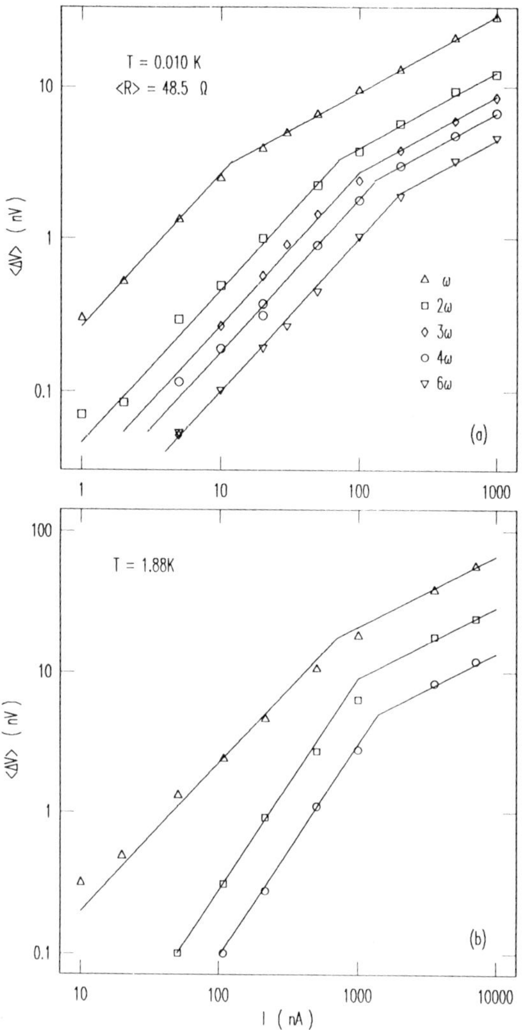

Figure 16. (a) Harmonic response from an Sb wire at $T = 0.01$ K as a function of drive current; (b) the response from the same wire at $T = 1.88$ K [65]. The solid lines illustrate the inferred dependence of the fluctuation amplitude on the drive current amplitude. For large currents, $\langle \Delta V \rangle \propto \sqrt{I}$ for all temperatures and all harmonic numbers. At low bias and low temperatures, $\langle \Delta V \rangle \propto I$, and at high temperatures, the same is true for fluctuations in the response at the fundamental ω, but the harmonic responses are $\propto \sqrt{I^3}$.

sion is evident in the experiment on GaAs-Al$_x$Ga$_{1-x}$As where (because of the long ℓ) the number of strong scattering events is surely small – although the number of small-angle scattering events probably is not. The tunnelling model reverts to linear response when eV is less than the level width for the tunnelling states. The corresponding energy in the metallic problem is E_C, which is larger than eV in *both* experiments discussed above.

The question remains then: is there a linear response regime for a metallic sample? If so, what is the energy scale that governs its range in voltage bias? Only two scales come to mind: (1) the disorder potential and (2) the level separation in the conduction band. The level separation is not an attractive candidate when it is much smaller than the level width E_C. In the GaAs-Al$_x$Ga$_{1-x}$As sample the disorder in the electrostatic potential is surely smaller than in the poly-crystalline metal films, and so at first this seems to be an unlikely candidate as well. The ripple on the Fermi surface, however, might be larger in the 2 D electron gas, because the screening is so much poorer in 2 D than in the 3 D where the screening is exponential after a length of order 1Å.

5 Summary

From measurements on small metallic wires at low temperatures, it is clear that the resistance of weakly disordered potentials is much more complicated than one expects from classical physics. The non-classical behavior results from quantum interference among the multitude of electron trajectories through the wires. Since the trajectories are stationary, random-walk paths (because the impurity potential is static and random), random fluctuations appear in the magnetoresistance. Any perturbation of the paths (AB effects, for instance) changes the interference condition, and so alters the resistance. The crossover to classical behavior is governed by ensemble averaging over energy or spatial coordinates. The non-local quality of the wave-function leads to resistances that are utterly forbidden by classical physics. Biasing the impurity potential leads to non-linear response that is in qualitative agreement with theoretical predictions for fluctuations in the current-voltage characteristic, but also to harmonic responses that are rather puzzling.

6 Acknowledgements

Helpful discussions with B. Al'tshuler, H. Baranger, A. Benoit, M. Büttiker, V. Chandrasekhar, D. diVincenzo, Y. Gefen, S. Hershfield, Y. Imry, C. Kane, R. Landauer. P. Lee, M. Ma, S. Maekawa, J. Rammer, P. Santhanam, D. Stone, C. Van Haesendonck, and R. Webb, which occurred over the years, are acknowledged with gratitude.

References

[1] G. Blonder, R. Dynes, and A. White, unpublished (1984)

[2] C.P. Umbach, S. Washburn, R.B. Laibowitz, and R.A. Webb, Phys. Rev. **B 30**, 4048 (1984); R.A. Webb, S. Washburn, C.P. Umbach, and R.B. Laibowitz, in: 'Localization, Interaction, and Transport Phenomena', G. Bergmann, Y. Bruynseraede, and B. Kramer, eds., Springer Ser. in Sol. St. Sc. **61**, 121 (1985)

[3] B.L. Al'tshuler, JETP Lett. **41**, 648 (1985)

[4] P. A. Lee, A.D. Stone, and J. Fukuyama, Phys. Rev. **B 35**, 1039 (1987)

[5] A.D. Stone, Phys. Rev. Lett. **54**, 2692 (1985)

[6] B.L. Al'tshuler and B.Z. Spivak, JETP Lett. **42**, 447 (1985)

[7] R. Landauer, Phil. Mag. **21**, 863 (1970); see also Z. Phys. **B 68**, 217 (1987), and references cited there

[8] P.A. Lee, Physica **140**, 169 (1986), and references cited there

[9] P.F. Bagwell and T.P. Orlando, Phys. Rev. **B 40**, 1456 (1989)

[10] N.D. Mermin and N.W. Ashcroft, 'Solid State Physics', 345ff, Holt, Rinehart and Winston, New York (1976)

[11] See A.G. Aronov and Yu.V. Sharvin, Rev. Mod. Phys. **59**, 744 (1987) and references to literature on weak localization and h/2e oscillations from coherent back-scattering cited there

[12] D.J. Thouless, Phys. Rev. Lett. **39**, 1167 (1977)

[13] B.L. Al'tshuler and B.I. Shklovskii, JETP **64**, 127 (1986)

[14] Y. Imry, Europhys. Lett. **1**, 249 (1986)

[15] Y. Imry, in: Directions on Condensed Matter Physics, G. Grinstein and E. Mazenko, eds., p. 101, World Publishing, Singapore (1986)

[16] B.L. Al'tshuler and D.E. Khmel'nitskii, JETP Lett. **42**, 291 (1985)

[17] W.F. Smith et al., unpublished

[18] S. Hershfield, Phys. Rev. **B 37**, 8557 (1988), and unpublished (1989)

[19] F.N.H. Robinson, Phys. Lett. **97 A**, 162 (1983); C. Van Haesendonck, unpublished (1985)

[20] S. Feng, P.A. Lee, and A.D. Stone, Phys. Rev. Lett. **56**, 1960 (1986)

[21] N.M. Zimmermann and W.W. Webb, Phys. Rev. Lett. **61**, 889 (1988); G.B. Alers, M.B. Weissmann, R.S. Averback, and A. Shyu, Phys. Rev. **B 40**, 900 (1989); N.B. Israeloff, M.B. Weissmann, G.J. Nieuwenhuys, and J. Kosiorowska, Phys. Rev. Lett. **63**, 794 (1989)

[22] N.O. Birge, B. Golding, and W.H. Haemmerle, Phys. Rev. Lett. **63**, 195 (1989)

[23] D. Mailly, M. Sanquer, J.-L. Pichard, and P. Pari, Europhys. Lett. **8**, 471 (1989)

[24] P. Debray, J.-L. Pichard, J. Vicente, and P.N. Tang, Phys. Rev. Lett. **63**, 2264 (1989)

[25] A.D. Stone, Phys. Rev. **B 39**, 10736 (1989)

[26] J.C. Licini, D.J. Bishop. M.A. Kastner, and J. Melngailis, Phys. Rev. Lett. **55**, 2987 (1985)

[27] Y. Aharonov and D. Bohm, Phys. Rev. **115**, 485 (1959)

[28] E. Merzbacher, Am. J. Phys. **30**, 237 (1962)

[29] R.G. Chambers, Phys. Rev. Lett. **5**, 3 (1960); A. Tonomura, N. Osakabe, T. Matsuda, T. Kawasaki, J. Endo, S. Yano, and H. Yamada, Phys. Rev. Lett. **56**, 792 (1986)

[30] B.S. Deaver and W.F. Fairbank, Phys. Rev. Lett. **7**, 43 (1961); R. Doll and M. Näbauer, ibid., 51 (1961)

[31] F. London, 'Superfluids I: Macroscopic Theory of Superconductivity', p. 152, Wiley, New York (1950); N. Byers and C.N. Yang, Phys. Rev. Lett. **7**, 46 (1961)

[32] N.B. Brandt, D.V. Gitsu, V.A. Dolma, and Ya.G. Ponomarev, JETP **92**, 515 (1987), and references to earlier work cited there

[33] D. Yu. Sharvin and Yu.V. Sharvin, JETP Lett. **34**, 272 (1981)

[34] B.L. Al'tshuler, A.G. Aronov, and B.Z. Spivak, JETP Lett. **33**, 101 (1981)

[35] S. Washburn and R.A. Webb, Adv. Phys. **35**, 375 (1986), and a forth-coming review article in Rep. Prog. Phys. (1990)

[36] G. Timp, A.M. Chang, P. Mankiewich, R. Behringer, J.E. Cunningham, T.Y. Chang, and R.E. Howard, Phys. Rev. Lett. **58**, 2814 (1987)

[37] B.L. Al'tshuler, V.E. Kravtsov, and I.V. Lerner, JETP Lett. **43**, 441 (1986), JETP **64**, 1352 (1986)

[38] G. Timp et al., Phys. Rev. **B 39**, 6227 (1989), and references cited there

[39] B. Mailly and M. Sanquer, this volume

[40] A.D. Stone and Y. Imry, Phys. Rev. Lett. **56**,189 (1986)

[41] Y. Gefen, Y. Imry, and M.Ya. Azbel, Surf. Sc. **142**, 203 (1984); Phys. Rev. Lett. **52**, 129 (1984); M. Büttiker, Y. Imry, R. Landauer, and S. Pinhas, Phys. Rev. **B 31**, 6207 (1985)

[42] D.J. Bishop and G.J. Dolan, Phys. Rev. Lett. **55**, 2911 (1985)

[43] R.A. Webb, S. Washburn, C.P. Umbach, and R.B. Laibowitz, Phys. Rev. Lett. **54**, 2696 (1985); V. Chandrasekhar, M.J. Rooks, S. Wind, and D.E. Prober, Phys. Rev. Lett. **55**, 1610 (1985); S. Datta, M.R. Melloch, S. Banyopadyay, R. Noren, M. Vaziri, M. Miller, and R. Reifenberger, Phys. Rev. Lett. **55**, 2344 (1985)

[44] C.P. Umbach, C. Van Haesendonck, R.B. Laibowitz, S. Washburn, and R.A. Webb, Phys. Rev. Lett. **56**, 386 (1986)

[45] Y. Gefen, Y. Imry, and M.Ya. Azbel, Phys. Rev. Lett. **52**, 129 (1984); Surf. Sc. **142**, 203 (1984)

[46] S. Washburn, C.P. Umbach, R.B. Laibowitz, and R.A. Webb, Phys. Rev. **B 32**, 4879 (1985), and unpublished

[47] A. Benoit, S. Washburn, R.A. Webb, D. Mailly, and L. Dumoulin, in: Anderson Localization, T. Ando and H. Fukuyama, eds., p. 346, Springer, Heidelberg (1988)

[48] M. Büttiker, Phys. Rev. Lett. **57**, 1761 (1986); IBM J. Res. Devel. **32**, 317 (1988), and references cited there

[49] S. Maekawa, Y. Isawa, H. Ebisawa, J. Phys. Soc. Jpn. **56**, 25 (1987); Y. Isawa, H. Ebisawa, and S. Maekawa, in: Anderson Localization, T. Ando and H. Fukuyama, eds., p. 329, Springer, Heidelberg (1988)

[50] M. Büttiker, Phys. Rev. M. **B 35**, 4123 (1987)

[51] A. Yu. Zyuzin and B.Z. Spivak, ZhETF **93**, 994 (1987) [JETP **93**, 560 (1987)]; Sol. St. Commun. **67**, 75 (1988)

[52] C.L. Kane, R.A. Serota, and P.A. Lee, Phys. Rev. **B 37**, 6701 (1988)

[53] C.L. Kane, P.A Lee, D.P. DiVinzenzo, Phys. Rev. **B 38**, 2995 (1988)

[54] S. Hershfield and V. Ambegaokar, Phys. Rev. **B 38**, 7909 (1988); S. Hershfield, Ann. Phys. **196**, 12 (1989)

[55] A. Benoit, S. Washburn, C.P. Umbach, R.B. Laibowitz, and R.A. Webb, Phys. Rev. Lett. **57**, 1765 (1986)

[56] A.D. Benoit, C.P. Umbach, R.B. Laibowitz, and R.A. Webb, ibid **58**, 2343 (1987); W.J. Skocpol, P.M. Mankiewich, R.E. Howard, L.D. Jackel, D.M. Tennant, and A.D. Stone, ibid **58**, 2347 (1987)

[57] H. Haucke, S. Washburn, A. Benoit, C.P. Umbach, and R.A. Webb, subm. to Phys. Rev. **B** (1990)

[58] C.P. Umbach, P. Santhanam, C. Van Haesendonck, R.A. Webb, Appl. Phys. Lett. **50**, 1289 (1987)

[59] V. Chandrasekhar, P. Santhanam, and D.E. Prober, Phys. Rev. Lett. **61**, 2253 (1988)

[60] C.W.J. Beenakker and H. van Houten, Phys. Rev. **B 37**, 6544 (1988)

[61] V. Chandrasekhar, P. Santhanam, and D.E. Prober, subm. to Phys. Rev. **B** (1990)

[62] D.P. DiVincenzo and C.L. Kane, Phys. Rev. **B 38**, 3006 (1988)

[63] H. Baranger, A.D. Stone, D.P. DiVincenzo, Phys. Rev. **B 37**, 6521 (1988)

[64] D.E. Khmel'nitskii and A.I. Larkin, Phys. Scripta **T 14**, 4 (1986)

[65] R.A. Webb, S. Washburn, and C.P. Umbach, Phys. Rev. **B 37**, 8544 (1988), and unpublished

[66] P. deVegvar, G. Timp, P.M. Mankiewich, R. Behringer, J.C. Cunningham, Phys. Rev. **B 40**, 3491 (1988). Some remarks here about the relations among energy scales in the metal experiments [65] are erroneous

[67] Y. Imry and S. Washburn, unpublished (1989)

[68] A.I. Larkin and K.A. Matveev, JETP **66**, 580 (1987); L.I. Glazman and K.A. Matveev, JETP **67**, 1276 (1987)

RANDOM TRANSFER MATRIX THEORY AND CONDUCTANCE FLUCTUATIONS

Jean-Louis Pichard

Service de Physique du Solide et de Résonance Magnétique
C.E.A. Saclay
91191 Gif sur Yvette Cedex, France

1 Random Matrix Theory: The Method and Some Physical Applications

Very often, in Physics and in particular in Quantum Mechanics, we are dealing with matrices. This can be the matrix H representing the Hamiltonian of the system, or its scattering matrix S, or, as studied in this work, its transfer matrix M at a given energy. In general, the matrix is huge and complex, especially if it is associated to a disordered system. An exact knowledge of this matrix, which corresponds to a particular system, on which measurements are made, is practically impossible and theoretically inadequate. On the one hand, the measurements are done on a real complex system, on the other hand the theory can only be based on a statistical ensemble of appropriate matrices. The definition of the matrix ensemble is an essential starting point. Very often, we consider *microscopic models*, e.g. an Anderson model [1] for disordered tight-binding Hamiltonians of non-interacting electrons

$$H = \sum_i V_i |i\rangle\langle i| + \sum_{i\ n\cdot n\ j} T_{ij} |i\rangle\langle j|. \tag{1}$$

In this one-orbital model, a lattice is assumed with nearest neighbor hopping and some more or less arbitrary distributions are taken for the diagonal and off-diagonal terms of H. One can hope that the real system is a particular member of such an ensemble, but we mostly hope that the physical properties, in mean or in distribution, do not crucially depend on the particular model and its microscopic distributions. There is an alternative way to introduce statistical ensembles of matrices, which is not based on microscopic considerations, but where matrix symmetry properties are crucial, and where the physical properties are introduced via *macroscopic constraints* and a *maximum entropy criterion*. While only recently applied to quantum transport, this method is not new, but has been used and tested in various domains. Section 1 of these notes could have been found in the sixties in a nuclear physics conference [2], in the seventies in the context of the physics of small metallic particles [3], or in the early eighties about

Quantum Coherence in Mesoscopic Systems
Edited by B. Kramer, Plenum Press, New York, 1991

classical and quantum chaos [4]. However, since 'Random Matrix Theory' is not yet very familiar in Solid State Physics, it appears appropriate to give as introduction a short review of its main ideas and results. We refer the reader to standard references [5, 6], in order to find either more rigorous derivations or more complete results.

Let us consider a Hamiltonian which can be represented by a $N \times N$-matrix i.e. an array of $N \times N$ complex numbers. This matrix is not an arbitrary set of $N \times N$ complex numbers, since it possesses symmetry properties. First of all, the space of definition of the matrix ensemble, which is not a $2(N \times N)$-dimensional Euklidian space, has to be determined.

1.1 Matrix Symmetry Properties and Dyson Threefold Way

We review the symmetry properties of Hamiltonians which are relevant for electronic quantum transport (spin 1/2 electrons). The symmetry properties of the S-matrix can be found in [7], and we shall give later the symmetry properties of the transfer matrix M.

Firstly, in the absence of sufficient spin-orbit scattering, the hopping elements of H are spin-independent and the total $2N \times 2N$-Hamiltonian can be separated into two uncoupled $N \times N$-Hamiltonians $H(\uparrow)$ and $H(\downarrow)$, corresponding to spin up and down electrons respectively. N is for instance the number of lattice sites in a one-orbital tight binding model for non-interacting electrons. In the absence of significant Zeeman splitting, we have in addition

$$H(\uparrow) = H(\downarrow) = H. \tag{2}$$

H must be equal to its Hermitian conjugate i.e. the complex conjugate of its transpose

$$H = H^\dagger = (H^T)^*. \tag{3}$$

This implies that H can always be diagonalized by a unitary transformation A

$$H = A \cdot H_D \cdot A^{-1} \tag{4}$$

$$A \cdot A^\dagger = A^\dagger \cdot A = \mathbf{1}, \tag{5}$$

where H_D is diagonal with elements $E_1, E_2, ..., E_N$. $\mathbf{1}$ denotes the unit matrix.

If the system is not invariant by time inversion (e.g. when a sufficient magnetic field is applied), H has no additional symmetry and the canonical group for A is that of the *unitary* transformations.

$$A^\dagger \cdot A = A \cdot A^\dagger = \mathbf{1}. \tag{6}$$

If one has time inversion symmetry, H must also be real

$$H = H^* \tag{7}$$

and the canonical group for A is that of *orthogonal* transformations

$$A \cdot A^T = A^T \cdot A = \mathbf{1} \tag{8}$$

In presence of strong spin-orbit scattering and in the absence of significant magnetic field, time inversion symmetry is conserved and it is convenient to write the $2N \times 2N$-Hamiltonian as a Hermitian quaternion-real $N \times N$-matrix. The quaternion elements of H are 2×2-matrices which satisfy a relation yielded by time inversion symmetry

$$\langle i, -\sigma|H|j, -\sigma'\rangle = \sigma\sigma'(\langle i, \sigma|H|j, \sigma'\rangle)^*, \tag{9}$$

in addition to hermicity

$$\langle i, \sigma|H|j, \sigma'\rangle = (\langle j, \sigma'|H|i, \sigma\rangle)^*. \tag{10}$$

Here, σ and σ' denote the spin variables ($\sigma, \sigma' = +1 \ or - 1$). Then, one can show [7] that any Hermitian quaternion-real $N \times N$-matrix can be written as

$$H = A \cdot H_D \cdot A^{-1}, \tag{11}$$

where A belongs in this third case to the canonical *symplectic* group and H_D is diagonal, real and scalar i.e. consists of N blocks of the form

$$\begin{pmatrix} E_a & 0 \\ 0 & E_a \end{pmatrix}. \tag{12}$$

Thus, H possesses a twofold degeneracy which is not due to spin degeneracy (removed by spin dependent hopping) but to time inversion symmetry (Kramers degeneracy).

This classification, obtained from time inversion symmetry, and spin rotation symmetry requirements was first introduced by Dyson [8], yielding three possible universality classes, related to the three types of division algebra with real coefficients, namely, the real numbers, complex numbers and quaternions. The *orthogonal, unitary and symplectic* classes, named as the canonical groups used for diagonalizing H, are now familiar in weak localization theory [9, 10]. Backscattering enhancement characterizes the orthogonal class, while being absent in the unitary class. The symplecticclass is characterized by weak anti-localization.

1.2 Matrix-Spaces and Measures in Eigenvalue-Eigenvector Coordinates

For simplicity, we continue to explain the method of Random Matrix Theory in the orthogonal case. Let H be a real symmetric $N \times N$-matrix. It contains $N(N + 1)/2$ real parameters. We associate a numerical volume dH with the matrix H. Its measure $\mu(dH)$ is defined by the product of the increments of the independent elements

$$\mu(dH) = \prod_{a=1}^{N} dH_{aa} \prod_{a>b}^{N} dH_{ab}. \tag{13}$$

We put a warning here: the rigorous definition of a volume element in a matrix space is in general more complex [11], mostly when the matrices do not form a group. The matrices M for instance require a more careful definition of a metric, and its associated metric tensor, and the measures associated with the numerical volume dS or $d(M^\dagger M)$ have to be shown to be unique. However, the simple definition eq. (13) is correct (except for an irrelevant factor).

The next step consists in expressing $\mu(dH)$ in terms of the eigenvalues and eigenvectors of H. Let us reproduce a simple argument [2] giving the eigenvalue dependence of $\mu(dH)$. From eq. (4), we have in component notation

$$H_{ab} = \sum_{c=1}^{N} E_c A_{ac} A_{bc}, \tag{14}$$

which defines the inverse transformation from the Hamiltonian matrix element variables to the eigenvalue-eigenvector variables. If we parametrize the matrices by their N eigenvalues $E_1, ..., E_N$ and a set of $N(N-1)/2$ parameters $\{x_a\}$ which are functions of the $\{A_{bc}\}$, we need to calculate the Jacobian $J_N(\{E_a\})$ of the change of coordinates

$$\prod_{a=1}^{N} dH_{aa} \prod_{a>b}^{N} dH_{ab} = J_N(E_1, ..., E_N, x_1, ..., x_{N(N-1)/2})\mu(dH_D)dx_1...dx_{N(N-1)/2}. \tag{15}$$

$\mu(dH_D)$ is the measure of the volume element associated with H_D

$$\mu(dH_D) = \prod_{c=1}^{N} dE_c. \tag{16}$$

The Jacobian matrix is given by:

$$\begin{pmatrix} \frac{\partial H_{11}}{\partial E_1} & \frac{\partial H_{11}}{\partial E_2} & \cdot & \cdot & \frac{\partial H_{11}}{\partial E_N} & \frac{\partial H_{11}}{\partial x_1} & \frac{\partial H_{11}}{\partial x_2} & \cdot & \cdot & \frac{\partial H_{11}}{\partial x_{N(N-1)/2}} \\ \cdot & \cdot & \cdot & \cdot & \cdot & \cdot & \cdot & \cdot & \cdot & \cdot \\ \cdot & \cdot & \cdot & \cdot & \cdot & \cdot & \cdot & \cdot & \cdot & \cdot \end{pmatrix}. \tag{17}$$

From eq. (14), we see that the first N columns do not depend on the eigenvalues, while the remaining columns are linear in the eigenvalues. Thus in the expansion of the determinant, each term will contain $N(N-1)/2$ factors, each of which is linear in the eigenvalues. Hence the Jacobian J_N is a polynomial of degree $N(N-1)/2$ in the eigenvalues. When two eigenvalues are equal, the corresponding eigenvectors are not unique, the inverse transformation is singular and $J_N(\{E_a\}) = 0$. This indicates that

$$J_N(\{E_a\}) \propto \prod_{a<b}^{N} |E_a - E_b|^{\beta}. \tag{18}$$

Since $J_N(\{E_a\})$ is of degree $N(N-1)/2$ in the orthogonal case, we must have $\beta = 1$. Thus, up to normalization constants, we have expressed the measure of the space of real-symmetric matrices in terms of the measure $\mu(dA)$ of the orthogonal group.

$$\mu(dH) = \prod_{a<b}^{N} |E_a - E_b|^{\beta} \mu(dH_D)\mu(dA). \tag{19}$$

The eigenvalue interaction coming from $J_N(\{E_a\})$ is the main result of any random matrix approach. It has only been derived from basic symmetry considerations and is related to the geometry of the considered matrix space. Similarly, one can obtain $J_N(\{E_a\})$ in the unitary and symplecticcases. Since the set of auxiliary parameters $\{x_a\}$ is larger with complex numbers or quaternions, one can show that relation (18) remains valid, but with other values for β. One gets $\beta = 1, 2, 4$ for the *orthogonal, unitary* and *symplectic* universality classes, respectively.

1.3 Matrix Ensembles of Maximum Information Entropy

The statistical ensemble will be defined by the probability $P(dH)$ that the matrix associated with our sample belongs to the volume element dH. This requires us to define a matrix density $\rho(H)$ in the matrix space

$$P(dH) = \rho(H)\mu(dH). \tag{20}$$

A general method has been proposed by Balian [12] for assigning a probability distribution to a matrix. Its principle is that the amount of information contained in the probability density should not exceed the minimum amount needed to satisfy the relevant physical properties of the sample. The problem is of course to know on the one hand what the relevant physical properties of the matrix are which have to be introduced via $\rho(H)$, and, on the other hand, to justify the use of a maximum entropy criterion. We shall discuss this point later for the case of the multiplicative transfer matrix. Here we just describe the general method.

As usual, the information entropy $S(\rho(H))$ associated to a density $\rho(H)$ is defined by

$$S(\rho(H)) = -\int \rho(H)\ln \rho(H)\mu(dH). \tag{21}$$

Then, the physical properties are introduced in specifying the possible measured expectation values C_a of a set of quantities $f_a(H)$ depending on the random matrix H,

$$C_a = \int f_a(H)\rho(H)\mu(dH). \tag{22}$$

A set of $p-1$ given expectation values C_a, and a normalization requirement

$$\int \rho(H)\mu(dH) = 1 \tag{23}$$

yield p constraints for $\rho(H)$. There is an infinite number of densities $\rho(H)$ which satisfy these p constraints, but only one which maximizes the entropy S. In order to maximize eq. (21) with the p constraints, one introduces p Lagrange multipliers L_a. One gets the equation

$$\int \delta\rho(H)(\log \rho(H) - \sum_{a=1}^{p} L_a f_a(H))\mu(dH) = 0 \tag{24}$$

for any variation $\delta\rho(H)$. The resulting "most random" density $\rho(H)$ is

$$\rho(H) = \exp\left(\sum_{a=1}^{p} L_a f_a(H)\right). \tag{25}$$

We give now two applications of this method. The first one gives the famous *Gaussian ensembles* [5], that have been extensively studied. It consists of specifying only the expectation value of the trace of $HH^\dagger$, which yields

$$\rho_G(H) = \exp\left(L_1 + L_2 \sum_{c=1}^{N} E_c^2\right). \tag{26}$$

The second example, that will be used later for the transfer matrix, consists in giving as input the level density $\rho(E)$ i.e. the expectation value $\langle trace\ \delta(H - E)\rangle$. The resulting density for H is

$$\rho(H) = \exp\left(\int L(E)\ trace\ (\delta(H - E))dE \right) = \exp(\ trace\ L(H)). \tag{27}$$

Note that the Gaussian ensembles have been defined by introducing a single "physical" constraint, in contrast to the second kind of ensembles where an a priori infinite number of constraints is assigned (the value of $\rho(E)$ at each point of the real axis). However the two definitions can be the same if the input density in the second kind corresponds to the level density which characterizes the first, the so-called Wigner semi-circle law [5].

1.4   Joint Probability Distribution and Coulomb Gas Analogy

The joint probability distribution $P(E_1, ..., E_N)$ of the eigenvalues is obtained after integration of eq. (20) over the eigenvector variables

$$P(E_1, ..., E_N) = \int_A \mu(dA) \left[\prod_{a<b}^{N} |E_a - E_b|^\beta\ \rho(H) \right]. \tag{28}$$

Note that we have just given as input in our two examples some spectral properties. This yields in the two cases for the joint probability distribution of the eigenvalues

$$P(E_1, ..., E_N) = C_{N,\beta} \prod_{a<b}^{N} |E_a - E_b|^\beta \prod_{c=1}^{N} F(E_c), \tag{29}$$

where the function F depends on the chosen constraints. This implies that the differential probabilities of the matrices A, which contain the eigenvectors, are the most random ones i.e. equal (up to a constant) to the invariant measure $\mu(dA)$ on the canonical orthogonal, unitary or symplectic groups respectively

$$P(dA) \propto \mu(dA). \tag{30}$$

The three Gaussian ensembles can also be obtained by requiring the invariance of the ensembles under the canonical transformations in addition to the statistical independence of the matrix elements $H_{ab}; a \leq b$ [5]. One gets for the three Gaussian ensembles

$$F_G(E) = exp - \left(\frac{\beta}{2} E^2 \right) \tag{31}$$

after a proper choice of the origin and the scale of energy. When the input is an arbitrary level density, the function F is given by

$$F(E) = exp(L(E)). \tag{32}$$

We can calculate the Lagrange multipliers $L(E)$ by approximate methods of statistical mechanics in the large-N limit, as explained in Wigner's derivation of the semi-circle-law [13]. We obtain

$$L(E) = - \int_{-\infty}^{\infty} \rho(E') \ln |E' - E|^\beta\ dE'. \tag{33}$$

In order to provide a precise and well-understood language that describes the statistical properties of the eigenvalues, we note that the resulting joint probability distribution can be formally written as the Boltzmann weight of a fictitious Hamiltonian

$$P\left(E_1,\ldots,E_N\right) = \exp{-\beta H\left(E_1,\ldots,E_N\right)} \tag{34}$$

$$H\left(E_1,\ldots,E_N\right) = -\sum_{a<b}^{N} \ln\left|E_a - E_b\right| + \int_{-\infty}^{+\infty} \rho(E')\ln\left|E - E'\right| dE'. \tag{35}$$

There is a mathematical identity between the level distribution of these matrix ensembles and the distribution of the N positions of classical negative point charges free to move on the real axis in the complex plane, in a state of thermodynamical equilibrium at a temperature $kT = \beta^{-1}$. These charges interact with a logarithmic Coulomb repulsion and feel a logarithmic Coulomb attraction from an input positive jellium of density $\rho(E')$. This gives us a very clear and simple "physical" picture of the distributions characterizing these maximum entropy ensembles. We note that most of the results (temperature and logarithmic form of the interactions) have a simple geometrical origin.

Using our physical intuition, we can guess that the local statistical properties, over an energy scale small compared to the scale of variation of $\rho(E)$, are governed by the logarithmic interactions and the value of β characterizing the universality class. Thus, we expect that the *local* statistical properties are *universal*, and we can understand without calculation that the fluctuations are larger in the orthogonal universality class ($kT = 1$) than in the unitary ($kT = 1/2$) or in the symplecticone ($kT = 1/4$).

Since the physical properties appear only in the confining potential yielded by the positive jellium, we see that only the *global* statistical properties, over an energy scale larger than the scale of variation of $\rho(E)$, are *non-universal*.

These predictions have received many numerical confirmations and have been examined more carefully by Dyson [14] within a Brownian motion model.

1.5   Wigner Surmises

The *level repulsion* of the Coulomb gas models yields universal statistics for the spacings s between consecutive levels, if s is measured in terms of the non-universal average spacing. Thus the statistics can be obtained from the Gaussian ensembles. Moreover, a very good approximation for the distribution $p(s)$ is given from a two-level matrix [2] ($N = 2$, Wigner surmises), giving

$$p(s) = \frac{\pi}{2} s^1 \exp{-\left(\frac{\pi}{4} s^2\right)} \tag{36}$$

$$p(s) = \frac{32}{\pi^2} s^2 \exp{-\left(\frac{4}{\pi} s^2\right)} \tag{37}$$

$$p(s) = \frac{2^{18}}{3^6 \pi^3} s^4 \exp{-\left(\frac{64}{9\pi} s^2\right)} \tag{38}$$

for $\beta = 1, 2$, and 4 respectively.

1.6   Two-Level Correlation Function

The spacing distribution mainly depends on the level repulsion. A more important distribution, which shows the *long range rigidity* of the series of eigenvalues, is the two-level correlation function obtained after integration of $P(E_1, ..., E_N)$ over $N-2$ variables. The integration is not trivial, but all the serious difficulties can be overcome by a beautiful method, due to Gaudin and Mehta [2,5]. The method is simpler for $\beta = 2$ where the joint probability distribution can be expressed as a Vandermonde determinant. If one knows the polynomials $P_i(E)$ of degree i which are orthogonal with respect to the weight $F(E)$ that appears in eq. (29) i.e.

$$\int_{-\infty}^{+\infty} P_i(E)P_j(E)F(E)dE = \delta_{ij}. \tag{39}$$

For the Gaussian unitary ensemble, these polynomials are the known Hermite polynomials. The calculation has also been done for non-Gaussian unitary ensembles, characterized by different sets of known orthogonal polynomials (Legendre, Laguerre...). The results give the same two-level function $p(E, E') = p(|E - E'|)$ on a scale where the average spacing is constant and scaled to one, in agreement with the idea that the local statistics is universal. The expression for $p(|E - E'|)$ for the unitary case can be found in [5], in addition to equations and asymptotic expressions for the orthogonal and symplectic Gaussian ensembles.

Integration over $N - 1$ variables gives the level density, a non-universal global function.

1.7   Linear Statistics and Δ_3-Statistics

When we have experimental or numerical series of levels, it is convenient, in order to test the validity of a maximum entropy model to define a "statistics" i.e. a number W that can be computed from the observed sequences of data. Let $\tilde{F}(x)$ a *smooth* function defined on an energy interval $[-E, +E]$. A linear statistic is defined by

$$W = \sum_{a=1}^{n} \tilde{F}(E_a), \tag{40}$$

where the sum is taken over the levels belonging to the interval $[-E, +E]$. Using the 2-point-correlation functions calculated from the Gaussian ensembles, one finds [15] that the variance of W is a pure number, *independent of the number of levels inside the interval*. Note however that this result supposes that W is evaluated on a local scale, if not non-universal corrections to W have to be taken into account, that can be weakly dependent on the global variation of the level density.

In order to measure the long range regularity of the series of levels another statistics is often used [2,15] . This statistic does not use a smooth function $\tilde{F}$ and exhibits a characteristic logarithmic divergence. It consists in the analysis of the deviation of the staircase graph $y = N(E')$, where y is equal to the number of levels in the interval $[-E, E']$, from a suitably chosen straight line $AE + B$. The Δ_3-statistic is the mean-square deviation of the staircase graph from the best linear fit i.e. it measures the deviation of the observed series from the best regularly spaced possible series in an energy interval of width $C = 2E$.

$$\Delta_3(C) = \min_{A,B} \frac{1}{C} \int_{-E}^{+E} (N(E') - AE' - B)^2 \, dE'. \tag{41}$$

From the 2-point correlation functions of the three Gaussian ensembles, one gets the approximate expressions

$$\Delta_3(C) = \frac{1}{1\pi^2} \ln C - 0.00695 \tag{42}$$

$$\Delta_3(C) = \frac{1}{2\pi^2} \ln C - 0.0590 \tag{43}$$

$$\Delta_3(C) = \frac{1}{4\pi^2} \ln(4\pi C) + \gamma + 1 + \frac{\pi^2 - 18}{8} \tag{44}$$

for $\beta = 1, 2$ and 4 respectively, $\gamma = 0.57721$ denotes the Euler constant. Note that these weak logarithmic divergences show that the long range regularity of the series of levels is extremely rigid, corresponding to a "wobbly crystal" rather than a classical plasma. If the levels were located at random in an interval of width C, the Δ_3-statistic of their uncorrelated positions will be linear as a function of C.

1.8   Main Physical Applications

As conclusion of this brief review, we give references where applications of *Random Matrix Theory* to physics can be found. As long ago as the early fifties, Wigner gave the basis of the theory and has proposed it as an appropriate conceptual tool in order to analyze the statistical distribution of the widths and the spacings of nuclear resonance levels. Dyson, Mehta, Gaudin and others developed the statistical theory of nuclear spectra that is now given in classical texts [2, 5] and review papers [6]. Numerous experimental data agree with the Wigner surmises and Δ_3-statistics. A further breakthrough of the theory occurred in the beginning of the sixties in the studies of the spectral fluctuations and thermodynamic properties of small metallic particles [3]. In the seventies, the statistical theory of nuclear reactions [16] was established using the same concepts. More recently, classical and quantum chaos studies [4] have again illustrated how physics can depend only on basic symmetry considerations. Random Matrix Theory has now penetrated mesoscopic physics and quantum transport theory with the works of Altshuler and Shklovskii [17], with Imry [18] and with maximum entropy approaches based on the multiplicative random transfer matrix that we shall develop in the two next sections.

2   Maximum Entropy Ensembles for the Random Transfer Matrix

Our analysis of the conductance fluctuations is based on the *multiplicative* transfer matrix and the two-probe Landauer formula [19]. But first, *let us point out that we are essentially dealing in the following with coherent wave diffusion through a random medium, so that our theory applies more generally to the transmission of waves* [20] *in a coherent multiple scattering regime (acoustic, electromagnetic waves...).* As usual [21], we divide

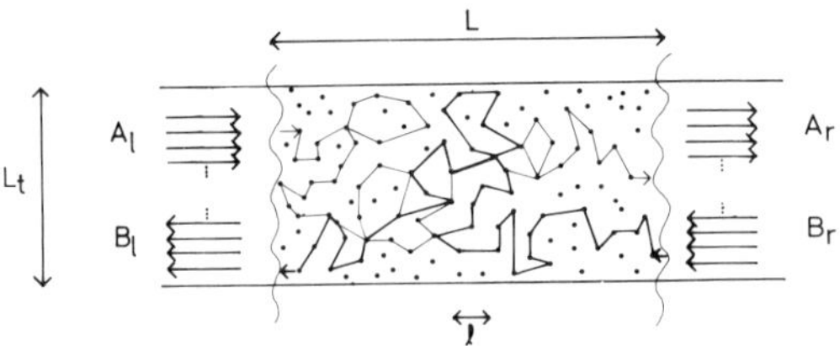

Figure 1. Scheme of a N-channel elastic scatterer with an elastic mean free path l.

a real sample into boxes of size L_ϕ, where L_ϕ is the phase coherence length, and we evaluate the conductance g of a coherent quantum box, assuming negligible thermal smearing of the Fermi surface ($kT < E_c$, E_c denoting the Thouless energy). Thus, g depends on the interference patterns of electronic waves on the Fermi surface. Let us concentrate our attention on a coherent quantum box of length L_ϕ (denote hereafter only by L) and of transverse width L_t. At the Fermi energy, if d denotes the dimensionality, we have typically $N = (k_F L_t)^{d-1}$ quantized channels for injecting the carriers into the box, both from left and right sides, and the system behaves as a N-channel elastic scatterer, as sketched in Fig. 1. We assume for N a non-specific value, which could be large $(5, 10, 10^3, 10^6, ...)$, or possibly infinite. We suppose also $L \gg l$, where l denotes the elastic mean free path. This means that electron transmission through the sample involves quantum interferences between a huge number of classical random walks. All the complex interference processes can be represented by a transfer matrix M, which relates at the Fermi energy the $2N$ quantized fluxes present at the right hand side of the box to those present at the left hand side

$$\begin{pmatrix} A_r \\ B_r \end{pmatrix} = M \begin{pmatrix} A_l \\ B_l \end{pmatrix}. \tag{45}$$

A_l, B_l, A_r, B_r are the complex N-component vectors ($2N$ if spins are taken into account) describing the flux amplitudes on left and right respectively. M has the same dimension as the scattering matrix S, defined by

$$\begin{pmatrix} A_r \\ B_l \end{pmatrix} = S \begin{pmatrix} A_l \\ B_r \end{pmatrix}, \tag{46}$$

where

$$S = \begin{pmatrix} \mathbf{t} & \mathbf{r'} \\ \mathbf{r} & \mathbf{t'} \end{pmatrix}, \tag{47}$$

$\mathbf{t}$ and $\mathbf{r}$ denoting the transmission and reflection matrix respectively.

2.1 Symmetries and Invariant Measures

We first review the symmetries properties of M, in order to define the available matrix space in which a point could correspond to a possible transfer matrix. We recover the Dyson Threefold Way introduced in section 1. Flux conservation implies that M is pseudo-unitary

$$M^\dagger \Sigma_z M = \Sigma_z, \tag{48}$$

where Σ_z is given by

$$\Sigma_z = \begin{pmatrix} 1 & 0 \\ 0 & -1 \end{pmatrix} \tag{49}$$

and $\mathbf{1}$ designating the $N \times N$ unit matrix.

Time reversal symmetry is removed if a magnetic field H is applied, and any matrix satisfying eq. (48) can represent an a priori possible system (unitary ensemble). For $H = 0$, one has to know if the electronic spin is conserved (orthogonal ensemble), or if spin flip can occur during the transfer due to the presence of spin-dependent hopping which however preserves time reversal symmetry (symplecticensemble).

The time reversal symmetry requirement for M is

$$M^* = KMK^T, \tag{50}$$

where

$$K = \begin{pmatrix} 0 & 1 \\ 1 & 0 \end{pmatrix} \tag{51}$$

in the absence of spin-dependent hopping [22], and where

$$K = \begin{pmatrix} 0 & \mathbf{k} \\ \mathbf{k} & 0 \end{pmatrix}. \tag{52}$$

in the presence of spin-dependent hopping [23]. K^T is the transpose of K. Note that M is a quaternion $2N \times 2N$-matrix when spins are included, in contrast to the orthogonal and unitary cases, and $\mathbf{k}$ is the quaternion N-matrix

$$\mathbf{k} = \begin{bmatrix} -i\sigma_y & & 0 \\ & \ddots & \\ 0 & & -i\sigma_y \end{bmatrix}, \tag{53}$$

where $-i\sigma_y = \begin{pmatrix} 0 & -1 \\ 1 & 0 \end{pmatrix}$ is a 2×2 matrix.

It is very convenient to introduce a matrix X for which we shall define a maximum entropy ensemble (global maximum entropy approach),

$$X = \frac{(M^\dagger M) + (M^\dagger M)^{-1} - 2 \cdot 1}{4}. \tag{54}$$

Using relations (45)(46) and (48), we can see that X is simply related to the transmission matrices $\mathbf{t}$ ant $\mathbf{t}'$ by

$$(X + 1)^{-1} = \begin{pmatrix} \mathbf{t}^\dagger \mathbf{t} & 0 \\ 0 & \mathbf{t}'\mathbf{t}'^\dagger \end{pmatrix}. \tag{55}$$

However, it is not necessary to introduce the X-matrix if a polar decomposition [24, 25] of the pseudo-unitary matrix M is used, requiring 4 unitary matrices $u^{(i)}$ and real diagonal matrix Λ, whose diagonal elements define the *radial parameters* $\{\lambda_a\}$ of M.

$$M = \begin{pmatrix} u^{(1)} & 0 \\ 0 & u^{(2)} \end{pmatrix} \begin{pmatrix} \sqrt{1+\Lambda} & \sqrt{\Lambda} \\ \sqrt{\Lambda} & \sqrt{1+\Lambda} \end{pmatrix} \begin{pmatrix} u^{(3)} & 0 \\ 0 & u^{(4)} \end{pmatrix} \equiv U\Gamma V. \tag{56}$$

With time inversion symmetry, we have the additional relations

$$u^{(2)} = \left(u^{(1)}\right)^*, \quad u^{(4)} = \left(u^{(3)}\right)^* \tag{57}$$

for the orthogonal universality class, and

$$u^{(2)} = \mathbf{k}\left(u^{(1)}\right)^* \mathbf{k}^T, \quad u^{(4)} = \mathbf{k}\left(u^{(3)}\right)^* \mathbf{k}^T \tag{58}$$

for the symplecticjuniversality class.

The $u^{(i)}$ characterize how the different incoming and outgoing channels are mixed by the diffusive sample, while the $\{\lambda_a\}$ describe the decay of the transmission coefficients of the "eigen-channels" of the disordered medium. It is useful for instance to realize that such a parametrization introduces only pure phases with the $u^{(i)}$, and an hyperbolic rotation with Γ, in the single channel case ($N = 1$). In a many channel sample, it is this generalized hyperbolic rotation which introduces the N radial parameters $\{\lambda_a\}$. When we diagonalize the Hermitian matrix

$$X = A X_D A^{-1}, \tag{59}$$

it is easy to see that the N *radial parameters of M are the eigenvalues of X*

$$X_D = \begin{pmatrix} \Lambda & 0 \\ 0 & \Lambda \end{pmatrix}. \tag{60}$$

Note that both the diagonalization of X and the polar decomposition of M are not unique in the unitary and symplecticicensembles.

The invariant measures of the volume element of the X-space and M-space have been calculated in [22,23] and in [24] respectively. We refer the reader to these published papers. These calculations are a little more complex than explained for the real-symmetric matrices. The calculation of $\mu(dX)$ has been inspired by Dyson's calculation of $\mu(dS)$, including the proof of the uniqueness and invariance of the chosen definition. The calculation of $\mu(dM)$ is based on an appropriate metric and associated metric tensor.

The invariant measure $\mu(dX)$, once expressed in eigenvalue-eigenvector coordinates, is equal to

$$\mu(dX) = \prod_{a>b}^{N} |\lambda_a - \lambda_b|^\beta \prod_{c=1}^{N} d\lambda_c \mu(dA) \tag{61}$$

and the invariant measure $\mu(dM)$, expressed in terms of the radial parameters and the unitary matrices $u^{(i)}$ of the polar decomposition eq. (56), is given by

$$\mu(dM) = \prod_{a>b}^{N} |\lambda_a - \lambda_b|^\beta \prod_{c=1}^{N} d\lambda_c \prod_i \mu\left(du^{(i)}\right), \tag{62}$$

which introduces the same interaction term between the $\{\lambda_a\}$. Here β depends only on the symmetry of the system and the same values than for the Hamiltonians. Conductors with no applied magnetic field H and no spin-orbit interaction are described by $\beta = 1$ (orthogonal ensemble), with magnetic field by $\beta = 2$ (unitary ensemble), and by $\beta = 4$ (symplecticicensemble) for $H = 0$ and strong spin-orbit interaction. $\mu(dA)$ is the invariant measure for the group of canonical transformations diagonalizing X (within an irrelevant factor due to the degree of freedom of the transformation eq. (59)) and $\mu(du^{(i)})$ is the invariant measure on the unitary group.

2.2 Advantages of a Transfer Matrix Description

2.2.1 The conductance as a linear statistic of the radial parameters $\{\lambda_a\}$ of M

The set of N real positive parameters $\{\lambda_a\}$ is simply related to the conductance g via a standard two-probe Landauer formula. Denoting by $|t_{ab}|^2$ the transmission coefficient between the channels a and b, this formula gives

$$g = 2 \sum_{a,b}^{N} |t_{ab}|^2 \equiv 2T, \tag{63}$$

when g is measured in units of e^2/h and when the voltage probes are separated by a length $L \geq L_\phi$. The factor 2 in eq. (63) is due to spin degeneracy for the orthogonal and unitary classes (in absence of a significant Zeeman splitting) and comes from Kramers degeneracy in presence of time reversal invariant spin-orbit scattering.

Using eq. (55), we obtain the exact relation

$$T = \sum_{a=1}^{N} \frac{1}{1 + \lambda_a}, \tag{64}$$

which implies that we do not need to know the eigenvectors statistics of X or the distribution of the $u^{(i)}$ in order to obtain the statistics of T and g, but only the joint probability distribution $P(\lambda_1, \lambda_2, ..., \lambda_N)$. This is in contrast to standard linear response formulations which express g in terms of the eigenvalues and eigenvectors of the system Hamiltonian, so that conductance fluctuations are yielded both by the fluctuations of the density of levels at the Fermi energy and by the fluctuations of their associated diffusion properties [17].

2.2.2 Transfer matrix and multiplicative combination law

We also point out that M is naturally generated by a multiplicative composition law. This can be seen by arbitrarily cutting the sample in L slices and realizing that the total transfer matrix is the product of the matrices M_k of each of the slices.

$$M = \prod_{k=1}^{L} M_k. \tag{65}$$

In contrast to matrices considered in section 1, this natural composition law generating the matrix ensemble is a major advantage at the heart of the maximum entropy description: any transfer matrix representing a sample is not only a complicated random matrix, but *results from the successive multiplication of many independent random transfer matrices.*

In order to give some intuition of the importance of this property, let us mention a much simpler example where the generating composition law has major implications: let us consider a random scalar generated by adding independent random scalars. *This information is essentially equivalent to specifying the main part of the distribution, if a standard law of large number and associated central limit theorem apply.* A law of large numbers is known for random matrix products: under weak conditions, Oseledec theorem [26] applied to the multiplicative matrix M guarantees that there is a limiting self-averaged matrix O

$$O = \lim_{L \to \infty} \left(M^\dagger M \right)^{\frac{1}{2L}} \tag{66}$$

and the N logarithms of its eigenvalues rigorously define the N inverse decay lengths $\{\alpha_a(\infty)\}$ in a many channel system. If we arrange the $\{\alpha_a(\infty)\}$ in order of increasing magnitude

$$\alpha_1 < \alpha_2 < \ldots < \alpha_N, \tag{67}$$

the localization length ξ is given by

$$\alpha_1(\infty) = \frac{1}{\xi} \tag{68}$$

and corresponds to the longest decay length [18, 27].

We are not aware of very relevant available results in the mathematical literature about the convergence in distribution of random matrix products, but *let us keep in mind that the knowledge of a generating composition law could give some insight into the distribution of M.*

2.2.3   "Logarithmic interaction and temperature"

A third very convenient property of M comes from the use of the radial parameters $\{\lambda_a\}$ for expressing the invariant measures $\mu(dX)$ or $\mu(dM)$, since they yield interaction terms identical to those obtained for the eigenvalues of Hamiltonian matrices. This allows us to apply the classical methods and known results of Random Matrix Theory. We shall have essentially the same Coulomb gas analogies at the same universal values for the temperature, and we can use for the $\{\lambda_a\}$ the same universal local distribution functions which have been calculated for particular Gaussian Models, as the spacing distributions, the two-level correlation functions, the Δ_3-statistics...

However, we have to mention two new features that we have not met in section 1.

First, the eigenvalues of Hamiltonian matrices are real and can go back and forth across the origin. The radial parameters are by definition real positive and cannot cross the origin. For a problem of "interacting particles", this gives a more subtle effect than cutting the range of variation of the $\{\lambda_a\}$ to the positive part of the real axis: this yields a "surface" two-point-correlation function which differs from the bulk one close to the origin [28].

Second, we have introduced the $\{\lambda_a\}$ because our symmetry analysis has shown that they are characterized by the classic Coulomb interaction. But let us define now the variables $\{\nu_a\}$ and $\{\alpha_a\}$ by

$$\lambda_a = \frac{\cosh(\nu_a) - 1}{2} \tag{69}$$

$$\nu_a = 2L \cdot \alpha_a. \tag{70}$$

Oseledec theorem states that the $\{\alpha_a(L)\}$ self-average in the large-L limit to a well defined set of "Lyapunov exponents" $\{\alpha_a(\infty)\}$. A *generic property* of this set [29], also observed in very different contexts [30, 31], is a tendency to have a more or less uniform density. This means that the $\{\lambda_a\}$ (as well as their spacings) diverge exponentially with L. Thus, if we start with a high density $\rho_L(\lambda)$ when L is small (corresponding to a good conductor), we will finish after many matrix multiplications to a very weak density when $L \gg \xi$ (corresponding to an insulator). This generic behavior, related to the multiplicative composition law of M, limits the application of the "universal local random matrix behaviors", as we shall see more precisely in section 3.

2.3 Global Maximum Entropy Model

In this approach [22, 23] developed with K. Muttalib, A. D. Stone and N. Zanon, we have used the X-matrix and applied the general method introduced in 1.3. Our model consists in taking for X the ensemble of maximum information entropy, given an eigenvalue density $\rho(\lambda)$. First note that the actual $\rho(\lambda)$ yields the actual averaged behavior of T in the diffusive and localized regime (and hence of g via Landauer formula)

$$\langle T \rangle = \int_0^\infty \frac{\rho(\lambda)d\lambda}{1+\lambda}. \tag{71}$$

At this stage, one can ask oneself two questions. Are there important features, mostly related to the eigenvectors of X (or to the matrices $u^{(i)}$), which are missing in this statistical ensemble? Or on the contrary, is it necessary to enter in the definition of this ensemble the whole function $\rho(\lambda)$, that is to say an a priori infinite number of parameters? (This point turns out to be the same as asking how many independent parameters enter into the definition of the chosen constraint, a question of importance concerning a possible underlying one-parameter scaling theory.)

Technically, the application of Balian's method is straightforward, giving for the density probability of the X-matrix

$$\rho_X(X) \propto \prod_{c=1}^{N} \exp\left[L(\lambda_c)\right], \tag{72}$$

and for the joint probability distribution, after the standard mean field approximation for calculating the Lagrange multipliers

$$P(\lambda_1, \ldots, \lambda_N) = \exp\left\{-\beta \cdot H(\lambda_1, \ldots, \lambda_N)\right\} \tag{73}$$

$$H(\lambda_1, \ldots, \lambda_N) = -\sum_{a>b}^{N} \ln|\lambda_a - \lambda_b| + \sum_{c=1}^{N} \int_0^\infty \rho(\lambda) \ln|\lambda_c - \lambda| \, d\lambda. \tag{74}$$

2.4 Applications to Conductance Fluctuations

2.4.1 Disordered conductors and universal conductance fluctuations

We assume the conductance and thus T to be large i.e. the input positive jellium to be mostly concentrated close to the origin on the real positive axis. In this regime, we know from microscopic diagrammatic calculations that $var(g)$ is a universal number (up to corrections of order $1/\langle g \rangle$) which weakly depends on the dimensionality d for uncorrelated disorder. We easily understand that the long range rigidity of the $\{\lambda_a\}$, obtained from

$\mu(dX)$, is the basic explanation of these "universal conductance fluctuations" [32, 33, 34]. As noted by Imry [18], g is a linear statistic of random matrix eigenvalues (see 1.7), and one can expect that

$$var(T) = var\left(\sum_{a=1}^{N_{\text{eff}}} \frac{1}{1+\lambda_a}\right) = \frac{constant}{\beta}. \tag{75}$$

The rigorous demonstration [28] of this result is long and technical and we prefer in these notes to skip it and to refer the reader to [18], which essentially gives the right argument and an estimate of the constant appearing in eq. (75). The difficulty in determining its exact value comes from the differences already mentioned in 2.2.3 between the transfer matrix M and the Hamiltonian H (which are not taken entirely into account in [18]). First the multiplicative character of M does not yield a uniform density $\rho(\lambda)$ even at a small "local" scale in the vicinity of the origin. This difficulty can be solved by defining from the $\{\lambda_a\}$ some rescaled variables (essentially the $\{\nu_a\}$) whose average positions are equally spaced, but which approximatively continue to keep locally the logarithmic interaction in order to use the known universal two-level correlation function (see section 3). We note that such a rescaling depends on the input density $\rho(\lambda)$. A second difficulty is the presence of the edge which introduces a surface term in the two-level correlation function $p(\lambda, \mu)$ close to the origin. This edge effect must also be considered in order to calculate $var(T)$. (As a matter of fact, ignoring the surface term in $p(\lambda, \mu)$ would give a $ln(N)$-behavior for $var(T)$.)

Since time inversion symmetry is broken by an applied magnetic field B, it is possible to observe for the variance of $g = 2T$ transitions between different universality classes, combined with the possible removal of a twofold degeneracy of the $\{\lambda_a\}$. A basic difference comes from the presence or the absence of a sufficient spin-orbit coupling i.e. sufficient for flipping the electronic spins during the transmission time through the quantum coherent box. In absence of spin-orbit coupling, time reversal symmetry is broken roughly when a flux quantum Φ_o is applied through a coherent area, increasing the dimensionality of the matrix-space available for M. This transition from the orthogonal to unitary universality class is accompanied by a reduction of $var(g)$ of a universal factor of two, (eq. (75), $\beta = 1 \rightarrow \beta = 2$). This reduction has been observed in Si-doped GaAs wires [35] and $Al_x Ga_{1-x} As/GaAs$ heterojunctions [36]. Increasing the field continues to split the substrate potential seen by the spin up and spin down electrons by an energy $E_Z = g_L \mu_B B$, where g_L denotes the Landé factor and μ_B the Bohr magneton. When this splitting exceeds some correlation energy, the contribution to g of spin up and down electrons are uncorrelated and must be added. The threshold splitting (kT or Thouless energy E_c) depends on the decay of energy correlation functions, which are dimensionality dependent [37]. This additional reduction about a factor of two has been observed in the variance of the conductance fluctuations of a $Al_x Ga_{1-x} As/GaAs$ heterojunction with a backgate [38]. In presence of a strong spin-orbit scattering, spin degeneracy is removed even in absence of an applied field B. However, time reversal symmetry yields a twofold degeneracy (Kramers degeneracy) which disappears when BL_ϕ^2 exceeds Φ_o. For the same cross-over field B_c, the matrix-space in which M is defined changes, and β decays from 4 to 2. Thus, a net factor-of-two reduction [39] accompanies also the breakdown of time inversion symmetry, but this effect cannot be followed by any further reduction, since spin degeneracy is removed even without Zeeman splitting. This effect has been observed in bismuth films where spin-orbit scattering is strong [40].

To our knowledge, the factor-of-four reduction yielded by the introduction of spin-orbit coupling in presence of time reversal symmetry has not been observed (transition from orthogonal to symplectic ensemble).

2.4.2   Anderson insulators

When $g \ll 1$, the input positive jellium is widely spread along the positive axis. The $\{\lambda_a\}$ and their spacings are exponentially large (Oseledec theorem) when we are deep in the localized regime ($L \gg \xi, \lambda_1 \gg 1$).

$$T \sim \lambda_1^{-1} \sim \exp - \left(\frac{2L}{\xi} \right). \tag{76}$$

T depends only on the location of the first λ and is no longer a linear statistic of the $\{\lambda_a\}$. Thus, we have no reason to expect universal conductance fluctuations.

2.5   Local Maximum Entropy Model

In contrast to the global approach, where the statistical ensemble is directly defined for the whole sample, in the approach developed [24] by Mello, Pereyra and Kumar, the sample is cut in a series of building blocks. A single constraint is imposed, in addition of a normalization requirement, for the maximum entropy statistical ensemble of the elementary blocks

$$\left\langle \operatorname{trace} M_{\delta L}^{\dagger} M_{\delta L} - \mathbf{1} \right\rangle \equiv \frac{\delta L}{l}. \tag{77}$$

Thus, the elastic mean free path l is the single physical parameter entering into the theory with L and N. The size δL of the elementary building block is assumed large enough in order to justify a maximum entropy description of its transfer matrix $M_{\delta L}$ (typically a size $\delta L > l$). One can check that the chosen constraint yields classical Ohm's Law for a length δL

$$g(\delta L) = \frac{Nl}{\delta L}. \tag{78}$$

Using eq. (25) and calculating the associated Lagrange multiplier, one gets for the probability density $\rho_{\delta L}(M')$ associated with the building block

$$\rho_{\delta L}(M') \propto \exp - \left[\frac{(N+1)l}{2\delta L} \operatorname{trace} \mathbf{\Lambda}' \right]. \tag{79}$$

The multiplicative combination law of M implies that the probability density associated with a sample of length L put in series with a building block of size δL is given by the convolution

$$\rho_{L+\delta L}(M) = \int \rho_L \left(M M'^{-1} \right) \rho_{\delta L}(M') \mu(dM'), \tag{80}$$

if the building blocks are statistically independent. Using perturbation theory and expanding in powers of δL, taking eventually the limit $\delta L \to 0$, Mello, Pereyra and Kumar have derived a diffusion equation for $P_S(\lambda_1, \lambda_2, ..., \lambda_N)$, which gives the evolution of the $\{\lambda_a\}$ when $S = L/l$ varies

$$\frac{\partial P_S(\{\lambda_a\})}{\partial S} = \frac{2}{\beta N + 2 - \beta} \sum_{a=1}^{N} \frac{\partial}{\partial \lambda_a} \left[\lambda_a(1 + \lambda_a) J_\beta(\{\lambda_a\}) \frac{\partial}{\partial \lambda_a} \frac{P_S(\{\lambda_a\})}{J_\beta(\{\lambda_a\})} \right]. \tag{81}$$

The difficulty of this "Fokker-Planck" equation is mostly due to the coupling factor $J_\beta(\{\lambda_a\})$ which comes from the measure $\mu(dM)$. However, after expanding in $1/\langle g \rangle$ the evolution equations for the expectation value of the pth moment of g, which are directly obtained from eq. (81), Mello [41, 42] has calculated at leading order in $1/\langle g \rangle$ the weak localization corrections

$$\langle g(L) \rangle = \frac{Nl}{L} - \frac{1}{3}, \tag{82}$$

when $\beta = 1$, and

$$\langle g(L) \rangle = \frac{Nl}{L}, \tag{83}$$

when $\beta = 2$; and the variance of the conductance fluctuations

$$var(g) = \frac{8}{15\beta}. \tag{84}$$

These results are very striking. They show that the local approach gives exactly the same values for $\langle g \rangle$ and $var(g)$ as the microscopic diagrammatic calculations [21] for *quasi-one-dimensional (1D) systems*. These results indeed are an indication that one gets the same statistical ensemble for M, either by a maximum entropy distribution for $M_{\delta L}$, or by multiplying the transfer matrices of microscopic slices whose the distribution are yielded by a plausible microscopic Hamiltonian distribution.

2.6   The Joint Probability Distribution of the Global Model as a Solution of the Diffusion Equation of the Local Model

We have defined two maximum entropy models for describing quantum transport in a disordered sample of length $S = \frac{L}{l}$. The *global* model could be summarized by an ansatz for the joint probability distribution of the $\{\lambda_a\}$ which is expressed by eqs. (72) and (73), where the input is the global density $\rho_S(\lambda)$. The *local* maximum entropy model, combined with the multiplicative combination law for M, yields a diffusion equation for $P_S(\lambda_1, \lambda_2, ..., \lambda_N)$, which describes its evolution with S. Due to the coupling term $J_\beta(\{\lambda_a\})$, one could believe this equation difficult to solve. However, we note that the maximum entropy distribution assumed in the local approach for the building block

$$P_{S\sim1}(\lambda_1, \lambda_2, ..., \lambda_N) = \prod_{a<b}^{N} |\lambda_a - \lambda_b|^\beta \prod_{c=1}^{N} F_{S\sim1}(\lambda_c), \tag{85}$$

where

$$F_{S\sim1}(\lambda) \propto \exp\left[-\frac{(N+1) \cdot l}{2 \, \delta L} \lambda \right] \tag{86}$$

is of the type assumed in the global approach i.e. corresponds to a Coulomb gas of N classical negative charges located on the real positive axis with logarithmic interactions at a temperature $kT = \beta^{-1}$. The corresponding input density $\rho_{S\sim 1}(\lambda)$ can be calculated from $F_{S\sim 1}(\lambda)$ and will give a positive jellium confining the negative charges near the origin. If we put in series S building blocks which represent S good conductors, the compatibility between the global and local models requires that this Coulomb gas analogy remains valid, the series of building blocks being just characterized by a positive jellium less confined close to the origin, since the conductance of the series is smaller than the conductance of its elements.

The proof of the compatibility between the two maximum entropy descriptions has been given in [43]. Let us take the global ansatz as a trial solution of the diffusion equation. After integrating over $\lambda_2, ..., \lambda_N$, we see that the input density $\rho_S(\lambda)$ which was not specified in the global approach must satisfy (in the large N-limit) the equation

$$\frac{\partial \rho_S(\lambda)}{\partial S} = 2\frac{\partial}{\partial \lambda}\left[\lambda(1 + \lambda)\rho_S(\lambda)\wp\left\{\int \frac{\rho_S(\lambda')d\lambda'}{\lambda - \lambda'}\right\}\right], \tag{87}$$

in order for the ansatz to be a solution of the diffusion equation. $\wp$ denotes a principal value integral and we underline that β does not appear in eq. (87).

On the other hand, integrating the diffusion eq. (81) over $\lambda_2, ..., \lambda_N$, we obtain the evolution of $\rho_S(\lambda)$ as a function of S, as implied by the local approach

$$\frac{\partial \rho_S(\lambda)}{\partial S} = \frac{2}{\beta N + 2 - \beta}\frac{\partial}{\partial \lambda}\left[\lambda(1 + \lambda)\left(\frac{\partial}{\partial \lambda}\rho_S(\lambda) - \beta(N - 1)\wp\left\{\int \frac{\rho_S(\lambda, \lambda')d\lambda'}{\lambda - \lambda'}\right\}\right)\right] \tag{88}$$

After the same mean-field approximation [13] ($\rho_S(\lambda, \lambda') \sim \rho_S(\lambda)\rho_S(\lambda')$) as the one made for the calculation of the Lagrange multipliers of the global ansatz, we see that the two approaches are identical in the large-N limit.

We underline the importance of this result: in contrast with the Hamiltonian ensembles where an ansatz "a la Balian" has no real justification, except to give both the correct local statistics and of course the right level density, we have shown here that such an ansatz for the $\{\lambda_a\}$ has in addition the very remarkable property that it remains valid when we consider a series of samples instead of a single one. *This implies that the validity of the two approaches cannot be restricted to good conductors, where $L \ll \xi$, but still holds after increasing the sample length, possibly crossing the localization length $(L \gg \xi)$ and then reaching an Anderson insulator.*

We note also that we have established the equivalence between the two models after having restricted the input density of the first to satisfy a given integro-differential equation implied by the second. This restricts our theory to the quasi-1 D systems and leaves open an important issue about the ability for the global ansatz to describe quantum transport in two- and three dimensions (2 D and 3 D). An element of an answer is contained in our numerical calculations which involve mostly strictly 2 D systems, indicating the relevance of the global approach outside quasi-1 D systems.

A complete description of quasi-1 D systems will be achieved with the solution of the integro-differential eq. (87). We have not been able to solve it, except in the limit $L \gg l$. As detailed in [43], it is convenient to use the set of variables $\{\alpha_a\}$ related to the $\{\lambda_a\}$ by the relations eqs. (68), (69), and one easily gets for the $\{\alpha_a\}$ a uniform density $\sigma(\alpha)$ in the large-N limit

$$\sigma(\alpha) = \begin{cases} Nl & , \quad 0 \leq \alpha < l^{-1} \\ 0 & , \quad \alpha > l^{-1} \end{cases}. \tag{89}$$

This means that the Lyapunov exponents of M are equally spaced in the local approach. We point out that a tendency for the Lyapunov exponents to be uniformly distributed is not specific to quantum transport and its related transfer matrix, but is characteristic of random matrix products. This is a generic property that is also observed in different contexts, e.g. dynamical systems [30, 31].

3 Comparison with Microscopic Models and Universal Relations for Anderson Insulators

The use of maximum entropy ensembles is quite recent in quantum transport theory. Microscopic tight binding Hamiltonian models are more familiar to most of us. First introduced by Anderson at the end of the fifties in his famous paper [1] "on the absence of quantum diffusion in certain random lattices", they have been widely studied later in numerical works, e.g. for numerical investigations of quasi-1 D localization, with extrapolation to higher dimensions via finite size scaling methods [27, 44, 45]. We have studied [22, 23, 46] the statistics of $\{\lambda_a\}$ generated from these microscopic models, first in order to check that their local statistical properties agree with our general predictions based on maximum entropy models, and second in order to determine their global densities. From this knowledge, we shall extend further the analysis of the global maximum entropy ensemble, assuming approximate input densities which essentially agree with these numerical results. This leads us to propose universal relations which characterize Anderson insulators.

3.1 Numerical Studies of Disordered Metallic Squares

The usual Anderson tight binding model [27] with constant nearest neighbor hopping and random site disorder has been considered for the orthogonal universality class. For

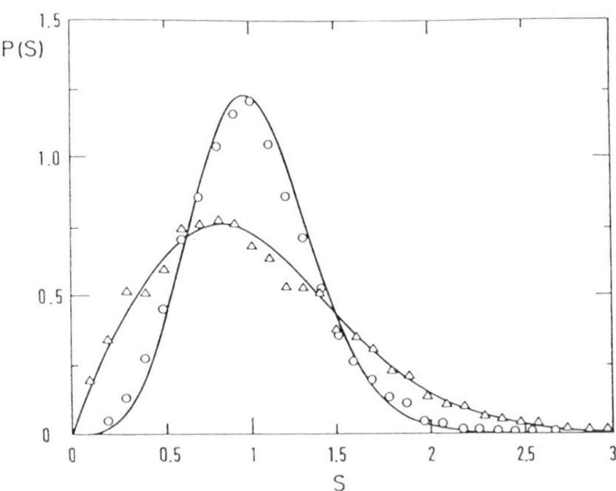

Figure 2. Spacing distributions between successive λ, calculated with a statistic of Anderson models (disordered metallic squares). The triangles correspond to $\beta = 1$ and the circles to $\beta = 4$. The continuous curves are the Wigner-surmises (eqs. (36) and (38)).

the unitary ensemble, we have studied a Hofstadter model [47] with site disorder. For the symplectic ensemble, the spin dependent nearest neighbor hopping term is random [23] and characterized by a disorder parameter W_{so}. In the three cases, our numerical studies have been done with a square lattice, having one orbital per site and where W is the width of the rectangular distribution of the uncorrelated site potential. The energy E is taken equal to zero (band center). Since we know that the quasi-1 D systems are correctly described by the local maximum entropy approach, it is very important to note that our systems are 2 D, as confirmed by the calculated value of $var(T)$ which accurately agrees [46] with the 2 D perturbative result in the metallic regime. Thus, we are checking the ability of the global approach to describe dimension two. From the microscopic Hamiltonian, the transfer matrix of the different columns of the studied squares are calculated and multiplied. The corresponding X-matrix is diagonalized, giving the set of N real positive parameters $\{\lambda_a\}$ associated with the disordered square. For low enough values of W and W_{so}, the simple matrix multiplication is numerically reliable in order to obtain X for $L_t = L \sim 100$-squares and $T \gg 1$ (metallic regime).

The sets of $\{\lambda_a\}$ obtained from typically 50×50-squares exhibit a very strong variation of their averaged spacings. In the Coulomb gas analogy, this means that N is not large enough in order to have a large number of charges interacting on a scale where the variation of the confining potential can be neglected. The classical method for getting rid of this non-universal variation is to unfold the obtained sets i.e. to rescale

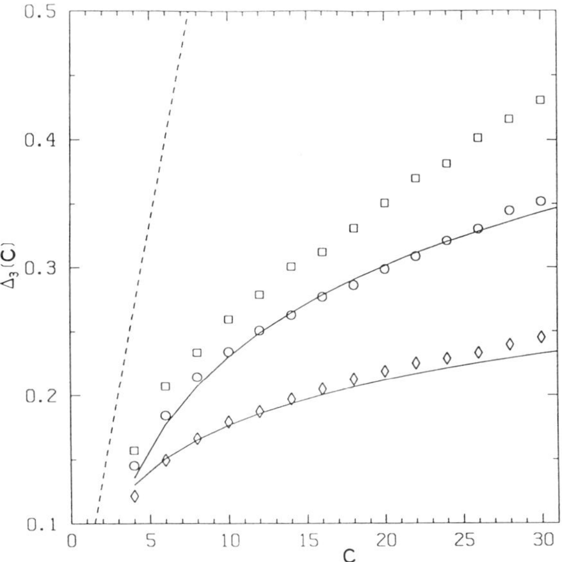

Figure 3. Δ_3-statistics as a function of the average number C of λ-parameters in the unfolded sequence calculated for metallic squares ($L < \xi$). The circles and the diamonds are calculated from Anderson models for $\beta = 1$ and 2, respectively. The continuous curves are the analytical predictions obtained from Random Matrix Theory (eqs. (42) and (43). The dashed line corresponds to uncorrelated sequences, and the squares to a size $L \approx \xi$.

numerically the sequences of $\{\lambda_a\}$ to sequences with constant average spacings. We do not need this unfolding in order to show that the spacing distributions are very well described by the Wigner surmises (Fig. 2), but it is only the unfolded sequences which present the expected universal 2-point correlation functions and Δ_3-statistics (Fig. 3). Level repulsion and long range rigidity of the microscopic models are clearly in agreement with the macroscopic approach of maximum information entropy for disordered metallic squares. When $g \sim 1$, the Δ_3-statistics are not at all in agreement with formula (41), as clear in Fig. 3. Does it mean that our maximum entropy description is only valid in the metallic regime? The answer is NO.

But, let us first show how the original sequence of $\{\lambda_a\}$ looks. Figure 4 is a good summary of the behavior actually observed on a given disordered metallic square compared to the ensemble averaged behavior. The variables plotted in Figure 4 are the set of $\{\alpha_a\}$ deduced from the $\{\lambda_a\}$ by relations (68) and (69). We note that their spacing are essentially uniform. To unfold the $\{\lambda_a\}$ is thus essentially equivalent to consider the set of variables $\{\alpha_a\}$ or $\{\nu_a = 2L\alpha_a\}$. The small fluctuations of these latter variables around their ensemble average positions is due to the interactions which have been obtained from symmetry considerations. The quasi-1 D limit of the $\{\alpha_a\}$, corresponding to the Lyapunov exponents of M, are also plotted. (This set cannot be reliably obtained by simple matrix multiplications, but by a different numerical algorithm [27]). We can see that $\lim_{L\to\infty} \alpha_a(L, L_t)$, which characterizes a very long quasi-1 D insulator, does not significantly differ from the ensemble average $\alpha_a(L = L_t)$ which characterizes a metallic square. We refer the reader to [46] for a detailed analysis of this small difference, *which is governed by finite size scaling laws.*

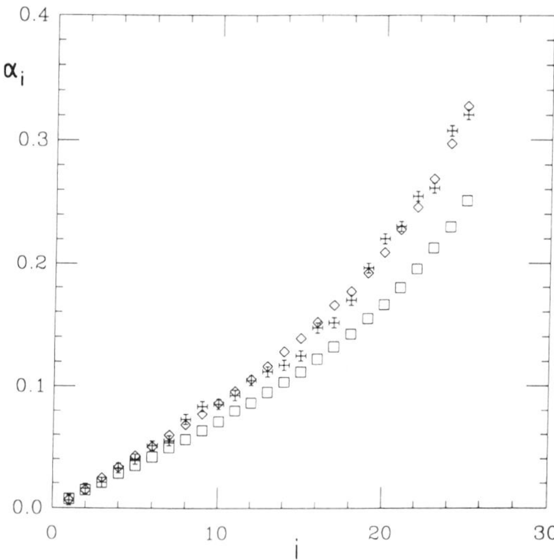

Figure 4. $\alpha_i(L = L_t)$ as a function of the index i for 25×25 metallic squares ($W = 2, \beta = 1$). The crosses correspond to a given disordered square, while the diamonds are the ensemble-averaged values. The squares are the inverse localization lengths $\alpha_i(L \to \infty, L_t)$.

3.2 Uniform Continuous Jellium Approximation for the ν-variables

These numerical studies of microscopically disordered metallic squares, former studies of Lyapunov spectrum of quasi-1 D strips and bars [44], with extrapolation to 2 D and 3 D systems in the large L_t-limit [29], lead us to approximate the input density $\sigma(\nu)$ of the global approach by a uniform density. We see in Fig. 4 that this is more correct for the best transmitting channels and partly wrong for the ν_a with $a \sim N$. We consider for simplicity the actual output density implied by the local approach in the large-N limit (eq. (89)) where the set of $\{\nu_a\}$ is uniformly spreaded along the real axis between 0 and $2L/Nl$. From (73), we get for the joint probability distribution for the $\{\nu_a\}$

$$\hat{P}(\{\nu_a\}) = \exp[-\beta \tilde{H}(\{\nu_a\})], \tag{90}$$

where the Hamiltonian $\tilde{H}$ is composed of three terms

$$\tilde{H}(\{\nu_a\}) = -\sum_{a>b}^{N} \ln|\cosh \nu_a - \cosh \nu_b| - \frac{1}{\beta}\sum_{c=1}^{N} \ln|\sinh \nu_c| + \sum_{c=1}^{N} V(\nu_c). \tag{91}$$

If one assumes eq. (89), the confining potential is given by

$$V(\nu) = \int_0^\infty \sigma(\nu') \ln|\cosh \nu - \cosh \nu'| \, d\nu' \sim \left(\frac{g_0}{4}\nu^2\right), \tag{92}$$

where $g_0 = Nl/L$ is the classical Ohmic conductance. Since the density (89) corresponds to quasi-1 D systems, we note in order to be more general that we just need to require a uniform density $\sigma(\nu)$ at the origin for getting $V(\nu) \sim A\nu^2$ for small values of ν, without the need of $A = g_0/4$.

The second term in eq. (91), coming from the change of variables $(\lambda \to \nu)$, is β-dependent.

As pointed out by Imry [18], $g \approx N_{\text{eff}}$ in the metallic regime, where N_{eff} is the number of open channels for which $(\cosh \nu + 1)^{-1} \approx 1$ i.e. $\nu \le 1$. Let us review the evolution of N_{eff} with the sample size L for different dimensionalities. We have $N \sim (k_F L_t)^{d-1}$ and $N_{\text{eff}} \sim Nl/L$. In the quasi-1 D limit, L_t fixed and $L \to \infty$, we spread the positive jellium out along the real positive axis, and N_{eff} goes to zero as a function of L. When $N_{\text{eff}} < 1$ (localized regime), $g \ne N_{\text{eff}}$, but depends only on the transmission property of the best transmitting closed channel (see 2.4.2). In 2 D, $L = L_t$, the uniform density and N_{eff} take a non-zero value $2/g_0 = 2L/Nl$ which does not depend on L. This agrees with the idea that $d = 2$ is the marginal dimensionality. In 3 D, the increase of the number of charges $N \propto L^2$ is stronger than their linear spreading when $L \to \infty$, yielding a linear increase with L of N_{eff}. Thus, our approximation guarantees a classical ohmic behavior for g in the thermodynamic limit for $d = 3$.

3.2.1 Conductors and standard random matrix behaviors

In the metallic regime, the uniform spacing $2/g_0 \ll 1$, and we have $N_{\text{eff}} \gg 1$ ν-levels in an interval of width 1. Expanding $\cosh \nu_a - \cosh \nu_b$ around a real positive value ν_0, we recover the Coulomb logarithmic interaction around ν_0 with a quadratic confining potential. This means that the variables $\{\nu_a\}$ have the same distributions as those calculated from the Gaussian ensembles on a scale of order 1, involving N_{eff} levels.

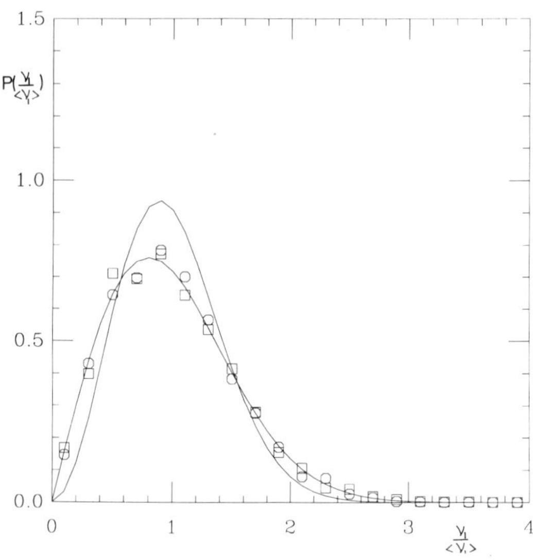

Figure 5. Distribution of the first level $\nu_1/\langle\nu_1\rangle$ calculated for metallic squares for $B = 0$ (squares) and $B \neq 0$ (circles). The continuous curves are the Wigner surmises (eqs. (36) and (37)).

Thus, we justify the starting point assumed by Imry in order to calculate $var(g)$ for a conductor, if one excepts the perturbation due to the rigid edge at the origin. One can see that the edge effect is equivalent to introduce symmetric charges located at $-\nu_a$ for $\nu_a \ll 1$ after expanding $\cosh \nu$ around 0

$$\ln|\cosh\nu_a - \cosh\nu_b| \approx \ln|\nu_a - \nu_b| + \ln|\nu_a - (-\nu_b)|. \tag{93}$$

The interaction of symmetrically located charges comes from the β-dependent second term of the R.H.S of eq. (93):

$$\frac{1}{\beta}ln|2\sinh\nu_a| \approx \frac{1}{\beta}\ln|\nu_a - (-\nu_a)|. \tag{94}$$

When $\beta = 1$, the symmetry around the origin is exact and we can guess that the spacing between ν_1 and $-\nu_1$ must have a distribution close to the one from the Wigner surmise. Figure 5 confirms this conjecture. When $\beta \neq 1$, the charge located at ν_1 is less repelled from the origin, since its symmetry has its charge divided by $1/\beta$. Then, we do not expect the Wigner surmise to describe the distribution of ν_1 outside the orthogonal universality class. This agrees with data calculated from microscopic models (Fig. 5).

3.2.2   Insulators and ν-lattice with quasi-independent Gaussian fluctuations

In the localized regime, as implied by the multiplicative composition law of M, the $\{\nu_a\}$ and their spacings diverge linearly with L. For a strongly insulating sample, we have typically

$$1 \ll \nu_1 \ll \nu_2 \ll \ldots \ll \nu_N \tag{95}$$

and the interaction term in (91) reduces to

$$\ln|\cosh\nu_a - \cosh\nu_b| \approx \nu_a \tag{96}$$

for any $\nu_b \ll \nu_a$. Then, the potential U felt by level ν_a reduces to

$$U(\nu_a) \approx -\left(a - 1 + \frac{1}{\beta}\right)\nu_a + \frac{g_0}{4}\nu_a^2. \tag{97}$$

This means that the $\{\nu_a\}$ have equilibrium lattice positions ν_a^0 given by

$$\nu_a^0 = \left(a - 1 + \frac{1}{\beta}\right)\frac{2}{g_0} \tag{98}$$

and quasi-independent Gaussian fluctuations around ν_a^0 with a variance

$$\langle \delta^2(\nu_a)\rangle = \frac{2}{\beta g_0} = \frac{2L}{\beta Nl}. \tag{99}$$

3.3 Universal Relations for Anderson Insulators

Let us concentrate our attention on the first level ν_1, since

$$\ln(g) \approx \ln(T) \approx -\nu_1, \tag{100}$$

if we neglect a $\ln 2$ term which comes from a possible twofold degeneracy. Deep in the localized regime, we have

$$\langle \nu_1\rangle = \nu_1^0 = \frac{2L}{\xi}. \tag{101}$$

Then, setting $a = 1$ in (98), we obtain a very simple relation for the localization length:

$$\xi(\beta) = \beta \cdot \xi(\beta = 1) \tag{102}$$

i.e. the breakdown of time inversion symmetry is accompanied by universal multiplication factors of the localization domains. The specific input density that we have assumed eq. (89), yields in addition

$$\xi(\beta = 1) = Nl. \tag{103}$$

This dependence can be obtained by noting that $g \approx 1$ when $L \approx \xi$ if one estimates g by a simple Ohm's Law ignoring any quantum corrections. This is only a rough estimate characteristic of quasi-1 D systems. These relations are in agreement with a previous work on a model of weakly coupled 1 D chains by Dorokhov [48], but we believe that the relation eq. (102) is more general and still applies for higher dimensionalities. Let us recall the necessary assumptions in order to obtain (102). First, we need a uniform density $\sigma(\nu)$ close to the origin in order to have a quadratic confining potential for the first ν-levels. Second, if we consider an experiment where the value of β will be changed with an applied magnetic field B, we neglect a possible modification of this density induced by B. This assumption is of course incorrect for a large applied field B, which could open gaps in $\sigma(\nu)$ if the disorder is too weak (quantized Hall regime), but seems correct if the cyclotron length $L_H \gg l$. The potential $U(\nu_1)$ felt by the first level will be

$$U(\nu_1) = -\frac{1}{\beta}\nu_1 + A\nu_1^2, \tag{104}$$

where A does not depend on B. Then, the application of a magnetic field will be accompanied by a universal effect only related to geometrical changes of the matrix-space of definition of the transfer matrix M. (Note that the corresponding effect in the metallic regime consists in the weakening of the symmetric charge repulsion exhibited in Fig. 5). The relation (102) could be very easily checked in the Mott variable range hopping regime [49] where the temperature dependence of the conductivity $\sigma_e(T)$ is given by

$$\sigma_e(T) \propto \exp\left[-C(\frac{T_o}{T})^{1/(d+1)}\right] \tag{105}$$

and where T_0 is related to ξ by the relation

$$T_o \sim \frac{1}{n(0)\xi^d}, \tag{106}$$

$n(0)$ being the density of states at the Fermi level.

When a time reversal symmetry breaking magnetic field B is applied, we predict that the localization length ξ is multiplied by a universal factor two in absence of spin-orbit coupling, or is divided by a universal factor two in presence of a strong spin-orbit coupling. Very recent measurements [50] confirm both predictions. The first experiment has been done with a Si-doped GaAs insulating sample (no spin-orbit) between 20 mK up to $1K$ and it has been found that $\sigma_e(T)$ behaves according to eq. (105). T_o *decreases as a function of the applied magnetic field and saturates* to a value which exactly agrees with the expected value for the localization length ($\beta = 1 \rightarrow \beta = 2$ and $\xi(B) = 2\xi(0)$ for $B > B^* \approx$ 2000 gauss). Here also, we neglect a B-dependence of $n(0)$ in relation (106) for these weak applied fields. The second experiment has been performed on a Yttrium-Silicon amorphous alloy which is known to have a strong spin-orbit scattering. *Here, T_o increases as a function of B and saturates* to a value which agrees as expected with $\xi(B) \approx \xi(0)/2$ for $B > 4\,\mathrm{T}$. Since the second sample is a better insulator than the first, we are not surprised to be forced to apply a larger magnetic field in order to remove time inversion symmetry. More complete measurements illustrating these universal effects will be published soon. We are also planning to try to detect these universal effects in measuring the complex wavevector-and-frequency-dependent dielectric function function which depends [47] on ξ as

$$\lim_{q\to 0}\lim_{\omega\to 0} \epsilon(q,\omega) \sim \xi^2. \tag{107}$$

In addition to the β-dependence of ξ, we get *a simple relation between the mean and the variance of* $\ln T$ *which does not depend of* β. Using eqs. (99) and (100), we obtain for $\ln(T)$ (and hence for $\ln(g)$) a Gaussian distribution with

$$\langle \delta^2(\ln T)\rangle \sim -\langle \ln T\rangle. \tag{108}$$

For 10×10 strongly disordered squares ($L \approx 6\xi$), our numerical data agree with a one-parameter Gaussian distribution satisfying eq. (108), as given in Fig. 6. In the same figure, we can see that 10×250 quasi-1 D strips have a different behavior: the calculated data for ν_1 have a distribution which can be very well fitted by a Gaussian satisfying

$$\langle \delta^2(\ln T)\rangle = -2\langle \ln(T)\rangle. \tag{109}$$

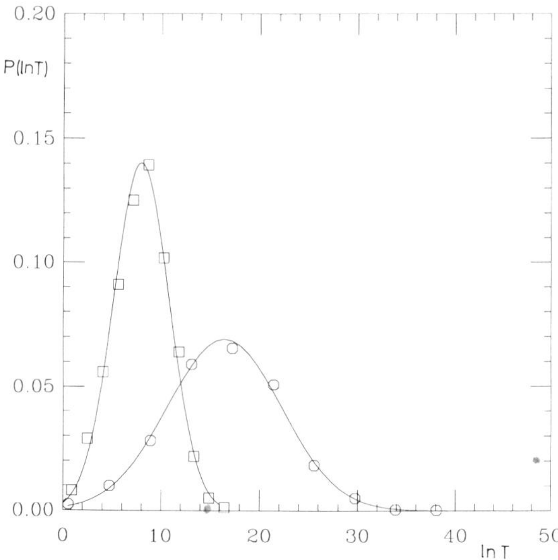

Figure 6. Distribution of $\ln(T)$ calculated for strongly disordered insulating squares ($L = L_t = 10, W = 12$, squares) and for quasi-1 D weakly disordered insulating strips ($L_t = 10, L = 250, W = 3$, circles). The continuous curves are Gaussian distributions with $var(\ln(T)) = -\langle \ln(T) \rangle$ for the square and $var(\ln(T)) = -2\langle \ln(T) \rangle$ for the strips.

Note that this factor of two, which is not present in the square geometry, has been obtained for purely 1 D chains [51] in the weak disorder limit. We suspect a coarse-grained approximation to be the origin of the difference: we have assumed as an input a continuous uniform $\sigma(\nu)$, and we have obtained as an output of our theory a density consisting in a sum of Gaussian peaks arranged on a lattice. This effect deserves further study but it appears that such a coarse-graining works for 2 D squares but not for quasi-1 D systems.

3.4 Quasi-One-Dimensional Anderson Insulators. Theory and Numerical Studies

Because of the presence of this factor 2 in relation (109), let us consider directly the diffusion eq. (81) in order to avoid approximations for $\sigma(\nu)$. We assume again that we are deep in the localized regime and taking eq. (95) into account, the diffusion eq. (81) of the local approach reduces to

$$\frac{\partial}{\partial S}\tilde{P}(\{\nu_a\}) = \frac{2}{\beta N + 2 - \beta}\sum_{a=1}^{N}\left[\frac{\partial}{\partial \nu_a}\tilde{P}(\{\nu_a\}) - (1 + \beta(a-1))\tilde{P}(\{\nu_a\})\right] \qquad (110)$$

for the ν-variables, whose the solution is

$$\tilde{P}(\{\nu_a\}) = \prod_{a=1}^{N}\left[\frac{1}{\sqrt{2\pi\sigma}}\right]\exp-\left[\frac{(\nu_a - \nu_a^0)^2}{2\sigma^2}\right]. \qquad (111)$$

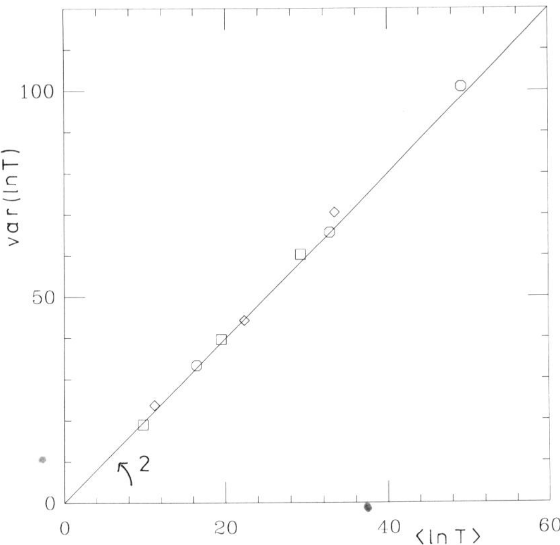

Figure 7. Variance of $\ln(T)$ as a function of $\langle \ln(T) \rangle$ for quasi -1 D strips of length $L = 250, 500, 750$. The applied magnetic field B in units of flux quanta per lattice cell is equal to 0 (circle), 0.03 (square) and 0.0003 (diamond).

The lattice position ν_a^0 are given by

$$\nu_a^0 = 2S \cdot \frac{1 + \beta(a-1)}{\beta N + 2 - \beta} \tag{112}$$

which agrees in the large N-limit with eq. (98) and the variance of the quasi-independent Gaussian fluctuations is given by

$$\sigma^2 = 2S \cdot \left(\frac{2}{\beta N + 2 - \beta} \right), \tag{113}$$

which differs in the large N-limit from eq. (99) by precisely a factor two! Setting $a = 1$ in the two previous relations, we get

$$-\langle \ln(T) \rangle = \nu_1^0 = \frac{2L}{(\beta N + 2 - \beta)l} \tag{114}$$

and

$$\langle \delta^2 \nu_1 \rangle = \langle \delta^2(\ln(T)) \rangle = \sigma^2 = 2\nu_1^0 = -2\langle \ln(T) \rangle. \tag{115}$$

We have checked the relation (115) in quasi-1 D microscopic disordered models. The results are shown in Fig. 7, confirming the presence of a factor two for various lengths and values of the applied magnetic field in the *universal β-independent relation relating the mean and the variance of* $\ln(g)$. We underline the interest of this relation: as in the metallic regime where g has an essentially one-parameter distribution (Gaussian with a universal variance), we obtain also for $\ln(g)$ in the localized regime a one-parameter distribution for an N-channel scatterer. We have not investigated for the moment if this

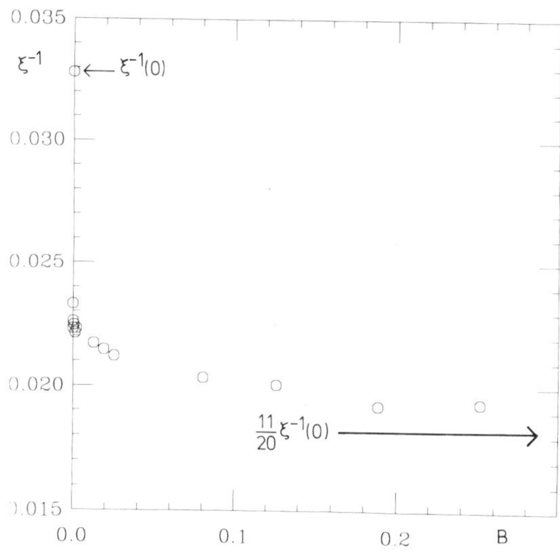

Figure 8. Inverse localization length ξ^{-1} of a strip as a function of the applied magnetic field B. $W = 3, L_t = 10$. B is measured in units of $1/(2\pi)$ flux quanta per lattice cell.

relation disappears for larger disorder as in purely 1 D-chains [51]. From eq. (111), we see that the localization length of quasi-1 D systems is given by

$$\xi = (\beta N + 2 - \beta)l, \tag{116}$$

which agrees with relation (102) in the large N-limit. We report in Fig. 8 numerical calculations of the localization length of a site-disordered Hofstadter model in a quasi-1 D geometry. We have taken periodic transverse boundary conditions in order to get rid of edge states. We just present calculations done for large enough disorder ($W = 3$) in order to approximately erase the complex underlying magnetic structure. For $N = 10$, we predict $\xi(H) = (20/11)\xi(H = 0)$ in approximate agreement with the calculated lengths. In addition to the mentioned experiments, this is an other indication of these simple and universal relations that we believe to be generic for the localization lengths. Nevertheless, we note in addition that Hofstadter's model is known to have a very rich and complex band structure, where the commensurability between the flux quantum ϕ_0 and the flux passing through a lattice cell Ba^2 plays a role [47]. Then, a strong magnetic field yields $\beta = 1$ or $4 \to \beta = 2$, but may also vary the number N of propagation channels without disorder via gap opening effects which are not considered in our theory. We have observed that these additional effects are washed out by a sufficiently large disorder, and the expected simple behavior of ξ to exactly occur for larger values of N, at particularly well-chosen values of B (commensurate situations). However, we believe that these additional difficulties are irrelevant for understanding the magnetoconductance behavior outside a quantized Hall effect context (as shown by the measurements of the activated conductivity in 3 D samples).

4 Summary

We have presented a macroscopic theory for quantum transport based on maximum entropy ensembles for the multiplicative transfer matrix. Symmetry considerations play a central role in these models where the physical properties are only introduced via global constraints, as the density $\rho(\lambda)$ in the global approach, or the elastic mean free path l in the local approach.

We have compared the statistics yielded by the global approach for the radial parameters of M with the ones calculated from detailed microscopic models. We have not found any disagreements (for the level spacing distributions, the Dyson-Mehta Δ_3-statistics or the correlation functions) between the macroscopic model and microscopically disordered metallic squares i.e. 2 D-systems. Though numerical studies remain to be done in 3 D, we do not think that any surprising results will restrict the validity of standard Random Matrix Theory for the $\{\lambda_a\}$, which describe quantum transport in disordered *conductors*. For a disordered *insulator*, we have seen that a generic spreading of their density along the positive real axis destroys the classical Wigner-Dyson statistics, nevertheless the global ansatz eq. (72) remains valid. Since our approach is non-perturbative, we have given novel and universal predictions for the localized regime: a simple β-dependence of the localization lengths which is confirmed by experiments performed on Si-doped GaAs and amorphous Yttrium-Silicon insulators, and a universal β-independent relation between the mean and the variance of the Gaussian fluctuations of $\ln(g)$, which remains to be observed.

What are the reasons of the success of these maximum entropy models? We have to some extent furnished elements of the answer by stressing the importance to have a natural composition law generating our ensembles. The question is to know why microscopic models, giving a huge amount of information via the distribution of a huge Hamiltonian, and maximum entropy models characterized only by a very few global constraints give basically the same distribution for the relatively small matrix M (compared to the Hamiltonian) describing the transfer at the Fermi surface. This point has been discussed in the context of the local approach [52], but we point out that the proof that the global ansatz eq. (72) remains valid when the composition law generating the ensemble is iterated furnish an important step towards the understanding of those issues. A better insight into the central limit theorems on matrix products will be extremely helpful. It will make precise how many of the details of the original microscopic distribution get lost after many multiplications and under what conditions one ends up with "simple and universal" maximum entropy models.

5 Acknowledgements

I gratefully acknowledge the collaboration of Y. Imry, K. Muttalib, A. D. Stone and N. Zanon with whom the "global approach" has been developed. I am also indebted to P. A. Mello and K. Slevin for a more recent collaboration including developments in the insulating regime, which have also being possible thanks to P. Le Doussal. I thank my colleagues at Saclay, M. Sanquer, P. Debray and D. Mailly, for a fruitful collaboration which provides me the opportunity to have contact with the real world.

References

[1] P.W. Anderson, Phys. Rev. **109**, 1492 (1958)

[2] Statistical Theories of Spectra: Fluctuations, C.P. Porter, ed, Academic Press, New York (1965)

[3] J.A.A.J. Perenboom, P. Wyder, and F. Meier, Phys. Rep. **78**, 175 (1981), and references cited therein

[4] Quantum Chaos and Statistical Nuclear Physics, T.H. Seligman and H. Nishioka, eds, Springer, Berlin, Heidelberg, New York (1986), and references cited therein

[5] M.L. Mehta M.L., Random Matrices, Academic Press, New York (1967)

[6] T.A. Brody et al, Rev. Mod. Phys. **53**, 385 (1981)

[7] F.J. Dyson, J. of Math. Phys. **3**, 140 (1962)

[8] F.J. Dyson, J. of Math. Phys. **3**, 1199 (1962)

[9] G. Bergmann G., Phys. Rep. **107**, 1 (1984)

[10] F. Wegner, Nucl. Phys. **B 316**, 663 (1989)

[11] L.K. Hua, Harmonic Analysis of Functions of Several Complex Variables in the Classical domains, Amer. Math. Soc., Providence, R.I. (1963)

[12] R. Balian, Nuevo Cimento **57**, 183 (1968)

[13] E.P. Wigner, Can. Math. Congr. Proc, p. 174, Univ. of Toronto Press, Toronto, Canada (1957)

[14] F.J. Dyson, J. of Math. Phys. **13**, 90 (1972)

[15] F.J. Dyson and M.L. Mehta, J. of Math. Phys. **4**, 701 (1963)

[16] P.A. Mello and T.H. Seligman, Nucl. Phys. **A 344**, 489 (1980), and references cited therein

[17] B.L. Altshuler and B.I. Shklovskii, Sov. Phys. JETP **64**, 127 (1986)

[18] Y. Imry, Europhys. Lett. **1**, 249 (1986)

[19] R. Landauer, I.B.M. J. Res. Dev. **1**, 223 (1957)

[20] E. Akkermans, P.E. Wolf, R. Maynard, and G. Maret, J. Phys. France **49**, 77 (1988)

[21] P.A. Lee, A.D. Stone, and H. Fukuyama, Phys. Rev. **B 35**, 1039 (1987)

[22] K. Muttalib, J.-L. Pichard, and A.D. Stone, Phys. Rev. Lett. **59**, 2475 (1987)

[23] N. Zanon and J.-L. Pichard, J. Phys. France **49**, 907 (1988)

[24] P.A. Mello, P. Pereyra, and N. Kumar N., Ann. Phys. **181**, 290 (1988)

[25] P.A. Mello and L.-L. Pichard, submitted to J. Phys. France

[26] V.I. Oseledec, Trans. Moscow Math. Soc. **19**, 197 (1968)

[27] J.-L. Pichard and G. Sarma G., J. Phys. **C 14**, L127 (1981)

[28] P.A. Mello, J.-L. Pichard, and K. Slevin, unpublished

[29] J.-L. Pichard and G. André, Europhys. Lett. **6**, 477 (1986)

[30] J.P. Eckmann and C.E. Wayne C.E., J. of Stat. Phys. **50**, 853 (1988)

[31] G. Paladin and A. Vulpiani A., J. Phys. **A 19**,1881 (1986)

[32] B.L. Altshuler, Sov. Phys. JETP Lett. **41**, 648 (1985)

[33] P.A. Lee and A.D. Stone A. D., Phys. Rev. Lett **54**, 1622 (1985)

[34] S. Washburn and R.A. Webb, Adv. Phys. **35**, 375 (1986)

[35] D. Mailly and M. Sanquer, these volume

[36] D. Mailly, M. Sanquer, J.-L. Pichard, and P. Pari P., Europhys. Lett. **8**, 471 (1989)

[37] A.D. Stone, Phys. Rev. **B 39**, 10736 (1989)

[38] P. Debray, J.-L. Pichard, J. Vicente, and P.N. Tung P.N., Phys. Rev. Lett **63**, 2264 (1989)

[39] S. Feng, Phys. Rev. **B 39**, 8722 (1989)

[40] N.O. Birge, B. Golding, and W.H. Hammerle W.H., Phys. Rev. Lett. **62**, 195 (1989)

[41] P.A. Mello, Phys. Rev. Lett. **60**, 1089 (1988)

[42] P.A. Mello and A.D. Stone, preprint

[43] P.A. Mello and J.-L. Pichard, Phys. Rev. Rapid Com. **B 40**, 5276 (1989)

[44] L.-L. Pichard and G. Sarma G., J. Phys. **C 14**, L617 (1981)

[45] A. MacKinnon and B. Kramer, Phys. Rev. Lett. **47**, 1546 (1981)

[46] J.-L. Pichard, N. Zanon, Y. Imry, and A.D. Stone, J. Phys. France **1**, 1 (1990)

[47] D. Hofstadter, Phys. Rev **B 14**, 2239 (1976)

[48] O.N. Dorokhov, Sov. Phys. JETP **58**, 606 (1983)

[49] e.g. Y. Imry, in: Percolation, Localization and Superconductivity. NATO ASI series **109**, Plenum Press, New York (1983)

[50] M. Sanquer, unpublished

[51] K. Slevin, these volume

[52] P.A. Mello and B. Shapiro, Phys. Rev. **B 37**, 5860 (1988)

SENSITIVITY OF QUANTUM CONDUCTANCE FLUCTUATIONS TO TIME REVERSAL SYMMETRY

Dominique Mailly and Marc Sanquer[†]

Laboratoire de Microstructure et Microelectronique, CNRS
196, avenue H. Ravera, 92220 Bagneux, France
[†]Service de Physique du Solide et de Résonance Magnétique
C.E.A. Saclay, 91191 Gif sur Yvette Cedex, France

In the last ten years the progresses of microtechnology has opened the field of mesoscopic systems, where quantum information is not lost at low temperature during the
travel of electrons through the sample. Beside the weak localization (WL) phenomena,
that has become the appropriate tool to determine the scale of inelastic processes in
macroscopic systems, mesoscopic samples exhibit a very unusual phenomena: the complex interference pattern resulting from the diffusive motion of electrons results in a very
complicated but reproducible magnetoconductance curve (MC curve), which is sensitive
to any microscopic disorder configuration change. The MC curve contains so a very
large amount of informations but it obeys to universal statistical rules, as predicted by
Universal Conductance Fluctuations (UCF) theory.

It has been predicted from the beginning of the theoretical studies that UCF are
sensitive to the time-reversal symmetry (TRS) in a universal way: the amplitude of
the UCF should decrease by a factor-of-two by application of a magnetic field H_c such
that $H_c L_\phi^2 = h/e$ where L_ϕ is the phase breaking length. This is also the typical field
necessary to suppress the WL contribution to the conductance. The tricky point for
an experimental confirmation is to obtain a sufficient statistics of conductances not
generated by varying the magnetic field as in a usual MC experiment. In fact, with the
use of the ergodic hypothesis it is easy to get from the MC curve the variance over a large
statistical ensemble of conductances, but it always corresponds to the situation where
there is no TRS. However at $H = 0\,T$ generation of UCF is only possible by Fermi level
or disorder configuration variations.

In semi-conducting devices it is convenient to deplete the conduction band by a gate
voltage, which induces UCF superimposed to a large variation of the mean conductance
due to depletion or to sample shape changes. These latter effects could be in principle
corrected in order to compare properly the variance of the conductance at different fixed
magnetic fields. On the other hand, in metallic wires, fluctuations of conductance with
the time are sometimes detected either as random telegraph noise (RTN) or $1/f$-noise.

Quantum Coherence in Mesoscopic Systems
Edited by B. Kramer, Plenum Press, New York, 1991

UCF theory predicts a surprisingly large sensitivity of the conductance at low temperature to any scatterer dynamics [1]. As an illustration, let us consider at $T = 0\,\mathrm{K}$ a very large two-dimensional conductor containing a number of scatterers as large as we want. Let this conductor to be highly disordered such that $k_F l \approx 1$. In that case the displacement of *one* scatterer induces typically a conductance change as large as $(25800\,\Omega)^{-1}$! A large number of experiments has been devoted to the interplay between UCF and noise in conductors, and with respect to the enormous complexity of interpretations of the noise problem, a very nice and simple test for quantum interferences detection is that, as the UCF, the noise amplitude should decrease by a factor-of-two when TRS is broken. This is in fact the case either for the $1/f$-noise detected below the nitrogen temperature in Bismuth wires by Birge et al. [2] or for the RTN induced by a voltage pulse in Si:GaAs wires and $\mathrm{Al}_x\mathrm{Ga}_{1-x}\mathrm{As/GaAs}$ heterojunctions at $T = 45\,\mathrm{mK}$ by Mailly et al. [3].

Birge et al. have obtained the full crossover curve for the magnetic field dependence of the noise amplitude in their Bismuth wires (a system with strong spin-orbit coupling). But a direct comparison with the WL contribution to the magnetoresistance is still lacking as also in analysis of the UCF induced by a gate voltage in a $\mathrm{Al}_x\mathrm{Ga}_{1-x}\mathrm{As/GaAs}$ heterojunction versus the magnetic field by Debray et al. [4].

The aim of this paper is precisely to compare *in a single sample* the magnetic field dependence of the weak localization and of the entire UCF, and to establish definitely – in a very clear and demonstrative experiment – how the reduction by a factor-of-two of the UCF amplitude takes place when the TRS is broken. For that purpose Si:GaAs doped at $10^{18}\,\mathrm{cm}^{-3}$ is chosen because

(1) there is no spin-orbit coupling and the weak localization correction is well known,

(2) the elastic mean free path is 40 nm i.e. small compared to the sample dimensions and to the phase breaking length at low temperature, as well as small compared to the magnetic length for field below 0.5 Teslas, so that the UCF theory is well established (the sample is 10 μm long and the cross section is $0.6 \times 0.3\,\mu\mathrm{m}^2$),

(3) the disorder parameter $k_F l = 11$, so that we are far enough from the metal-insulator transition.

To produce a large number of displacements of scatterers we use the most simple way by applying thermal cyclings to our wire. In this material at $T = 45\,\mathrm{mK}$ there is no intrinsic scatterers dynamics contrarily to metals, but if the sample is warmed up to room temperature another MC curve is obtained without any change of macroscopic parameter as the carrier density of sample shape. The MC curve before and after the thermal cycling are completely uncorrelated. It is not possible to achieve this in a noise experiment at low temperature because the disorder dynamics is too slow, even in metals. With many thermal cyclings, we obtain a statistical series of samples at $T = 45\,\mathrm{mK}$, such that no thermal smearing of the Fermi surface occurs and consequently the thermal length $L_T = \sqrt{hD/k_B T}$ does not enter the problem. Essentially, L_ϕ is the only parameter in the theory. Moreover it can be quantitatively determined by the WL fit of the mean behavior of our statistical series.

To obtain a series of 50 samples we have repeated only 8 thermal cyclings up to room temperature and completed the experiment by applying long (1 second) current pulses to the wire at $T = 45\,\mathrm{mK}$. The effect of those pulses is to dissipate high power

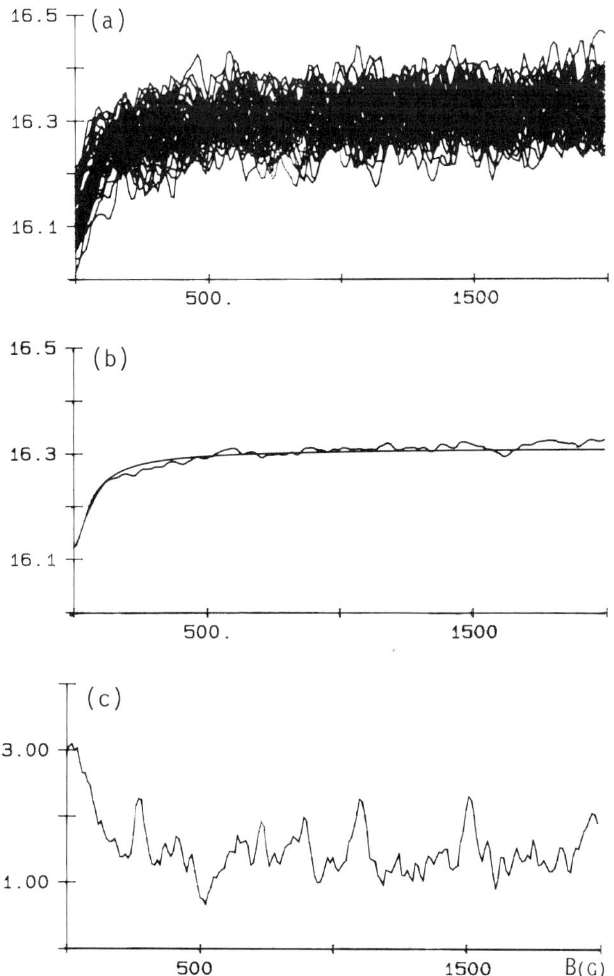

Figure 1. (a) 46 reproducible magnetoconductance curves $T = 45\,\mathrm{mK}$ obtained by thermal cyclings on the *same* SiGaAs wire in units of e^2/h. (b) The mean value of the 46 MC curves and the 1 D weak localization fit which gives $L_\phi = 3.0\,\mu\mathrm{m}$. (c) The variance over 46 samples in units of $10^{-3} \times (e^2/h)^2$. A quantum of magnetic flux passes through $L_\phi \times W$ (W is the width of the sample) for a magnetic field of 23 Gauss. The horizontal scale is the magnetic field in Gauss and is the same for the 3 plots.

directly in the wire and to decorrelate the MC curves. It is not equivalent to room temperature thermal cyclings (i.e. simply a disorder configuration change), because the pulses reduce the mean conductance of the wire. We have evidence that their principal effect is to deplete the conduction band and so to decrease the Fermi level, but we are able to correct the small variation of the mean conductance induced by each pulse. Our final result is plotted in Fig. 1a. The mean behavior is represented in Fig. 1b and the variance over the 50 samples as a function of the magnetic field in Fig. 1c. A very important advantage of our method is that we can fit accurately the WL effect and obtain a good estimation of L_ϕ in the wire used in the experiment. This is not possible with only one MC curve – the error could be as large as a factor two –, and it is always questionable for the determination of L_ϕ to use a sample with a larger geometry, because the etching could greatly change the defect nature, number and localization. L_ϕ is found to be $3.0\,\mu$m with the use of the standard one-dimensional (1 D) correction [5]. We note that because the length between the voltage probes is $10\,\mu$m, L_ϕ could be underestimated by about 10 percent [6].

A good estimation of L_ϕ is crucial because the predicted amplitude of the UCF in our quasi-1 D wire at $T = 45\,$mK is deduced from [7] to be:

$$var(g) = \frac{4}{15}\left(\frac{e^2}{h}\right)^2 \left(\frac{L_\phi}{L}\right)^3 , \tag{1}$$

when TRS is absent. By using $L_\phi = 3.0\,\mu$m in this equation we find $var(g) = 7.2 10^{-3}(\frac{e^2}{h})^2$, that is 4.8 times larger than the observed one (Fig. 1c). To recover the right value it is enough to reduce L_ϕ to $1.8\,\mu$m, which is not incompatible with the WL correction. Our principal result is obtained by comparison of Figs. 1b and 1c. *The UCF amplitude decreases by a factor-of-two when TRS is broken at exactly the same magnetic field where the weak localization is suppressed.*

We have performed the experiment up to $H = 1\,T$ and the result is not modified. The same experiment has also been performed at $T = 4.2\,$K and the same result is obtained i.e. the UCF decreases also by a factor-of-two when the WL correction to the MC disappears (at $T = 4.2\,$K): L_T does not enter the problem except to average the UCF.

References

[1] S. Feng, P.A. Lee, and A.D. Stone, Phys. Rev. Lett. **56**, 1960 (1986); ibid **56**, 2772 (1986)

[2] N.O. Birge, B. Golding, and W.H. Haemmerle, Phys. Rev. Lett. **62**, 195 (1989)

[3] D. Mailly, M. Sanquer, J.-L. Pichard, and P. Pari, Europhys. Lett. **8**, 471 (1989); D. Mailly and M. Sanquer, Surf. Sc. **229**, 260 (1990)

[4] P. Debray, J.-L. Pichard, J. Vicente, and P.N. Tung, Phys. Rev. Lett. **63**, 2264 (1989)

[5] B.L. Al'tshuler and A.G. Aronov, JETP Lett. **33**, 499 (1981)

[6] V. Chandrasekhar, D.E. Prober, and P. Santhanam, Phys. Rev. Lett. **61**, 2253 (1988)

[7] P.A. Mello, Phys. Rev. Lett. **60**, 1089 (1988)

UNIVERSAL CONDUCTANCE FLUCTUATIONS IN NARROW CHANNEL MOSFETs

Jacob Caro, Jianrong Gao, Adrianus H. Verbruggen,
Sybrand Radelaar, Jan Middelhoek* †

Delft Institute for Microelectronics and Submicron Technology
Delft University of Technology, Delft, The Netherlands
*Vakgroep IC-Technology and Electronics, Twente University
Enschede, The Netherlands

1　Introduction

Mesoscopic normal-metal wires and rings exhibit random, reproducible fluctuations in
the magnetoconductance (MC) $G(B)$ due to quantum interference of electron waves [1,2].
These fluctuations are called universal conductance fluctuations (UCF) because at zero
temperature their magnitude is of order of e^2/h, regardless of the degree of disorder and
the sample size. The fluctuation pattern is sensitive to the microscopic distribution of
elastic scatterers for electrons and therefore the fluctuation traces are sometimes referred
to as magneto-fingerprints. At finite temperature the size of the sample usually exceeds
the inelastic diffusion length L_{in} of the electrons, and due to classical self-averaging the
magnitude of the fluctuations is reduced.

The amplitude and the magnetic correlation length of the fluctuations are predicted
[2] to depend on temperature via L_{in}. Thus, by varying the temperature of the sample an
important experimental test of the theory can be made. So far, such detailed experiments
on small Si-MOSFETs have not been reported.

In this paper we present new data on UCF in narrow ($0.14\,\mu$m and $0.43\,\mu$m) channel
MOSFETs. The emphasis of the work is on the *temperature dependence* of UCF in the
MC. A more extensive account of this research can be found in [3].

2　Temperature Dependence of Universal Conductance Fluctuations

The rms magnitude of the UCF is defined as $rms[G/(e^2/h)] = rms(g) = \langle (g - \langle g \rangle)^2 \rangle^{1/2} = \langle \delta g^2 \rangle^{1/2}$. Here angular brackets denote averaging over the magnetic field. The magnetic
correlation length B_c is a typical scale of the spacing between peaks and the valleys in
$g(B)$. B_c is obtained from the half-width of the correlation function in magnetic field,
defined as $F(\Delta B) = \langle \delta g(B) \cdot \delta g(B + \Delta B) \rangle$.

Quantum Coherence in Mesoscopic Systems
Edited by B. Kramer, Plenum Press, New York, 1991

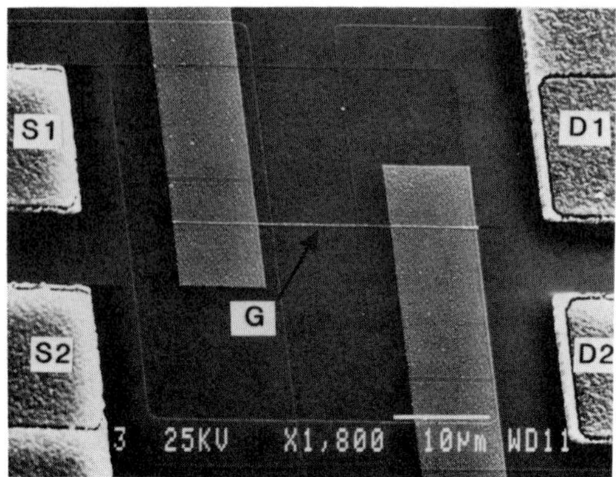

Figure 1. Narrow channel MOSFET for quantum interference studies. The Figure shows source and drain areas (with contacts S_1, S_2 and D_1, D_2) and in the centre the area of the gate oxide. G denotes the narrow metal gate.

The temperature dependence of $rms(g)$ and B_c depends on the size of the conductor in relation to L_{in} and L_T. L_{in} is defined as $(D\tau_{in})^{1/2}$, D being the diffusion constant and τ_{in} the inelastic scattering time, while the thermal length L_T is given by $(hD/kT)^{1/2}$. In case $L_{in} < L_T < L$ (which applies to this work) $rms[g(B)]$ for the one-dimensional (1 D) and 2 D geometry is given by [2]

$$rms(g) = \left(\frac{L_{in}}{L}\right)^{3/2} \begin{cases} 0.52 & L_{in} > W \\ 0.61\left(\frac{W}{L_{in}}\right)^{1/2} & L_{in} < W \end{cases} \tag{1}$$

The magnetic correlation length B_c for the 1 D and 2 D geometry is given by [2]

$$B_c = 1.2\frac{h}{e}\frac{1}{L_{in}} \begin{cases} W^{-1} & L_{in} > W \\ L_{in}^{-1} & L_{in} < W \end{cases} \tag{2}$$

The temperature dependence of $rms(g)$ and B_c in these equations arises from the dependence of L_{in} on $\tau_{in}(T)$.

3   Devices

The devices used are n-channel Si-MOSFETs with a narrow gate. The gate length L is $4.3\,\mu$m and the gate width W is 0.14 and $4.3\,\mu$m. After the growth of a 25 nm thick gate oxide the gate pattern is written in a double layer of PMMA resists [4] with different molecular weight. Then NiCr gates are produced by evaporation and lift-off. More details of the fabrication process can be found elsewhere [5]. We fabricated several types of devices with different W/L ratios, the smallest gate width realized being 140 nm. An SEM photograph of a narrow channel device is shown in Fig. 1. At helium temperatures the devices are normally off and the maximum mobility of the narrowest devices is about $8000\,\mathrm{cm^2/Vs}$.

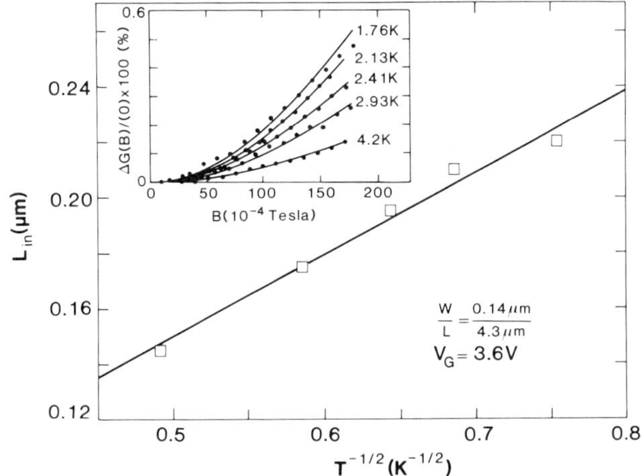

Figure 2. The inelastic length L_{in} in narrow device N 7 obtained from the LFMC is plotted against $T^{-1/2}$. $V_G = 3.6$ V. MC data and fits are displayed in inset.

4   Experimental Results

4.1   Low Field Magnetoconductance

In Fig. 2 results are given of low field MC measurements. We carried out such experiments on a number of devices also used for UCF studies, in order to determine L_{in}. Figure 2 shows a plot of L_{in} versus $T^{-1/2}$ for device N 7, which has a 0.14 μm wide gate. The values of L_{in} were derived from fits of the theoretical 1 D-expression [6] for the low field MC (LFMC) to the experimental data. The width of the channel was also obtained as a result of the fits and agreed within 5% with the lithographic width of the gate. Experimental data for temperatures between 1.76 and 4.2 K and corresponding fits are shown in the inset of Fig. 2. As may be seen the fits describe the data very well. We conclude that the MOSFET is 1 D with respect to weak localization, although marginally at 4.2 K. Further, L_{in} depends linearly on $T^{-1/2}$, the best fit for the experimental conditions of Fig. 2 being $L_{in} = 0.3 \, \mu$m $(T/1 \, \text{K})^{-1/2}$. Apparently in our devices $\tau_{in} \propto L_{in}^2$ is proportional to $1/T$. This indicates [7] that electron-electron interaction is the dominant inelastic scattering mechanism.

4.2   Universal Conductance Fluctuations

Traces of fluctuations in the MC of device N 25 ($W/L = 0.14 \, \mu$m$/4.3 \, \mu$m) are shown in Fig. 3. The gate voltage was 4.0 V, giving an electron density of 3.4×10^{12} cm^{-2}. As may clearly be seen the amplitude of the fluctuations gradually decreases with increasing temperature. At 4.2 K the fluctuations are small but still visible. Figure 3 shows two traces taken at 0.5 K. These were obtained under identical experimental conditions. The lower one was obtained first and the time interval between the measurements was 15 h. During this time the sample was kept below 1 K. Apparently the fluctuations reproduce very well in this device. To demonstrate the gate voltage dependence of the MC Fig. 4 shows traces taken from the same device, in the same experimental run, but at a gate voltage

                                           J. CARO et al.

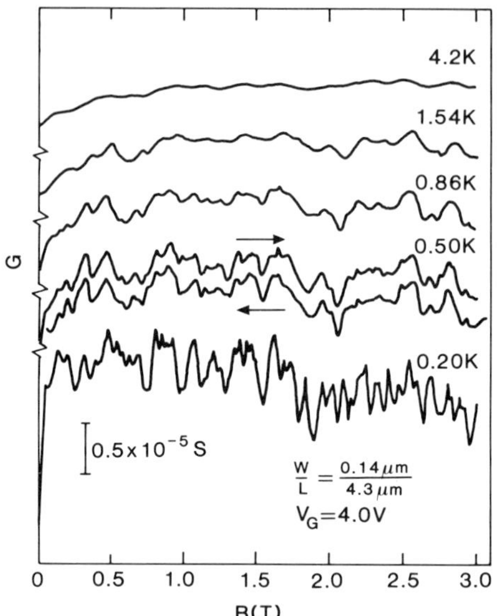

Figure 3. UCF in the MC of narrow device N 25 for a
number of temperatures. $V_G = 4.0\,$V. The time interval
between the two traces at $0.5\,$K was $15\,$h and they were
taken with opposite sweep direction. For each of the six
curves $\langle G \rangle$ amounts to about $108\,\mu$S.

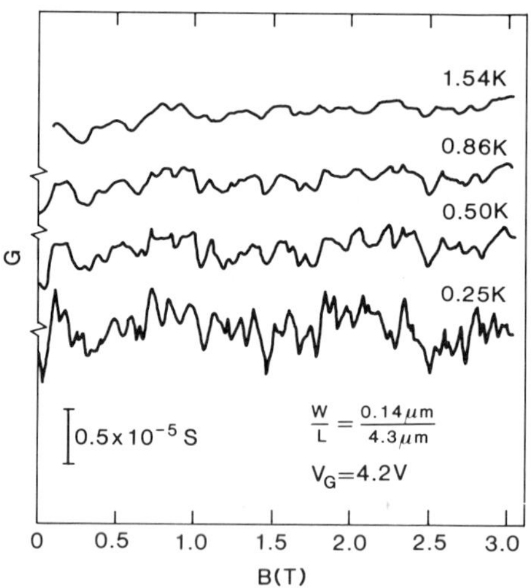

Figure 4. UCF in the MC of narrow device N 25 for a
number of temperatures. $V_G = 4.2\,$V. For each of the four
curves $\langle G \rangle$ amounts to about $113\,\mu$S.

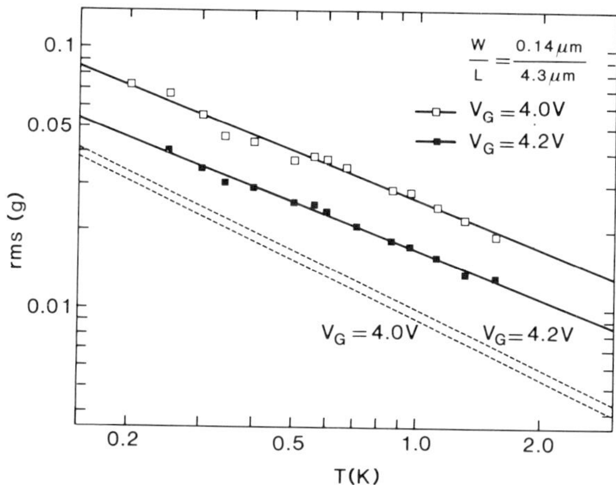

Figure 5. Temperature dependence of $rms(g)$ of the fluctuations on a log-log scale for $V_G = 4.0\,\text{V}$ and $V_G = 4.2\,\text{V}$. Dashed lines are theoretical relations corresponding to the experimental data.

of 4.2 V. In comparing Figs. 3 and 4 one concludes that the two magneto-fingerprints are completely different, which is due to different interference conditions at a different Fermi energy.

In order to calculate $rms(g)$ as a function of temperature a background was subtracted from each curve. To eliminate the influence of weak localization, calculations were limited to $B > 0.4\,T$. This procedure was followed for two data sets that include the curves of Figs. 3 and 4. Results are plotted in Fig. 5, which also shows least square fits to the data points.

In order to compare with the theory, we first determined the dimensionality of the device from L_{in} (derived from LFMC measurements) and L_T (calculated from D, which follows from the average conductance $\langle G \rangle$). We found $L_{in}(T = 1.54\,\text{K}) = 0.23\,\mu\text{m}$ and $L_T(1.54\,\text{K}) = 0.63\,\mu\text{m}$, both for $V_G = 4.0\,\text{V}$. Since L_T, $L_{in}(1.54) > W$ the 1 D limit applies to Fig. 3 (at 4.2 K the device is not in this limit, so we excluded the 4.2 K trace). The 1 D limit holds for $V_G = 4.2\,\text{V}$ as well, since [8] $L_{in} \propto g$ and $L_T \propto g^{1/2}$. Now, by using the temperature dependence of $L_{in}(\propto T^{-1/2})$ and the value of W derived from the LFMC measurements, eq. (1) predicts the values of $rms(g)$. The theoretical predictions are given as dashed lines in Fig. 5. The slopes of the experimental lines are strongly indicative for a $T^{-1/2}$ behaviour of L_{in}.

For the two sets of UCF data the correlation functions were calculated to yield $B_c(T)$. In Fig. 6 resulting B_c values are plotted as a function of temperature. The lines again are best fits. The corresponding theoretical relations were calculated from eq. (2) using the same data for L_{in} as in calculating $rms(g)$. The results are displayed as dashed lines in Fig. 6. We conclude that there is a good agreement between the magnitudes of B_c^{expt} and B_c^{theo}. Again, the slopes of the experimental lines are strongly indicative for a $T^{-1/2}$ behaviour of L_{in}. The values of the parameters that describe the relations for $rms(g)$ and B_c just discussed are given in Table 1.

In Fig. 7 traces of fluctuations in the MC of device N 19 at temperatures between 0.35 and 4.2 K are given. This device is relatively wide ($W/L = 0.43\,\mu\text{m}/4.3\,\mu\text{m}$) and

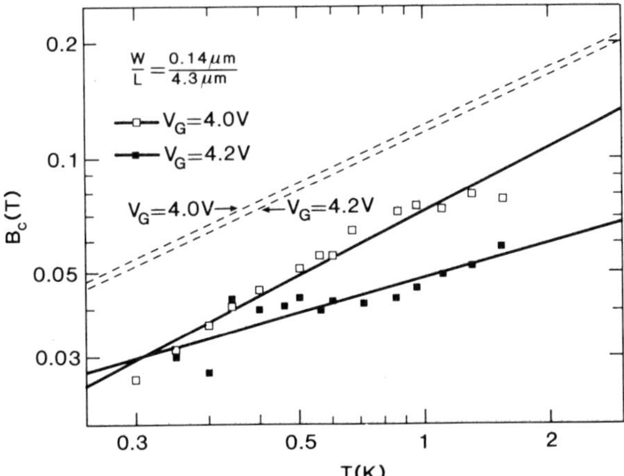

Figure 6. Correlation length B_c of the fluctuations as a function of temperature on a log-log scale for $V_G = 4.0\,\text{V}$ and $V_G = 4.2\,\text{V}$. Dashed lines are corresponding theoretical relations.

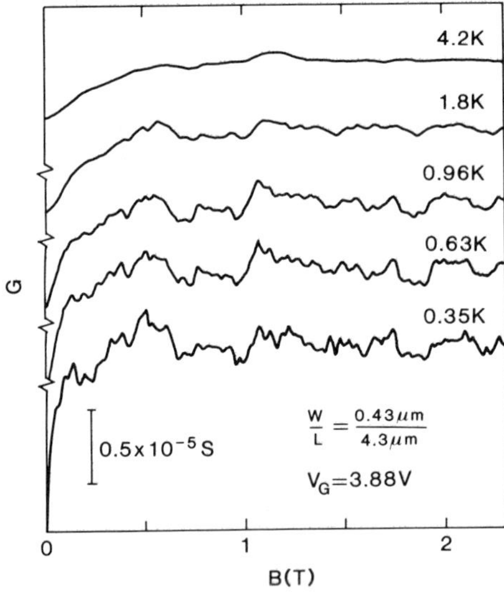

Figure 7. UCF in the MC of 2 D device N 19 for a number of temperatures, $V_G = 3.88\,\text{V}$, $\langle G \rangle$ amounts to $106\,\mu\text{S}$.

Table 1. Values of the parameters a,b,c,d that describe $rms(g)^{\text{expt,theo}} = a(T/1\text{ K})^b$ and $B_c^{\text{expt,theo}} = c(T/1\text{ K})^d$.

		$V_G(V)$	a	b	$c(mT)$	d
1 D	expt	4.0	$(2.7 \pm 0.2) \times 10^{-2}$	-0.63 ± 0.06	74 ± 5	0.56 ± 0.07
		4.2	$(1.7 \pm 0.1) \times 10^{-2}$	-0.61 ± 0.07	48 ± 3	0.3 ± 0.1
	theo	4.0	9.1×10^{-3}	-0.75	122	0.5
		4.2	9.8×10^{-3}	-0.75	117	0.5
2 D	expt	3.88	$(1.5 \pm 0.3) \times 10^{-2}$	-0.45 ± 0.07	58 ± 4	0.23 ± 0.1
	theo	3.88	7.6×10^{-3}	-0.5	170	1.0

was shown to operate in 2 D limit. The gate voltage used was 3.88 V, giving an electron density of $1.8 \times 10^{12}\,\text{cm}^{-2}$ and an average conductance $\langle G \rangle$ of $106\,\mu\text{S}$. For the data set that includes the traces of Fig. 7 $rms(g)$ was calculated. Further, from the curves correlation functions were derived from which B_c values were obtained. For $V_G = 3.88$ V, also using LFMC experiments, it can be derived that $L_{in} = 0.17\,\mu\text{m}\ (T/1\text{ K})^{-1/2}$ and $L_T = 0.45\,\mu\text{m}\ (T/1\text{ K})^{-1/2}$, so that the criterion $L_{in} < L_T < L$ is satisfied. Therefore eqs. (1) and (2) are applicable. From these we derived theoretical relations for $rms(g)$ and B_c. The values of the parameters that describe the relations for $rms(g)^{\text{expt,theo}}$ and $B_c^{\text{expt,theo}}$ for device N 19 are also given in Table 1. We conclude that the magnitudes of $rms(g)^{\text{expt}}$ and $rms(g)^{\text{theo}}$ agree to within a factor of two. For the slope of the $rms(g)$ lines the agreement is excellent, also strongly indicating a $T^{-1/2}$ behaviour of L_{in} for the 2 D regime. For B_c the situation is different. Although experimental values have the proper order of magnitude, the temperature dependence of B_c is smaller than predicted by the theory.

5   Conclusions

We have investigated the temperature dependence of fluctuations in the MC of narrow Si inversion layers in a wide temperature range $(0.2 - 4.2\text{ K})$. Weak localization effect measurements have been used for an independent determination of the inelastic diffusion length of the electrons. For the 1 D geometry we found good agreement with the UCF theory for the magnitude of $rms(g)$ and B_c of the fluctuations. The temperature dependence gives a strong indication that τ_{in} is proportional to $1/T$, as expected for electron-electron scattering. For the 2 D geometry the magnitude of $rms(g)$ is in accord with the UCF theory and the temperature dependence is in excellent agreement with $\tau_{in} \propto 1/T$. B_c has the proper order of magnitude, but the temperature dependence strongly deviates from the theoretical prediction. At this point new experiments are prompted.

6   Acknowledgments

The authors wish to express their gratitude to A. Kooy and G. Boom for their advice and for MOS-processing of the wafers. We are very grateful to the Solid State Physics Group of the Faculty of Applied Physics for the use of their dilution refrigerator and to

L.J. Geerligs for technical assistance. This work is part of the research program of the
"Stichting Fundamenteel Onderzoek der Materie (FOM)", which is financially supported
by the "Nederlandse Organisatie voor Wetenschappelijk Onderzoek (NWO)".

† Deceased. The first four authors gratefully acknowledge the special contributions of
Jan Middelhoek to this work.

References

[1] Several authors in IBM J. Res. Develop. **32** (3) (1988), and references therein

[2] P.A. Lee and A.D. Stone, Phys. Rev. Lett. **55**, 1622 (1985); P.A. Lee, A.D. Stone,
and H. Fukuyama, Phys. Rev. **B 35**, 1039 (1987)

[3] J.R. Gao, J. Caro, A.H. Verbruggen, S. Radelaar, and J. Middelhoek, Phys. Rev.
B 40, 11676 (1989)

[4] A.H. Verbruggen, B.M.G. de Lange, B.A.C. Rousseeuw, H.A.H. Billiet, and
S.Radelaar, Microelectronic Engineering **11**, 561 (1990)

[5] J.R. Gao, J. Caro, A.H. Verbruggen, S. Radelaar, and J. Middelhoek, Microelectronic Engineering **9**, 373 (1989)

[6] B.L. Al'tshuler and A.G. Aronov, JETP Lett. **33**, 499 (1981)

[7] D.J. Bishop, R.C. Dynes, and D.C. Tsui, Phys. Rev. **B 26**, 773 (1982)

[8] W.J. Skocpol, L.D. Jackel, E.L. Hu, R.E. Howard, and L.A. Fetter, Phys. Rev. Lett.
49, 951 (1982)

CHAPTER 8

LOCALIZATION

LOCALIZATION AND FLUCTUATIONS

Angus MacKinnon

Blackett Laboratory
Imperial College
London SW7 2BZ, UK

1 Introduction

The subject of fluctuations in the transport properties of disordered solids is often presented as an effect that was first observed in the mid 80's. It may well be that such effects were first observed experimentally at that time, but they have been dogging the theory and especially those doing computer simulations from much earlier. In this paper I shall present a personal view of the subject. I offer no apology for the large dose of hindsight which the discussion will contain.

The problem of fluctuations is intrinsic to the behaviour of waves in a disordered medium. This is true whether we are dealing with electrons in a disordered solid, microwaves in a waveguide containing polystyrene balls, light in milk, even gravity (water) waves scattered from the islands of an archipelago. Indeed the non-electronic manifestations are often much easier to observe than the electronic ones. Electrons are, in the main, subject to many scattering processes with different characteristics, whereas other wave phenomena offer much cleaner experimental conditions.

1.1 Self-Averaging

Before starting the discussion it is worthwhile to define some terminology. Consider an extrinsic property $\Gamma(L)$ of a particular random system of size L. Γ could, for example, represent the conductance $G = 1/R$. By looking at a large number of statistically different realizations of the random system it would be possible to calculate the mean value of $\Gamma(L)$, $\langle \Gamma(L) \rangle$ or the variance $\langle \Delta\Gamma(L)^2 \rangle = \langle \Gamma(L)^2 \rangle - \langle \Gamma(L) \rangle^2$ or any other moment.

In statistical mechanics it is often assumed that a large system, of size L, can be considered to be made up of a large number of statistically independent smaller systems, of size ℓ. In such a case the $\Gamma(L)$ behaves statistically as

$$\langle \Gamma(L) \rangle \sim \langle \Gamma(\ell) \rangle (L/\ell) \sim L, \tag{1}$$

$$\langle \Delta\Gamma(L)^2 \rangle \sim \langle \Delta\Gamma(\ell)^2 \rangle (L/\ell) \sim L, \tag{2}$$

Quantum Coherence in Mesoscopic Systems
Edited by B. Kramer, Plenum Press, New York, 1991

$$\Delta_{\text{rel}}\Gamma(L) = \sqrt{\langle\Delta\Gamma(L)^2\rangle/\langle\Gamma(L)\rangle^2} \sim L^{-1/2}, \tag{3}$$

where $\Delta_{\text{rel}}x$ represents the *relative* error in x. A quantity whose relative error falls with increasing system size is said to be *self-averaging*. In such cases the limit

$$\lim_{L\to\infty}\Gamma(L) = \lim_{L\to\infty}\langle\Gamma(L)\rangle \tag{4}$$

is true for (almost) all realizations of the random system.

The assumption of self-averaging is fundamental to the development of statistical mechanics. Without it the whole edifice of thermodynamics would collapse. Thus it is not surprising that for many years the assumption of self-averaging lay at the heart of any theory of disordered systems (e.g. [1]). It was a brave man indeed who dared to question it.

2   Calculating the Conductivity of One-Dimensional Chains

2.1   History

Ever since the work of Mott and Twose [2] and even earlier it had been known that all states are localized in 1 D disordered systems. This of course implies the absence of metallic conductivity in 1 D chains. In the late seventies attempts were made to apply the Kubo-Greenwood formula to the problem [3]. This proved rather difficult, but the difficulties were eventually overcome by the development of an algorithm which could calculate the conductivity for any length of chain without running into computer storage limits [4, 5].

The reason for the difficulties is easily understood [6]. In a finite system the spectrum is discrete, only particular energies are eigenenergies of the system. If the Fermi energy falls in a gap (almost always the case) the $T = 0$ conductivity will be zero. If the Fermi energy happens to coincide with an eigenenergy the Kubo-Greenwood formula may yield an infinite conductivity. In order to overcome this problem it is necessary to define an energy broadening function, either by considering a finite temperature or by introducing an imaginary part of the energy into the formalism. In practice it is found that the number of eigenenergies contained within the broadened Fermi energy, namely

$$N_{\text{eigen}} \sim \eta L, \tag{5}$$

where η is the energy broadening, controls the statistical behaviour (Fig. 1). The relative error in the calculated conductivity is found to obey

$$\Delta_{\text{rel}}\sigma \sim N_{\text{eigen}}^{-1/2} \sim (\eta L)^{-1/2}. \tag{6}$$

The introduction of η has unfortunate side effects. η^{-1} is a phenomenological lifetime which introduces inelastic scattering. For large η or weak disorder this may constitute the dominant scattering process. Even small η allows hopping between localized states so that the conductivity goes as

$$\sigma \sim \eta\lambda^2 \tag{7}$$

where λ is the localization length [4].

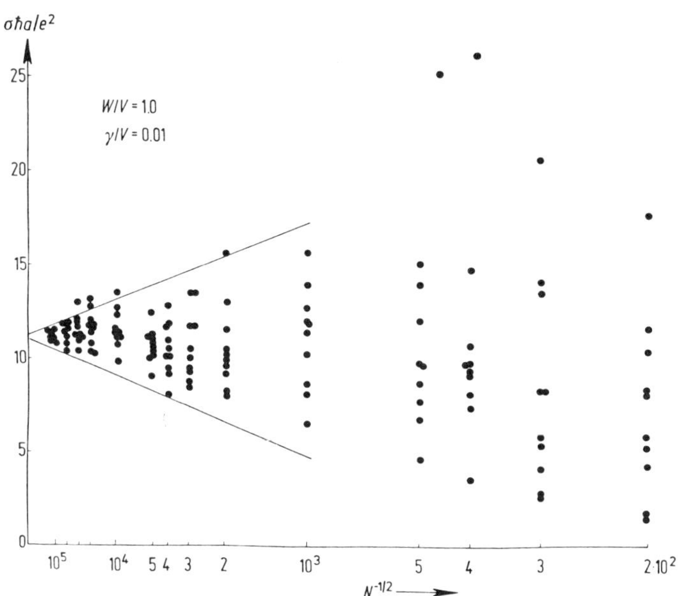

Figure 1. Conductivity σ of a 1 D chain of length N versus $N^{-1/2}$ for various samples described by the Anderson Hamiltonian (12) with rectangular disorder $-W/2 < \epsilon < W/2$, with $W/V = 1.0$ and $\eta/V = 0.01$ [4].

We are thus faced with 2 contradictory conditions. In order to achieve $\sigma = 0$ we must take the limit $\eta \to 0$. On the other hand the relative error in σ diverges in this limit.

All is not lost however. There is another way of overcoming the problem of the discrete spectrum: the system can be embedded in infinite perfect leads, the so-called Landauer [7] geometry. Then the spectrum becomes continuous and the problem vanishes ...or does it? Surely the effects of the leads will only be apparent within about a localization length λ of the ends. When $L \gg \lambda$ the leads must be all but irrelevant. In that case we are still faced with an error which diverges when η is switched off, $\sigma = 0\pm\infty$.

2.2   Transmission Coefficients

The Landauer geometry suggests an alternative formulation of the problem in terms of the transmission coefficient T, the probability that an electron which enters one end of the sample is emitted from the other end. This can be related to the Kubo-Greenwood formula [8] by

$$G = \frac{e^2}{h}T \tag{8}$$

or through the Landauer [7] formula

$$G = \frac{e^2}{h}\frac{T}{1 - T}, \tag{9}$$

where $G = \sigma L^{d-2}$ is the conductance. The validity of each of these formulæ has been the subject of considerable dispute [9] which need not concern us here.

If we ignore all back scattered electrons we can arrive at a simple picture of the behaviour of T. The total transmission coefficient $T(L)$ is then simply the product of the transmission coefficients of individual segments of the system i.e.

$$T(L) = \prod_{i=1}^{L/\ell} T_i(\ell) = \exp\left(\sum_{i=1}^{L/\ell} \ln T_i(\ell)\right). \tag{10}$$

If we assume that the $T_i(\ell)$ are statistically independent then $\ln T(L)$ is a self-averaging quantity of the sort discussed earlier in eq. (1)-(3). Note however that

$$\ln\{T \pm \Delta T\} = \ln\{T(1 \pm \Delta T/T)\} \approx \ln T \pm \Delta T/T, \tag{11}$$

so that the *absolute* error in $\ln T$ is approximately the *relative* error of T, and

$$\langle \Delta \ln T(L)^2 \rangle \sim L, \tag{12}$$

so that the $T(L)$ itself is *not self-averaging*.

This simple model has another important property, namely that the distribution of $\ln T(L)$ is asymptotically normal. This is a simple consequence of the central limit theorem, as long as $\langle \Delta \ln T_i(\ell)^2 \rangle$ is finite.

It is very important to note however that this does not imply that the distribution of $T(L)$ is normal. Indeed its properties can be summarised as follows

$$0 \leq T(L) \leq 1, \tag{13}$$

$$\langle T(L) \rangle \sim \exp(-\alpha L), \tag{14}$$

$$\Delta_{\mathrm{rel}} T(L) \sim L^{1/2}. \tag{15}$$

An alternative interpretation was suggested by Azbel [10] in which he considered localized states of length λ distributed at random positions in the disordered chain. Then the probability that an electron will be transmitted from one end to the other depends on the overlap of the states with both ends. The transmission coefficient due to a state centred at position r will be

$$T \sim \exp\left[-r/\lambda\right]\exp\left[-(L-r)/\lambda\right] = \exp\left[-(L-2r)/\lambda\right] \tag{16}$$

which has its maximum value when $r = 1/2L$. The details of this argument have been disputed by Pendry [11] as it contains an over simplistic description of the localized states and ignores some important correlations. Nevertheless it provides a useful insight into the origin of the fluctuations.

2.3   Transfer Matrices

The above discussion involved a drastic approximation in that we ignored all back-scattered electrons. Now let us try to do the calculation correctly. For this we shall use the traditional Anderson model [12]

$$\epsilon_i a_i + V_{i,i+1} a_{i+1} + V_{i,i-1} a_{i-1} = E a_i, \tag{17}$$

where the diagonal matrix elements ϵ are random and the off-diagonal elements V are usually defined to be unity. The a_i are then the coefficients of the wave function on the different sites. Using $V = 1$ eq. (17) can be rewritten in the form

$$\begin{aligned}
\begin{pmatrix} a_{i+1} \\ a_i \end{pmatrix} &= \begin{pmatrix} E - \epsilon_i & -1 \\ 1 & 0 \end{pmatrix} \begin{pmatrix} a_i \\ a_{i-1} \end{pmatrix} \\
&= \prod_{j=1}^{i} \begin{pmatrix} E - \epsilon_j & -1 \\ 1 & 0 \end{pmatrix} \begin{pmatrix} a_1 \\ a_0 \end{pmatrix} \\
&= \prod_{j=1}^{i} \mathsf{M}_i \begin{pmatrix} a_1 \\ a_0 \end{pmatrix},
\end{aligned} \tag{18}$$

where M_i are called *transfer matrices*.

By comparing eq. (10) with eq. (18) we see that they have the same form except that M_i are matrices whereas T_i are scalars. This problem has been studied by a large number of authors [13, 14, 15] who used numerical methods and analytically by Abrikosov [16, 17].

By choosing the initial vector $[a_1, a_0]$ to represent a transmitted wave the transfer matrix M can be related to the transmission coefficient t for the *amplitude* of plane waves incident on the disordered system. Kirkman and Pendry [18] showed how to calculate all the moments of t and some aspects of $\ln t$ and of $tt^\dagger$. We thus have a more or less complete picture of the behaviour of a 1 D system. The results of an exact calculation agree with those of the very approximate arguments given earlier.

2.4   Resistances

What has all this got to do with experiment? Usually experimentalists measure resistances $R = 1/G$. By combining the definitions of eqs. (8), (9) or eq. (10) with our results for the statistical behaviour of the transmission coefficient we obtain the result that the mean resistance *diverges* exponentially with the length and the variance of the resistance diverges even faster than the mean. Clearly, since we implied the behaviour of T and hence G from that of $\ln T$ the same conclusions must apply to R.

Some caution is required in the interpretation of this result. The resistance is usually measured by measuring the voltage between two contacts on a sample carrying a known current, $R_{12} = V_{12}/I$. It has occasionally been argued that if the current I is fixed then $R \sim \exp(\alpha L)$ implies $R_{13} > R_{12} + R_{23}$ and hence $V_{13} > V_{12} + V_{23}$. This latter statement is, of course, false (conservation of energy). The exponential increase in resistance is valid only in the absence of inelastic scattering, that is when the phase of the wave function is well defined. It is impossible to measure a voltage without introducing some inelastic scattering. This and a number of related questions have been discussed in a series of papers by Büttiker et al. [19].

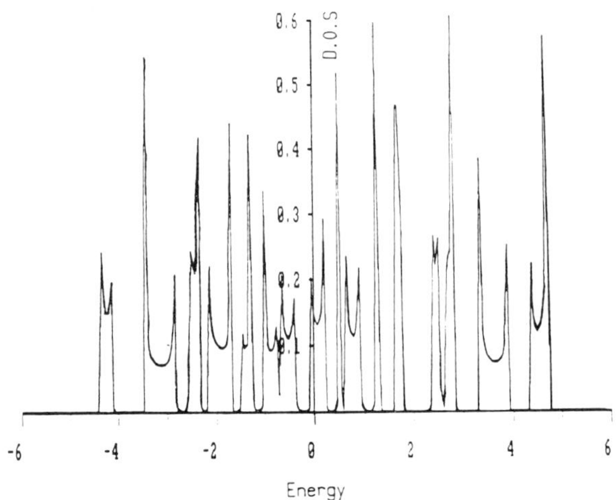

Figure 2. The Density of States of a system consisting of a disordered unit cell periodically continued to infinity in one direction [20].

2.5 Maximal Fluctuations

In order to understand a bit more about the statistical behaviour of the conductance consider an alternative set of boundary conditions for our disordered system. If we link identical systems end to end we can form an infinite 1 D crystal in which our disordered system is a unit cell. Since we now have a crystal we can apply Bloch's theorem and get bands and gaps. In fact we get as many bands as there are sites in the unit cell. Since there is no further symmetry the bands are unlikely to cross and there will probably be gaps between all consecutive bands (Fig. 2).

The model discussed here should not be confused with that which gives rise to persistent currents (see paper by Imry in this volume). In that case the band structure arises by varying the magnetic flux through a single ring. In this case we are considering an infinite crystal composed of identical unit cells.

When the Fermi level lies in a band the transmission coefficient of the complete system must be unity, but when E_F is in a gap T is zero. Thus the statistics of the conductance (using eq. (8)) are particularly simple and can be characterised by a probability P that E_F is in a band. In this case all moments of the transmission coefficient $\langle T^n \rangle$ are equal, so that

$$\frac{\langle T^n \rangle}{\langle T \rangle} = 1 \qquad \text{for all } n. \tag{19}$$

The surprising aspect of this result is that it survives even when we consider the Landauer geometry (Fig. 3) [21]. In this case we have the result

$$\lim_{L \to \infty} \frac{\langle (tt^\dagger)^n \rangle}{\langle tt^\dagger \rangle} = C_n. \tag{20}$$

Since C_n is less than unity and weakly dependent on n the simple picture of open and shut channels is not quite true. The distribution of $tt^\dagger$ is such that for large L it is usually

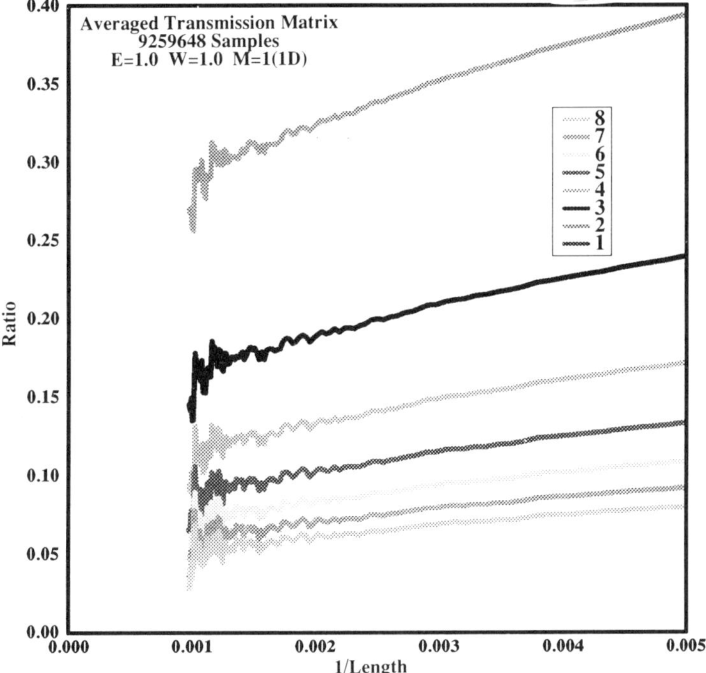

Figure 3. $\langle(tt^\dagger)^n\rangle/\langle tt^\dagger\rangle$ versus $1/L$ for a 1 D chain of up to $L = 1024$. The values of n are as in the insert. The averages were taken over $9,259,648$ samples with $E = 1.0$, $W = 1.0$.

of order zero, but occasionally *of order* unity. Nevertheless the approximation of open und shut channels with a probability $P = \langle tt^\dagger\rangle$ is a remarkably accurate description of the true behaviour.

One other interesting aspect of this result is that the fluctuations have the *maximum* value they could have for a given mean. This suggests that a description in terms of a *maximum entropy* might be appropriate.

3 Two- and Three-Dimensional Systems

The extension of the above considerations to 2 D and 3 D constitutes a much more difficult problem. Nevertheless considerable progress has been made.

3.1 The Metallic Limit

In recent years experimental and theoretical investigations have revealed a most surprising feature of the transport properties of metallic systems [22, 23]. At very low temperatures inelastic scattering processes are frozen out. If as a result of this (almost) no phase randomisation takes place within a finite sample, the statistical fluctuations of the conductance as a function of the Fermi level, an applied magnetic field, or the configuration of the impurities are larger than expected.

$$\Delta G = \left(\frac{e^2}{h}\right) f \qquad f \approx 1. \tag{21}$$

Within certain limits f is found to be universal, although there is a dependence on the dimensionality and shape of the sample. This finding is well supported by numerical and analytical calculations [24, 25, 26]. The conductivity σ is related to the conductance by the classical relation $G = \sigma L^{d-2}$, therefore the relative fluctuations vanish only according to $\Delta_{\rm rel} G \sim L^{2-d}$ in contrast to the classical result, $\Delta_{\rm rel} G \sim L^{-d/2}$. This means that even in 3 D metallic systems the zero-temperature conductance and resistance (and hence conductivity and resistivity) are only just self-averaging, and we should expect to be able to observe significant fluctuations.

3.2   The Bethe Lattice

The Bethe lattice [27] or Cayley tree [28] has properties which are more like a 3 or higher dimensional system while maintaining the mathematical tractability of 1 D systems. It is defined as a system in which each site is connected to 2 D neighbours, as in a normal lattice, but where there are no rings. This lack of a ring structure eliminates many of the terms which are known to contribute to *weak localization*. Nevertheless the properties of the Bethe lattice form a useful indicator of what might be expected from a real disordered lattice. A transmission coefficient may be defined by considering the probability that an electron starting at the centre will eventually reach somewhere on the surface.

The results of such a calculation based on the Anderson model eq. (17) are summarised in Fig. 4 where the logarithms of the arithmetic, geometric and harmonic means of T, $\ln\langle T\rangle$, $\langle \ln T\rangle$ and $-\ln\langle T^{-1}\rangle$ are plotted as a function of disorder on the metallic side of the metal insulator transition. The most startling aspect of this result is the step-like behaviour of $\ln\langle T\rangle$ in contrast to the smooth transition shown by the geometric and harmonic means.

What does this imply for the distribution of T? The interpretation is suggested by our previous discussion of *maximal fluctuations*. The distribution of T is such that it is usually very small but occasionally of order unity. $\langle T\rangle$ is dominated by the few large values whereas $\langle \ln T\rangle$ and $\langle T^{-1}\rangle$ are dominated by the smaller ones.

3.3   Spatial Fluctuations

In a 2 D or 3 D system the formulation in terms of transfer and transmission matrices is formally similar to that in 1 D except that the matrices become larger, of order the cross section of the system. This allows for another form of fluctuations. Do the electrons transmitted through a disordered 2 D or 3 D system emerge uniformly from the exit surface or only from a few points. A useful measure of this is provided by the participation ratio integrated over the exit surface. This is defined as

$$\mathcal{P} = \frac{\left[\int_S |\psi|^2 \, dS\right]^2}{\int_S |\psi|^4 \, dS \int_S dS} \tag{22}$$

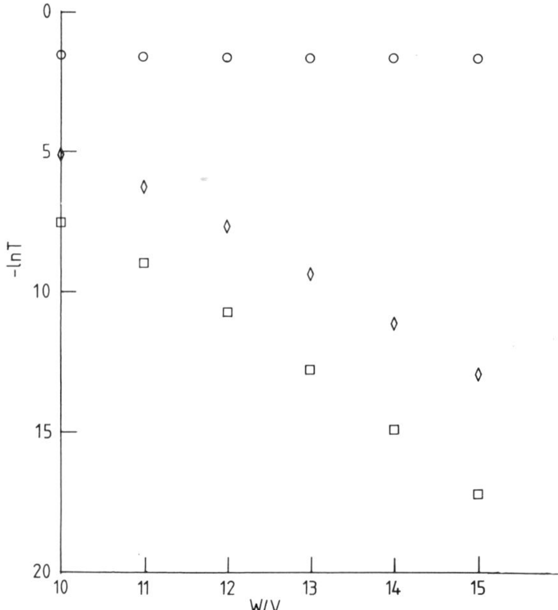

Figure 4. The arithmetic, geometric, and harmonic means of the transmission coefficient of a 3-fold coordinated Bethe lattice on the metallic side of the metal-insulator transition. $\ln\langle T\rangle$ (○), $\langle \ln T\rangle$ (◇), and $-\ln\langle T^{-1}\rangle$ (□) versus disorder W/V. The metal insulator is just beyond $W/V = 15$ where all means become zero.

and may be interpreted as the fraction of the surface from which the electrons emerge. For electrons emerging from a surface of cross section M^2 this will be

$$\lim_{M\to\infty} \mathcal{P} \sim \begin{cases} \text{const.} & \text{for extended states} \\ \lambda^2/M^2 & \text{for localized states} \\ M^{-1} & \text{at the transition} \end{cases} \tag{23}$$

The behaviour at the transition is confirmed by numerical simulation (Fig. 5). Hence at the transition the electrons emerge in a *fractal* distribution with an effective dimension of 1, rather like laser speckle, in which the intensity of the current is high on a small fraction of the surface.

Since this spatially fractal behaviour must be part of a continuous development as a function of disorder, energy etc. some element of it must be apparent on both sides of the transition. We should expect, therefore, that the *short range* behaviour of both localized and extended states will also be fractal. For a fuller discussion of this aspect see the paper by Schreiber in this volume.

3.4   Maximal Fluctuations

In introducing the idea of maximal fluctuations in 1 D systems we considered our disordered sample repeated infinitely often to form a 1 D crystal with a large unit cell. The same argument can of course be extended to 2 D and 3 D with essentially the same

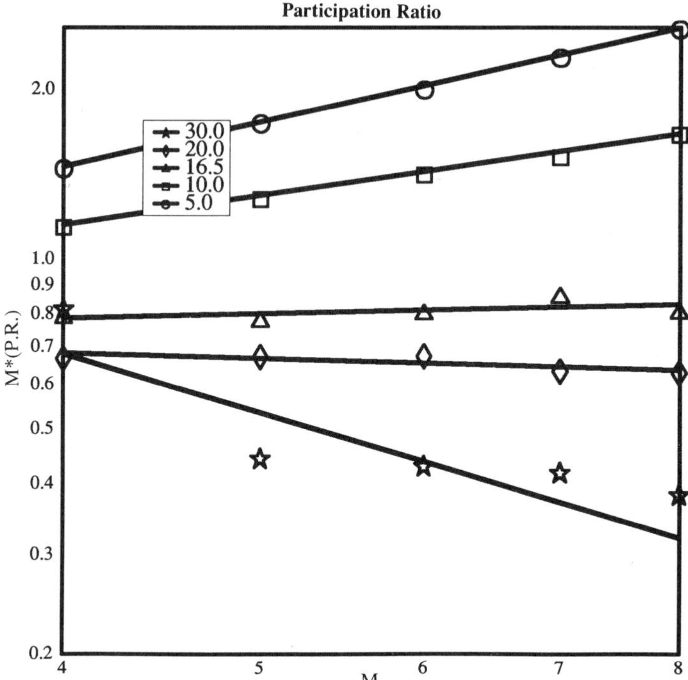

Figure 5. A log-log plot of $M*\mathcal{P}$ versus M for a 3 D Anderson model for various values of disorder W/V as marked in the inset. $\mathcal{P}$ is defined by eq. (22) for the electrons emerging from the surface and M is the width of the system perpendicular to the direction of the current. N.B. The cross-section is M^2. Note the horizontal line at the metal-insulator transition $(W/V = 16.5)$.

results [26]. The probability $\mathcal{P}$ of being in a band now exhibits a fixed point behaviour around $\mathcal{P}_c \approx 0.3$. For $\mathcal{P} > \mathcal{P}_c$, $\mathcal{P}$ increases with increasing system size, and for $\mathcal{P} < \mathcal{P}_c$ it decreases in a similar way to the conductance in conventional scaling theories [29]. This is not surprising given the close relationship between $\mathcal{P}$ and $\langle T \rangle$.

Now we revert to Landauer geometry. Again the picture of open and shut channels is essentially correct. Even more surprisingly, it works for extended states (Fig. 6). In fact the C_n's in eq. (20) are larger in this case than in 1 D so that the approximation is even better. This is true as long as the sample geometry is such that the length L is not larger than the width M and that both are larger than the mean free path.

4   Conclusions

Fluctuations in the conductance are an intrinsic property of all disordered systems at low temperatures. They manifest themselves in a number of ways. Extreme but reproducible dependence on Fermi level and magnetic field as well as the details of the disordered potential are essential characteristics. This effect is fundamentally due to interference and can be destroyed by any phase randomising process such as inelastic scattering from phonons, other electrons etc. Hence in solids the effect is limited to low temperatures

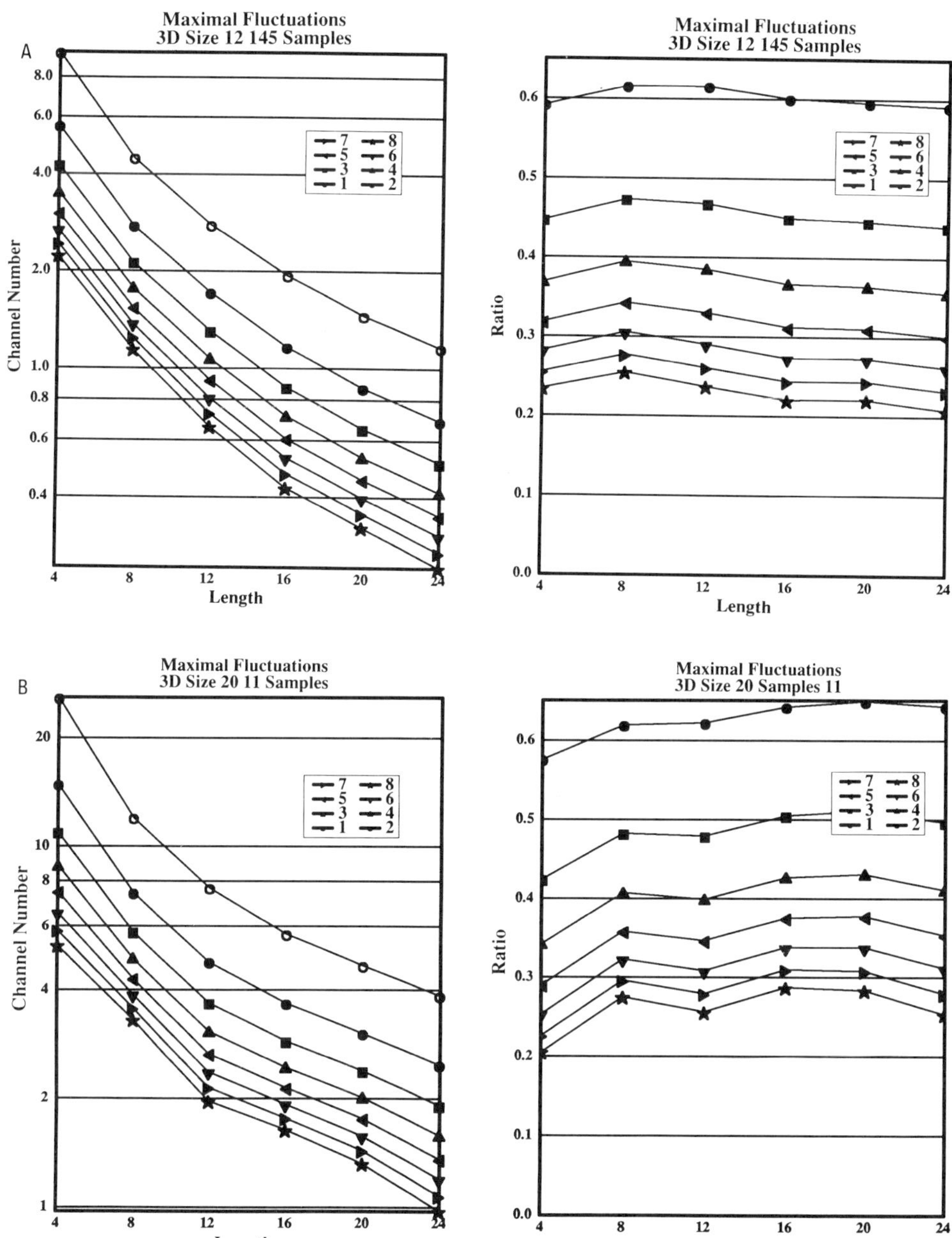

Figure 6. Channel Number ($\langle (tt^\dagger)^n \rangle$) versus length and Ratio ($\langle (tt^\dagger)^n \rangle / \langle tt^\dagger \rangle$) versus length for 3 D systems of cross section 12×12 and 20×20 and various values of n (see insert). The size 12 systems have been averaged over 145 samples and the size 20 ones over 11.

and small samples, but the analogous effects in optics, acoustics etc. may well be more easily observed.

In this paper we started with the onset of the fluctuations as the lifetime of the electrons increases, albeit unrecognised as such when first encountered. We have seen that much of the statistics can be understood by very simple models: the product of random transmission probabilities, or randomly open and shut channels. The fluctuations, in fact, have close to their maximum possible extent.

The fluctuations are not only a function of energy or magnetic field but also of position. This latter aspect still awaits experimental confirmation.

To sum up: *If it can fluctuate it will, as much as possible.*

5   Acknowledgements

I should like to acknowledge many useful discussions with J.B. Pendry, A.B. Prêtre, K.S. Chase, A.D.L. Liebert, J.P.G. Taylor and B. Kramer.

References

[1] V.L. Bonch-Bruevich, Fundamental Physics of Amorphous Semiconductors, F. Yonezawa, ed., p. 145, Springer Verlag, Berlin, Heidelberg (1981)

[2] N.F. Mott and W.D. Twose, Adv. Phys. **10**, 107 (1961)

[3] G. Czycholl and B. Kramer, Z. Phys. **B 39**, 193 (1980)

[4] A. MacKinnon, J. Phys. C: Sol. St. Phys. **13**, L1031 (1980)

[5] G. Czycholl, B. Kramer, and A. MacKinnon, Z. Phys. **B 43**, 5 (1981)

[6] B. Kramer, A. MacKinnon, and D. Weaire, Phys. Rev. **B 23**, 6357 (1981)

[7] R. Landauer, Phil. Mag. **21**, 863 (1970)

[8] E.N. Economou and C.M. Soukoulis, Phys. Rev. Lett. **46**, 618 (1981)

[9] D.C. Langreth and E. Abrahams, Phys. Rev. **B 31**, 2978 (1981)

[10] M.Ya. Azbel, Phys. Rev. **B 28**, 4106 (1983)

[11] J.B. Pendry, J. Phys. C: Sol. St. Phys. **20**, 733 (1987)

[12] P.W. Anderson, Phys. Rev. **109**, 1492 (1958)

[13] B. Andereck, E. Abrahams, J. Phys. C: Sol. St. Phys. **13**,L383 (1980)

[14] J. Sak, B. Kramer, Phys. Rev. **24**, 1761 (1981)

[15] Y. Kantor, A. Kapitulnik, Sol. St. Commun. **42**, 161 (1982)

[16] A.A. Abrikosov, Sol. St. Commun. **37**, 997 (1981)

[17] A.J. O'Connor, Commun. Math. Phys. **45**, 63 (1975)

[18] P.D. Kirkman and J.B. Pendry, J. Phys. C: Sol. St. Phys. **17**, 4327 (1984); J. Phys. C: Sol. St. Phys.17, 5707 (1984)

[19] M. Büttiker, Y. Imry, R. Landauer, and S. Pinhas, Phys. Rev. **B 31**, 6207 (1985)

[20] K.S. Chase, PhD Thesis, University of London (1987)

[21] J.B. Pendry, A. MacKinnon, and A.B. Prêtre, Physica **A 168**, 400 (1990)

[22] R.G. Wheeler, K.K. Choi, A. Goel, R. Wisnieff, and D.E. Prober, Phys. Rev. Lett. **49**, 1574 (1982)

[23] S. Washburn and R.A. Webb, Adv. Phys. **35**, 375 (1986)

[24] P.A. Lee and A.D. Stone, Phys. Rev. Lett. **55**, 1622 (1985)

[25] B.L. Altshuler, JETP Lett. **41**, 648 (1985)

[26] K.S. Chase and A. MacKinnon, J. Phys. C: Sol. St. Phys. **20**, 6189 (1987)

[27] H.H. Bethe, Proc. R. Soc. **A 216**, 45 (1935)

[28] A. Cayley, Collected Mathematical Papers, Vol **3** (1889)

[29] E. Abrahams, P.W. Anderson, D.C. Licciardello, and T.V. Ramakrishnan, Phys. Rev. Lett. **42**, 673 (1979)

LOCALIZATION AND STRING THEORY

Shinobu Hikami

Department of Pure and Applied Sciences
University of Tokyo
Meguro-ku, Komaba 3-8-1, Tokyo, 153 Japan

1 Introduction

Anderson localization has been investigated by the renormalization group method [1-3]. The diffusion-diffusion interaction of a particle in a random potential becomes important in two-dimensions (2 D), and eventually, the localized state is realized in 2 D when the random potential has time reversal invariance and space-inversion symmetry (orthogonal case). For the strong spin-orbit case and for the magnetic impurity or the magnetic field case, the localization behavior belongs to a different universality class (symplectic or unitary case) [4]. These universal localization behaviors are described by an effective Hamiltonian, which predicts the long-distance behavior of a diffusion mode. The diffusion constant is proportional to the conductivity. The effective Hamiltonian becomes equivalent to the non-linear σ-model near 2 D from the point of view of the renormalization group analysis for the diffusion constant. The non-linear σ-model [5] has a particular symmetry, and the field variable is restricted to $O(N)/O(p) \times O(N-p)$ manifold with the replica limit $p = N \to 0$ due to the random impurity average. We denote this manifold by $O(0)/O(0) \times O(0)$, hereafter.

Recently, the relation between the non-linear σ-model and a string theory has attracted attention in a field theory. String theory has a long history in high energy physics and recently, superstring theory has been intensively developed [6]. The condition of the conformal invariance of the string theory is shown to be equivalent to the vanishing of the β-function, which is related to the derivative of the effective action by a metric [7,8]. This effective action should be consistent with a string scattering amplitude. The n-point string amplitude can be evaluated by an operator expansion and it can be expanded in powers of the coupling constant (string tension). Thus, the effective action can be constructed and the β-function can be derived from it in principle up to any desired order. We will see later that the four-loop β-function of Anderson localization is easily obtained by a string effective action. However, there remains a subtle point in this procedure. The renormalization group β-function depends on the subtraction scheme. We use the minimal subtraction scheme with a dimensional regularization. In this scheme, the critical exponents are directly related to the derivative of the β-function.

Quantum Coherence in Mesoscopic Systems
Edited by B. Kramer, Plenum Press, New York, 1991

In this article, we explain the relation between the β-function and the string theory, in particular with minimal subtraction scheme. We also present our recent study of five-loop order for Anderson localization.

Anderson localization is the problem of particle localization in a random potential. Recently, the localization problem due to dissipation, which has been studied by Caldeira and Leggett [9], is shown to be related to the open string theory [10]. Since the Anderson localization problem is formulated in a closed string theory, it is of interest to consider the crossover from Anderson localization to Caldeira-Leggett's localization in terms of a string theory. We briefly mention this problem.

2 β-Function of a Supersymmetric Non-Linear σ-Model

The non-linear σ-model is extended to the supersymmetric non-linear σ-model with an introduction of Fermion field ψ coupled to the bosonic field ϕ in a supersymmetric way. The usual non-linear σ-model has this supersymmetric generalization. For example the famous $O(N)/O(N-1)$ non-linear σ-model with N-component, has a following supersymmetric generalization

$$H = \frac{1}{t} \int d^2x \left[\frac{1}{2} (\partial_\mu \phi)^2 + \frac{1}{2} i \bar{\psi} \gamma_\mu \partial_\mu \psi + \frac{1}{8} (\bar{\psi}\psi)^2 \right], \tag{1}$$

with $\phi^2 = 1$ and $\phi^a \psi^a = 0$. More generally, the bosonic (non-supersymmetric) non-linear σ-model is written as

$$A = \frac{1}{2t} \int d^2x g_{ik}(\phi) \partial_\mu \phi^i \partial_\mu \phi^k, \tag{2}$$

and we extend this action for the supersymmetric case as

$$A = \frac{1}{2t} \int d^2x d^2\theta g_{ik}(\Phi)(\bar{D}\Phi^i)(D\Phi^k), \tag{3}$$

where

$$D = \frac{\partial}{\partial\bar{\theta}} + \bar{\theta}\partial_\mu, \qquad \bar{D} = \frac{\partial}{\partial\theta} + \theta\frac{\partial}{\partial\mu}$$

$$\Phi^j(x,\theta) \quad = \quad \phi^j(x) + \bar{\theta}\psi^j + \frac{1}{2}\bar{\theta}\theta F^j. \tag{4}$$

Supersymmetry means here the full space-time supersymmetry in 2 D. For the supersymmetric non-linear σ-model, the bosonic part cancels with the fermionic part and hence the divergence becomes less than in the bosonic non-linear σ-model. The β-function of the supersymmetric non-linear σ-model becomes simpler than in the bosonic case for the same reason.

We have the following supersymmetric β-function [11-13],

$$\beta_{ij}(t) = \epsilon t - R_{ij}t^2 - \frac{1}{2}\zeta(3)[a_3 - a_6]t^5 + O(t^6), \tag{5}$$

where $\epsilon = d - 2$ and R_{ij} is the Ricci tensor [14]. $\zeta(3)$ is a transcendental number and is equal to 1.202. In the expression for the β-function eq. (5), there are no two- and three-loop terms, which are order t^3 and t^4 terms, respectively. For the $O(N)/O(p) \times O(N-p)$ manifold, we have $R_{ij} = (N-2)g_{ij}$ and a_3, a_6 become polynominals of p and N. The

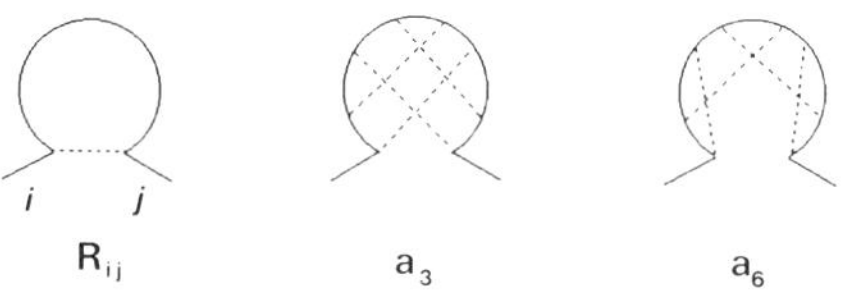

Figure 1.

terms R_{ij}, a_3 and a_6 are represented in Fig. 1. In Fig. 1, the external lines mean i and j of Riemann tensor R_{ikjl} of the present manifold. In the limit $p = N \to 0$, we have $a_3 = 24, a_6 = 0$, and the β-function becomes

$$\beta(t) = \epsilon t + 2t^2 - 12\zeta(3)t^5 + O(t^6). \tag{6}$$

We remark that this β-function coincides with the bosonic one, which has been evaluated by Wegner [15] for Anderson localization. The bosonic β-function has many terms and they become vanishingly small in the limit $p = N \to 0$ since they have factors of p and N. For Anderson localization, there are no two- and three-loop order terms [3]. The four-loop term has a transcendental factor $\zeta(3)$, and this $\zeta(3)$ does not appear in lower order of β-function.Therefore, the $\zeta(3)$ term is related to a new divergence and this term is shown to be common both for the supersymmetric case and for the bosonic cases.

The supersymmetric β-function of Kähler manifolds, which involve $O(N)/O(2) \times O(N-2), U(N)/U(p) \times U(N-p)$, and $O(N)/U(N)$ cases have a remarkable property. The β-function is vanishes except for the first-loop order term. In eq. (5), indeed $a_3 - a_6$ becomes zero for the Kähler manifold. The unitary non-linear σ-model $U(N)/U(p) \times U(N-p)$ has no $\zeta(3)$ factor in four-loop order [15]. The reason for this is that the manifold is Kähler. For the unitary case of Anderson localization, with magnetic impurities or with a magnetic field, the non-linear σ-model is described by $U(0)/U(0) \times U(0)$ manifold. In this case, the β-function does not correspond to the supersymmetric expression since it can be written by [5,11]

$$\beta(t) = \epsilon t - 2t^3 - 6t^5 + O(t^7). \tag{7}$$

For the unitary case, the supersymmetry is not realized. But it is Kähler, and hence the common factor $\zeta(3)$ term vanishes.

3   Five-Loop and Higher Order Terms

We consider the higher-loop terms of the β-function. The determination of the higher-loop order needs detailed studies. Here we discuss our recent studies about five-loop order [16].

In five-loop order, the supersymmetric non-linear σ-model β-function is expressed by

$$\beta(t) = \epsilon t - \sum_{n=1}^{\infty} \beta^{(n)} t^{n+1}, \tag{8}$$

$$\beta^{(5)}(t) = \lambda_1(A - B) + \lambda_2(E - F), \tag{9}$$

where A, B, E and F represent the five Riemann tensor products, which are given by Fig. 2. This form is consistent with a recent result by Gracey [17].

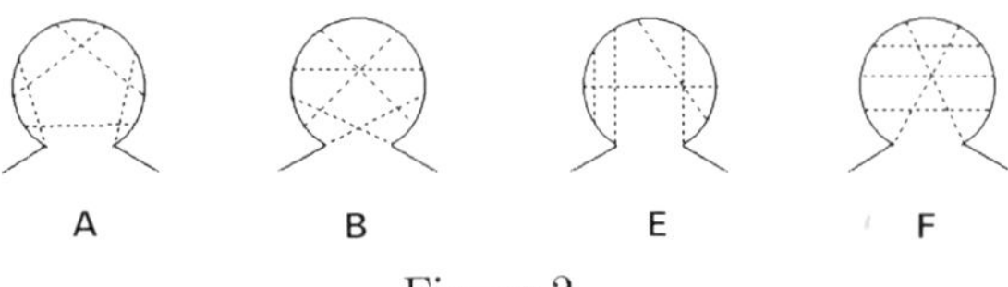

Figure 2.

There are equivalent classes among various terms of the five Riemann tensor products due to the following two identities for the symmetric space, which gives $\nabla_i R_{klmn} = 0$ by definition

$$R_{klmn} + R_{kmnl} + R_{knlm} = 0 \tag{10}$$

$$[\nabla_i, \nabla_j]R_{klmn} = R^{\rho}_{kij}R_{\rho lmn} + R^{\rho}_{lij}R_{k\rho mn} + R^{\rho}_{mij}R_{kl\rho n} + R^{\rho}_{nij}R_{klm\rho} = 0. \tag{11}$$

For the case of $O(0)/O(0) \times O(0)$, we have $A - B = 0, E = 0, F = 32$.

For the bosonic model, we have similarly [16]

$$\beta^{(5)}(t) = \lambda_1'(A - B) + \lambda_2'(E - F)$$

$$+ \lambda_3 C + \lambda_4 I + \lambda_5 R_1 + \lambda_6 R_6 + \lambda_7 R_8 + \lambda_8 R_9 + \lambda_9 R_{11}, \tag{12}$$

where C, I, R_1, R_6, R_8, R_9 and R_{11} consist of five Riemann tensor products, and they become zero for $O(0)/O(0) \times O(0)$ cases. λ_1' and λ_2' are believed to be same as λ_1 and λ_2 since they are common terms for the supersymmetric and for the bosonic case. Equation (12) is valid for the general manifold, we consider the famous $O(N)/O(N-1)$ model for eq. (12). In this case, the $1/N$ expansion has been studied up to order $1/N^2$ for the critical exponent [18-20]. From this result, we have

$$\lambda_9 = -\frac{2\zeta(3) - 1}{32}, \tag{13}$$

and

$$2\lambda_2' - \lambda_8 = \frac{35}{96} - 2\zeta(3) - \frac{27}{32}\zeta(4). \tag{14}$$

Thus a new transcendental number $\zeta(4)$ appears in five-loop order and it should be included in λ_2' since the R_9 term is related to the lower order term of β-function and the $E - F$ term is the term related to a new divergence. We are able to calculate the $\zeta(4)$ factor in five-loop order for orthogonal Anderson localization case by using $1/N$ results,

$$\beta(t) = \varepsilon t - 2t^2 - 12\zeta(3)t^5 + \frac{27}{2}\zeta(4)t^6 + O(t^7). \tag{15}$$

Also, the string scattering amplitude gives the coefficients of eq. (12). In eq. (12), the terms of R_i depend upon the subtraction scheme since they are expressed by the lower order β-functions. The β-function derived from the string effective action is also subtraction scheme dependent. However, there are subtraction independent terms, which should be evaluated for the determination of the β-function. The terms of $a_3 - d_6$ in eq. (2) and $(A-B), (E-F)$ in eq. (12) are subtraction independent and their coefficients are calculable from the string scattering amplitudes.

In the large N limit of the β-function of the $O(N)/O(N-1)$ manifold [21], transcendental numbers $\zeta(n)$ appear in $(n+1)$ loop order. This suggests that the β-function of the Anderson localization involves such $\zeta(n)$ terms in higher orders. It may be necessary to consider the large order behaviour of the perturbational series.

4   Discussion

It is of interest to investigate the structure of the β-function from the point of view of a string theory. The β-function is related to tree scattering amplitudes, which have no loop divergences. Recent string theory studies multi-loop divergences corresponding to the existence of genus. We can speculate that the non-perturbative term of the β-function for Anderson localization is somehow related to such a loop corrected β-function. Callan and Thorlacius [10] have pointed out that the open string vacua are related to the localization-delocalization transition due to the dissipation studied by Caldeira and Leggett [9]. For example, the problem of the extended state of the Landau level in the quantum Hall effect needs a treatment beyond the perturbative renormalization group analysis. We believe that this non-perturbative effect should be related to the boundary effect or a non-local interaction. Therefore, it is of interest to consider the situation of the crossover from the closed string problem to the open string problem This study is under way.

References

[1] E. Abrahams, P.W. Anderson, D.C. Licciardello, and T.V. Ramakrishnan, Phys. Rev. Lett. **42**, 673 (1979)

[2] F.J. Wegner, Z. Phys. **B 35**, 207 (1979)

[3] S. Hikami, Phys. Rev. **B 24**, 2671 (1981)

[4] S. Hikami, A.I. Larkin, and Y. Nagaoka, Prog. Theor. Phys. **63**, 707 (1980)

[5] E. Brézin, S. Hikami, and J. Zinn-Justin, Nucl. Phys. **B 165**, 528 (1980)

[6] Superstring Theory, Vol. **1**, M.B. Green, J.H. Schwarz, and E. Witten, eds., Cambridge University Press, Cambridge (1987)

[7] C.G. Callan, D. Friedan, E.J. Martinec, and M.J. Perry, Nucl. Phys. **B 262**, 593 (1985)

[8] R.R. Metsaev and A.A. Tseytlin, Phys. Rev. Lett. **B 185**, 52 (1987)

[9] A.O. Caldeira and A.J. Leggett, Ann. Phys. (N.Y.) **149**, 374 (1983); Phys. Rev. Lett. **46**, 211 (1981)

[10] C.G. Callan and L. Thorlacius, Nucl. Phys. **B 329**, 117 (1990)

[11] S. Hikami, Physica **A 167**, 149 (1990)

[12] D.J. Gross and E. Witten, Nucl. Phys. **B 277**, 1 (1986)

[13] M.T. Grisaru, A.E.M. Van de Ven, and D. Zanon, Nucl. Phys. **B 287**, 189 (1987)

[14] D. Friedan, Phys. Rev. Lett. **45**, 1957 (1980)

[15] F.J. Wegner, Nucl. Phys. **B 316**, 663 (1989)

[16] S. Hikami and A. Fujita, preprint

[17] J.A. Gracey, J. Phys. **A 23**, 2183 (1990)

[18] A.N. Vassil'ev, Yu. M. Pis'mak, and Yu. R. Honkonon, Theor. Math. Phys. **46**, 104 (1981)

[19] W. Bernreuther and F.J. Wegner, Phys. Rev. Lett. **57**, 1383 (1986)

[20] S. Hikami, Nucl. Phys. **B 215**, 555 (1983)

[21] S. Hikami and E. Brézin, J. Phys. **A 11**, 1141 (1978)

SCALING EXPONENTS AT A MOBILITY EDGE IN TWO DIMENSIONS

Spiros Evangelou

Physics Department, Division of Theoretical Physics
University of Ioannina
Ioannina 451 10, Greece

It is now widely recognized that electrons in very dirty metals localize [1] and an Anderson metal-insulator transition occurs in disordered lattice systems when the Fermi level crosses a mobility edge which separates extended from localized states in the energy spectra. Such viewpoint of localization relies on a simplified non-interacting electron picture. In this context a one- parameter scaling theory [2] offers the theoretical framework to understand the results for the associated critical behavior within a few universality classes. In fact, three universality classes are distinguished [3], depending on symmetry: the orthogonal in the case of a random potential, the unitary when a magnetic field is added and the symplectic when spin-orbit coupling is also present. A firm prediction exists for the presence of a phase transition for the orthogonal universality class only in three or higher dimensions. Two-dimensional (2 D) disordered systems, however, are believed to display mobility edges and extended states when spin-orbit coupling or a strong magnetic field are present, that is for the symplectic and unitary universality classes, respectively. Our theme is precisely the study of a mobility edge in 2 D and the details of the associated critical behaviour which are largely open questions. 2 D is also an especially promising area to study conformal invariance [4]. Emphasis is placed on the following aspects: Firstly, the evaluation of the localization length exponent ν and the correlation exponent η by scaling the dominant Lyapunov exponent near the transition. Secondly, a complete numerical analysis of the spectral distributions for all the Lyapunov exponents in connection with the predictions of weak disorder theories. Finally, a short discussion follows on a statistical description of the fluctuations and correlation of wavefunction amplitudes and the energy levels at the mobility edge.

Wegner [5] has recently shown that a fixed point exists in the scaling beta function for the symplectic ensemble (SE) in $d = 2$ from appropriate results in $d = 2 + \varepsilon$ dimensions. It takes the form (a is a constant and $\zeta(3) \approx 1.202$)

$$\beta_{SE}(g) = \varepsilon + a \cdot \frac{1}{g} - 3\frac{\zeta(3)a^4}{4} \cdot \frac{1}{g^4} + 0\left(\frac{1}{g^5}\right). \tag{1}$$

In $d = 2(\varepsilon = 0)$ for large conductance g the term of $O(1/g)$ is positive and dominates while for spinless electrons the corresponding term is negative leading to localized states

Quantum Coherence in Mesoscopic Systems
Edited by B. Kramer, Plenum Press, New York, 1991

for a sufficiently large system. In the symplectic case the conductance g does not decrease but increases instead logarithmically with a linear system size L. If the negative four-loop term is included (eq. (1)) the beta function bends over and permits an Anderson transition to appear even in $d = 2$. The existence of this mobility edge was also numerically established [6-8]. Constructive interference and dephasing effects which lead to an increase of the conductivity are experimentally observed in metallic films (e.g. of Mg or Al with a few percent of Au or Pd atoms which add spin-orbit scattering) [9].

We have considered a general spin-1/2 Schrödinger equation with a random potential and spin-orbit coupling at each lattice site in 2 D using an orthogonalized site basis representation. The corresponding statistical ensemble consists of symplectic matrices, it is spin-dependent and preserves the time- reversal invariance. The equation written for the 2 D strip of length L and width M takes the form

$$\mathbf{V}_{lm,l-1,m}\Psi_{l-1,m} + \sum_{m'=m\pm 1}\mathbf{V}_{lm,lm'}\Psi_{lm'} + \mathbf{V}_{lm,l+1,m}\Psi_{l+1,m} = (E - \varepsilon_{lm})\Psi_{lm}, \qquad (2)$$

where a site is denoted by the pair (lm), with $l = 1, 2, \ldots, L$, $m = 1, 2, 3, \ldots, M$ and $\Psi_{lm} = (\Psi_{lm}^{+}, \Psi_{lm}^{-})$ is a vector in spinor space associated with every site. The diagonal elements ε_{lm} are independently distributed random variables with density

$$P(\varepsilon_{lm}) = 1/W, \qquad \varepsilon_{lm} \in [-W/2, W/2]. \qquad (3)$$

The hopping matrix elements in spinor space are represented by 2×2 complex matrices which connect every pair of nearest- neighbor sites that is

$$\mathbf{V}_{lm,l'm'} = \begin{pmatrix} 1 + i\mu t^z & -t^y + i\mu t^x \\ t^y + i\mu t^x & 1 - i\mu t^z \end{pmatrix}_{lm,l'm'} \qquad (4)$$

where by time reversal-invariance the t^x, t^y, t^z associated to every lattice bond are real and independent random variables chosen from a uniform probability distribution defined on the interval $[-1/2, 1/2]$. The strength of the random spin-orbit coupling is given by μ. The parameter W describes the disorder strength and the metal-insulator transition occurs at $W_c \approx 7.0$ [7] when $\mu = 1$ in $d = 2$. For $W > W_c$ all states are localized and the conductivity is zero, while for $W < W_c$ mobility edges appear in the band separating localized states near the edges from extended states near the band center.

The most successful method [10,11] for computations in finite samples utilizes the one-dimensional (1 D) transfer matrix formulation of the problem. The linear lattice size L is taken exceedingly large (by orders of magnitude) in comparison with the size M in the perpendicular direction. Therefore, one considers a very long strip in 2 D. The disordered material is further connected to two ideal leads on its left and right so that current properties over the entire sample can be studied. In this representation the full transmission properties are expressed in terms of the eigenvalues of certain transfer matrix products [10] and the conductance g and the transport coefficient t can be directly evaluated in terms of these eigenvalues. Scaling is made possible by varying the perpendicular system size M. The method is outlined in [11] and is applied in the rest of the paper to a mobility edge in $d = 2$.

The Lyapunov exponents for the random strips of very long length L and finite width M define a hierarchy of length scales ξ_j, $j = 1, 2, \ldots, M$ can be computed from the inverses of the positive exponents $\gamma_1 \leq \gamma_2 \leq \gamma_3 \leq \ldots \leq \gamma_M$. In fact, because all states should be always localized in the adopted very long length strip ($L \to \infty$, i.e. the

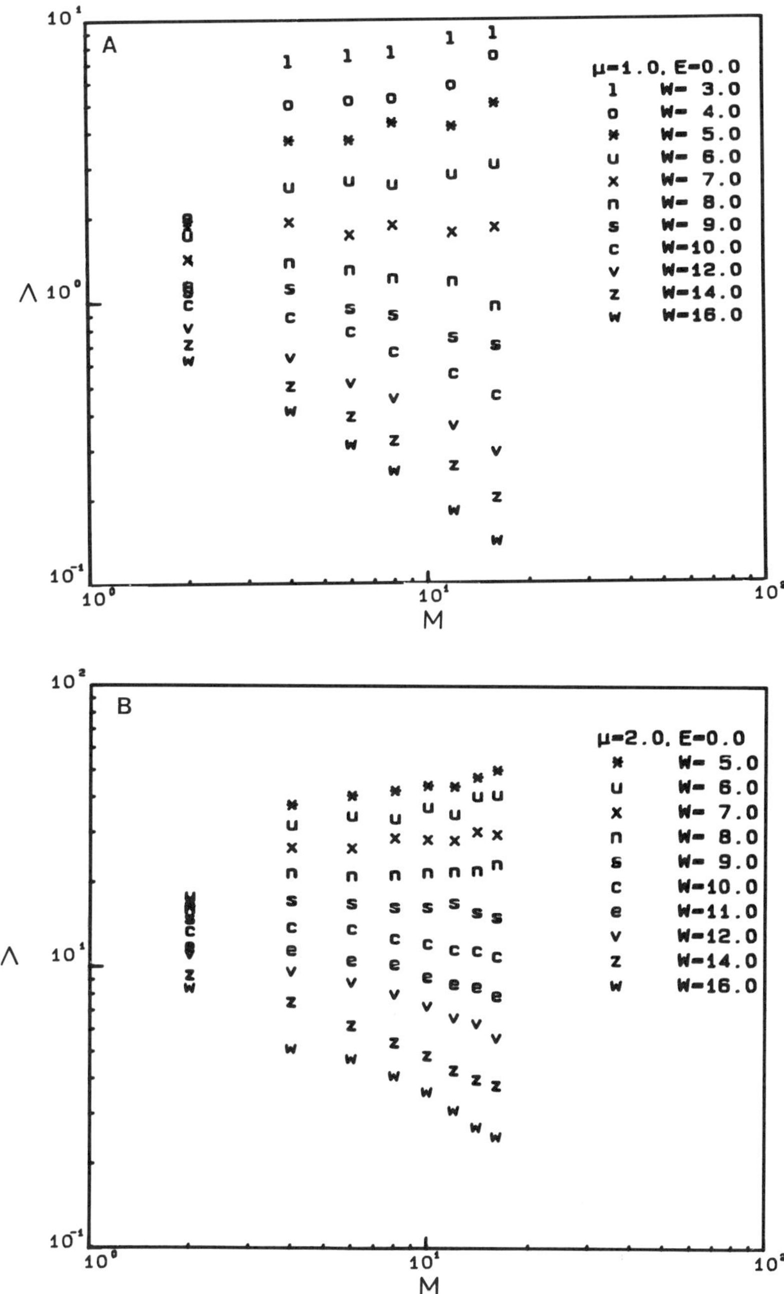

Figure 1. The renormalized localization length Λ versus the circumference of the cylinder M in a log-log plot at the centre of the band ($E = 0.0$) for $\mu = 1.0, 2.0$.

quasi-1 D limit) the γ_j's are always non-zero, different and come in opposite pairs. The smallest positive exponent, γ_1, gives the dominant transmission channel. In the limit of the transverse size M approaching infinity ($M \to \infty$) this defines the inverse localization length as $\xi = \lim_{M \to \infty} \gamma_1^{-1}$. For extended states $\gamma_1 \to 0$ and the localization length ξ is infinite while for localized states $\gamma_1 \to 1/\xi \neq 0$ and ξ is finite. In the one-parameter scaling theory other scales apart from γ_1 do not enter.

Firstly we focus on the largest length scale $\gamma_1^{-1} = \xi_M$ varying the width M following the technique of finite-size scaling [10,11]. The results of the calculations which were all done at $E = 0$, are shown in Fig. 1 for $\mu = 1$ and 2 and various values for W. In practice our calculations are restricted up to $M = 16$ and a relative statistical accuracy of a few percent was obtained in most cases by taking L up to 10^5. From these collected raw data it can be observed that a transition exists in 2 D. The re-scaled length $\Lambda = \xi_M/M$ obeys a scaling relationship of the form

$$\Lambda = \xi_M/M = f(M/\xi), \tag{5}$$

where ξ is the localization length of the infinite 2 D system. For a given M the length Λ is a function of W, E and μ. For low Λ the scaling behaviour reduces to $\Lambda = \xi/M$; in general the data approximately fit to a scaling function f and length ξ such that $\Lambda = f(\xi/M)$ that is, the data for different parameters reduce to a single scaling curve f versus M/ξ which is displayed in Fig. 2. The function f has now two branches [11]. From eq. (5) we are able to calculate ξ as a function of disorder and μ. The values of ξ for many Ws are also plotted so that the exponent ν can be deduced from the relation $\xi \approx |W - W_c|^{-\nu}$. For $\mu = 1$ the value $\nu \approx 1.60$ is approximately obtained and the value $\nu \approx 1.48$ for $\mu = 2$. The results of Fig. 2 support the validity of the one-parameter scaling theory. Our values for ν are smaller than the estimates of [2] who found $\nu \approx 2$. From [5] the obtained values for ν to four-loop order are such that the inequality $\nu \geq 2/d$ is violated when it should generally hold for disordered systems [13]. We have also estimated the power-law of the correlations to be $\eta = (1/\pi) \cdot (1/\Lambda^*)$ where Λ^* is the fixed- point value of Λ by assuming conformal invariance of correlations [4,14] at the transition. We evaluated an estimate for the correlation exponent $\eta \approx 0.17$ and we find that this remains approximately the same for different values of μ. From the data of [7] who used a different model we extracted a value very close to our η. Unfortunately, there are no reliable analytical estimates for ν and η to compare our results. Our results favour of the one-parameter scaling theory [2]. We have also carried out detailed calculations of all the γ_j's and considered their scaling properties when M varies. For a fixed M the Lyapunov exponents γ_j are distributed with a more or less uniform density in the extended regime [15]. The spectrum of the γ_j's in the localized regime shows a gap at its centre, that is between positive and negative eigenvalues, which equals $2/\xi$ when $M \to \infty$. The results for the distribution are shown in Fig. 3. The one-parameter scaling theory can be again approximately recovered from the data of Fig. 3 as shown in [15].

In conclusion we have shown that the one-parameter scaling theory remains, at least approximately, valid for the SE in $d = 2$ and the exponents ν and η are estimated. The distribution of the Lyapunov exponents is regular and not incompatible with the one- parameter scaling theory. Our numerical data were contrasted versus already known perturbational and other field-theoretic predictions [5]. By analogy with the Wigner-Dyson statistics we expect for the symplectic universality class that the universal symmetry index $\beta = 4$ will characterize the spectral fluctuations in the metallic phase [16] as well as the magnitude of the universal mesoscopic conductance fluctuations.

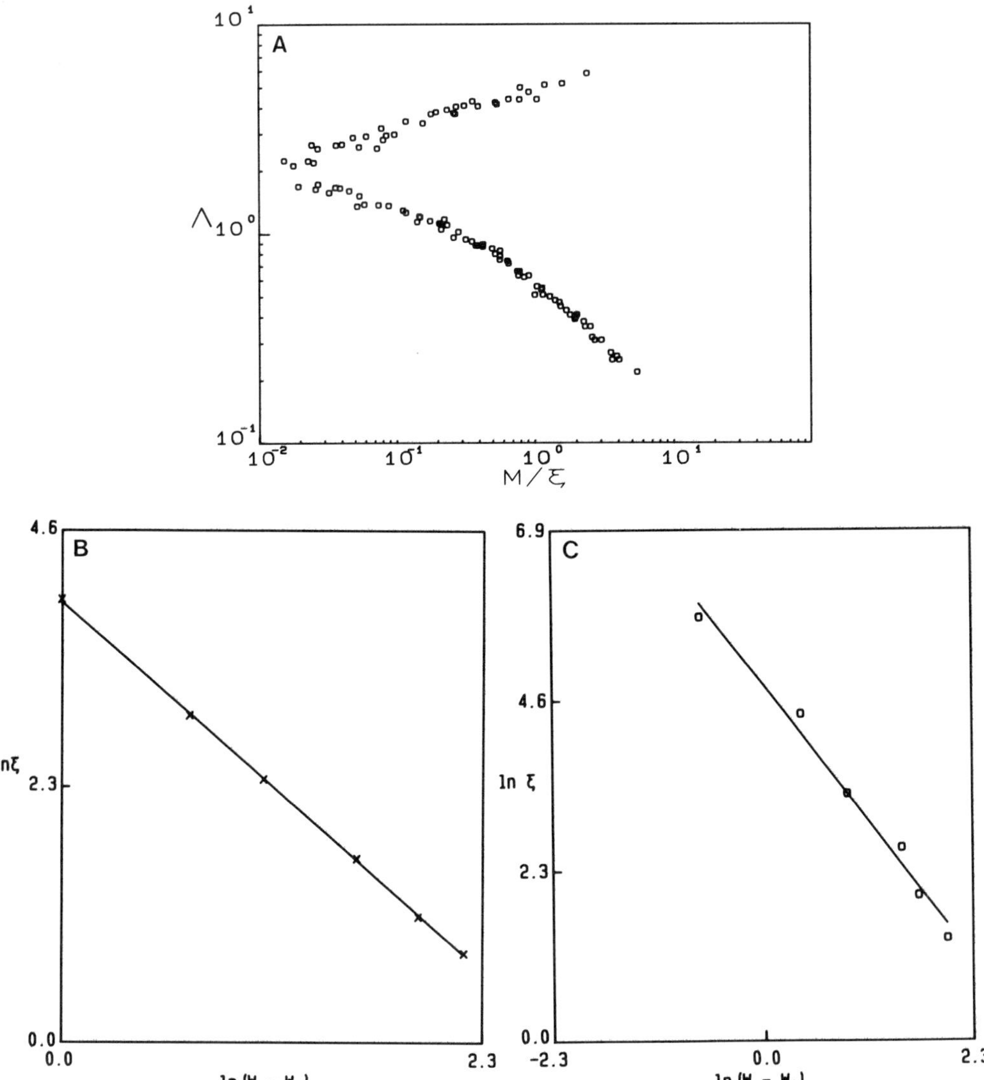

Figure 2. (a) The scaling function $\Lambda = f(\xi/M)$ for the SE in $d = 2$. In (b) ξ is plotted versus $|W - W_c|$ in a double-log plot and the exponent ν is extracted when $\mu \approx 1$, $W_c = 7.0$ and (c) when $\mu \approx 2, W_c = 8.5$.

The characterization of the fluctuations and correlations at the critical point is in many respects an open question, see however [16-18].

Acknowledgments

This work was supported in part by a Π.EN.E.Δ. Research Grant of the Greek Secretariat of Science and Technology and from an EEC grant for problems in Statistical Mechanics.

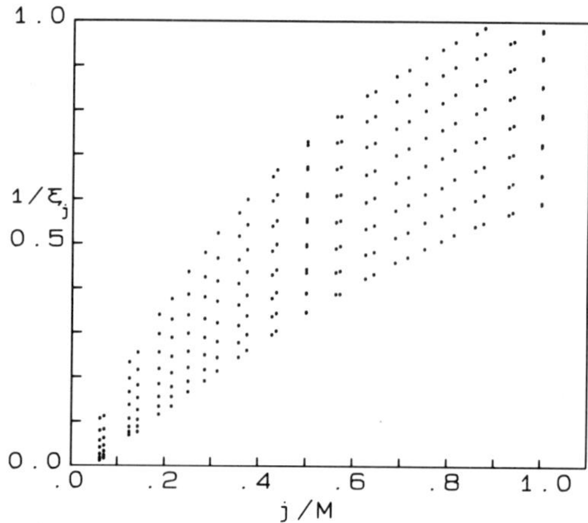

Figure 3. The distribution of the calculated exponents is shown for $M = 14, 16$ and $\mu = 2$.

References

[1] P.W. Anderson, Phys. Rev. **109**, 1492 (1958)

[2] E. Abrahams, P.W. Anderson, L.C. Licciardello, and T.V. Ramakrishnan, Phys. Rev. Lett. **42**, 673 (1979)

[3] S. Hikami, A.I. Larkin, and Y. Nagaoka, Prog. Theor. Phys. **63**, 707 (1980)

[4] J.L. Cardy, J. Phys. A: Math. Gen **17**, L385 (1984)

[5] F. Wegner, Phys. Rep. **67**, 15 (1980); ibid Nucl. Phys. **B 270**, 1 (1980). A negative correction for the conductivity exponent s is found in Nucl. Phys **B 316**, 663 (1989), i.e. $s = 1 - 9\zeta(3)\varepsilon^3/4 + O(\varepsilon^4)$ which gives $\nu = s/\varepsilon$ violating the inequality $\nu \geq 2/d$ (see [13])

[6] S.N. Evangelou and T.A.L. Ziman, J. Phys. C: Sol. St. Phys. **20**, L235 (1987)

[7] T. Ando, Phys. Rev. **B 40**, 5325 (1989)

[8] A.MacKinnon, in: Localization Interaction and Transport Phenomena, Springer Ser. in Sol. St. Sc. **61**, 90 (1985)

[9] G. Bergmann, Phys. Rep. **107**, 1 (1984)

[10] J.-L. Pichard and G.J. Sarma, J. Phys.: Sol. St. Phys. **14** L127, 617 (1981)

[11] A.MacKinnon and B. Kramer, Z. Phys. **B 53**, 1 (1983)

[12] J.-L. Pichard, J .Phys.: Sol. St. Phys. **14** L127, 617 (1981)

[13] J.T. Chayes, L.Chayes, D.S. Fisher, and T. Spencer, Phys. Rev. Lett. **57**, 2999 (1986)

[14] J.-L. Pichard and G. J. Sarma , J. Phys.: Sol. St. Phys. **18**, 3457 (1985)

[15] J.-L. Pichard and G. J. Andre , Europhys. Lett. **2**, 477 (1986)

[16] S.N. Evangelou, Phys. Rev. **B 39**, 12895 (1989)

[17] J.T. Chalker and G.J. Daniel, S.N. Evangelou, and B. Nahm (to be published)

[18] S.N. Evangelou (J. Phys. **A**, to appear)

ONE-PARAMETER SCALING OF THE LOCALIZATION LENGTH IN HIGH MAGNETIC FIELDS

Bodo Huckestein

Physikalisch-Technische Bundesanstalt
Bundesallee 100, W-3300 Braunschweig, F. R. Germany

1 Introduction

A disordered two-dimensional system in the presence of a strong perpendicular magnetic field is of considerable theoretical interest. Most theoretical tools fail in dealing with this situation. The only quantity accessible to exact calculations is the density of states which consists of a series of disorder broadened Landau bands [1]. For other physical quantities analytical treatments have to rely on resummations of diverging series [2] or lead to models that seem to be too complicated to solve [3]. Only in the limit of a long-range correlated random potential are semi-classical arguments applicable and analytical results can be obtained [4, 5, 6]. In such a situation numerical calculations might help to understand the properties of the system under consideration. Besides the density of states the most easily calculated quantity is the localization length ξ. If ξ is finite then the DC conductivity $\sigma_{ii}(\omega = 0)$ vanishes and the system is an insulator. Furthermore, the localization length acts as a scaling parameter that governs the critical behaviour of physical quantities near the center of the Landau bands [7]. This behaviour can be obtained by investigating the exponential decay length in a finite system as a function of the size of the system. It can further be argued that at finite temperatures this system size dependence is replaced by a dependence on a temperature dependent phase coherence length. As a consequence the critical behaviour of the localization length shows up in the temperature dependence of physical quantities, like the conductivity, when the Fermi energy lies near the center of a Landau band.

In the following section the numerical methods used to obtain the critical behaviour are outlined. In section 3 the numerical results and in section 4 the connection to the experimental observations are discussed.

Quantum Coherence in Mesoscopic Systems
Edited by B. Kramer, Plenum Press, New York, 1991

2   Numerical Methods

The localization length can be defined by the asymptotic exponential decay of the modulus of the one-particle Green's function $G(E; \mathbf{r}, \mathbf{r}')$

$$\xi^{-1}(E) = - \lim_{|\mathbf{r}-\mathbf{r}'|\to\infty} \frac{1}{2|\mathbf{r}-\mathbf{r}'|} \ln |G(E; \mathbf{r}, \mathbf{r}')|. \tag{1}$$

This definition holds for an infinite two-dimensional system. In order to introduce a finite length into the problem the system treated numerically has periodic boundary conditions in the y-direction with a periodicity length M. The Hamiltonian H for a particle in the presence of a vector potential $\mathbf{A}$ and a random scalar potential $V(\mathbf{r})$

$$H = \frac{1}{2m}(\mathbf{p} - e\mathbf{A})^2 + V(\mathbf{r}) \tag{2}$$

transforms for these boundary conditions and a Landau gauge of the vector potential, $\mathbf{A} = Bx\mathbf{e}_y$, into

$$H = \sum_{nk}\sum_{n'k'} |nk\rangle\langle nk|H|n'k'\rangle\langle n'k'|, \tag{3}$$

$$\langle nk|H|n'k'\rangle = (n + 1/2)\,\hbar\omega_c\,\delta_{n,n'}\delta_{k,k'} + \langle nk|V|n'k'\rangle. \tag{4}$$

The Landau states $|nk\rangle$ are given by

$$\langle \mathbf{r}|nk\rangle = \frac{1}{\sqrt{Ml_c}} e^{-iky}\chi_n\left(\frac{x - kl_c^2}{l_c}\right), \tag{5}$$

where

$$\chi_n(x) = \left(2^n n! \sqrt{\pi}\right)^{-1/2} H_n(x)e^{-x^2/2} \tag{6}$$

is the eigenfunction of a linear harmonic oscillator, $H_n(x)$ is Hermite's polynomial, $l_c = (\hbar/eB)^{1/2}$ is the magnetic length and $\omega_c = eB/m$ the cyclotron frequency. Due to the periodicity in y-direction k is a discrete quantum number and takes on values of integer multiples of $2\pi/M$. In the following it will be assumed that the magnetic field is sufficiently high, such that $|\langle nk|V|n'k'\rangle| \ll \hbar\omega_c$, and the projection of the Hamiltonian onto the lowest Landau level can be used. The Landau level index $n = 0$ will be dropped from the further equations and the zero of energy will be shifted to $\hbar\omega_c/2$.

For the system with a finite periodicity length M an exponential decay length $\lambda_M(E)$ along the x-direction can be defined by introducing a width-averaged Green's function

$$G^2(E; x, x') = \frac{1}{M^2} \int_0^M dy \int_0^M dy' |G(E; \mathbf{r}, \mathbf{r}')|^2, \tag{7}$$

$$\lambda_M^{-1}(E) = - \lim_{|x-x'|\to\infty} \frac{1}{|x - x'|} \ln G^2(E; x, x'). \tag{8}$$

In the limit $M \to \infty$, $\lambda_M(E)$ tends to the localization length of the infinite system $\xi(E)$. It is most conveniently calculated from the Green's function in Landau state representation $G(E; k, k') = \langle k|(E - H)^{-1}|k'\rangle$. Rewriting eq. (7) in terms of $G(E; k, k')$

$$G^2(E; x, x') = \frac{1}{M^2} \int_0^M dy \int_0^M dy' \sum_{k_1,\ldots,k_4} \langle \mathbf{r}|k_1\rangle \langle k_1|(E - H)^{-1}|k_2\rangle \langle k_2|\mathbf{r}'\rangle$$
$$\langle \mathbf{r}'|k_3\rangle \langle k_3|(E - H)^{-1}|k_4\rangle \langle k_4|\mathbf{r}\rangle, \tag{9}$$

$$\propto \sum_{k_1,k_2} \chi^2 \left(\frac{x - k_1 l_c^2}{l_c}\right) \chi^2 \left(\frac{x' - k_2 l_c^2}{l_c}\right) |G(E; k_1, k_2)|^2, \tag{10}$$

it follows that

$$\lambda_M^{-1}(E) = -\lim_{|k-k'|\to\infty} \frac{1}{2|k - k'| l_c^2} \ln |G(E; k, k')|. \tag{11}$$

The Green's function $G(E; k, k')$ can be calculated using a generalization of the recursive method of MacKinnon [8]. Consider the Hamiltonian $H^{(K)}$ of eq. (3) in the lowest Landau level taking into account the states $k = \tilde{k} 2\pi/M$, $\tilde{k} = 1, \ldots, \tilde{K}$ and its decomposition

$$H^{(K)} = \sum_{\tilde{k},\tilde{k}'=1}^{\tilde{K}} |k\rangle V_{k,k'} \langle k'| \tag{12}$$

$$= H_0 + H', \tag{13}$$

$$H_0 = H^{(K-1)} + |K\rangle V_{K,K} \langle K|, \tag{14}$$

$$H' = \sum_{\tilde{k}=1}^{\tilde{K}-1} (|k\rangle V_{k,K} \langle K| + \text{h.c.}), \tag{15}$$

$V_{k,k'} = \langle k|V|k'\rangle$. The related Green's functions are $G^{(K)} = (E - H^{(K)})^{-1}$ and $G_0 = (E - H_0)^{-1}$. Expressing $G^{(K)}$ in terms of H_0 and H' yields the recurrence relations

$$G^{(K)} = G_0 + G_0 H' G^{(K)}, \tag{16}$$

$$G_0 = G^{(K-1)} + \frac{|K\rangle\langle K|}{E - V_{K,K}}. \tag{17}$$

The disorder potential enters these equations via the matrix elements $V_{k,k'}$. They can be calculated from an assumed real space potential $V(\mathbf{r})$ by using

$$V_{k,k'} = \int d^2r \langle k|\mathbf{r}\rangle V(\mathbf{r})\langle \mathbf{r}|k'\rangle \tag{18}$$

as has been done in previous calculations [9, 10]. However, this procedure is due to the double integral involved very computer time consuming. A much more efficient approach is to generate random matrix elements $V_{k,k'}$ directly in such a way as to model a certain behaviour of the real space potential [11]. The precise nature of the disorder potential in a real sample is unknown and the critical behaviour of the theory should not depend on such details. Therefore we specify only the two lowest order correlation functions of the disorder potential, e.g.

$$\overline{V(\mathbf{r})} = 0, \tag{19}$$

$$\overline{V(\mathbf{r})V(\mathbf{r} + \mathbf{d})} = V^2 \delta(\mathbf{r} - \mathbf{r}'), \tag{20}$$

i.e. a white noise real space potential. From these equations the two lowest order correlation functions of the matrix elements of the Hamiltonian eq. (3) can be calculated

$$\overline{V_{k,k'}} = 0, \tag{21}$$

$$\overline{V_{k,k+q}V_{k'+q',k'}} = \frac{V^2}{\sqrt{2\pi}l_c M}\delta_{q,q'}\exp\left(-\frac{q^2 l_c^2}{2} - \frac{(k-k')^2 l_c^2}{2}\right). \tag{22}$$

Equation (22) shows that the $V_{k,k'}$ form a band-diagonal matrix with correlations along the diagonals. $|V_{k,k+q}|^2$ falls off on the average as $\exp(-q^2 l_c^2/2)$ so that the matrix elements $V_{k,k+q}$ are essentially zero for $ql_c \gg 1$. This means that of the order of $\tilde{q} = qM/2\pi = M/2\pi l_c$ diagonals of the matrix contain non-zero elements.

Gaussian correlations among random numbers as given by eq. (22) can be generated from uncorrelated random numbers by summing them with gaussian weights. The following matrix elements obey the correlation function given by eq. (22)

$$V_{k,k+q} = \exp\left(-q^2 l_c^2/4\right)\sum_{j=-\infty}^{\infty} U_{2k+j2\pi/M,q}\exp\left(-j^2(2\pi l_c)^2/4M^2\right), \tag{23}$$

$$\overline{U_{k,k'}} = 0, \tag{24}$$

$$\overline{U_{k,q}U_{k',q'}} = \frac{V^2}{\sqrt{2\pi}l_c MA}\delta_{k,k'}\delta_{q,q'}, \tag{25}$$

$$A = \sum_{j=-\infty}^{\infty}\exp\left(-j^2(2\pi l_c)^2/2M^2\right). \tag{26}$$

$U_{x,q}$ is related to the Fourier transform of $V(\mathbf{r})$ with respect to y [11]. These matrix elements have thus a correlation function corresponding to a δ-correlated real space potential. Their calculation involves a single sum that can be truncated due to the gaussian factor in eq. (23). In the actual calculation the summation was truncated for $|j| > M/\sqrt{2\pi}l_c$. The storage requirements are of the order of $4M/2\pi l_c^2$ matrix elements even in the limit $k \to \infty$.

Using the matrix elements given by eq. (23) and the recurrence relations eqs. (16) and (17) the exponential decay length $\lambda_M(E)$ can be calculated according to eq. (11) as a function of system width M and Fermi energy E. In order to obtain the localization length $\xi(E)$ these data have to be extrapolated for $M \to \infty$. This was done by assuming the existence of a one-parameter scaling of the quantity $\lambda_M(E)/M$ near the center of the Landau band, i.e.

$$\Lambda \equiv \frac{\lambda_M(E)}{M} = f\left(\frac{M}{\xi(E)}\right), \tag{27}$$

where $\xi(E)$ is a scaling parameter that diverges at a critical energy E_c according to

$$\xi(E) = \xi_0|E - E_c|^{-\nu} \tag{28}$$

and that can be identified with the localization length. The motivation for this assumption stems from the observation that at the band center the localization length seems indeed to diverge. In the study of disordered Anderson models such scaling behaviour was observed near a mobility edge [12]. However, the existence of a one-parameter scaling in the Quantum Hall regime has been questioned [3, 9, 10]. In the present approach a statistical test procedure was employed that allows to check the validity of the assumption a posteriori [13]. In this way it is possible to show that the numerical data are compatible with the assumed one-parameter scaling law. Thus, the extrapolation to infinite system size can be performed on a justifiable basis.

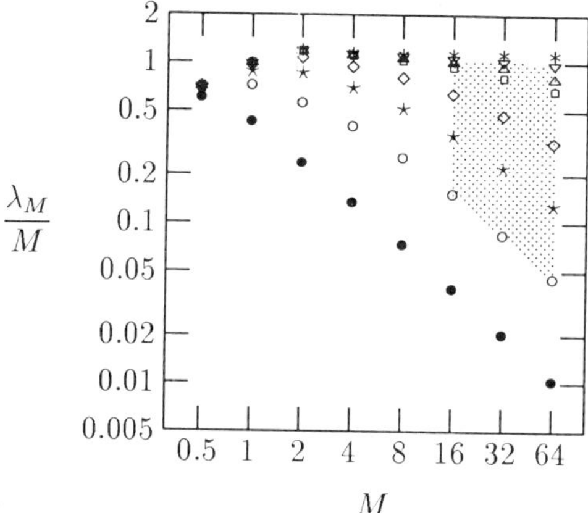

Figure 1. The renormalized exponential decay length λ_M/M as a function of system width M (in units of $\sqrt{2\pi}\,l_c$) for energies $E = 0.01$ (*), 0.05 ($\triangledown$), 0.07 ($\triangle$), 0.1 ($\square$), 0.18 ($\diamond$), 0.3 ($\star$), 0.5 ($\circ$) and 1 ($\bullet$) (in units of Γ). Data in the shaded area show scaling behaviour (Fig. 2).

3 Numerical Results

The numerical calculations were performed for 8 different energies E between 0.01 and 1 in units of the bandwidth Γ in the self-consistent Born approximation [14], $\Gamma^2 = 8V^2/2\pi l_c^2$. The periodicity length M was varied between $1/2$ and 64 in units of $\sqrt{2\pi}\,l_c$. The difference $|k - k'|l_c^2$ in eq. (11) was taken to be $10^5\sqrt{2\pi}\,l_c$ which is a macroscopic distance of about $2\,\text{mm}$ in a magnetic field of $10\,\text{T}$. Figure 1 shows the results of the calculations. For $E = 0.01\,\Gamma$ the normalized exponential decay length is constant as a function of M. This means that the localization length $\xi(E)$ which is the limiting value of λ_M as $M \to \infty$ diverges in the band center. For $E = \Gamma$ the exponential decay length converges rapidly to a finite value for $\xi(E)$. For intermediate values of the energy λ_M grows but not as fast as M.

The data shown in fig. (1) can be interpreted in terms of a single parameter scaling of $\lambda_M(E)/M$. It will not be possible to write all data as a function of $M/\xi(E)$ since for $M \to 0$ the renormalized exponential decay length $\lambda_M(E)/M$ tends to $2/\pi$ independent of energy [10]. However, one would expect scaling behaviour only as the asymptotic behaviour for $M \to \infty$. Furthermore, critical behaviour can only be expected near the phase transition so that data for large E have to be discarded in obtaining the scaling behaviour from the raw data. A statistical test shows that for $16 \leq M \leq 64$ and $0.05 \leq E \leq 0.5$ (shaded region in fig. 1) the data are compatible within the error bars with the scaling behaviour described by eqs. (27) and (28). Figure 2 shows the resulting scaling function $\Lambda(M/\xi(E))$. The critical exponent of the localization length is found to be $\nu = 2.34 \pm 0.04$. The critical energy E_c was taken to be zero in the fit and can be at most about $1\,\%$ of the band width Γ.

The obtained value for the critical exponent is in excellent agreement with the analytical result of $7/3$ obtained by Mil'nikov and Sokolov in the approximation of slowly varying random potential [6]. Since the present result was obtained in the opposite limit of δ-correlated potential it seems likely that the critical exponent does not depend on the

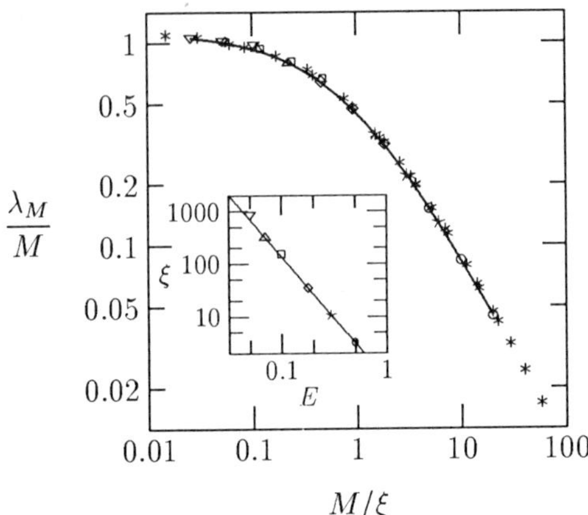

Figure 2. The renormalized exponential decay length λ_M/M as a function of the scaling variable $M/\xi(E)$. The data plotted are taken from the shaded area of Fig. 1. The solid curve approximates the scaling function Eq. (27). For comparison data from the network model [15] are shown ($*$). The inset shows the localization length $\xi(E)$.

correlation length of the random potential. Furthermore, the data obtained by Chalker and Coddington for the network model fit to the same scaling curve as is shown in fig. 2 [15].

4 Relation to Experiment

The scaling behaviour of the localization length is a consequence of the phase coherence of the wave function. At finite temperature the phase coherence is broken by inelastic processes. The wave function will then show phase coherence only over a finite range. The range is described by a phase coherence length L_{in} that is a function of temperature $L_{\text{in}} \propto T^{-p/2}$. In a first approximation this length replaces M as an effective system size. The scaling function eq. (27) then becomes a function of temperature $\Lambda = f^*(T^{-\kappa}|E - E_c|)$ with $\kappa = p/2\nu$. This peculiar temperature dependence shows up in the temperature dependence of the derivatives of Λ with respect to the Fermi energy $\mathrm{d}^n\Lambda/\mathrm{d}E^n \propto T^{-n\kappa}$ at $E = E_c$. This temperature behaviour is precisely the temperature behaviour observed by Wei et al. in measurements of ρ_{xx} and ρ_{xy} in heterostructures [16]. If these quantities depend on the same variable as Λ then the measured value for $\kappa = 0.42 \pm 0.04$ allows to extract a value for the exponent of the phase coherence length L_{in} of $p/2 = 1.0 \pm 0.1$. This does agree with the usual result of Fermi liquid theory but disagrees with $p = 1$ for dirty metals in two dimensions in the absence of a magnetic field [17].

5 Summary

We have presented a method to generate a random matrix representing a disordered twodimensional system under the influence of a high perpendicular magnetic field. This allows to calculate the exponential decay length of the modulus of the one-particle

Green's function on a strip for sufficiently large system widths to observe a single parameter scaling behaviour of the renormalized decay length. The localization length of the infinite system appears as a scaling parameter that diverges in the band center with a critical exponent $\nu = 2.34$. This scaling behaviour is consistent with measurements of the temperature dependent Quantum Hall effect.

References

[1] F. J. Wegner, Z. Phys. B **51**, 279 (1983)

[2] S. Hikami, Prog. Theor. Phys. **76**, 1210 (1986)

[3] H. Levine, S. B. Libby, and A. M. M. Pruisken, Phys. Rev. Lett. **51**, 1915 (1983)

[4] M. Tsukada, J. Phys. Soc. Jpn. **41**, 1466 (1976)

[5] S. A. Trugman, Phys. Rev. B **27**, 7539 (1983)

[6] G. V. Mil'nikov and I. M. Sokolov, JETP Lett. **48**, 536 (1988)

[7] B. Huckestein and B. Kramer, Phys. Rev. Lett. **64**, 1437 (1990)

[8] A. MacKinnon, J. Phys. C **13**, L1031 (1980)

[9] H. Aoki and T. Ando, Phys. Rev. Lett. **54**, 831 (1985)

[10] T. Ando and H. Aoki, J. Phys. Soc. Jpn. **54**, 2238 (1985)

[11] B. Huckestein and B. Kramer, Solid State Comm. **71**, 445 (1989)

[12] A. MacKinnon and B. Kramer, Z. Phys. B **53**, 1 (1983); B. Kramer, K. Broderix, A. MacKinnon, and M. Schreiber, Physica A **167**, 163 (1990)

[13] B. Huckestein, Physica A **167**, 175 (1990)

[14] T. Ando and Y. Uemura, J. Phys. Soc. Jpn. **36**, 959 (1974)

[15] J. T. Chalker and P. D. Coddington, J. Phys. C **21**, 2665 (1988)

[16] H. P. Wei, D. C. Tsui, M. Paalanen, and A. M. M. Pruisken, Phys. Rev. Lett. **61**, 1294 (1988); H. P. Wei, S. W. Hwang, D. C. Tsui, and A. M. M. Pruisken, Surf. Sci. **229**, 34 (1990)

[17] B. L. Altshuler, A. G. Aronov, and D. E. Khmel'nitskii, J. Phys. C **15**, 7367 (1982)

PROBABILITY AND SCALING IN ONE-DIMENSIONAL DISORDERED SYSTEMS

Keith Slevin

Service de Physique du Solide et de Résonance Magnétique
CEA Saclay
91191 Gif-sur-Yvette, France

The single parameter scaling theory of localization [1] is central to much of our present understanding of transport in disordered systems predicting as it does the existence of a metal-insulator transition in dimension $d = 3$ and the localization of electron states for $d = 2$ and $d = 1$. The central assumption of the theory, embodied in the definition of the beta function [1], is a smooth one parameter scaling behavior of the conductance g with system size. However as is now widely appreciated this assumption is not correct since it ignores the existence of fluctuation phenomena such as the universal conductance fluctuations (UCF). The basic problem is that in the absence of inelastic effects the conductance of a disordered system does not self average as the system size increases but remains sensitive to the exact microscopic arrangement of impurities. In the localized regime the fluctuations in g diverge exponentially with increasing system size, and even in the metallic regime they are not zero (the UCF mentioned above). The question then arises of how to reconcile the scaling theory with the fluctuating behavior of g. One solution is to propose that the distribution $p(g)$ of g obeys a single parameter scaling, for example

$$p_L(\ln g) = p\left(\ln g, \langle \ln g \rangle_L\right) \tag{1}$$

$$\frac{\partial \langle \ln g \rangle_L}{\partial \ln L} = \beta\left(\langle \ln g \rangle_L\right). \tag{2}$$

This proposal is most easily checked in one dimension (1 D). It is found that the variable $-\ln T$ is normally distributed in the long length limit and that both its mean and variance obey scaling relations of the form eq. (2). Here T is the transmission coefficient of the system and can be related to the conductance through the Landauer formula. In general this implies a two parameter scaling of the distribution unless a relation exists between the mean and variance. For a random phase model, where the transmission and reflection phases are assumed uniformly distributed and uncorrelated with the transmission amplitude, just such a relationship exists in the weak disorder limit. It is found that [2,3] for weak disorder

$$var(-\ln T) = 2 \cdot \langle -\ln T \rangle. \tag{3}$$

Quantum Coherence in Mesoscopic Systems
Edited by B. Kramer, Plenum Press, New York, 1991

We shall see below, using a transfer matrix method, that this result also holds for the 1 D Anderson model[1] where the transmissions and reflections are non-uniformly distributed and are also correlated with the transmission amplitude. We shall also see that in general the transmissionj phase distribution obeys a two parameter scaling.

Here we present only the outline of the calculation preferring to concentrate on the results, for the details the reader is referred to [4]. We suppose that the transfer matrix for the system is composed of a product of a large number of individual transfer matrices each describing a separate "slice" of the system. We assume that these matrices are identically and independently distributed. It has been shown in previous work [5] how to define a generalized transfer matrix in terms of which an expression for the positive integer moments of $1/T$ can be obtained. This expression is analytically continued to non-integer moments and the statistical cumulants of $-\ln T$ obtained by differentiation

$$c_n(-\ln T) = \frac{d^n}{dN^n} \ln\langle T^{-N}\rangle. \tag{4}$$

Here n refers to the order of the cumulant and the derivative is evaluated at $N = 0$. In the long length limit this yields a simple result for the cumulants provided that the eigenvalue spectrum of the generalized transfer matrix satisfies certain conditions, here again we refer the reader to a more detailed discussion in [4]. We find subject to these conditions that

$$c_n(-\ln T) \to L \cdot a_n \qquad L \to \infty, \tag{5}$$

where the a's are coefficients independent of the system size L. We see that all the cumulants depend linearly on L, a situation which is analogous to the random phase model where the cumulants scale exactly as L. The difference here is that the cumulants scale only asymptotically as L with corrections to this behavior at finite L. From eq. (5) we deduce that the distribution of $-\ln T$ approaches a normal distribution in the long length limit.

$$p(-\ln T) = \frac{1}{\sqrt{2\pi L a_2}} \exp\left[\frac{-(-\ln T - L a_1)^2}{2 L a_2}\right]. \tag{6}$$

Evaluating the coefficients in eq. (5) explicitly for the 1 D Anderson model we find

$$\langle -\ln T\rangle \;=\; L a_1 = L \cdot \frac{var(\epsilon)}{4 - E^2}$$

$$var(-\ln T) \;=\; L a_2 = 2L \cdot \frac{var(\epsilon)}{4 - E^2}, \tag{7}$$

confirming eq. (3). Here $var(\epsilon)$ is the variance of the site energy distribution in the Anderson Hamiltonian. Since we are calculating only to first order in the variance our result implies a single parameter scaling only for weak disorder. To investigate the range of validity of this result we present in Fig. 1 the results of a numerical simulation of the 1 D Anderson model with a rectangular site energy distribution of width W. In the Figure we plot the mean versus the variance for various values of energy E and disorder W. For strong disorder we find that there are clear deviations from eq. (3). Note that such deviations alone are not sufficient to imply a crossover to two parameter scaling.

---

[1] Provided the length of the system is much longer than the localization length.

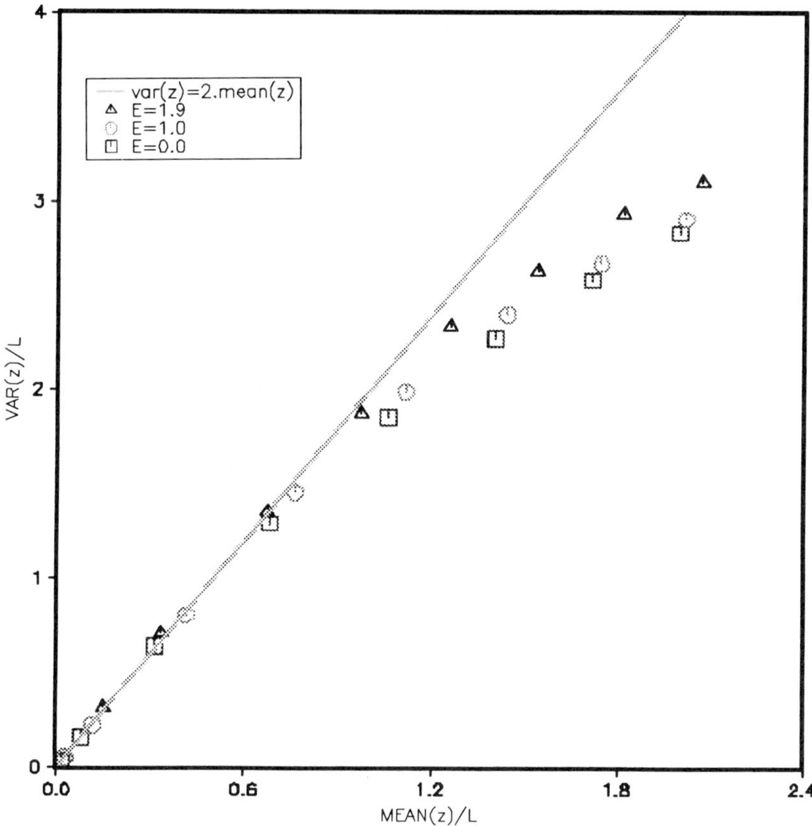

Figure 1. Relation between the mean and variance of $-\ln T$ for various energies E (square $E = 1.9$, cross $E = 1.0$ and octagon $E = 0.0$) and disorders W in the range $1 < W < 14$.

For single parameter scaling it would be sufficient that all the simulation results fell on a common curve and it is the departure from this behavior for large W which implies a crossover to two parameter scaling.

The utility of the approach explained here becomes apparent when we examine the higher cumulants of $-\ln T$ since this allows to determine whether the (positive integer) moments of T or $1/T$ are consistent with the normal approximation for $-\ln T$. Again calculating to first order in $var(\epsilon)$ we find that all higher cumulants are zero from which we conclude that the moments of $1/T$, effectively the resistance moments, are consistent with the asymptotic form for $p(-\ln T)$ for weak disorder. On the other hand, the moments of T, which under certain conditions can be identified with the conductance moments, are never consistent with a log-normal distribution for T. This point is confirmed by a direct calculation of the positive moments of T [6] and is a consequence of the fact that the positive moments of T are dominated by atypical resonance or "necklace" states [7] which are not well described by the scaling theory.

A similar analysis for the distribution of the phase φ of the (amplitude) transmission coefficient yields to first order in the site energy variance

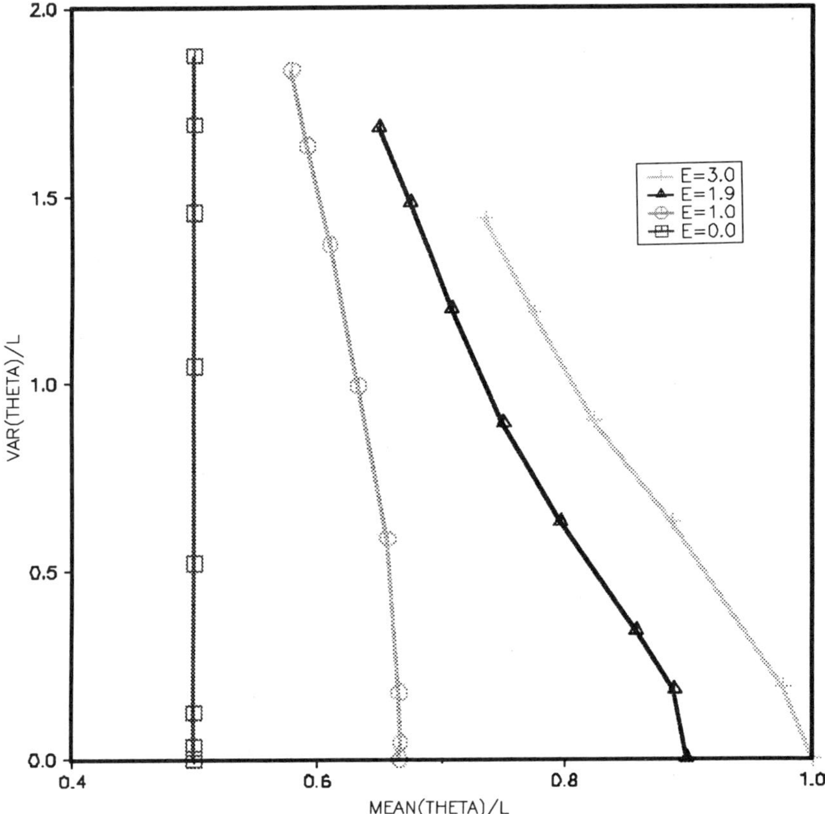

Figure 2. Relation between mean and variance of the transmission phase for various energies E (octagon $E = 0.0$, square $E = 1.0$, plus $E = 1.9$ and cross $E = 3.0$ and disorder W in the range $1 < W < 14$. Solid lines are guides to the eye only.

$$\langle \varphi \rangle \; = \; L \cdot \cos^{-1}\left(\frac{E}{2}\right)$$

$$var(\varphi) \; = \; \frac{3L}{2} \cdot \frac{var(\epsilon)}{4 - E^2}. \tag{8}$$

This implies a two parameter scaling for the phase distribution even for weak disorder. In Fig. 2 we plot the results of a numerical simulation for the phase. Contrary to eq. (8) we find that the mean phase is not independent of disorder. This is a consequence of the approximation to first order in the site energy variance. Nevertheless Fig. 2 does confirm the main point of the perturbation calculation that there is no one to one relation between the mean and variance.

We conclude that the distribution of $-\ln T$ for the 1 D Anderson model obeys in general a two parameter scaling, which in the weak disorder limit this reduces to a single parameter scaling. The phase distribution however obeys a two parameter scaling even in this limit.

References

[1] P.W. Anderson, D. Thouless, E. Abrahams, D. Fisher, Phys. Rev. Lett. **B 22**, 3519 (1979)

[2] P. Mello, J. of Mathem. Phys. **27**, 2876 (1986)

[3] B. Shapiro, Phil. Mag. **56** (6), 1031 (1987)

[4] K. Slevin and J.B. Pendry, to appear in J. of Phys.: Cond. Matter **2**, 2821 (1990)

[5] P. Kirkmann and J.B. Pendry, J. Phys. **C 17**, 4327 (1984)

[6] P. Kirkmann and J.B. Pendry, J. Phys. **C 17**, 5707 (1984)

[7] J.B. Pendry, J. Phys. **C 20**, 733 (1987)

CHAPTER 9

SUPERCONDUCTING SYSTEMS

SUPERCONDUCTING WIRE NETWORKS

Bernard Pannetier

Centre National de la Recherche Scientifique
Centre de Recherches sur les Très Basses Temperatures
Laboratoire associé á l'Université J. Fourier
B.P. 166X, 38042 Grenoble Cedex, France

1 Introduction

Superconductivity offers an unique example in condensed matter physics where the electrons occupy a single coherent quantum state over macroscopic distances. In homogeneous superconductors energy considerations usually ensure that, in most physical situations, the superconducting pair wavefunction is nearly constant in both phase and amplitude. The development of modern microfabrication techniques has made possible recently to produce a wide variety of complex artificial structures where, to a certain extent, the physicist is able to constrain, at will, the wavefunction, for example by enforcing the boundary conditions (BC) or by applying an external magnetic field [1,2].

This lecture focuses on some amazing properties of discrete artificial superconducting systems. The system consists of basic superconducting filaments which are coupled together to form a higher dimensional structure. Being of reduced dimensionality (0 D or 1 D) is not a severe constraint for the individual filaments because of the large coherence length typical of superconductivity. In fact the characteristic dimensions of the elements need only be less than the Ginzburg-Landau coherence length which becomes very large in the vicinity of T_c. Fig. 1 shows an example of a square aluminum wire network. Elementary strands have typical cross section 80×250 nm and length $3\,\mu$m. In a significant temperature range near the phase transition line, the superconducting properties of these structures are controlled entirely by the network topology. Their detailed analysis relies upon the remarkable fact that all Cooper pairs occupy coherently the same quantum state. For a given geometry, the allowed quantum states are those determined by the quantum interference effects at the macroscopic scale of the network.

The first section is a short review of the superconducting properties near the mean field phase transition line. We will follow the classical approach of Abrikosov, extended to include more general geometries than that of simple bulk type II materials. As will be shown there is a direct correlation between the equilibrium superconducting properties and the energy spectrum of charged particle in the underlying lattice. The most illustra-

Quantum Coherence in Mesoscopic Systems
Edited by B. Kramer, Plenum Press, New York, 1991

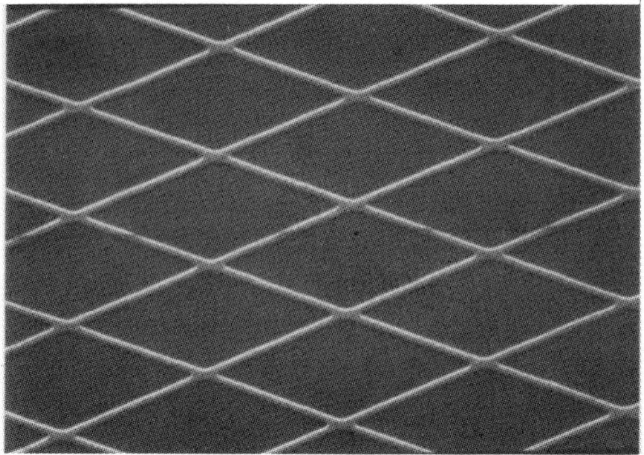

Figure 1. Small part of a square superconducting network: SEM oblique view. The filaments are 80 nm thick, 250 nm wide aluminum wires made by e-beam direct writing followed by a lift-off process. The lattice parameter is $a = 3.0\,\mu$m.

tive example is certainly the infinite square network. In the second part we examine with some detail the extremely rich Landau level spectrum of electrons in a square periodic lattice, and the implications of the subtle quantization effects on the lattice for the superconducting properties of the network. Such concepts as frustration phenomena induced by the magnetic field, rationality and commensurability effects become experimentally accessible in such systems through critical temperature, critical current or magnetization measurements. In the last part we consider systems of lower dimensionality and more exotic geometries. An example is given by a superconducting dot, which is a simple realization of a 0 D system. Structures of intermediate dimensionality such as self similar arrays or quasi-crystalline networks are also considered. Those structures represent in some sense an elementary step in the route to random structures. Examples of percolation networks will be discussed. We will see that even in such simple artificial random structures, our understanding of the effect of disorder is still extremely modest.

2   Superconductivity Near the Phase Boundary

It is well accepted that the nucleation of superconductivity in bulk sample is accurately described by the mean field Ginzburg-Landau (GL) theory. This mean field approach is based upon simple assumptions on the superconducting phase. It is supposed that it can be described by a complex order parameter $\Psi e^{i\theta}$ which goes smoothly to zero at the second order phase transition. In addition the fluctuations of both the order parameter amplitude and phase are neglected.

According to the GL theory, the free energy density can be expanded as function of the slowly varying order parameter $\Psi(r)$ as

$$F_s = F_n + \int d^3 r \left\{ \alpha |\Psi|^2 + \frac{\beta}{2}|\Psi|^4 + \frac{\hbar^2}{2m^*}\left| \left(-i\nabla - \frac{2\pi}{\phi_o}A\right)\Psi \right|^2 + \frac{b^2}{2\mu_0} \right\} \tag{1}$$

where F_n is the free energy in the normal state, $\hbar$ the Planck constant and $\phi_o = h/2e$

the superconducting flux quantum. Note that the mass m^* and the charge $2e$ are taken to be twice that of the unpaired electrons in the solid. The material parameters are the temperature dependent coherence length $\xi(T) = \xi(0)/(1 - T/T_0)^{1/2}$ and the coefficient of the fourth order term β. T_o is the zero field critical temperature and $b = |curl\,A|$ is the local magnetic induction. The temperature dependence is included in the linear coefficient $\alpha = -\hbar^2/2m^*\xi^2 \sim (T - T_0)$.

Minimizing the free energy with respect to the order parameter and the vector potential leads to the non-linear coupled differential GL equations which describe the spatial variations of the order parameter and the supercurrents. Except in very simple geometries, the complete non-linear equations cannot be solved exactly. In complex geometries one must make some simplifications to treat the non-linear term. Close enough to the second order phase boundary, a perturbative approach inspired from the famous Abrikosov theory [5] for the mixed state of type II superconductors has been proposed by Y.Y. Wang et al. [6] in the general case of wire networks of arbitrary topology. The essential ingredient is the assumption that the actual order parameter can be constructed by combination of the solutions of the linear GL equations.

It is convenient to formulate the linearized GL equations as a Schrödinger equation for a particle having the charge $(2e)$ and the mass m^* of the Cooper pairs. This analogy provides a useful correspondence between superconductivity and more general electronic properties that we will exploit now. In bulk superconductors, the solutions of the linear problem are the Landau orbits of the free electron pairs. In networks of arbitrary geometry the Schrödinger equation must be solved with appropriate BC [4]

$$\frac{\hbar^2}{2m^*}\left(-i\nabla - \frac{2\pi}{\phi_o}\mathbf{A}\right)^2 \Psi_\sigma = E_\sigma \Psi_\sigma \tag{2}$$

$$n\left(-i\nabla - \frac{2\pi}{\phi_o}A\right)\Psi_\sigma\bigg|_b = -\frac{i}{b}\Psi_\sigma. \tag{3}$$

The set of eigenstates Ψ_σ represent all the possible states which are local minima of the quadratic gradient term in the free energy. The subscript σ indexes the different possible quantum numbers (band index, wavevector, etc. ...). The BC (eq. (3)) lead to two relations, the first one (real part) is just the condition of conservation of supercurrent, the second one (the imaginary part), more specific to superconductivity, $d\Psi/dx = -\Psi/b$ rules the depression of Ψ near the boundary. According to the size of the characteristic length b, either Ψ is zero at the interface (interface $S - N$ or low ξ materials), or Ψ has an horizontal slope (regular S-vacuum interface).

Following [6], we calculate the free energy for the state (Ψ_σ, E_σ), in an external magnetic field H. For simplicity, it is convenient to assume that the superconducting specimen is cylindrical so that all relevant quantities (order parameter, current density, local magnetic field) are independent upon the z-coordinate along which the external field is applied. Such specimen represents for example the cylindrical extension of the square network shown in Fig. 1. The demagnetization factor is zero in this case. Demagnetization effects will be accounted for in a second step. It is also useful to introduce the induced field $b_s = b - \mu_0 H$ (parallel to H) defined as the difference between the local microscopic induction and the external field. b_s is of order of $|\Psi|^2$ and can therefore be assumed very small. On the other hand, since Ψ_σ is an eigenstate of eq. (2), the gradient term in the free energy can be replaced by $E_\sigma|\Psi_\sigma|^2$. By incorporating the correction due to b_s in $E_\sigma(H)$, the Gibbs energy difference is

$$G_s(H) - G_n(0) = \int d^3r \left\{ \left(\alpha + E_\sigma + \mu_0 b_s \frac{dE_\sigma}{dH} \right) |\Psi_\sigma|^2 + \frac{\beta}{2}|\Psi_\sigma|^4 + \frac{b_s^2}{2\mu_0} - \frac{1}{2}\mu_0 H^2 \right\} \quad (4)$$

which depends explicitly on two free parameters, b_s and the amplitude of Ψ_σ (not fixed in the homogeneous equation eq. (2)). Upon minimization of ΔG with respect to these two parameters, the following expression is obtained for the average order parameter amplitude of state σ

$$\langle|\Psi_\sigma|^2\rangle = \frac{m^*}{2\mu_0 e^2 \beta_A(2\kappa^2 - \mu^2)} \left\{ \frac{1}{\xi^2} - \frac{2m^* E_\sigma}{\hbar^2} \right\}. \quad (5)$$

κ is the GL parameter, defined by $\beta = \mu_0 2\kappa^2(e\hbar/m^*)^2$ and $\mu = (m^*/e\hbar)dE_\sigma/dH$ is a dimensionless parameter.

The expressions for the Gibbs free energy and the magnetization for state σ are obtained from eqs. (4) and (5)

$$G_s(H) - G_n(0) = \frac{m^*}{2\mu_0 \beta_A(2\kappa^2 - \mu^2)} \left(\frac{\phi_0}{2\pi} \right)^2 \left\{ \frac{1}{\xi^2} - \frac{2m^* E_\sigma}{\hbar^2} \right\}^2 \quad (6)$$

$$M = \langle|\Psi_\sigma|^2\rangle \frac{dE_\sigma}{dH}. \quad (7)$$

Remarkably, these general expressions for the thermodynamic quantities near T_c are obtained without specifying explicitly the network geometry. In fact, the influence of its geometry is incorporated in two averaged quantities both related to the state σ:

a) the energy $E_\sigma(H)$ coming from the gradient term as a result of the external magnetic field and the BC. On the second order phase boundary, superconductivity nucleates into the state $\Psi_\sigma = \Psi_{\min}$ which corresponds to the lowest energy $E_{\min}$. Accordingly, the mean field critical temperature is determined by the lower edge ($\alpha = -E_{\min}$) of the spectrum of eq. (2).

$$T_c(H) = T_0 \left\{ 1 - \frac{2m^*}{\hbar^2}\xi^2(0)E_{\min}(H) \right\}. \quad (8)$$

Below T_c the order parameter square amplitude $|\Psi_\sigma|^2$ grows linearly as $T_c - T$. Its amplitude depends upon the fourth order term of the free energy expansion through κ and β_A,

b) the parameter β_A which is defined as

$$\beta_A = \frac{\langle\Psi^4\rangle}{\langle\Psi^2\rangle^2} \quad (9)$$

is the generalized Abrikosov parameter first introduced in the theory of the vortex state in type II superconductors [5]. β_A is the inverse participation ratio of the wavefunction Ψ and characterizes its degree of localization. The statement that β_A must be minimum allows to determine, among the degenerate solutions of the linearized equation (eq. (2)), the most stable configuration of the order parameter. A well known example is the triangular lattice of vortices in type II superconductors. An example for the square network will be discussed in section 3.

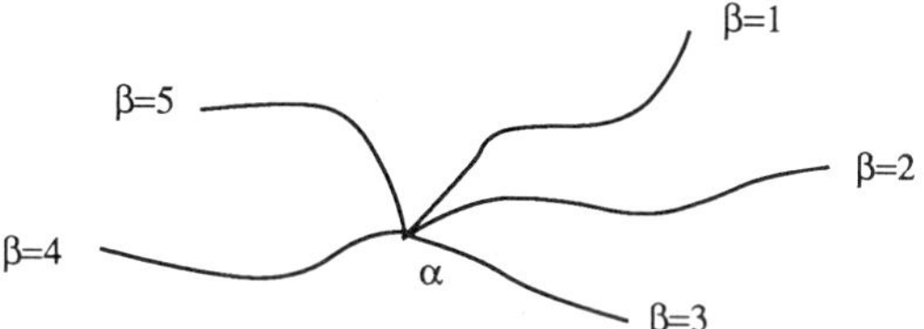

Figure 2. Wire network: node α and nearest neighbors β (length $\ell_{\alpha\beta}$).

As a last step, one must include the demagnetization effect for real two-dimensional (2 D) networks. A very crude approximation consists in assuming that the specimen has a flat ellipsoidal contour. This is the condition for the internal field to be uniform in the sample ($N_x = N_y = 0, N_z = 1 - \pi e/4R$ for an ellipsoid of thickness e and radius R). The field inside the specimen is $H = H_{\text{applied}} - N_z M$. As far as the magnetization remains small and as one can neglect the second order variations of the field, the effect of the demagnetization field can be approximated by replacing $2\kappa^2 - \mu^2$ by $2\kappa^2 - \mu^2(1 - N_z/\beta_A)$ in the expressions eqs. (5), (6) and (7). As a final result, at equilibrium, the order parameter, free energy and magnetization are obtained by including the demagnetization effects and substituting $E_\sigma \to E_{\text{min}}$ in eqs. (5) and (7). Using eq. (8), a convenient formulation may be given in terms of the $T_c(H)$ function

$$\langle|\Psi_{\text{min}}|^2\rangle = \frac{m^*}{2\mu_0 e^2 \beta_A(2\kappa^2 - \mu^2(1 - N_z/\beta_A))} \left\{ \frac{T_c(H) - T}{T_0 \xi^2(0)} \right\} \tag{10}$$

$$M = \frac{\hbar^2}{2m^* T_0 \xi^2(0)} \langle|\Psi_{\text{min}}|^2\rangle \frac{dT_c}{dH}. \tag{11}$$

It is worth noticing that the above formulation leads to the well known magnetic properties of the ideal bulk type II superconductors in the continuous limit, $E_{\text{min}} = \frac{1}{2}\hbar\omega_c = e\hbar H/m^*, \mu = 1$, and $\beta_A = 1.16$ [4,5].

3   Wire Network Formalism

3.1   Node Equations

The formalism introduced by de Gennes and Alexander [7,8] in their pioneer work on superconducting networks allows to consider networks of arbitrary geometry, provided the elementary strands can be assumed 1 D. Here 1 D means that the wire diameter is smaller than both the GL coherence length and penetration depth.

Let s be the curvilinear coordinate on the strand linking node α to node β (Fig. 2). Along strand $\alpha\beta$, eq. (2) writes

$$\left(-i\frac{\delta}{\delta s} - \frac{2\pi}{\phi_0}A_\parallel\right)^2 \Psi(s) = \frac{2m^* E}{\hbar^2}\Psi(s) \tag{12}$$

with S-vacuum BC:

$$\sum_\beta \left\{-i\frac{\delta}{\delta s} - \frac{2\pi}{\phi_0}A_\parallel\right\} \Psi(s = 0) = 0 \tag{13}$$

where $A_\parallel$ is the component of the vector potential parallel to the wire. Solving the first equation we obtain the order parameter along the wire $\alpha\beta$

$$\Psi(s) = \frac{\exp i\gamma_{\alpha s}}{\sin u_{\alpha\beta}}\left\{\Psi_\alpha \sin u_{s\beta} + \Psi_\beta \exp i\gamma_{\alpha\beta} \sin u_{\alpha s}\right\} \tag{14}$$

where $\gamma_{\alpha\beta}$ is $2\pi/\phi_0$ times the line integral of the vector potential along strand $\alpha\beta$ and $u_{\alpha\beta} = \ell_{\alpha\beta}\sqrt{2m^*E}/\hbar$ is the normalized length of strand $\alpha\beta$.

The supercurrent in strand $\alpha\beta$ follows immediately from the GL pair current, and depends sinusoidally, as the Josephson current, on the phase difference between nodes α and β

$$J_{\alpha\beta} = \frac{2e\hbar}{m^*\xi \sin u_{\alpha\beta}}|\Psi_\alpha\Psi_\beta| \sin(\phi_\alpha - \phi_\beta - \gamma_{\alpha\beta}). \tag{15}$$

Applying the BC (eq. (13)) at node α we obtain the Alexander node equations

$$\sum_\beta -\Psi_\alpha \cot u_{\alpha\beta} + \Psi_\beta \frac{\exp i\gamma_{\alpha\beta}}{\sin u_{\alpha\beta}} = 0. \tag{16}$$

This set of linear equations (one for each node of the network), is the reduced form of eq. 2 for a discrete lattice. Replacing $u_{\alpha\beta}$ by $\ell_{\alpha\beta}/\xi(T)$ restores the original presentation of the Alexander equations for the along the phase transition line.

Consider as an example, a ring of radius R made of a 1 D filament, and assume a fictitious node α. The summation over nearest neighbors involves two symmetric paths linking α to itself, clockwise and counterclockwise. Since $\ell_{\alpha\alpha} = 2\pi R$ and $\gamma_{\alpha\alpha} = 2\pi\phi/\phi_0$, eq. (16) becomes $\cos u = \cos\gamma$. The solution leads to the well known energy spectrum of paired electrons on a ring

$$E = \frac{\hbar^2}{2m^*R^2}\left\{\frac{\phi}{\phi_o} - n\right\}^2. \tag{17}$$

Combined with eq. (8), this relation leads to the famous Little-Parks curve [9] which provided the first demonstration of fluxoid quantization in superconductor in 1962. Little and Parks found that the phase transition line of a hollow cylinder $T_c(H)$ was an oscillatory function of the applied flux with period given by the superconducting quantum ϕ_0.

$$1 - \frac{T_c(H)}{T_c(0)} = \frac{\xi^2(0)}{R^2}\left\{\frac{\phi}{\phi_0} - n\right\}^2. \tag{18}$$

The application of the de Gennes-Alexander's formalism to micro-networks [10,11] of more complex geometries (lasso, ladder) is a new example of basic calculations on quantum circuits in solid state physics, which are now accessible to experimental tests.

3.2   Regular Networks

Of interest are the regular networks where every strand has the same length, $\ell_{\alpha\beta} = a$. The node equations are equivalent to a tight binding Schrödinger equation with hopping integral $t_{\alpha\beta} = \exp -i\gamma_{\alpha\beta}$, and tight binding energy $\epsilon = z\cos u = z\cos a\sqrt{2m^*E}/\hbar$

$$\epsilon\Psi_\alpha = \sum_\beta t_{\alpha\beta}\Psi_\beta. \tag{19}$$

This equivalence assumes that every node has the same number of nearest neighbors $z_\alpha = z$. This assumption is no more valid in finite systems since z_α is smaller at the edge of the lattice (for example $z = 3$ instead of 4 for the 2 D square lattice) except in the following edge configuration: the missing number of branches is added to the edge nodes (one for the square network) and is connected to a normal metal or magnetic material which ensures that Ψ vanishes at the dead end. This situation corresponds to the unusual s-normal BC: $\Psi = 0$ at the boundary. The usual s-vacuum BC is recovered by including the variable coordination number into the hopping integral $t_{\alpha\beta} = \exp -i\gamma_{\alpha\beta}/z_\alpha$. This has some interesting implications in the comparison between edge states in superconducting disks and quantum dots (see section 5.3)

3.3 A few Remarks about Notations

The link between the energy E and the tight binding energy ϵ is made through the equation

$$E = \frac{\hbar^2 u^2}{2m^* a^2} = \frac{\hbar^2}{2m^* a^2} \left[\pm \arccos \frac{\epsilon}{z} + 2\pi N \right]^2 \tag{20}$$

where $z = 4$ for the square network. The band index N is generally taken as $N = 0$.

In real networks the wires have a finite width w. If w is small enough, the transverse modes within the wire are not excited. Only some extra energy, due to the penetration of the magnetic field in the volume of the wires, must be included. For uniform width w this additional energy is quadratic in w and H

$$E = E_{\text{network}} + E_{\text{wire}} = \frac{\hbar^2}{2m^* a^2} \left[u^2 + \frac{\pi^2}{3} \frac{H^2 w^2}{\phi_0^2} \right]. \tag{21}$$

In the following section we focus on the infinite square lattice on which has been the object of the major attention in this context.

4 Square Network

Study of the infinite square network has permitted considerable progress in the understanding of the frustration effect induced by the magnetic flux on the superconducting properties of wire networks. Since, as emphasized in the first section, these are controlled by the electron energy spectrum of the underlying lattice, it is useful to discuss in some details the features of Landau levels in a square periodic lattice.

4.1 Landau Levels

Consider an infinite square network of period a, exposed to a perpendicular field $\mathbf{H} = curl\mathbf{A}$, with (arbitrary) gauge $A = Hx$ along y axis (Fig. 3). Let us call $\phi(\phi = Ha^2)$ the magnetic flux through one elementary cell.

Using translational invariance properties, one can write the wavefunction $\Psi_{m,n}$ at node (m, n) as a planewave $\Psi_{m,n} = \Psi_m e^{ink}$ along y direction. The Alexander equations (eq. (19)) reduce to

$$\epsilon f_m = f_{m+1} + f_{m-1} + 2\cos(k - m\gamma) f_m \tag{22}$$

which is a 1 D tight binding equation with a periodic potential $V_m = 2\cos(\gamma m - k)$.

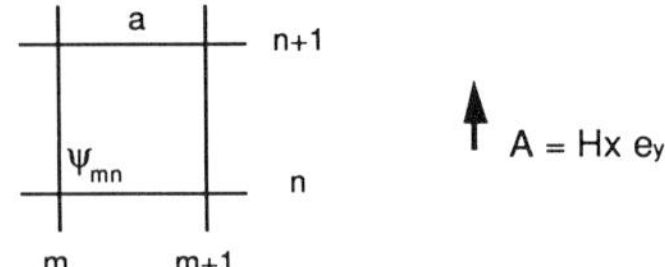

Figure 3. The elementary cell of a square network.

Figure 4. The Landau levels structure of electrons in an
infinite square lattice. Vertical axis is the tight binding
energy $\epsilon = 4\cos u$ in the interval (-4, +4). Horizontal
axis is the reduced magnetic flux ϕ/ϕ_0 in the first period
(0,1). Correspondence with the usual energy units through
eq. (20) [14].

This equation appears in different problems of statistical physics such as electrons
in incommensurate crystals [12], Bloch electrons in ultra high magnetic fields, and more
recently the flux phase in high Tc superconductors [13]. Here the period is controlled by
the external magnetic field through $\gamma = 2\pi\phi/\phi_0$. The properties of its fractal spectrum
(Fig. 4) were originally investigated by Azbel, and later in more detail by Hofstadter [14].
The high complexity of the spectrum arises from the competition of two spatial periods.
The first one is the lattice spacing a of the square network and the second one is the
magnetic period $\ell_H = \sqrt{\phi_0/\pi H}$ which can be tuned continuously by the field strength.
The genesis of this spectrum has been discussed in [10]. As far as superconductivity
is concerned, the significant part of this spectrum is in the vicinity of the ground state
$E_{\min}$. Higher energy levels correspond to unfavorable metastable states not occupied in
practice, except in presence of thermal fluctuations ignored in this discussion or under
non-equilibrium situations (strong current injection for example). We will come back to
this point in the discussion of critical currents in section 4.5.

In the low field region ($\gamma \to 0$ and $\epsilon \to 4$) the wavefunction varies slowly at the scale of the lattice parameter a and can be expanded according to the continuous approximation $\Psi_{m+1} = \Psi_m + a d\Psi_m/dx + \ldots$. The node equation then becomes, with $V(x) = 2 - 2\cos\gamma x/a$, a periodic potential,

$$\left\{ -a^2 \frac{d^2}{dx^2} + V(x) \right\} \Psi = (4 - \epsilon)\Psi. \tag{23}$$

To leading order, the potential is quadratic with Landau level solutions as in free space and the electron orbits are giant orbits which extend over a distance $a/\sqrt{\gamma}$ much larger than the lattice parameter.

$$4 - \epsilon = \pm(1 + 2n)\gamma. \tag{24}$$

Although in this continuous limit ϵ is independent of the lattice parameter, the discreteness of the lattice does influence the low field limit of the spectrum. Equation (24) with $n = 0$ leads to an energy $E = \pi\hbar^2 H/2m^*\phi_0$ which is half the energy of the first Landau level in free space. The origin of this factor 2 is the 1 D character of the internode coupling. Its physical significance has been checked experimentally [15].

As the field is increased, the overlap between Landau orbits centered at successive minima of the potential $V(x)$, becomes larger and the electrons can jump from one orbit to the other. As results the discrete Landau levels turn into energy bands. Higher order terms in the expansion provide an estimate of the bandwidth of the lowest Landau level, which grows exponentially in $1/H$

$$4 - \epsilon = \gamma + \frac{\gamma^2}{16} + 2\exp{-\frac{\pi^2}{\gamma}} \cos\frac{2\pi k}{\gamma} + \ldots \tag{25}$$

We consider now the rational magnetic flux, $\phi/\phi_0 = p/q$ (p, q integers). The node equation is invariant under translation $m \to m + q$. Therefore q linear equations are sufficient to describe the system. The solution are Bloch states constructed on a basic superunit cell of size $qa \times qa$. The dispersion relation is given by the roots of a polynomial of degree q

$$P_q(\epsilon) = 2\cos qk_x + 2\cos qk_y. \tag{26}$$

For low q we have $P_1 = \epsilon, P_2 = \epsilon^2 - 4$, etc. $\ldots$. The energy spectrum at $\phi/\phi_0 = p/q$ consists of q bands separated by energy gaps (Fig. 4). The ground state ϵ_m is obtained for $P_q(\epsilon) = 4(k = 0)$ and corresponds to a cusp-like minimum in the $\epsilon(\phi/\phi_0)$ curve. Clearly the particular organization of the wavefunction at commensurate flux (ϕ/ϕ_0 rational) allows for some obvious optimization of the energy. For example, one finds $\epsilon_m(0) = 4, \epsilon_m(1/2) = 2\sqrt{2}, \epsilon_m(1/3) = \ldots$.

Close to a rational ($\gamma = 2\pi p/q + \delta\gamma$), a continuous expansion similar to the above low field continuous limit can be performed [16]. Now the wavefunction may have strong variations within one superunit cell, but is assumed to vary slowly from one node to the equivalent node on the adjacent superunit cell $\Psi_{m+q} = \Psi_m + qa d\Psi_m/dx + \ldots$.

As in the vicinity of zero field, one finds, in the $\delta\gamma \to 0$ limit, a harmonic oscillator equation with Landau levels solutions

$$4 - P(\epsilon) = P'(\epsilon)\frac{d\epsilon}{d\gamma} = \pm(2n + 1)\delta\gamma + A\delta\gamma. \tag{27}$$

Two important differences must be noted with respect to the zero field case. (i) The slope $d\epsilon/d\gamma$, $1/P'(\epsilon)$, depends on the p/q ratio and increases with q. (ii) There is an additional linear term $A\delta\gamma$ which makes, in general, Landau levels asymmetric (left hand slope different from right hand slope).

The expansion near rational ϕ/ϕ_0 is equivalent to assuming that the the phase shift a between two equivalent nodes in adjacent super unit cells is a slowly varying function of space. The new term $A\delta\gamma$ can also be viewed as a correction to the Bohr-Sommerfeld quantization rule [17] for a closed cycle in phase space (x, α). These remarks are relevant for the discussion of experiment on the magnetization of superconducting networks.

4.2   The Phase Transition Line

The first experimental study of the phase transition line of an artificial 2 D superconducting wire network was reported by Pannetier et al. [18] in 1983 on a honeycomb In network. Shortly after, a similar experiment performed on a square network [18] revealed in great detail the subtle effect of the magnetic flux discussed above. The sample studied in this work was prepared by reactive ion etching of a pure Al film 80 nm thick into a grid shape containing $2.7 \cdot 10^6$ identical cells of size $6 \times 6\mu m^2$. The overall size of the network was $1 \times 1\,cm^2$. The Al resistivity was 0.51 $\mu\Omega cm$ at 4.2 K. The resistance of each strand was 0.3 Ω. Fig. 5 shows the field dependence of the critical temperature. Well-defined oscillations are observed on top of a parabolic background. This parabolic envelope reflects the effect of finite width of the Al strands (here $2\mu m$). In the more recent samples made by electron-beam lithography, the width is small enough ($w \approx 0.2\mu m$) that this parabolic envelope is no more observed (see paragraph 5.3). The observed oscillations are the result of fluxoid quantization in the smallest cell of the network (Little-Parks oscillations, see [9]). The period of these oscillations, 0.606 Oe, corresponds to one flux quantum $\phi_0 = 2.07 \cdot 10^{-7}\,gauss\,cm^2$ per unit cell within the accuracy of the lattice size measurement ($a = 6.01\mu m$).

The most noticeable features of the $T_c(H)$ curve are shown in the magnified view of the first period in the lower part of Fig. 5. There are finite slopes at $\phi/\phi_0 = 0$ and 1,

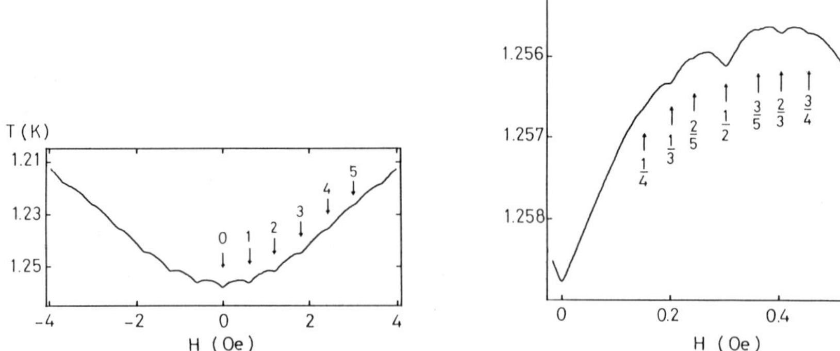

Figure 5. Critical temperature vs magnetic field for a square network made of superconducting Al (left). The arrows indicate the magnetic field values corresponding to the integral number of flux quanta per unit cell of the network. Magnified view of the first period showing secondary dips at rational ϕ/ϕ_0 (indicated by the arrows, right) [17].

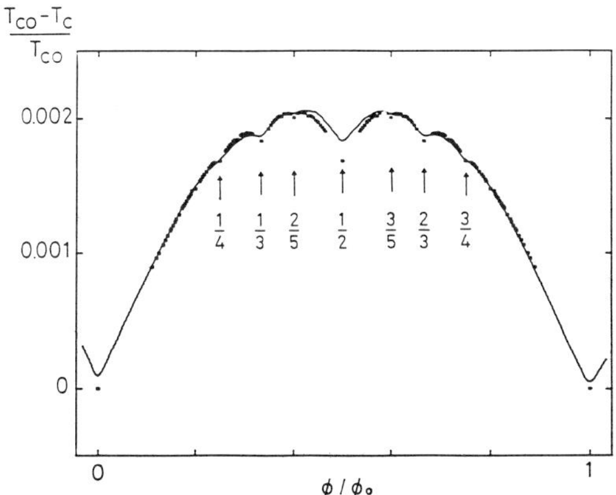

Figure 6. Solid curve: same as in Fig. 5 (right) in reduced units and with parabolic contribution from the bulk of the wires being subtracted. Dots: theoretical values calculated from the edge of the Landau level spectrum (Fig. 4) for rational $\phi/\phi_0 = p/q (q < 30)$ [18].

downwards concavity of the modulation, well-defined dips at $\phi/\phi_0 = 1/4, 1/2, 2/5, 1/3, \ldots$, in addition to the fundamental dips at 0 and 1.

All of these features, which are absent in the one loop case, establish a very accurate test of the theoretical prediction discussed in the previous section. A comparison between the theory and the experiment is shown in Fig. 6. The continuous curve is the critical line (same data as in Fig. 5 (lower part)) plotted in reduced units $\Delta T_c/T_c$ vs ϕ/ϕ_0 after subtracting the parabolic background. The theoretical values

$$\frac{\Delta T_c}{T_c} = \frac{\xi(0)^2}{a^2} \arccos^2 \frac{\epsilon}{4} \tag{28}$$

as computed numerically for ϕ/ϕ_0 rational (with $q < 30$) using the characteristic equation $Pq(\epsilon) = 4$, are shown on the same figure. The best fit is obtained from $T_c(0) = 1.259\,K$ and $\xi(0) = 0.305\mu$. The obtained value of $\xi(0)$ is in reasonable agreement with that expected for dirty Al if one takes the mean free path $l = 80\,\text{nm}$ and the intrinsic coherence length $\xi_0 = 1.6\mu\text{m}$.

The variation of the lowest energy state of the Landau level spectrum as a function of ϕ determines the effect of frustration induced by the magnetic field on the superconducting transition. The frustration parameter is simply the reduced flux ϕ/ϕ_0. Each rational $\phi/\phi_0 = p/q$ corresponds to a certain amount of frustration in the network giving rise to a particular periodic organization of the order parameter and supercurrents. Physically this corresponds to a fluxoid quantization phenomenon through superunit cells resulting in a commensurate structure at rational ϕ/ϕ_0. This quantization effect is at the origin of the reduction of T_c without any normal core.

4.3 Configuration of the Equilibrium Order Parameter at Rational ϕ/ϕ_0

The configuration of the order parameter as well as the distribution of supercurrents in the mixed state of a square network have been studied by Y.Y. Wang et al. [6]. When $\phi/\phi_0 = p/q$, the solutions of the linearized equation having minimum energy are represented by a wavefunction which is q-fold degenerate. Following Abrikosov, one assumes that the actual order parameter can be constructed from a linear combination of these degenerate states. The general expression for Ψ_{mn} at node (m,n) is expressed as function of these q degenerate states

$$\Psi_{mn} = \sum_{\ell=0}^{q-1} C_\ell \Psi_{m-\ell,n} \exp(2i\pi n\ell\phi/\phi_0) \tag{29}$$

where the complex coefficients C_ℓ have to be determined from the requirement that the Abrikosov parameter β_A must be minimum.

The determination of C_ℓ requires an explicit calculation of $\langle\Psi^2\rangle$ and $\langle\Psi^4\rangle$ over the q^2 strands of the basic $q \times q$ superunit cell using the network formalism (eqs. (14) and (19)). This calculation has been carried out in [6] for low q values $q = 1$ to 5, and provides the amplitude of the order parameter and the supercurrent density at every point of the network. The results are illustrated in Fig. 7 for $\phi/\phi_0 = 1/2$ and $1/3$. The equilibrium configuration consists of a periodic arrangement of basic superunit cells containing 2×2 and 3×3 elementary cells respectively. In the first case the configuration is a checkerboard arrangement, which consists of an alternate sequence of clockwise and counterclockwise current loops. The amplitude of the order parameter is uniform across the network. At higher q values, the order parameter is non-uniform and takes maximum values in strands where no supercurrent is flowing. On the other hand, Ψ remains almost constant along the current lines (see arrows). This property reminds of the Abrikosov vortex state of type II superconductors where the current lines coincide rigorously with

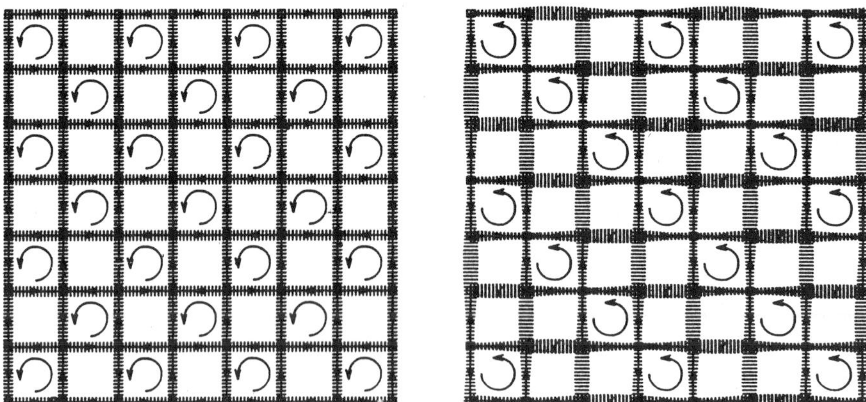

Figure 7. The equilibrium superconducting phase for $\phi/\phi_0 = 1/2$ (left graph) and $1/3$ (right graph). The order parameter amplitude is represented by the thickness of the lines. The circular arrows show the location of the current loops (i.e. vortex location) [6].

the lines of constant order parameter amplitude. Here the vortices have no core, nowhere does the order parameter vanish. The singularity only takes place in the phase of the order parameter, and this is best illustrated in the $\phi/\phi_0 = 1/2$ case.

The above pictures show similarities with the zero temperature ground state of Josephson junction arrays [19,20]. Actually the equilibrium state obtained from the present analysis has the same phase configuration at least for low rational ($q < 5$) as that obtained from the XY model. To date there is no direct experimental evidence of such a configuration.

4.4   Magnetization Effects

Using accurate ac-modulation techniques, P. Gandit et al. [21] have measured the low field susceptibility and the derivative dM/dT of extended square superconducting networks containing $4 \cdot 10^6$ square indium cells of side $a = 3.2\mu$m, made by electron beam lithography.

The zero field ac susceptibility dM/dH, measured under field modulation dH from $100\,\mu$Oe to $10\,m$Oe, was found to drop from zero in the normal state to a large negative value independent of the magnetic field in the superconducting state. From the susceptibility at saturation the demagnetization coefficient N_z of the network could be determined. The obtained N_z is very close to 1, as expected in this geometry ($1 - N_z = 5 \cdot 10^{-5}$). Within the experimental accuracy, the susceptibility follows the GL linear temperature dependence near the critical temperature. In the intermediate region, however, it exhibits sharp peaks at both integer and rational values of the reduced flux ϕ/ϕ_0. This behaviour reflects the oscillation of the order parameter amplitude which is expected to vary as $T_c(H) - T$. The phase transition line as derived from these susceptibility measurements confirms the oscillatory effect obtained from resistive measurements. The main features of the phase boundary are: periodicity with respect to the magnetic flux, dips at integer and rational $\phi/\phi_0 = p/q$ (up to $q = 3$). However the precise shape of the $T_c(H)$ curve, in particular at very low field, seems to differ from the expected variation. This discrepancy may indicate the limits of the mean field treatment which is based upon equilibrium properties of the superconducting phase and ignores the dynamics of vortices.

In contrast to dM/dH which determines the effect of diamagnetic screening currents, the dM/dT measurements are performed at constant field (without disturbing the fragile arrangement of the flux superlattice) and therefore provide a direct and sensitive probe of the properties the equilibrium magnetization. The details of the experimental set up were presented in [21]. The dM/dT curves are reversible in a range of temperature of ~ 100 mK below T_c ($T_c = 3.15\,K$). As the temperature is decreased from the normal state, dM/dT presents a maximum near $T = 3.09\,K$ either positive or negative according to the magnetic field value. This is illustrated in Fig. 8 which shows an enlarged view of the low field part of the dM/dT curve measured at fixed temperature $T = 3.11K$ as a function of the external magnetic field under field cooling cycles.

One notices the sharp magnetization jump as the field is scanned from negative to positive small field. The same jump is observed at integer ϕ/ϕ_0 and similar but smoother jumps are also seen at rationals 1/2, 1/3, 2/3. These discontinuities which result from sudden changes in the directions of diamagnetic currents are not seen in the $T_c(H)$ line which is sensitive only to the energy of the diamagnetic currents. Actually the

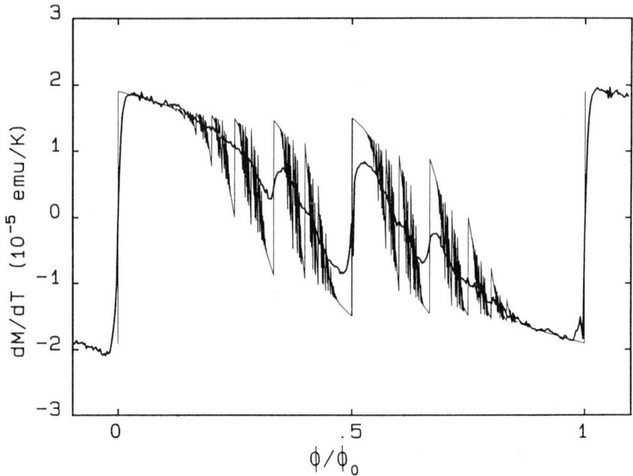

Figure 8. dM/dT as a function of ϕ/ϕ_0 for the first period. Experimental curve (heavy line) obtained at $T = 3.11\,K$ with a temperature modulation $dT = 12.7\,\mathrm{mK}$. Theoretical curve (thin line) from the derivative $d\epsilon/d\gamma$. Magnetization jumps are observed on the experimental curve up to $q = 5$ [21].

magnetization jumps denote discontinuous modifications of the set of quantum numbers which rules the ground state as the magnetic flux is scanned between $\phi/\phi_0 = 0$ and 1.

Most features of the experimental curves can be understood in the framework of the mean field theory. As introduced in the first section of this paper, the equilibrium magnetization of a superconducting network is the product of the superconducting electron density $|\Psi_m|^2$ by the field derivative of $T_c(H)$. Since in the network geometries, both N_z and β_A are close to 1, it is justified to drop the factor $\mu^2(1 - N_z/\beta_A)$. The derivative dM/dT assumes a simple expression as a function of the slope of the lowest Landau level obtained from the network formalism

$$\mu_0 \frac{dM}{dT} = \frac{\phi_0}{2\pi\xi(0)^2 T_c} \frac{1}{2\kappa^2\beta_A} \frac{u}{2\sin u} \frac{d\epsilon}{d\gamma} \tag{30}$$

The steplike curve shown in Fig. 8 is the theoretical expectation value for dM/dT. The only free parameter is the amplitude which is mainly controlled by the GL parameter κ.

Coming back to the energy spectrum shown in Fig. 4, one notices that the aspect of the spectrum in the vicinity of rationals (see for example $\phi/\phi_0 = 1/3$), is a replica of the whole energy spectrum. More specifically, as discussed in the theoretical section, this spectrum forms in the vicinity of rationals a set of discrete energy levels linear in the field which is reminiscent of the Landau level structure in free space. Investigation of the continuum limit near a rational reduced flux (see section 3.1), and advanced theoretical treatments [16] have shown that these levels have a cusp-like structure with different slopes at left and right hand. This new Landau level structure, associated with giant quantized orbits formed by coupling many superunit cells $q \times q$, also reflects an interesting modification of the Bohr-Sommerfeld quantization rule [16,17]. The asymmetric jumps are apparent on the experimental curve. One notices that the fine structure becomes less and less pronounced as q increases. A possible explanation for this loss of coherence could be the effect of superconducting fluctuations or the presence of topological disorder.

4.5 Critical Currents

Resistive T_c measurements as well as magnetization experiments probe the properties of the equilibrium superconducting state. The ability of a superconducting network to carry a transport current relates to different properties of the superconducting state. Different approaches have to be considered according to the strength of the internode coupling. In weakly coupled networks, i.e. when the coupling energy is smaller than $k_B T$, the fluctuations in the phase of the order parameter are essential, and control the transition to the resistive state [22, 23]. The mechanisms for the appearance of dissipation involve time dependent voltages given by the Josephson relation. Since we focus here on strongly coupled wire networks, the amplitude and phase of the order parameter are well established below T_c, and therefore only the supercurrent carried by the whole superconducting condensate has to be considered. Again this property is related to the energy band spectrum of the paired electrons on the underlying lattice.

Several attempts have been made to determine the critical current of wire networks from the GL theory. H. Fink et al. [24] have solved the non-linear GL equations in simple network geometries. Treating the non-linear term of the GL energy as a perturbation, Y.Y. Wang et al. [25] have been able to reduce the calculation of the critical current to an energy band problem. The magnetic field effects predicted in these different approaches have many qualitative similarities with that obtained in the original study of the frustrated XY model ground state by Teitel and Jayaprakash [19]. Detailed measurements of the critical current of wire networks have been performed only very recently. O. Buisson

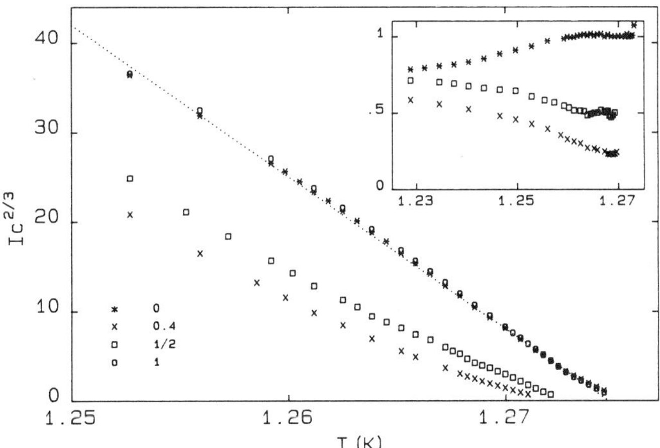

Figure 9. Plot of the 2/3 power of the critical current of a square network as a function of temperature at different reduced fields: $(*)$ $\phi/\phi_0 = 0$, $(\times)$ $\phi/\phi_0 = 0.4$, $(\square)$ $\phi/\phi_0 = 1/2$ and $(\circ)$ $\phi/\phi_0 = 1$. Inset shows the T-dependence of the prefactor $C(H)$ for the first three field values [26].

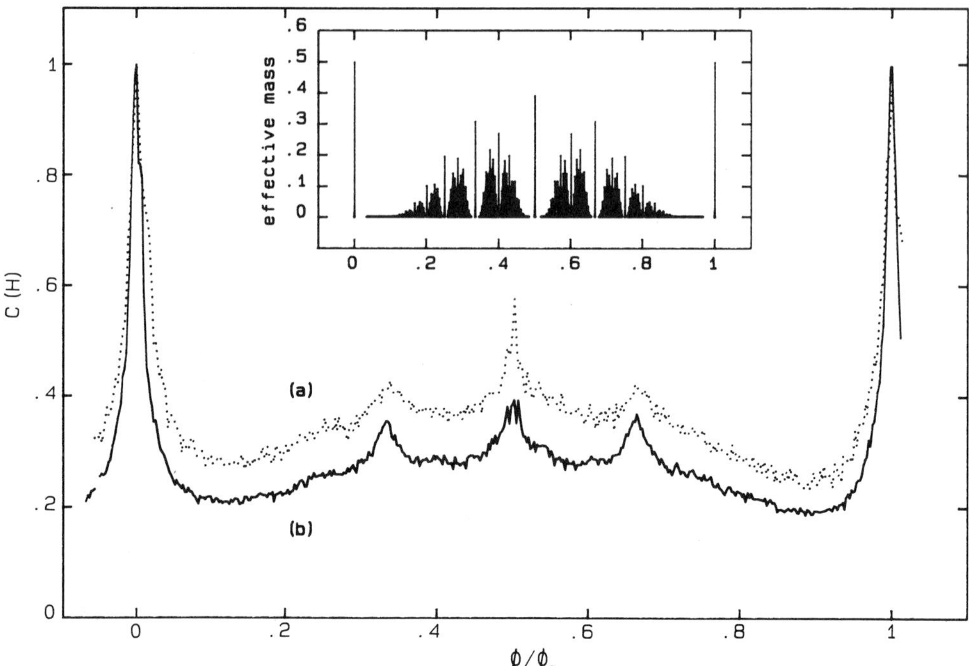

Figure 10. Plot of the prefactor $C(H)$ as measured in a square network of width 30 cells voltage criteria for I_c determination are respectively 50 nV (dotted line) and 20 nV (continuous line). In the inset the vertical bar indicate the theoretical variation of the effective mass term $\sqrt{m^*/m_{\mathrm{eff}}}$ (see eq. (32)) at some rational values $\phi/\phi_0 = p/q$ [26].

et al. [26] have investigated the critical current of various Al networks patterned in the shape of strips of finite width: 30 cells, 10 cells and 1 cell (ladder). Experiments were performed within a few mK below the mean field transition temperature ($\xi > a$ and effective penetration depth λ_{eff} larger than the width of the sample). At zero magnetic field, the current-voltage characteristic exhibits discontinuous voltage jump at a well-defined critical current value with no evidence of fluctuation induced voltage at the foot of the transition. In small magnetic fields ($\phi/\phi_0 \sim$ a few percent), the voltage discontinuity is preceded by a smooth voltage increase in the nV range, strongly field dependent, which indicates some precursor dissipation induced by the presence of vortices. This low voltage tail makes the critical current determination sensitive to the voltage criterion.

The main experimental results of [26] are:

a) The critical current follows the $T^{3/2}$ dependence of the depairing limit. This is illustrated in Fig. 9 which show a plot of $I_c^{2/3}$ vs temperature. For example the zero field curve shows a linear variation from which a critical current density of $J_c(0) = 8.8 \cdot 10^{10}\,\mathrm{A/m^2}$ is deduced at zero temperature. This density is close to the depairing limit of a single filament $J_c(0) = 9.0 \cdot 10^{10}\,\mathrm{A/m^2}$, as estimated for the actual material parameters of the Al films. A linear dependence is also observed at finite magnetic fields. The line crosses the horizontal axis at a lower temperature which is identified as the field dependent critical temperature $T_c(H)$. In addition the slope is strongly reduced and exhibits a very singular field dependence which is shown in Fig. 10. Deviations from linearity are discussed in [26].

b) The parameter $C(H) = I_c/I_c(0)(1-T/T_c(H))^{3/2}$ as deduced from the experimental critical current curves is strongly dependent on the magnetic field. Its variations with respect to ϕ/ϕ_0, shown in Fig. 10, consist of sharp peaks at integer and rational reduced fluxes on top of a uniform background.

Close enough to T_c a simple argument gives the depairing critical current of a periodic wire network. Since the order parameter is a Bloch state Ψ_k, the GL pair current density carried by the superconducting phase, $J = 2e\Psi^2 v_s$, can be written in terms of the group velocity $v_s = dE_k/d(\hbar k)$ of state k.

Accordingly, the corresponding pair current density can be written as

$$J_k = \frac{J_n}{\beta_A}\left(1 - \frac{2m^* E_k}{\hbar^2}\xi^2\right)\frac{m^*}{\hbar^2}\xi\frac{dE_k}{dk} \tag{31}$$

where $\mu_0 J_n = \phi_0/2\pi\xi\lambda^2$ is the critical current of the London theory. This expression is very similar to the pair current of a single 1 D filament. 2 D network effects are included in the dispersion equation $E(k)$. In the above expression, β_A plays the role of the inverse participation ratio of state Ψ_k. High β_A values correspond to a wavefunction having a strongly localized character and therefore a poor ability to carrying current.

Near the phase boundary, one can define, as in usual band theory, an effective mass m_{eff} from the quadratic expansion of $E(k)$ near its minimum value E_m. The depairing critical current, as defined as the maximum pair current, is obtained in the same way as in a single wire as function of the band parameters E_m and m_{eff}

$$J_c = \frac{2}{3\sqrt{3}}\frac{J_n}{\beta_A}\sqrt{\frac{m^*}{m_{\text{eff}}}}(1 - T/T_c(H))^{3/2} \tag{32}$$

with $T_c(H)$ given by the mean field expression (eq. (8)).

This approach explains the main experimental observations, in particular the $T^{3/2}$ dependence. The most interesting result is the prefactor $C(H)$. It turns out that the external magnetic flux not only induces singularities on the $T_c(H)$ curve but it also strongly affects the critical current slope through the effective mass at the band edge in the following way.

An interesting property of the Landau levels spectrum is the emergence of bands at commensurate fluxes (rational ϕ/ϕ_0). Strictly speaking, in a perfect infinite lattice, the bandwidth is non-zero at rational fluxes $\phi/\phi_0 = p/q$ only (see eq. (26)), i.e. when the wavefunction is an extended state. It is expected to cancel at irrational fluxes where electronic states are localized. Since the effective mass is proportional to the reciprocal of the bandwidth, one expects, from eq. (32), a highly singular variation of the critical current as function of the magnetic flux. The inset of Fig. 10 shows the theoretical effective mass ratio obtained from the dispersion equation for the infinite lattice $P_q(\epsilon) = 2 + 2\cos qk_x$. The resulting critical current curve consists of a discrete set of δ-functions located at rational fluxes. The asymptotic low field limit $\sim \exp(-\pi\phi_0/2\phi)$ is exponentially small (eq. (25)), as well as the asymptotic limit near rationals. It is interesting to note that the field dependence of the critical current is very similar to that predicted from the models of Josephson junctions arrays [19,20]. In fact the physical arguments are very similar. To a certain extent, the phase fluctuations of the order parameter can be ignored in these strong coupled systems.

This model which describes the critical current in terms of band properties of the lattice explains reasonably well the experimental observations on a square network.

Indeed the critical current measures the ability of the superconducting wavefunction to carry a current. Following this idea, the introduction of disorder should depress the critical current by turning extended states into localized states. A more refined theory should include finite size and disorder effects.

5   Non-Periodic Networks

The need for understanding the role of disorder in granular and composite superconductors was the original motivation of de Gennes [7] and Alexander [8] to introduce the wire network models. This approach was one of the simplest ways to model the influence of disorder the emphasis being on geometrical inhomogeneities in the superconducting volume. In the last few years, a wide diversity of network geometries has been explored in order to test different approaches for randomness. We focus on three examples where the dimensionality plays the dominant role. Up to now the study of these exotic geometries was restricted to the phase transition line only.

5.1   Fractal and Percolative Networks: the Role of Disorder

Self similar superconducting networks were first considered as a model for investigating the diamagnetic properties of random superconducting systems [8] near the percolation threshold. Such systems have a natural fractal structure which cannot be represented by periodic models. The Sierpinski Gasket (SG), which has a built in scale invariance has been widely used as a simple representation of this class of fractal structures. Recently, several groups [27, 28] have studied the phase transition line of artificial wire networks having the shape of a Sierpinski Gasket in order to find evidence for the expected scaling properties of these structures.

Figure 11 shows an example of superconducting SG. At each iteration, the linear size is increased by a factor 2 while the total wire length increases by a factor 3. Such construction has a fractal dimension $\bar{d} = \ln 3/\ln 2 = 1.585$. It was pointed out by Alexander and Orbach [29] that this reduced dimension was not sufficient to characterize the physical properties of fractal objects. At least one more dimension, the spectral (or fracton) dimension $\tilde{d} = \bar{d}/(1 + \Theta/2)$, where Θ is the anomalous diffusion exponent, is necessary to describe harmonic properties (diffusion, vibration, ...). The spectral dimension can be calculated exactly for the SG, $\tilde{d} = 2\ln 3/\ln 5 = 1.365$.

The $T_c(H)$ curve obtained experimentally in 10 stages [27] and 6 stages [28] artificial wire gaskets was found to exhibit an extremely rich fine structure (Fig. 12) which extends over four orders of magnitude in the magnetic field (1mOe to 10 Oe for a gasket with elementary triangle of side $a = 3.2\mu$m). These structures are indicative of quantization effects occurring in the loops of various sizes which form the different stage of the fractal structure. The useful field parameter is the reduced flux $\alpha = \phi/\phi_0$ in the smallest cell of the network, i.e. stage 0 of the gasket. It controls the magnetic length $\ell_H \sim a/\sqrt{\alpha}$, which can be tuned over the different length scales of the structure. Let us summarize the novel features of these self similar objects (Fig. 12).

The high field regime ($\phi/\phi_0 > 1$) is dominated by the single loop effects as in regular arrays. The $T_c(H)$ curve is a periodic function of ϕ/ϕ_0, with period 1. The magnetic length probes the structure at length scales below (a) and therefore cannot feel the self-similarity.

More interesting is the low field regime ($\phi/\phi_0 \ll 1$) which clearly exhibits the self-similar properties of the gasket. The scale invariance, which is obvious on the geometrical

Figure 11. Micrograph of part of an Al network having the shape of a Sierpinski Gasket. The smallest wire is 3.2 μm length. The linear size is increased by a factor 2 at each iteration while the total wire length increases by a factor 3.

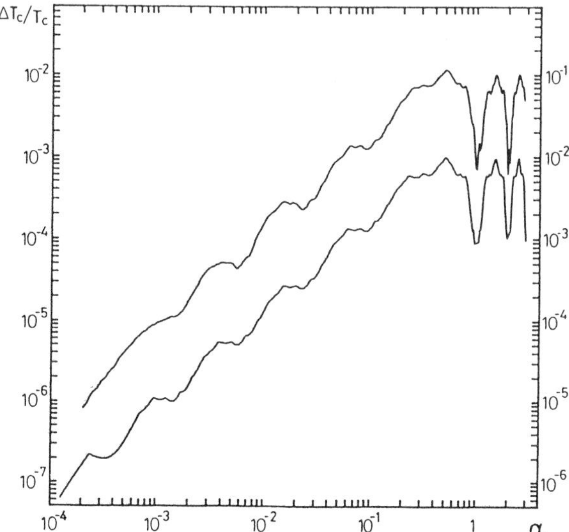

Figure 12. Phase transition line of a 6-stages SG in reduced units. Upper curve experiments (fitting parameters $T_c = 1.230$ and $\xi(0) = 0.34\mu$m). Lower trace theory for stage 6. The two curves have been shifted for clarity [28].

construction, is also observed over several levels of hierarchy from stage 0 ($\alpha = 1$) to stage 5 ($\alpha = 4^{-5}$) on the superconducting phase transition line of the network (Fig. 12). Using nesting properties for the gasket it is possible to show [8] that the sequence of corresponding points $\Delta T_c(\alpha_n)$ for $\alpha_n = 1/2 \times 4^n$ falls on a simple curve defined by $\Delta T_c = \alpha^\gamma$ where $\gamma = \bar{d}/\tilde{d}$ is the ratio of the fractal to the spectral dimensions introduced above. The exponent γ obtained from the experimental curve coincides with the expected value for the Sierpinski Gasket $\gamma = \ln 5/\ln 4$ within a few percent. Oscillations superposed onto the power law are a particular manifestation of a very general property: the periodic correction to scaling. In a physical problem where renormalization group transformations are exact (this is the case here, at each iteration $L \to 2L, \phi \to 4\phi, \ldots$), periodic corrections to power law are expected. A discussion of this effect is given in [16, 28] in the more general situation of the weak localization problem. The details of the fine structure of the $T_c(H)$ curve can be described very accurately by the lower edge of the electronic spectrum of the Schrödinger equation on the gasket [28, 30] (Fig. 12 lower trace).

By measuring the phase boundary of a triangular array of third order SG's, Gordon et al. [31] have modeled the dimensionality crossover from fractal behaviour at small length scales to 2 D behaviour at larger scales. The array adds length scale inhomogeneity to the pure fractal networks. At small scale, i.e. when the magnetic length is smaller than the size of the third order gasket, the SG arrays is inhomogeneous and exhibits self-similarity (Fig. 13 high field) as in the infinite SG. As the magnetic probe length is increased, the experiment shows a net crossover to a 2 D behaviour characterized by a linear critical field slope (Fig. 13a low field limit). In this ideal system, the crossover length measures the size of the largest voids in the structure, which is the size of the third order gasket. This length plays the role of the percolation correlation length (here $\xi_p = 8a$).

The percolation networks provides a more realistic model for real inhomogeneous structures with the addition of randomness and dead-end bonds [31]. The existence of a single characteristic length scale is a simplifying feature common to all aspects of the

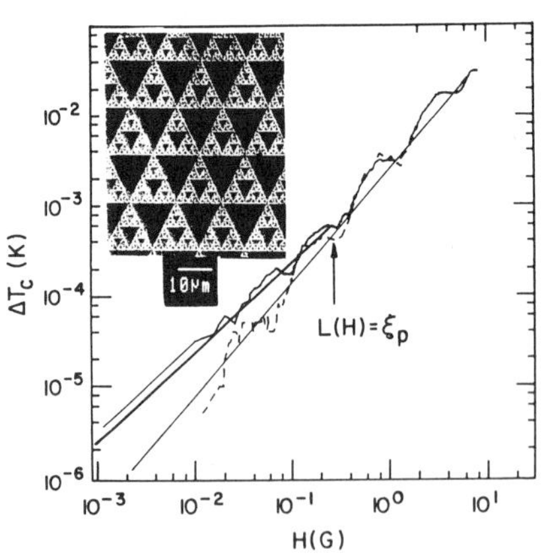

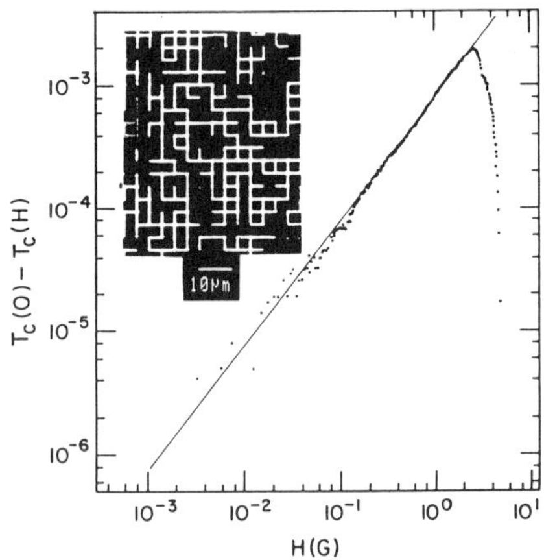

Figure 13. Left: phase transition line for a periodic array of 3 stages SG (solid curve) in comparison to the pure SG (dashed curve). The straight line has a slope $m = 1$ for the array and $m = 1.161$ for the SG. Right: same for a percolation network with $p = 0.56$. Slope $m = 1$ is observed over the whole field range [31].

percolation problem. This percolation correlation length $\xi_p = a(p - p_c)^{-\nu}$ diverges near the percolation threshold p_c with a well characterized exponent ($\nu = 4/3$ in 2 D), and measures the length scale at which the infinite cluster becomes uniform. There has been quite an extensive literature on the superconducting properties of random systems and their relation to the percolation problem in the early eighties (see for example [32]). The comparison with the SG array allows a didactic view of the main features of these systems. In particular it is natural to expect, in the critical line, a crossover comparable to that of the SG array with an additional control parameter given by p. At large length scale, $\ell_H \gg \xi_p$, the network is seen as an homogeneous 2 D object with some microscopic disorder. The T_c line must be linear in field with a reduced slope as result of the weakening of the apparent coherence length near p_c. The slope should vary according to $\Delta T_c \sim H(p - p_c)^k$ where the positive exponent k depends on the scaling models [33].

At short length scale, i.e. as the field is larger than ϕ_0/ξ_p^2 one expects a fractal behaviour similar to that observed in the SG, $\Delta T_c \sim H^{1+\theta/2}$. In the percolation case, the anomalous diffusion exponent $\Theta_{perc} = 0.8$ is much larger than in the SG, and therefore a power law exponent of 1.4 should be observed. In addition, the SG, no fine structure is expected since the different loop sizes are random. Here the scaling is not exact, the self-similarity is only a statistical property.

By using electron beam lithography, Gordon et al. [31] have fabricated artificial percolation networks of size 800×800 cells. The wire bond are present with probability p. Since $p_c = 0.5$ for 2 D bond percolation, $p_c = 0.5$, the estimated correlation length in the studied samples (0.54, 0.56, 0.60), is quite long, bounded by $\xi_p(0.6) = 37\mu$m.

Surprisingly, the fractal behaviour expected over most of the experimental range of field (0.01 to 10 Oe, which corresponds to a probe length ℓ_H varying from 40 to 1 μm), is not observed (Fig. 13). Instead, a very smooth linear T_c variation is obtained up to the field ($H \approx \phi_0/a^2 \approx 2\mathrm{Oe}$) where fluxoid quantization effects in the smallest loop become dominant. It seems that the sample remains in the homogeneous regime over the whole field range. The slope of the linear field dependence is characterized by the parameter A defined by $A = H\xi^2(T)/\phi_0$. The plot of the measured A values, as function of p, leads to an exponent $k \approx 1$, close to the prediction of [31], $k = \nu\theta = 1.06$, and in agreement with results of numerical simulation on site percolation networks [34]. The observed slope A shows no evidence of saturation as p approaches p_c. This is consistent with the absence of crossover mentioned above.

Another evidence of the failure of the fractal model in artificial percolation networks has been given by Steinmann et al. [35], who observed both the low field homogeneous regime and the high field periodic regime (ℓ_H smaller than the lattice cell), but not the intermediate self similar regime. Rather they observed giant reproducible oscillations in the phase transition line (Fig. 14) which were indicative of fluxoid quantization effects at an intermediate characteristic loop size close to the estimated percolation length. Such oscillations are not expected in large arrays. Presently, there is no theoretical interpretation of this effect.

Recently, Soukoulis et al. [36] have examined the implication of the extended or localized nature of the eigenstates in site diluted square networks for the $T_c(H)$ curve. A small amount of randomness introduces a mobility edge that separates the domain where the superconductivity nucleates in small isolated regions from the one where it spreads macroscopically over the whole specimen. They compared the experimental data of [31] with their determination of the parameter $A \sim (p - p_c)^k$ as obtained respectively for the band edge ($k = 2.5$) and for the mobility edge ($k = 0.84$). They found that the experimental curve lies in between the two numerical predictions which only give bounds

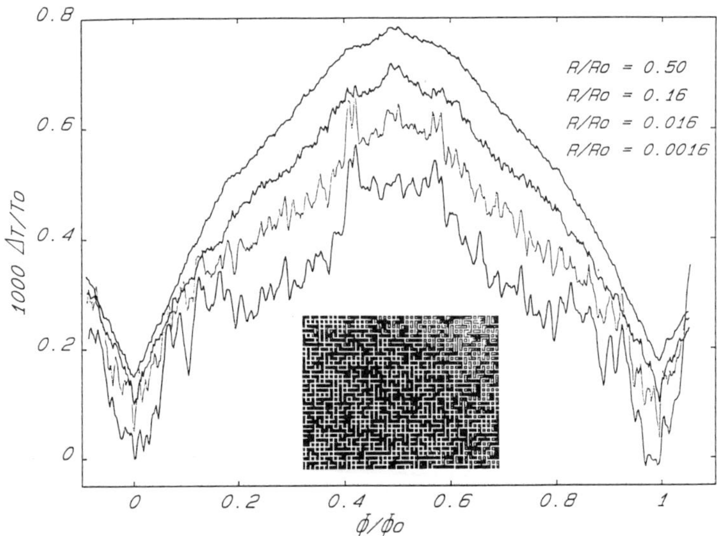

Figure 14. Critical temperature for a percolation network ($p = 0.6$) as measured at different levels of the resistive transition (R_0 normal state resistance). Giant reproducible oscillations are observed on the low resistance traces [35].

to the real T_c curve. In fact, the non-linear terms of the GL equations, which are neglected in this mean field treatment, have to be taken into account in the determination of the $T_c(H)$ curve. Soukoulis et al. suggest that the breakdown of the mean field theory may explain the lack of dimensionality crossover in percolating superconducting networks.

On the other hand the observation of unexpectedly large reproducible T_c oscillations raises the question of fluctuation effects in these objects. Looking at the structures of the backbone near the percolation threshold makes it obvious that superconductivity of the whole structure is controlled by a very small number of small "strategic" bridges whose properties probably cannot be explained by the usual scaling theories. The problem of superconducting percolation networks is still open and calls for a more systematic study.

Other realizations of weak geometrical disorder were made by Benz et al. who introduced deliberate positional disorder in square Josephson junction arrays [37]. The main oscillations (ϕ/ϕ_0 integer or fractional) in samples with disorder were found to decay with increasing field at a rate depending on the amount of disorder. For example the peak to peak amplitude of the principal oscillation at integer ϕ/ϕ_0, which measures short length scale correlations, decayed with a linear envelope, disappearing at a critical reduced flux given by $\phi/\phi_0 \approx 0.95/\Delta^*$ where Δ^* is the width of the probability distribution of site displacements. Mean field simulations for different kinds of lattice distortion (gaussian or truncated uniform distribution) are in good agreement with experimental magnetoresistance measurements on Josephson arrays. Granato and Kosterlitz have considered the effect of positional disorder on long range order [38].

5.2 Quasi-Periodic Networks

The study of quasi-periodic superconducting networks was stimulated by the discovery of quasi-crystalline alloys and by the need for understanding the electronic structure of this new class of lattice which has neither translational invariance, nor scale invari-

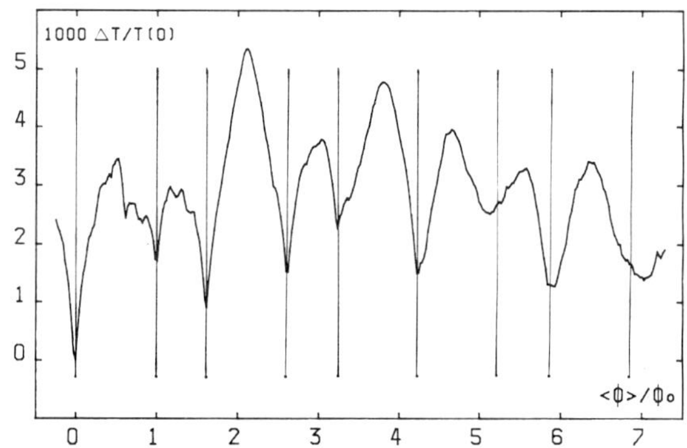

Figure 15. Phase transition line (experimental) for a Penrose network. The horizontal axis is the ratio $\langle \phi \rangle / \phi_0$ of the average flux per cell to the flux quantum. The vertical bars mark the quasi-periodic sequence $1, \tau, \tau^2, \ldots$ [40].

ance. Various quasi-periodic networks have been considered, most have been made by the repetition of two cells with incommensurate areas as, for instance 2 D Fibonnacci sequences, eightfold symmetric Penrose lattices [39], fivefold Penrose tiling [40], etc. In these systems, due to the incommensurability of the basic cell area there is no way to satisfy fluxoid quantization simultaneously in all closed path, and this contrasts with the simple periodic networks or regular fractal lattices discussed earlier. Experiments on quasi-periodic networks have revealed a very complex phase boundary which greatly differ from the regular arrays (Fig. 15). The relevant field parameter is the average flux per cell. The main dips can be indexed by considering the rational approximants of the cell area. In the 5-fold Penrose network for instance, the golden mean $\tau = (\sqrt{5} + 1)/2$ is approximated by the Fibonacci sequence $2/1$, $3/2$, $5/3$, $8/5$, $13/8$ etc. One finds, instead of the single main period ϕ_0, a quasi-periodic sequence of two periods which correspond to the successive increment of the fluxoid numbers $((2,1), (3,2), (5,3) \ldots)$ in both type of cells. A series of sharp dips is observed in most quasi-periodic structures at low fields (less than one flux quantum in every cells of the networks). These dips, predicted by the mean field GL-theory, are the equivalent of the rational peaks $\phi/\phi_0 = p/q$ of the periodic structures. They are indicative of the effect of long range order and must correspond to some favorable arrangement of the flux lattice on the underlying quasi-periodic network. This is a very complex situation, as compared to the regular networks, since two more parameters enter: the cell area ratio (which controls the short scale behaviour) and the tile arrangement (which controls the long range order). However, the mean field theory has shown itself extremely accurate in fitting experimental data in aperiodic networks in general [41]. An intermediate situation was discussed by Santhanam et al. [42] who have considered periodic networks with complex unit cells made of a mixture of squares and equilateral triangles. For a more complete discussion on these structures we refer to the recent reviews given in [1].

5.3 Singly Connected Superconducting Dots

One of the simplest elementary superconducting circuits one can think of is a small 2 D superconducting disk. In some sense it may be viewed as the analogy of the quantum dots, or the quantum billiard, with specific properties given the nature of the super-

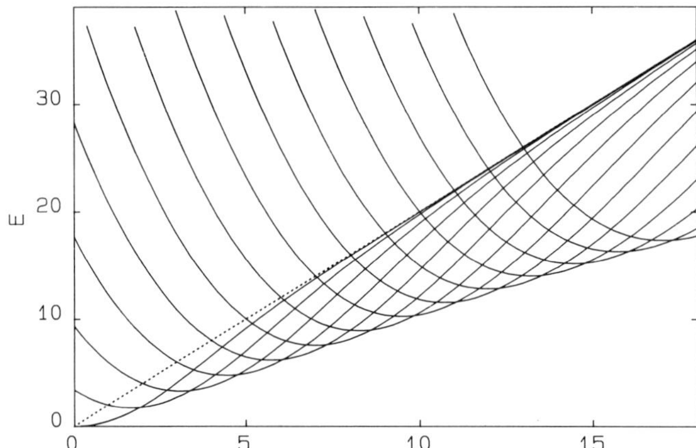

Figure 16. Lower energy levels of a charged particle confined in an isolated circular disk vs reduced applied magnetic flux $\phi/\phi_0 = \pi R^2 H/\phi_0$. The dashed line shows the first Landau level. BC are those for a superconducting to insulator interface.

conducting wavefunction. Recent experiments on finite Al disks [43] have revealed new interesting features of the edge states in such geometries.

Consider a perfect circular disk exposed to a perpendicular magnetic field. The energy spectrum of eq. (2) consists of Landau levels with additional discrete energy levels corresponding to the edge states. For the usual BC ($\Psi = 0$ at the edge) these energy levels are enhanced as compared to the bulk Landau levels. Here, instead, the specific BC at the superconducting interface, lead to a lowering of the edge state energy levels. Fig. 16 shows the prominence of these discrete levels below the lowest Landau level (dashed line). Two main regions of the magnetic field can be distinguished according to the ratio of the disk radius to the magnetic length.

a) In the low field limit ($\ell_H = \sqrt{\phi_0/\pi H} > R$), the superconducting wavefunction is in the $0\,$D state and spreads almost uniformly over the whole disk area. The pair velocity varies linearly as $A(r) = \frac{1}{2}H \times r$ where r is the radial distance leading to a kinetic energy quadratic in magnetic field.

b) In the high field limit ($\ell_H < R$), the wavefunction is localized within a distance ℓ_H near the edge of the disk. From the point of view of superconductivity, two distinct effects are superposed. Firstly the existence of edge states is well-known at the contact between a superconducting volume and an insulator. In the simple case of a half infinite plane (in this example the second effect is absent) two opposite pair currents develop near the interface at an approximate distance ℓ_H of each other. It is known [4] that the energy of these edge states is linear in field and smaller by a factor 0.59 than that of the first Landau level. Hence the boundary at the outer radius of the disk produces an average linear field dependence of the lower edge of the energy spectrum as shown in Fig. 16. The second effect follows from the ring shape of the edge state. Bohr-Sommerfeld quantization rules impose an additional condition to the closed orbits which results in an alternate unbalance of the two edge currents and therefore a periodic modulation superimposed to the

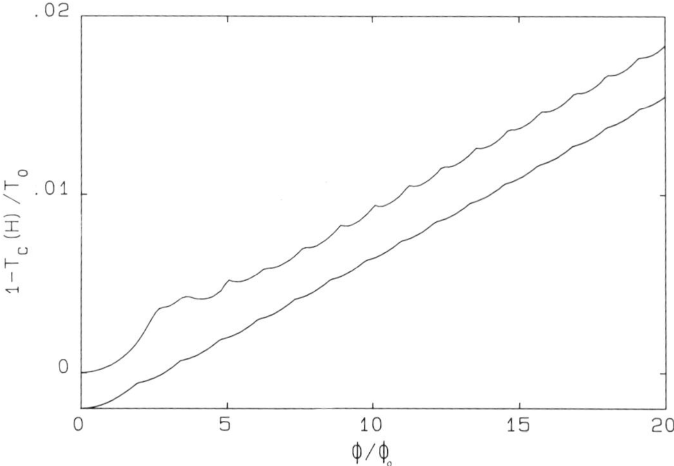

Figure 17. Variation of the transition temperature of a 14.6μm diameter superconducting disk. The horizontal axis unit corresponds to a magnetic field of 0.124 Oe. Upper curve theory, lower curve: experiment [43].

global linear field dependence. At high fields the period of this modulation is expected to follow the asymptotic limit

$$\Delta H = \frac{\phi_o}{\pi R^2}\left(1 + \sqrt{0.592}\frac{\ell_H}{R}\right) \tag{33}$$

which corresponds to a change of one flux quantum through a surface slightly smaller than the disk area. This effect is similar to the oscillation of the T_c line and magnetization of multiply-connected specimens (rings, periodic networks...). It is observed here (see Fig. 17) on a 1 D state within a bulk sample that results from BC at the disk circumference.

One must emphasize that the occurrence of oscillations at the edge of the spectrum is far from being trivial. In fact this part of the spectrum is extremely sensitive to the actual BC. For example, the BC used by Sivan et al. [44] in the quantum dot problem, or by Nakamura et al. [45] drive these states into the energy gaps between Landau levels and therefore suppressed the oscillatory effect of the lowest level. These BC are in fact similar to those between a superconductor and a normal metal.

A detailed study of the behaviour of superconducting disks of various shapes has been performed by O. Buisson et al. [43]. Fig. 17 shows typical results on the phase transition line of a circular disk in magnetic fields up to 2.5 Oe. Both the quadratic low field regime ($\phi/\phi_0 < 3$) and the high field linear regime are clearly observed with a unique fitting parameter $\xi(0) = 0.2\mu$m. The slope of the high field linear part is 0.59 times that of the first Landau level, as expected. Over the whole field range (0 to 6 Oe), about 50 periods of oscillation are observed on top of the linear asymptotic line. The period is found to decrease in good agreement with eq. (8) down to 0.133 Oe at $\phi/\phi_0 = 50$. This limit compares well to that expected from the disk area ($\phi_0/\pi R^2 = 0.124$ Oe) using the measured diameter $2R = 14.6\mu$m. The intermediate field regime is not so well understood and seems to be very sensitive to the BC, presence of contact leads, disk shape, etc.

6 Conclusions

A main line of thought in this paper has been to illustrate how the superconducting properties of artificial networks are governed by quantum interference effects in the underlying lattice structure. Dimensionality can be scanned easily from 0 D (superconducting dots) to fractal dimensionality and 2 D systems. In some sense superconductivity allows to do mesoscopic physics at macroscopic scale, on well characterized man made objects. For the various lattice geometries which have been explored, the mean field GL theory discussed in this paper, and the mapping of the superconducting properties to a band structure problem, explain quite well the fine structure and the frustration effect in the presence of a magnetic field, at least in regular lattices. However, the theory seems to fail in systems with a small amount of disorder.

The question of disorder in superconducting systems remains a major challenge. Intensification of the systematic study of artificial networks with intentional disorder is certainly one of the most promising way to address this very difficult question, both from the experimental and the theoretical point of view.

Up to now, much effort has been devoted to the investigation of the equilibrium properties and transport properties near the phase boundary. There are still many experiments to do, particularly in the area of dynamical properties (noise, microwave studies, ...).

The prodigious capabilities of modern microfabrication techniques provides the physicist with almost unlimited possibilities of realizing small superconducting, semiconducting, or composite coherent objects. The role of thermal and quantum fluctuations in very weak superconducting arrays and the role of charging effects in small tunnel junctions (see Mooij's lecture in this volume) is one of the most exciting open questions in this field.

7 Acknowledgements

Much of the work described here was carried out in CRTBT Grenoble by O. Buisson, J. Chaussy, B. Douçot, P. Gandit, R. Steinmann, and Y.Y. Wang. Special thanks are due to my colleagues in Centre National d'Etudes des Telecommunications Meylan, for their valuable contribution in realizing the electron beam lithography on the samples. Stimulating conversations with M. Giroud, H. Fink and correspondence with F. Nori are also acknowledged.

References

[1] See for a recent review: Coherence in Superconducting networks, J.E. Mooij and G.B. Schön, eds., Physica **B 152** (1988)

[2] Proceedings of the Electronic properties of 2 D systems, to appear in Surf. Sc. (1990), and references therein

[3] V.L. Ginzburg and L. Landau, Zh Eksp. Teor. Fiz. **20**, 1064 (1950)

[4] P.G. de Gennes, Superconductivity of metals and alloys, W.A. Benjamin, ed., New York (1966)

[5] A.A. Abrikosov, Zh. Eksp. Teor. Fiz. **32**, 1442 (1957) [Sov. Phys. JETP **5**, 1174 (1957)]

[6] Y.Y. Wang, R. Rammal, and B. Pannetier, J. Low Temp. Phys. **68**, 301 (1987)

[7] P.G. de Gennes, C.R. Ac. Sci. **B 292**, 9 and 279 (1981)

[8] S. Alexander, Phys. Rev. **B 27**, 1541 (1983)

[9] W.A. Little and R. Parks, Phys. Rev. **A 13**, 97 (1964)

[10] H.J. Fink, A. Lopez, and R. Maynard, Phys. Rev. **B 25**, 5237 (1982); J. Simonin, D. Rodriguez, and A. Lopez, Phys. Rev. Lett. **49**, 944 (1982); J.Riess, J. Phys. Lett. **43**, L277 (1982)

[11] R. Rammal, T.C. Lubensy, and G. Toulouse, Phys. Rev. **B 27**, 2820 (1983)

[12] see for example J.B. Sokolov, Phys. Rep. **126**, 189 (1985)

[13] P.W. Anderson, Phys. Scr. **T 27**, 660 (1989); P.B. Wiegmann, ibid, 160; P. Lederer, D. Poilblanc, and T.M. Rice, Phys. Rev. Lett. **63**, 1519 (1989); R. Rammal and J. Bellissard, Phys. Rev. Lett., to be published

[14] M.Y. Azbel, Zh. Eksp. Teo. Fiz. **46**, 929 (1964) [Sov. Phys. JETP **19**, 634 (1964)]; D.R. Hofstadter, Phys. Rev. **B 14**, 2239 (1976)

[15] Y.Y. Wang, B. Doucot, R. Rammal, and B. Pannetier, Phys. Lett. **A 145** (1986)

[16] Y.Y. Wang, B. Pannetier, and R. Rammal, J. de Phys. **48**, 2067 (1987)

[17] see R. Rammal, in Ref. [1]

[18] B. Pannetier, J. Chaussy, and R. Rammal, J. Phys. Lett. **44**, L853 (1983); B. Pannetier, J. Chaussy, R.Rammal, and J.C. Villégier, Phys. Rev. Lett. **53**, 1845 (1984)

[19] S. Teitel and C. Jayaprakash, Phys. Rev. Lett. **51** (1983); see also S. Teitel, in Ref. [1]

[20] T.C. Halsey, Phys. Rev. **B 31**, 5728 (1985); T.C. Halsey, Phys. Rev. Lett. **55**, 1018 (1985)

[21] P. Gandit J. Chaussy, B. Pannetier, A. Vareille, and A. Tissier, Europhys. Lett., 623 (1987); and in Ref. [1]

[22] B. Jeanneret, Ph. Fluckiger, J.L. Gavilano, Ch. Leemann, and P. Martinoli, Phys. Rev. **B 40**, 11374 (1989)

[23] see H.J. Mooij, this volume

[24] H.J. Fink, D. Rodriguez, and A. Lopez, Phys. Rev. **38**, 8767 (1988); H.J. Fink and V. Grundfeld, Phys. Rev. **B 31**, 600 (1985)

[25] Y.Y. Wang, B. Pannetier, and R.Rammal, J. Physique **49**, 2045 (1988)

[26] O. Buisson, submitted to Europhys. Lett. (March 1990)

[27] J.M. Gordon, A.M. Goldman, J.Maps, D.Costello, R.Tiberio, and B. Whitehead, Phys. Rev. Lett. **56**, 2280 (1986)

[28] B. Douçot, Y.Y. Wang, J. Chaussy, B. Pannetier, A.Vareille, and D. Henry, Phys. Rev. Lett. **57**, 1235 (1986); Y.Y. Wang, R. Steinmann, J. Chaussy, R. Rammal, and B. Pannetier, Jpn. J. Appl. Phys. **26**, Sup. 26-3 (1987)

[29] S. Alexander and R. Orbach, J. Phys. Lett. **43**, L625 (1982)

[30] R.Rammal and G. Toulouse, Phys. Rev. Lett. **49**, 1194 (1982); J.M. Ghez, Y.Y. Wang, R. Rammal, B. Pannetier, and J. Bellissard, Sol. St. Commun. **64**, 1291 (1987)

[31] J.M. Gordon, A. Goldman, and B. Whitehead, Phys. Rev. Lett. **59**, 2311 (1987); and in Ref. [1]

[32] see for example G. Deutscher, in: Applications of percolation, in Chance and Matter, J. Souletie, J. Vannimenus, and R. Stora, eds., North Holland (1987)

[33] R. Rammal, T.C. Lubensky, and G. Toulouse, J. Phys. Lett. **44**, L65 (1983); P.G. de Gennes, in: Percolation, Localization and Superconductivity, A.M. Goldman and S.A. Wolf, eds., Plenum Press, New York (1984); see also Refs. [31] and [32]

[34] J. Simonin and A. Lopez, Phys. Rev. Lett. **56**, 2649 (1986)

[35] R. Steinman and B. Pannetier, Europh. Lett. **5**, 559 (1988); Physica **C 153**, 1487 (1988)

[36] C.M. Soukoulis, G.G. Grest, and Q. Li, Phys. Rev. **B 38**, 12000 (1988)

[37] M.G. Forester, H.J. Lee, M. Tinkham, and C. Lobb, Phys. Rev. **B 37**, 5966 (1988); S. Benz, M.G. Forester, M. Tinkham, and C.J. Lobb, Phys. Rev. **B 38**, 2869 (1988)

[38] E. Granato and J.M. Kosterlitz, Phys. Rev. Lett **62**, 823 (1989); see also E. Granato and J.M. Kosterlitz, in Ref. [1]

[39] A. Behrooz, M.J. Burns, H. Deckam, D. Levine, B. Whitehead, and P.M. Chaikin, Phys. Rev. Lett. **57**, 368 (1986)

[40] Y.Y. Wang, R. Steinman, J. Chaussy, R. Rammal, and B. Pannetier, Jpn. J. of Appl. Phys. **26**, 1415 (1987); see also D.J. Van Harlingen, K.N. Springer, G.C. Hilton, and I. Tien, in Ref. [1]

[41] F. Nori and Q. Niu, in Ref. [1]

[42] P. Santhanam, C.C. Chi, and W.W. Molsen, Phys. Rev. **B 37**, 2360 (1988)

[43] O. Buisson et al., to be published

[44] U. Sivan, Y. Imry, and C. Hartzstein, Phys. Rev. **B 39**, 1242 (1989)

[45] K. Nakamura and H. Thomas,Phys. Rev. Lett. **61**, 247 (1988)

PHASE COHERENCE AND CRITICAL RESISTANCE IN SUPERCONDUCTING GRANULAR FILMS

Shun-ichi Kobayashi

Faculty of Science, University of Tokyo
7-3-1 Hongo, Bunkyoku
Tokyo 113, Japan

A granular superconductor is an ensemble of small superconducting grains, the sizes and shapes of which are randomly distributed to some extent. The grains are interacting with each other through the Josephson coupling which is also random. The physical properties of granular superconductors depend largely on the mean size of grains, d, and the mean Josephson coupling energy, E_J. Usually, d is smaller than the coherence length. Therefore each grain is a so-called zero dimensional superconductor. In the limit of large size and weak coupling, the system can be regarded as a random Josephson network, while in the opposite limit, it is a homogeneously disordered superconductor. Many interesting phenomena are expected between these two limits. In this paper, the experiments of two-dimensional (2 D) thin films of granular superconductors are reviewed.

In 2 D systems, the sheet resistance is independent of the system size, as long as the size is larger than the coherence length of any type. This peculiar feature allows us to investigate directly the physical properties on a microscopic scale, because the sheet resistance of the whole system is essentially the same as that of the constituent element. For 2 D networks of junctions, the sheet resistance is equal to that of the individual junctions, provided that the distribution of the junction resistance is narrow enough and that the network is sufficiently uniform.

The first attention was paid on the competition between the thermal and the quantum fluctuation of the order parameters in small Josephson junction networks [1]. This effect arises from the number-phase uncertainty relation for the quantum condensate. In a very small junction with very small capacitance, the charging energy E_c associated with a tunnelling of a single cooper pair can be comparable with the thermal energy $k_B T$. When $k_B T$ is larger than E_c but smaller than E_J, the network can establish a global coherence, because the thermal fluctuation of the number of the pairs allows the phase of the pairs to be well defined. On the contrary, at low temperatures where $k_B T \ll E_c$, the number of pairs cannot fluctuate and inevitably the phase fluctuates. Then $E_J \cos \theta$

Quantum Coherence in Mesoscopic Systems
Edited by B. Kramer, Plenum Press, New York, 1991

averages out and the coherence disappears. Therefore, a re-entrance might appear in the phase diagram of the temperature and the Josephson-coupling-strength plane.

Resistivity measurements [2] where performed on a series of samples which were prepared by repeating vacuum evaporation of small amount of tin and very light oxidation. By this method, we get small particles making contact with neighbors via thin layers of insulating oxide. This type of granular films (GF) have been extensively used by the author's group. The average diameter d of the particles was 30 nm with which particles have certainly bulk BCS properties. The temperature dependence of the sheet resistance of these films showed minima below T_c when their normal sheet resistance fell in a certain range. At the higher side of this range, the films behaved as semiconductors, and at the lower side, they underwent a superconducting transition with decreasing temperature. This behavior was naturally assigned to the expected re-entrance. Later, however, the theory of re-entrance of this sort has been reexamined by several authors [3]. At present, the theoretical necessity of the re-entrance is still not clear.

In the limit of small d, there will be no difference between a GF and microscopically disordered continuous films. For the latter, the Anderson localization phenomenon is important. The theory predicted that the superconducting transition temperature will be suppressed by the random potential and the electron-electron interaction [4]. This effect was first observed in thin films of Zn [5]. It was also confirmed in samples prepared by various methods [6, 7]. The improved theory [8] can reproduce the full profile of T_c as a function of the sheet resistance. The sheet resistance at which T_c vanished is not universal but depends on material parameters.

Meanwhile, a new aspect of a small single junction has been revealed. Let us consider a very small junction in which E_c, E_J and the BCS energy gap are comparable. For a crude estimation, let all these three be equal to 1 K. To satisfy this, the capacitance should be of the order of 10^{-15} F, requiring the junction area as small as 50×50 nm^2 for the barrier thickness of several Å. The critical Josephson current should be of the order of 10^{-8} A. This is equivalent to the normal resistance of the junction being about $R_Q = h/4e^2 \ (= 6.45\,k\Omega)$. Here we used the relation between the critical current and the normal resistance given by Ambegaokar et al. [9]. It is important to keep in mind that this normal resistance has nothing to do with the ohmic dissipation of the junction but represents the inverse probability of the electron tunnelling. E_J induces zero voltage Josephson current in the junction, while E_c tends to prevent the tunnelling. Therefore, when the size of the junction is further reduced, it is naturally expected that the junction would not carry a supercurrent.

The role of normal resistance was, however, studied from somewhat different points of view [10, 11, 12]. The model considered in these theories was an ideal non-dissipative junction shunted by an ohmic resistor. The dissipation in this resistor is represented by an ensemble of harmonic oscillators coupled to a simple system of the Josephson junction, and it was shown that this additional system can be considered as the effective capacitance of the junction. When the intrinsic capacitance is very small (as it is in very small junctions), the additional effective capacitance becomes dominant and determines whether the phase difference between two superconductors of the junction stays constant or fluctuates quantum mechanically. In the former case, a zero voltage Josephson current can flow, while in the latter, the junction is always resistive. The critical resistance was shown to have a universal value, R_Q. It was also shown that in a 2 D regular network of small Josephson junctions, the critical sheet resistance that separates the superconducting from the resistive phase is also R_Q, although there appears a numerical factor of the order of one [12]. There exists another type of theory, in which the dissipation associated with the quasi-particle tunnelling is considered. This theory also gives the result that the critical sheet resistance for the phase coherence in small junctions is R_Q [13].

The first report which claimed the observation of R_Q was of the experiment on the films of Sn prepared by depositing metal vapour onto a liquid He cooled substrate (QCF, quench condensed film) [14]. The thickness was increased step by step and the temperature dependence of the resistance was measured et each step keeping the system always lower than 17 K. With increasing thickness, the system showed successively three phases: semiconducting, normal-metallic and superconducting, as shown in Fig. 1a. The phase boundary from normal-metallic to superconducting behavior appeared at a sheet resistance very close to R_Q. The resistance minima similar to those in ref. 2 were also observed in semiconducting region.

The old data of the GF taken to see the existence of the re-entrance [2] was re-examined, and it was found that the boundary between the superconducting and semi-conducting phases falls at R_Q. The substantial difference between the results of GF and QCF is that in the former no normal-metallic phase exists.

After then, more data have been accumulated for both types of samples. Ref. [15] reconfirmed that the critical sheet resistanceindexcritical sheet resistance in GF of Sn is about $4\,k\Omega$ which is close enough to R_Q (Fig. 1b). For GF of Al [16] the same critical sheet resistance was obtained as shown in Fig. 1c. The particle size d is smaller than d of the GF of Sn in Fig. 1b. For QCF, Ga, Pb [17] and Nb [18] were examined and the same results with R_Q as that for QCF of Sn were obtained. A new preparation technique, which used very thin film of Ge to underlie QCF (we shall call it QCF/Ge), was applied to Pb and Bi [19]. QCF/Ge is known to provide more continuous film than simple QCF. QCF/Ge of Pb and Bi both showed a smooth decrease of T_c, while the critical resistances were different; for Bi close to R_Q, and for Pb much higher.

The experimental results of T_c so far observed can be classified into three types, which are schematically shown in Fig. 2. The results of GF (Sn [2, 15] Al [16]) and QCF (Sn [14], Pb, Ga [17]) belong to (a). Type (b) applies only for QCF/Ge (Bi [19]). Many of continuous dirty films, QCF/Ge (Pb [19]), and GF (Zn with very small d [5]) behave as (c).

Among the structures of GF, QCF, QCF/Ge and dirty continuous films, that of GF is most suitable to be identified as the Josephson junction array, because it consists of well defined grains of superconductor, the size of which is large enough to be free from order parameter fluctuations and the quantum size effect, and the controlled oxide barriers. For GF, the parameters such as E_J and the capacitance of the junctions can be easily estimated. For example, the parameters for the samples in Fig. 1b are about 3 K and 10^{-16}F, respectively. The result that all the GF's have the critical resistance close to R_Q suggests the validity of the theories for R_Q, although it is not yet possible to judge which is the real underlying mechanism.

The QCF of Sn, Pb and Ga show the behavior of type (a) with R_Q. It is conjectured [17] that the QCF is a thin film with microscopic cracks which serve as the tunnelling junctions, and therefore, the system is a random Josephson network. However, because the structure changes when the film is warmed up, direct observation by microscope is difficult and has not been done. The most clear difference between QCF and GF is that the normal-metallic phase exists in the former but not in the later. In this aspect, QCF resembles the continuous disorder films which, however, do not show universal R_Q. There

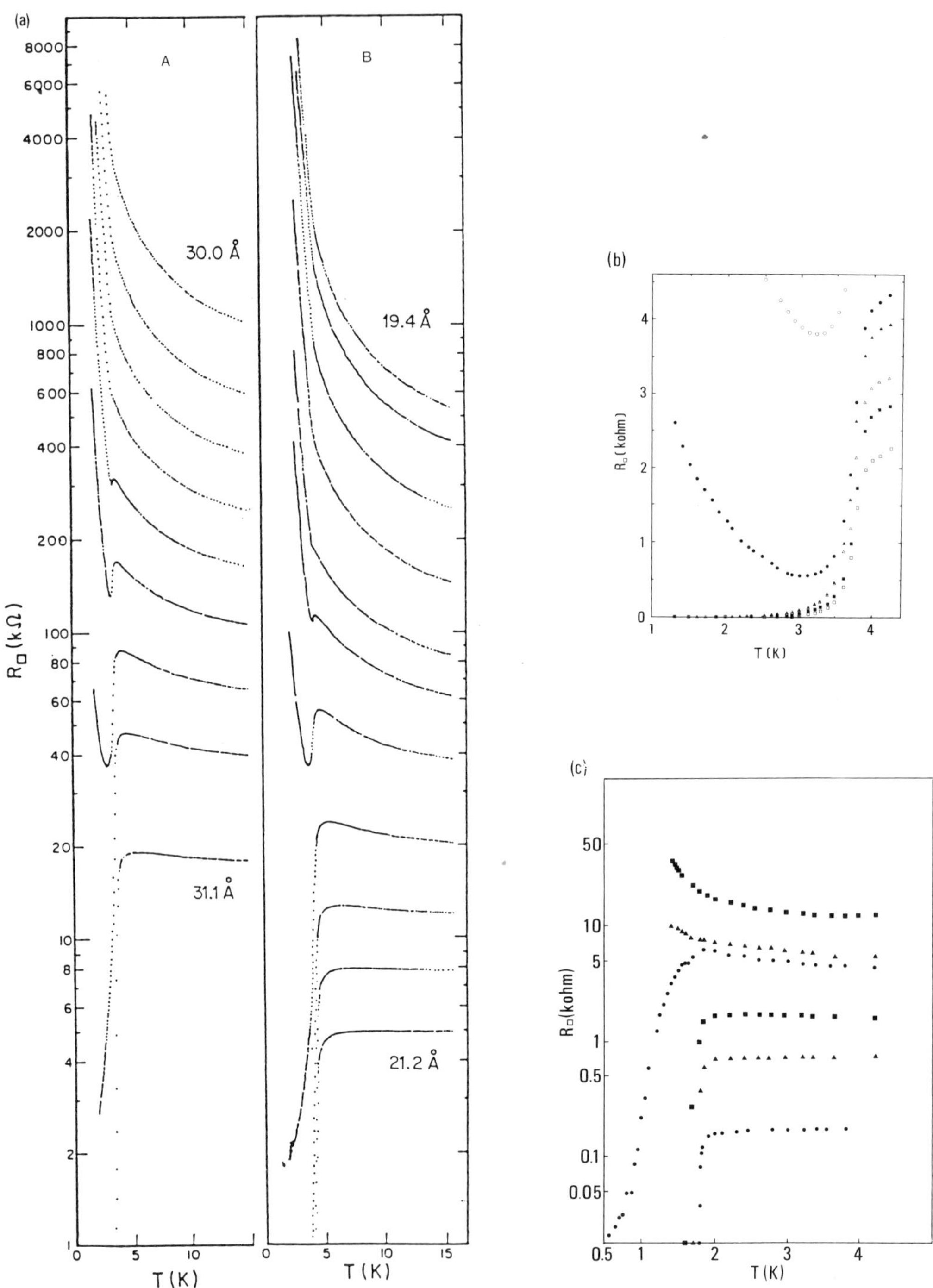

Figure 1. (a) Sheet resistance vs temperature for two sets of QCF of Sn [14], (b) sheet resistance versus temperature for GF of Sn [15], (c) sheet resistance vs temperature for GF of Al [16].

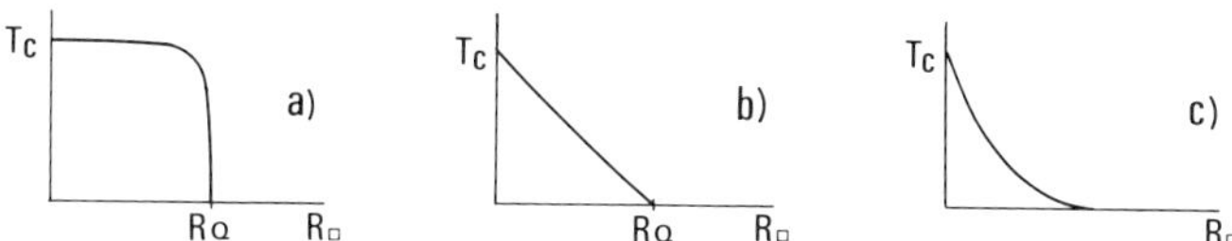

Figure 2. Three types of the dependence of T_c on the sheet resistance. For type (c) there is no universal resistance at which T_c vanishes.

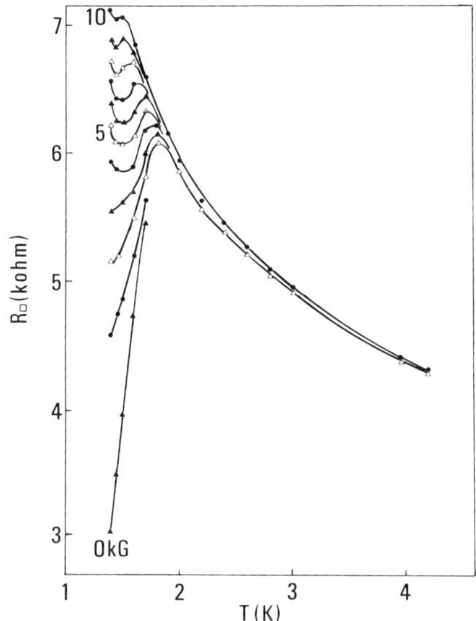

Figure 3. Sheet resistance versus temperature of GF of Al in magnetic field ranging between $0\,\mathrm{kG}$ ($\triangle$) and $10\,\mathrm{kG}$ ($\bullet$) in steps of $1\,\mathrm{kG}$ [16].

is a theory [20] which suggests a normal-metal-like resistive behavior for the granular films, but it does not explain the absence of the normal-metallic phase in GF.

Only QCF/Ge of Bi [19] belongs to type (b). If QCF/Ge of Bi is a continuous film as the authors claim, it is rather natural to consider that the result of QCF/Ge of Bi is of type (c) and R_Q is a coincidence, because type (c) in the case T_c vanishes at a sheet resistance either higher or lower than R_Q. There is a theory which suggests that the superconductivity of continuous films also vanishes at R_Q [21]. But it does not give any idea how to explain many results of the type (c) which do not have the universal R_Q.

In Fig. 3, the temperature dependence of sheet resistance of the GF of Al in magnetic field is shown [16]. The sample is the last one which shows superconductivity with increasing sheet resistance in Fig. 1c. The sample turns out to be semiconducting when the sheet resistance just above T_c exceeds $6\,k\Omega$, which is close to R_Q, with increasing magnetic field. The minimum field necessary to switch the phase is about $4\,kG$. This field produces a flux in a junction four orders of magnitude smaller than the flux quantum, and in the smallest area enclosed by the junctions three orders. Therefore, this phase transition with magnetic field is a cooperative phenomenon involving many junctions. It is very interesting that here again the transition is associated with R_Q.

In summary, the sharp disappearance of global superconductivity at the sheet resistance near R_Q was observed in the GF and in some of the QCF. The structure of the former is the 2 D random Josephson network, and that of the latter is also expected to be similar. In magnetic fields, the sheet resistance increases with increasing field, and the transition was observed when the sheet resistance just above T_c exceeds R_Q. This means that the transition at R_Q is not a single junction property but it involves many junctions. In the continuous films, T_c smoothly decreases with increasing sheet resistance and the results are well explained by localization theories, while one example is reported to have a critical sheet resistance equal to R_Q. More elaborate studies are necessary to understand the physics of R_Q.

The author thanks F. Komori, S. Katsumoto and W. Zwerger for stimulating discussions.

References

[1] E. Simanek, Sol. St. Commun. **31**, 419 (1979)

[2] S. Kobayashi, Y. Tada, and W. Sasaki, J. Phys. Soc. Jpn. **49**, 2075 (1980)

[3] K.B. Efetov, Sov. Phys. JETP **51**, 1015 (1980); P. Fazekas, Z. Phys. **B 45**, 2836 (1982); J.V. Jose, Phys. Rev. **B 29**, 2836 (1984)

[4] S. Maekawa and H. Fukuyama, J. Phys. Soc. Jpn. **51**, 1380 (1982)

[5] S. Okuma, F. Komori, Y. Ootuka, and S. Kobayashi, J. Phys. Soc. Jpn. **52**, 2639 (1983)

[6] J.M. Geball and M.R. Beasley, Phys. Rev. **B 29**, 4167 (1984); M.R. Beasley, Jpn. J. Appl. Phys. **26**, 1967 (1987)

[7] A.F. Hebard and M.A. Paalanen, Phys. Rev. Lett. **54**, 2155 (1985)

[8] A.M. Finkel'stein, JETP Lett. **45**, 46 (1987)

[9] V. Ambegaokar and A. Baratoff, Phys. Rev. Lett. **10**, 486 (1963)

[10] A. Schmid, Phys. Rev. Lett. **51**, 1506 (1983)

[11] M.P.A. Fisher and W. Zwerger, Phys. Rev. **B 32**, 6190 (1985)

[12] S. Chakravarty, G.L. Ingold, S. Kivelson, and A. Luther, Phys. Rev. Lett. **56**, 2303 (1986)

[13] V. Ambegaokar, U. Eckern, and G. Schön, Phys. Rev. Lett. **48**, 1745 (1982)

[14] B.G. Orr, H.M. Jeager, and A.M. Goldman, Phys. Rev. **B 32**, 7586 (1985)

[15] S. Kobayashi and F. Komori, J. Phys. Soc. Jpn. **57**, 1884 (1988)

[16] S. Kobayashi, A. Nakamura, and F. Komori, in preparation

[17] H.M. Jeager, D.B. Haviland, A.M. Goldman, and B.G. Orr, Phys. Rev. **B 32**, 4920 (1985); H.M. Heager, D.B. Haviland, B.G. Orr, and A.M. Goldman, Phys. Rev. **B 40**, 182 (1989)

[18] A. Asamitsu, S. Okuma, and N. Nishida, private communication

[19] D.B. Haviland, Y. Liu, and A.M. Goldman, Phys. Rev. Lett. **62**, 2180 (1989)

[20] A.I. Larkin, Yu.N. Ovchinnikov, and A. Schmid, Physica **B 152** 266 (1988)

[21] M.P.A. Fisher, G. Grinstein, and S.M. Girvin, Phys. Rev. Lett. **64**, 587 (1990)

QUANTUM EFFECTS IN GRANULAR SUPERCONDUCTORS

Wilhelm Zwerger

Institut für Theoretische Physik
Universität Göttingen
W-3400 Göttingen, F. R. Germany

1 Introduction

Granular superconductors are strongly inhomogeneous systems consisting of locally superconducting grains in an insulating matrix. In these lectures we will discuss the consequences of the Coulomb interaction on the mesoscopic scale of grains and of dissipative effects in these systems. Throughout we assume that the granular system may be modelled as a network of tunnel junctions which are normal junctions above and Josephson junctions below the transition to local superconductivity at T_c^ℓ. The lectures are organized as follows: section 2 starts with a discussion of tunneling in non-superconducting granular systems and in particular the behavior of the conductance due to the effects of finite size and charging in small grains. In section 3 we discuss the appearance of local superconductivity within the grains and the Josephson coupling between them. It is shown that both thermal fluctuations and Coulomb effects may lead to a destruction of global superconductivity. The influence of dissipative effects associated with normal electron tunneling is treated in section 4 for single junctions. In section 5 we determine the phase diagram at zero temperature of one and two-dimensional (2 D) networks of quantum Josephson junctions including dissipation. Finally in section 6 we discuss corresponding experiments on artificially created Josephson junction arrays and very thin granular films.

For previous brief reviews of granular superconductors see for instance the articles by Deutscher, Doniach, Imry and Rosenblatt in the NATO ASI series 1983 [1] and the proceedings of a conference on inhomogeneous superconductors [2].

2 Non-Superconducting Granular Systems and Coulomb Effects

Granular structures both in two and three dimensions are usually prepared by evaporation of metals on a substrate for instance in the presence of oxygen or together with an

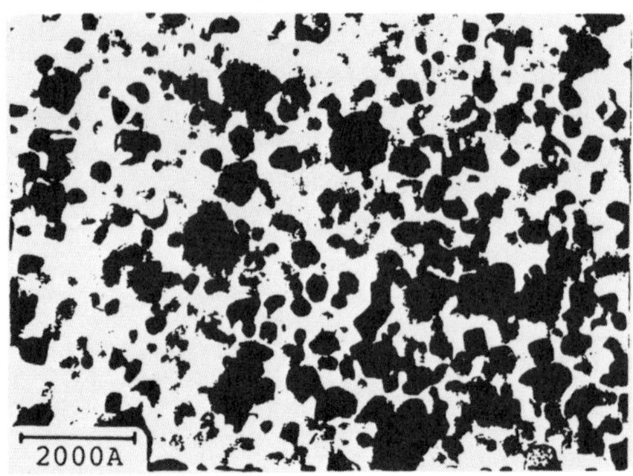

Figure 1: Electron micrograph of a granular tin film ([3]).

insulating material to account for the non-conducting matrix. Alternatively the evaporation of very thin films on a substrate at low temperatures results in a structure of disconnected islands whose average separation decreases by adding more material. In any case one obtains a collection of isolated metallic grains with typical size L_0 which is usually in the range of several hundred Å (see Fig. 1 taken from [3]). As long as tunneling between neighboring grains is negligible, the overall conductance of such a system vanishes. Tunneling, however, eliminates the classical percolation transition and leads to a finite conductance even in a disconnected system. The granular system may then be considered as a network of tunnel junctions between neighboring grains. To estimate the conductance of a single junction, we use the golden rule result

$$1/\tau = \frac{2\pi}{\hbar}|t|^2\, N(\epsilon_F) \tag{1}$$

for the lifetime of an electron at the Fermi energy ϵ_F due to tunneling to a neighboring grain. Here $|t|^2$ is the square of the corresponding tunneling matrix element averaged over the Fermi surface and $N(\epsilon_F) \approx n(\epsilon_F)L_0^3$ the density of final states ($n(\epsilon_F)$ is the density of states per energy and volume and is equal to $3n/2\epsilon_F$ for a free Fermi gas). Assuming a symmetrical junction in the presence of a finite voltage V, there are $eV\, N(\epsilon_F)$ states available and the associated current is $I = e^2 V N(\epsilon_F)/\tau$. The dimensionless conductance $g_0 = R_Q/R_0$ of a single junction is then given by

$$g_0 = \pi^2 |t|^2 N^2(\epsilon_F). \tag{2}$$

Here the quantum unit of resistance $R_Q = h/4e^2 = 6.45\,k\Omega$ has been chosen for later convenience and the index $_0$ indicates that g_0 is the conductance on the scale L_0 of single grains. The result eq. (2) can be interpreted in terms of Thouless' [4] picture of conductance being the ratio $g_0 = v/w$ between the coupling energy $v \approx \hbar/\tau$ of neighboring grains and the average level spacing

$$w \approx N(\epsilon_F)^{-1} \approx \epsilon_F/N \tag{3}$$

at the Fermi energy for a grain with N electrons. Typical values of w are of the order of a few Kelvin for grain sizes $L_0 \approx 100$ Å and $N \approx 10^5$ (we assume throughout that $k_B = 1$ and thus energies are measured in degree Kelvin). In eq. (3) we have implicitly assumed that the irregularity in grain surfaces removes all degeneracies which would be present for instance in a sphere where w would scale as $N^{-2/3}$ (see Mühlschlegel in [1]). Due to the exponential dependence of $|t|^2$ on grain separation s even a narrow say Gaussian distribution in s leads to a very broad log-normal distribution of conductances g_0. According to an argument by Ambegaokar, Halperin and Langer [5] the actually measured overall conductance of the network is then determined by the condition that the subset of all junctions with conductances larger than the measured one forms a percolating network.

Within the picture of a normal granular system as a network of tunnel junctions it will behave like a standard metal with a finite conductance at zero temperature. However apart from a possible superconducting transition to be discussed below, there are two different reasons which lead to a breakdown of metallic behavior at low temperatures. First of all it is clear that the electronic states in a granular system are localized on the scale L_0 of the grain size. As a result, at zero temperature, we expect to have an insulator behaving in a similar way than a disordered *continuous* system with localization length equal to L_0. At $T \neq 0$ tunneling to neighboring localized states then requires a finite activation energy w, since the levels are discrete with an average spacing as given in eq. (3). The conductance will thus vanish exponentially like $\exp -w/T$ for temperatures $T \ll w$ and the system is indeed insulating at $T = 0$. This conclusion remains valid also in the case of variable range hopping, where the conductance will vanish like $\exp -(T_0/T)^{\frac{1}{d+1}}$ (for granular systems a recent discussion is given by Németh and Mühlschlegel [6]). Now in our derivation of a finite tunnel conductance g_0 above, the finite size of the grains has in fact been neglected by assuming a continuum of final states. This is justified, however, only if the level discreteness is effectively smeared out either by a thermal energy $T \gg w$ or, as pointed out by Kawabata [7], a sufficient broadening of the levels due to strong tunneling (see below). The condition for observing metallic behavior may also be formulated in terms of length rather than energy scales. Defining L_φ as the length beyond which electrons have lost their phase coherence [4], a granular system behaves as a network of tunnel junctions as long as L_φ is smaller or equal to the grain size L_0.

A second effect which destroys metallic behavior at low temperatures is related to the effects of the Coulomb interaction on the mesoscopic scale of grains. Consider the transport of a single electron between two originally neutral neighboring grains. Then after the tunneling process we have a configuration which may be viewed as a tiny capacitor being charged by a single electron. With U as the associated increase in electrostatic energy we thus expect that at temperatures $T \ll U$ the conductance will again be thermally activated and proportional to $\exp -U/T$ as first pointed out by Neugebauer and Webb [8]. More precisely, the true activation energy is one half of the energy U_0 necessary to create a *fully* dissociated pair. Since simple estimates, however, give $U_0 \approx 2U_1$ [9] the apparent activation energy is essentially the Coulomb energy U_1 associated with a pair of oppositely charged neighboring grains. Therefore, at low enough temperatures, a granular system will always display semiconducting behavior due to the finite size of the grains which is responsible both for the level spacing w as well as the Coulomb energy U. In order to determine U we consider a sphere of radius L_0 embedded in a non-conducting medium with dielectric constant ε which gives $U = e^2/\varepsilon L_0$ as an order of magnitude estimate. Using $\varepsilon = 10$ and $L_0 = 100$ Å for instance, this leads to $U = 167\,°K$ showing that charging energies may be rather large. Indeed the ratio between the level spacing and the Coulomb energy is of order

$$w/U \approx \pi^2 \varepsilon \frac{a_B}{k_F L_0^2} \tag{4}$$

with $a_B \approx 0.5$ Å the Bohr radius. For reasonable grain sizes this is small compared to one unless the dielectric constant ε is very large. Thus the Coulomb gap U is usually much larger than the level spacing w. Unfortunately, quantitative estimates of the charging energy are very difficult to obtain since the effective dielectric constant is affected by the presence of neighboring grains. In addition, measurable tunnel conductances require that the systems are relatively close to the threshold for metallic connectedness where ε diverges.

As in the case of non-interacting electrons the simple exponential temperature dependence of the conductance is changed if there is variable range hopping. With the assumption that the ratio between grain radius and spacing is constant it was shown by Abeles et.al. [9] that the corresponding temperature dependence is $\exp -(\tilde{T}_0/T)^{1/2}$. The same behavior has been found by Efros and Shklovskii [10] for hopping between localized *single* electronic states with Coulomb interactions. However, in contrast to [9], the latter result is due to a many body effect which also leads to a vanishing density of states at the Fermi energy. The associated $(\tilde{T}_0/T)^{1/2}$ - law is thus not tied to specific properties of the localized states as in the model of Abeles et.al. [9]. For a recent discussion of the crossover between Mott and Coulomb variable range hopping see [6].

According to the arguments above, non-superconducting granular systems will always exhibit semiconducting behavior at temperatures lower than the Coulomb gap U. However in the very thin films to be discussed in section 6 below, it is found that there is a crossover between semiconducting and metallic behavior at constant temperature when the film thickness d is increased. In view of the extremely tiny changes in d which are necessary to induce this crossover, it is clear that the effect cannot be due to a geometrical reduction of U with increasing grain size. It is rather the conductance which, due to the tunneling nature of transport, depends very sensitively on the film thickness. A simple qualitative argument which is able to explain the semiconducting to metallic crossover with increasing conductance is due to Abeles and Sheng [11] and Imry and Strongin [12]. Consider two adjacent grains and a single electron tunneling between them. Then according to simple electric circuit theory the corresponding charged configuration relaxes back to the original one within a time $\tilde{\tau} = R_0 C$ with C the effective capacitance and R_0 the usual tunnel resistance. The quantum mechanical uncertainty associated with this relaxation is given by $\hbar/\tilde{\tau} \approx g_0 U$. It is now plausible that the charging effects are smeared out when this uncertainty becomes larger than the Coulomb energy itself, i.e.

$$\frac{\hbar}{\tilde{\tau}} > U. \tag{5}$$

Remarkably the value of U cancels in this relation and we find that charging effects become irrelevant when the dimensionless tunnel conductance g_0 is larger than a numerical constant of order one. A more quantitative theory of this effect on the level of a single normal tunnel junction was given by Brown and Simanek [13,14] (see section 4). They found that for strong tunneling $g_0 \gg 1$ the Coulomb gap is smeared out, leading to metallic behavior $g(T) \approx g_0$ even at zero temperature. Experimentally this has been verified in recent measurements by Geerligs et.al. [15] who found a crossover resistance of about 10 $k\Omega$. We mention that a different criterion for a crossover between semiconducting and metallic behavior was used by Kawabata [7]. In the spirit of a Hubbard model description he took $w > U$ instead of eq. (5) as the condition for obtaining metallic behavior. Since $v \approx g_0 w$, however, this leads to a crossover value of the tunnel conductance $g_0 \approx U/w$ which is always large compared to one in contradiction with the experimental results.

3 Superconductivity and Josephson Coupling

We now turn to granular systems in which isolated metallic grains become locally superconducting below a temperature T_c^ℓ. A theory of superconductivity in small grains is due to Mühlschlegel, Denton and Scalapino [16]. Assume that the corresponding bulk material is superconducting below T_c with a value Δ for the gap at $T = 0$. Clearly T_c depends on the Coulomb interaction and the amount of disorder. However we will not be concerned with such effects here but assume Δ to be a given parameter. Now in spite of the absence of a true phase transition in any finite system, in practice even a small grain exhibits a more or less rounded transition into a superconducting state below a certain local condensation temperature T_c^ℓ. Usually one expects that T_c^ℓ is smaller than the bulk T_c and is decreasing with grain size. However this is not necessarily the case. Indeed for quench condensed films the inhomogeneous system often has a higher transition temperature than the bulk [17]. As shown in [16], a superconducting state can be established only as long as the level spacing at the Fermi energy is smaller than the gap, i.e.

$$w < \Delta. \tag{6}$$

Taking a spherical grain this condition leads to a minimum grain size

$$L_s = \left(\Delta n(\epsilon_F)\right)^{-1/3} \tag{7}$$

which is of the order $L_s \approx 50 - 100$ Å for Δ being several Kelvin and typical density of state values. We thus find that only grains above a certain minimum size become locally superconducting while in smaller ones superconductivity is quenched by finite size effects. Usually there is also a maximum value of L_0 which is set by the Ginzburg-Landau coherence length ξ. Indeed for $L_0 < \xi$, which we assume to be valid throughout, the spatial dependence of the order parameter is negligible. The Ginzburg-Landau free energy is then a functional only of the individual values ψ_ℓ of the order parameter in each grain [18]. The relation eq. (6) may be understood as a condition for the validity of Andersons theorem which guarantees the existence of superconductivity for an arbitrary form of the time reversed paired states as long as the sum over the discrete levels may be replaced by an integral with a given finite density of states [19,20]. In particular this shows that even a system with only localized states at the Fermi energy may still become superconducting. Within a mean field approximation [20] superconductivity would persist for arbitrarily small localization lengths $L_0 < L_s$. Quantum fluctuations of the order parameter, however, will set a finite lower bound on the size of the localized states below which superconductivity is completely quenched [21].

Obviously local superconductivity within the grains does not necessarily imply global superconductivity in the whole sample. Indeed in many granular superconductors the resistance only gradually goes to zero below T_c^ℓ over a range of temperatures and in some cases two clearly separated transitions may be observed [22]. The first one is associated with the occurrence of local superconductivity in individual grains, while the second one is related to the ordering of the phases between different grains. It is only the latter transition which leads to a vanishing resistance since supercurrents now are able to flow also in the insulating regions between the grains. This is possible through the Josephson effect in which pairs tunnel coherently through a non-superconducting barrier. As a result there is a coupling energy between two close enough grains which depends on the relative phase of their corresponding order parameters

$$E(\varphi) = -E_J \cos \varphi. \tag{8}$$

Here the Josephson coupling energy E_J is the tunneling amplitude for the transfer of a single pair in units of energy. Following an argument due to Ferrell and Prange [23], eq. (8) is the energy of a Bloch state $|\varphi> = \sum_n e^{in\varphi}|n>$ where $|n>$ is a state with $n = 0, \pm 1, \ldots$ transferred pairs. The phase φ is thus the conjugate variable to the number n of pairs. The supercurrent associated with a certain value of φ is given by $I = I_c \sin \varphi$ with a critical current $I_c = 2eE_J/\hbar$. An external current smaller than I_c can thus be transported by adjusting the phase to a corresponding constant value. By virtue of the second Josephson relation $\dot\varphi = 2eV/\hbar$ which relates $\dot\varphi$ to the difference V in electrochemical potential across the junction, there is no potential drop below I_c and thus the junction is superconducting. The amplitude of E_J can be calculated from the microscopic theory by using a tunneling Hamiltonian [24] and is given by

$$E_J = \frac{1}{2}g_0\Delta(T)\tanh\frac{\Delta(T)}{2T}. \tag{9}$$

The proportionality of E_J to the normal state conductance g_0 is simply a reflection of the fact that the amplitude for pair tunneling is proportional to the probability for tunneling of a single electron. As a result of the usually very wide distribution of conductances g_0 in a granular system there is then a corresponding spread in the strength of the Josephson coupling E_J. Clearly large values of E_J do not play a role since very strongly coupled grains can be considered as an effective single one. Indeed, as pointed out by Doniach in [1], a granular structure with scale L_0 is present only if the Josephson coupling E_J is smaller than the superconducting condensation energy $L_0^3 n(\epsilon_F)\Delta^2$ per grain at $T = 0$. Using eqs. (7) and (9), this is equivalent to $g_0 < (L_0/L_s)^3$. Considering junctions with small values of E_J, however, it may be possible that the Josephson junctions which can carry a current equal or larger than the measuring current without dissipation do not form a percolating network. In this case there is no global superconductivity even in the absence of any fluctuations and a percolation type transition may occur [25]. In the following we will restrict ourselves to cases in which such effects are absent and where the granular system may be modelled by a network of simple Josephson junctions. Thus we assume that the temperature is well below the local transition temperature T_c^ℓ such that fluctuations in the magnitude of the order parameter are negligible and essentially all electrons are condensed into pairs. Moreover we disregard the effects of disorder and replace the granular system by a regular array of Josephson junctions with a fixed average value of E_J. For systems like those shown in Fig. 1, the latter approximation may seem rather questionable. However, as we will see in section 5, for certain aspects like the crossover between metallic and superconducting behavior as $T \to 0$ this may be justified. With these approximations the Josephson coupling energy then takes the form of an XY model

$$H_\varphi = -E_J \sum_{\langle \ell\ell'\rangle} \cos(\varphi_\ell - \varphi_{\ell'}). \tag{10}$$

Here $\langle \ell\ell' \rangle$ denotes a sum over nearest neighbor pairs with no double counting and φ_ℓ is the phase of the order parameter in grain ℓ. In order to take into account the Coulomb interaction between pairs on the mesoscopic scale of grains, the electrostatic energy is expressed in terms of an effective (inverse) capacitance matrix $(C^{-1})_{\ell\ell'}$. With n_ℓ as the deviation of the pair number from its average value in grain ℓ the Coulomb energy is then given by

$$H_n = \frac{(2e)^2}{2} \sum_{\ell\ell'} (C^{-1})_{\ell\ell'} n_\ell n_{\ell'}. \tag{11}$$

In the absence of screening the inverse capacitance matrix would asymptotically decay as $|\ell - \ell'|^{-1}$. For any non-vanishing conductance, however, the long range part of the Coulomb interaction is screened out (note that the short screening length *within* a grain is not relevant for this argument). As a result, the capacitance matrix is usually approximated by a diagonal one and the Coulomb effects are thus reduced to a Hubbard like effective charging energy U of a single grain. More generally, however, depending on the spatial distribution of the electric field, both self and – at least – nearest neighbor charging energies play a role. In cases where the latter are dominant one may even have a situation in which the effective Coulomb interaction is logarithmic over a substantial range leading to a charge unbinding transition (see the contribution by R. Fazio in this volume).

The introduction of Coulomb effects now crucially changes the physics compared to the standard XY-type models eq. (10). Indeed, as shown above, phase and pair number are conjugate variables. Thus, as first pointed out by Anderson [26], in a quantum mechanical treatment they become operators with the canonical commutation relation [27]

$$[\varphi_\ell, n_{\ell'}] = i\delta_{\ell\ell'}. \tag{12}$$

Within the self charging approximation which is expected to apply in most granular systems one thus arrives at a quantum mechanical Hamiltonian

$$H = \frac{U}{2} \sum_\ell n_\ell^2 - E_J \sum_{\ell\ell'} \cos(\varphi_\ell - \varphi_{\ell'}). \tag{13}$$

As pointed out by Efetov [19] this is essentially a Hubbard model for the pairs which are bosons in this approximation. At zero temperature one thus expects a kind of Mott transition in which the pairs are delocalized when the Josephson coupling dominates the charging energy. For $U/E_J \gg 1$ on the other hand, the pairs will be localized and transport is possible only by thermal activation leading to insulating behavior at $T = 0$. Thus in a similar way in which Coulomb effects destroy the metallic behavior in normal granular systems they may suppress the appearance of global superconductivity in a network of locally superconducting grains even at $T = 0$. Physically this is a result of the strong quantum fluctuations of the phases when $U \gg E_J$ which prevents the formation of long range phase coherence necessary for true superconducting behavior. At finite temperature also thermal fluctuations contribute to destroy a superconducting state with ordered phases. In the limit of negligible charging energy $U \to 0$ the model reduces to a classical XY model with a critical temperature T_c of order E_J. With increasing U, T_c will decrease monotonically going to zero at a critical value of E_J/U of order one. The complete phase diagram calculated on the basis of a mean field approximation [12,19,28]

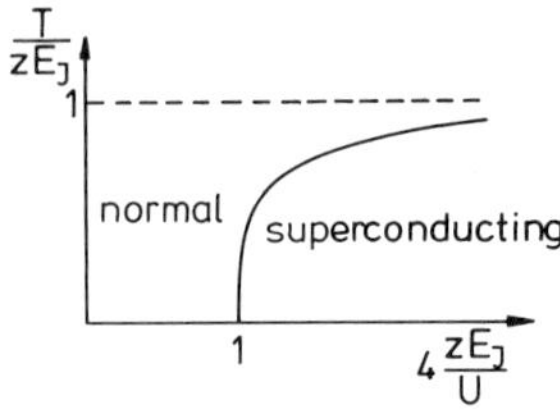

Figure 2: Mean field phase diagram of a Josephson junction
array with Coulomb effects ([12]).

is shown in Fig. 2 for a regular array with $2z$ the number of nearest neighbors. In the
vicinity of the critical value $4zE_J/U = 1$ the transition temperature to superconductivity
very rapidly goes to zero. In order for the model to be applicable up to the classical
i.e. $U \to 0$ value of the transition temperature $T_c^{c\ell} \approx E_J$ it is necessary that $T_c^{c\ell}$ is
still smaller than the local condensation temperature T_c^{ℓ}, i.e. the ordering of the phase
and the occurrence of superconductivity within the grains are separated. Since T_c^{ℓ} is of
order Δ this leads, using eq. (9), to the condition $g_0 < 1$. Thus, as found by Ebner
and Stroud [25], it is only for high enough inter-grain resistances that phase ordering
is well separated from T_c^{ℓ}. Now in the previous section we have found that a similar
condition determines the crossover between metallic and semiconducting behavior for
normal granular systems. Thus whenever the conductance g_0 is large enough to suppress
the semiconducting behavior through Coulomb effects *above* T_c^{ℓ}, the superconducting
transition at T_c^{ℓ} itself and the phase ordering appear together.

4   Dissipation in Single Junctions

As was mentioned in section 2 the crossover between semiconducting and metallic behavior in normal granular systems is due to the smearing of Coulomb effects by tunneling
between the grains. For a theoretical description of this problem we define a phase
variable θ by $\dot{\theta} = eV/\hbar$ like in the Josephson case. Then, as shown in [29] (see also the
contribution by G. Schön in this volume), the partition function for a tunnel Hamiltonian
model of a normal tunnel junction including Coulomb effects can be written in terms of
a functional integral $Z = \int D\theta \, \exp -S_\theta$ with dimensionless action

$$S_\theta = \frac{\hbar}{2U} \int_0^{\beta\hbar} d\tau \left(\frac{d\theta}{d\tau}\right)^2 + 2 \int_0^{\beta\hbar} d\tau \int_0^{\beta\hbar} d\tau' \alpha(\tau - \tau') \sin^2\left(\frac{\theta(\tau) - \theta(\tau')}{2}\right). \qquad (14)$$

Here the function $\alpha(\tau)$ is given by

$$\alpha(\tau) = \frac{g_0}{\pi^2} \frac{(\pi T/\hbar)^2}{\sin^2(\pi T\tau/\hbar)}. \qquad (15)$$

Formally the non-local term in eq. (14) arises from integrating out the tunneling electrons
while the kinetic energy part in S_θ represents the contribution of the charging energy
$\frac{C}{2}V^2$ with capacity $C = e^2/U$. The periodicity in θ of the non-local term is a reflection of
the discrete nature of charge transfer. The action eq. (14) is essentially a one-dimensional
(1 D) classical XY model with long range interaction decaying as τ^{-2} (this is valid at
$T = 0$ while at $T \neq 0$ the classical system has a finite length proportional to $\beta\hbar$). Within
a spin wave calculation Šimánek [14] showed that fluctuations in θ are frozen for $g_0 > 1$

although no true phase transition with long range order exists. Thus the free quantum rotor behavior at $g_0 = 0$ whose discrete spectrum is responsible for the Coulomb gap in the conductance [29]

$$g(T) = g_0 \frac{U}{2T} \exp -U/2T. \tag{16}$$

is quenched for $g_0 > 1$. In the latter regime it is the phase θ rather than the conjugate charge which has a well defined value. The Coulomb gap is then smeared out leading to metallic behavior $g(T) \approx g_0$ of the conductance even at $T = 0$ [13] in agreement with the qualitative argument eq. (5).

For a single Josephson junction the relevant variable is the Josephson phase φ and a functional description of the corresponding partition function was developed by Ambegaokar, Eckern and Schön [30]. In this case the tunneling of pairs is clearly non-dissipative. Thus in a single Josephson junction even an arbitrary strong tunneling does not smear out the Coulomb effects as was the case in a normal tunnel junction. Indeed the associated action

$$S_\varphi = \frac{\hbar}{2U} \int_0^{\beta\hbar} d\tau \left(\frac{d\varphi}{d\tau}\right)^2 - \frac{E_J}{\hbar} \int_0^{\beta\hbar} d\tau \cos\varphi(\tau) \tag{17}$$

essentially describes a quantum mechanical particle with mass proportional to U^{-1} in a periodic potential whose eigenstates are extended. Thus, in principle, for any non-zero Coulomb energy the phase is delocalized and the Josephson effect, which requires a definite value of φ, is quenched by quantum fluctuations. In practice, the delocalization of the phase is apparent only for times larger than the inverse bandwidth which becomes exponentially large for $E_J \gg U$ and thus may be unobservable. In addition there are often dissipative effects which screen the Coulomb interaction and prevent a complete delocalization of the phase. In order to incorporate such effects consider first an ideal tunnel junction between two identical superconductors. Elimination of the electronic degrees of freedom then leads to eq. (17) plus an additional contribution [30]

$$\delta S_\varphi = \int_0^{\beta\hbar} d\tau \int_0^{\beta\hbar} d\tau' k(\tau - \tau') \sin^2\left(\frac{\varphi(\tau) - \varphi(\tau')}{4}\right). \tag{18}$$

Here the function $k(\tau)$ at $T = 0$ is given by

$$k(\tau) = 2\frac{g_0}{\pi^2}\left(\frac{\Delta}{\hbar}\right)^2 K_1^2(\Delta\tau/\hbar) \tag{19}$$

with K_1 the modified Bessel function. At zero temperature and for a finite value of the gap Δ no quasi-particles are present. Thus at $T = 0$ the contribution eq. (18) does not describe dissipative processes and can only give a renormalization of the properties of a bare Josephson junction due to *virtual* excitation of quasi-particles. Indeed approximating

$$\varphi(\tau) - \varphi(\tau') \approx (\tau - \tau')\frac{d\varphi(\tau)}{d\tau} \tag{20}$$

and linearizing the sin, eq. (18) becomes equivalent to a reduction of the Coulomb energy [30]

$$1/U \rightarrow 1/U + 3g_0/64\Delta \tag{21}$$

with increasing normal state conductance g_0. Thus quantum phase fluctuations are reduced, however the Josephson effect still remains quenched. A strictly well defined phase at $T = 0$, however, is possible if there is some normal electron tunneling even at zero temperature for instance due to pair breaking effects. Formally this is described by keeping the Josephson coupling E_J finite but taking the limit $\Delta \to 0$ in eq. (19) (generalized to $T \neq 0$). Then, as it must, eq. (18) reduces to the non-local dissipative contribution of a normal tunnel junction eq. (14) up to a factor of 4 which is due to the fact that φ and θ differ by a factor of 2 arising from the pair versus single electron charge. Together with eq. (17) one then obtains a Josephson junction with a finite subgap conductance g_0. This model has been studied in some detail by Guinea and Schön and exhibits a rather rich structure [31,32]. In a simplified version we neglect the discreteness in the transfer of normal electrons, i.e. the sin is replaced by its argument. The model then reduces to a particle in a periodic potential which is subject to ohmic dissipation [33]. Similar to the normal junction case discussed above it then turns out that a well defined phase at $T = 0$ exists for large enough conductance $g_0 > 1$ [34]. For $g_0 < 1$, on the other hand, the Josephson junction shows a finite resistance, even though the expected Coulomb gap behavior is lost in the approximation. In conclusion, we find, that both in normal tunnel junctions as well as in Josephson junctions the Coulomb effects are smeared out when the normal state conductance g_0 is larger than a critical value of order one. In the Josephson case, though, it is necessary to have a finite subgap resistance while for an ideal tunnel junction an increasing g_0 only leads to a gradual decrease in the charging energy.

5   Josephson Junction Networks

We will now generalize the above discussion of quantum effects and dissipation to regular networks of Josephson junctions in 1 D and 2 D following [35]. For simplicity we will discuss the 1 D model

$$H = \frac{U}{2} \sum_\ell n_\ell^2 - E_J \sum_\ell \cos(\varphi_{\ell+1} - \varphi_\ell) \tag{22}$$

introduced by Bradley and Doniach [36] (see also [37]). Within a functional integral formulation on the basis of number eigenstates and using the Villain approximation for $\exp -\varepsilon E_J(1 - \cos \varphi)$ the partition function turns out to be [35]

$$Z = \sum_{h_{\ell,j}=0,\pm 1,\ldots} \exp -\left(\frac{\varepsilon U}{2} \sum_{\ell,j}(h_{\ell+1,j} - h_{\ell,j})^2 + \frac{1}{2\varepsilon E_J} \sum_{\ell,j}(h_{\ell,j+1} - h_{\ell,j})^2 \right). \tag{23}$$

Here $\varepsilon = \beta/N_\tau \to 0$ is the discretization unit for steps $j = 1,\ldots,N_\tau$ in imaginary time $\tau \in (0,\beta\hbar)$ and $h_{\ell,j}$ is an integer valued variable. In the form eq. (23), the partition function corresponding to the Josephson network Hamiltonian eq. (22) is a classical discrete Gaussian model for roughening of a 2 D interface. One of the dimensions (ℓ) is the actual physical dimension of the network while the other (j) is related to the imaginary time $\tau = \varepsilon\hbar j$ describing the quantum fluctuations. By a standard method [38] the model eq. (23) may be transformed into a 2 D neutral Coulomb gas with logarithmic interaction

$$v(r \to \infty) = 2\pi(E_J/U)^{1/2}\ln(r). \tag{24}$$

Here $r = (\ell^2 + (\omega_0\tau)^2)^{1/2}$ is the dimensionless distance between the charges and $\hbar\omega_0 = (E_J U)^{1/2}$ the energy scale for pair density oscillations in the chain of Josephson junctions. Since the effective Coulomb interaction is short ranged they have a linear dispersion $\omega_q = \omega_0 q$ in the limit of vanishing wavevector $q \to 0$. The charges of the Coulomb gas behave like classical vortices in 2 D but are in fact quantum mechanical excitations of the 1 D system. As is well known, the 2 D Coulomb gas with interaction eq. (24) has a metal to insulator transition at a critical value $(E_J/U)_c \approx 1.2$ [36]. In terms of the original Josephson junction problem this corresponds to an insulator to superconductor transition. Indeed for E_J/U larger than its critical value the frequency dependent conductivity $Re\,\sigma(\omega)$ has a singularity at zero frequency proportional to $n_s\delta(\omega)$ which is characteristic of a superconducting state with a finite superfluid density n_s [35]. It is important to note that this result can be directly obtained from the partition function or, more precisely, the corresponding imaginary time current correlations. In fact it is a unique feature of quantum phase transitions that unlike in classical statistical mechanics the static and the dynamical behavior are non-separable and time dependent correlation functions are contained in the equilibrium theory. For values of E_J/U smaller than the critical one the Coulomb gas is metallic, corresponding to a smooth interface with a finite step free energy [39]. In this regime the pairs are localized and the chain of Josephson junctions is insulating at $T = 0$. At finite temperature one expects a non-zero conductance due to thermally activated hopping of pairs with an activation energy which is equal to the step free energy. The activation energy will vanish as E_J/U approaches its critical value from below and thus the semiconducting behavior gradually goes away before the system becomes superconducting.

Let us now assume that the Josephson junctions in eq. (22) have a finite subgap conductance g_0 which is approximated by a shunt resistor with continuous charge transfer. Within the functional integral formulation it is straightforward to incorporate such a dissipative coupling. Using the Coulomb gas description the only change is that the interaction eq. (24) acquires an additional logarithmic contribution [35]

$$\delta v(\ell, \tau \to \infty) = 2g_0 \delta_{\ell,0} \ln(\tau/\tau_0) \tag{25}$$

which acts only along the τ-direction at fixed ℓ. This interaction has two effects: First it enhances the insulating nature of the Coulomb gas and thus the critical coupling $(E_J/U)_c$ necessary for superconductivity will decrease with increasing g_0. Quantitatively, however, this effect is small. More importantly, the contribution eq. (25) leads to an anisotropic ordering for $g_0 > 1$ but arbitrarily small E_J/U in which the Coulomb gas charges are bound in τ-direction but free in ℓ-direction. Physically this describes a state in which the chain of Josephson junctions has a well defined local phase on each individual grain but no global phase coherence. The broken symmetry in this state is the local discrete symmetry $\varphi_\ell \to \varphi_\ell + 2\pi$ for each separate ℓ. The complete phase diagram at zero temperature is shown in Fig. 3. There are three different phases: At large values of E_J/U there is a globally ordered superconducting state with long range phase coherence (algebraic in 1 D and $T = 0$). It is characterized by a $\delta(\omega)$ singularity in the real part of the conductivity and thus a full Meissner effect. At small values of E_J/U but conductances $g_0 > 1$ there is a phase with vanishing resistance but no long range order in the phases. Finally for small values of E_J/U and $g_0 < 1$ there is a normal state which becomes an insulator in the limit $g_0 \to 0$. The transition at $g_0 \approx 1$ is sharp only at $T = 0$ and is due to a *local* freezing of quantum phase fluctuations. It is therefore essentially independent of dimensionality and the geometric structure of the network. In fact a

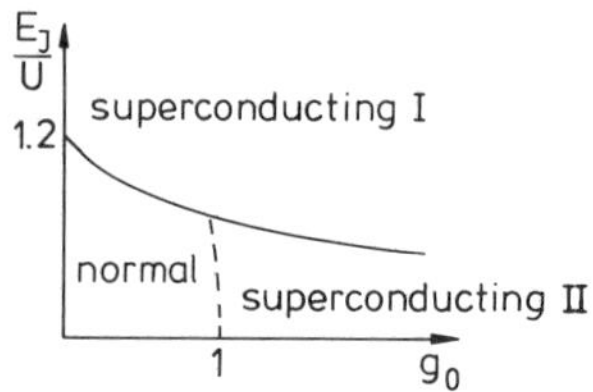

Figure 3: Phase diagram at zero temperature for a dissipative chain of Josephson junctions ([35]).

similar analysis for the more complicated 2 D array of Josephson junctions [35] shows a phase diagram which is only quantitatively different from the one shown in Fig. 3. In particular the critical value of the conductance g_0 is again close to one though probably somewhat smaller. The transition to true long range phase coherence now occurs at $(E_J/U)_c \approx 0.11$ at $g_0 = T = 0$. In contrast to the 1 D case, however, this transition is also present at finite temperature.

6   Experiments and Conclusion

An experiment in which the competition between Coulomb effects on a mesoscopic scale and superconductivity by Josephson tunneling can directly be investigated was recently performed by Geerligs et.al. [40]. By lithographical methods regular arrays of ultra-small Josephson junctions were built (see Fig. 4) which had values of E_J/U ranging between 0.06 and 0.5 (in regular arrays the self charging effect is small and the effective U is thus renormalized by the nearest neighbor contribution). There is no dissipative element in these systems at low temperature and thus the model eq. (13) is expected to provide a reasonable description of the experiments. The resistance of five of these arrays is shown in Fig. 5 as a function of temperature well below the value $T_c^\ell \approx 1.4\,K$. Clearly

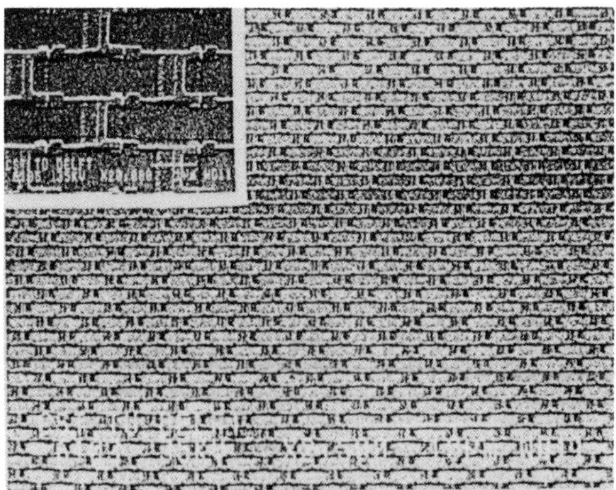

Figure 4: 2 D array of ultra-small Josephson junctions (Centre for Submicron technology, Delft, Research report 1988).

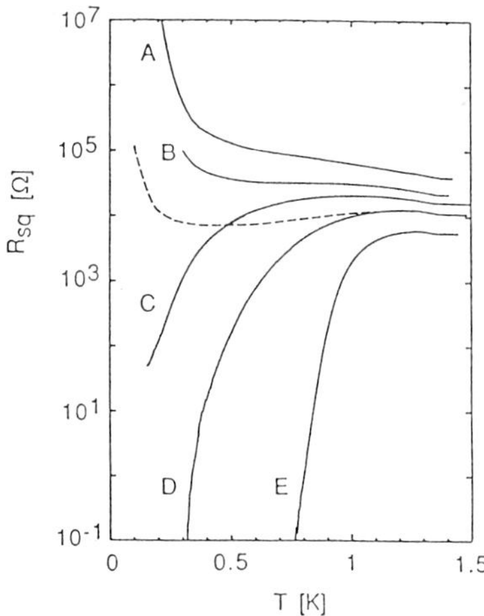

Figure 5: Resistance as a function of temperature for five different arrays as shown in Fig. 4 ([40], the dashed line is at finite magnetic field).

the arrays A and B which have small values of E_J/U show semiconducting behavior at low temperatures with no sign of superconductivity. Instead, their current voltage characteristics display a Coulomb gap which allows to estimate U. On the other hand the arrays D and E with larger values of E_J/U appear to become truly superconducting at low enough temperatures. Array C finally appears to be near the boundary between insulating and superconducting behavior. It has a value $E_J/U \approx 0.16$ which is reasonably close to the theoretical result $(E_J/U)_c \approx 0.11$ mentioned above, thus supporting the model used.

The experiments which originally did initiate much of the theoretical activity in this subject were performed by Goldman and coworkers [41]. In these experiments ultra-thin granular films of bulk superconducting materials were deposited on a substrate at low temperatures. It was thus possible to study the onset of superconductivity as a function of film thickness d ranging between 10 and 50 Å. The corresponding resistances as a function of temperature for two different series of depositions are shown in Fig. 6. Obviously three qualitatively different regimes may be distinguished:

1. For very small film thicknesses the resistance increases exponentially with decreasing temperature. Very thin films thus exhibit semiconducting behavior due to Coulomb effects. However superconductivity still exists on the local level of grains. Indeed in all but the very thinnest samples the activation energy suddenly *increases* at a temperature T_c^ℓ which is close to the bulk superconducting transition temperature. This increase is due to the fact that below T_c^ℓ there is a superconducting gap in addition to the Coulomb gap. As a result, the transition to local superconductivity within the grains even enhances the tendency towards an insulating state.

2. For slightly larger thicknesses there is a very pronounced drop in the resistance at T_c^ℓ, usually by several orders of magnitude. At low temperatures, however, the resistance remains finite ending either in temperature independent flat tails

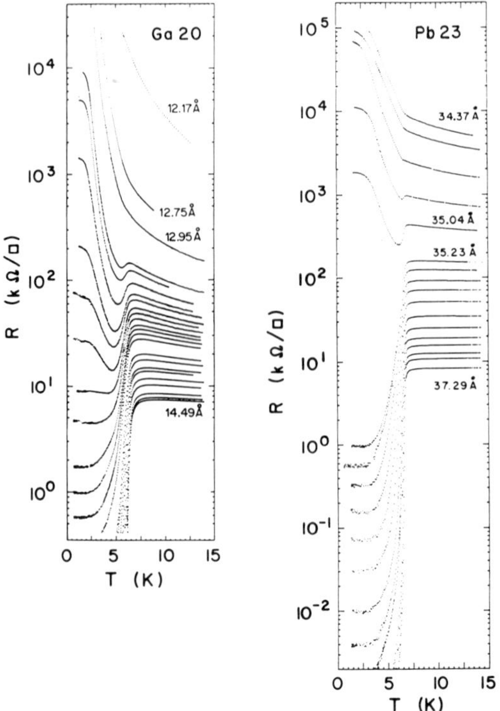

Figure 6: Resistance as a function of temperature for a series of Ga or Pb films of varying thickness ([41]).

or even rising again after going through a minimum. In this regime the films therefore exhibit metallic behavior both above and below T_c^ℓ, however with very different values of the resistance.

3. For still larger thicknesses finally the drop at T_c^ℓ indeed leads to a superconducting state with vanishing resistance. In practice the resistance falls below the experimental resolution of around $10^{-3}\Omega$ which is more than six orders of magnitude smaller than the corresponding normal state values.

Compared to the above experiments on regular arrays, the common feature is that superconductivity is established for small values of the normal state resistance R_n while Coulomb effects prevail for large R_n. One rather simple reason for this is related to the fact that via eq. (9) the strength of the Josephson coupling on each junction is proportional to the value of the corresponding normal state conductance g_0. To the extent that g_0 may be identified with the actual *global* conductance of the 2 D system, this shows that the characteristic energy E_J which favors superconductivity by ordering of the phases increases with increasing normal state conductance. At $T = 0$ the simple criterion $E_J/U \approx 1$ can then be rewritten as $R_Q/R_n \approx U/\Delta$. For the particular values of U and

Δ in the above experiments on regular Josephson junction arrays this leads to a critical value of the normal state resistance of order 13 $k\Omega$ as observed. Now the surprising observation made in the context of the experiments [41] on granular films was that *independent* of material or film thickness a true superconducting state always occurred at the same value of the normal state resistance R_n which was close to the quantum unit for pairs $R_Q = 6.5\,k\Omega$. Indeed this is consistent with other experiments on quench condensed films [42,43], but the apparently universal nature of this threshold resistance was only realized through the systematic work by Goldman and coworkers. While there is still no detailed understanding of these experiments, in particular the behavior of the resistance as a function of temperature, there are at least simple arguments showing that in 2 D granular films the resistance threshold may be independent of material parameters. They are essentially based on the qualitative considerations discussed above for the smearing of the Coulomb effects with increasing normal state conductance. Specifically, a simple model which is hoped to capture the essential physics consists of a network of Josephson junctions which have a subgap conductance g_0. A possible origin of such ohmic resistive channels in real granular systems may be grains with sizes $L_0 < L_s$ which do not become superconducting below T_c^ℓ. Thus in parallel to Josephson tunneling between larger grains there is also normal electron tunneling between smaller ones (see Fig. 1). Estimates of the ratio E_J/U in the considered experiments indicate that E_J/U is smaller than the critical value necessary for long range phase coherence. Thus, according to the results discussed above, one expects a transition to a superconducting state at a critical value of the conductance of order one. In agreement with the observations, this criterion is independent of material parameters like U or Δ. The possibility of this explanation however crucially relies on the identification of the *local* inter-grain conductance g_0 with the actually measured *global* normal state conductance. Now, as was shown in section 2, for $g_0 > 1$ Coulomb effects are smeared out in the normal state. Thus in this regime the phase coherence length L_φ is essentially equal to the grain size L_0 and by the unique scale independence of the resistance in 2 D the identification is possible. For three-dimensional granular systems corresponding arguments lead to a threshold in *resistivity* $\rho = R_Q a$ with a as a typical inter-grain dimension. As an order of magnitude one obtains $\rho \approx 10^{-2}\Omega cm$ which is in fact not far from corresponding experimental values [44].

We mention that an alternative argument for the threshold $R_n \approx R_Q$ in 2 D granular films which is based on the reduction eq. (21) of the nearest neighbor charging energy with increasing normal state conductance has been put forward by Ferrell and Mirhashem [45] and by Chakravarty et.al. [46]. While this renormalization will certainly be important for obtaining a quantitative estimate of where the transition between insulating and superconducting behavior occurs, a theory based on this effect only seems not to be sufficient in explaining an apparently universal resistance threshold beyond the mean field approximation [35]. Finally we mention that also in *continuous* films, which show a reduction of T_c with increasing normal state resistance R_n due to microscopic disorder, superconductivity is usually lost for values R_n of order or larger than the quantum unit R_Q. In this case, however, no universal threshold value is observed. Indeed experiments give values of R_n^c between 3.5 $k\Omega$ [47] and 9.5 $k\Omega$ [48] or even larger [49]. However, there is a recent argument [50] which shows that the *zero* temperature value of the resistance at the transition has the universal value R_Q. Clearly more work is needed to understand the crossover between continuous and granular films and in particular to calculate the dependence of the resistance on temperature in the different regimes.

References

[1] Percolation, Localization and Superconductivity, A.M. Goldman and S.A. Wolf, eds., Nato ASI series, Plenum Press, New York (1984)

[2] Inhomogeneous Superconductors, T.L. Francavilla, D.V. Gubser, J.R. Leibowitz, and S.A. Wolf, eds., AIP Conference Proceedings No. **58**, AIP, New York 1979

[3] S. Kobayashi and F. Komori, J. Phys. Soc. Jpn. **57**, 1884 (1988); see also the contribution by S. Kobayashi in this volume

[4] D.J. Thouless, Phys. Rev. Lett. **39**, 1167 (1977)

[5] V. Ambegaokar, B.I. Halperin, and J.S. Langer, Phys. Rev. **B 4**, 2612 (1971)

[6] R. Németh and B. Mühlschlegel, Z. Phys. B - Condensed Matter **70**, 159 (1988)

[7] A. Kawabata, J. Phys. Soc. Jpn. **43**, 1491 (1977)

[8] C.G. Neugebauer and M.B. Webb, J. Appl. Phys. **33**, 74 (1962)

[9] B. Abeles, P. Sheng, M.D. Coutts, and Y. Arie, Adv. Phys. **24**, 407 (1975)

[10] A.L. Efros and B.I. Shklovskii, J. Phys. **C 8**, L49 (1975)

[11] B. Abeles and P. Sheng, in Electrical Transport and Optical Properties of Inhomogeneous Media, J.C. Garland and D.B. Tanner, eds., AIP Conference Proceedings No. **40**, AIP, New York (1978)

[12] Y. Imry and M. Strongin, Phys. Rev. **B24**, 6353 (1981)

[13] R. Brown and E. Simanek, Phys. Rev. **B34**, 2957 (1986)

[14] E. Simanek, Phys. Lett. **119 A**, 477 (1987)

[15] L.J. Geerligs, V.F. Anderegg, C.A. van der Jeugd, J. Romijn, and J.E. Mooij, Europhys. Lett. **10**, 79 (1989)

[16] B. Mühlschlegel, D.J. Scalapino, and R. Denton, Phys. Rev. **B 6**, 1767 (1972)

[17] W. Buckel and R. Hilsch, Z. Phys. **138**, 109 (1954)

[18] G. Deutscher, Y. Imry, and L. Gunther, Phys. Rev. **B 10**, 4598 (1974)

[19] K.B. Efetov, Sov. Phys. JETP **51**, 1015 (1980)

[20] M. Ma and P.A. Lee, Phys. Rev. **B 32**, 5658 (1985)

[21] M. Ma, B.I. Halperin, and P.A. Lee, Phys. Rev. **B 34**, 3136 (1986)

[22] Such separated transitions are found for instance in arrays of SNS weak links, see D.W. Abraham, C.J. Lobb, M. Tinkham, and T.M. Klapwijk, Phys. Rev. **B 26**, 5268 (1982)

[23] R.A. Ferrell and R.E. Prange, Phys. Rev. Lett. **10**, 479 (1963)

[24] V. Ambegaokar and A. Baratoff, Phys. Rev. Lett. **10**, 486 (1963) and ibid. **11**, 104 (Erratum) (1963)

[25] C. Ebner and D. Stroud, Phys. Rev. **B 25**, 5711 (1982)

[26] P.W. Anderson in Lectures on the Many Body Problem, E. Caianello, ed., Academic Press, New York (1964)

[27] Note that φ is periodic and n is discrete and thus the commutation relation should more properly be written as $[e^{i\varphi}, n] = -e^{i\varphi}$

[28] E. Simanek, Sol. St. Commun. **31**, 419 (1979)

[29] E. Ben-Jacob, E. Mottola, and G. Schön, Phys. Rev. Lett. **51**, 2064 (1983); for a different approach see T.L. Ho, Phys. Rev. Lett. **51**, 2060 (1983)

[30] V. Ambegaokar, U. Eckern, and G. Schön, Phys. Rev. Lett. **48**, 1745 (1982), and Phys. Rev. **B 30**, 6419 (1984)

[31] F. Guinea and G. Schön, J. Low Temp. Phys. **69**, 219 (1987)

[32] For a comprehensive recent review see G. Schön and A.D. Zaikin, Phys. Rep. to be published (1990)

[33] See for instance A.J. Leggett in Chance and Matter, J. Souletie, J. Vannimenus, and R. Stora, eds., Les Houches 1986, North Holland, Amsterdam (1987)

[34] A. Schmid, Phys. Rev. Lett. **51**, 1506 (1983); M.P.A. Fisher and W. Zwerger, Phys. Rev. **B 32**, 6190 (1985); W. Zwerger, Phys. Rev. **B 35**, 4737 (1987)

[35] W. Zwerger, Europhys. Lett. **9**, 421 (1989), and Z. Phys. B - Condensed Matter **78**, 111 (1990)

[36] R.M. Bradley and S. Doniach, Phys. Rev. **B 30**, 1138 (1984)

[37] S.E. Korshunov, Europhys. Lett. **9**, 107 (1989); P. Bobbert, R. Fazio, G. Schön, and G.T. Zimanyi, Phys. Rev. **B 41**, 4009 (1990)

[38] S.T. Chui and J.D. Weeks, Phys. Rev. **B 14**, 4978 (1976)

[39] D.S. Fisher and J.D. Weeks, Phys. Rev. Lett. **50**, 1077 (1983)

[40] L.J. Geerligs, M. Peters, L.E.M. de Groot, A. Verbruggen, and J.E. Mooij, Phys. Rev. Lett. **63**, 326 (1989)

[41] H.M. Jäger, D.B. Haviland, A.M. Goldman, and B.G. Orr, Phys. Rev. **B 34**, 4920 (1986)

[42] R.C. Dynes, J.P. Garno, and J.M. Rowell, Phys. Rev. Lett. **40**, 479 (1978)

[43] A.E. White, R.C. Dynes, and J.P. Garno, Phys. Rev. **B 33**, 3549 (1986)

[44] M. Kunchur, P. Lindenfeld, W.L. McLean, and J.S. Brooks, Phys. Rev. Lett. **59**, 1232 (1987)

[45] R.A. Ferrell and B. Mirhashem, Phys, Rev. **B 37**, 648 (1988)

[46] S. Chakravarty, S. Kivelson, G.T. Zimanyi, and B.I. Halperin, Phys. Rev. **B 35**, 7256 (1987)

[47] A.F. Hebard and M.A. Paalanen, Phys. Rev. **B 30**, 4063 (1984)

[48] D.B. Haviland, Y. Liu, and A.M. Goldman, Phys. Rev. Lett. **62**, 2180 (1989)

[49] M. Strongin, R.S. Thompson, O.F. Kammerer, and J.E. Crow, Phys. Rev. **B 1**, 1078 (1970)

[50] M.P.A. Fisher, G. Grinstein, and S.M. Girvin, Phys. Rev. Lett. **64**, 587 (1990)

CHAPTER 10

TUNNEL JUNCTIONS

VORTICES AND CHARGES IN TUNNEL JUNCTION NETWORKS

Herre S.J. van der Zant, Lambert J. Geerligs, Johan E. Mooij

Department of Applied Physics
Delft University of Technology
P.O. Box 5046, 2600 GA Delft, The Netherlands

1 Introduction

Films of granular materials have long been modeled as two- (2 D) or three- (3 D) dimensional networks of grains, coupled by tunnel junctions [1]. The electrical resistance of such films is supposed to be concentrated in the junctions, whereas the grains themselves are relatively pure. It has only recently become possible to fabricate such tunnel junction networks artificially, using the lithographic techniques developed for microelectronics. So far only 2 D networks have been made. Now parameters can, to a certain extent, be varied independently and theoretical models can be verified quantitatively against experimental results. The networks in 'natural' films are irregular, which in the percolative limit is an essential aspect. However, many of the important properties are well represented in regular arrays of metallic islands, coupled by identical tunnel junctions. We only consider such arrays here. The interest in tunnel junction arrays is certainly not only prompted by the analogy with the films. The regularity of fabricated tunnel junction arrays also introduces special effects, with their own special physics. In these lectures, we will concentrate on two main topics: the resistive behavior of superconducting arrays and the charging effects associated with the transfer of a single electron or a single Cooper pair. Resistance in 2 D superconductors is usually associated with flow of vortices and, as the title implies, this will play a prominent role here as well. However, in junction arrays we also find a different dissipative regime, better described by coherent phase slip. When vortices are present, they may not move in the familiar viscous way. As for the charging effects, they lead to a suppressed conduction in the normal state as well as in the superconducting one. Only since a few years are these effects studied experimentally in fabricated structures. We discuss effects of charge quantization in small networks with only a few junctions and in large arrays. It turns out that there is a remarkable analogy between the resistance due to vortex flow in superconducting arrays and the conductance due to flow of single charges in charging effects dominated arrays of superconducting and normal metal tunnel junctions.

Quantum Coherence in Mesoscopic Systems
Edited by B. Kramer, Plenum Press, New York, 1991

Tunnel junction arrays are not only model systems for granular films, in the superconducting state they are also practical realizations of the well-known XY-model of statistical mechanics in 2 D. Spin directions are replaced by the phase of the order parameter. Experimental results on arrays can be compared with results from theoretical calculations using renormalization techniques or from computer simulations.

We consider small tunnel junctions for which the spatial dependence in the barrier plane can be ignored. A junction is characterized by four parameters: the ideal tunnel resistance R_t, the resistance R_Ω of a possible shunt channel with ohmic conduction, the capacitance C and the superconducting critical current I_0. Obviously, in the normal state I_0 is zero. For superconductors with energy gap Δ, I_0 and R_t are connected by the Ambegaokar-Baratoff relation

$$I_0(T) = (\pi\Delta/2eR_t)\tanh(\Delta/2k_BT) \tag{1}$$

The relevant energy scale for superconducting phase coherence is

$$E_j = (\Phi_0/2\pi)I_0 \tag{2}$$

with $\Phi_0 = h/2e$. For charging effects the energy scale is

$$E_C = e^2/2C. \tag{3}$$

When $k_BT \gg E_J$, superconductivity is suppressed. When $k_BT \gg E_C$, charging effects are not important. At low temperatures for normal metal junctions, the charging effects may lead to a strong reduction of the conduction. For superconducting junctions, the behavior at low temperatures depends on the ratio

$$x = E_C/E_J \tag{4}$$

Theory predicts and experiments show a superconductor/insulator transition when x is varied from low to high values. In practice charging effects have to be considered when the junction size is $0.01 \ \mu\text{m}^2$ or smaller. That area corresponds to a capacitance of about $10^{-15} F$ or $E_C = 0.9 \ K$, given the usual barriers.

The values of R_t and R_Ω are also very important. R_t determines the mixing of the electron or Cooper pair states on both sides of the barrier. Localized charges, essential for charging effects, exist only when R_t and R_Ω are much larger than the quantum resistance $R_c = h/(4e^2) = 6.5k\Omega$. The strength of the superconducting phase coherence is not suppressed by parallel quasi-particle conduction, but the dynamic resistive behavior is strongly influenced. The relevant parameter, called after McCumber, is

$$\beta_c = 2\pi\Phi_0^{-1}I_0R_{qp}^2C \tag{5}$$

Here R_{qp} includes both the quasi-particle tunneling and ohmic channels. At low temperatures the tunneling conductance disappears exponentially for voltages below the gap, the ohmic component is often relatively insensitive to temperature. For the overdamped case, $\beta_c \ll 1$, single junctions are non-hysteric and vortex flow in arrays is controlled by viscous forces. That regime has been studied extensively. Arrays of superconductor-normal metal-superconductor (SNS) junctions, which are in the extreme overdamped limit, have been used as well as tunnel junctions. For the underdamped case with $\beta_c \gg 1$, single junctions and arrays show a hysteric current-voltage characteristic. Similar as the dynamics of a single junction can be described with the picture of a particle

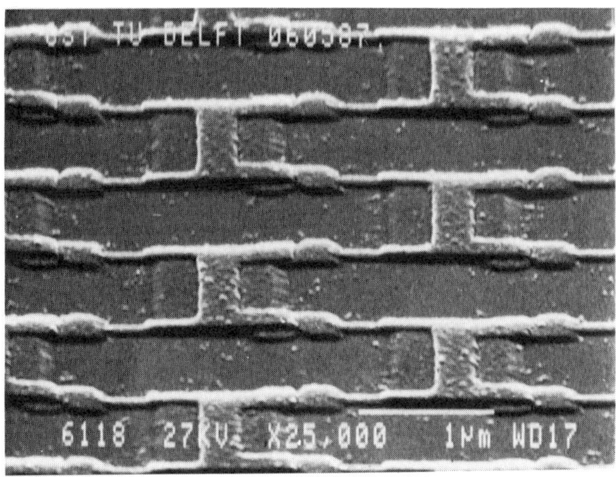

Figure 1: Scanning electron microscope (SEM) photograph of 2 D Aluminum array. The junctions are the small (100 x 100 nm) overlapping regions of the narrow lines, about 400 nm to the left of the broad cross-lines of the H-like islands.

with mass in a washboard potential, it is possible to attribute a mass to the vortices in an underdamped array.

In the regular superconducting arrays, frustration can be introduced by a perpendicular magnetic field. The penetration depth is usually very large with respect to the cell size. When each elementary cell contains a flux equal to a rational fraction of a flux quantum, a special arrangement of phase differences is possible with a relatively low energy. The clearest example is the 'fully frustrated' array with half a flux quantum per cell. In the regime where charging effects dominate, similar frustration can in principle be introduced by inducing a fractional charge on each island by capacitive coupling. However, in practice it is difficult to obtain this kind of frustration homogeneously.

Our tunnel junction arrays are fabricated using electron beam lithography and shadow evaporation. A mask is fabricated, containing holes through which the electrodes are deposited. The mask is supported at a distance of about 500 nm from the substrate. After evaporation of the first electrode at a certain angle, the barrier is made and the second electrode is evaporated from a different angle. With Aluminum, very high quality tunnel junctions are obtained without leakage, because oxidation renders a good barrier over the whole surface. With Niobium, the barrier has to be fabricated with evaporation of Silicon. As a result the edges are more susceptible to leakage. The photograph of Fig. 1 shows an Aluminum array with junctions of $(100\,\text{nm})^2$ and a period of $2\mu\text{m}$.

These lecture notes are organized as follows: After this introduction, in section 2 we discuss 2 D arrays of overdamped ($\beta_c \ll 1$) superconducting tunnel junctions with small E_C. This is the classical limit, where the superconducting phase is well defined. We consider the 2 D phase transition and the influence of a perpendicular magnetic field (frustration). Some first results on underdamped junction arrays are described in section 3. The discussion of charging effects begins in section 4 with single and double junctions and small linear arrays. Section 5 is about single charges in 2 D arrays.

2 Classical Vortex Regime

In this section we consider superconducting tunnel junction arrays where charging effects can be ignored. We may assume that the phase of the superconducting order parameter has a well-defined value ϕ_i for island i. The Hamiltonian for this system is the sum of the Josephson coupling energy terms

$$H = -E_J(T) \sum_{\langle ij \rangle} \cos(\phi_i - \phi_j - A_{ij}) \tag{6}$$

The sum is over nearest neighbor pairs only. A_{ij} is the line integral of the vector potential from the center of island i to the center of island j. In this form the expression is gauge invariant. In the absence of a magnetic field we can choose the vector potential to be zero, and obtain the same Hamiltonian as for the 2 D XY-model of spins with one continuous degree of freedom. It may contain vortices, excitations that have the topological property that the phase changes by 2π when a circuit is completed around them. There are vortices with positive and with negative circulation. The relevant temperature is the real temperature normalized to the coupling energy prefactor, which for our junction array is

$$\tau = k_B T / E_J \tag{7}$$

At very low temperatures, no vortices are present. At somewhat higher temperatures, there appear vortex-anti-vortex pairs, two vortices of opposite sign at a certain mutual distance r. The energy of such a pair is limited, as far away the phase pattern is not distorted. The free energy of a single isolated pair increases logarithmically with r (a is the lattice constant of the array)

$$U(r) = \Phi_0 I_0 \ln(r/a) + 2\mu_c \tag{8}$$

Here $2\mu_c$ is the free energy of a pair at a separation r equal to one lattice cell. μ_c contains an entropy contribution. With increasing temperature, more and more pairs are generated in thermal equilibrium with larger and larger separation. Pairs of small separation modify the interaction of larger pairs, and renormalization methods are required to calculate the properties of the system. A comparison is made with a 2 D Coulomb gas, where the charges interact logarithmically as well. The renormalization of the interaction by smaller pairs is then introduced by a scale and temperature dependent dielectric constant. The theory was developed by Kosterlitz and Thouless [2] and by Berezinskii [3]. A review of the application to superconducting films is given in [4]. The treatment for junction arrays was first developed by Lobb et al. [5]. A phase transition occurs, where at the transition temperature in thermal equilibrium the first free vortex (a pair of infinite separation) appears. The transition temperature is

$$\tau_c = (\pi/2)\epsilon_c^{-1} \tag{9}$$

where ϵ_c is a constant, slightly larger than 1. ϵ_c is the 'dielectric constant' for screening of the 2 D Coulomb gas at infinite scale at the transition temperature. The value is non-universal. Above the transition temperature the density of free vortices grows in a specific way

$$n_f = C \exp\{-[b(\tau - \tau_c)^{-1}]^{1/2}\}. \tag{10}$$

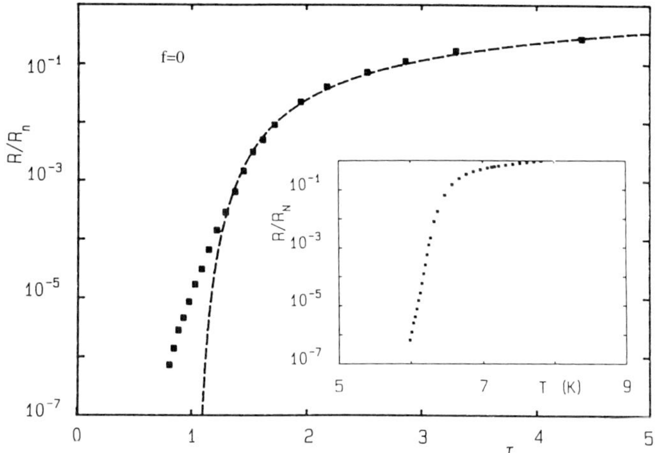

Figure 2: Resistive transition of an array in zero field. τ is the reduced temperature $k_B T / E_j(T)$. The array is 383 cells long and 63 cells wide and has $R_{qp} = 170\,\Omega$. The dashed line is the Kosterlitz Thouless prediction. Inset: resistance versus real temperature.

In a superconducting array it is not a priori clear that the vector potential can be set to zero, even in zero external field. A phase difference leads to a supercurrent, which in turn generates a field. In fact these currents lead to screening. The relevant length is the penetration depth for perpendicular fields

$$\Lambda = \Phi_0 / (4\pi^2 \mu_0 I_0). \tag{11}$$

Within this length, the logarithmic potential for pairs holds, outside it falls off faster. In arrays at the transition temperature the value of L is about $2\,\mathrm{cm}$, divided by the transition temperature in K [5]. This means that in a wide temperature range, for most arrays, the size is well within this penetration length.

As mobile vortices in a superconducting sample lead to resistance, the phase transition can directly be studied. It has clearly been observed. It is not very prominently visible in the resistance measured with a small excitation current, because the number of free vortices increases extremely slowly above T_c. Although the resistive transition can well be fitted to the predicted specific temperature dependence eq. (10), as shown in Fig. 2, the theoretical equation contains too many unknown fitting parameters to make the correspondence very significant. In the non-linear resistance a much clearer fingerprint is found. When the voltage V is measured as a function of the current I, it is found that over a certain temperature range near the transition the dependence is well described with a proportionality of V with I^a. Coming from low temperatures, the exponent a increases till it reaches the universal value 3 at the transition temperature, then jumps to 1 for higher temperatures. Figure 3 shows this kind of behavior for an array of Niobium junctions [6].

The phase transition as discussed so far refers to infinite systems. Clearly, even in large arrays the finite penetration depth makes it impossible to obtain the infinite scale to which theory refers. The finite size will round off the transition somewhat. There are,

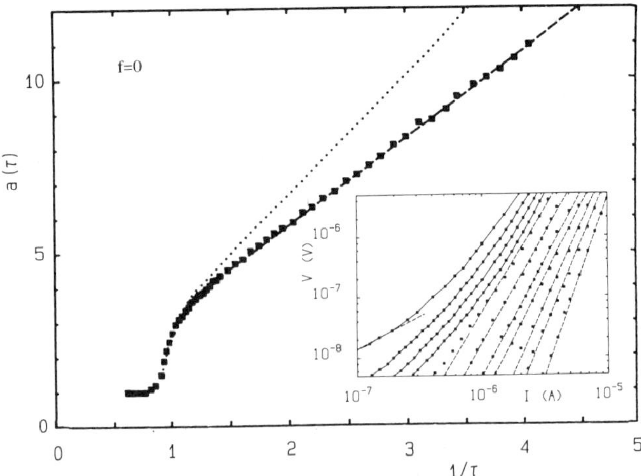

Figure 3: Exponent of non-linear resistance, defined from $V \alpha I^a$, versus inverse reduced temperature. Inset shows $\log V$ versus $\log I$ at different temperatures (high to low T from left to right). At the critical temperature the exponent jumps from 3 to 1. Dashed line: result from Monte Carlo simulation.

however more consequences to the presence of boundaries. A practical superconducting array has superconducting banks at the two opposite edges where the current is applied. Along the other edges, there is vacuum. In computational simulations, also necessarily performed on limited sizes, usually periodic boundary conditions are used at all edges. A vortex in a real array is repelled by the superconducting banks, it sees a mirror vortex of the same sign there. Along the vacuum sides on the other hand, the energy is lower. The vortex sees a mirror vortex of opposite sign. The energy surface of a single vortex is rather flat in the middle of the array. For arrays which are longer than wide, the energy in the middle is $\pi E_J \ln(2W/\pi)$, where the width W is expressed in unit cells. A magnetic field lowers the energy in the middle, leaving higher energy values nearer the edges. The field is expressed in the flux per unit cell, divided by the flux quantum. This is called the frustration f

$$f = \Phi/\Phi_0 \tag{12}$$

It should be remembered that for the junctions under discussion here, the field penetrates homogeneously over the whole array.

In Fig. 4 the magnetoresistance is shown of a Niobium array at three different temperatures. Clearly an oscillation is present, with a period of one flux quantum per cell ($\Delta f = 1$). This is the manifestation of flux quantization in these 2 D networks. At the lowest temperature one also sees the beginning of additional structure. In Fig. 5 such structure is shown on a finer scale. The symmetry around $f = 0$ is very high, indicating that the noise-like variations are in fact completely reproducible. A very deep dip is found at $f = 1/2$ (and other values $f = n + 1/2$), next the dips at $1/3$ and $2/3$ are very prominent. It is possible to assign values n/m to all dips. Their relative strength varies with temperature and also with measuring current. We have systematically investigated

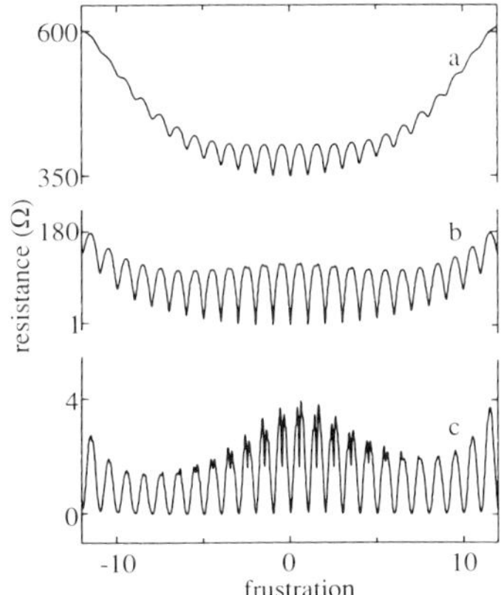

Figure 4: magnetoresistance of Niobium array at different temperatures. (a) $\tau = 5.2$; (b) $\tau = 1.35$; (c) $\tau = 0.45$.

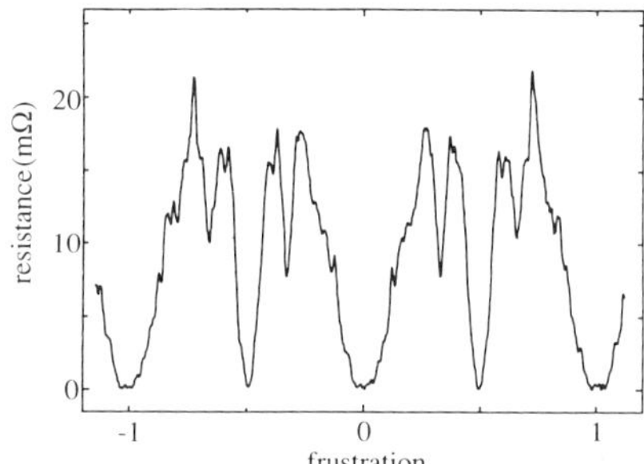

Figure 5: The magnetoresistance on expanded scale at $\tau = 0.22$.

the magnetoresistance of long, narrow arrays of varying width and different edges, which yields a general picture of complexity that increases with width. No clear, simple rules for the prominent occurrence of certain dips m/n have yet been found. A comparison of these results is being made with results of numerical dynamic simulations by Eikmans and Van Himbergen.

The superconducting transition of arrays at $f = 1/2$ has been determined and compared with the transition at $f = 0$. Figure 6 shows results. The linear resistance shows a similar shape as at $f = 0$, and can also well be fitted to results of the Kosterlitz-Thouless theory. The transition temperature is considerably lower. The fact that there is zero resistance below the transition proves the 'fully frustrated' state at $f = 1/2$ to be an ordered state. Although enough flux is present in the array for one vortex per two cells, no flux flow resistance develops. In the ordered state all junctions have a phase difference of $\pi/4$, around each cell a circulating current is flowing with a magnitude of $I_0 \sin(\pi/4)$. The current direction alternates from cell to cell in a checker-board pattern. Vortices may occur as deviations from that ordered pattern. Also domain walls may occur as distortions of the checkerboard pattern where black and white fields have a neighbor of the same color. In Fig. 6 also the exponent of the non-linear resistance is shown. One clearly sees the jump indicative of the Kosterlitz-Thouless-Berezinskii transition.

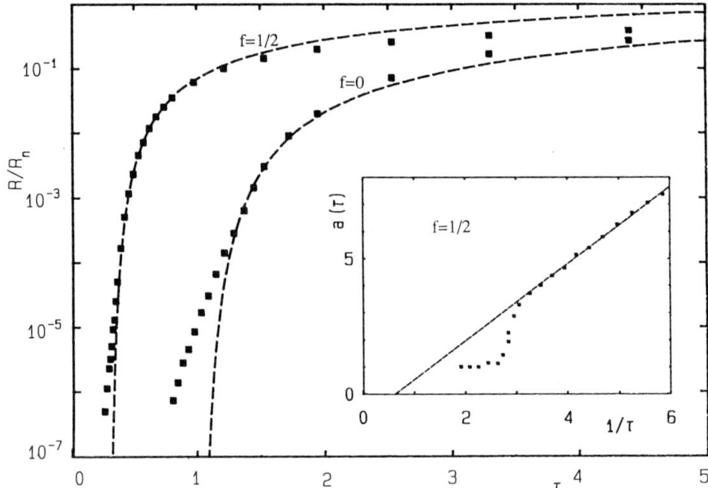

Figure 6: Resistance versus reduced temperature for fully frustrated array ($f = 1/2$) compared with $f = 0$. Inset: exponent of non-linear resistance.

3   Ballistic Vortex Flow

In the previous section we considered arrays of overdamped junctions, i.e. with a shunt conductance in parallel to the Josephson channel of sufficient magnitude to suppress hysteresis in the basic junction characteristic. If the capacitance is large enough and the shunt conductance R_{qp} small enough, the junctions are underdamped. The McCumber parameter of eq. (5) is determining in this respect. We now turn our attention to junctions with $\beta_c > 1$.

When a vortex moves, the phase difference across junctions in the neighborhood changes. As a consequence, there is a voltage difference across these junctions and their capacitance. The capacitive energy or the electric field energy is larger as the vortex velocity is higher. Eckern and Schmid [7] discuss this regime. The electric energy term is proportional to the square of the velocity v. If it is written as a kinetic energy $M_v v^2/2$, where v is expressed in cell constants per second, the effective 'mass' of the vortex is

$$M_v = (h/2e)^2 C/2 \tag{13}$$

We ignore a small second contribution to the mass connected with the self capacitance of islands. When the driving force for vortex flow is cut off, the kinetic (electric) energy leads to a continued movement of the vortices. It depends on the damping how long this motion will continue. At very low temperature, with high quality unshunted junctions, the resistance for voltages below the gap can be extremely high. Consequently, the viscous drag force on a vortex is small and the vortex should be accelerated by a transport current to very high velocity. In that regime vortex flow is ballistic rather than diffusive/viscous. It is not completely clear what should be expected. The maximum velocity is probably determined by the sudden increase in damping due to quasi-particle generation when the rate of phase change exceeds $2\Delta/h$. It may, for other parameter values also happen when the kinetic energy is larger than 2Δ and sufficient to break pairs. It may be that vortices that reach the 'terminal' velocity are stopped suddenly, or that an equilibrium between driving force and drag force is found at a velocity just below the gap value. Another

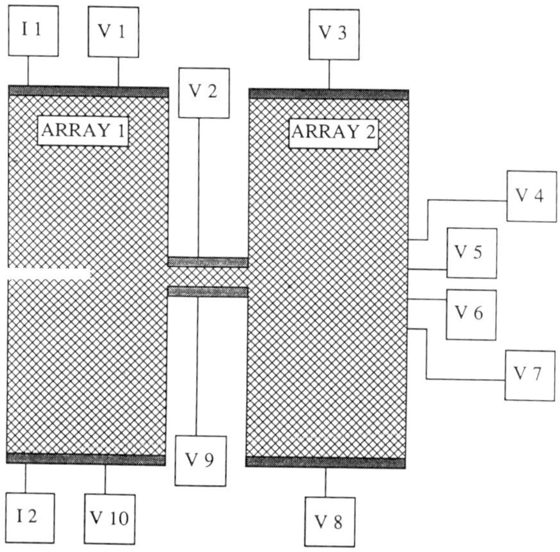

Figure 7: Sample layout for measurement of ballistic vortex flow. Vortices generated in left-hand section are accelerated by current and detected in right-hand section without current.

unclear issue is the velocity component parallel to the transport current, which leads to a Hall voltage. Even for films of dirty superconductors no consistent picture of these aspects has yet been made up, for arrays the force balance determining the direction of vortex flow is even less clear at this stage.

In Fig. 7 we show the layout of a sample that has been designed to try and detect ballistic vortices. It has been designed with T.P. Orlando. It consists of two 2 D arrays, connected by a narrow channel which is also a junction array. Only in array 1 a transport current is applied. Vortices are generated by the current and/or a field and accelerated by the current. Some of them will go through the channel into the second array. If the vortices behave in a ballistic manner, they will keep going over a certain distance even if there is no driving current. If the flow is viscous, they will pile up near the channel exit, but ballistic vortices will travel on to the opposite side where they are to be detected by local voltage probes. We have performed initial measurements on this sample and indeed saw some indications of the type of behavior expected.

However, in the underdamped regime the array shows a very strong tendency to a different mode, which we indicate as switching rows. In Fig. 8 a typical I-V characteristic is shown. Plotted is the voltage between contacts $V2$ and $V9$ of Fig. 7, as a function of the current between $I1$ and $I2$. At each step, a row of junctions switches into the voltage carrying state, here immediately to the gap voltage for our Aluminum junctions. We have strong indications from measurements with local probes that each step is associated with a single row. When the row carries a voltage, the junctions in the row perform a mutually coherent phase slip process. In particular in zero field when vortices are generated by a high transport current, the steps are completely dominating the characteristic. In a field, for a certain range of junction parameters, temperature and current, we also sometimes find clear flux flow behavior. It is too early to evaluate the results. Earlier,

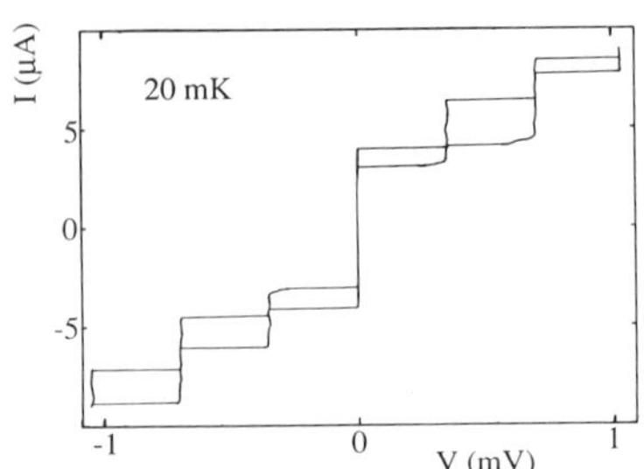

Figure 8: Discontinuous step-wise increase of voltage (row-switching) in underdamped Aluminum array.

in 2 D arrays of just underdamped Niobium junctions, van der Zant et al. [8] found a similar discontinuous behavior with switching of single rows. There, however, the voltage reached was still considerably below the gap and the pieces of the I-V characteristic had a finite slope. Clearly, phase and vortex dynamics in 2 D arrays of underdamped junctions require a thorough study of theory as well as a detailed experimental investigation. Ballistic vortices in the highly underdamped regime will be macroscopic quantum objects.

4   Charging Effects in Small Networks

In this section we will discuss phenomena in short arrays of tunnel junctions that arise due to small capacitance of the junctions. These are usually indicated by the general term charging effects. It will provide the background to understand charging phenomena in e.g. 2 D arrays. We will focus on the experimental aspects, a thorough theoretical treatment of the charging effects, both in these small arrays and in 2 D systems is given by other authors in this volume. For junctions in the normal state the charging effects have recently become well understood. It is now possible to think of fabricating devices such as a single electron transistor or a single electron turnstile. In this section we will first discuss the single electron physics. In the superconducting state the physics changes in several ways. Cooper pairs are present in addition to quasi-particles, and the Josephson coupling can cause coherent Cooper pair tunneling, dependent on the interaction with charging energy. We will present some experimental results and give our present interpretation, which is less established than the theory for single electron effects.

We first discuss normal metal junctions and arrays. The basic reason for the occurrence of charging effects is the discreteness of charge transfer in tunnel junctions. The prerequisite for the following simple theory of charging effects is that the charge states of a junction are well distinguished. With charge state we mean a junction with a given charge (number of charge carriers) and the accompanying charge distribution in the environment. A well defined charge state requires that the junction resistance is high enough ($R_t > h/e2$) that virtual electron tunneling is not important. In that case the tunneling rate is simply related to the energy difference $\Delta E = E_f - E_i$ of those charge states [9]

$$G = \Delta E/e^2 R_t[\exp(\Delta E/kT) - 1] \approx -\Delta E/e^2 R_t \qquad kT \ll \Delta E \qquad (14)$$

We have found that the behavior of the devices we have examined so far can be very accurately described with this equation for the tunneling rate. However, the calculation

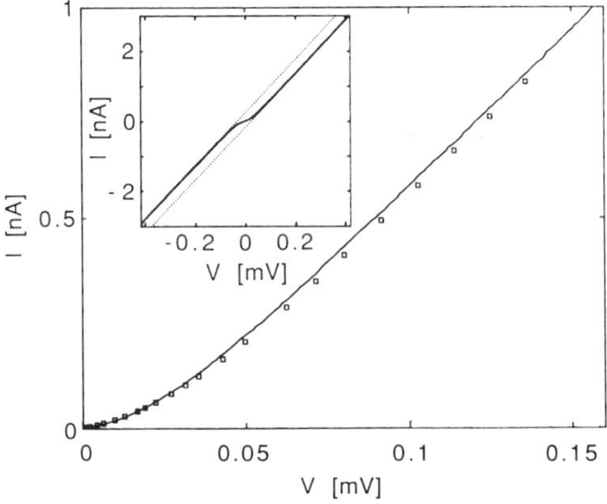

Figure 9: I-V characteristic of a single junction in the normal state, current-biased through junction arrays. A Coulomb gap is clearly observed. The temperature is 55 mK, $R_t = 132\,\mathrm{kOhm}$, $C = 3fF$.

of the energy difference needs some further explanation. Only since recently can it be considered established that for usual experimental circuits this change is the sum of the change in energy of the bias sources and the capacitances present in the system. This is true for systems that are voltage biased, which is usually the case because the impedance of the junctions is quite high. Without special precautions the stray capacitances that are present in the on-chip circuit provide a voltage source of relatively low impedance. By applying Thevenin's rule on the environment of one junction, a simple equation for the energy change in a tunneling event across the junction can be obtained:

$$\Delta E = -e(Q - Q_c)/C. \tag{15}$$

This means that there is a critical charge for electron tunneling given by

$$Q_c = e/[2(1 + C_e/C]. \tag{16}$$

Here Q is the junction charge, C the junction capacitance and C_e the equivalent capacitance of the circuit shunting the junction. In a linear array of n equal junctions, C_e is approximately equal to C/n.

In Fig. 9 we show the I-V characteristic of a single junction, current biased by including 2 D arrays in the leads to the junction. It is clear from this figure that the Coulomb gap can be observed for a single junction. The value of the voltage offset corresponds well with the known capacitance of the junction. In Fig. 10 we give more examples of the charging effects. It shows I-V curves of a double Aluminum junction at two different gate voltages. For two equal junctions in series the critical charge $Q_c = e/4$. This means that the junction charge has to be larger than $e/4$ to obtain conduction. If the metal island between the junctions is neutral, the junctions have equal charge and the voltage over the double junction must therefore be larger than $e/2C$. This is visible as a voltage gap in Fig. 10 (solid curve). If the island is charged by an amount $e/2$, both junctions can tunnel at zero bias voltage across the double junction. This results in conduction at arbitrary non-zero voltage. This suppression of the Coulomb gap is also visible in Fig. 10 (dashed curve). Alternative to a true charging of the island a gate

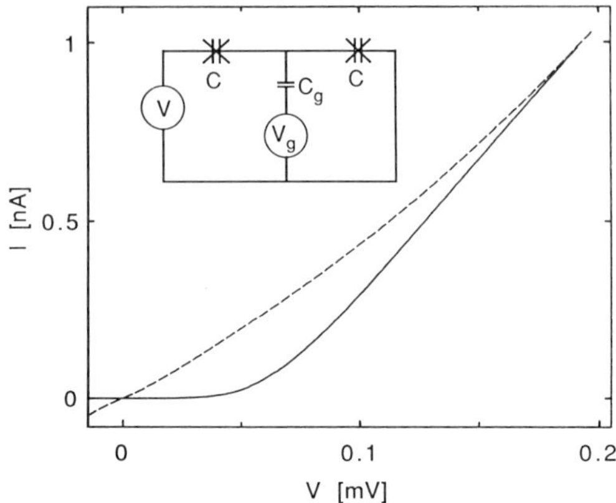

Figure 10: Coulomb gap for two junctions in series ($R_t =$ $58k\Omega$). Drawn line: gate voltage adjusted for maximum gap, dashed line: gate voltage adjusted for minimum gap. The inset shows the measurement configuration.

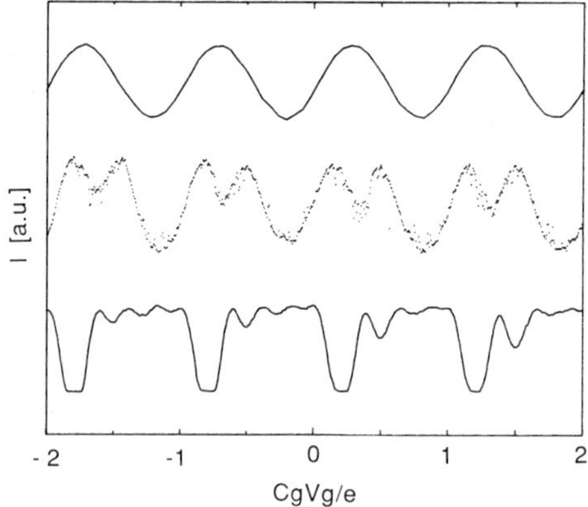

Figure 11: Gate voltage modulation for series arrays of 2 junctions (upper trace), 3 junctions (middle trace) and 5 junctions (lower trace). The bias voltage is fixed below the maximum gap, the current recorded versus gate voltage.

voltage V_g can be applied to the island via a capacitance C_g (see the inset of Fig. 10). For the junction charges this has the same result as an island charge C_gV_g.

The effect of applying a gate voltage can be displayed with the current versus the gate voltage, recorded for a fixed bias voltage. Examples are shown in Fig. 11 for 2,3 and 5 junctions in series. One clear feature is the periodicity of the curves. This periodicity corresponds to the equivalence of charges C_gV_g and $C_gV_g + ne$ for the conductive properties, due to the possibility of tunneling of single electrons. The number of current dips in one period equals the number of central metal islands. This is similar to the resistance of an array of superconducting junctions frustrated by a magnetic field. These show also periodicity in the frustration (the applied magnetic field) with

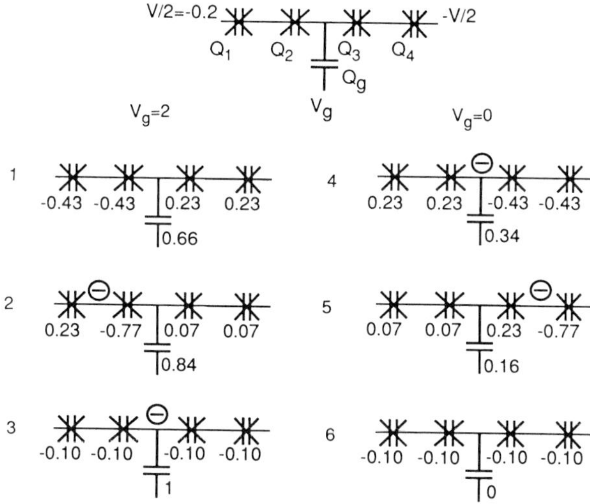

Figure 12: Schematic of operation for turnstile device. Sequence runs from 1 to 6. For $C_g = C/2$, tunneling only occurs when the absolute value of the charge, normalized to e, is larger than $1/3$. Charges on junctions at various moments are indicated.

fine structure dependent on the number of parallel unit cells. There are more parallels between arrays of normal and superconducting junctions, on which we will return later.

One other important result can be deduced from Fig. 11. The curves show varying offset in V_g-direction, indicating that the metal islands are polarized even without applying on purpose a gate voltage. We think the polarizing fields arise from trapped charges, e.g. near the junctions. A result that supports this idea is the observation of telegraph noise in the current for several devices. The current jumps between two I-V_g curves that are offset in gate voltage with respect to each other. Apparently in such a device a trapped charge can jump between two positions. This topic is worth some consideration because it can affect device behavior and disturb experiments. For devices consisting of only a few junctions these offset gate charges can be tuned away by using individually adjustable gates. For larger systems they will have to be avoided or the device behavior should be insensitive to their presence.

We describe one device, fabricated of these small normal metal junctions, that exemplifies the possibilities of these junctions for practical purposes. We call it a turnstile for single electrons. It was designed and the experiments were performed with H. Pothier, D. Esteve, C. Urbina and M.H. Devoret. Fig. 12 shows schematically an array of 4 junctions in which single electrons are transferred in a controlled way with an rf gate voltage. Per cycle one electron is passed, resulting in a current $I = ef$. The working principle is to exceed the critical charge for electron tunneling only for the junctions in the left arm during the first half of the cycle, and only in the right arm during the second half. This can be realized because the central island has a gate capacitance of the order of the junction capacitance (in contrast to the other islands with a much smaller capacitance). Therefore an electron that has passed junction 1 and 2, polarizes this

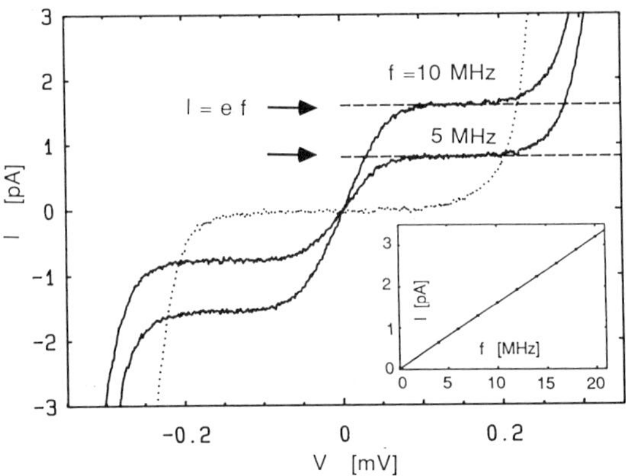

Figure 13: I-V characteristic for turnstile device. Dotted curve: no RF modulation. Drawn lines: gate modulation with frequency $f = 5$ and 10 MHz. Current plateaus are observed at exactly $I = ef$. Inset: plateau levels versus frequency. Slope is e within measuring accuracy.

capacitor with an amount of charge that brings the junction charges below the critical charge for tunneling. The electron is then trapped on the central island until the gate voltage is decreased sufficiently to let it tunnel out through the other arm. Fig. 13 shows the resulting I-V curves. The Coulomb gap that arises without rf gate voltage is lifted to a current plateau $I = ef$ when an rf gate voltage is applied. To obtain a wide current plateau, it is necessary to adjust the DC gate voltage as well as the rf amplitude. We have found that the relation $I = ef$ holds within the experimental accuracy of about 0.3%. It is expected that the intrinsic accuracy of the device is much higher and that it could be used as a current standard [10]. It would be very advantageous to put several devices in parallel to increase the current to a more practical level than the present pA's. However, this is an example of a situation where the offset charges are very disadvantageous since they affect the stability of the trapping of an electron on the central island. In the experiment it was useful to tune the other metal islands with auxiliary DC gate voltages via small gate capacitances. Of course this will be hard to realize if a large number of these turnstile devices are put in parallel.

In the superconducting state at low temperature the density of quasi-particles is low, so that conduction through a junction by single electron tunneling is almost negligible. Cooper pair tunneling can provide conduction under certain conditions [11,12]. The Josephson coupling energy E_J is proportional to the matrix element in the junction Hamiltonian that couples states differing in junction charge by one Cooper pair. It causes the mixing of these states when they are not too different in energy compared with E_J. When two such states are equal in energy Cooper pair tunneling at a rate $E_J/\hbar$ can occur. When they differ in energy by an amount ΔE, a state with a Cooper pair on one side of the junction will start to oscillate, yielding an AC supercurrent. Since Cooper pairs can not gain kinetic energy, we expect that dc conduction by Cooper pair tunneling is only possible for $\Delta E = 0$. Note that the above description is very similar to the Josephson effects in large capacitance (classical) junctions. The difference is that for large junctions charge states differing by many Cooper pairs are strongly mixed. This results in a description with the junction phase as the c-number, since the phase is the quantum number conjugate to the junction charge. The crossover for small capacitance to a description in junction charge has been extensively described, mostly theoretically, as macroscopic quantum effects in a Josephson junction [13].

In an array of junctions the charge transfer occurs via intermediate states, with the charge residing on an island between the junctions. These states differ typically by an amount of order E_C in energy. Single electrons can only tunnel incoherently between neighboring positions with a rate dependent on the energy change. Cooper pairs can also tunnel between neighboring positions, but with a rate dependent on the coupling. If $E_J \ll E_C$ only tunneling for which $\Delta E = 0$ will occur, across individual junctions. If E_J is comparable to E_C, or if the intermediate states differ less than E_C in energy (a situation that can be accomplished by e.g. applying a gate voltage), the coupling can be large enough that coherent Cooper pair tunneling across more than one junction occurs at a significant rate. The array can in that case be described as a single junction. Cooper pair tunneling between neighboring positions has been described by Fulton et al. [11], in experiments on a double junction. They observed conduction by alternating quasi-particle tunneling across one junction and Cooper pair tunneling across the other. Conductance was only significant if bias voltage and gate voltage were applied in such a way that the Cooper pair had zero energy gain and the quasi-particle energy gain was larger than 2Δ (2Δ is the BCS quasi-particle excitation gap). This latter condition ensured that the quasi-particle tunneling rate was not proportional to the (very low) subgap conductance but instead to the normal state conductance.

In Fig. 14 we show I-V curves for arrays of 5 superconducting junctions that illustrate the two conduction mechanisms outlined in the previous section. For an array with $E_J \approx E_C$ at zero voltage coherent Cooper pair transfer through the 5 junctions is possible, resulting in a supercurrent. For an array with $E_J \ll E_C$ the coupling for Cooper pair tunneling across the complete array is negligible (it is proportional to $(E_J/E_C)^4$) so that no supercurrent is observed. The I-V curve for this latter array shows a voltage gap of about $150\,\mu\mathrm{V}$. This gap is twice the normal state gap. Therefore the gap is probably a result of Coulomb blockade of Cooper pair tunneling. At the gap voltage Cooper pair tunneling across one junction takes place with zero energy gain. We think conduction in this case is a process of alternating Cooper pair and quasi-particle tunneling, like the process described by Fulton et al. The current is very low because now the quasi-particle tunneling rate is determined by the subgap conductance.

Confirmation of the ideas outlined above must come from quantitative agreement between experiments and theory. Unfortunately there are some problems for such a comparison. The measurements in the superconducting state are very sensitive to external interference. Very careful filtering of the bias leads at low temperature will be necessary. The current at the various bias voltages is strongly dependent on the quasi-particle conductance, which is very low at low temperatures but difficult to measure directly.

5   Charging Effects in 2 D Arrays

When in 2 D arrays the ratio $x = E_c/E_j$ is increased, a transition occurs from the regime where the superconducting phase is well-defined to the regime where charging effects dominate and the superconducting metal turns into an insulator. For a discussion of the theoretical aspects of this transition we refer to the papers of Schön and Zwerger in this volume. The metal-insulator transition has been seen in quench-condensed granular

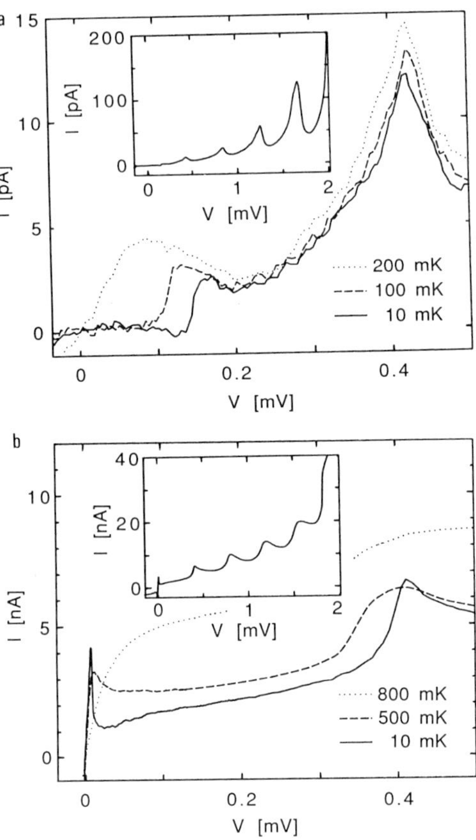

Figure 14: I-V characteristics of series array of 5 junction in superconducting state. (a) $E_j/k_B = 0.13\,$K and $E_c/k_B = 0.45\,$K. A Cooper pair gap is clearly visible at $10\,$mK near $0.14\,$mV. (b) $E_j/k_B = 1.4\,$K and $E_c/k_B = 0.9\,$K. At $10\,$mK a supercurrent is present. (Finite slope due to 2 point measurement). The inset shows the larger scale I-V curves.

films [14] but for these systems the parameters are not well known. It is now possible to repeat these experiments with fabricated large 2 D arrays of junctions with small capacitance. We describe here the experimental results of Geerligs et al. [15]. They studied arrays of Aluminum junctions with $C = 1fF$, and varying R_t. I_0 and E_j are inversely proportional to this R_t. The phase boundary between superconducting and insulating phases not only depends on x, but also on the tunneling conductance which reduces the effects of charge quantization by renormalizing the effective capacitance. Figure 15 shows a series of resistive transitions for arrays with fixed E_c and increasing R_t. It is clearly seen that at low temperatures the low resistance samples E and D become superconducting, while the resistance of samples B and A increases strongly below $300\,$mK. For array C, the voltage at low bias current switches fast between zero and a finite value, making it impossible to assign a resistance below $0.1\,$K. The transition occurs between samples B and D, for a value of x close to 1.5. This is in very good

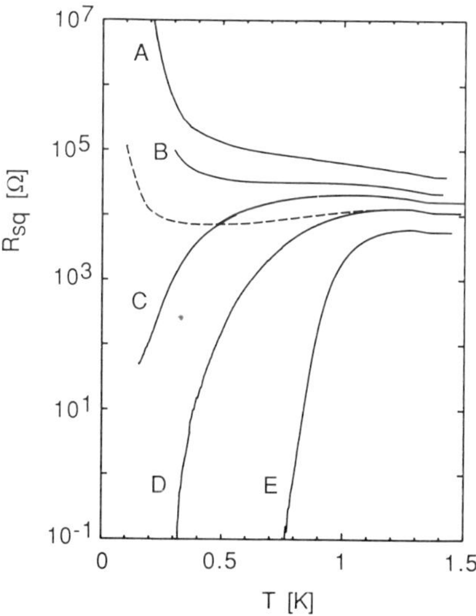

Figure 15: Resistive transition for 2 D arrays of small capacitance Aluminum junctions with varying R_n (from 36 kΩ for A to 4.8 kΩ for E) and E_j/E_c ratio (from 3.9 to 0.53). Dashed line: $R(T)$ for array D ($R_t = 9.7$ kΩ) with application of magnetic field with $f = 1/2$.

agreement with theoretical predictions which show a certain scatter, with the value 1.5 well in their range. In particular, the prediction of Chakravarty et al. [16] which includes the capacitance renormalization predicts a transition for a junction resistance of 13 $k\Omega$, given the geometry and material, in excellent agreement with the experimental data.

We will now consider charges in arrays of normal metal tunnel junctions. Averin and Likharev [9] have discussed charges in 1D arrays. When the array consists of islands, with a self-capacitance C_0 and a nearest neighbor capacitance C, the electrical potential of island j due to a charge e on island k is equal to:

$$\Phi_j = \Phi_k \exp(|j - k|/\Lambda) \tag{17}$$

where $\Omega_k = e/(C^2 + 4CC_0)^{1/2}$ and $\Lambda^{-1} = \mathrm{arc\,cosh}(1 + C_0/2C)$.

This solution has the character of a soliton. The soliton can move if tunneling is possible. It is repulsed by a soliton or an edge voltage of the same sign and also by an open edge, but attracted to a grounded edge. No systematic experimental study of solitons in a 1D chains has yet been performed.

In 2 D arrays a similar situation arises. Here there is no simple analytic solution for the potential on the discrete lattice sites, but in a quasi-continuous approximation it is easy to obtain expressions. We consider a 2 D array as depicted at the top of Fig. 16. When a charge is present on one of the islands, the potential at a distance r is:

$$\Phi(r) = \alpha K_0(r/\Lambda). \tag{18}$$

Where $\Lambda = (C/C_0)^{1/2}$ and α is a constant. K_0 is the modified Bessel function. When the screening length Λ is shorter than the sample size, solitonic solutions again exist. We concentrate here on the case that Λ is larger than the sample size. In that case the Bessel

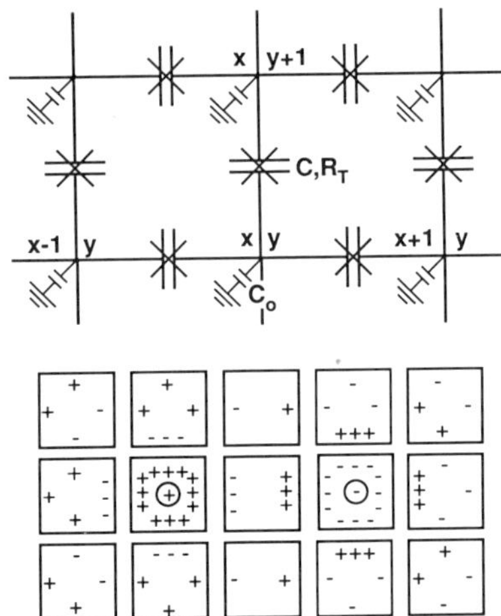

Figure 16: Top: Equivalent circuit for 2 D array with junctions characterized by their capacitance C and tunneling resistance R_t and with self-capacitance C_0 of islands to ground. Bottom: Charge-anti-charge pair, in 2 D array. Squares indicate islands. Only islands which contain circles have net charge. Other islands are polarized.

function can be expanded and one finds approximately that $\Phi(r) = -a \ln(r/\Lambda)$. This is the same potential as for the Coulomb gas or the 2 D XY-model of statistical mechanics. These systems exhibit a Kosterlitz-Thouless-Berezinskii phase transition, which for the charge-array will be at [18]

$$k_B T_c = E_C / 4\pi\epsilon_c \tag{19}$$

Similar as for the vortex-pair-unbinding transition, ϵ_c is a non-universal constant. We expect that below T_c no free charges will be present but only dipoles, above T_c the density will initially grow with a square-root-cusp dependence on temperature. As calculated by Fazio and Schön [17], the critical temperature will be depressed when the tunneling conductance is large, compared with the inverse quantum resistance.

For the vortex transition in 'classical' arrays the resistance is zero below T_c, for the charging transition in small junction arrays the conductance is zero below T_c. Above T_c, the presence of free charges leads to increasing conductance.

The previous discussion only applies for $r \ll \Lambda$. In practice it is quite difficult to make Λ larger than about 20-50. This may lead to significant rounding of the transition in an experiment. In Fig. 17 we show the experimental 'conductive transition' for the same array as sample B in Fig. 15, in the normal state (N, strong magnetic field applied) as well as in the superconducting state (S). For this array Λ is about 17. Clearly, the overall shape is not too different from the shape of $R(T)$ for the vortex pair unbinding transition. However, for the N curve fitting to the square-root cusp equation is not very satisfactory. In the superconducting state such a fit is better possible. It is remarkable how the $R(T)$ curve for the superconducting array is shifted to lower temperatures. Over a considerable temperature range, the conductance of the superconducting sample is far below that in the normal state, an unusual result. This by itself is a clear demonstration

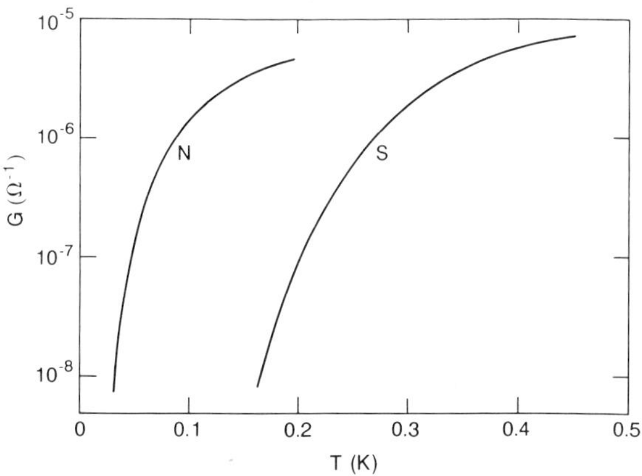

Figure 17: Conductance of array with $15k\Omega$ junctions in normal state (N) and superconducting state (S) as a function of temperature.

that charging effects are dominating. The double charge of the superconducting charge carriers leads to four times higher charging energies. Indeed, the temperature shift is by about a factor of 4. Further investigation in samples with longer screening length is needed. The influence of random fractional charge induced by defects may also lead to disorder smearing the transition.

We thank all colleagues for help and discussions, as well as the Foundation for Fundamental Research on Matter FOM for financial support.

References

[1] The field is reviewed in the Proceedings of the NATO Advanced Research Workshop on Coherence in Superconducting Networks, J.E. Mooij and G.Schön, eds., Physica **B 152**, 1-302 (1988)

[2] J.M. Kosterlitz and D.J. Thouless, J. Phys. **C 6**, 1181 (1973)

[3] V.L. Berezinskii, Zh. Eksp.Teor. Fiz. **59**, 907 (1970) [Sov. Phys. JETP **32**, 493 (1971)]

[4] J.E. Mooij, in: Percolation, Localization and Superconductivity, A.M. Goldman and S.A. Wolf, eds., p. 325, Plenum, New York (1983)

[5] C.J. Lobb, D.W. Abraham and M. Tinkham, Phys.Rev. **B 27**, 150 (1983)

[6] H.S.J. van der Zant, H.A. Rijken and J.E.Mooij, J. Low Temp. Phys. **79**, 289 (1990)

[7] U. Eckern and A. Schmid, Phys. Rev. **B 39**, 6441 (1989)

[8] H.S.J. van der Zant, C.J. Muller, L.J. Geerligs, C.J.P.M. Harmans and J.E. Mooij, Phys. Rev. **B 38**, 5154 (1988)

[9] D.V. Averin and K.K. Likharev, Single Electronics, in: Quantum Effects in Small Disordered Systems, B. Al'tshuler, P. Lee and R. Webb, eds., Elsevier, Amsterdam, to be published

[10] L.J. Geerligs, V.F. Anderegg, P.A.M Holweg, J.E.Mooij, H. Pothier, D. Esteve, C. Urbina and M.H. Devoret, Phys. Rev. Lett. **64**, 2691 (1990)

[11] T.A. Fulton, P.L. Gammel, D.J. Bishop, L.N. Dunkleberger and G.J. Dolan, Phys. Rev. Lett. **63**, 1307 (1989)

[12] D.V. Averin and V.Ya.Aleshkin, JETP Lett. **50**, 367 (1989)

[13] G. Schön and A.D. Zaikin, Quantum coherent effects phase transitions and the dissipative dynamics of ultra small tunnel junctions, submitted to Phys. Reports

[14] see e.g. B.G. Orr, H.M. Jaeger and A.M. Goldman, Phys. Rev. **B 32**, 7586 (1985)

[15] L.J. Geerligs, M. Peters, L.E.M. de Groot, A. Verbruggen and J.E. Mooij, Phys. Rev. Lett. **63**, 326 (1989)

[16] S. Chakravarty, S. Kivelson, G. Zimanyi and B.I. Halperin, Phys. Rev. **B 35**, 7256 (1987)

[17] R. Fazio and G. Schön, this volume and submitted to LT 19

[18] J.E. Mooij, B.J. van Wees, L.J. Geerligs, M. Peters, R. Fazio and G. Schön, Phys. Rev. Lett. **65**, 645 (1990)

SINGLE ELECTRON EFFECTS IN SMALL TUNNEL JUNCTIONS

Dmitri V. Averin* and Gerd Schön

Department of Applied Physics
Delft University of Technology
Lorentzweg 1, 2628 CJ Delft, The Netherlands

1 Introduction

With modern lithographic techniques tunnel junctions with areas smaller than $10^{-2}\mu\text{m}^2$ can be fabricated in a controlled fashion corresponding to capacitances as small as $C \approx 10^{-16}F$ (see Geerligs and Mooij, this volume). Charging effects in these systems strongly influence the transport properties at low temperatures. The basic theoretical description of these effects is by now well established. A comprehensive summary of this description is presented in section 2. In section 3 we discuss some selected extensions of the basic theory, still using a perturbative approach. In many experiments the influence of the external circuit on the charging effects is not negligible, and recent developments of the theory concentrated on these effects. Frequently it requires a non-perturbative approach. We, therefore, summarize in section 4 the formulation of the macroscopic quantum dynamics with arbitrary dissipation and apply it to several selected problems in section 5. For more extensive reviews of the field we refer to two recent articles [1, 2].

2 Background

2.1 Single Electron Tunneling in Normal Junctions

2.1.1 The tunneling rate

We first consider a normal tunnel junction consisting of two metallic electrodes separated by an insulating barrier. If the capacitance C of the junction is small the voltage change $\Delta V \approx e/C$ and the energy change $\Delta E \approx E_C \equiv e^2/2C$ associated with a single electron tunneling are large. The two most prominent consequences are the Coulomb blockade of tunneling and the correlation between different single electron tunneling events. The essence of the Coulomb blockade is that tunneling is prohibited as long as the voltage is small since a tunneling process would increase the charging energy of the junction as

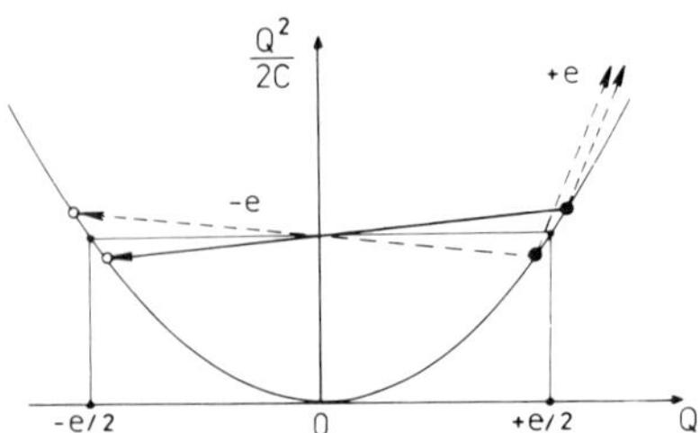

Figure 1: Charging energy of a tunnel junction and single electron transitions. The rate depends on the energy difference in the tunneling process. At zero temperature transitions with $\Delta E^\pm > 0$ are forbidden. The correspond to $|Q| < e/2$ (dashed arrow). Transitions with $\Delta E^\pm < 0$ are allowed (full arrow).

visualized in Fig. 1. This leads to a Coulomb gap in the DC current-voltage characteristics. The correlations in the tunneling arise since the voltage change resulting from one tunneling event changes the rate of subsequent tunneling events. They can lead to coherent single electron tunneling (SET) oscillations of the voltage across the junction.

Let us discuss these single electron effects in more detail. As long as the tunneling conductance is small $1/R_t \equiv G_t \ll 1/R_Q$, where

$$R_Q \equiv h/4e^2 \approx 6.45k\Omega \tag{1}$$

is the quantum resistance, the quantum fluctuations of the charge of the junction are negligible, and one can describe the charge dynamics as a classical stochastic process resulting from random electron tunneling events. The tunneling probabilities for the transitions which increase or decrease the charge from Q to $Q \pm e$ are [3], see also [4]

$$\Gamma^\pm(Q,T) = \frac{\Delta E^\pm(Q)}{e^2 R_t} \left[\exp\left(\frac{\Delta E^\pm(Q)}{k_B T} \right) - 1 \right]^{-1}. \tag{2}$$

Here ΔE is the difference in the junction energy before and after the tunneling process. The transitions are visualized in Fig. 1. At zero temperature the rates reduce to

$$\Gamma^\pm(Q,T=0) = \begin{cases} (1/e)I_t(|\Delta E^\pm|/e) & \text{for } \Delta E^\pm < 0 \\ 0 & \text{for } \Delta E^\pm > 0 . \end{cases} \tag{3}$$

This means that tunneling occurs only if during the process the charging energy decreases. For $|Q| < e/2$ the transitions would cost energy $\Delta E^\pm > 0$, and both $\Gamma^+(Q)$ and $\Gamma^-(Q)$ vanish. This is the 'Coulomb blockade' mentioned above. At finite temperatures transitions which are unfavorable with regard to energy can occur by thermal excitation.

An external current increases the charge continuously, and the charge evolves as

$$\frac{dQ}{dt} = I_x + I_t. \tag{4}$$

Here I_t is the stochastic tunneling current. Alternatively we can describe the stochastic evolution of Q by a master equation for $\sigma(Q,t)$, which is the probability that at time t

Figure 2: Two equivalent circuits of externally biased junctions are shown: (a) A current biased junction with a parallel Ohmic shunt and (b) a voltage biased junction with a series shunt.

the junction capacitance is charged with a charge Q [3]

$$\frac{\partial \sigma(Q,t)}{\partial t} = -I_x(t)\frac{\partial \sigma(Q,t)}{\partial Q} - [\Gamma^+(Q) + \Gamma^-(Q)]\sigma(Q,t)$$

$$+\Gamma^-(Q+e)\sigma(Q+e,t) + \Gamma^+(Q-e)\sigma(Q-e,t). \tag{5}$$

If the junction is shunted by a Ohmic resistor the external current flows in part through this shunt. The shunt also causes temperature dependent current fluctuations, which means the charge evolves for $1/R_s \equiv G_s \ll 1/R_Q$ according to a Langevin equation

$$\frac{dQ}{dt} = I_x(t) - \frac{1}{R_s}\frac{Q(t)}{C} + \tilde{I}(t) + I_t. \tag{6}$$

The spectral density of the fluctuating current $\tilde{I}(t)$ is given by Nyquist's formula

$$S_I(\omega) = \frac{1}{2\pi}\int dt e^{i\omega t}\langle \tilde{I}(t)\tilde{I}(0)\rangle_\omega = \frac{k_B T}{\pi R_s}, \tag{7}$$

and in the master equation (5) we have to add the usual diffusion terms to the right hand side

$$\frac{1}{R_s}\frac{\partial}{\partial Q}[\sigma V] + \frac{k_B T}{R_s}\frac{\partial^2 \sigma}{\partial Q^2}. \tag{8}$$

Although we discuss in this section mostly the current biased (I_x) shunted junction (see Fig. 2a), one should notice that this is equivalent to a voltage biased (V_x) junction with a series resistance, shown in Fig. 2b. All we have to do is to identify $V_x = I_x R_s$. The current through the latter circuit is the tunneling current I_t. Therefore, in the following we frequently concentrate on this current.

2.1.2 Properties of a single junction

SET oscillations

It is straightforward to simulate the stochastic equations or to solve numerically the master equation (5) for arbitrary I_x. The solution of the latter, $\sigma(Q,t)$ is illustrated in Fig. 3 for a situation where an unshunted junction is biased by a small DC current I_x at low temperatures. In this case the junction shows SET oscillations. In the Coulomb blockade interval $-e/2 \leq Q \leq e/2$ the charge changes only due to the external current, but as soon as Q exceeds the value $e/2$ tunneling can occur and lower the charge by e. This takes the system back to the vicinity of $-e/2$, inside the range of Coulomb

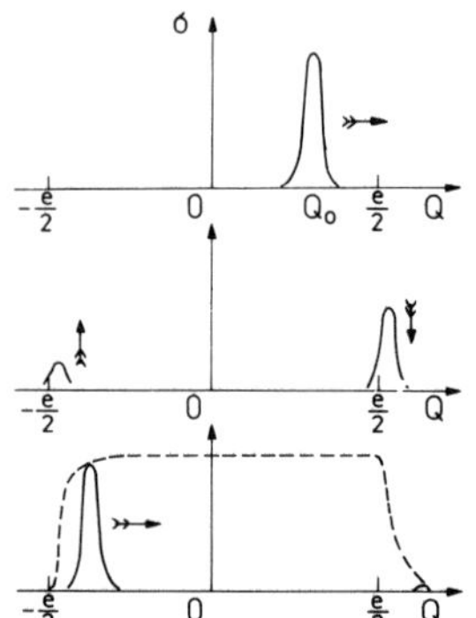

Figure 3: Qualitative picture of the time evolution of the probability $\sigma(Q,t)$ in the parameter regime where SET oscillations can be found.

blockade. The process repeats, which leads to the oscillations of the voltage across the junction with frequency

$$f_{\text{SET}} = I_x/e. \tag{9}$$

Although the tunneling process is stochastic, the Coulomb energy, and the fact that eventually all added charge has to tunnel introduces correlations most pronounced at low temperatures and small external currents. A shunt destroys these correlations, and the SET oscillations acquire a finite linewidth. For weak shunts $G_s < I_x C/e < G_t$ it is

$$W_{\text{SET}} \approx \frac{1}{R_s C} \left(\frac{I_x R_t C}{e} \right)^{1/2}.$$

For stronger shunts $G_s \gtrsim G_t$ the SET oscillations become completely washed out.

The Coulomb gap

In the regime of SET oscillations, $I_x R_t C/e \ll 1$, in the absence of a shunt and at low temperature the Coulomb blockade leads to a more than linear increase of the voltage with increasing current

$$\bar{V} = [\pi I_x R_t e/2C]^{1/2}. \tag{10}$$

For large currents, $I_x R_t C/e \ll 1$, when the SET oscillations are suppressed, the DC current-voltage (I-V) characteristic becomes linear but shifted relatively to the Ohmic line

$$\bar{V} = I_x R_t + \frac{e}{2C}\text{sign}\, I_x. \tag{11}$$

This means that the DC voltage extrapolated to $I_x = 0$ gives a non-zero value $V_g = e/2C$, which is denoted as the 'Coulomb gap'.

The influence of finite temperatures on the DC I-V characteristic is displayed in Fig. 4. At high temperatures $k_B T \gtrsim E_C$ the SET oscillations are destroyed and the I-V curve approaches a linear form also for small currents. The Coulomb gap (defined as the

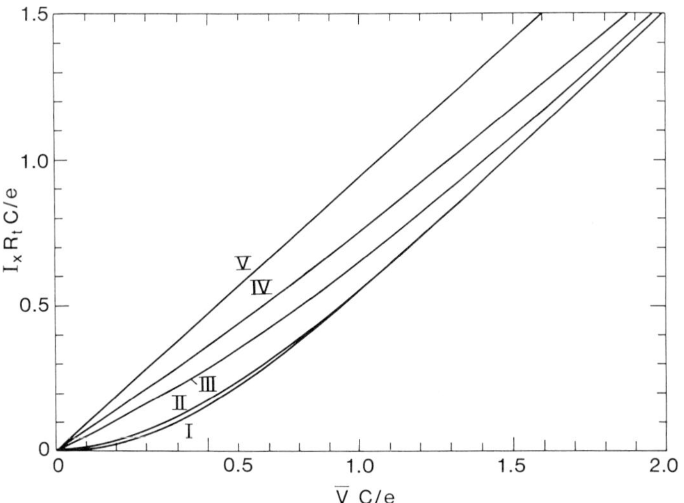

Figure 4: The I-V characteristic of a current biased normal tunnel junction (without shunt) at different temperatures displaying the effect of the Coulomb blockade and the Coulomb gap.

offset at large currents) does not get destroyed. However, it requires larger currents to see it.

The influence of a shunt R_s on the I-V characteristic is displayed in Fig. 5. We display here, for reasons discussed above, the DC value of the current through the junction, $I_t = I_x - V/R_s$. In contrast to the SET oscillations the Coulomb gap in the $I_t - V$ curve is not affected, even in the regime $G_s \gg G_t$ where SET oscillations are completely suppressed. For large currents the DC $I_t - V$ characteristic is still given by the relation eq. (11) with I_x replaced by I_t. At small currents the form of the $I_t - V$ characteristic

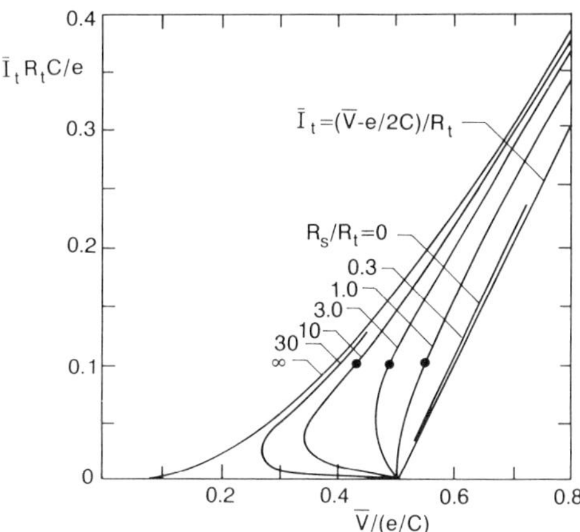

Figure 5: The I-V characteristic of a current biased normal tunnel junction with different shunt conductances at $T = 0$. Displayed is the tunneling current through the junction [8].

depends on G_s. It is interesting to notice that for large $G_s \gg G_t$ it approaches for all currents a linear form shifted by V_g. Quantum fluctuations associated with the shunt for $G_s \gtrsim 1/R_Q$ suppress charging effects at small voltages and further modify the I-V curve. But still the Coulomb gap is recovered at large voltages. See section 5 for a discussion of this limit.

The Coulomb gap has been observed in several experiments. The necessary decoupling from the external circuit and the realization of a current source can be achieved by considering a junction in series with other junctions [5, 6, 7]. The array acts as a resistor and by proper choice of the setup can inject charge in a quasi-continuous fashion into the junction of interest.

2.1.3   Multi-junction systems

Many of the complications caused by the external circuit do not arise if we consider instead of a single junction a network of junctions. In this case the presence of the other junctions with high resistance provides the necessary weak coupling to the external circuit. The dynamics of the network is governed again by the tunneling of electrons across individual junctions. A new feature arises, namely after a tunneling process the charges on all islands redistribute according to the rules of electrostatics. Hence the voltages across each of the junctions of the system change. This gives rise to correlations between tunneling events in different junctions in addition to correlations between subsequent tunneling events in one junction.

The rates of the electron tunneling in the multi-junction systems are still given by eqs. (2) and (3) as for a single junction. However the question arises what is the relevant energy change $\Delta E^{\pm}$. For the junction networks considered in most of the experiments the energy change is the change of the electrostatic energy of the *whole* network (global rule). But if one would achieve a sufficient weak coupling of the junctions, e.g. by inserting high resistors between them, the relevant energy change would be that of the individual junction where the tunneling occurred (local rule). In addition to specific features associated with the correlation between the tunneling events the DC current-voltage characteristics of all junction networks demonstrate the Coulomb blockade of tunneling at small voltages and the Coulomb gap at large voltages [9, 10, 11]. We consider now in more detail some specific examples.

Double-junction system and the Coulomb staircase

The simplest non-trivial multi-junction system consists of two junction in series displayed in Fig. 6a. The charging energy of this system is

$$U = \frac{Q^2}{2C_\Sigma} - \frac{eV}{C_\Sigma}(C_1 n_2 + C_2 n_1), \qquad C_\Sigma \equiv C_1 + C_2, \tag{12}$$

$$Q = e(n_1 - n_2) + Q_0. \tag{13}$$

Here V is the voltage across the whole system, C_i is the capacitance of the i-th junction, and n_i is the number of electrons that have tunneled through it. Furthermore, $Q_0 = \delta V C_\Sigma$ is the non-integer part of charge on the central electrode, where δV is the difference between the Fermi levels of the central and external electrodes of the system at $V = 0$.

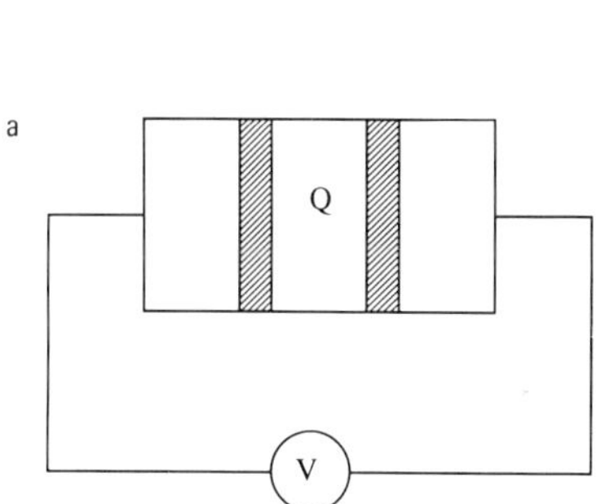 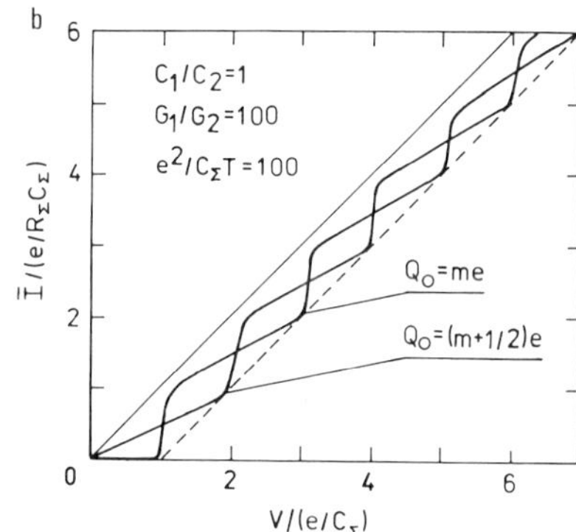

Figure 6: (a) A double junction geometry; (b) the current voltage characteristic of the double junction displaying the Coulomb staircase.

The tunneling rate depends on the charge on the central electrode [12, 13, 14, 15]. The tunneling of an electron through one of the junction increases the voltage across the other one. This increases the tunneling rate in the latter and thus gives rise to correlations between tunneling in both junctions. Such correlations lead to a periodic modulation of the resistance of the system or equivalently to steps in the DC I-V characteristic [16, 17]. The position of these steps depends on Q_0. An example, calculated numerically is shown in Fig. 6b. The steps are most pronounced when there exists a strong asymmetry between the two junction conductances G_i.

The origin of the steps can be understood as follows: a rare tunneling event in the low conducting junction (say junction 2) is rapidly followed by a tunneling through the better conducting junction restoring the equilibrium value of the charge $Q = Q_0 + e\text{Int}\{(VC_2 - Q_0 + e/2)/e\}$ where $\text{Int}(x)$ is the integer part of x. The current I flowing through the system is determined by the rate of the tunneling in the weak junction. The voltage drop across this junction, $\Delta V = (VC_1 + Q)/C_\Sigma$, and the current $I = G_2 \Delta V$, increase stepwise with increasing V, each step corresponding to an increase by e of the charge Q on the central electrodes of the system. The resulting I-V characteristic was called "Coulomb staircase". It has been observed in several experiments (see e.g. [18, 19].

1 D junction array. SET oscillation

In the voltage biased double-junction system considered above only the correlations between tunneling events in different junctions exist. The tunneling through the whole system is uncorrelated in time. In contrast to that in a 1 D array with a larger number of junctions $N > 2$ (see Fig. 7a) both types of correlations, i.e. "space" correlations between tunneling events in different junctions and time correlations of tunneling in one junction occur simultaneously even under the conditions of voltage bias. The reason for that is that an electron inserted in one internal, say the k-th, electrode of the array polarizes all other electrodes and thus induces a finite charge on all junctions of the

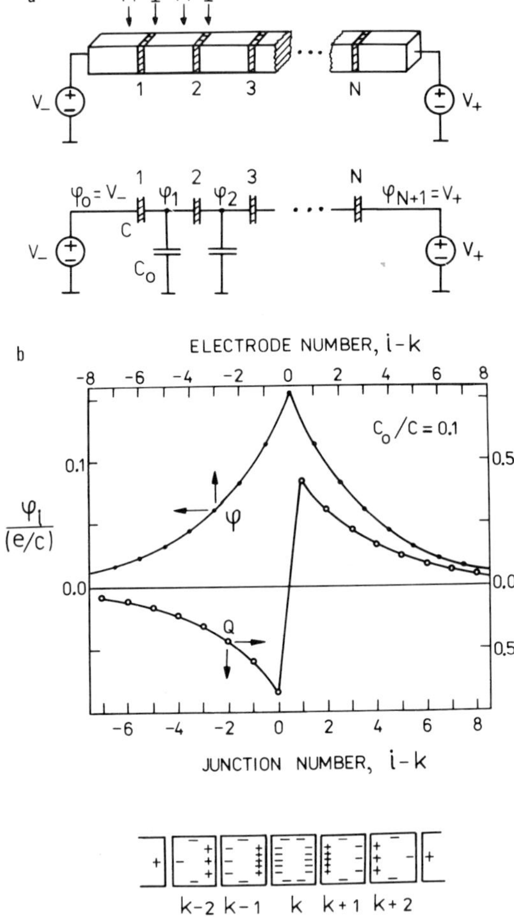

Figure 7: (a) 1 D multi-junction array; (b) polarization of
the array by an electron inserted on the k-th electrode [10].

system and a finite electrostatic potential φ_i on all electrodes (Fig. 7b). In 1 D it is given
by [10]

$$\varphi_i = \frac{e}{C_{\text{eff}}} e^{\lambda|i-k|}, C_{\text{eff}} = \left(C_0^2 + 4C_0C\right)^{1/2}, \quad \lambda = \text{Arch}\left(1 + \frac{C_0}{2C}\right), \tag{14}$$

where C is the junction capacitance and C_0 is the capacitance of the electrodes of the
array with respect to some conducting background. The functional dependence eq. (14)
is unchanged when the electron tunnels along the array, even in the presence of other
electrons. For this reason it can be called a 'single electron soliton' [10].

The polarization results in a repulsion of the charges. As a result they tend to form
a regular structure which can be viewed as a Wigner lattice. When a voltage larger than
a threshold voltage V_{th}

$$V_{th} \approx \frac{e}{C(e^\lambda - 1)}$$

is applied across the array electrons are inserted into the array with some rate. In spite of the randomness of the single electron tunneling, the Wigner lattice persists. But it becomes distorted and moves along the array. This motion gives rise to periodic SET oscillations of the charge and the potential of the electrodes of the array with the frequency I/e. Notice that for any given junction of the array all other junctions act as a large resistor injecting charge in a quasi-continuous into it (i.e. fixing an external current I_x for it), in spite of the fact that the whole array is voltage biased.

The SET oscillations in the array can be phase locked in the usual way by external ac signal with the frequency f_{ac}. This results in the periodically spaced voltage steps on the DC I-V curve of the array, or equivalently in differential resistance peaks at currents $I = enf_{ac}$. Such a behavior have been observed in experiment by Delsing et al. [20] providing an experimental support for the notion of time correlated single electron tunneling in small tunnel junctions.

2 D junction networks

Also 2 D networks (with N_s islands in series and N_p in parallel) have been studied along the same lines as indicated above [9, 11]. For large currents $I_x \gg e/R_tC$ we can find an approximate expression for the voltage across the whole array [11]

$$\bar{V} = I_x R(N_s/N_p) + \frac{e}{2} \sum_{j=1}^{N_s} \left[(\mathbf{C}^{-1})_{j+1,j+1} - 2(\mathbf{C}^{-1})_{j+1,j} + (\mathbf{C}^{-1})_{j,j} \right]. \tag{15}$$

Here $\mathbf{C}^{-1}$ is the inverse capacitance matrix. The indices $j = 1, \ldots N_s$ refer to one chain through the array. We assumed translational invariance in the perpendicular direction. For $N_s, N_p \gg 1$ in a square array the Coulomb gap approaches the limiting value

$$V_g = (1/4)N_s e/2C$$

which agrees well with the numerical simulations [9].

The 2 D arrays are also interesting from a different point of view [21]. The charges on the islands (or solitons) interact via a potential $\Phi(r) \propto K_0(r/\Lambda)$ where $\Lambda = (C/C_0)^{1/2}$ and K_0 is a Bessel function. It is possible to fabricate arrays with small capacitance to the ground $C_0 \ll C$. In these systems for not too large distances $r/\Lambda \ll 1$ the charges interact with a logarithmic rule $\Phi \propto (e/2\pi C) \ln(r/\Lambda)$. Thus the junction array provides a physical realization of a Coulomb gas, which should show a Kosterlitz-Thouless-Berezinskii type transition at a critical temperature $T_c \approx E_C/4\pi$. Below the transition temperature the charges are bound in dipoles and the array is insulating, above free charges allow a finite conductance. The influence of the single electron tunneling on the transition is to lower the transition temperature [22]. For a further discussion of this effect, as well as the competition between the charge ordering and the phase ordering in a superconducting array (i.e. the vortex unbinding transition) see the article of Fazio in this volume.

2.2   Josephson Junctions

2.2.1   The Hamiltonian and the band structure

In contrast to the single electron tunneling the Cooper pair tunneling current is non-dissipative and can be accounted for by a simple Hamiltonian. In terms of the phase difference φ across the junction and the charge Q on the electrodes it is

$$H_0 = \frac{Q^2}{2C} + U(\varphi). \tag{16}$$

The first term is the charging energy, the potential

$$U(\varphi) = -E_J \cos\varphi - I_x \hbar\varphi(2e) \tag{17}$$

accounts for the Cooper pair tunneling and the external current. For junctions with large capacitance this description reduces to a classical equation of motion, which is in essence the balance of supercurrent, external current and capacitive displacement current. In general the charge Q and the phase difference $\hbar\varphi/2e$, although both are macroscopic degrees of freedom, should be considered as quantum mechanical conjugate variables (for extensive discussions see e.g. [23])

$$Q = \frac{\hbar}{i}\frac{d}{d(\hbar\varphi/2e)}. \tag{18}$$

In an unbiased Josephson junction the potential accounts only for the Josephson coupling and hence is 2π-periodic in φ. Hence the eigenstates of the Hamiltonian H_0 eq. (16) are Bloch states

$$\Psi_{n,Q_x}(\varphi + 2\pi) = e^{i2\pi Q_x/2e}\Psi_{n,Q_x}(\varphi). \tag{19}$$

They depend on a parameter Q_x, which we denote as 'quasi-charge' in analogy to the quasi-momentum of a Bloch state in a periodic lattice. The energy levels E_n develop into Q_x-dependent bands $E_n(Q_x)$ [24]. The first Brillouin zone extends over the range $-e \le Q_x \le e$. The detailed form of $E_n(Q_x)$ depends on the ratio between the charging energy scale E_C and the Josephson coupling energy E_J. In the 'nearly free electron' limit $E_J \ll E_C$ the energy bands in the extended zone scheme are approximately given by a parabola

$$E_n(Q_x) \approx Q_x^2/2C \tag{20}$$

(with appropriate assignment of band indices n to different ranges of Q_x) and with band gaps at the boundary of the Brillouin zones $|Q_x| = qe, q = \pm1, \pm2, \ldots$. The energy gap between the two lowest bands is

$$\delta E_1 = E_J. \tag{21}$$

From the properties of the Mathieu functions we know that the higher band gaps δE_n between the band n and $n-1$ decrease quickly, asymptotically as $\delta E_n \approx E_C(E_J/2E_C)^n/n^{n-1}$.

In the tight binding limit $E_J \gg E_C$ the ground state energy is

$$E_0(Q_x) = -\Delta_0 \cos\frac{2\pi Q_x}{2e}, \tag{22}$$

where the bandwidth $2\Delta_0$ follows from properties of the Mathieu function

$$\Delta_0 = 16\left(\frac{E_J E_C}{\pi}\right)^{1/2}\left(\frac{E_J}{2E_C}\right)^{1/4}\exp\left[\left(8\frac{E_J}{E_C}\right)^{1/2}\right] \tag{23}$$

and the energy gap to the first excited band is

$$\delta E_1 = \hbar\omega_0 = (8E_J E_C)^{1/2}. \tag{24}$$

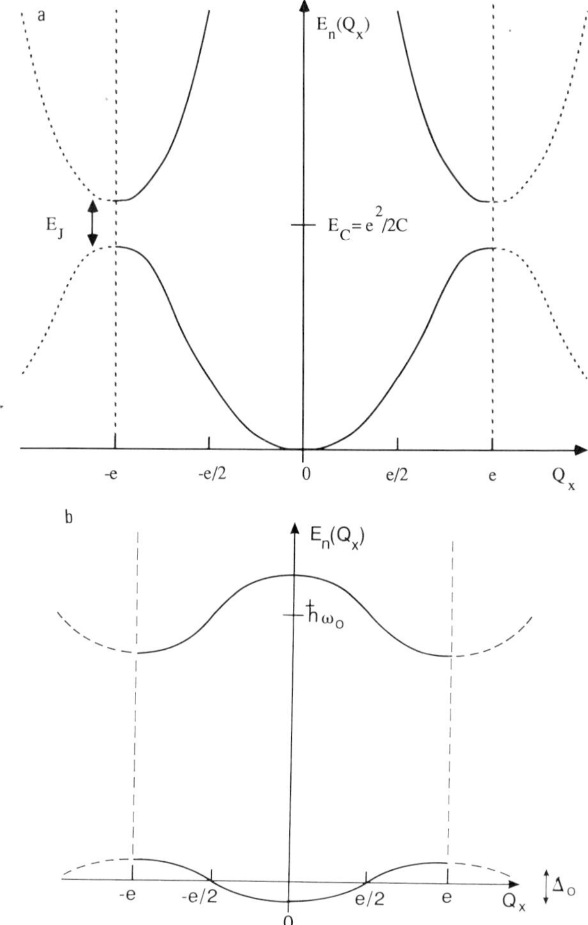

Figure 8: Band energy versus the quasi-charge of a Josephson junction in two limits (a) the 'nearly free electron' limit
$E_J \ll E_C$; (b) the 'tight binding' limit $E_J \gg E_C$.

The band structure for both limits is displayed in Fig. 8a and b, respectively.

If at low temperatures the external charge Q_x is increased adiabatically, such that $d(Q_x/2e)/dt \ll \delta E_1 \hbar$, the system stays in its ground state. The energy as well as other observables such as the voltage

$$V_0 = \langle 0|V|0 \rangle = dE_0(Q_x)/dQ_x \tag{25}$$

vary in time. If $I_x = dQ_x/dt$ is constant they oscillate with frequency [25, 24]

$$f_B = I_x/2e. \tag{26}$$

These oscillations are analogue to the 'Bloch oscillations' of a crystal electron driven by an electric field, and they are named accordingly. Both in the case of Bloch and SET oscillations the voltage oscillates in time due to a coherent charge transfer either of a Cooper pair or of a single electron. Bloch oscillations are deterministic due to the

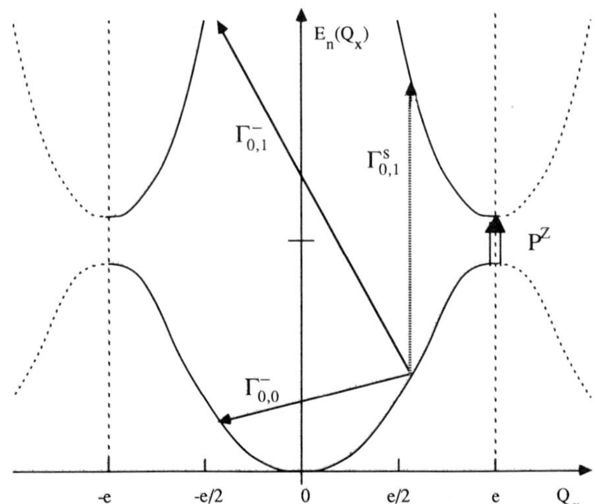

Figure 9: Transitions in the band structure due to a shunt (Γ^s), single electron tunneling ($\Gamma^\pm$) and Zener tunneling (P^Z).

coherent nature of the Cooper pair tunneling. SET oscillations have a stochastic origin but acquire a coherence due to the Coulomb interaction.

2.2.2   Transitions in the energy bands

Effect of an Ohmic shunt

In a Josephson junction the effect of a weak Ohmic resistor (with $1/R_s \ll 1/R_Q = 4e^2/h$) can be described in a similar fashion as discussed in the normal case. If the system is in a Bloch state $\Psi_n(Q_x)$ characterized by Q_x and band index n, the quasi-charge evolves according to a Langevin equation (for the moment we ignore quasi-particle tunneling)

$$\frac{dQ_x}{dt} = T_x(t) - \frac{V_n}{R_s} + \tilde{I}(t) \qquad \text{for} - e \leq Q_x \leq e. \tag{27}$$

The current through the shunt depends on the voltage, which is now $V_n(Q_x) = dE_n(Q_x)/dQ_x$. The fluctuations cause a diffusion of Q_x within one band, but now they also lead to stochastic transitions between different bands. The rate for these interband transitions are

$$\Gamma^s_{n,n'}(Q_x) = \frac{\Delta E_{n,n'}(Q_x)}{4e^2 R_s} |\varphi_{n,n'}(Q_x)|^2 \left[\exp\left(\frac{\Delta E_{n,n'}(Q_x)}{k_B T}\right) - 1\right]^{-1}. \tag{28}$$

The rate can be derived from the Hamiltonian with Ohmic dissipation (see below) by Golden Rule arguments. Here $\Delta E_{n,n'}(Q_x) = E_{n'}(Q_x) - E_n(Q_x)$ is the energy difference in the transition, and $\varphi_{n,n'}(Q_x) = \langle \Psi_{n'}(Q_x)|\varphi|\Psi_n(Q_x)\rangle$ is the relevant matrix element between the states involved. This matrix element is diagonal in Q_x. Hence the stochastic transitions are vertical in the band picture (see Fig. 9).

Quasi-particle tunneling in Josephson junctions

We turn now to the effect of quasi-particle tunneling in superconducting junctions [3, 26, 27, 28, 29, 30, 31]. As long as the tunneling is weak it can be taken into account perturbatively. It allows stochastic transitions between states differing in Q_x by $\pm e$. The probabilities for the transitions are

$$\Gamma^{\pm}_{n,n'}(Q_x) = \frac{1}{e}\left|s^{\pm}_{n,n'}\right|^2 I_{qp}\left(\frac{\Delta E^{\pm}_{n,n'}(Q_x)}{e}\right)\left[\exp\left(\frac{\Delta E^{\pm}_{n,n'}(Q_x)}{k_B T}\right) - 1\right]^{-1}, \qquad (29)$$

which is a straightforward extension of eq. (1) with the following modifications

1. the energies are the band energies (see Fig. 8), i.e. $\Delta E^{\pm}_{n,n'}/Q_x) = E_{n'}(Q_x \pm e) - E_n(Q_x)$

2. the matrix elements for the transitions are

$$s^{\pm}_{n,n'} = \langle \Psi_{n'}(Q_x \pm e)\,|\exp(\pm i\varphi/2)|\,\Psi_n(Q_x)\rangle. \qquad (30)$$

   In the limit of small $E_J \ll E_C$ they are unity for $n' = n \pm 1$, and for transitions within the lowest band $n = 0$ and vanish otherwise.

3. The quasi-particle tunneling current $I_{qp}(V)$ depends in the familiar fashion on the superconducting energy gap Δ. In much of our discussion we will assume for simplicity that the quasi-particle current is characterized by subgap conductance $I_{qp}(V) \approx G_t V$.

Zener tunneling

In addition if the driving current I_x is not adiabatically small the junction undergoes transitions between different states in different bands n but characterized by the same Q_x. This process is denoted as 'Zener tunneling'. A simple picture emerges in the limit $E_J \ll E_C$. In this case the band gaps are narrow and we can assign a probability for such a transition whenever we pass a boundary of a Brillouin zone. The probability for a Zener transition, say at $Q_x = e$ from band $n - 1$ to the adjacent band n is

$$P^Z_{n-1 \leftrightarrow n} = \exp\left[-\frac{\pi}{8}\frac{\delta E_n^2}{n E_C}\frac{e}{\hbar I_x}\right] \equiv \exp\left[-\frac{I_Z}{I_x}\right]. \qquad (31)$$

The second relation defines the Zener current I_Z. Ben-Jacob et al. [32] where the first not to be mislead by Ziman [33] and to write down the correct factors. We ignore interference effects between subsequent Zener transitions, which is justified if there exists a 'phase breaking' time τ_ϕ shorter than $2e/I_x$. [1] The Zener transitions are indicated in Fig. 9 by an arrow. For large E_J/E_C it cannot be accounted for by a simple transition probability

---

[1] Actually here an independent argument applies: Due to the decrease in the band gaps the Zener tunneling probability in a Josephson junction quickly approaches unity with increasing n. Hence the system will be driven to higher energy states after the first Zener transition across the first gap, and interference effects do not arise.

but at the same time the mixing of different quantum states is less important. Zener tunneling drives the junction to higher energy states. This is balanced by the relaxation processes induced by an Ohmic shunt or the quasi-particle tunneling.

The master equation

The rules given above specify the stochastic time evolution of the system. It can be described by combining the Langevin eq. (27) and the transitions between different bands and charges as described by the rates eqs. (28), (29) and (31). Alternatively we can consider the time evolution of $\sigma_n(Q_x, t)$, i.e. the probability to find the system at time t in a state $\Psi_n(Q_x)$. Expectation values of physical quantities are taken with $\sigma n(Q_x, t)$, e.g. $\langle V \rangle = \sum_n V_n \sigma_n$. The probability obeys by the master equation [3]

$$\frac{\partial \sigma_n(Q_x)}{\partial t} = -I_x(t)\frac{\partial \sigma_n}{\partial Q_x} + \frac{1}{R_s}\frac{\partial}{\partial Q_x}[\sigma_n V_n] + \frac{k_B T}{R_s}\frac{\partial^2 \sigma_n}{\partial Q_x^2}$$

$$+ \sum_{n'=0}^{\infty} \left\{ -\left[\Gamma_{n,n'}^s(Q_x) + \Gamma_{n,n'}^+(Q_x) + \Gamma_{n,n'}^-(Q_x) + \Gamma_{n,n'}^Z(Q_x) \right] \sigma_n(Q_x) \right.$$

$$+ \left[\Gamma_{n',n}^s(Q_x) + \Gamma_{n',n}^Z(Q_x) \right] \sigma_{n'}(Q_x)$$

$$+ \left. \Gamma_{n,n'}^-(Q_x + e)\sigma_{n'}(Q_x + e, t) + \Gamma_{n,n'}^+(Q_x - e)\sigma_{n'}(Q_x - e, t) \right\}. \tag{32}$$

The terms on the right describe the effect of the external current I_x and the diffusion of the quasi-charge Q_x within one band due to the current noise of the shunt. Furthermore, the shunt and the quasi-particle tunneling cause interband transitions with rate $\Gamma_{n,n'}^s(Q_x)$ and $\Gamma_{n,n'}^\pm(Q_x)$, respectively. In the limit $E_J \ll E_C$ we can account for the Zener tunneling between adjacent bands n and $n \pm 1$ by the rate [34]

$$\Gamma_{n-1\leftrightarrow n}^Z(Q_x) = \frac{\pi}{8\hbar}\frac{\delta E_n^2}{nE_C}e\delta(|Q_x| - e). \tag{33}$$

Integration of the *rate* eq. (33) across the boundary of the Brillouin zone reproduces the Zener tunneling *probability* eq. (31). Effectively it leads to boundary conditions for the master equation. For instance if upon increasing Q_x the bands n and n' meet at $Q_x = e$ the boundary conditions (in the reduced band scheme) are $\sigma_{n'}(-e + 0) = P_{n\leftrightarrow n'}^Z \sigma_n(e - 0) + \left(1 - P_{n\leftrightarrow n'}^Z\right)\sigma_{n'}(e - 0)$.

2.2.3   Properties of single Josephson junctions

Josephson junction shunted by an Ohmic shunt

Consider a Josephson junction shunted by an Ohmic shunt which is driven by an external DC current I_x. If this external current is weak $I_x \ll I_Z$ and provided that the temperature is low $T \approx 0$ the junction remains in the lowest band. Below a threshold $I_x < I_{th}$

$$I_{th} = \begin{cases} e/R_s C & \text{for } E_J \ll E_C \\ \pi \Delta_0/eR_s & \text{for } E_J \gg E_C \end{cases} \tag{34}$$

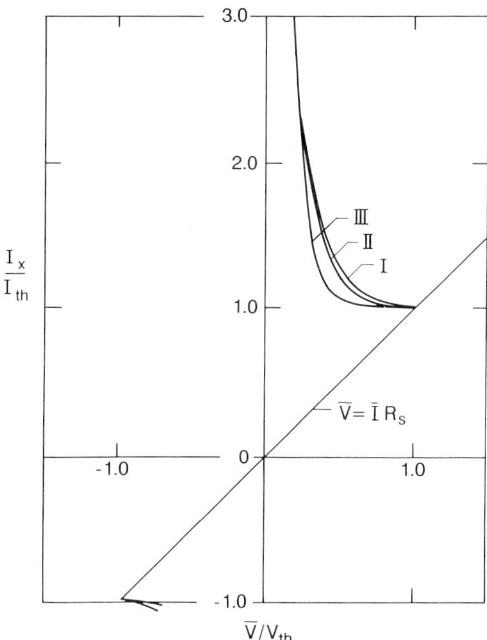

Figure 10: The I-V characteristic of a shunted Josephson junction. The current scale I_{th} is defined in eq. (35), the voltage scale is $V_{th} = R_s I_{th}$. The three curves correspond to $E_J \ll E_C (I), E_J \approx E_C$ (II), $E_J \gg E_C$ (III).

the current flows completely through the shunt and the voltage is constant $\langle V \rangle = I_x/R_s$. For larger currents the junction is driven to the boundary of the Brillouin zone and we find a cross-over to Bloch oscillations. As a consequence the DC voltage decreases with increasing current. Above the threshold $I_x > I_{th}$ the DC current-voltage characteristic becomes [24]

$$\bar{V}/R_s = \begin{cases} I_x - 2I_{th}/\ln\left(\dfrac{I_x+I_{th}}{I_x-I_{th}}\right) & \text{for } E_J \ll E_C \\[2ex] I_x - (I_x^2 - I_{th}^2)^{1/2} & \text{for } E_J \gg E_C \end{cases} . \tag{35}$$

Here $\bar{V}$ is the time average of $\langle V(t) \rangle$. This result is shown in Fig. 10. For large currents $I_x \gg I_{th}$ (but still $I_x \ll I_Z$) it reduces to $\bar{V} = I_{th}^2 R_s/2I_x$. The Bloch oscillation frequency depends on the current through the junction and in comparison to eq. (26) is reduced to $f_B = I_t/2e$.

At finite temperatures the current noise in eq. (27) gives rise to intra-band fluctuations, which lead to a finite linewidth $2W_B$ of the Bloch resonances. In the classical regime, as long as $\hbar W_B \ll k_B T \ll \delta E_1$, for currents just above I_{th}, this is

$$W_B = (\pi/e)^2 k_B T/R_s. \tag{36}$$

At still larger temperatures $k_B T \gtrsim \delta E_1$ (which is E_J in the limit $E_J \ll E_C$) and for larger currents $I_x \gtrsim I_Z$ the interband transitions due to thermal activation eqs. (28), (29) or due to Zener tunneling across the bandgaps are frequent. Thus the superconducting

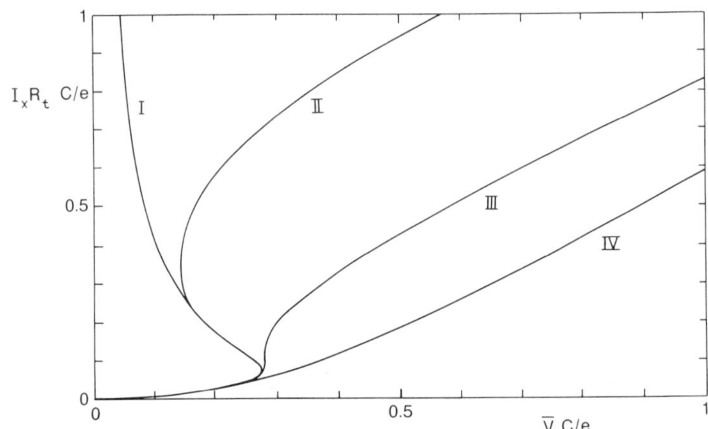

Figure 11: The I-V characteristic of a Josephson junction with quasi-particle tunneling in the limit $E_J \ll E_C$ for different strength of the Zener tunneling controlled by the parameter α_t. The curves I to IV correspond to $\alpha_t \approx 0, \alpha_t = 0.01, 0.05$, and 0.5.

properties are washed out and the junction behaves much like a normal junction. These results apply as long as the shunt conductance is small $1/_Rs \ll 1/R_Q$. The general description will be presented in later sections.

Quasi-particle tunneling in Josephson junctions

We now study the effect of single electron tunneling processes on the properties of the Josephson junction. Again we increase the quasi-charge Q_x adiabatically by an imposed dc current I_x. For small currents $I_x \ll e/R_tC$, i.e. small sweeping rates compared to the rate of tunneling, Q_x rarely reaches the boundary of the Brillouin zone. Therefore, Bloch oscillations cannot develop, but under suitable conditions we see SET oscillations. The situation is very similar to normal junctions. For instance for $F_J \ll E_C$ the DC I-V characteristic is the same as in a normal junction eq. (10), except for quantitative details due to the structure in $I_t(V)$.

For larger currents $I_x \gtrsim e/R_tC$ the quasi-charge Q_x is frequently driven to the boundary of the Brillouin zone. Provided that the Zener tunneling between the bands is weak Bloch oscillations develop, while single electron tunneling becomes relatively less important. Plots of the resulting DC I-V characteristics are shown in Fig. 11 for the limit $E_J \ll E_C$. If we ignore Zener tunneling the Coulomb gap of the normal junction is replaced by a 'nose'-like structure, which arises as a crossover from single electron tunneling dominated behavior to Bloch oscillations. The current and voltage scales are given by $e/(R_tC)$ and the bandwidth E_C/e, respectively. At large currents the I-V characteristic reduces to

$$\bar{V} = \frac{E_C}{12I_xR_tC} \qquad \text{for} \qquad e/R_tC \ll I_x \ll I_Z. \tag{37}$$

Also in the opposite limit $E_J \gg E_C$ we find a nose-like structure in the I-V curve, on a much smaller current scale $I_x \approx 0.1\Delta_0/(eR_t)$ and much smaller voltage scale given by the bandwidth Δ_0 eq. (23). Asymptotic results are

$$
\bar{V} = \begin{cases} \pi\sqrt{\Delta_0 R_t I_x/2e} & \text{for} \qquad I_x \ll \Delta_0/eR_t \\[12pt] 2\Delta_0^2/e^2 R_t I_x & \text{for} \qquad I_x \gg \Delta_0/eR_t \end{cases}
\tag{38}
$$

We want to emphasize this result: a Josephson junction driven by a DC current shows a finite voltage, i.e. it is not superconducting at all. This property is a consequence of the quantum mechanical nature of the junction. We note, however, that in the limit of classical junctions $E_J \gg E_C$ the voltage and current scales of the nose-like structure become exponentially small. We further point out that in the limit $E_J \leq E_C \leq \Delta_{BCS}$ the current scale, which depends on the subgap conductance $1/R_t$, may be very small.

At higher temperatures thermal activation allows transitions to the higher bands. In the case $E_J/E_C < 1$ these transitions involve energy differences ΔE of the order of E_C, and *not* $\Delta E \approx E_J$. Therefore, the nose-like structure persists for temperatures $k_B T \lesssim E_C$ and only vanishes beyond.

Interband transitions and single electron tunneling

If the current I_x driving the junction is not adiabatically small transitions between neighboring bands are also caused by Zener tunneling with the probability given in eq. (31). Since for the structures discussed the relevant scale for the currents is $e/R_t C$ we rewrite the Zener tunneling probability as

$$
P_{0\leftrightarrow 1}^Z = \exp\left[-\frac{1}{8}\left(\frac{E_J}{E_C}\right)^2 \frac{e}{I_x R_t C}\frac{\pi e^2 R_t}{2\hbar} \right].
\tag{39}
$$

In this form we see that for currents $I_x R_t C/e \approx 0.1$, i.e. on the scale of the nose, Zener tunneling is negligible only if the resistance R_t is large, i.e. if the dissipation is weak $\alpha_t \ll (E_J/E_C)^2$, where the dimensionless tunneling conductance is

$$
\alpha_t \equiv h/(\pi^2 e^2 R_t).
\tag{40}
$$

For larger currents the transitions to higher bands occur with higher probability. As a result the current-voltage characteristic approaches that of a normal junction. In Fig. 11 we have also plotted the I-V characteristica of a Josephson junction with fixed value of E_J/E_C for different strength of α_t. As the probability of Zener tunneling increases we observe a transition from the nose-like I-V characteristic, typical for the system where bandgaps prevent transitions to higher bands, to an I-V characteristic of the shape as encountered in a normal junction, where the current can drive the system to high energies. At intermediate strength of the Zener tunneling the Coulomb gap appears both as a 'nose' as well as a shift of the high voltage asymptote.

In the limit of small at Zener tunneling sets in sharply at a crossover current I_{cr} [34, 35]

$$
I_{cr} = \frac{\pi}{8} I_c \left(\frac{2\alpha_t}{a}\right)^{1/2} = \left(\frac{I_Z e}{a R_t C}\right)^{1/2},
\tag{41}
$$

where $I_c = 2eE_J/\hbar$ and the numerical coefficient is $a = 1.645$. The crossover current I_{cr} is much smaller than the Zener breakdown current $I_Z \sim eE_J^2/\hbar E_C$. Thus already a relatively small current $I_{cr} < I_x < I_Z$ is sufficient to destroy the Bloch oscillations, although the probability of Zener tunneling between the two lowest bands for each traversal through the band may still be exponentially small.

We summarize these important results. In the limit $\alpha_t \ll (1/4)(E_J/E_C)^2 \ll 1$ the I-V characteristic consists of four distinct pieces, characterized by the tunneling current $e/R_t C$ and I_{cr}.

1. The region of small currents $I_x \lesssim 0.1 e/R_t C$ is dominated by relaxation by single electron tunneling. The voltage increases $V(I_x) \approx (\pi I_x R_t e/2C)^{1/2}$ with I_x, qualitatively as in a normal junction.

2. At larger currents I_x, but still smaller than the crossover current $0.1 e/R_t C \lesssim I_x \lesssim I_{cr}$, Bloch oscillations gain importance. Hence the DC voltage decreases $V(I_x) \approx E_C/(12 I_x R_t C)$. This leads to the nose-like structure in $I(V)$. This regime is not affected by the Zener tunneling.

3. At the crossover current $I_x \approx I_{cr}$ in a narrow regime of width $\delta I_x \approx e/R_t C \ll I_{cr}$ Zener tunneling rapidly gains importance. The voltage changes from a value near zero to a value $I_{cr} R_t$.

4. For larger currents we have a linear behavior with Coulomb gap $V = I_x R_t + e/2C$ as in a normal junction.

For larger values of at the different regions merge as shown in Fig. 11.

2.2.4   Arrays of Josephson junctions

Next we consider networks of Josephson junctions. Unless some mechanism breaks the superconducting phase coherence such a network forms a multi-degree of freedom quantum system described by a multi-degree of freedom Hamiltonian. If we ignore dissipation the Hamiltonian in terms of the phases φ_j of the islands is

$$H_0 = \sum_{ij} \frac{1}{2} Q_i C_{ij}^{-1} Q_j - \sum_{\langle ij \rangle} E_J \cos(\varphi_i - \varphi_j) \quad , \quad Q_j - \frac{\hbar}{i} \frac{\partial}{\partial(\hbar\varphi_j/2e)}. \tag{42}$$

The invariance of the Hamiltonian under the translation $\varphi_j \to \varphi_j + 2\pi$ implies that the eigenstates are Bloch states [36]

$$\Psi_{n\mathbf{q}}(\varphi) = \exp(i\varphi \cdot \mathbf{q}/2e) u_{n\mathbf{q}}(\varphi). \tag{43}$$

Here $\mathbf{q} = \{q_1, q_2, \ldots\}$, and q_j is the quasi-charge on the island j. The functions $u_{n\mathbf{q}}(\varphi)$ are 2π periodic and orthonormal, and n denotes the band index. The reciprocal lattice is hypercubic with lattice constant $2e$. In the limit $E_C \gg E_J$ the Bloch states are plane waves in the extended zone scheme, and the ground state energy is given by

$$E_{\mathbf{q}}^{(0)} = \min_{p_i=0,\pm1,\ldots} \left\{ \frac{1}{2}(\mathbf{q} - \mathbf{p}2e) \cdot \mathbf{C}^{-1} \cdot (\mathbf{q} - \mathbf{p}2e) \right\}. \tag{44}$$

At the boundaries of the Brillouin zones the Cooper pair tunneling , described by the potential in eq. (42), mixes the degenerate states and yields a band splitting. This is a straightforward, though tedious extension of the single junction analysis. For details see [36].

Again the single electron tunneling, shunt and Zener tunneling lead to transitions in the multi-dimensional bandstructure. The resulting I-V characteristics for small arrays are shown in the article [36]. The overall shape is similar to the single junction result but the voltage scales with the number of junction in series. This supports that the I-V characteristic of a Josephson array found by Geerligs et al. [7], which has the shape of the Bloch nose and a crossover to a resistive branch, can be interpreted as evidence of the bandstructure and the Bloch oscillations.

3 Density Matrix Approach

3.1 General Discussion

In the previous section we considered the semi-classical "Golden-rule" type approach to the single electron effects. It is based on the concept of a stochastic transitions which can be described by the master equation (33) for the probability density $\sigma(q, t)$. In the present section we shall discuss a slight generalization of this approach to cases where one is interested not only in the evolution of probabilities, which are diagonal elements of the junction density matrix ρ, but also in the evolution of the off-diagonal elements of the density matrix.

To achieve such a generalization one can consider an equation for the density matrix of the whole system, which will be traced over the internal degrees of freedom of the junction electrodes, i.e. the tunneling degrees of freedom which will be denoted by t, and those of the shunt s. Since for large resistances, $R_t, R_s \gg R_Q$, the couplings H_t and H_s between these degrees of freedom and the junction macroscopic degrees of freedom, φ and Q, are weak, one can treat $H_{t,s}$ as a perturbation. In lowest non-vanishing order of standard perturbation theory (see e.g. [37]) one gets

$$\dot{\rho}(t) = F_t + F_s, \quad F_{t,s} = -\hbar^{-2} \int_{-\infty}^{t} dt' Tr_{t,s} \left\{ [H_{t,s}(t), [H_{t,s}(t'), \rho(t')\rho_{th}]] \right\}, \tag{45}$$

where ρ_{th} is the equilibrium density matrix of the junction electrodes (excluding the principle degrees of freedom φ and Q) or of the shunt.

The coupling operator H_t is the standard tunnel Hamiltonian, which can be written in the following form

$$H_t = e^{-i\varphi(t)/2} H^+ + e^{i\varphi(t)/2} H^- \,, \quad H^+ = \sum_{k_L, k_R} T_{k_L k_R} c_{k_L}^+ c_{k_R} \,, \quad H^- = (H^+)^+, \tag{46}$$

where $c_{k_i}^+$ and c_{k_i} are the creation and annihilation operators of electrons in the left and right electrode. By writing the phase factors we made explicit the time dependence which accounts for a voltage difference across the junction. The phase should be considered itself as a quantum degree of freedom. The coupling to the shunt is described by

$$H_s = \frac{\hbar}{2e} I_s \varphi, \tag{47}$$

where I_s is the operator of the current through the shunt which can be specified, for example, within the Caldeira-Leggett model (see section 4.1).

Since the perturbations $H_{t,s}$ are weak, the density matrix ρ evolves slowly in time and we use a Markoffian approximation. This amounts to replacing $\rho(t')$ on the right hand side of eq. (45) by $\rho(t)$. After that we can express the integrals in eq. (45) by the usual linear response functions which describe the shunt and the quasi-particle tunnel current. We assume that both currents are linear functions of the voltage, so that

$$\frac{2e^2}{\hbar^2} Re \int_0^\infty dt e^{-i\omega t} \langle H^+(t) H^- \rangle = \frac{\hbar\omega}{R_t} \left[\exp\left(-\frac{\hbar\omega}{T}\right) - 1 \right]^{-1}, \tag{48}$$

$$Re \int_0^\infty dt e^{-i\omega t} \langle I_s(t) I_s \rangle = \frac{\hbar\omega}{R_s} \left[\exp\left(-\frac{\hbar\omega}{T}\right) - 1 \right]^{-1}. \tag{49}$$

In this way one can show that the diagonal part of eq. (45), in the basis of eigenstates of the junction Hamiltonian eq. (10), reproduces the corresponding part of the master equation (33). Below we shall consider several specific examples where the off-diagonal elements of ρ are also of importance.

3.2   Bloch Oscillations in the Tight-Binding Limit

Let consider a single junction with strong Josephson coupling, $E_J \gg E_C$, biased by a sufficiently large current $I_x \gtrsim e\Delta_0/\hbar$, where $2\Delta_0$ is the width of the lowest energy band eq. (23). In this case one cannot treat the current as a perturbation. It modifies considerably the eigenstates of the junction Hamiltonian eqs. (16), (17), i.e. they are no longer pure Bloch states. We therefore include the term describing the current into the basic Hamiltonian of the junction, together with the charging energy and Josephson coupling energy. If the current is still small compared to the separation between energy bands, $I_x/e \ll \omega_0$, and the temperature is also small, $T \ll \hbar\omega_0$, a one-band approximation holds and the Hamiltonian has the form

$$H_0 = \sum_\nu -\hbar\omega_B |\mu\rangle\langle\mu| - \frac{\Delta_0}{2} \left(|\mu\rangle\langle\mu+1| + |\mu+1\rangle\langle\mu| \right). \tag{50}$$

Here we used the Bloch oscillation frequency $h I_x/2e \equiv \hbar\omega_B$ and $|\mu\rangle$ is the lowest energy state in the μ-th minimum of the Josephson potential. The eigenstates $|\mu\rangle$ of the Hamiltonian eq. (50) are known to form the "Stark ladder" (see e.g. [38])

$$E_m = -m\hbar\omega_B, \qquad |m\rangle = \sum_\mu J_{m-\mu}\left(\frac{\Delta_0}{\hbar\omega_B}\right) |\mu\rangle, \tag{51}$$

where J_p are the Bessel functions. The result shows that in the limit $\hbar\omega_B \gg \Delta_0$ the states $|m\rangle$ are close to localized in one minimum of the potential.

It is straightforward to calculate the matrix elements describing the quasi-particle current and the shunt current in the m-representation

$$\langle m'| e^{\pm i\varphi/2} |m\rangle = (-1)^m J_{m'-m}\left(\frac{2\Delta_0}{\hbar\omega_B}\right), \tag{52}$$

$$\langle m'|\varphi|m\rangle = 2\pi \left(m\delta_{m,m'} - \frac{\Delta_0}{2\hbar\omega_B} \sum_\pm \delta_{m,m\pm 1} \right). \tag{53}$$

Using this matrix element one can derive from the general expression eq. (45) the following master equation for the density matrix in the m-representation

$$\dot{\rho}(m,m') \;=\; \frac{2}{e^2 R_t}\sum_p\left\{\frac{p\hbar\omega_B}{\exp(p\hbar\omega_B/T)-1}J_p^2\left(\frac{2\Delta_0}{\hbar\omega_B}\right)\right.$$

$$\left.\Big[(-1)^{m-m'}\rho(m+p,m'+p)-\rho(m,m')\Big]\right\}-\frac{\pi^2 T}{e^2 R_s}(m-m')^2\rho(m,m')$$

$$+\;\frac{V_{th}^2}{2\hbar\omega_B R_s}\sum_{\pm}\frac{\pm 1}{\exp(\pm\hbar\omega_B/T)-1}\left[\rho(m\pm 1,m'\pm 1)-\rho(m,m')\right]. \quad (54)$$

Here we defined the threshold voltage (compare eq. (29)) $V_{th}=\pi\Delta_0/e$.

One can evaluate the DC voltage across the junction making use only of the equation for the diagonal of the density matrix (extending the work [39])

$$\langle V\rangle=\frac{\hbar}{2e}\langle\dot{\varphi}\rangle=\frac{\pi\hbar}{e}\sum_m m\dot{\rho}(m,m)=\frac{V_{th}^2}{2I_x}\left[\frac{1}{R_s}+\frac{4}{\pi^2 R_t}\right]. \quad (55)$$

However, since the operator V is non-diagonal in the m-representation,

$$V=\frac{V_{th}}{2}\sum_{\pm}|m\rangle\langle m\pm 1|, \quad (56)$$

one needs also off-diagonal elements of the density matrix to calculate the voltage correlation function $\langle V(0)V(t)\rangle$. The result is [38]

$$\langle V(0)V(t)\rangle=\frac{V_{th}^2}{2}\cos(\omega_B t)\rho(m+1,m;t), \quad (57)$$

where

$$\rho(m+1,m;t)\;=\;\exp(-W_B t)\,,$$

$$W_B\;=\;\frac{\pi^2 T}{e^2 R_s}+\frac{2\hbar\omega_B}{e^2 R_t}\sum_p pJ_p^2\left(\frac{2\Delta_0}{\hbar\omega_B}\right)\coth\left(\frac{p\hbar\omega_B}{2T}\right), \quad (58)$$

is the solution of eq. (54) with the initial condition corresponding to a voltage eigenstate. This results shows that the system considered exhibits narrow-band Bloch oscillations in wide range of parameters, with a half-width W_B of the spectral line given above, even when the stationary states of the junction are nearly localized in one minimum of the Josephson potential.

3.3 Resonant Tunneling of Cooper Pairs

Let us study now the influence of a weak Josephson coupling on the properties of the system of two tunnel junctions connected in series. This extends the analysis of a normal double junctions presented in section 2.1.3 [40, 41]. This system can be studied conveniently by means of the density matrix equation outlined above. In contrast to quasi-particle tunneling the dissipation-free Cooper pair tunneling is strong only if the difference δ in electrostatic energy U of the system before and after the tunneling of a Cooper pair is small, $|\delta| \ll E_C$. If the voltage drop across the total system is small $eV \ll E_C$, a resonance condition can be satisfied for tunneling through the whole system. Furthermore, if the impedance Z_e of external circuit is small, $|Ze| \ll R_Q$, (which is the usual experimental condition) the total phase difference across both junctions behaves classically and the system exhibits ordinary Josephson oscillations. The critical current, however, depends on the charge Q of the central electrode of the system [42].

At larger voltages, $V \sim E_C/e$ at most one junction can be in resonance at a given voltage and central charge Q. In order to describe the Cooper pair tunneling in this case it is convenient to write down an equation for the density matrix $\rho(Q, Q')$ in the basis of the Q eigenstates. Assuming that the condition for resonance $U(Q) \approx U(Q + 2e)$ is satisfied for transitions in one (the j-th) junction, and retaining only the resonant terms we get the following equation for ρ in the Schrödinger representation

$$\dot{\sigma}(Q) = -(E_j/\hbar)\mathrm{Im}[\rho(Q, Q + 2e)] + F_t\{\sigma(Q)\}, \tag{59}$$

$$\dot{\sigma}(Q + 2e) = (E_j/\hbar)\mathrm{Im}[\rho(Q, Q + 2e)] + F_t\{\sigma(Q + 2e)\}, \tag{60}$$

$$\dot{\rho}(Q, Q + 2e) = i(E_j/2\hbar)[\sigma(Q) - \sigma(Q + 2e)] - (i\delta/\hbar + \Gamma)\rho(Q, Q + 2eh), \tag{61}$$

where $\sigma(Q) \equiv \rho(Q, Q)$, E_j is the Josephson coupling energy of the j-th junction, and $\delta = U(Q+2e) - U(Q)$ is the difference in electrostatic energy of the total system eqs. (12), (13). With F_t we denoted the contribution of single electron tunneling to the evolution of diagonal elements of the density matrix

$$F_t\{\sigma(Q)\} = \sigma(Q + e)\Gamma_j^-(Q + e) + \sigma(Q - e)\Gamma_j^+(Q - e) - \sigma(Q)\left[\Gamma_j^-(Q) + \Gamma_j^+(q)\right], \tag{62}$$

while Γ is the decay rate of the off-diagonal elements of the density matrix due to this tunneling

$$\Gamma = \frac{1}{2}\sum_{j,\pm}\left[\Gamma_j^\pm(Q) + \Gamma_j^\pm(Q + 2e)\right]. \tag{63}$$

Here $\Gamma_j^\pm$ are the rates of single-electron tunneling given by eqs. (2), (3).

Solving eq. (61) for $\rho(Q, Q+2e)$ and substituting it into eqs. (59), (60) one can see that they describe, in addition to single-electron tunneling, tunneling of Cooper pairs through the j-th junction between the state Q and the state $Q+2e$. Due to the relaxation in eq. (61) a stationary state can be reached where the rate of Cooper pair tunneling is [41]

$$\gamma_j(Q) = \frac{\Gamma E_j^2}{2\left[\delta^2 + (\hbar\Gamma)^2\right]}. \tag{64}$$

Thus, the contribution of the Cooper pair tunneling to the DC current flowing through the junctions is essential only at resonances. Because of the discreteness of the charge Q eq. (13) the conditions for the resonance can be satisfied only at certain voltages,$V = (en + Q_0)/C_2$ for resonance in the first junction, and $V = -(en + Q_0)/C_1$ for the second junction. As a result, the Cooper pair tunneling leads to periodically spaced resonance current spikes in the I-V characteristic of the system. Examples of numerically computed I-V characteristics which contain resonance current spikes are shown in Fig. 12. Such spikes have been observed in recent experiment [40].

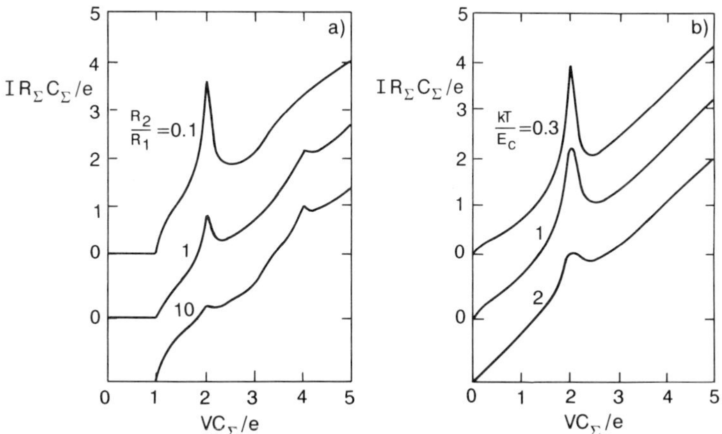

Figure 12: I-V characteristic of a double Josephson junction for the following parameters: $Q_0 = 0, C_1 = C_2, E_1 = 0.1E_C, E_2 = 0$, and $R_Q/R_\Sigma = 0.01$ where $1/R_\Sigma = 1/R_1 + 1/R_2$ [43]. (a) dependence on the ratio G_1/G_2 (for the middle curve $R_Q/R_\Sigma = 0.03$); (b) suppression of the current peak with increasing temperature.

3.4 Tunneling of the Charge via Virtual States in Multi–Junction Systems

In this section we extend the theory to higher order in the tunneling and thus find limitations of the simple description discussed in section 2. One of the main features of single electron charging effects in small junctions or multi-junction systems is the Coulomb blockade of tunneling: for small voltages across the system, $V < V_{th}$, where V_{th} is some threshold voltage, the tunneling is suppressed. For vanishing temperature, low conductance of the external circuit $1/R_s$ and low tunnel conductance $1/R_t$ of the junctions this suppression is absolute. But thermal fluctuations of the charge or quantum fluctuations due to finite conductances give rise to a non-zero tunneling rate.

At low temperatures only the quantum tunneling is important. In order to calculate the rate of tunneling of the charge in one junction in general one should use the non-perturbative description of the tunnel and shunt conductances of the sections 4 and 5. However, perturbatively we can calculate the rate of the tunneling also in multi-junction systems by means of the "Golden Rule"

$$\gamma = \frac{2\pi}{\hbar}|\langle i|M|f\rangle|^2\delta(E_i - E_f), \tag{65}$$

where E_i and E_f are the initial and final energy of the system, and $\langle M \rangle$ is the matrix element of tunneling.

Outside the Coulomb blockade region the dynamic of the system is governed by the "one- junction" tunneling with rate eq. (2), (3), which is of first order in the small junction conductances. In the Coulomb blockade regime the tunneling through any one of the junctions in the system is energetically unfavorable and hence suppressed. The only favorable tunneling event is the transfer of an electron through the whole system, which, however, is of higher order in the junction conductances. Following standard prescriptions [44] one can write down the transition amplitude for such a tunneling. For instance in the case of a double junction the transition through the whole system consists

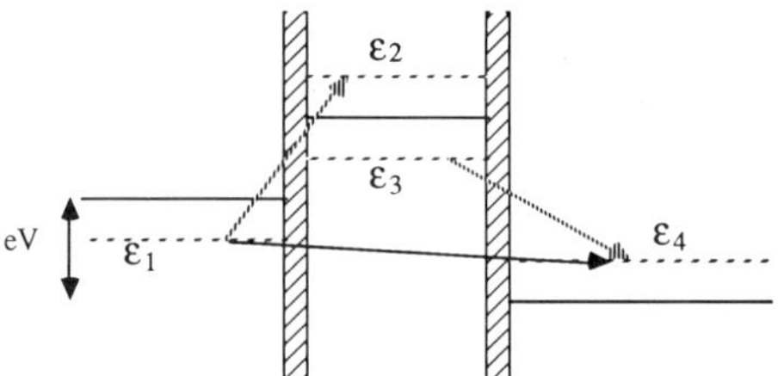

Figure 13: The virtual transitions involved in the 'macroscopic quantum tunneling of the charge' through a double junction.

of two virtual transitions involving intermediate states with bare energies ϵ_i and ϵ_k (see Fig. 13). In addition we have to take into account the electrostatic energy E_j, which depends on the order of tunneling, i.e. whether the tunneling occurs first in the junction $j = 1$ or $j = 2$. As a result we have for the transition amplitude through the total system

$$M(\epsilon_1, \epsilon_2, \epsilon_3, \epsilon_4) = T^{(1)}_{\epsilon_1\epsilon_2} T^{(2)}_{\epsilon_3\epsilon_4} \left(\frac{1}{\epsilon_1 + \epsilon_2 + E_1} + \frac{1}{\epsilon_3 + \epsilon_4 + E_2} \right), \tag{66}$$

where $T^{(j)}_{\epsilon_i\epsilon_k}$ is the matrix element of the tunneling in the j-th junction. The electrostatic energies E_j follow from eqs. (12), (13). They are

$$E_1 = \frac{e}{C_\Sigma,} \left(\frac{e}{2} + Q_0 - VC_2 \right), \qquad E_2 = \frac{e}{C_\Sigma,} \left(\frac{e}{2} - Q_0 - VC_1 \right). \tag{67}$$

Assuming that the tunneling events which involve states with different energies ϵ_i are uncorrelated, which is valid for a sufficiently large central electrode of the system, we can sum up all the transition *rates* and obtain

$$\gamma = \frac{2\pi}{\hbar} \int d\epsilon_1 d\epsilon_2 d\epsilon_3 d\epsilon_4 f(\epsilon_1) f(\epsilon_2) f(\epsilon_3) f(\epsilon_4) M^2(\epsilon_1, \epsilon_2, \epsilon_3, \epsilon_4) \delta(\epsilon_1 + \epsilon_2 + \epsilon_3 + \epsilon_4 - eV). \tag{68}$$

Here $f(\epsilon)$ is the Fermi distribution function, and we made use of $1 - f(\epsilon) = f(-\epsilon)$. At $T = 0$ the integrals in eq. (68) can be calculated explicitly and one gets [45]

$$\gamma = \frac{\hbar}{2\pi e^4 R_{t1} R_{t2}} \left\{ \left(1 + \frac{2}{eV} \frac{E_1 E_2}{E_1 + E_2 + eV} \right) \left[\sum_{j=1,2} \ln \left(1 + \frac{eV}{E_j} \right) \right] - 2 \right\} eV. \tag{69}$$

For small voltages this expression reduces to

$$\gamma = \frac{\hbar}{12\pi e R_{t1} R_{t2}} \left(\frac{E_1 + E_2}{E_1 E_2} \right)^2 V^3. \tag{70}$$

It follows that the current I flowing through the system $I = e\gamma$ is non-linear for all voltages.

At finite temperatures the I-V characteristic becomes linear for $eV \ll T$, and the conductance of the system, up to a constant of order unity, is given by

$$G = (dI/dV)_{V=0} \approx \frac{\hbar}{2\pi e^2 R_{t1} R_{t2}} \left(\frac{E_1 + E_2}{E_1 E_2} T \right)^2. \tag{71}$$

One can easily generalize these considerations to the case of an N-junction array and show that for small voltages, $V \ll V_{th}$, the MQT rate γ decreases rapidly with increasing N. For $T = 0$ it is [38]

$$\gamma \propto \prod_j \left(\frac{R_Q}{\pi^2 R_{tj}}\right) \left(\frac{V}{V_{th}}\right)^{2N-1} . \tag{72}$$

The macroscopic quantum tunneling of the charge limits the accuracy of the performance of the Langevin equation devices based on single-electron tunneling, in particular the accuracy of current quantization in the electron turnstile device discussed in this volume by Geerligs and Mooij.

4 Theory of Dissipation in Quantum Mechanics and Quantum Dynamics of JosephsonJunctions

In general a Josephson junction is subject to dissipation. The question how dissipation can be described on a quantum mechanical level was addressed in the paper of Caldeira and Leggett [46, 23]. Their approach is based on a phenomenological model, which we will summarize first. We then will review the derivation of the description of dissipation in Josephson junctions starting from the microscopic many body theory. In section 5 we will make use of this description in order to study the influence of an arbitrary linear environment on the single electron effects.

4.1 The Caldeira-Leggett Model for Ohmic Dissipation

The model Hamiltonian

The current balance of a Josephson junction with Ohmic dissipation is

$$C\frac{\hbar\ddot{\varphi}}{2e} + \frac{1}{R_s}\frac{\hbar\dot{\varphi}}{2e} + \frac{d}{d\hbar\varphi/2e}U(\varphi) = \tilde{I}(t). \tag{73}$$

The potential is given by eq. (17). It accounts for the supercurrent and the external current. The current noise $\tilde{I}(t)$ of the shunt resistor has a Gaussian distribution and spectrum given by eq. (7). It is straightforward to generalize the description to arbitrary frequency dependent dissipation and corresponding noise. This is relevant for junctions shunted by an element with a general impedance $Z(\omega)$.

Linear dissipation can be described by coupling the principal variable φ to a bath of harmonic oscillators [46] $H = H_0 + H_{\text{bath}}$

$$H_0 = \frac{Q^2}{2C} + U(\varphi) \; ; \quad H_{\text{bath}} = \sum_j^\infty \left[\frac{P_j^2}{2M_j} + \frac{M_j}{2}\Omega_j^2 \left(X_j - \frac{c_j}{M_j\Omega_j^2}\frac{\hbar\varphi}{2e}\right)^2\right] . \tag{74}$$

The distribution of frequencies of the bath and the coupling strengths define a spectral density $J(\omega) \equiv (\pi/2) \sum_j^\infty \left(c_j^2/M_j\Omega_j\right) \delta\left(\omega - \Omega_j\right)$. If $J(\omega)$ is chosen as

$$J(\omega) = \frac{1}{R_s}\omega \qquad 0 \le \omega \le \Omega_c. \tag{75}$$

Ohmic dissipation is reproduced in the classical limit. A general frequency dependence of the impedance $Z(\omega)$ can be modelled by a corresponding choice for $J(\omega)$.

The classical equation of motion eq. (73) is derived by eliminating the bath coordinates from the infinite number of coupled equations of motion for φ and X_j. On the other hand, the Hamiltonian eq. (74) can also be analyzed on a quantum mechanical level. Key quantities of interest are propagators in imaginary times or the partition function, which can be written as path integrals over all degrees of freedom. Since the bath is harmonic its degrees of freedom can be integrated out exactly. Thus the partition function can be reduced to a path integral over the principal variable $\varphi(\tau)$ only $Z = \int D\varphi(\tau)\exp\{-S_{\text{eff}}[\varphi(\tau)]/\hbar\}$. It involves a non-trivial effective action

$$S[\varphi] = \int_0^{\hbar\beta} d\tau \frac{C}{2}\left(\frac{\hbar}{2e}\frac{\partial\varphi}{\partial\tau}\right)^2 + U(\varphi) + \int_0^{\hbar\beta} d\tau \int_0^{\hbar\beta} d\tau' \frac{1}{8}\alpha_s(\tau-\tau')[\varphi(\tau)-\varphi(\tau')]^2. \tag{76}$$

Although S_{eff} does not contain explicitly the bath degrees of freedom it still accounts for the bath and the dissipation by the non-local last term. For Ohmic dissipation, i.e. if $J(\omega)$ is given by eq. (75), the kernel $\alpha_s(\tau)$ is

$$\alpha_s(\tau) = \alpha_s \frac{1/\hbar\beta^2}{\sin^2/(\pi\tau/\hbar\beta)} \qquad \text{with} \qquad \alpha_s = \frac{h}{4e^2 R_s}. \tag{77}$$

Its Fourier transformed is $\alpha(\omega_n) = -\alpha_s\hbar|\omega_n|/\pi$ (with Matsubara frequencies $\omega_n = 2\pi n/\hbar\beta, |\omega_n| \leq \Omega_c$). From the imaginary time formulation we obtain the equilibrium properties of the system, and after analytic continuation also some information about the dynamics in real times.

Real time description

For a direct study of the real time dynamics of the system it is useful to consider the time evolution of a density matrix. It involves a forward and a backward time evolution operator. Starting from the Hamiltonian eq. (74), after tracing out the bath degrees of freedom one arrives at a reduced density matrix

$$\langle\varphi_{1f}|\rho_{red}(t_f)|\varphi_{2f}\rangle = \int d\varphi_{1i} \int d\varphi_{2i} K\left(\varphi_{1f},\varphi_{2f},t_f ; \varphi_{1i},\varphi_{2i},t_i\right)\langle\varphi_{1i}|\rho_s(t_i)|\varphi_{2i}\rangle.$$

Here we assumed that at time t_i the density matrix of the system factorizes $\rho(t_i) = \rho_s(t_i)\rho_{\text{bath}}(t_i)$ and that at t_i the bath is in thermal equilibrium. The time evolution of the system is described by the "influence functional" K. It contains two propagators from the forward and backward time evolution operators, which can be expressed as path integrals over $\varphi_1(t)$ and $\varphi_2(t)$. Both get coupled when the bath is traced out. After introducing 'center of mass' and 'relative' variables $\phi(t) = [\varphi_1(t) + \varphi_2(t)]/2$ and $\chi(t) = \varphi_1(t) - \varphi_2(t)$ one can write K as

$$K\left(\varphi_{1f},\varphi_{2f},t_f ; \varphi_{1i},\varphi_{2i},t_i\right) = \int_{\phi_i}^{\phi_f} D\phi(t) \int_{\chi_i}^{\chi_f} D\chi(t) e^{\{iS_1[\phi,\chi]-S_2[\chi]\}/\hbar}. \tag{78}$$

The effective action is complex $S_{\text{eff}}[\phi,\chi] = S_1[\phi,\chi] + iS_2[\chi]$ with

$$S_1[\phi,\chi] = -\int_{t_i}^{t_f} dt \left[\frac{\hbar\chi(t)}{2e}C\frac{\hbar\ddot\phi(t)}{2e} + U\left(\phi+\frac{\chi}{2}\right) - U\left(\phi-\frac{\chi}{2}\right)\right]$$

$$+ \int_{t_i}^{t_f} dt \int_{t_i}^{t} dt'\alpha_s^I(t-t')\phi(t')\chi(t) \tag{79}$$

$$S_2[\phi, \chi] = \frac{1}{2} \int_{t_i}^{t_f} dt \int_{t_i}^{t_f} dt' \chi(t) \alpha_s^R(t - t') \chi(t'). \tag{80}$$

The kernels $\alpha_s^R(t)$ and $\alpha_s^I(t)$ are the real time equivalent of the kernel $\alpha_s(\tau)$. More precisely, analytic continuation of the imaginary time kernel $\alpha_s(\tau)$ from the upper (lower) half plane yields the functions $\alpha_s^{>(<)}$. The real and imaginary parts of these functions determine the kernels $\alpha_s^{I(R)}$ [47, 23]

$$\alpha_s^{>(<)}(t) \;=\; \alpha_s^R(t)_{(\pm)} i \alpha_s^I(t)$$

$$= \frac{\hbar^2}{2e^2} \sum_j \frac{c_j^2}{2M_j \Omega_j} \left[\coth \frac{\hbar \Omega_j}{2k_B T} \cos(\Omega_j t)_{(\pm)} i \sin(\Omega_j t) \right]. \tag{81}$$

If the spectral density $J(\omega)$ is chosen to produce Ohmic damping, i.e. if it is given by eq. (75), then $\alpha_s^{(I,R)}$ is given by

$$\alpha_s^I(\omega) \;=\; \frac{\hbar^2}{2e^2} \frac{i\omega}{R_s}$$

$$\alpha_s^R(\omega) \;=\; \frac{\hbar^2}{2e^2} \frac{\omega}{R_s} \coth \frac{\hbar \omega}{2k_B T}. \tag{82}$$

For a general frequency dependent impedance $Z(\omega)$ one has

$$\alpha_s^I(\omega) \;=\; \frac{\hbar^2}{2e^2} i Z^{-1}(\omega) \omega$$

$$\alpha_s^R(\omega) \;=\; \frac{\hbar^2}{2e^2} Re Z^{-1}(\omega) \omega \coth \frac{\hbar \omega}{2k_B T}. \tag{83}$$

The imaginary part of the action, S_2, can be interpreted as describing the effect of a quantum noise. Under certain conditions the real time description outlined above reduces to a quantum Langevin description with a Gaussian noise [47].

The effective action given above can also be derived from a different model, namely the microscopic many body description of conduction electrons in a metal with impurities. This has been shown recently by Schön and Zaikin [48] along the lines to be presented in the following section.

4.2 From the Microscopic Theory of Josephson Junctions to Macroscopic Quantum Mechanics with dissipation

In Josephson junctions or weak links we can start from the microscopic theory. At the outset we have a large number of Fermions, and we first have to introduce the important collective variables. In this way we can derive an effective description of a Josephson junction with dissipation [49, 50]. For a review and extensions see also Schön and Zaikin [2]. The starting point is the grand canonical Hamiltonian $H = H_0 + H_{int}$

$$H_0 = \int d^3r \Psi_\sigma^+(\mathbf{r}) \left[-\frac{1}{2m}(\hbar\nabla - ie\mathbf{A})^2 - \mu + U(\mathbf{r}) \right] \Psi_\sigma(\mathbf{r})$$

$$H_{int} = \int d^3r \int d^3r' \Psi_\sigma^+(\mathbf{r})\Psi_{\sigma'}^+(\mathbf{r}') \left[-\frac{1}{2}g(\mathbf{r})\delta(\mathbf{r}-\mathbf{r}')\delta_{\sigma-\sigma'} \right.$$

$$\left. + e^2 w(\mathbf{r}-\mathbf{r}') \right] \Psi_{\sigma'}(\mathbf{r}')\Psi_\sigma(\mathbf{r}). \tag{84}$$

It describes conduction electrons in a potential $U(\mathbf{r})$ which interact via a BCS attraction of strength $g(\mathbf{r})$ and the Coulomb interaction $w(\mathbf{r}-\mathbf{r}')$. The model also describes normal conducting parts where $g(\mathbf{r}) = 0$. For simplicity we did not write here the magnetic field energy.

Imaginary time analysis

The partition function of the system contains a trace over the Fermion fields Ψ. In order to handle the BCS and Coulomb interactions we introduce by a Hubbard-Stratonovich transformations the order parameter field $\Delta(\mathbf{r},\tau) = |\Delta(\mathbf{r},\tau)|e^{-1\varphi(\mathbf{r},\tau)}$ and a potential $V(\mathbf{r},\tau)$, respectively. This makes the Hamiltonian quadratic in the Fermi fields, which allows us to perform the trace over Ψ. The partition function then becomes $Z = \int D^2\Delta(\mathbf{r},\tau) \int DV(\mathbf{r},\tau) \int D^3\mathbf{A}(\mathbf{r},\tau) \exp\{-S[\Delta,V,\mathbf{A}]/\hbar\}$ with effective action

$$S[\Delta,V,\mathbf{A}] = \hbar \operatorname{tr} \ln \mathbf{G}^{-1}[\Delta,V,\mathbf{A}] + \int_0^{\hbar\beta} d\tau \int d^3r\, g^{-1}|\Delta(\mathbf{r},\tau)|^2$$

$$+ \frac{1}{8\pi} \int_0^{\hbar\beta} d\tau \int d^3r\, [\nabla V(\mathbf{r},\tau)]^2. \tag{85}$$

Here $\mathbf{G}$ is the Green's function of the system and is a matrix in Nambu space. A gauge transformation removes the phase from Δ and makes it appear explicitly in combination with the scalar and vector potential. Defining the superfluid velocity $\mathbf{v}_s = -\frac{1}{2m}(\hbar\nabla\varphi + 2e\mathbf{A})$ we find

$$\mathbf{G}^{-1}[\Delta,V,\mathbf{A}] = \left\{ -\hbar\frac{\partial}{\partial\tau}\mathbf{1} + i\hbar(\mathbf{v}_s\nabla)\mathbf{1} + \left[\frac{\hbar^2\nabla^2}{2m} + \mu - \frac{m}{2}\mathbf{v}_s^2 \right.\right.$$

$$\left.\left. + i\left(\frac{\hbar}{2}\frac{\partial\varphi}{\partial\tau} - eV\right) - U(\mathbf{r}) \right]\tau_3 - |\Delta|\tau_1 \right\} \delta(\mathbf{r}-\mathbf{r}')\delta(\tau-\tau'). \tag{86}$$

In principle $|\Delta(\mathbf{r},\tau)|$, $\mathbf{A}(\mathbf{r},\tau)$, $V(\mathbf{r},\tau)$ and $\varphi(\mathbf{r},\tau)$ are all independent and fluctuating collective variables. However, for our purposes the integrations over the first three can be performed with sufficient accuracy in a semiclassical approximation. The least action condition $\delta S/\delta|\Delta(\mathbf{r},\tau)| = 0$ yields the classical BCS theory. The deviations from this least action solution are strongly suppressed [51]. Hence we replace in the following $|\Delta|$ by its BCS value. If we combine Maxwell's equations with the least action condition $\delta S \Delta \mathbf{A}(\mathbf{r},\tau) = 0$ we find the Meissner effect. The action is a minimum if Josephson's relation is satisfied

$$\hbar\partial\varphi/\partial\tau = 2eV(\mathbf{r},\tau). \tag{87}$$

Deviations from eq. (87) are also suppressed by bulk energies, and we will ignore them.

Within the electrodes forming the junction the electric field is vanishing small. However, it may exist in the region of the barrier. Thus the Coulomb interaction can be described by the geometric capacitance C_0 and the last term in eq. (85) can be rewritten in terms of the voltage difference across the junction and the geometric capacitance $\int d\tau \frac{1}{2} C_0 V(\tau)^2$.

In a tunnel junction tr $\ln \mathbf{G}^{-1}$ can be evaluated further by expanding it in the tunneling matrix elements. This leads to the following effective action for a tunnel junction (excluding terms referring to the magnetic field and the rest of the circuit)

$$S[\varphi] = \int_0^{\hbar\beta} d\tau \frac{C_0}{2} \left(\frac{\hbar}{2e} \frac{\partial\varphi}{\partial\tau} \right)^2 - \int_0^{\hbar\beta} d\tau \int_0^{\hbar\beta} d\tau' \left[\alpha_{qp}(\tau - \tau') \cos \frac{\varphi(\tau) - \varphi(\tau')}{2} \right.$$

$$\left. - \beta(\tau - \tau') \cos \frac{\varphi(\tau) + \varphi(\tau')}{2} \right]. \tag{88}$$

The kernels α and β are products of two diagonal or two off-diagonal Greens functions (in Nambu space) of the two electrodes (for details see Eckern et al. [52]). The kernel $\alpha_{qp}(\tau)$ is related to the quasi-particle tunneling current $I_{qp}(V)$ by

$$\alpha_{qp}(\omega_n) = \frac{\hbar}{e} \int \frac{d\omega}{2\pi} \frac{\omega}{\omega^2 + \omega_n^2} I_{qp} \left(\frac{\hbar\omega}{e} \right). \tag{89}$$

Some important limits are

1. If the junction is formed by ideal BCS superconductors, then at $T = 0$ the kernel $\alpha_{qp}(\tau)$ decays proportional to $\exp(-2\Delta|\tau|/\hbar)$. If the phase varies slowly in time on the scale $\hbar/\Delta$ the α_{qp}-term of the action renormalizes the capacitance

$$C = C_0 + \delta C , \qquad \delta C = 3\pi\hbar/32\Delta R_n. \tag{90}$$

   Here we introduced the conductance of the junction in the normal state $\hbar/R_n = 4\pi e^2 |T|^2 N^2(0)$ where $N(0)$ is the normal state density of states, and we assumed that the junction to be symmetric.

2. In a normal junction, i.e. for $\Delta = 0$, the kernel $\alpha_{qp}(\tau)$ coincides with the kernel found for Ohmic dissipation eq. (77) with strength characterized by the normal state conductance $1/R_n$.

3. Many tunnel junctions are not ideal. Then the junction even at $T = 0$ has a finite subgap conductance $1/R_t \neq 0$. This can be accounted for by using the actual I_{qp} in eq. (89). Approximately we can set $\alpha_{qp}(\tau) = \alpha_0(\tau) + \alpha_t(\tau)$ where $\alpha_0(\tau)$ leads to the capacitance renormalization eq. (90) and $\alpha_t(\tau)$ is given by eq. (77) with $1/R_s$ replaced by the subgap conductance $1/R_t$.

$$\alpha_t(\tau) = \alpha_t \frac{\pi^2}{4} \frac{1/\hbar\beta^2}{\sin^2(\pi\tau/\hbar\beta)} \qquad \alpha_t = \frac{4}{\pi^2} \frac{h}{4e^2 R_t}. \tag{91}$$

The kernel $\beta(\tau)$ appears in combination with $\cos[\varphi(\tau) + \varphi(\tau')]/2$. This emphasizes its zero frequency component $\beta(\omega_n = 0) = -I_c/2e$. Thus the action contains a term

$S_\beta = -\int d\tau (I_c/2e)\cos\varphi(\tau)$. The non-zero frequency components of β produce the so-called $'\cos\varphi'$ term [53, 54]. In order to keep the discussion transparent, we ignore this term here. Then the action reduces to

$$S[\varphi] = \int_0^{\hbar\beta} d\tau \frac{C_0}{2}\left(\frac{\hbar}{2e}\frac{\partial\varphi}{\partial\tau}\right)^2 + U(\varphi) + \int_0^{\hbar\beta} d\tau \int_0^{\hbar\beta} d\tau' \left\{ \alpha_{qp}(\tau - \tau') \right.$$

$$\left. \left[1 - \cos\frac{\varphi(\tau) - \varphi(\tau')}{2}\right] + \frac{1}{8}\alpha_s(\tau - \tau')[\varphi(\tau) - \varphi(\tau')]^2 \right\}. \tag{92}$$

The potential $U(\varphi)$ contains the Josephson coupling $E_J\cos\varphi$ and other terms, see e.g. eq. (17).

In eq. (91) we also account for the Ohmic dissipation. The corresponding kernel $\alpha_s(\tau)$ is given by eq. (77). For convenience we defined α_s and α_t slightly differently.

In the limit where the phase deviates only weakly from an equilibrium value the dissipation caused by quasi-particle tunneling and that due to an Ohmic shunt agree. But for larger deviations of the phase, of the order of π, distinct differences arise reflecting the difference between a discrete and a continuous charge transport.

Real time analysis

We also consider the real time evolution of the density matrix of the junction which is governed by the influence functional. The effective action is again complex $S_{\text{eff}}[\phi, \chi] = S_1[\phi, \chi] + iS_2[\phi, \chi]$. For a tunnel junction it is [50, 52]

$$S_1[\phi, \chi] = -\int_{t_i}^{t_f} dt \frac{\hbar\chi(t)}{2e} C_0 \frac{\hbar\ddot\phi(t)}{2e} + 16 \int_{t_i}^{t_f} dt \int_{t_i}^{t} dt' \left[\alpha_{qp}^I(t - t')\sin\frac{\phi(t) - \phi(t')}{2}\right.$$

$$\left. -\beta^I(t - t')\sin\frac{\phi(t) + \phi(t')}{2}\right] \sin\frac{\chi(t)}{4}\cos\frac{\chi(t')}{4}. \tag{93}$$

$$S_2[\phi, \chi] = 4 \int_{t_i}^{t_f} dt \int_{t_i}^{t_f} dt' \left[\alpha_{qp}^R(t - t')\cos\frac{\phi(t) - \phi(t')}{2}\right.$$

$$\left. +\beta^R(t - t')\cos\frac{\phi(t) + \phi(t')}{2}\right] \sin\frac{\chi(t)}{4}\sin\frac{\chi(t')}{4}. \tag{94}$$

On one hand, the kernels $\alpha_{qp}^{R(A)}$ are obtained by analytic continuation of the imaginary time kernel $\alpha_{qp}(\tau)$, on the other they are related to the tunneling current voltage characteristic. Following Werthamer [53] and Larkin and Ovchinnikov [54] we define the quantities

$$\alpha_{qp}^{<(>)}(t) = (\pm)\frac{\hbar}{e} \int_{-\infty}^{\infty} \frac{d\omega}{2\pi} \exp(-i\omega t)\frac{1}{\exp\left[(\pm)\beta\hbar\omega\right] - 1} I_{qp}(\omega) \tag{95}$$

$$\beta^{<(>)}(t) = (\mp)\frac{\hbar}{e} \int_{-\infty}^{\infty} \frac{d\omega}{2\pi} \exp(-i\omega t)\frac{1}{\exp\left[(\pm)\beta\hbar\omega\right] - 1} I_c(\omega) \tag{96}$$

where

$$\left.\begin{array}{c} I_{qp}(\omega) \\ I_c(\omega) \end{array}\right\} = \frac{1}{2eR_n} \int_{-\infty}^{\infty} dE\, N\left(E + \frac{\hbar\omega}{2}\right) N\left(E - \frac{\hbar\omega}{2}\right)$$

$$\left[\tanh\left(\frac{E + \hbar\omega/2}{2k_B T}\right) - \tanh\left(\frac{E + \hbar\omega/2}{2k_B T}\right)\right] \left\{\begin{array}{c} 1 \\ \frac{\Delta^2}{[(E+\hbar\omega/2)(E-\hbar\omega/2)]} \end{array}\right. \quad (97)$$

and $N(E)$ is the normalized density of states of the superconductor. Clearly $I_{qp}(\omega = eV/\hbar)$ is the quasi-particle tunneling current of a voltage biased junction. Finally we obtain α_{qp}^R and α_{qp}^I as the real and imaginary part of $\alpha_{qp}^{<(>)}(t)$

$$\alpha_{qp}^{<(>)}(t) = \alpha_{qp}^R(t)(\pm)i\alpha_{qp}^I(t). \quad (98)$$

Both satisfy the fluctuation-dissipation theorem $\alpha_{qp}^R(\omega) = Im\,\alpha_{qp}^I(\omega)\coth(\hbar\omega/2k_B T)$. In the limit $\Delta = 0$ they simply reduce to the same form as in the Ohmic case eq. (82) with R_s replaced by R_n. In general they depend on the details of the quasi-particle current voltage characteristic $I_{qp}(V)$.

An equivalent definition applies for $\beta^{R(I)}$. For transparency we treat the β-terms in the approximation which lead to the simple form eq. (92). On this level the β-term of S_1 eq. (93) becomes

$$-\int_{t_i}^{t_f} dt\, E_J[\cos(\phi + \chi/2) - \cos(\phi - \chi/2)]$$

whereas it drops from S_2 eq. (94). The imaginary part of the action S_2 can be interpreted as a noise term similarly as we did at the end of section 3.1. However, now we have to generalize to state-dependent noise sources. See Eckern et al. [52] for details.

5 Applications and some Selected Problems Beyond the Perturbative Regime

From the microscopic description presented in the preceding section the perturbative treatment of sections 2 and 3 can be justified. For example the transition rate due to an Ohmic shunt which we quoted in section 2 has been derived by Golden Rule arguments from the Hamiltonian eq. (74). This leads naturally to the matrix element involving j quoted there. The transition rate to single electron tunneling has been derived from the microscopic theory. Also the macroscopic quantum properties of a Josephson junction, which in the simplest case are expressed by the Hamiltonian eq. (16) find their justification from the analysis of section 4. In this section we will discuss a number of selected problems based on the full effective action. Much more work has been done on this level by different techniques. For a rather detailed review we refer to the article by Schön and Zaikin [2].

5.1 The Influence of the Environment on the Charging Effects in Normal Junctions

When discussing the charging effects in section 2 we basically ignored the influence of the environment, assuming that the coupling between the junction and the external circuit can be made weak. This condition is difficult to achieve in experiments with single junctions. On one hand, this is one of the reasons why most of the experiments dealing with charging effects are conducted on junction networks, on the other hand by suitable experimental means one can approach the ideal situation [55, 56]. It is the purpose of this section to study in more detail the influence of an arbitrary environment on charging effects such as the Coulomb blockade showing up in the tunnel I-V characteristics.

If we consider a tunnel junction which is coupled to an external circuit it is important to consider the back-influence of the electromagnetic field propagating in the geometry given by the circuit onto the tunneling itself. This problem had been studied first by Nazarov [57]. He started from the microscopic Hamiltonian of the many electron system and effectively repeated the steps outlined in section 4 which lead to a reduction of the many electron problem to one described by one degree of freedom. Furthermore, he studied in detail the electromagnetic field propagation in some given geometries. The essence of this analysis is that the back-influence of the electromagnetic field can be accounted for completely by giving the impedance $Z(\omega)$ of the environment seen by the tunneling electron [58]. After this observation and after we have already reduced the problem in section 4 to the important degrees of freedom we can present now a compact re-derivation of Nazarov's formulation and summarize some results.

Weak tunneling in an arbitrary environment

A systematic way to analyze expectation values of physical quantities is to consider a suitable generating functional. We start in the imaginary time representation which is more transparent. Later we will perform the analytic continuation to real times. Of particular interest are the expectation value and fluctuations of the tunneling current which we obtain from

$$Z[\xi] = tr \left\{ T_\tau \exp \left[-\int_0^{\hbar\beta} d\tau [H - I_t \xi(\tau)]/\hbar \right] \right\}. \tag{99}$$

As in the previous section we can express the generating functional by a path integral. The only modification is that the tunneling matrix element is replaced by [4] $T[\xi] = T[1 + ie\xi(\tau)]\delta(\tau - \tau')$. The path integral is over the phase of the order parameter in the superconducting case. In a normal junction we also introduce a phase as the integral of the voltage $\hbar d\varphi/d\tau = 2eV$ (the factor 2 is included for convenience to allow a parallel discussion of the normal and superconducting case). It is now straightforward to show that the expectation value of the (quasi-particle) tunneling current is

$$\langle I_t(\tau) \rangle = \frac{1}{Z[0]} \int D\varphi I_t[\varphi, \tau] e^{-S[\varphi]/\hbar}, \tag{100}$$

where $S[\varphi]$ is given by eq. (92) and

$$I_t[\varphi, \tau] = -2e \int_0^{\hbar\beta} d\tau' \alpha_t(\tau - \tau') \sin \frac{\varphi(\tau) - \varphi(\tau')}{2}. \tag{101}$$

Similarly we can express higher moments of the current. We note that for fixed, classical $\varphi(\tau)$ the expression eq. (101) reproduces, after analytic continuation to real times, the known expression for the tunneling current (which is trivial in a normal junction, but is the Werthamer functional in the superconducting state). The formulation of the problem indicated by eq. (100) had been applied for the evaluation of the tunnel current fluctuations of mesoscopic normal junctions in the limit of weak tunneling [4, 59]. In this case the action $S[\varphi]$ reduces to the charging energy and the necessary integrals are all Gaussian. Here we will consider several extensions.

We consider a normal tunnel junction, assuming that the tunneling conductance is small $\alpha_t \ll 1$ but allow for an environment characterized by an impedance $Z(\omega)$ with arbitrary strength and frequency dependence. In this limit we can expand in the tunneling matrix element. This means that the expectation value of the tunnel current is given by eq. (100), but the action $S[\varphi]$ eq. (92) can be taken in the limit $\alpha_t = E_J = 0$, but arbitrary $\alpha_s(\omega_\nu) \neq 0$. This limit will be denoted by $S_0[\varphi]$. We have in mind a junction which is voltage biased. In order to account for this we split the phase as $\varphi(\tau) = \varphi_0(\tau) + \varphi_1(\tau)$, where after analytic continuation to real times $\hbar\varphi_0(t) = 2eVt$, whereas φ_1 denotes the fluctuating part. This means we can write the expectation value of the tunneling current in the form

$$\langle I_t(\tau)\rangle = -2e \int_0^{\hbar\beta} d\tau' \alpha_t(\tau - \tau') \left\langle \cos \frac{\varphi_1(\tau) - \varphi_1(\tau')}{2} \right\rangle_0 \sin \frac{\varphi_0(\tau) - \varphi_0(\tau')}{2}. \qquad (102)$$

The expectation value is taken with the action $S_0[\varphi_1]$. Since this action is quadratic in the phase the necessary expectation value can be written as

$$\left\langle \cos \frac{\varphi_1(\tau) - \varphi_1(\tau')}{2} \right\rangle_0 = \exp\left\{ -\frac{1}{8} \left\langle [\varphi_1(\tau) - \varphi_1(\tau')]^2 \right\rangle_0 \right\} \qquad (103)$$

and can be evaluated exactly. It should be noted that expectation values of products evaluated by path integrals always refer to the time ordered products. We now can perform the analytic continuation to real times $\tau \to it$. This involves a deformation of the contour of the time integration forth and back along the real time axis. The time ordering refers to this contour. We make the order of the time arguments explicit and write

$$\langle I_t(t)\rangle = -2ie \int_{-\infty}^{t} dt' \left\{ \alpha_t^>(t - t') e^{\frac{1}{4}\langle[\varphi_1(t) - \varphi_1(t')]\varphi_1(t')\rangle_0} \right.$$

$$\left. -\alpha_t^<(t - t') e^{\frac{1}{4}\langle[\varphi_1(t') - \varphi_1(t)]\varphi_1(t)\rangle_0} \right\} \sin \frac{\varphi_0(t) - \varphi_0(t')}{2}. \qquad (104)$$

The kernels $\alpha_t^>(t)$ and $\alpha_t^<(t)$ which are obtained by the analytic continuation of $\alpha_t(\tau)$ are related to the response and correlation functions by eq. (98). With the help of the definitions eqs. (95), (97), and (98) they can be expressed explicitly. For a constant voltage bias $\varphi_0(t) = 2eVt/\hbar$ we can rewrite the expression eq. (104) in the from presented by Devoret et al. [58]

$$\langle I_t\rangle = \frac{1}{eR_t} \int_{-\infty}^{\infty} dE \int_{-\infty}^{\infty} dE' \quad \{ f(E)[1 - f(E')]P(E - E' + eV)$$

$$-[1 - f(E)]f(E')P(E' - E - eV)\}, \qquad (105)$$

where

$$P(E) = \frac{1}{2\pi} \int_{-\infty}^{\infty} dt \exp\{K(t) + iEt\} \qquad (106)$$

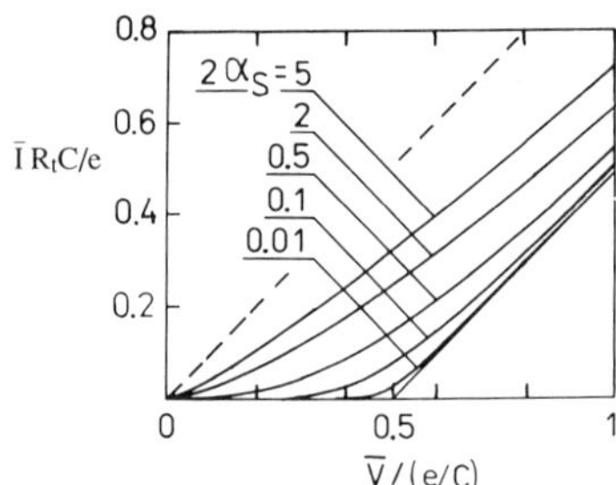

Figure 14: The tunnel current at $T = 0$ of a normal junction with an Ohmic shunt of strength α_s (with permission from Odintsov [60, 61], see also [1]).

and the correlator $K(t)$ is

$$K(t) = \frac{1}{4} \langle [\varphi(t) - \varphi(0)]\varphi(0)\rangle_0 . \tag{107}$$

Notice that the operators in $K(t)$ have a certain order and are not symmetrized as in a ordinary correlation function. Since the action S_0 is quadratic the function $K(t)$ can be evaluated explicitly, e.g by means of the real time description eqs. (78) - (82). Consider a circuit with the junction capacitance and an arbitrary impedance $Z(\omega)$ which is characterized by

$$Z_t(\omega) = [i\omega C + Z^{-1}(\omega)]^{-1}. \tag{108}$$

In this case

$$K(t) = 2e^2 \int_0^\infty \frac{d\omega}{2\pi} \frac{Re Z_t(\omega)}{\hbar\omega} \left\{ \coth \frac{\hbar\omega}{2kT}[\cos(\omega t) - 1] - i\sin(\omega t)\right\}. \tag{109}$$

The problem is thus reduced to elementary integrations.

Ohmic shunt of arbitrary strength

As an explicit example we consider a junction shunted with an Ohmic resistor i.e. $Z(\omega) = R_s$. By evaluating eqs. (105) – (109) one gets the the I-V characteristic shown in Fig. 14. For $\alpha_s \ll 1$ one recovers the perfect Coulomb blockade discussed before. For stronger shunt conductance, however, the Coulomb blockade is washed out, i.e. the I-V characteristic acquires a finite slope at small V even at $T = 0$. This process can also be considered as a 'macroscopic quantum tunneling of the charge' which we discussed in section 3.4. For large voltages the Coulomb gap is recovered. However, the cross-over to a shifted I-V curve occurs at larger voltages than in the limit $\alpha_s = 0$. For small voltages one gets an analytic result [31, 60, 61, 45, 58]

$$\langle I_t\rangle = \frac{1}{R_t}V\frac{1}{\Gamma(2 + 1/2\alpha_s)}\left(e^{-\gamma}\frac{\pi VC}{2e\alpha_s}\right)^{1/2\alpha_s} \quad \text{for } V \ll \alpha_s e/C. \tag{110}$$

As can be seen from the references this formula has some history, which deserves some comment. The result eq. (110) is equivalent to a result obtained by Panyukov and Zaikin [31]. They considered a current (I_x) biased, resistively shunted junction. However, the two problems are equivalent, as we discussed in section 2. In order to describe the voltage biased problem one merely has to replace I_x of the current biased problem by V/R_s and separate the tunneling current from the Ohmic part. Odintsov derived this result as an approximation for a slightly different problem (see below), but it was realized later [45] that his analysis actually describes the problem considered here, a weak tunneling barrier with an arbitrary Ohmic shunt. Technically his analysis was equivalent to the one presented here. Finally, the result eq. (110) was re-derived by Devoret et al. [58]. They find the numerical coefficient $e^{-\gamma}$, where $\gamma = 0.5772\ldots$ is Euler's constant, which is related to the high frequency part of the environment [62].

General frequency dependent impedance

The analysis presented above allows for an arbitrary frequency dependence of the impedance $Z(w)$. For instance, one can study the influence of a spatially distributed capacitance [63]. This question is crucial for the understanding how strongly stray capacitances influence the charging effects. It had been argued before that only the capacitance within a 'relativistic horizon' $\Delta r = c\tau$, where c is the velocity of light and τ some typical time scale of the tunneling process, could influence the tunneling rate. It was conjectured [18, 64] that the relevant time is the 'tunneling time' of Büttiker and Landauer [65], which would be very short for the tunneling barriers, inverse proportional to the barrier height of order $1\,\mathrm{eV}$. However there exists another time, proportional to the inverse energy difference in the tunneling process, which is of order $E_C \approx 1\,\mathrm{K}$. The most important result of Nazarov's analysis is that the longer time, i.e. in tunnel barriers the one related to the energy difference is the relevant scale. One consequence is that in the experiments on tunnel junction networks the energy change of the whole network determines the tunneling rate. For lower barriers both times are important [66].

These conclusions can be illustrated by an example evaluated by Devoret et al. [58]. Considered a junction in a circuit with an inductance L without dissipation. In this case the correlator is

$$K(t) = \rho \left\{ \coth\left(\frac{1}{2}\beta\hbar\omega_L\right) [\cos(\omega_L t) - 1] - i\sin(\omega_L t) \right\}, \tag{111}$$

where $\omega_L = (LC)^{1/2}$ is the oscillation frequency of the environmental mode and $\rho = E_C/\hbar\omega_L$ is the ratio of the charging energy scale and the mode excitation energy $\hbar\omega_L$. For zero temperature the current becomes for $V > 0$

$$\langle I_t \rangle = \frac{1}{eR_t} e^{-\rho} \sum_{n=0}^{n_{\max}} \frac{\rho^n}{n!} (eV - n\hbar\omega_L), \tag{112}$$

where $n_{\max}$ is the largest integer below $eV/\hbar\omega_L$. The example demonstrates that the tunneling is an inelastic process. The tunneling electron excites modes of the environment depending on the energy difference in the tunneling process.

5.2   Strong Single Electron Tunneling

In section 2 we described the single electron tunneling perturbatively, based on a transition rate obtained by Golden Rule arguments. This leads among other things to the phenomenon of Coulomb blockade. The perturbative description, however, is valid only provided that the tunneling is weak, more precisely that dimensionless parameter α_t is small, $\alpha_t \ll 1$. Brown and Simanek [67] analyzed the effect of single electron tunneling in a normal junction with strong tunneling. The result is that the Coulomb blockade is weakened, again an effect which we may call 'macroscopic quantum tunneling of the charge'. Schön [68] and Guinea and Schön [26, 27] showed that strong virtual quasi-particle tunneling processes influence the quantum mechanical eigenstates and energy bands of Josephson junctions and studied the influence on the Coulomb blockade in this case. Both problems require handling the non-Gaussian action given in eq. (92) with $\alpha_t \neq 0$.

Let us consider the normal junction problem further: the idea of Brown and Simanek was to replace the action eq. (92) with $\alpha_t \neq 0$ (but $\alpha_s = E_J = 0$) by a Gaussian action with an effective Ohmic dissipation of strength α_{var}. They used a variational principle to fix the strength of this effective dissipation, then used the variational action to evaluate via Kubo's formula the conductance of the junction. It is clear that this refers only to the linear regime where I and V approach zero. The result compares well with the experiments of Geerligs et al. [6]. The experimental shape for strong at looks similar to what we had shown in Fig. 14 for strong α_s. Also Odintsov [60, 61] described a junction with strong quasi-particle dissipation. In an attempt to handle the complicated action he replaced – ad hoc – the non-harmonic action eq. (92) with $\alpha_t \neq 0$ by a harmonic one with an effective Ohmic dissipation with strength given by α_t. While this is an uncontrolled approximation, he performed the same type of analysis as outlined above for the problem of a junction with arbitrary Ohmic shunt. His conclusions coincide largely with those mentioned above.

Strong single electron tunneling also modifies the properties of normal and superconducting junction arrays. In normal arrays with dominant capacitances between the islands a charge unbinding Kosterlitz-Thouless-Berezinskii transition can occur separating an insulating from a resistive region [21]. The transition temperature is of order of E_C for weak tunneling. For stronger tunneling it decreases and eventually vanishes. In superconducting junctions the charge unbinding and the vortex unbinding transitions compete and both are influenced by the dissipation. See the article by R. Fazio (this volume) for a further discussion.

5.3   Influence of the Environment on the Cooper Pair Tunneling

In a similar fashion as outlined above we can describe how the Cooper pair tunneling is influenced by quantum fluctuations in a shunt. We consider the limit of weak Josephson coupling, $E_j \ll E_C$, ignore for the moment the quasi-particle tunneling, but allow for a shunt with arbitrary strength. [2] In the limit considered the DC current through a voltage biased Josephson junction is

$$I = 2e \left(\frac{E_J}{\hbar}\right)^2 \int_0^{\hbar\beta} d\tau \, \langle \sin\varphi \cos\varphi(\tau)\rangle_0 =$$

---

[2] For weak voltage bias the junction undergoes at $T = 0$ a phase transition at a critical strength of the dissipation $\alpha_s \geq 1$ [72]. However, here we consider a system with a large voltage bias where this does not show up.

$$= 2e \left(\frac{E_J}{\hbar}\right)^2 \int_0^{\hbar\beta} d\tau \sin\varphi_0(\tau) \exp\left\{-\frac{1}{2}\left\langle[\varphi_1(\tau)-\varphi_1(0)]^2\right\rangle_0\right\}. \tag{113}$$

The expectation value is evaluated again with an action S_0 containing the charging energy and the effect of the Ohmic dissipation. As before the analytical continuation involves a deformation of the contour of integration. The result is [69]

$$I = 2e \left(\frac{E_J}{\hbar}\right)^2 \int_0^{\infty} d\tau \sin\left(\frac{2eVt}{\hbar}\right) Im\left\{e^{-4K(t)}\right\}, \tag{114}$$

where $K(t)$ is given by eqs. (107)-(109). Limiting results can be derived from this expression. For instance at $T = 0$ if the junction is shunted by an Ohmic resistor one finds for small voltages [70, 71]

$$I = \frac{V}{R_s} \frac{\pi^2(E_J R_s C)^2}{\alpha_s \Gamma(2/\alpha_s)} \left(\frac{\pi V C}{e\alpha_s}\right)^{2/\alpha_s - 2}. \tag{115}$$

On the other hand, for larger voltages the current shows a resonant structure [38]

$$I = \frac{\pi^{1/2} e E_J^2}{2\hbar^2\gamma} \exp\left\{-\left[\frac{e(V-e/C)}{\hbar\gamma}\right]^2\right\}$$

$$\text{where}\quad \gamma^2 = \max\left\{\frac{2e^2 T}{\hbar^2}, \left(\frac{1}{R_s C}\right)^2 \frac{1}{2\alpha_s}\ln\left(\frac{1}{\alpha_s}\right)\right\}.$$

The peak arises due to resonant Cooper pair tunneling at voltages $V \approx e/C$ when the energies before and after the tunneling process are nearly equal.

6 Conclusion

We did not attempt in this article to give a review of the whole field. Rather we summarized the necessary background information and concentrated on a few recent developments. The basic principles can be discussed in single junctions. However, the coupling to the external circuit makes the ideal situation difficult to achieve. In order to reduce the complications due to the external circuit one can consider junction networks. Alternatively one can try to study the influence of the environment in more detail. The applications discussed in this article aimed in these two directions.

7 Acknowledgements

We would like to conclude by acknowledging the large number of stimulating and pleasant discussion which we had with our colleagues and coworkers on the issues covered in this article. They are V. Ya. Aleshkin, V. Ambegaokar, E. Ben-Jacob, U. Eckern, L.J. Geerligs, U. Geigenmüller, F. Guinea, K.K. Likharev, J.E. Mooij, K. Mullen, Yu. N. Nazarov, A.A. Odintsov, A. Schmid, and A.D. Zaikin. One of us D.V.A. also acknowledges the hospitality of the Delft University of Technology.

* permanent address: Department of Physics, Moscow State University, Moscow 119899 GSP, U.S.S.R.

References

[1] D.V. Averin and K.K. Likharev, to be published in 'Quantum Effects in Small Disordered Systems', B.L. Altshuler, P.A. Lee, and R.A. Webb, eds. (1990)

[2] G. Schön and A.D. Zaikin, to be published in Physics Reports (1990)

[3] D.V. Averin and K.K. Likharev, J. Low Temp. Phys. **62**, 345 (1986); Zh. Eksp. Teor. Fiz. **90**, 733 (1986) [Sov. Phys. JETP **63**, 427, (1986)]

[4] E. Ben-Jacob, E. Mottola, and G. Schön, Phys. Rev. Lett. **51**, 2064 (1983)

[5] P. Delsing, T. Claeson, K.K. Likharev, and L.S. Kuzmin, Phys. Rev. Lett. **63**, 1180 (1989)

[6] L.J. Geerligs, V.F. Anderegg, C.A. van der Jeugd, J. Romijn, and J.E. Mooij, Europhys. Lett. **10**, 79 (1989)

[7] L.J. Geerligs, M. Peters, L.E.M. de Groot, A. Verbruggen, and J.E. Mooij, Phys. Rev. Lett. **63**, 326 (1989)

[8] K.K. Likharev, IBM J. Res. Develop. **32**, 144 (1988)

[9] U. Geigenmüller and G. Schön, Europhys. Lett. **10**, 765 (1989)

[10] K.K. Likharev, N.S. Bakhvalov, G.S. Kazacha, and S.I. Serdyukova, IEEE Trans. Magn. **25**, 1436 (1989)

[11] N.S. Bakhvalov, G.S. Kazacha, K.K. Likharev, and S.I. Serdyukova, submitted to Physica **B** (1990)

[12] H.R. Zeller and I. Giaever, Phys. Rev. **181**, 789 (1969)

[13] J. Lambe and R.C. Jaklevic, Phys. Rev. Lett. **22**, 1371 (1969)

[14] I.O. Kulik and R.I. Shekhter, Zh. Eksp. Teor. Fiz. **68**, 623 (1975) [Sov. Phys. JETP **41**, 308, (1975)]

[15] J.B. Barner and S.T. Ruggiero, Phys. Rev. Lett. **59**, 807 (1987)

[16] K.K. Likharev, IEEE Trans. Magn. **23**, 1142 (1987)

[17] K. Mullen, E. Ben-Jacob, R.C. Jaklevic, and Z. Schuss, Phys. Rev. **B 37**, 98 (1988)

[18] P.J.M. van Bentum, R.T.M. Smokers, and H. van Kempen, Phys. Rev. Lett. **60**, 2543 (1988)

[19] R. Wilkins, E. Ben-Jacob, and R.C. Jaklevic, Phys. Rev. Lett. **63**, 801 (1989)

[20] P. Delsing, K.K. Likharev, L.S. Kuzmin, and T. Claeson, Phys. Rev. Lett. **63**, 1861 (1989)

[21] J.E. Mooij, B.J. van Wees, L.J. Geerligs, M. Peters, R. Fazio, and G. Schön, Phys. Rev. Lett. **65**, 645 (1990)

[22] R. Fazio and G. Schön, to be published in Physica B; and the Proc. of the ICTPS'90, Rieo de Janeiro (1990)

[23] A.O. Caldeira and A.J. Leggett, Physica **121 A**, 587 (1983)

[24] K.K. Likharev and A.B. Zorin, J. Low Temp. Phys. **59**, 347 (1985); D.V. Averin, A.B. Zorin, and K.K. Likharev, Zh. Eksp. Teor. Fiz. **88**, 692 (1985) [Sov. Phys. JETP **61**, 407 (1985)]

[25] A. Widom, G. Megaloudis, T.D. Clark, H.Prance, and R.J. Prance, J. Phys **A 15**, 3877 (1982)

[26] F. Guinea and G. Schön, Europhys. Lett. **1**, 585 (1986)

[27] F. Guinea and G. Schön, J. Low Temp. Phys. **69**, 219 (1987)

[28] M. Büttiker, Physica Scripta **T14**, 82 (1986); Phys. Rev. **B 36**, 3548 (1987)

[29] D.V. Averin, Fiz. Nizk. Temp. **13**, 364 (1987) [Sov. J. Low Temp. Phys. **13**, 208 (1987)]

[30] U. Geigenmüller and G. Schön, Physica **152 B**, 186 (1988)

[31] S.V. Panyukov and A.D. Zaikin, J. Low Temp. Phys. **73**, 1 (1988)

[32] E. Ben-Jacob, K. Mullen, Y. Gefen, and Z. Schuss, Phys. Rev. **B 37**, 7400 (1988)

[33] J.M. Ziman, Principles of the Theory of Solids, section **6**, Cambridge (1969)

[34] A.D. Zaikin and I.N. Kosarev, Phys. Lett. **131 A**, 125 (1988)

[35] A. Maassen van der Brink, U. Geigenmüller, and A.D. Zaikin, submitted to Physica **B** (1990)

[36] U. Geigenmüller and G. Schön, in: Proc. of the 3rd Intl. Symposium Foundations of Quantum Mechanics, p. 276, Tokyo (1989)

[37] M. Lax M., Phys. Rev. **145**, 110 (1966)

[38] D.V. Averin and A.A. Odintsov, 1990a, Fiz. Nizk. Temp. **16**, 16 (1990) [Sov. J. Low Temp. Phys., to be published]

[39] Yu.N. Ovchinnikov and B.I. Ivlev, Phys. Rev. **B 39**, 9000 (1989)

[40] T.A. Fulton, P.L. Gammel, D.J. Bishop, L.N. Dunkleberger, and G.J. Dolan, Phys. Rev. Lett. **63**, 1307 (1989)

[41] D.V. Averin and V.Ya. Aleshkin, Pis'ma Zh. Eksp. Teor. Fiz. **50**, 331 (1990) [JETP Lett. 50, 367 (1990)]

[42] K.K. Likharev and A.B. Zorin, Jpn. J. Appl. Phys. **26** (suppl. 3), 1407 (1987)

[43] V.Ya. Aleshkin and D.V. Averin, Resonant tunneling of Cooper pairs in a double Josephson junction system, submitted to Physica **B** (1990)

[44] R.P. Feynman and A.R. Hibbs, Quantum Mechanics and Path Integrals, section 6, McGraw-Hill, New York (1965)

[45] D.V. Averin and A.A. Odintsov, Phys. Lett. **A 140**, 251 (1989)

[46] A.O. Caldeira and A.J. Leggett, Phys. Rev. Lett. **46**, 211 (1981); Ann. Phys. **149**, 374 (1983)

[47] A. Schmid, J. Low Temp. Phys. **49**, 609 (1982)

[48] G. Schön and A.D. Zaikin, Phys. Rev. **B 40**, 5231 (1989)

[49] V. Ambegaokar, U. Eckern, and G. Schön, Phys. Rev. Lett. **48**, 1745 (1982)

[50] V. Ambegaokar, in: NATO ASI Series **109** Percolation, Localization, and Superconductivity, A.M. Goldman and S.S. Wolf, eds., Plenum Press (1984)

[51] V.L. Ginzburg, Sov. Physics-Solid State, **2**, 1824 (1960)

[52] U. Eckern, G. Schön, and V. Ambegaokar, Phys. Rev. **B 30**, 6419 (1984)

[53] N.R. Werthamer, Phys. Rev. **147**, 255 (1966)

[54] A.I. Larkin and Yu. N. Ovchinnikov, Zh. Eksp. Teor. Fiz. **51**, 1535 (1966) [Sov. Phys. JETP**24**, 1035 (1967)]

[55] J.M. Martinis and R.L. Kautz, Phys. Rev. Lett. **63**, 1507 (1989)

[56] A.N. Cleland, J.M. Schmidt, and J. Clarke, Phys. Rev. Lett. **64**, 1565 (1990)

[57] Yu.V. Nazarov, Zh. Eksp. Teor. Fiz. **96**, 240 (1989) [Sov. Phys. JETP ....]

[58] M.H. Devoret, D. Esteve, H. Grabert, G.-L. Ingold, H. Pothier, and C. Urbina, preprint (1990)

[59] G. Schön, Phys. Rev. **B 32**, 4469 (1985)

[60] A.A. Odintsov, Zh. Eksp. Teor. Fiz. **94**, 312 (1988) [Sov. Phys. JETP **67**, 1265 (1988)]

[61] A.A. Odintsov, Fiz. Nizk. Temp. **14**, 1038 (1988)

[62] H. Grabert, private communication (1990)

[63] Yu.V. Nazarov, Pis'ma Zh. Eksp. Teor. Fiz. **49**, 105 (1989) [JETP Lett. ....]

[64] P.J.M. van Bentum, H. van Kempen, L.E.C. van de Lemput, and P.A.A. Teunissen, Phys. Rev. Lett. **60**, 369 (1988)

[65] M. Büttiker and R. Landauer, Phys. Rev. Lett. **49**, 1739 (1982)

[66] Yu.V. Nazarov, preprint (1990)

[67] R. Brown and E. Simnek, Phys. Rev. **B 34**, 2957 (1986)

[68] G. Schön, in SQUID '85, H.D. Hahlbohm and H. Lbbig eds., Walter de Gruyter & Co (1985)

[69] D.V. Averin, Yu. V. Nazarov, A.A. Odintsov, Incoherent tunneling of Cooper pairs and magnetic flux quanta in ultrasmall Josephson junctions, submitted to Physica **B** (1990)

[70] M.P.A. Fisher and W. Zwerger, Phys. Rev. **B 32**, 6190 (1985)

[71] A.D. Zaikin and S.V. Panyukov, Phys. Lett. **120 A**, 306 (1987)

[72] A. Schmid, Phys. Rev. Lett. **51**, 1506 (1983)

CHARGE KOSTERLITZ-THOULESS-BEREZINSKII TRANSITION IN ARRAYS OF TUNNEL JUNCTIONS

Rosario Fazio*

Department of Applied Physics
Delft University of Technology
Lorentzweg 1, 2628 CJ Delft, The Netherlands

1 Introduction

In recent years much attention has been devoted to the study of junction arrays [1]. When the islands become superconducting a phase transition, in which the system reaches a state of phase coherence, takes place. In two-dimensional (2 D) arrays the transition is driven by the dissociation of vortex-anti-vortex pairs. When the dimensions of the islands become such that it is not possible to disregard charging energy quantum fluctuations become important. The main effect is the suppression of the vortex unbinding transition temperature [2] and also at $T = 0$ it is possible to have a disordered phase. The tunneling of quasi-particles or Ohmic shunts introduce dissipation [3]. The most striking consequence is the existence of a critical value of the normal sheet resistance below which the superconductivity is recovered even when the charging energy exceeds the Josephson term [4]. Experiments of fabricated 2 D networks show a similar transition between an insulating and a superconductive state as the resistance of the sample is varied [5].

Very recently Mooij et al. [6] (see also van der Zant et al., this volume) pointed out that interesting collective effects can take place in the charge gas. In particular it was observed that in same cases the electrostatic interaction between charges depends logarithmically on the distance. This implies the possibility of a Kosterlitz-Thouless-Berezinskii (KTB) [7] transition in which real charge anti-charge pairs dissociate. The experiments of [6] hint at this hypothesis. In the case of *charge*-KTB, both normal and superconducting arrays will show the transition but at a different temperature due to the different charge unit involved (Cooper pairs in the superconducting case). The analogy with the familiar classical 2 D Coulomb gas holds as long as tunneling of quasi-particle and Cooper pairs is disregarded, except that it is needed to establish thermal equilibrium. In the opposite case quantum dynamic enters into the problem and it will strongly affect the phase diagram.

Quantum Coherence in Mesoscopic Systems
Edited by B. Kramer, Plenum Press, New York, 1991

In the following I will discuss the influence of single-electron and Cooper pair tunneling on this phase transition. In the superconducting case, beside the charge transition in itself, a detailed understanding of the interplay between the *charge*-KTB and the *vortex*-KTB it is important for the complete phase diagram.

2 Arrays of Normal Junctions

The model we will consider in order to describe the properties of an array of normal tunnel junctions was derived from a microscopic approach in [8,9] (see also Averin and Schön, this volume). The partition function is expressed as a path integral over a macroscopic field φ ($\hbar = 1$)

$$Z = \int D\varphi \exp\{-A[\varphi]\} \tag{1}$$

where the action is defined as

$$
\begin{aligned}
A[\varphi] &= A_{ch}[\varphi] + A_D[\varphi] \\[2mm]
&= \frac{1}{16E_0}\sum_i \int_0^\beta d\tau \left(\frac{d\varphi_i}{d\tau}\right)^2 + \frac{1}{16E_C}\sum_{\langle ij\rangle} \int_0^\beta d\tau \left(\frac{d\varphi_{ij}}{d\tau}\right)^2 \\[2mm]
&\quad - \sum_{\langle ij\rangle}\int_0^\beta d\tau \int_0^\beta d\tau'\, \alpha(\tau-\tau') \cos\left[\frac{\varphi_{ij}(\tau) - \varphi_{ij}(\tau')}{2}\right].
\end{aligned}
\tag{2}
$$

The first two terms in eq. (2) represent the charging contribution due to the self-capacitance C_0 ($E_0 = e^2/2C_0$) to the ground and the nearest neighbors capacitance C ($E_C = e^2/2C$) respectively. As it will be discussed in details later, in order to see the *charge*-KTB transition we should further assume that the self capacitance is negligible compared with C ($C_0 \ll C$). The third term describes the dissipation due to the quasi-particle tunnelling. The subscripts i,j label the islands on the lattice, $\varphi_{ij} = \varphi_i - \varphi_j$, the brackets mean that the lattice summation is restricted to nearest neighbors. The dissipative kernel is defined

$$\alpha(\tau) = \alpha_T / \left(\beta^2 \sin^2(\pi\tau/\beta)\right)^2 \tag{3}$$

where $\alpha_T = h/(4e^2 R_T)$ ($\pi/2e^2 R_T$ in the units we chose).

As discussed in detail in [10], the limits of the path integration over the field φ depend on the allowed charge states of the system; in the case under consideration we assume that no continuous charge transport (i.e. Ohmic dissipation) takes place so that the symmetry of the action requires that $\varphi_i(\beta) = \varphi_i(0) + 4\pi n_i$. Hence, the integral in eq. (1) includes also a summation over the winding numbers. We remind that in this formalism the generalized momentum conjugated to the phase φ_i is the charge in the i-th island.

In order to study the winding number contribution to the action we will employ the following decomposition of the phase [11]

$$\varphi_i(\tau) = \varphi_{i0} + \frac{4\pi n_i}{\beta}\tau + \vartheta_i(\tau), \qquad \vartheta_i(0) = \vartheta_i(\beta) = 0. \tag{4}$$

The use of eq. (4) shows in a transparent way the winding number contribution to the charging part of the action. The charging part of the action decouples into two terms, one, $A_{ch}[\vartheta]$, depending on $\vartheta_i(\tau)$, the other depending on the winding number.

$$A_{ch}[n] = \frac{\pi^2}{\beta E_0} \sum_i n_i^2 + \frac{\pi^2}{\beta E_C} \sum_{\langle ij \rangle} (n_i - n_j)^2 \tag{5}$$

which resembles the so called Discrete Gaussian Model (DGM) [12] was studied extensively in connection with the roughening transition of solid-solid interfaces. The DGM can be mapped onto a 2 D Coulomb gas which, as it is well known, shows the KTB transition. In the case considered in eq. (5) the critical temperature is

$$T_{cn}^{(0)} = \frac{1}{4\pi} E_C. \tag{6}$$

The critical temperature eq. (6) separates the system in two regions. In low temperature phase the charges form dipoles while in the high temperature phase only free charges are present in the system. At this stage it is possible to borrow all of the results known for KTB transition, and to translate to coupling constants (self-capacitance and mutual capacitance) of the array of normal junctions. In the charge representation the nearest neighbor capacitance provides the logarithmic interaction between charges in distant islands, while the self-capacitance introduces an infrared cutoff to the interaction. A finite value of C_0 rounds the transition introducing "finite size" effects in the problem. The Coulomb gas with finite cut-off was studied extensively in connection with the KTB transition in superconducting films [13].

In the limit in which α_T is very small we can include the dissipation perturbatively [14]. To the first order in α_T the partition function is

$$Z \approx Z_{ch,\vartheta} \sum_{\{n\}} \left\{ 1 + \sum_{\langle ij \rangle} \int_0^\beta d\tau \int_0^\beta d\tau' \alpha(\tau - \tau') \left\langle \cos \left[\frac{\vartheta_{ij}(\tau) - \vartheta_{ij}(\tau')}{2} \right] \right\rangle_{0,\vartheta} \right.$$

$$\left. \cdot \cos \left[\frac{2\pi}{\beta} n_{ij}(\tau - \tau') \right] \right\} \exp\{-A_{ch}[n]\} \approx Z_{0,\vartheta} \sum_{\{n\}} \exp\{-A_{\text{eff}}[n]\} \tag{7}$$

where $n_{ij} = n_i - n_j$. $Z_{ch,\vartheta}$ is the partition function associated to the charging contribution to the action from the ϑ field. The effective action for the winding numbers is defined

$$A_{\text{eff}}[n] = A_{ch}[n] - \sum_{\langle ij \rangle} \int_0^\beta d\tau d\tau' \alpha(\tau - \tau') g_{ij}(\tau - \tau') \cos \left[\frac{2\pi}{\beta} n_{ij}(\tau - \tau') \right] \tag{8}$$

and

$$g_{ij}(\tau - \tau') = \exp \left\{ -\frac{1}{8} \langle [\vartheta_{ij}(\tau) - \vartheta_{ij}(\tau')]^2 \rangle_{ch,\vartheta} \right\}. \tag{9}$$

Due to the exponential decay in time of the correlator we can consider a "gradient expansion" of the last term in eq. (8). The first non-zero term renormalizes the nearest neighbors capacitance so that we recover the same model as defined in eq. (5) but with an effective mutual capacitance. In the limit of vanishing self-capacitance it is then straightforward to calculate the first order correction of the KTB transition temperature due to the quasi-particle dissipation.

$$\frac{T_{cn}}{T_{cn}^{(0)}} = 1 - 0.1\alpha_T. \tag{10}$$

As expected, the effect of dissipation is to decrease the KTB transition. The 'capacitance renormalization' that is obtained in the limit of small dissipation is expression of the fact that when the tunneling is small, the classical picture of charges sitting on the island is still meaningful and quantum dynamics merely renormalizes the coupling constants.

For strong dissipation it is more appropriate to perform an expansion of the ϑ field. Truncating the expansion to the harmonic terms the dissipative term in the action reduces to

$$A_D[\vartheta, n] \approx 2\alpha_T \sum_{\langle ij \rangle} |n_i - n_j| + \frac{1}{8} \sum_{\langle ij \rangle} \int_0^\beta d\tau d\tau' \alpha(\tau - \tau')$$

$$\cdot \, [\vartheta_{ij}(\tau) - \vartheta_{ij}(\tau')]^2 \cos\left[\frac{2\pi}{\beta} n_{ij}(\tau - \tau')\right]. \tag{11}$$

As the dissipation increases the fluctuation in the phase are suppressed so that it is sufficient to consider only the first term in eq. (11). One is left with the study of the following classical statistical mechanical problem

$$A_{\text{eff}}[n] = \frac{\pi^2}{\beta E_0} \sum_i n_i^2 + \frac{\pi^2}{\beta E_C} \sum_{\langle ij \rangle} (n_i - n_j)^2 + 2\alpha_T \sum_{\langle ij \rangle} |n_i - n_j|. \tag{12}$$

The above effective action is the the sum of two part. The first has been already recognized as the DGM model and the last has also been studied (ASOS model [15]) by itself in connection with the roughening transition or at interfaces. As the temperature approaches zero only the third term survives; the effective action then reduces to the ASOS model which last undergoes a transition of the KTB type at the critical value of coupling constant

$$\alpha_T \approx 0.45 \tag{13}$$

as determined from Monte Carlo calculations [15]. It corresponds to a resistance of $\approx 15 k\Omega$.

At this point some comments seem appropriate in order to discuss the validity of the above results. The effective action eq. (12) was obtained under the assumption of large dissipation; this implies that, strictly speaking, we can use it only to discuss the charge system only in its plasma state. However the possibility to establish that for large dissipation we are in the disordered phase of a system showing a KTB transition allows us to conclude the presence of a KTB point at a critical value of α_T. Moreover all these step models (i.e. models in which at each lattice site is attached an integer valued variable running form $-\infty$ to ∞) belong to the same universality class so that we expect that employing eq. (12) also for not too large value of dissipation will *only* give quantitative changes. The very existence of the phase transition and its characteristics are not affected by these approximations.

The effect of the temperature is to decrease the critical value of dissipation. It is possible to evaluate the phase boundary by mapping eq. (12) onto a generalized XY model. The two models are dual in the sense that the weak coupling region is mapped into the strong coupling region of the XY action. In the following I apply the method discussed in [16] to the model defined in eq. (12). For this purpose it is convenient to treat as independent the bond variables $n_{ij} = n_i - n_j$. This can be done introducing N (N = number of lattice sites) constraints which demand that the sum of n_{ij} over each elementary plaquette is zero. This set of constraints can be written as

$$\delta\left(\sum_p n_{ij}, 0\right) = (2\pi)^{-1} \int_0^{2\pi} d\gamma_I \exp\left\{i\gamma_I \sum_p n_{ij}\right\}$$

where the auxiliary variables γ_I are defined on the dual lattice sites labelled with the capital letters and the symbol $\sum_p$ means the summation around the plaquette. It is possible to interchange the sum over the integer variables and the integration over γ with the result

$$Z_{XY} = \int_0^{2\pi} \prod_I d\gamma_I \exp\left\{-\sum_{\langle IJ\rangle} W(\gamma_I - \gamma_J)\right\} . \tag{14}$$

where the interaction energy $W(\gamma)$ for a generic bond in the dual lattice is given by the following expression

$$-W(\gamma_I - \gamma_J) = \ln\left\{\sum_{n=-\infty}^{\infty} \exp\left\{-A_{\text{eff}}[n] + \text{in}(\gamma_I - \gamma_J)\right\}\right\} . \tag{15}$$

In the case in which we consider eq. (12) we obtain

$$\exp\{-W(\gamma_I - \gamma_J)\} = \left(\frac{4\pi^3}{\beta E_C}\right)^{-1/2} \tanh(2\alpha_T) \int_{-\infty}^{\infty} dx \exp\left\{-\frac{\beta E_C x^2}{4\pi^2}\right\}$$

$$\cdot \frac{1}{1 - \cos(\gamma_I - \gamma_J + x)/\cosh(2\alpha_T)} . \tag{16}$$

The Boltzmann factor eq. (16) of the generalized XY model contains terms proportional to $\cos k\gamma$ ($k = 1, 2, \ldots$). The first harmonic ($k = 1$) contributes to the pure XY model with a coupling constant given by

$$W(\gamma) = \text{const} - \frac{\exp\left\{-\pi^2/\beta E_C\right\}}{\cosh(2\alpha_T)} \cos\gamma . \tag{17}$$

At $T \approx 0$ we recover the standard result for the ASOS model. At finite small temperatures the coupling constant of the XY model is renormalized so that the critical line bends towards smaller values of the dissipation. The shift of the critical value of dissipation, at low temperatures is

$$\alpha_T \approx 0.45\left(1 - \frac{\pi}{4}\frac{T}{T_{cn}^{(0)}}\right) . \tag{18}$$

The phase diagram for an array of normal junctions is shown in Fig. 1. In the region below the transition line the charges are bound in dipoles so that it is an insulating state, while above dipoles unbind and the charge gas is in a plasma phase.

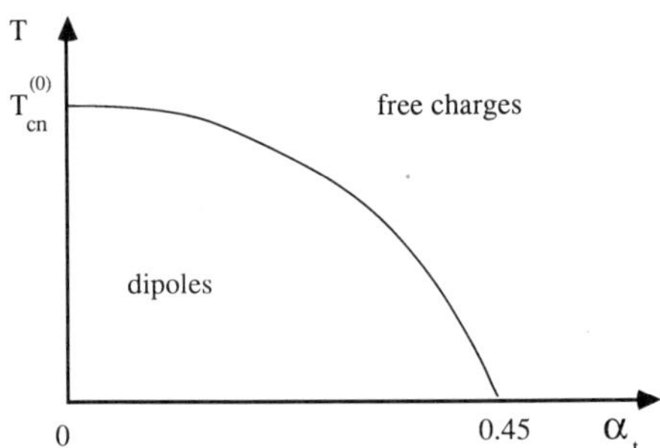

Figure 1: The KTB transition temperature is plotted versus the dissipation strength for an array of normal junctions. The critical temperature suddenly drops when the resistance of the sample is of the order of $15k\Omega$.

3 Josephson Arrays

In the case of superconducting arrays we should consider also the Josephson term and the action reads

$$A_s[\varphi] = A[\varphi] - E_J \sum_{\langle ij \rangle} \int_0^\beta d\tau \cos \varphi_{ij}(\tau) \tag{19}$$

where $A[\varphi]$ was defined in eq. (1) and E_J is the Josephson coupling energy. If we disregard for the moment the contribution of the quasi-particles ($\alpha_T \approx 0$) the unit of charges is $2e$ (Cooper pair) and correspondingly the boundary condition for the phase is $\varphi_i(\beta) = \varphi_i(0) + 2\pi n_i$. As a result the charging contribution will be of the same form as eq. (5) but with coupling constants that are 4 times smaller. The corresponding transition temperature is increased to the value [6]

$$T_{cs}^{(0)} = \frac{1}{\pi} E_C. \tag{20}$$

The experiments reported [6] give strong indication of this behavior. The possibility of such a transition in the superconducting case in the limit of (E_J and $\alpha_T \approx 0$) was also suggested in [17]. The effect of small Josephson tunneling can be treated perturbatively along the same lines to obtain eq. (10) and the resulting shift is [14]

$$\frac{T_{cs}}{T_{cs}^{(0)}} = 1 - 0.98 \left(\frac{E_J}{E_C} \right)^2. \tag{21}$$

If the superconducting gap Δ is large compared with the charging energy then only virtual quasi-particle transitions which renormalize the nearest neighbors capacitance contribute

$$\frac{E_C^{ren}}{E_C} = \left[1 + \frac{3}{8} \frac{E_C}{\Delta} \alpha_T \right]^{-1}, \tag{22}$$

and the transition temperature eqs. (21) (22) changes in a trivial way. On the other hand, when the superconducting gap is small compared with E_C, quasi-particle tunneling dominates and the transition temperature goes back to the value given by eq. (6). A crossover between the two limits is expected when $\Delta \approx E_C$.

As already anticipated in the introduction, in Josephson arrays also a transition to a phase-coherent state takes place. In the 2 D case the transition, at least at finite temperature, is driven by the dissociation of vortex-anti-vortex pairs. The transition temperature is of the order of E_J. Both charging effects and dissipation strongly affect the transition [2,3]. Different types of order can be present in the phase diagram. In the limit in which one of the energies (charging or Josephson) dominates the other the system undergoes a *charge*-KTB or a *vortex*-KTB respectively. Both transition temperatures are lowered by quantum fluctuations. In the intermediate region the analysis is much more complicated.

In order to study the crossover between both we map the problem onto one with two coupled Coulomb gases representing the charge and vortex degrees of freedom, respectively, which allows us to establish a duality relation between the two gases [18,19].

The partition function for an array of Josephson junctions (without quasi-particle tunneling) can be written in a representation involving charge trajectories $Q_i(\tau)$ replacing the velocities $\dot{\varphi}_i(\tau)$ [20]

$$Z = \int_{Q_{i0}}^{Q_{i0}} \prod_i DQ_i \sum_{\{n\}} \int_{\varphi_{i0}}^{\varphi_{i0}+2\pi n_i} \prod_i D\varphi_i \exp\left\{ -\frac{1}{2} \int_0^\beta d\tau \left[\sum_{ij} Q_i(\tau) U_{ij} Q_j(\tau) \right.\right.$$

$$\left.\left. + i \sum_i Q_i(\tau)\dot{\varphi}_i(\tau) + \sum_{ij} E_J \cos\varphi_{ij}(\tau) \right] \right\} . \tag{23}$$

The charging matrix $U_{ij} = 4e^2/C_{ij}^{-1}$ will be specified below. The summation over the winding numbers fixes the charges to be integer.

Vortex degrees of freedom can be introduced by means of the Villain transformation [21]. It allows us to integrate out the phases at the expense of introducing a set of integers $\vec{m} = (m_x, m_y)$ at each (space) bond. We introduce a lattice, with spacing ϵ, also in the time direction but keep denoting it by τ. The integration over the φ fields can be performed with the final result

$$Z = \sum_{\{Q_{i\tau}\}} \sum_{\{\vec{m}_{i\tau}\}} \exp\left\{ -\frac{\epsilon}{2} \sum_{ij\tau} Q_{i\tau} U_{ij} Q_{j\tau} - \frac{1}{2\epsilon E_J} \sum_{ij\tau} \vec{m}_{i\tau} \vec{m}_{j\tau} \right\} . \tag{24}$$

The summations are constraint by the 'continuity' equations at each lattice point

$$\vec{\nabla} \cdot \vec{m}_{ij} - \partial_\tau Q_{i\tau} = 0. \tag{25}$$

The constraint can be solved by [22]

$$m_{i\tau}^{(\mu)} = n^{(\mu)}(\vec{n} \cdot \vec{\nabla})^{-1}\partial_\tau Q_{i\tau} + \epsilon_{\mu\nu}\nabla_\nu A_{i\tau}. \tag{26}$$

The operator $(\vec{n} \cdot \vec{\nabla})^{-1}$ is the line integral on the lattice, $\epsilon_{\mu\nu}$ is the anti-symmetric tensor and $A_{i\tau}$ is a discrete scalar filed. There are other ways to solve the constraint eq. (25), but eq. (26), which corresponds to the choice of the axial gauge, allows us to express the partition function in terms of real charges and vortices.

After a Poisson resummation the partition function becomes

$$Z = \sum_{\{Q_{i\tau}\}} \sum_{\{P_{i\tau}\}} \exp\left\{-H_{CCG}(\{Q_{i\tau}\}, \{P_{i\tau}\})\right\} \tag{27}$$

where $Q_{i\tau}$ and $P_{i\tau}$ represent the charge and vorticity in the space-time lattice, and the Hamiltonian of the coupled Coulomb gas is

$$H_{CCG} = \sum_{ij\tau}\left\{ (2/\pi)\epsilon E_C Q_{i\tau}G(i-j)Q_{i\tau} + \pi\epsilon E_J P_{i\tau}G(i-j)P_{j\tau} \right.$$

$$\left. + \frac{1}{2\epsilon E_J}\partial_\tau Q_{i\tau}G(i-j)\partial_\tau Q_{j\tau} + 2\pi i Q_{i\tau}\Theta(i-j)\partial_\tau P_{j\tau} \right\}. \tag{28}$$

with kernels

$$G(i-j) = -\ln|i-j| + c \tag{29}$$

$$\Theta(i-j) = \tan^{-1}(y_i/x_i) \tag{30}$$

Obviously for $E_J = 0$ or $E_C = 0$ the only contribution to the partition function comes from the fields which are constant in time τ and, therefore, the classical Coulomb gases of charges or vortices, respectively are recovered.

As it stands the Hamiltonian eq. (28) is not symmetric with respect to the exchange $Q \leftrightarrow P$. This is due to the non-local kinetic contribution for the charge, which arises from the spin-wave contribution of phases φ.

Related coupled Coulomb gas models were discussed in [23] in connection with transitions in frustrated XY systems. In this case it is convenient to see the different time slices as different replicas of the Coulomb gas which are coupled by the time derivatives present in eq. (28). The renormalization group analysis of [24] can be extended to this case. At this stage we can say that it leads to a fixed point dominated by quantum fluctuations, at which charges and vortices are self-dual. This corresponds to the point

$$T = 0 \qquad (E_j/E_C)_c = 2/\pi^2. \tag{31}$$

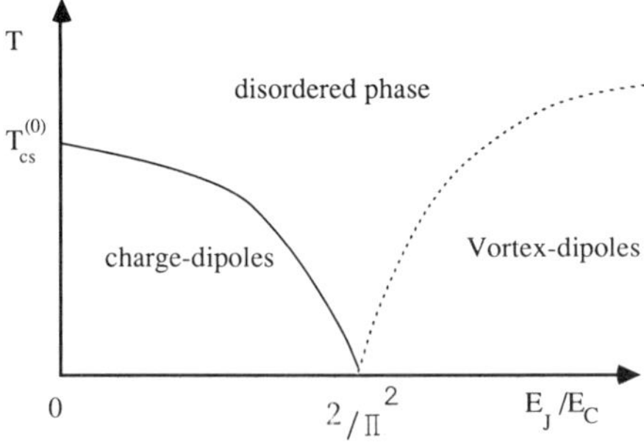

Figure 2: Qualitative phase diagram for a Josephson array. The dashed and the continuous lines represent respectively the *vortex*-KTB and *charge*-KTB transition temperatures. The existence of the transition temperature at $T = 0$ can be derived from duality arguments.

This result was also obtained by van Wees [25]. Actually the result eq. (27) should be corrected by a ratio of dielectric constants, which differs from one due the lack of duality away from the fixed point. The RG equations also suggest that there exists a regime where quantum fluctuations are dominant. Hence at non-zero temperature with increasing values of E_J/E_C the system undergoes a transition from a charge ordered to a disordered to a phase ordered state. The qualitative phase diagram is also shown in Fig. 2 (for a more detailed discussion of the interplay between *charge* and *vortex* KTB transitions see also [26]).

4 Conclusion

2 D arrays of tunnel junctions can undergo a KTB type phase transition (if the mutual capacitance is much larger than the capacitance to ground). In the ordered phase the charges on different islands are bound in dipoles. I briefly reviewed the properties of the transition with emphasis on the influence of single-electron and Cooper pair tunneling on this charge-anti-charge unbinding. Both dissipation and Josephson tunneling lower the transition temperature. Dissipative effects act in a different way in the normal and superconductive case due to the different excitation spectrum. In the superconducting case the additional phase ordering rises the possibility of a more complex phase diagram in which charge unbinding and vortex unbinding act in a complementary way.

5 Acknowledgements

I reviewed here work that has been done, and that is still in progress, with Gerd Schön; I am also indebted to him for many useful advices in the preparation of the manuscript. I also acknowledge fruitful discussions with J.E. Mooij, B.J. van Wees, U. Geigenmüller, H.M. Jaeger, G. Falci and G. Giaquinta. This work was supported by a CNR-NATO grant.

* On leave from: Istituto di Fisica, Facoltà di Ingegneria, Università di Catania, viale A. Doria 6, 95129 Catania, Italy

References

[1] Proc. of the NATO Advanced Research Workshop on Coherence in Superconducting Networks, J.E. Mooij and G. Schön, eds. Physica **B 152** (1988)

[2] L. Jacobs, J.V. Jos, and M.A. Novotny, Phys. Rev. Lett. **53**, 2177 (1984)

[3] S. Chakravarty, G.L. Ingold, S. Kivelson, and A. Luther, Phys. Rev. Lett. **56**, 2303 (1986); A. Kampf and G. Schön, Phys. Rev. **B 36**, 3651 (1987); R. Fazio, G. Falci, and G. Giaquinta, Sol. St. Commun. **71**, 275 (1989); W. Zwerger, J. Low Temp. Phys. **72**, 291 (1988); G. Falci, R. Fazio, V. Scalia, and G. Giaquinta, to appear in LT19, Physica **B** (1990); see also ref. [1]

[4] H.M. Jaeger et al., Phys. Rev. **B 34**, 4920 (1986); Phys. Rev. **B 40**, 182 (1989);

[5] L.G. Geerligs and J.E. Mooij, Physica **B 152**, 212 (1988); L.J. Geerligs, M. Peters, L.E.M. de Groot, A. Verbruggen, and J.E. Mooij, Phys. Rev. Lett. **63**, 326 (1989)

[6] J.E. Mooij, B.J. van Wees, L.J. Geerligs , M. Peters, R. Fazio, and G. Schön, subm. to Phys. Rev. Lett.

[7] J.M. Kosterlitz and D.J. Thouless, J. Phys. **C 6**, 1181 (1973) ; V.L. Berezinskii, Zh. Eksp. Teor. Fiz. **59**, 907 (1970) [Sov. Phys. JETP **32**, 493 (1971)]

[8] V. Ambegaokar, U. Eckern, and G. Schön, Phys. Rev. Lett. **48**, 1745 (1982)

[9] E. Ben-Jacob, E. Mottola, and G. Schön, Phys. Rev. Lett. **51**, 2064 (1983)

[10] G. Schön and A.D. Zaikin, Physica **B 152**, 203 (1988)

[11] S. Chakravarty, S. Kivelson, G.T. Zimanyi, and B.I. Halperin, Phys. Rev. **B 37**, 7256 (1988)

[12] S.T. Chui and J.D. Weeks, Phys. Rev. **B 14**, 4978 (1976)

[13] P.Minnhagen, Rev. Mod. Phys. **59**, 1001 (1987)

[14] R. Fazio and G. Schön, to appear in LT19, Physica **B** (1990)

[15] Y. Saito and H. Müller-Krumbhaar, in: Applications of the Monte Carlo Method, K. Binder, ed., Topics in Current Physics **36**, Springer-Verlag (1984)

[16] H.J.E. Knops, Phys. Rev. Lett. **39**, 766 (1977); J.D. Weeks, in: Ordering in Strongly Fluctuating Condensed Matter Systems, T. Riste, ed., NATO-ASI series **50**, Plenum Press, New York (1980)

[17] A. Widom and S. Badjou, Phys. Rev. **B 37**, 7915 (1988)

[18] L.P. Kadanoff, J. Phys. **A 11**, 1399 (1978)

[19] D.R. Nelson and B.I. Halperin, Phys. Rev. **B 21**, 5312 (1980)

[20] F. Guinea and G. Schön, J. Low Temp. Phys **69**, 219 (1987)

[21] J.V. Josè, L.P. Kadanoff, S. Kirkpatrick, and D.R. Nelson, Phys. Rev. **B 16**, 1217 (1977); M.P.A. Fischer and D.H. Lee, Phys. Rev. **B 39**, 2756 (1989)

[22] S. Elitzur, R. Pearson, and J. Shigemitzu, Phys. Rev. **D 19**, 3638 (1979)

[23] E. Granato and J.M. Kosterlitz, Phys. Rev. **B 33**, 4767 (1986); M. Yosefin and E. Domany, Phys. Rev. **B 32**, 1778 (1985)

[24] J. Cardy and S. Ostlund, Phys. Rev. **B 25**, 6899 (1982)

[25] B.J. van Wees, preprint

[26] R. Fazio and G. Schön, Proc. of Intl. Conference on Transport Properties of Superconductors, Rio de Janeiro (1990)

CHAPTER 11

QUANTUM CHAOS

QUANTUM SIGNATURES OF CHAOS

Fritz Haake

Fachbereich Physik
Universität-Gesamthochschule Essen
W-4300 Essen 1, F. R. Germany

1 Quantum Chaos as a Game with Quadratic Equations

The distinction between level clustering and level repulsion is one of the quantum analogues of the classical distinction between globally regular and predominantly chaotic motion (see Figs. 1, 2, 3). In order to reveal level repulsion under conditions of global classical chaos special care may be necessary: (i) subspectra referring to different values of the quantum numbers related to symmetries must be dealt with separately and (ii) for systems with quantum localization only levels whose wavefunctions have overlapping support must be admitted. A "level" may either be an energy eigenvalue E in the case of autonomous systems or, for periodically driven systems, a quasi-energy φ, i.e. an eigenphase of the unitary Floquet operator transporting the wavevector from period to period.

I shall here be concerned with three varieties of quantum systems displaying level repulsion. They differ in the degree β of level repulsion which is also referred to as the universality class index. The degree in question describes the limiting behavior of the distribution $P(S)$ of nearest-neighbor spacings S for small S,

$$P_\beta(S) \sim S^\beta \qquad \text{for} \qquad \beta \to 0. \tag{1}$$

The so-called orthogonal, unitary, and symplectic universality classes of dynamical systems have $\beta = 1, 2$, and 4, respectively (see Fig. 2). Which universality class a given dynamical system may belong to is decided by the set of unitary and anti-unitary symmetries. Anti-unitary symmetries are related to (generalized) time reversal invariance [1,2,3].

For a given spectrum consisting of many levels the level spacing distribution takes the form of a histogram (see Fig. 2). As is well known, such histograms generically are faithful to the smooth distributions predicted by Random Matrix Theory for ensembles of $N \times N$ matrices in the limit $N \to \infty$. Rather than aiming at the construction of the full functions $P_\beta(S)$ for $0 \leq S < \infty$ I shall confine myself to the discussion of the small-S behavior eq. (1). This rather modest goal can be reached by using quite elementary methods.

Quantum Coherence in Mesoscopic Systems
Edited by B. Kramer, Plenum Press, New York, 1991

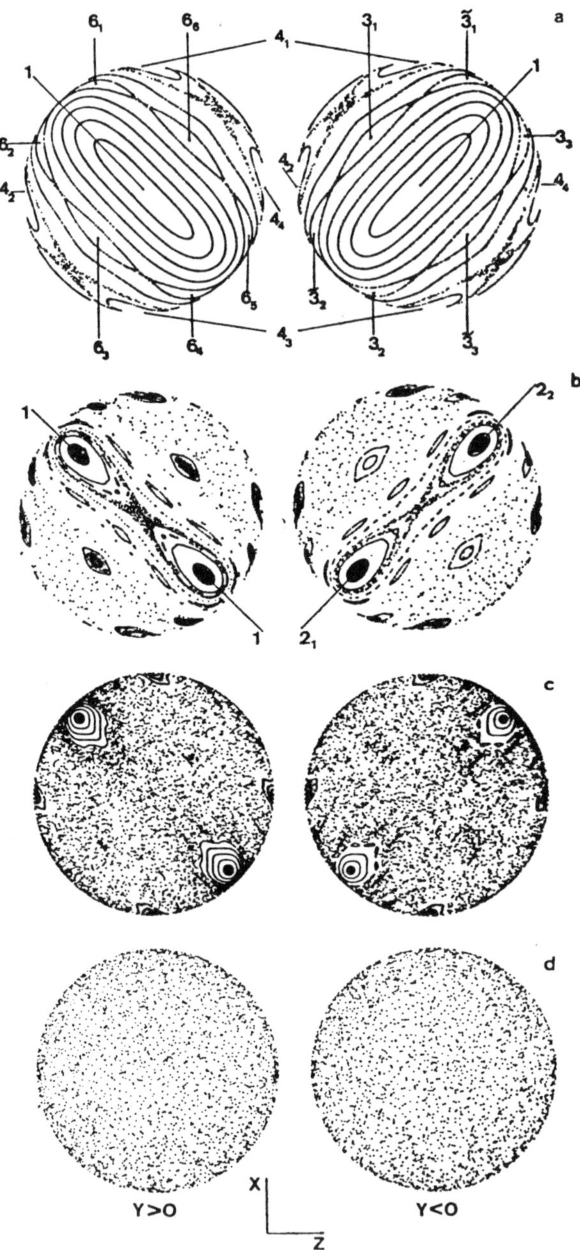

Figure 1: Sequence of four classical phase space portraits for a periodically kicked top. As a coupling constant is increased from (a) to (b) and so forth, chaos is seen to expand from small islands (a) to full coverage of the phase space (d).

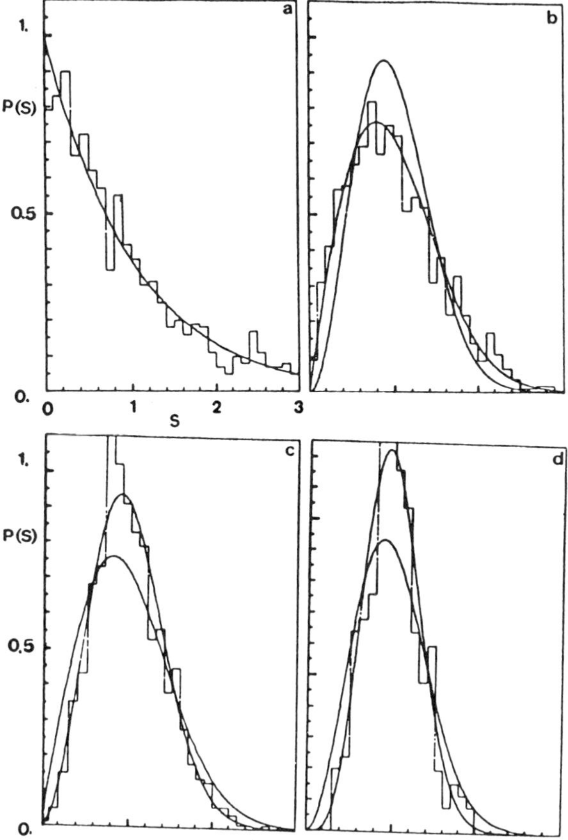

Figure 2: Level spacing distributions for periodically kicked tops. The histogram in (a) refers to a case with predominantly regular motion in the classical limit, while (b), (c) and (d) correspond to conditions of global classical chaos. The latter three cases differ by the degree of level repulsion which is linear (b), quadratic (c), and quartic (d); the smooth curves describe the predictions of Random Matrix Theory for the circular orthogonal, unitary, and symplectic ensembles.

More specifically, I propose to show that (i) $\beta = 1$ if there is a (generalized) time reversal invariance T squaring to $+1$, (ii) $\beta = 2$ if there is no such invariance, and (iii) $\beta = 4$ if there is an invariance with $T = -1$, the latter case being characteristic of systems with Kramer's degeneracy[1,2,3].

Some remarks about time reversal are in order now. I shall consider autonomous systems; the generalization to periodically driven systems may be looked up in the literature [3]. Schrödinger's equation

$$i\hbar\dot{\psi}(t) = H\psi(t) \tag{2}$$

is said to be time reversal invariant if for every solution $\psi(t)$ there is another one,

$$\widetilde{\psi}(t) = T\psi(-t), \quad i\hbar\dot{\widetilde{\psi}}(t) = H\widetilde{\psi}(t) \tag{3}$$

with T an anti-unitary operator squaring to either $+1$ or -1,

$$T^2 = \pm 1. \tag{4}$$

By demanding that $\widetilde{\psi}(t)$ obey the same Schrödinger equation as $\psi(t)$ one easily finds that the Hamiltonian must commute with T,

$$THT^{-1} = H. \tag{5}$$

The reader may recall that a unitary operator leaves the scalar product of two arbitrary state vectors unchanged in the sense $\langle\psi|\chi\rangle = \langle U\psi|U\chi\rangle$. An anti-unitary operator T, on the other hand, does change a scalar product into its complex conjugate,

$$\langle T\psi|T\chi\rangle = \langle\psi|\chi\rangle^* = \langle\chi|\psi\rangle. \tag{6}$$

The squared modulus of a scalar product, i.e. probabilities, are left invariant by both unitary and anti-unitary operations. The additional requirement eq. (4) distinguishes the special anti-unitary operators deserving the name time reversal operations [4]: any wave function should be reproduced, up to a constant factor c, when operated upon twice by T, $T^2\psi = c\psi$; it is easy to see that ± 1 are the only possible values for c.

It may be well to give a few examples. The most well known one is *conventional time reversal*, which leaves all coordinates x unchanged, reverses the sign of all momenta and angular momenta, and replaces wavefunctions (in the position representation) by their complex conjugates

$$x \;\rightarrow\; T_0 x T_0^{-1} = x$$

$$p \;\rightarrow\; T_0 p T_0^{-1} = -p$$

$$L \;\rightarrow\; T_0 L T_0^{-1} = -L$$

$$\psi(x) \;\rightarrow\; T_0\psi(x) = \psi^*(x). \tag{7}$$

This yields an invariance for a free particle with the Hamiltonian $H = p^2/2m$ but not for a charged particle in a magnetic field since $T(p - eA)^2 T^{-1} = (p + eA)^2$.

However, a charged particle in a homogeneous magnetic field does have another anti-unitary symmetry. Let $\vec{B} = (0,0,B)$ point in the z-direction and the vector potential be $\vec{A} = (-1/2)\vec{x} \times \vec{B}$ such that the Hamiltonian reads

$$H\frac{1}{2m}\left\{(p_x + By/2)^2 + (p_y - Bx/2) + p_z^2\right\}. \tag{8}$$

Clearly, this H commutes with $T_1 = UT_0$ where T_0 is conventional time reversal and U a rotation by π around any axis perpendicular to $\vec{B}$. There is nothing "false" about the non-conventional time reversal operation T_1: it obeys eqs. (2), (3), (4), (5).

To break the invariance under T_1 it suffices to switch on a homogeneous electric field $\vec{E}$ tilted arbitrarily against the magnetic field $\vec{B}$. Even then, another non-conventional time reversal invariance remains which is easily checked to be the product of T_0 with a reflection about the plane spanned by $\vec{B}$ and $\vec{E}$. To verify that symmetry it must be realized that $\vec{B}$ and $\vec{E}$ behave differently under spatial reflections.

Breaking any kind to time reversal invariance in an atom or nucleus obviously requires inhomogeneous external fields. While sizeable inhomogeneities across a nucleus are certainly inaccessible to the human experimenter, the most highly excited Rydberg atoms may not make for hopeless tasks forever.

One more ingredient will be needed to prove eq. (1), nearly degenerate perturbation theory. Imagine the Hamiltonian of some dynamical systems to have the form $H = H_0 + \lambda V$ and consider the fate of two neighboring eigenvalues $E_1(\lambda)$ and $E_2(\lambda)$ when the control parameter λ is varied. Clearly, once the distance $|E_1(\lambda) - E_2(\lambda)|$ is smaller than the distance of either level to any third one, the close encounter in consideration can be studied by nearly degenerate perturbation theory. One may assume the two eigenvalues and the corresponding eigenvectors known for a particular value λ_0 of λ and diagonalize H in that basis. If the two levels are non-degenerate the basis is two-dimensional and the eigenvalues of the two by two matrix have the difference

$$E_1 - E_2 = \sqrt{(H_{11} - H_{22})^2 + 4|H_{12}|^2} \tag{9}$$

At this point it becomes obvious that by varying a single parameter like λ the distance of neighboring levels can be steered through a minimum but not, in general, made to vanish. The discriminant in eq. (9) being a sum of non-negative squares the number of parameters needed to enforce a crossing is two or three, depending on whether the matrix H_{ij} is real symmetric or complex Hermitian.

The density of level spacings may be defined as

$$P(S) = \langle \delta(S - E_1 + E_2) \rangle \tag{10}$$

with the average to be performed over all neighboring levels of the Hamiltonian. Alternatively, one may average over the set of two by two matrices H associated with close encounters of level pairs, provided a suitable distribution function for the matrix elements is available. The latter average takes the form [5]

$$
\begin{aligned}
P(S) &= \int dx\, dy\, dz W(x,y,z)\delta\left[S - \sqrt{x^2 + y^2 + z^2}\right] \\
&= S^2 \int dx\, dy\, dz W(Sx, Sy, Sz)\delta\left[1 - \sqrt{x^2 + y^2 + z^2}\right],
\end{aligned}
\tag{11}
$$

where I have assumed that no symmetry restricts the Hamiltonian to be a real matrix. If the density $W(x,y,z)$ exists as a reasonably smooth function and neither diverges nor vanishes at $x = y = z \to 0$, the spacing distribution approaches the power-law behavior eq. (1) with $\beta = 2$. This degree is typical of the so-called unitary (Gaussian or circular) ensembles of random matrices [1-3].

An anti-unitary symmetry of H with $T^2 = 1$ modifies the situation just described in a small but consequential detail: the Hamiltonian can then be represented as a real symmetric matrix, $H_{ij} = H_{ji}$ by employing T invariant orthonormal basis vectors, $\psi_i = T\psi_i$, $\langle \psi_i | \psi_j \rangle = \delta_{ij}$. An arbitrary basis $\{\varphi_i\}$ is easily turned into a T invariant one in the following way. Let $\psi_1 \sim \varphi_1 + T\varphi_1$ such that $T\psi_1 = \psi_1$ and normalize, $\langle \psi_1 | \psi_1 \rangle = 1$; then take a combination $\widetilde{\psi}_2$ of the φ_i orthogonal to $\psi_1 \langle \psi_1 | \widetilde{\psi}_2 \rangle = 0$, choose $\psi_2 \sim \widetilde{\psi}_2 + T\widetilde{\psi}_2$, and again normalize etc. The reality of H then follows from the anti-unitary of T and

$[H, T] = 0$: indeed, $\langle \psi_1 | H | \psi_2 \rangle = \langle T\psi_1 | TH | \psi_2 \rangle^* = \langle \psi_1 | H | \psi_2 \rangle^*$; due to its Hermiticity the Hamiltonian matrix is then also symmetric. It follows that the discriminant in eq. (9) is a sum of only two non-negative reals. The number of integrations in eq. (11) is thus reduced to two and the degree of level repulsion to $\beta = 1$ which also characterizes the so-called orthogonal (Gaussian or circular) ensembles of random matrices.

Finally, when a (generalized) time reversal invariance holds with $T^2 = -1$ an interesting complication arises due to Kramer's degeneracy, a double degeneracy of each level. To check on this statement, imagine ψ to be an eigenvector of H; then $T\psi$ is also an eigenvector pertaining to the same eigenvalue due to $[H, T] = 0$; but ψ and $T\psi$ are orthogonal: $\langle \psi | T\psi \rangle = \langle T\psi | T^2\psi \rangle^* = -\langle T\psi | \psi \rangle^* = -\langle \psi | T\psi \rangle = 0$. Therefore, in treating a near miss of two levels by nearly degenerate perturbation theory I must diagonalize a four by four matrix. That matrix is restricted by Hermiticity and the symmetry $[H, T] = 0$ in precisely such a way that its secular determinant takes the form of the square of a two by two determinant. It is easy to verify that [6]

$$H = \begin{pmatrix} \alpha + \beta & 0 & \gamma - i\sigma & -\epsilon - i\delta \\ 0 & \alpha - \beta & \epsilon - i\delta & \gamma + i\sigma \\ \gamma + i\sigma & \epsilon + i\delta & \alpha - \beta & 0 \\ -\epsilon + i\delta & \gamma - i\sigma & 0 & \alpha - \beta \end{pmatrix}, \tag{12}$$

where $\alpha, \beta, \gamma, \delta, \epsilon, \sigma$ are six real parameters; indeed, after arranging the basis vectors in the order $|1\rangle, T|1\rangle, |2\rangle, T|2\rangle$ consider, e.g. $\langle 1 | HT | 1 \rangle = \langle T1 | THT | 1 \rangle^* = -\langle T1 | H | 1 \rangle^* = -\langle 1 | HT | 1 \rangle = 0$. The two eigenvalues have the difference

$$E_1 - E_2 = 2\sqrt{\beta^2 + \gamma^2 + \delta^2 + \epsilon^2 + \delta^2}. \tag{13}$$

When determining the small-S behavior of $P(S)$ in analogy to eq. (11) I now encounter a fivefold integral and thus the quartic degree of level repulsion, $\beta = 4$, otherwise known to be characteristic of the so-called symplectic (Gaussian or circular) ensembles of random matrices [1-3].

2   Random Matrix Theory as Statistical Mechanics

Random Matrix Theory is well known to be quite successful in describing fluctuations in the quantum spectra of dynamical system with global chaos in the classical limit. That theory models Hamiltonians as random Hermitian matrices and Floquet operators as random unitary matrices. For the sake of a change I shall deal with Floquet operators, F, $FF^\dagger = 1$, in this section; the transcription to Hamiltonians is mostly a matter of notation, though.

Dyson's circular ensembles [1-3] of unitary $N \times N$ random matrices imply the following joint probability densities for the N eigenphases φ_i

$$P(\varphi) = \frac{1}{N!(2\pi)^N} \prod_{1<j}^{1...N} \left| e^{-i\varphi_i} - e^{-i\varphi_j} \right|^\beta. \tag{14}$$

The degree of level repulsion β appears again here and distinguishes the three universality classes as

$$\beta = \begin{cases} 1 & \text{orthogonal} \\ 2 & \text{unitary} \\ 4 & \text{symplectic} \end{cases}. \tag{15}$$

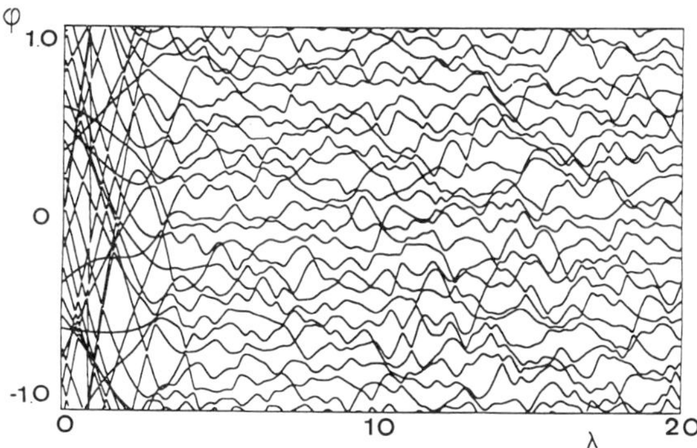

Figure 3: Quasi-energy spectrum of a periodically kicked top versus some coupling constant λ; top and coupling constant the same as in Fig. 1; Fig. 1a, b, c, and d correspond to $\lambda = 1$, $\lambda = 2.5$, $\lambda = 3$, and $\lambda = 6$, respectively. Note that the spectrum changes its character from level clustering to level repulsion in the same range of λ, $\lambda \approx 3 \pm 0.5$, in which the classical transition from predominance of regular motion to predominance of chaos takes place. The range $\lambda \gtrsim 3$ has λ independent spectral fluctuations; this is the range for equilibrium statistics treated in section 2. As λ increases from 0 to 3 a relaxation into equilibrium takes place such as the one treated in lecture 3.

From the distribution eqs. (14), (15) all measures of fluctuations in the spectrum can be derived by suitable integrations.

Rather than describing the amazing degree to which the Floquet spectra of classically chaotic systems are faithful to the predictions of Random Matrix Theory, I shall here focus on the question as to why that should be the case. To that end I shall point out that Random Matrix Theory can be reinterpreted as equilibrium statistical mechanics of a certain fictitious classical gas.

When contemplating Fig. 3 the reader will realize the analogy of the bundle of N level curves $\varphi(\lambda)$ with a bundle of trajectories of N particles in a one-dimensional configuration space with λ as a time; the particle-particle interaction appears as repulsive. In fact, on the basis of a discovery of Pechukas [7] the analogy can be extended into a rigorous one-to-one relation. Consider a Floquet operator of the form

$$F(\lambda) = e^{-i\lambda V} e^{-iH_0} \tag{16}$$

with two Hermitian operators H_0 and V, both independent of the control parameter λ; such Floquet operators arise for periodically kicked systems for which the perturbation λV is switched on periodically and impulsively. The control parameter dependence of the eigenvalue problem

$$F(\lambda)|m, \lambda\rangle = e^{-i\varphi_m(\lambda)}|m, \lambda\rangle \qquad m = 1, 2, \ldots N \tag{17}$$

can be shown to be equivalent to the classical motion generated by the Hamiltonian [8,3]

$$\mathcal{H} = \frac{1}{2}\sum_m p_m^2 + \frac{1}{8}\sum_{m \neq n} \frac{|\ell_{mn}|^2}{\sin^2\left((\varphi_m - \varphi_n)/2\right)}. \tag{18}$$

The correspondence between eqs. (17) and (18) is as follows

$$\lambda \;\leftrightarrow\; \text{time}$$

$$\varphi_m \;\leftrightarrow\; \text{particle coordinates}$$

$$p_m = \langle m|V|m\rangle \;\leftrightarrow\; \text{momenta}$$

$$\ell_{mn} = -2\langle m|V|n\rangle \left(e^{i(\varphi_m-\varphi_n)} - 1\right) \;\leftrightarrow\; \text{angular momenta.} \qquad (19)$$

The matrix elements of V occurring here are meant in the Floquet basis eq. (17). The only unusual feature of the dynamics in question is that the strengths of the pair interaction involve dynamical variables ℓ_{mn} rather than being constants; consequently, the phase space encountered has the dimension

$$\mathcal{N}_\beta = 2N + \beta N(N-1)/2, \qquad (20)$$

the universality class index β enters here in counting the number of independent parameters in the off-diagonal elements of the matrix V. To establish the Hamiltonian equations of motion the usual Poisson brackets for coordinates and momenta must be employed as well as independent such brackets for the angular momenta; the latter, looking different for the three universality classes [3,9], constitute Lie algebras of generators of rotations in appropriate (βN) dimensional vector spaces – hence the name "angular momenta".

Previous to the discovery of its relation to the quantum mechanical eigenvalue problem eqs. (16) and (17) the classical N particle system with the Hamiltonian eq. (18) had been known in the mathematical literature as a variant of the Calogero-Moser dynamics [10,11,12]. As a byproduct of the equivalence in question the integrability of the fictitious-particle dynamics is worth being noted. (The reader may recall that even the harmonic chain, a prototypical integrable N particle system, requires diagonalization of an $N \times N$ matrix.) In fact, by employing for a moment the eigenbasis of V it is easy to see that the λ dependence of the matrix F in eq. (16) can be built with N harmonic oscillations; in other words, the classical trajectories implied by the Hamiltonian $\mathcal{H}$ are confined to an N torus within the $\mathcal{N}_\beta$ dimensional phase space; moreover, the N frequencies involved have no reason to be commensurate and thus in general give rise to an ergodic motion in the N torus [13].

An infinity of constants of the motion of the fictitious-particle dynamics,

$$C_\mu(\varphi,p,\ell) = tr\left(\ell^{\mu_1} V^{\mu_2} \ell^{\mu_3} V^{\mu_4} \dots\right), \; \mu_i = 1,2,3,\dots, \qquad (21)$$

becomes accessible once it is recognized that the Hamiltonian equations can be written in the so-called Lax from [14], $\dot V = [g, V]$ and $\dot\ell = [g, V]$ with a suitable "generating matrix" g. Due to the relation between V_{mn} and ℓ_{mn} given in eq. (19) the phase space functions C_μ in general diverge as two coordinates coincide, $\varphi_m - \varphi_n \to 0$ at $\ell_{mn} \neq 0$; this fact may either be interpreted in quantum parlance (avoided crossings, level repulsion) or in the classical-particle picture (repulsive pair interaction). Unfortunately, it is not known at present whether or not the set of constants of the motion eq. (21) contains a subset of $\mathcal{N}_\beta - N$ independent ones allowing to nail down the N torus.

In the limit $N \to \infty$ the fictitious gas invites a statistical description [15,16,3]. The appropriate equilibrium ensemble would be the generalized microcanonical one which specifies the N torus by assigning sharp values to suitable $\mathcal{N}_\beta - N$ constants of the motion. Technically easier to work with is the canonical ensemble

$$\rho(\varphi, p, \ell) = Z^{-1} \exp\left\{ - \sum_\mu \alpha_\mu C_\mu(\varphi, p, \ell) \right\} \tag{22}$$

with Lagrange parameters α_μ. The argument to follow assumes all C_μ in eq. (22) to be of the form eq. (21), real valued, and bounded from below. Of special interest is the reduced distribution of all N coordinates, obtained by integrating out all N momenta and the $\mathcal{N}_\beta - N$ independent angular momenta

$$P(\varphi) = \int d^N p \left[d^{\mathcal{N}_\beta - 2N} \ell \right] \rho(\varphi, p, \ell). \tag{23}$$

Clearly, this distribution vanishes at particle crossings, due to the divergence of the C_μ mentioned above. To ascertain the behavior near crossings I abandon the angular momenta ℓ_{mn} in favour of the V_{mn} as integration variables,

$$P(\varphi) = \left(\prod_{m<n} \left| e^{-i\varphi_m} - e^{-i\varphi_n} \right|^\beta \right) \widetilde{P}(\varphi)$$

$$\widetilde{P}(\varphi) = \int d^N p \left[d^{\mathcal{N}_\beta - 2N} V \right] \rho \left(\left\{ \varphi, p \left(i V_{mn} e^{-i\varphi_m} - e^{-i\varphi_n} \right) \right\} \right). \tag{24}$$

Note that for fixed finite V_{mn} the function $\widetilde{P}(\varphi)$ has no reason vanish at a particle crossing. The degree to which the fictitious particles avoid crossings is therefore solely determined by the Jacobian of the transformation $\ell \to V, \det(\partial \ell / \partial V) \sim P(\varphi)/\widetilde{P}(\varphi)$, which actually is nothing but the joint eigenvalue density eq. (14) predicted by Random Matrix Theory.

Any noticeable φ dependence of the function $\widetilde{P}$ would make for a difference between the eigenvalue statistics implied by Random Matrix Theory and equilibrium statistical mechanics of the fictitious gas. Interestingly, could I exclude from the ensemble eq. (22) all constants of the motion eq. (21) with $\mu_1 = \mu_3 = \mu_5 = \ldots = 0$, i.e. admit only $tr V^\mu$, I would face a $\widetilde{P}$ strictly independent of the φ and thus rigorously recover Random Matrix Theory, even more interestingly, the usual microcanonical ensemble $\exp(-\beta \mathcal{H})$ does have this nice property since $\mathcal{H} \sim tr V^2$, unfortunately, however, of the $tr V^\mu$ only N are independent and these are definitely insufficient in number to nail down the N torus. It follows that there can be no rigorous equivalence, with respect to the distribution $P(\varphi)$ of all N coordinates, between ergodic motion of the N torus and Random Matrix Theory for fixed finite N. What does hold instead is a weaker, asymptotic correspondence: the level spacing distribution [16,3] (as well as certain other quantities which typically involve correlations between only a few levels) come out the same in the two approaches, provided the limit $N \to \infty$ is taken. For the proof of that asymptotic equivalence I refer the reader to the original literature. The proof is based on representing the φ dependence of $\widetilde{P}(\varphi)$ by an N fold Fourier series; none of the finite-frequency terms can compete, for $N \to \infty$, with the zero-frequency one in its effect on the level-spacing distribution.

Some non-trivial modifications of the foregoing consideration, become necessary for the level dynamics of quantum Hamiltonians

$$H(\lambda) = H_0 + \lambda V \tag{25}$$

of autonomous systems. The equivalent fictitious gas was identified by Pechukas [7] as governed by the Hamiltonian function

$$\mathcal{H} = \frac{1}{2}\sum_m p_m^2 + \frac{1}{2}\sum_{m\neq n}\frac{|\ell_{mn}|^2}{(E_m - E_n)^2}.$$ (26)

Here, the particle coordinates E_m correspond to the eigenvalues of H, in contrast to the eigenphases φ_m of Floquet operators the E_m are not confined to a finite interval but rather to $0 < E_m < \infty$. The diagonal elements V_{mn} of the perturbation V in the eigenrepresentation of $H(\lambda)$ again play the role of moments while $\ell_{mn} = -2V_{mn}(E_m - E_n)$.

Unfortunately, due to the infinite range of the E_m and the absence of any confining potential in eq. (26), the repulsive pair interaction causes indefinite expansion of the Pechukas gas and thus defines the application of equilibrium statistical mechanics. The instability of the gas corresponds to the structure of the original quantum Hamiltonian $H = H_0 + \lambda V$: for $\lambda \to \infty$ the eigenenergies must fly apart as $E_m(\lambda) \sim \lambda$; their spectral density thins out correspondingly. In as much as only fluctuations in the spectrum of $H(\lambda)$ are of interest it is intuitive to rescale as $\widetilde{H}(\lambda) = g(\lambda)(H_0 + \lambda V)$ such that the mean density of levels remains independent of λ. For the fictitious gas that rescaling $H \to \widetilde{H}$ means a non-canonical transformation with new coordinates $\widetilde{E}_m = gE_m$, a new time $\tau, \lambda \to \lambda(\tau), g = g(\tau)$. By requiring the transformed dynamics to be again Hamiltonian in character with a Hamiltonian function $\widetilde{\mathcal{H}}$ independent of τ, two differential equations for the two functions $g(\tau)$ and $\lambda(\tau)$ are obtained. The resulting Hamiltonian function [17,3]

$$\widetilde{\mathcal{H}} = \mathcal{H} + \frac{1}{2}\sum_m \widetilde{E}_m^2$$ (27)

differs from Pechuka's one by a harmonic binding potential which prevents the modified gas from exploding as $\tau \to \infty$.

Integrability is not destroyed by the confining potential in eq. (27); the Hamiltonian equations of motion can again be written in the Lax form and the expression eq. (21) reappear as constants of the motion. As an additional conserved quantity the Hamiltonian $\widetilde{\mathcal{H}}$ itself must be listed since now $\widetilde{\mathcal{H}}$ is not of the form eq. (21).

Equilibrium statistical mechanics does make sense for the confined gas. A restricted canonical ensemble involving $\widetilde{\mathcal{H}}$ as well as at most N different constants of the motion trV^μ now rigorously implies the densities $P(\widetilde{E})$ of all N levels predicted by the Gaussian ensembles of random Hermitian matrices [1,2,3],

$$P(\widetilde{E}) \sim \left(\prod_{i<j}\left|\widetilde{E}_i - \widetilde{E}_j\right|^\beta\right) e^{-\sum_m \widetilde{E}_m^2},$$ (28)

here the product of eigenvalue differences arises as a Jacobian like in eq. (24) while the Gaussian factor is due to the confining potential $1/2\sum_m \widetilde{E}_m^2$ in the additional constant of the motion $\widetilde{\mathcal{H}}$; in writing down eq. (28) I have absorbed the Lagrange parameter associated with $\widetilde{\mathcal{H}}$ in the energy scale.

No proof is available right now for the conjecture that the complete equilibrium ensemble (which specifies the submanifold of the $\widetilde{\mathcal{N}}_\beta$ dimensional phase space accessible to a trajectory) also implies the eigenvalue density eq. (28) in the limit $N \to \infty$.

3 Relaxation of Spectra into Equilibrium

Fluctuations within quantum spectra often change their character when a perturbation is switched on. Transitions from non-generic to generic behavior for either integrable or non-integrable systems, crossover from level clustering to level repulsing upon destroying integrability, change of universality class when breaking or restoring an anti-unitary symmetry are examples frequently met with.

The level dynamics discussed in the previous section allows to interpret such transitions as relaxations into equilibrium. (Needless to say, the apparent contradiction between classical Hamiltonian dynamics of the fictitious gas and effective relaxation of certain "macroscopic" quantities is the same here as for real many-particle systems and so is its resolution.)

For the sake of concreteness I shall now describe a particular relaxation, the change of the level spacing distribution $P(S, \lambda)$ accompanying the breaking of time reversal invariance in a Hamiltonian $H(\lambda) = H_0 + \lambda V$. To that end I assume H_0 invariant under some anti-unitary T but V not so restricted and $H(\lambda)$ classically non-integrable for any value of λ; moreover, with respect to their separate spectra H_0 and V are taken to be typical numbers of the Gaussian orthogonal and unitary ensembles (GOE and GUE), respectively. Fluctuations in the spectrum should thus reflect a transition from the orthogonal to the unitary universality class.

As a joint density of all matrix elements of $H(\lambda)$ I may consider [18,3]

$$P(H, \lambda) = \langle \delta \left(H - (H_0 + \lambda V) \right) \rangle, \tag{29}$$

where the brackets denote an average over the GOE with respect to H_0 and over the GUE with respect to V. After performing the N^2 fold integration over the elements of V I obtain the convolution

$$P(H, \lambda) = \int dH_0 P_{\text{GOE}}(H_0) P_{\text{GUE}} \left(\frac{H - H_0}{\lambda} \right) \lambda^{-N^2}. \tag{30}$$

At this point it is important to recall that $P_{\text{GUE}}(H)$ is a product of N^2 Gaussians with a separate factor for each matrix element. It follows that $P_{\text{GUE}}((H - H_0)/\lambda)\lambda^{-N^2}$ can be interpreted as the Green's function of an N^2 dimensional free diffusion process with λ^2 as the time; indeed, for $\lambda \to 0$ that function shrinks in width to the N^2 dimensional delta function $\delta(H - H_0)$ and then correctly reproduces $P(H, 0) = P_{\text{GOE}}(H)$ in eq. (30); for $\lambda \to \infty$, like the Green's function of any free diffusion it shrinks to zero height and expands to infinite width, keeping constant global weight with respect to all of its independent variables. Similarly "sick" is the behavior of the convoluted distribution at large λ, $P(H, \lambda) \to \lambda^{-N^2} P_{\text{GUE}}(H/\lambda)$. Of course, $P(H, \lambda)$ must fail to approach a stationary limit as $\lambda \to \infty$ since the Hamiltonian $H = H_0 + \lambda V \to \lambda V$ itself grows indefinitely with the coupling constant λ; that problem was already considered at the end of the previous section.

With the aim of keeping the mean density of levels constant as λ is varied I again rescale the energy by a function $g(\lambda)$ such that eqs. (29) and (30) are replaced by

$$\begin{aligned} P(H, \lambda) &= \langle \delta \left[H - g(H_0 + \lambda V) \right] \rangle \\[2mm] &= \int dH_0 P_{\text{GOE}}(H_0)(\lambda g)^{-N^2} P_{\text{GUE}} \left(\frac{H - g H_0}{\lambda g} \right). \end{aligned} \tag{31}$$

By imposing the boundary conditions

$$\lambda g \to \begin{cases} 0 & \text{for} \quad \lambda \to 0 \\ 1 & \text{for} \quad \lambda \to \infty \end{cases} \tag{32}$$

I secure the desired limiting behavior for the interpolating distribution

$$P(H,\lambda) \to \begin{cases} P_{\text{GOE}}(H) & \text{for} \quad \lambda \to 0 \\ P_{\text{GUE}}(H) & \text{for} \quad \lambda \to \infty \end{cases} \tag{33}$$

As already indicated above the rescaling function g should be fixed so as to enforce Wigner's semi-circle law [1,2,3] for the mean density of levels, which holds at $\lambda = 0$ and $\lambda \to \infty$, in the whole interval $0 \le \lambda < \infty$. A slightly indirect way of implementing that goal is to allow for a new "time" τ, i.e. for two functions $\lambda(\tau)$ and $g(\tau)$, and to require that $(\lambda g)^{-N^2} P_{\text{GUE}}((H - gH_0)/\lambda g) \equiv G(H, \tau | H_0)$ remain the Green's function of some Markovian random process, i.e. obey the Chapman-Kolmogorov equation [19]. In contrast to the free diffusion encountered for $g \equiv 1$ the new Markovian process respecting the boundary conditions eq. (32) must necessarily involve generalized forces preventing the levels of the family of rescaled Hamiltonian $g(H_0 + \lambda V)$ from flying apart. In fact, the Chapman-Kolmogorov equation yields two functional equations for $\lambda(\tau)$ and $G(\tau)$ which, together with the boundary conditions eq. (32) have the unique solution $\lambda(\tau)^2 = e^{2\tau} - 1, g(\tau) = e^{-\tau}$. The resulting random process is the N^2 dimensional Ornstein-Uhlenbeck process [19]: each matrix element H_{ij} behaves like the amplitude of an overdamped harmonic oscillator driven by Gaussian white noise. The difference to the free diffusion encountered before the rescaling lies in the systematic linear restoring force effective for each matrix element; that restoring force is analogous both in origin and in its effects to the one encountered for the modified Pechukas gas in the previous section.

Within Random Matrix Theory the matrix-valued Ornstein-Uhlenbeck process has long been known as Dyson's-Brownian-motion model [20,2]. Originally introduced by Dyson as an ad hoc dynamization of the Gaussian ensemble, it now reappears with its status slightly elevated: since the time τ us uniquely related to the control parameter λ in the quantum Hamiltonian the model amounts to a rigorous description of the dependence of fluctuations in the spectrum of $g(H_0 + \lambda V)$ on λ. From the point of view of the fictitious gas dynamics discussed in the previous section the Ornstein-Uhlenbeck process appears due to (i) a particular non-linear reparametrization of the "time" and (ii), more importantly, a "teleological" average: the Green's function in eq. (31) ows its Gaussian character to the average of the delta distribution over the final Gaussian unitary ensemble. One may think of $\delta(H - g(H_0 + \lambda V))$ as tracing out a single trajectory in an N^2 dimensional space while the Green's function refers to a bundle of such trajectories, the bundle being designed faithful to the asymptotic ensemble to be approached as $\lambda \to \infty$.

Incidentally, the Gaussian nature of both P_{GOE} and P_{GUE} in eq. (31) is responsible for Wigner's semi-circle law to hold for $0 \le \lambda < \infty$ or, equivalently, $0 \le \tau < \infty$; more specifically, the width of the convoluted Gaussian for each off-diagonal matrix element, given by the mean value of $(ReH_{ij})^2 + (ImH_{ij})^2$, is independent of τ and therefore the traditional construction of the semi-circle law goes through unchanged [3].

It is interesting to note that the Ornstein-Uhlenbeck process for the matrix $H = g(H_0 + \lambda V)$ contains a closed subdynamics for the eigenvalues $E_i(\tau)$ which is also Markovian in character [19]. That process may be described by the Fokker-Planck equation

$$\frac{\partial}{\partial \tau} W(E, \tau) = \sum_{i=1}^{N} \left\{ \frac{\partial}{\partial E_i} \left(E_i + \frac{1}{2} \sum_{j(\neq i)} \frac{1}{E_i - E_j} \right) + \frac{\partial^2}{\partial E_i^2} \right\} W(E, \tau). \tag{34}$$

The confining effect of the rescaling is manifest in the linear restoring force while the pair interaction force takes care of level repulsion. In view of that latter interaction the model in consideration is often referred to as to Dysons's Coulomb gas.

References

[1] Statistical Theory of Spectra, C.E. Porter, ed., Academic Press, New York (1965)

[2] M.L. Mehta, Random Matrices and the Statistical Theory of Spectra, Academic Press, New York (1965)

[3] F. Haake, 'Quantum Signatures of Chaos', Springer-Verlag, Berlin, to appear (1990)

[4] E.P. Wigner, Group Theory and its Applications to the Quantum Mechanics of Atomic Spectra, Academic Press, New York (1959)

[5] M.V. Berry, in: Chaotic Behavior of Deterministic Systems, G. Iooss, R.H.G. Hellemann, and R. Stora, eds., Les Houches Session XXXVI, 1981, North Holland, Amsterdam (1983)

[6] R. Scharf, B. Dietz, M. Kús, F. Haake, and M.V. Berry, Europhys. Lett. **5**, 383 (1988)

[7] P. Pechukas, Phys. Rev. Lett. **51**, 943 (1983)

[8] F. Haake, M. Kús, and R. Scharf, Lecture Notes in Phys. **282**, F. Ehlotzky, ed., Springer, Berlin (1987)

[9] M. Kús, to be published

[10] F. Calogero and C. Marchioro, J. Math. Phys. **15**, 1425 (1974)

[11] B. Sutherland, Phys. Rev. **A 5**, 1372 (1972)

[12] J. Moser, Adv. Math. **16**, 1 (1975)

[13] M. Kús, Europhys. Lett. **5**, 1 (1988)

[14] P.D. Lax, Comm. Pure Appl. Math. **21**, 467 (1968)

[15] T. Yukawa, Phys. Rev. Lett. **54**, 1883 (1985); Phys. Lett. **116 A**, 227 (1986)

[16] B. Dietz and F. Haake, Europhys. Lett. **9**, 1 (1989); Z. Phys., to be published

[17] F. Haake and G. Lenz, to be published

[18] G. Lenz and F. Haake, to be published

[19] H. Risken, The Fokker Planck Equation, Springer, Berlin (1984)

[20] F. Dyson, J. Math. Phys. **3**, 140 (1982)

NON-SEPARABLE VARIABLES: HIERARCHICAL QUANTIZATION AND TUNNELING RESONANCES

Mark Ya. Azbel

Raymond and Beverly Sackler Faculty of Exact Sciences
School of Physics and Astronomy, Tel Aviv University
Ramat Aviv, Tel Aviv 69978, Israel

1 Elliptic Shell in a Magnetic Field

The study of an electron in a co-focal elliptic shell is very instructive. In classical mechanics its motion is chaotic. In quantum mechanics variables are separable in the absence of magnetic field. Then the energy spectrum is readily determined. When the interfocal distance d infinitely increases, while the shell width is kept finite, the spectrum above a mobility edge energy E_n consists of intersecting branches (n is the branch number). Lower energy states $E < E_n$ are localized in the widest region [1].

Non-zero magnetic field, H, normal to the shell, makes the variables non-separable. A narrow shell allows for a hierarchical quantization, which is readily generalized to an arbitrary shape in any dimensionality [1]. First one establishes the (one-dimensional (1 D)) quantization of the energy levels with respect to the smallest length scale, keeping at this stage other coordinates as parameters. These are the "first stage classical orbits". Then the "second stage" quantization with respect to the next smallest stage is again 1 D. In the WKB approximation it readily reduces to the corresponding Bohr-Sommerfeld quantization of the effective Hamiltonian provided by the previous stage, etc. The resulting spectrum and states are given in terms of sets of integers, in analogy with separable problems. (This procedure was numerically verified in [2]). However, non-separability of variables leads to the interaction of previously intersecting branches, to level repulsion and to the formation of irregular energy gaps [2]. Numerically it is convenient to study the elliptic shell $x = a_1 r \cos \theta$, $y = a_2 r \cos \theta$, with the boundaries $r = r_0$, $r = 1$. When $a_1 = a_2 = a$, the shell is circular, the variables separate, and the energy spectrum $E = E(B_1)$, $B_1 = ea^2 H/\hbar c$, consists of intersecting branches, see Fig. 1 from [2]. A non-circular shell yields localized states (whose energies depend little on magnetic field) and lifts degeneracy, see Fig. 2 from [2]. Physically localization quenches the Aharonov-Bohm effect, while irregular gaps yield irregular de Haas-van Alphen oscillations.

Quantum Coherence in Mesoscopic Systems
Edited by B. Kramer, Plenum Press, New York, 1991

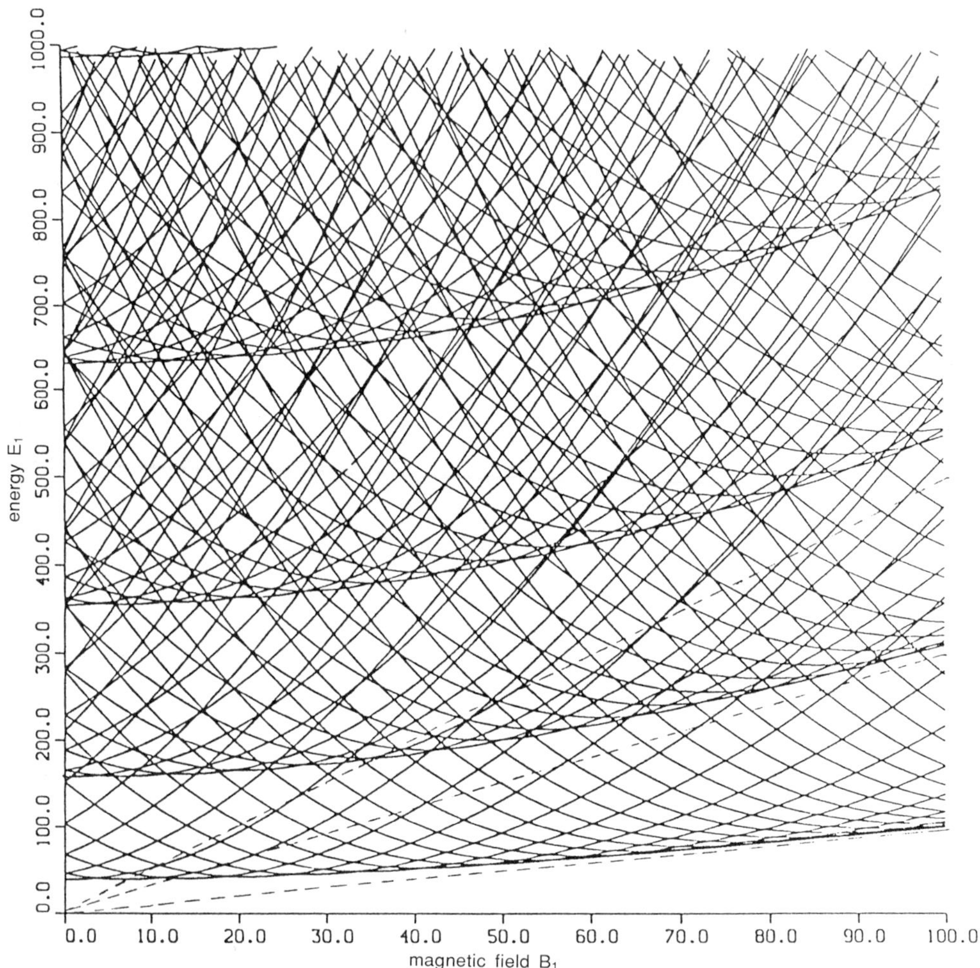

Figure 1: Energy spectrum $E_1 = 2ma^2E/\hbar^2$ as a function of magnetic field $B_1 = ea^2H/\hbar c$ for a circular shell with $r_0 = 0.5$.

Another effect of magnetic field H is the emergence of skipping orbits, which at least temporarily escape the collision with the opposite ellipse edge. As a result, the corresponding localization region and length increase with H. Strong enough magnetic field allows the first stage classical orbit to escape even the narrowest region, leads to *delocalization* and *insulator-metal* transition.

Consider this effect in more detail. Suppose in elliptic coordinates $x = d\cos\theta\cos\phi$, $y = d\sinh\theta\sin\phi$ the shell edges are $\theta = \theta_0$, $\theta = \theta_0 + \eta$, where $\eta \ll 1$. The first stage classical Hamiltonian in small magnetic fields is [1]

$$\epsilon = \left[p^2 + (n\pi)^2\right] / \left[f(\phi) + \frac{1}{3}(B/n\pi)^2 f^3(\phi)\right], \quad f(\phi) = \cosh^2\theta_0 - \cos^2\phi. \tag{1}$$

Here $\epsilon = 2mEd^2\eta^2/\hbar^2$ is the dimensionless energy, n is an integer, $B = (\eta d/L_0)^2$ is the dimensionless magnetic field, $2B\epsilon^{1/2}f^{3/2}(\phi) \ll (n\pi)^2$, and $L_0 = (2\hbar c/eH)^{1/2}$ is the

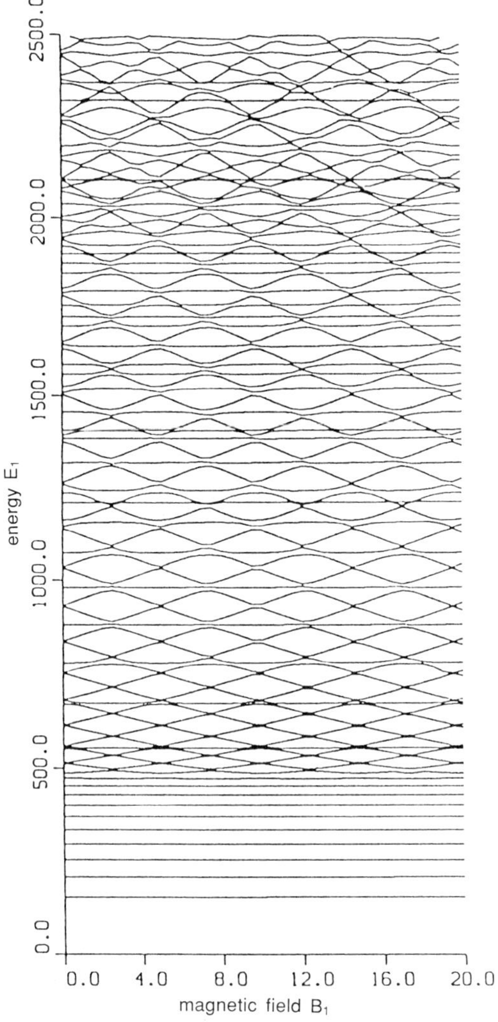

Figure 2: Energy spectrum $E_1 = E(B_1)$, $E_1 = 2ma_1a_2E/\hbar^2$, $B_1 = ea_1a_2H/\hbar c$ for an elliptic shell with $r_0 = 0.8$ and $a_2 = 0.8a_1$.

magnetic length. Equation (1) presents the orbit in the phase space of ϕ and its conjugate momentum $p(\phi)$. In the second stage quantization $p \to \hat{p} \equiv i^{-1}\eta d/d\phi$, i.e. $\eta \ll 1$ plays the role of the "dimensionless Planck constant". The second stage WKB quantization is $\int p(\phi)d\phi = (\ell + 1/2)\pi\eta$, ℓ being an integer. It yields [1] $\epsilon_{n\ell}$ and the second stage Schrödinger equation

$$\hat{p}^2 + (n\pi)^2 \left\{ 1 - \left[1 + (2\ell + 1)\eta/\pi n \sqrt{f} \right] \left[1 + (B/n\pi)^2 f^2/3 \right]^2 \right\} = 0. \tag{2}$$

The effective potential energy decreases with magnetic field, thus leading (for localized states) to the increase in the classically available region $\Delta\phi$ (which is determined by $\hat{p} \to 0$ in eq. (2)) and $\Delta x = d\cosh\theta_0\Delta\phi$; and to the decrease in the localization length, corresponding to the WKB probability density ρ in the classically forbidden region: $\rho\alpha\exp(-2\int|p|d\phi/\eta)$. Note that Δx first increases αH^2, then αH, and finally, at the mobility edge, embraces the whole shell with the extended state. Classical eq. (2), with $\hat{p}\to p$, determines "initial" $p(0)$ for given n,ℓ and allows to study the corresponding first stage classical orbit.

2 Orbital Variable Range Hopping Magnetoresistance

Until recently theory predicted [3] only negative orbital magnetoconductance in the variable range hopping (VRH) regime. Then experiments [4] stimulated the Nguyen-Spivak-Shklovskii (NSS) model [5] with positive magnetoconductance [6]. In this section I discuss when the model is applicable, why the conductance increases, what are the other options and relate one of them to physics of the previous section. For simplicity, I consider a 2 D case in a Corbino disc geometry, where the potential energy V is infinite at the edges $r=R_1$ and $r=R_2$, and is random at $R_1\leq r\leq R_2$. A hint at the situation is provided by infinite magnetic field $H\to\infty$, when an electron moves along equipotential lines [7] $V=$ const. When V is larger than a certain critical V_c (i.e. the corresponding Fermi energy is high enough), then the curves $V=$ const $>V_c$ are located [7,8] in the vicinity of $r\sim R_1$ and $r\sim R_2$, embrace $r=0$ and lead [9] (at zero temperature $T=0$ and $R_2-R_1\to\infty$) to quantized Hall effect (QHE), with zero conductance and zero resistance. When $V<V_c$, then the equipotential lines do not encircle $r=0$ and yield an insulator with infinite resistance and zero conductance. In a finite Corbino disc equipotential lines in the vicinity of V_c come from one disc edge to the other, and yield a metal.

In a general case the Mott VRH reduces [3] to the calculation of eigenenergies and localization of eigenstates. First I consider a model of [10], which allows one to accurately reduce the magnetoresistance problem to 1 D. In this model the potential energy of randomly situated "metal impurities" equals $-\sum_n(\hbar^2/2m)V_n(y-y_n)\delta(x-x_n)$; $V_n(-y)=V_n(y)$. Since x-size of impurities is zero, one may order x_n's and chose $x_{n+1}>x_n$, $n=0,1,\ldots,N-1$; $x_N\equiv+\infty$. In the Landau gauge the Schrödinger equation reads

$$\partial^2\psi/\partial x^2+(\partial/\partial y+ibx)^2\psi+\left[-K^2+\sum_n V_n(y-y_n)\delta(x-x_n)\right]\psi=0, \qquad (3)$$

where $K^2=-2mE/\hbar^2<0$, $b=eH/c\hbar$. When $x_{n-1}<x<x_n$, the solution to eq. (3) may be presented as

$$\psi=\int\exp(iky)\vec{A}_k^n\vec{g}_k(x)dk,\quad \vec{g}_k''(x)=\left[K^2+(k+bx)^2\right]\vec{g}_k(x);\quad g_k^\pm(\mp\infty)=0. \qquad (4)$$

The matching conditions at $x=x_n$ are $\delta\psi(x_n,y)\equiv\psi(x_n+0,y)-\psi(x_n-0,y)=0$; $\delta\partial\psi/\partial x_n=-V_n(y-y_n)\psi(x_n,y)$. They yield the recurrence relation for $\vec{A}_k^n$

$$\vec{A}_k^{n+1}=\vec{A}_k^n+\int\hat{R}_{k\chi}^n\vec{A}_\chi^n d\chi$$

$$\vec{R}_{k\chi}^{n,\alpha\beta}=-\alpha D_{kn}\frac{g_{\chi n}^\beta}{g_{kn}^\alpha}\exp\left[-i(k-\chi)y_n\right]W_n(k-\chi). \qquad (5)$$

Here $g_{kn}\equiv g_k(x_n)$; $D_{kn}=1/[\ell n(g_{kn}^+/q_{kn}^-)]'$ and $W_n(k)=\int\exp(-iky)V_n(y)dy$.

By eq. (4), $g_k^{\pm}(\pm\infty) \to \infty$. So, $A_k^{0-} = A_k^{N+} = 0$, and eq. (5) allows one to express $A^{0-} = 0$ via A^{N-} and to reduce the calculation of the spectrum to the eigenvalues of one homogeneous linear integral equation (for A^{N-}). Then eqs. (4) and (5) provide the eigenstates ψ. Suppose, following the Lifshitz model [11], that the wave function changes slowly compared to $V_n(y)$. Then in eq. (3) one may replace $V_n(y - y_n)\psi$ by $V_n(y - y_n)\psi_n$, $\psi_n \equiv \psi(x_n, y_n)$. This is formally equivalent to replacing (a relatively slow) $W_n(k - \chi)$ by $W_n(k)$ in eq. (5). The resulting degenerate equations reduce to N equations for $\psi_n = \int \exp(iky_n)\vec{g}_{kn}\vec{A}_k^n dk = \int \exp(iky_n)\vec{g}_{kn}\vec{A}_k^{n+1}dk$, $n = 0, 1, \ldots, N - 1$, which determine the eigenenergies and, by eqs. (4) and (5) with the known ψ_n, the eigenstates in the magnetic field

$$\psi_n = \sum_{m=0}^{N-1} S_{nm}\psi_m, \; S_{nm} = \int \exp\left[ik(y_n - y_m)\right] (g_{kn}^{\alpha}/g_{km}^{\alpha}) \, D_{km}W_m(k)dk, \qquad (6)$$

where $\alpha = -$ if $m \leq n$ and $\alpha = +$ if $m > n$. Since x_n increases with n, S_{nm} by eq. (6) depends on x_m (via g, D and, randomly, via W_m) and (regularly) on $(\vec{r}_n - \vec{r}_m)$

$$\psi(\vec{r}_n) = \sum_{m=0}^{N-1} S(\vec{r}_n - \vec{r}_m, x_m)\psi(\vec{r}_m). \qquad (7)$$

If the average $Kr_{nm} \equiv K|\vec{r}_n - \vec{r}_m| \gg 1$ for $n \neq m$, while $Hr_{nm} \ll K$, then WKB approximation yields

$$S(\vec{r}_{nm}, x_m) \approx W_m(0)\exp\left[-Kr_{nm} - iH(x_n + x_m)\frac{y_{nm}}{2} - \frac{H^2 r_{nm}^3}{24\,K}\right], n \neq m$$

$$S(0, x_m) \approx \int W_m(k)dk/2\sqrt{K^2 + k^2}. \qquad (8)$$

This equation is valid for any potential wells and may be considered as the refined (by the exponential H^2 decay) NSS model. By eqs. (7) and (8), in the leading WKB approximation the spectrum is independent of magnetic field [12] and is the same as for a single well at this eigenenergy location. This is natural for a very narrow well with a very small magnetic flux through it. By eqs. (7) and (8), when eigenfunctions of different wells overlap exponentially little, the VRH magnetoresistance exponentially *increases* with $H^2 r_M^3/24\,K$, where r_M is the Mott hopping distance, c.f. [3]. When eigenfunctions of different wells overlap, one must use eq. (7). Suppose $Kr_{nm} \ll 1$. Then

$$\psi(\vec{r}) \approx \int S(r - r', x')P(|x - x'|)P(|y - y'|)\psi(r')dr', \qquad (9)$$

where P is the probability density for independent $|x - x'|$, $|y - y'|$. The Fourier transformation with respect to y reduces the problem to the 1 D integral equation. Then localization depends on the relation between H and $|\ell n W(0)|$ and may *decrease* with H, c.f. [5,6].

Special consideration may be given to (exponentially little probable here) $Kr_{mn} \gtrsim 1$ and the $|\psi|$ oscillations with H (which may be accounted for by, e.g., the Poisson formula for eq. (7)).

The magnetoresistance is very different in the (new) revised version of the well-known Miller-Abrahams random resistor network [13]. To explicitly account for localization and (which is even more important) for magnetic field effects, I ascribe to each "metallic resistor" the varying width (of order of the "true" localization length ξ), the length of order of the Mott hopping distance r_M, the shape of the WKB tunneling path, and the potential energy $(-V_0)$ inside and zero outside it. These resistors account for percolating potential wells. The latter go from one edge of the sample to the other only when [14] $V > V_c$. Otherwise there must be random gaps $\sim r_M$ between resistor clusters [14]. Since VRH is essentially [15] 1 D, the same is assumed about the network. Of course, such a model becomes invalid in a very strong magnetic field, when magnetic length $L_0 < \xi$. Then the potential inside resistors must also be considered random (which leads to QHE or to insulator).

When $H = 0$, such a model, according to section 1, leads to localization in the vicinity of the local width maxima. It provides three types of magnetoresistance change. The resistance related to the barrier between metallic clusters exponentially *increases* due to the magnetic field induced localization of eq. (8). The resistance of the slowly varying width regions exponentially decreases (αH^2 and αH) according to the magnetic field induced *delocalization* of section 1. When the width changes "too rapidly" for a given magnetic field, or when H is "too weak" for a given width change, then the corresponding localization and thus magnetoresistance almost do not change.

Thus, there may be three types of the VRH magnetoresistance behaviour. In a "Mott insulator" with quasi-independent potential wells magnetoresistance is positive. In a "Mott metal" ($V > V_c$) magnetoresistance is negative, with certain almost H-independent regions. In the "intermediate" material, in a magnetic field below a certain value, the interplay between positive barrier magnetoresistance, quasi-H-independent regions and negative magnetoresistance delocalizing regions may lead to oscillations on the top of the negative magnetoresistance envelope. In sufficiently high H, when all clusters are delocalized, positive barrier magnetoresistance always wins. The transition from negative to positive magnetoresistance envelope occurs at temperature $T \sim E_F/(k_F\xi)^4$ where E_F denotes the Fermi energy.

A detailed study of each of the cases, as well as their "phase diagram", will be given in more detail elsewhere.

3   Time-Space Oscillations and Giant Resonances in Tunneling

An interesting case of non-separable variables is related to time response in tunneling. It is also related to the problem of a tunneling traversal time. The fact that this problem remains a subject of controversy after almost six decades [16,17], is a hint to its more profound nature of the intrinsic question: what is the tunneling traversal time, *how* one may *measure* it? In classical mechanics a precise measurement of traversal time is straightforward. One may probe a 1 D stationary flow of particles with the weak perturbative force $f = \delta(x)F(t)$, where $F(t) = F(-t)$ starts at $t = -t_0$, and $F(t) = 0$ when $t < -t_0$. If the probability density at x starts changing at $t = -t_0 + t_r$, then the retardation time t_r is the traversal time, and $v_r = dx/dt_r$ is the traversal velocity of particles. In quantum mechanics a perturbative potential $\delta U = \delta(x)F(t)$ yields an energy uncertainty $\delta E \sim h/T$, where T is the characteristic time of $F(t)$. So, $T \sim h/\delta E$ is the time uncertainty δt. To little change the velocity, δE must be $\ll E$, i.e. the impulse must be *slow*, $T \gg h/E$. To allow for the relatively accurate measurement, the traversal time t_r must be $\gg \delta t$, i.e. the detector must be sufficiently far off from the perturbation. Under these conditions

$$t_r \gg T \gg h/E \tag{10}$$

one may define the time t_r. Suppose the wave function $\psi(t, x)$ at $t = -t_0$ is a stationary wave function with the energy E. Then

$$i\hbar\dot{\psi} = -(\hbar^2/2m)\psi'' + [U + F(t)\delta(x)]\psi, \tag{11}$$

$$\psi(-t_0, x) = \psi_E(x)\exp(iEt_0/\hbar), \quad \psi_E'' + (2m/\hbar^2)(E - U)\psi_E = 0. \tag{12}$$

The relative change in probability density is

$$\Delta(t, x) = |\psi(t, x)/\psi(-t_0, x)|^2 - 1. \tag{13}$$

The perturbation at $x = 0$ is maximal, together with $|F(t)|$, at $t = 0$. It is natural to define the time t_r, when $|\Delta|$ reaches a maximum at x, as a retardation time t_r

$$\max_t |\Delta(t, x)| = |\Delta(t_r, x)|. \tag{14}$$

Indeed, when $E > U$, then the WKB solution to eqs. (11)-(12) yields

$$\Delta(t, x) = -\frac{g^2(t - t_r)}{1 + g^2(t - t_x)} \; ; \; g(t) = F(t)/\hbar|v(0)|, \tag{15}$$

where $v(x) = \sqrt{2m[E - U(x)]}$, and $t_r = \int_0^x dx/v$ is the classical traversal time. Clearly, by eq. (15), $|\Delta|$ is maximal at $t = t_r$, as it should be.

Now consider tunneling with $E < U_0 = \max U(x)$. If $F(t)$ starts at a finite $t = -t_0$, i.e. $F = 0$ at $t < -t_0$, then F must be discontinuous at $t = -t_0$. Suppose $F^{(n)}(-t_0)$ has a jump; then the probability to increase the energy by $\hbar\omega$ is $\alpha(\omega t_0)^{-n}$, while the resulting increase in the transmission from $x = 0$ to a given x is $\alpha\exp(\omega t_{BL})$, where $t_{BL} = \int_0^x dx/|v|$ is the Büttiker-Landauer traversal time [17]. In virtue of eq. (10) (for t_r replaced by the only characteristic time t_{BL}), the exponent always wins, and the perturbation *always* propagates *above* the barrier! The only possibility for the perturbation to tunnel is to be related to an entire function $F(t)$, starting at $t = -t_0 = -\infty$, i.e. to be related to an *infinitely long* impulse, with the only "chosen" time being $t = 0$, when $F(t)$ is maximal. (The same reasoning is applicable to essential singularity.) This is the main point in the choice of a "probing impulse": tunneling must win the competition with activation. It determines how "long ago" the impulse must start, what are the other "real experiment" conditions, etc. [18].

Note that the difference from the tunneling *rate* situation (such as a decay of an unstable nuclei) is crucial: "consider a large box with a particle that can escape through a thin tube. The particle spends a long time bouncing around the box until it escapes and then a shorter time in the tube, while escaping" (Landauer [17]). The WKB solution to eqs. (11)-(12) with $t_0 \to \infty$ (under the conditions $t_{BL} \gg T \gg h/E$ yields

$$\Delta(t, x) = |1 + g(t + it_{BL})|^{-2} - 1, \tag{16}$$

i.e. as one should expect in tunneling, the retardation time is imaginary: $t_r = -it_{BL}$. The response (which may be studied for, e.g., $F(t) = F_0\exp(-t^2/T^2)$) is (see Fig. 3) (i) strongly non-linear in g, (ii) oscillatory in time and space, (iii) symmetric with respect to time inversion, (iv) exhibits advanced and retarded resonant maxima and advanced,

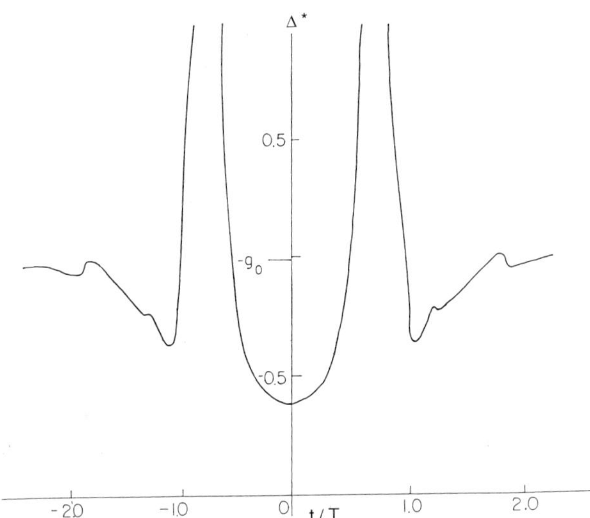

Figure 3:   Tunneling   response   close   to   resonance $(t_{BL}/T = 2.225)$ for $g = 0.01 \exp(-t^2/T^2)$; $\Delta^* = |\psi(t,x)/\psi(-\infty,t)| - 1$.

retarded and instantaneous (at $t = 0$) anti-resonant minima in time. All of them scale with the Büttiker-Landauer time t_{BL} (but t_r/t_{BL} may vary from $-\infty$ to $+\infty$); their $|\Delta|$ may be large even when $-t/T \gg 1$ and the perturbation is still exponentially weak. (v) When

$$g\left(t + it_{BL}\right) = -1, \tag{17}$$

giant resonances (with $\Delta = \infty$) occur, they correspond to local (in time and space) eigenvalues of the perturbation. Since t_{BL} depends on x and the energy E, resonances occur at certain energies and times at any x. When, e.g., $g = g_0 \exp(-t^2/T^2)$, $g_0 > 0$, then the resonance t_{BL}/T and t_r/T are

$$t_{BL}/T = \sqrt{(Q_n - \ell)/2} \qquad t_r/T = \pm\sqrt{(Q_n + \ell)/2}, \tag{18}$$

where $Q_n = \sqrt{\ell^2 + (2n+1)^2\pi^2}$, $\ell = \ln(1/g_0)$. When $n \to \infty$, then $t_r \simeq \pm t_{BL}$.

Thus, imaginary retardation time is equivalent to the ensemble of real advanced, retarded and instantaneous times. Of course, advanced responses are possible due to the analyticity of $F(T)$; any of its intervals determines its whole "future".

The presented approach is readily applicable to any dimensionality and type of waves (electromagnetic, acoustic, hydrodynamic etc.). It always leads to the response enhancement, which is advanced, retarded and instantaneous. This may affect a time-dependent interaction in a barrier; a one-particle escape from a condensed phase (then its effective potential is time-dependent due to the condensed phase adjustment to tunneling); wave propagation in a region, forbidden in geometrical optics. The latter may be important in the VRH regime.

References

[1] M.Ya. Azbel and O. Entin-Wohlman, J. Phys. **A 22**, L957 (1989); Phys. Rev. **B 41**, 395 (1990)

[2] D. Bergman, O. Entin-Wohlman, and M.Ya. Azbel, to be published

[3] B.B. Suprapto and P.B. Butcher, J. Phys. **C 8**, L517 (1975); B.I. Shklovskii and A.L. Efros, Electronic properties of doped semiconductors **45**, 210, Springer-Verlag, New York (1984)

[4] A. Hartstein, A.B. Fowler, and K.C. Woo, Physica **117-118 B**, 655 (1983); for later experiments see Y. Shapir and Z. Ovadyahu, Phys. Rev. **B 40**, 12441 (1989) and refs. therein

[5] V.I. Nguyen, B.Z. Spivak, B.I. Shklovskii, JETP Lett. **41**, 42 (1985); Sov. Phys. JETP **62**, 1021 (1982)

[6] O. Entin-Wohlman, Y. Imry, and V. Sivan, Phys. Rev. **B 40**, 8342 (1989) and refs. therein

[7] R.F. Kazarinov and S. Luryi, Phys. Rev. **B 25**, 7626 (1982); Phys. Rev. **B 27**, 1386 (1983)

[8] M.Ya. Azbel, Sol. St. Commun. **54**, 127 (1985); M.Ya. Azbel and O. Entin-Wohlman, Phys. Rev. **B 32**, 562 (1985)

[9] B.I. Halperin, Helv. Phys. Acta **56**, 75 (1983); Phys. Rev. **B 25**, 2185 (1982); M.Ya. Azbel and M.H. Brodsky, Phil. Mag. **50**, 237 (1984)

[10] M.Ya. Azbel, Phys. Rev. Lett. **47**, 1015 (1981)

[11] I.M. Lifshitz, Usp. Fiz. Nauk **83**, 617 (1964)

[12] It is convenient to chose $V(y) = (V_0/y)\sin(y/y^*)$. Then $W(k) = V_0$ for $|k| < 1/y^*$; $W(k) = 0$ for $|k| > 1/y^*$, and a single impurity yields $K = (y^* \sinh V_0^{-1})^{-1}$. The Lifshitz model is valid ($K \ll 1/y^*$) in a shallow well: $V_0 \ll 1$.

[13] A. Miller and E. Abrahams, Phys. Rev. **120**, 745 (1960); V. Ambegaokar, B.I. Halperin, and J.S. Langer, Phys. Rev. **B 4**, 2612 (1971), and ref. [3]

[14] See on their nature R. Meir and A. Aharony, Phys. Rev. **B 37**, 6349 (1989), and refs. therein

[15] See, e.g., M.Ya. Azbel, Sol. St. Commun. **43**, 515 (1982); Topics in Sol. St. Phys. **61**, 162 (1985)

[16] See, e.g., MacColl, Phys. Rev. **40**, 621 (1932), E.P. Wigner, Phys. Rev. **98**, 145 (1955); A.J. Leggett, Prog. Theor. Phys. Suppl. **69**, 80 (1980), E.H. Hauge and J.A. Støvneng, Rev. Mod. Phys. **61**, 917 (1989), and refs. therein

[17] M. Büttiker and R. Landauer, Phys. Rev. Lett. **49**, 1739 (1982), Phys. Scripta **32**, 429 (1987); IBM J. of Res. and Devel. **30**, 451 (1986); R. Landauer, Nature **341**, 567 (1989). See a detailed discussion on tunneling traversal time and refs. in M. Büttiker, to be published

[18] Note that an arbitrarily large (albeit exponentially weak) energy increase allows for the arbitrarily quick (albeit exponentially weak) response at any distance x

DYNAMICAL LOCALIZATION, DISSIPATION, AND REPEATED MEASUREMENTS

Robert Graham

Fachbereich Physik
Universität Essen GHS
W-4300 Essen, F. R. Germany

The study of the quantum signatures of classical chaos [1-3] has brought to light a number of new quantum effects, which may be combined under the label "dynamical localization". It has been seen in the excitation and (or) ionization of Rydberg atoms in a strong microwave field [4-6], however the phenomenon is more general and there is a whole class of such experiments, some of them already performed, others that may be conceived, which can be fruitfully considered from a common point of view: (i) one is dealing with quantum systems somewhere near the border between the quantum and the classical regime, which are (ii) driven by a periodic external electromagnetic field of a frequency larger compared to a neighboring level spacing, and of an intensity big enough to indirectly couple many unperturbed levels via multi-photon processes.

Examples are chaotic molecular vibrations excited in the field of an infrared laser [7], or periodically driven Josephson junctions sufficiently small to behave like single quantum systems similar to single atoms or molecules [8]. Or one may consider mesoscopic normal metal rings at low temperature of radius so small that the continuum of Bloch states of the electrons split into discrete quantum states. A strong direct coupling causing transitions between neighboring quantum states is achieved by the induction of a constant electric field E along the circumference of the ring [9] causing the effective coupling of many states by multiple transitions. Due to the periodicity of the ring with circumference L there arises an effective periodicity of the external perturbation in time with frequency ω, where $\hbar\omega = eEL$. Further examples are single ions or atoms in traps [10] whose motion is driven by external fields or micromasers driven by the repeated injection of excited atoms [11].

In a classical description a prominent role is played by "resonances" [12], the stable neighborhoods of stable periodic states of the system. The primary resonances are around periodic orbits where the external field performs m cycles, while the system performs 1 cycle. In between them there are Kolmogorov-Arnold-Moser (KAM)-tori for which m is fractional and irrational. For sufficiently high intensity of the external field the resonances "overlap" [12,13], i.e. they are mutually coupled and tend to destabilize each other, destroying, with increasing intensity, the KAM-tori with increasingly irrational m. Once resonance overlap is complete, i.e. the last KAM-torus has been destroyed, the resulting chaotic motion of the system connecting and surrounding the resonances has an essentially random phase relative to the external field and the absorption or emission of electromagnetic energy becomes a random process with zero average but finite mean square, defining a classical diffusion process for the absorbed energy [12,13].

The quantum description more or less agrees with the classical description for times short enough to make the discrete nature of the energy levels ineffective by the uncertainty principle [14]. Thus, an energy eigenstate of the undriven system starts to spread diffusively in energy after the external driving field is suddenly switched on. However, for sufficiently long times the quantum description need not be close to the classical description, the only requirement of the correspondence principle being that the time after which both descriptions disagree diverges to infinity in the classical limit [15]. In fact for the type of systems considered here the classical and the quantum description for long times differ radically. Where classical theory predicts diffusive spreading of energy, quantum theory, under the conditions described above, predicts localization with respect to the absorbed energy due to a class of novel but typical quantum mechanical coherence effects, dynamical localization [14-22,25]. Theoretically one may distinguish at least three types of dynamical localization according to their main causes, but in practice dynamical localization seems to be realized not as one of these pure types but as a mixture of two or all three of them.

The first is Anderson-type localization [16,17] or localization by the random interference of the superposition of all the many different multi-photon transition amplitudes between two given unperturbed states sufficiently far apart. To form an intuitive picture and also for a theoretical analysis [16,17] it is useful to think about the multi-photon transitions along the energy ladder, caused by the direct local coupling of each state to states in its neighborhood, as the coherent propagation of the probability amplitude along a one-dimensional (1 D) chain (the energy ladder) of random scatterers (the levels on the ladder), the randomness being due to the effective randomness of the phases of the states on the energy ladder [16,17]. In this picture, intuition gained from the study of Anderson localization in 1 D random media with short-ranged correlations can be brought to bear on the problem, and one predicts localization to occur under the circumstances described, provided the phase factors $\exp(-iE_n t/\hbar)$ of the states E_n on the energy ladder are sufficiently random and their coupling is sufficiently short-ranged [16,17]. Mathematically localization manifests itself in the discreteness of the spectrum of the Floquet operator (acting in an infinite-dimensional Hilbert space), which is the unitary operator which propagates the quantum system over one period of the external field.

A second type of dynamical localization [18-20] can be defined semi-classically as being caused by the presence, in classical phase space, of just barely broken KAM-tori which may still exist as Cantor sets between the classical resonances. Classically such "Cantori" are penetrable by phase-points and a characteristic action W may be associated with each of them equal to the classical phase area passing them with one period. Semi-classically a Cantorus is expected to act as an impenetrable barrier, giving rise to localization, if the area W is to small to support a quantum state $W < \hbar$, where $\hbar$ is Planck's constant. This quantum effect is just an extension, due to the finite size of $\hbar$ of the "trivial" localization by unbroken KAM-tori which already exists classically. One should expect that localization by random interference continuously goes over to this case either if the intensity of the external field is reduced sufficiently so that Cantori with W sufficiently small appear, or if states are considered corresponding to classical manifolds

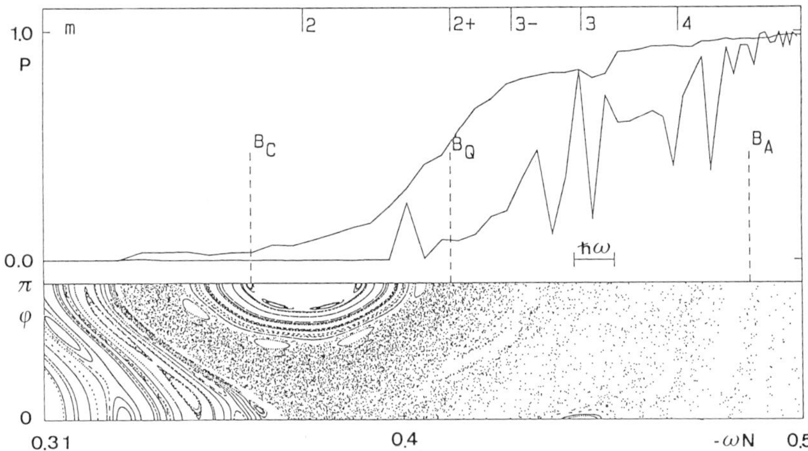

Figure 1: Dissociation probability of periodically driven molecular vibrations in a Morse potential as a function of the molecular energy in the initial state and the corresponding structure in phase space [24].

sufficiently close to the classical chaos border in phase space where such Cantori always exist.

A third type of dynamical localization is associated with unstable periodic orbits of the classical system and their stable and unstable manifolds in phase space [21-23]. Eigenstates of the Hamiltonian or, in the present context, the Floquet operator, may develop "scars" [21], i.e. they are peaked around unstable periodic orbits, in the sense that their overlap with coherent states centered anywhere along these manifolds becomes large [23]. Unstable periodic orbits whose period is a multiple of the external period always appear together with the stable periodic orbits giving rise to the primary non-linear resonances [12,13]. The stable and unstable manifolds of these unstable periodic orbits are known as "wild separatices" which intersect transversally infinitely often and form a "homoclinic tangle" in the phase space between two neighboring resonances. In recent work on the microwave excitation of Rydberg atoms the homoclinic tangle of some unstable periodic orbits was found to lead to strong scars [22].

As an example where all versions of dynamical localization can be seen at work we consider the excitation of molecular vibrations and dissociation in high-frequency fields, which has been analyzed theoretically in [24]. Figure 1, taken from [24], shows in its upper part the dissociation probability after 100 vibration periods versus the initial molecular energy (here denoted by $-\omega N$ where $-\omega N = 0$ in the ground state and $-\omega N = 1/2$ at the dissociation threshold), both classically (upper curve) and quantum mechanically (lower curve). In the lower part of the Figure a classical Poincare surface of section of the system is shown over the same energy-axis; the vertical axis here gives the phase variable φ canonically conjugate to the molecular energy. The Poincare surface of section is taken for constant total energy of field and molecule (put equal to zero arbitrarily) and at the minimum of the nuclear distance during each period of vibration. The classical phase-space is symmetrical under $\varphi \to -\varphi$, therefore only the interval $0 \leq \varphi \leq \pi$ is shown. The integers $m = 2, 3, 4$ indicate the position of the primary resonances.

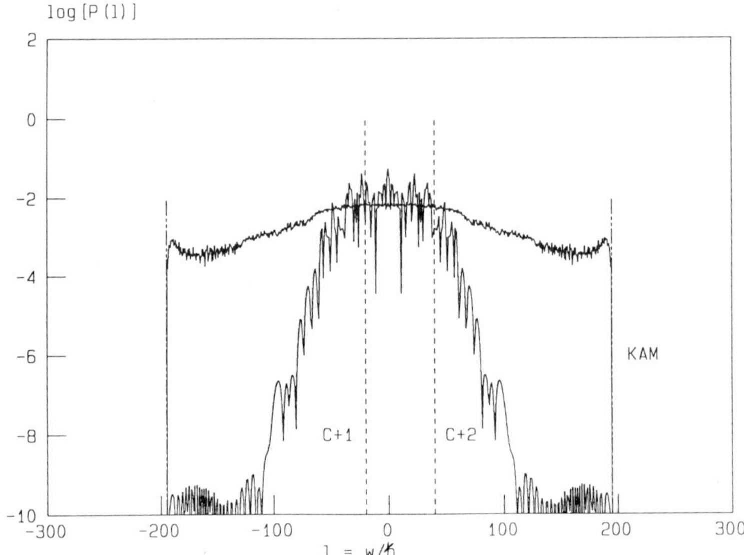

Figure 2: Distribution of the action variable $w = \hbar l$ in a stochastic separatrix layer described by the separatrix map after 10^4 iterations classical (curve denoted by KAM, extending to the first unbroken KAM-torus) and quantized (localized) [26].

The symbols 2+ and 3− denote the estimated phase-averaged positions of the two least penetrable Cantori between the resonances $m = 2,3$ with m equal to 2 plus or 3 minus the golden mean, respectively, where the criterion for Cantorus localization is satisfied for 2+ but not for 3−. The estimate is based on the positions of the corresponding KAM-tori in the undriven system. B_C indicates the phase-averaged position of the classical chaos border according to a simple analytical estimate [12,13], but both the classical phase portrait and the classical dissociation threshold show that the actual classical chaos border is somewhat lower. B_Q is the quantum dissociation border derived from the Cantorus 2+, while B_A indicates the distance from the dissociation threshold given by the localization length, again according to a simple analytical estimate [25]. It can be seen that, apart from resonance effects leading to peaks in the quantum dissociation curve, B_Q defines a border where quantum dissociation starts (evidence for Cantorus localization), but it remains below the classical result roughly up to the border defined by B_A (evidence for Anderson-type localization). It should be noted that the number of molecular vibrational eigenstates is finite in real two-atomic molecules and in the Morse potential for which Fig. 1 was obtained. Therefore Anderson-type localization, while reducing the dissociation rate, cannot be expected to suppress dissociation completely, whereas the Cantorus border should be very effective unless it is close to the dissociation border. Finally we note that the dissociation curve shows a pronounced resonance structure of states which dissociate comparatively easily and states which are much more stable than their neighbors. Comparatively stable states can be seen to exist immediately above and below the classical resonances $m = 3$ and $m = 4$ (evidence for localization by scarring on the homoclinic tangles around resonances).

Another example of dynamical localization on a chaotic separatrix layer [26] is considered in Fig. 2. A chaotic separatrix layer appears e.g. in a current-driven Josephson junction, whose equations of motion are isomorphic to those of a driven pendulum [13]. Its vibrational and librational motions in the Josephson junction correspond to states of zero and non-zero average voltage, respectively. In phase space these states are on different sides of the separatrix, which splits up to form a homoclinic tangle if a periodic driving current is applied [13]. For states close to the separatrix the change over one vibration or libration period is well described by the separatrix map [12,13] which has been quantized [26] in order to study the quantum transport across the homoclinic tangle. In this description the relevant variables are the action variable of the unperturbed system whose origin is chosen on the unperturbed separatrix, and the canonically conjugate angle variable. In Fig. 2 we plot the classical and quantum probability distribution of the action variable after 10^4 iterations of the map for an initial state (or, classically, an ensemble of initial states with equi-distributed phases) at the origin (i.e. on the classical separatrix). We also indicate the positions of the two least penetrable of a whole series of Cantori, which pile up towards the separatrix (i.e. the origin) with rapidly increasing penetrability [26]. The outer and least penetrable one is seen to act as a localizing barrier for the quantum system in agreement with the findings of [20] for other systems. Figure 2 seems to indicate that localization on separatrix layers at least in some cases may be related to localization by the Cantori surrounding the separatrix layer. While the separatrix-map used in Fig. 2 applies in the limit of a classically narrow stochastic separatrix layer further numerical results [27] show that dynamical localization can also be seen, under appropriate conditions, in current-driven Josephson junctions with stochastic separatrix layers sufficiently broad to destabilize the junctions's ground state. In addition other signatures of quantum chaos are predicted to show up in such junctions such as repulsion of quasi-energy levels which tends to suppress the low-frequency $1/f$-noise present in the classical regime [27].

The dynamical localization phenomena described are quantum mechanical coherence effects and very sensitive to all perturbations which tend to reduce coherence like external noise [28], dissipation [29,30], and any other coupling to the outside world such as the repeated or continuous interaction of a quantum system with a measuring device [31]. For instance, Rydberg atoms readily interact with thermal fields unless one cools to low temperatures, driven molecules can possess many degrees of freedom, only some of which couple to the external field, while others serve as a reservoir which can take up energy. Josephson junctions contain an effective resistive shunt leading to dissipation and measurements induce additional losses. Electrons in mesoscopic metal rings dissipate and take up energy when scattering inelastically from phonons, and again, suffer additional perturbations due to measurements [32]. Perturbations by repeated continuous measurements confront us with a new situation of fundamental importance [33], now that experimenters are able to isolate single quantum systems (rather than ensembles) which can be observed continuously in time.

Dissipation and measurements performed on a small quantum system continuously in time both amount to a coupling of the system to another macroscopic system or reservoir. Eliminating the variables of the latter one is left with the reduced dynamics of the small quantum system, which now, however, must be described by a statistical mixture, accounting for the partial loss of coherence due to dissipation and (or) at least partial collapse of the wavefunction due to the measurements performed. If the reservoir taking up the dissipated energy and (or) the measuring device have memory times which are short on the time-scale of the small quantum system the reduced dynamics may be treated as Markovian. Then periodically driven systems can still be described by

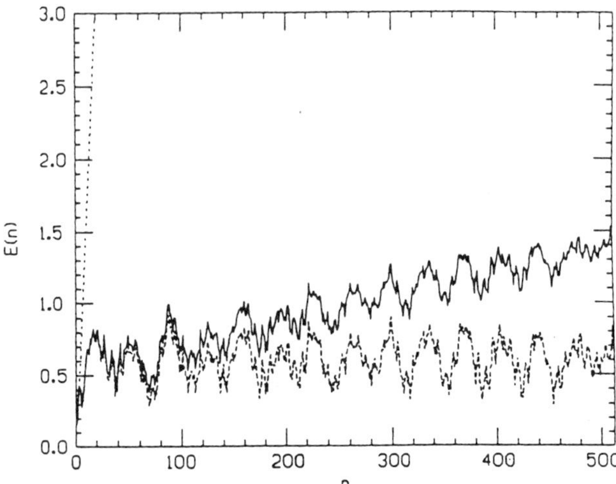

Figure 3: Mean energy of the quantum kicked rotor unperturbed (lower curve), perturbed by a continuous measurement of the angular momentum distribution (upper curve),
and classical (dashed curve) [31].

quantum maps $\rho_n \to \rho_{n+1}$, where ρ_n denotes the statistical operator of the small quantum system after n periods, and these maps can be numerically iterated [29,30,34]. In Fig. 3 we present a result for a model system, the kicked rotor, whose distribution over angular momentum states is monitored continuously in time [31]. Very similar results hold if the same rotor is subject to some dissipation [30]. The rotor is in the state of vanishing angular momentum initially. Its mean squared angular momentum is plotted against the number of kicking periods without continuous measurement (lower curve), and with continuous measurement summed over all possible outcomes (upper curve; increasing the accuracy of the measurement by increasing the coupling between the system and the meter increases also the average slope of the curve). The initial rise of the curve is described by classical random diffusion (dashed line) which is not noticeably disturbed by measurements, for larger times the curve without measurement displays localization by random interference, while the curve with continuous measurement shows a regime of measurement-induced diffusion instead.

As another closely related example we consider in Fig. 4 the mean energy of an electron on top of the Fermi sea in a tiny metal ring penetrated by a magnetic flux increasing linearly in time with (upper curve) and without (lower curve) taking into account dissipation due to the electron-phonon coupling [34]. The mean energy as a function of the time displays localization by random interference [35] or a dissipation induced diffusion process [34] in the cases without or with the electron-phonon coupling, respectively, in complete analogy to the results for the kicked rotor [30].

Finally, Fig. 5 shows the result [36] of recent experimental work [6,36] on a beam of Rydberg atoms (Rb $84P_{3/2}$) passing a wave-guide supporting a strong microwave field and some artificial broad-band noise of controlled and variable strength. The interaction time could be controlled and varied in this experiment in the range $10 - 5000\,n\,\mathrm{sec}$. After the atoms emerge from the wave-guide the width of the distribution of the resulting wave-

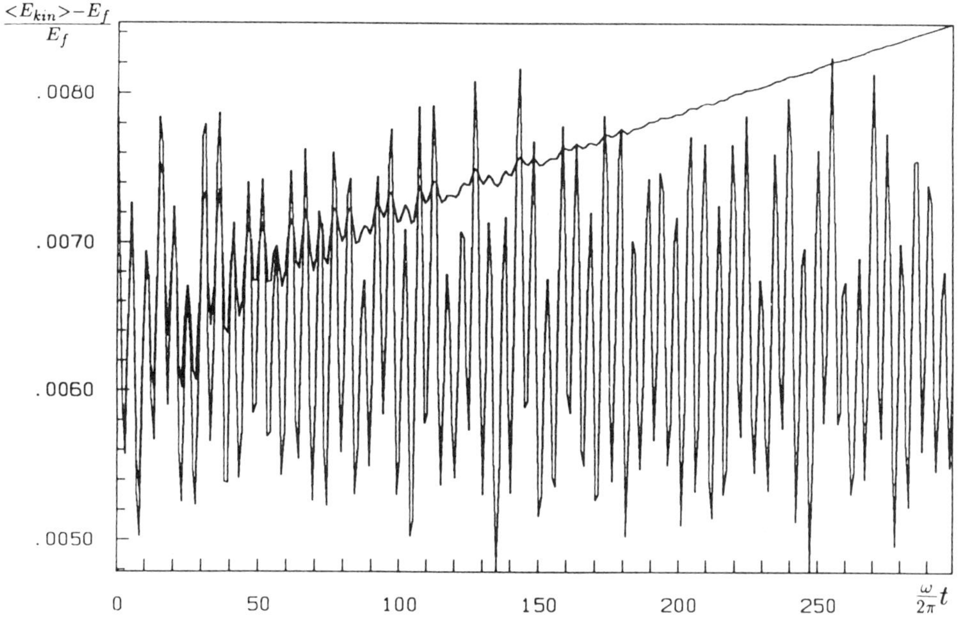

Figure 4: Mean energy of an electron in a normal metal ring in a magnetic field increasing linearly in time with (upper curve) and without (lower curve) electron-phonon coupling [34].

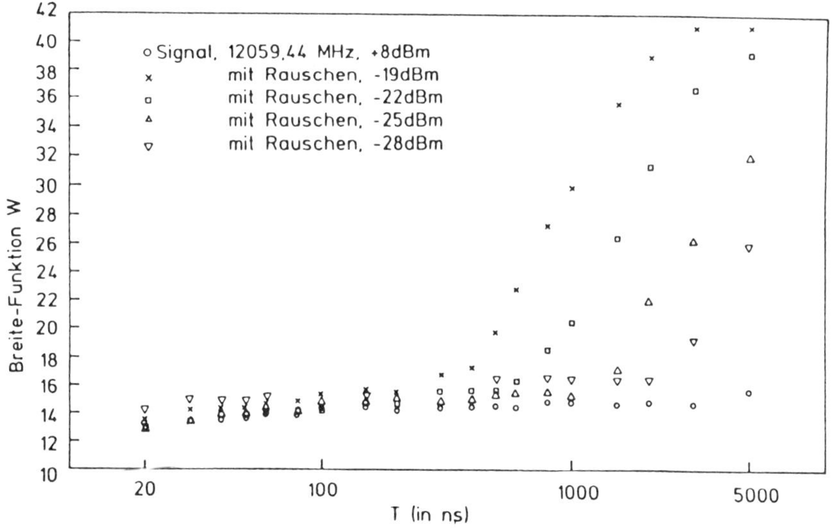

Figure 5: Width function $W = \exp(-\sum_n P_n \ln P_n)$ of the distribution P_n over Rydberg states in Rb 84 after interacting with a coherent microwave field plus noise for the time T ([36]).

packet over the principal quantum numbers was measured by a field-ionization technique. Without and also with sufficiently small noise dynamical localization of the wave-packet over about 14 Rydberg states is seen to occur (the initial spreading from 1 to 14 occurs in less than $10\,n\,\sec$) which is replaced, for stronger noise, by a noise-induced diffusion process whose initiation time and diffusion rate scale with the noise intensity in agreement with theory [6], demonstrating the presence of non-trivial dynamical localization and its crucial dependence on coherence (sensitivity to noise).

Acknowledgements

Some of the results reported here have been obtained in collaborations with Michael Höhnerbach, Nikolaus Bubner, Jürgen Keymer, Thomas Dittrich, and Roland Hübner whom I wish to thank for their various contributions. The support of this work by the Deutsche Forschungsgemeinschaft through the Sonderforschungsbereich 237 "Unordnung und große Fluktuationen" is gratefully acknowledged.

References

[1] G. Casati and L. Molinari, Suppl. to Progress in Theoretical Physics, to appear

[2] B. Eckhardt, Phys. Rep. **163**, 205 (1988)

[3] F. Haake, 'Quantum Signatures of Chaos', Springer-Verlag, Berlin (1990)

[4] E.J. Galvez, B.E. Sauer, L. Moorman, P.M. Koch, and D. Richards, Phys. Rev. Lett. **61**, 2011 (1988)

[5] J.E. Bayfield and D.W. Sokol, Phys. Rev. Lett. **61**, 2007 (1988); J.E. Bayfield, G. Casati, I. Guarneri, and D.W. Sokol, Phys. Rev. Lett. **63**, 364 (1989)

[6] R. Blümel, R. Graham, L. Sirko, U. Smilansky, H. Walther, and K. Yamada, Phys. Rev. Lett. **62**, 341 (1989)

[7] R.V. Ambartsumyan, Y.A. Gorokhov, V.S. Letokov, and G.N. Makarov, Sov. Phys. JETP **42**, 993 (1976)

[8] J.M. Martinis, M.H. Devoret, and J. Clarke, Phys. Rev. Lett. **55**, 1543 (1985)

[9] W. Washburn and R.A. Webb, Adv. in Phys. **35**, 375 (1986)

[10] W. Nagourney, J. Sandberg, and H.G. Dehmelt, Phys. Rev. Lett. **56**, 2797 (1986); Th. Sauter, R. Blatt, W. Neuhauser, and P.E. Toschek, Phys. Rev. **57**, 1696 (1986); F. Diedrich, E. Peik, J.M. Chen, W. Quint, and H. Walther, Phys. Rev. Lett. **59**, 2931 (1987)

[11] D. Meschede, H. Walther, and G. Müller, Phys. Rev. Lett. **54**, 551 (1985)

[12] B.V. Chirikov, Phys. Rep. **52**, 263 (1979)

[13] A.J. Lichtenberg and M.A. Lieberman, 'Regular and Stochastic Motion', Springer-Verlag, Berlin (1983)

[14] B.V. Chirikov, F.M. Izrailev, and D.L. Shepelyansky, Sov. Sc. Rev. C **2**, 209 (1981)

[15] G. Casati, B.V. Chirikov, F.M. Izrailev, and J. Ford, Lecture Notes in Phys. **93**, 334 (1979)

[16] S. Fishman, D.R. Grempel, and R.E. Prange, Phys. Rev. Lett. **49**, 509 (1982); Phys. Rev. **A 29**, 1639 (1984)

[17] D.L. Shepelyansky, Phys. Rev. Lett. **56**, 677 (1986)

[18] R.S. MacKay, J.D. Meiss, and I.C. Percival, Physica **13 D**, 55 (1984)

[19] R.S. MacKay and J.D. Meiss, Phys. Rev. **A 37**, 4702 (1988)

[20] T. Geisel, G. Radons, and J. Rubner, Phys. Rev. Lett. **57**, 2883 (1986); R.C. Brown and R.E. Wyatt, Phys. Rev. Lett. **57**, 1 (1986)

[21] E.J. Heller, Phys. Rev. Lett. **53**, 1515 (1984)

[22] R.V. Jensen, M.M. Sanders, M. Saraceno, and B. Sundaram, Phys. Rev. Lett. **63**, 2771 (1989)

[23] M.C. Gutzwiller, J. Math. Phys. **12**, 343 (1971); M.V. Berry, Proc. Roy. Soc. **A 423**, 219 (1989); E.B. Bogomolny, Physica **31 D**, 169 (1988)

[24] R. Graham and M. Höhnerbach, Phys. Rev. Lett. **64**, 637 (1990); to be published

[25] G. Casati, B.V. Chirikov, I. Guarneri, and D.L. Shepelyansky , Phys. Rep. **154**, 77 (1987); G. Casati, I. Guarneri, and D.L. Shepelyansky, IEEE J. Quantum Electronics **24**, 1420 (1988)

[26] N. Bubner and R. Graham, to be published

[27] R. Graham, M. Schlautmann, and D.L. Shepelyansky, to be published; R. Graham and J. Keymer, to be published

[28] J.D. Hanson, E. Ott, and M. Antonsen, Phys. Rev. **A 29**, 1819 (1984); S. Adachi, M. Toda, and K. Ikeda, Phys. Rev. Lett. **61**, 659 (1988)

[29] R. Graham, Z. Phys. **B 59**, 75 (1985); R. Graham and T. Tel, Z. Phys. **B 60**, 127 (1985); R. Graham, S. Isermann, and T. Tel, Z. Phys. **B 71**, 237 (1988)

[30] T. Dittrich and R. Graham, Z. Phys. **B 62**, 515 (1986); Europhys. Lett. **4**, 263 (1987); **7**, 287 (1988); Ann. Phys. (N.Y.), to appear

[31] T. Dittrich and R. Graham, in: Instabilities and Nonequilibrium Structures II, E. Tirapegui and D. Villarroel, eds., 145, Kluewer, Dordrecht (1989); Europhys. Lett. **11**, 589 (1990); to be published

[32] M. Büttiker, Y. Imry, and R. Landauer, Phys. Lett. **96 A**, 365 (1983); R. Landauer and M. Büttiker, Phys. Rev. Lett. **54**, 2049 (1985); R. Landauer, Phys. Rev. **B 33**, 6497 (1986)

[33] M.D. Srinivas and E.B. Davies, Opt. Acta **28**, 981 (1981)

[34] R. Graham and R. Hübner, to be published

[35] Y. Gefen and D.J. Thouless, Phys. Rev. Lett. **59**, 1752 (1987); G. Blatter and D.A. Brownse, Phys. Rev. **B 37**, 3856 (1988)

[36] A. Buchleitner, 'Chaos bei Rydberg-Atomen', Diplomarbeit , Ludwig-Maximilians-Universität, München (1989); R. Blümel, A. Buchleitner, R. Graham, L. Sirko, U. Smilansky, and H. Walther, to be published

HAMILTONIAN CHAOS IN LATERAL SURFACE SUPERLATTICES

Theo Geisel, Josef Wagenhuber

Institut für Theoretische Physik, Universität Regensburg
Universitätsstr. 31, W-8400 Regensburg, F. R. Germany, and
Institut für Theoretische Physik, Universität Frankfurt
W-6000 Frankfurt, F. R. Germany

Semiconductor physics and technology have reached a stage, where the creation of diverse artificial microstructures can be realized. High-purity $Al_xGa_{1-x}As/GaAs$ heterojunctions permit a ballistic motion of charge carriers in two dimensions (2 D) with elastic mean free paths of the order of $10\,\mu$m. In addition, lateral structures such as quantum dots, quantum wires, or lateral surface superlattices [1-3] (LSSL) are imposed in search of novel electronic properties for future devices. These lateral structures represent a 2 D potential for the charge carriers. We demonstrate in this article that the ballistic motion of charge carriers typically exhibits chaotic behavior. At present the spatial scales of the structures are still larger ($\geq$ a factor of 10 for LSSLs) than the Fermi wavelength. The particle dynamics can thus be described by classical approximations. We have shown some time ago that these approximations exhibit chaotic particle motions associated with $1/f$-noise [4]. On the other hand, as the spatial scales will be reduced further, these systems will also become interesting objects and a testing field for studies in quantum chaos.

In the present article we study the case of a ballistic charge carrier in an integrable superlattice potential subject to a magnetic field [5]. This problem is also of interest as the classical limit of the problem of a Bloch electron in an incommensurate magnetic field studied by Hofstadter and others [6]. In order to circumvent the unaccessibly high magnetic fields required to observe his self-similar electronic band structure, Hofstadter already suggested to use artificial 2 D superlattices with much larger lattice spacing than in natural crystals. Thereby, however, the electron dynamics also approaches the classical limit. Meanwhile such superlattices are available for experiments. Lateral superlattices with 1 D modulations [1,2] and 2 D modulations [3] have been realized with lattice parameters down to about 200 nm. This scale is still large enough to explain experimental observations on the basis of classical approximations for the dynamics of wave packets [7]. We report various types of chaotic diffusion (1 D and 2 D normal and anomalous diffusion) and point out the relevance of Kolmogorov-Arnold-Moser-theory (KAM) [5]. The mechanisms for the onset of 1 D-diffusion and 2 D-diffusion are explained

Quantum Coherence in Mesoscopic Systems
Edited by B. Kramer, Plenum Press, New York, 1991

by homoclinic intersections and the break-up of invariant KAM-tori, respectively. We conclude that chaotic diffusion may lead to a reduction of the elastic mean free path.

We consider a classical particle with charge e, mass m, and energy E moving in a 2 D periodic potential under the influence of a homogeneous magnetic field $\mathbf{B} = B\mathbf{z}$, described by the Hamiltonian

$$H(x, y, p_x, p_y) = (1/2m) \left[(p_x + eBy/2)^2 + (p_y - eBx/2)^2 \right] + V(x, y), \tag{1}$$

where $V(x, y) = V_0 \left[2 + \cos{(2\pi x/a)} + \cos{(2\pi y/a)} \right]$ is an isotropic (superlattice) potential represented by the lowest Fourier components. In the case of a semiconductor heterostructure, m represents the effective mass m^* and E the Fermi energy E_F. Measuring energy in units of V_0, lengths in units of the lattice constant a, and time in units of the inverse harmonic frequency $\omega_0 = (4\pi^2 V_0/a^2 m)^{1/2}$ leads to scaled variables

$$\widetilde{H} = H/V_0, \qquad \widetilde{x} = 2\pi x/a, \qquad \widetilde{y} = 2\pi y/a, \qquad \widetilde{t} = \omega_0 t. \tag{2}$$

The equations of motion then read (omitting the tildes for convenience)

$$\dot{x} = v_x \qquad\qquad \dot{v}_x = \sin x + 2\lambda v_y$$

$$\dot{y} = v_y \qquad\qquad \dot{v}_y = \sin y - 2\lambda v_x \tag{3}$$

corresponding to the Hamiltonian

$$H(x, y, p_x, p_y) = (p_x + \lambda y)^2 / 2 + (p_y - \lambda x)^2 / 2 + V(x, y) \tag{4}$$

with

$$V(x, y) = 2 + \cos x + \cos y. \tag{5}$$

The dimensionless quantity

$$\lambda = eBa/(16\pi^2 m V_0)^{1/2} = \omega_c/(2\omega_0) \tag{6}$$

proportional to the applied magnetic field B describes the non-integrable coupling between the two degrees of freedom and is related to the bare cyclotron frequency ω_c. Note that there are two integrable limits in this model, that is $\lambda \to 0$ and $\lambda \to \infty$.

The potential V of eq. (5) has minima at the energy $E = 0$, saddle points at $E = 2$, and maxima at $E = 4$. Thus in the regime $E \leq 2$ all orbits are restricted to one unit cell for all values of λ. For $E > 2$, localized and delocalized orbits may coexist. In this article we concentrate on the intermediate energy range $2 < E < 4$, where qualitatively the results do not depend on the value of E. Due to space limitations we present numerical results only for the special choice $E = 2.92$. In this energy range we find delocalized periodic and drifting quasi-periodic orbits coexisting with localized motions in phase space. The most interesting feature, however, is the occurrence of a variety of (deterministic) diffusive motions, which we have analyzed by means of the velocity power spectrum

$$S_\alpha(\omega) = \frac{1}{2\pi} \int_{-\infty}^{\infty} \langle v_\alpha(t) v_\alpha(0) \rangle \, e^{i\omega t} dt, \quad (\alpha \in \{x, y\}). \tag{7}$$

We have determined $S_y(\omega)$ using the Wiener-Khinchin theorem for a set of initial conditions located near the saddle point of the potential. With increasing λ starting with

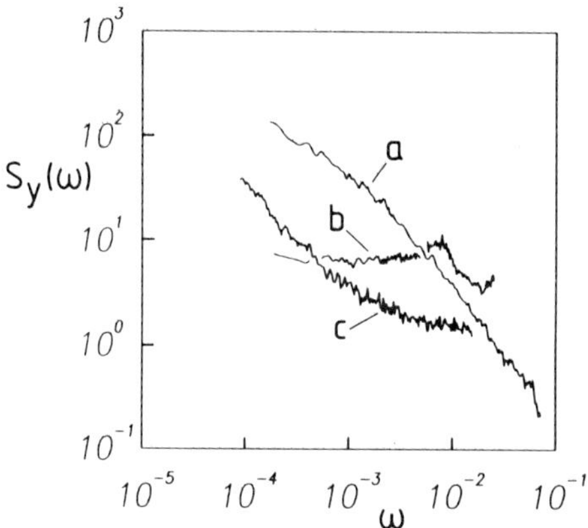

Figure 1: Velocity power spectra $S_y(\omega)$ for various magnetic field strengths λ. For $\lambda = 10^{-3}$ (a) and $\lambda = 0.07$ (c) $S_y(\omega)$ diverges as $\omega^{-\beta}$ with $\beta \approx 1$ corresponding to anomalous diffusion. For $\lambda = 10^{-2}$ (b) $S_y(\omega)$ remains finite for $\omega \to 0$ corresponding to normal diffusion. In each case the initial conditions were chosen in the stochastic layers (e.g. Fig. 3).

$\lambda = 10^{-4}$ we obtained the following results. For small magnetic fields ($\lambda \leq 10^{-3}$) the velocity fluctuations show $1/f$-noise (see Fig. 1), i.e. a low-frequency divergence of the spectral density $S_y(\omega) \sim \omega^{-\beta}$ with $\beta \lesssim 1$ (Fig. 1a). As explained in previous articles [4,8], this must be associated with anomalous diffusion, where the mean-square displacement $\sigma_y^2(t) = \langle y^2(t) \rangle$ grows faster than linearly in time, $\sigma_y^2(t) \sim t^{1+\beta}$. Above $\lambda = 10^{-3}$ there is a transition region where the character of the spectrum changes and near $\lambda = 10^{-2}$ the low-frequency divergence has disappeared (Fig. 1b). We thus can conclude that there is a transition to normal diffusion characterized by a finite value of the diffusion coefficient $D = \pi S(\omega = 0)$ and by a linear growth of the mean-square displacement. Inspection of the orbits in the above two cases shows that the particle diffuses along one axis only and remains confined in the perpendicular direction. At a critical field strength $\lambda_c \approx 0.014$ there is a transition from 1 D diffusion to 2 D diffusion, where the particle performs a random walk in the x-y-plane. With increasing values of λ the 2 D diffusion becomes anomalous (see Fig. 1c) and a normal regime is reached again above $\lambda \approx 0.3$.

We can understand the mechanisms for diffusion as well as the transitions between the regimes by analyzing the Poincaré surfaces of section [9] (Figs. 2, 3, 4). All sections show the y-v_y-plane at $x = \pi$ (mod 2π), i.e. in the potential minima. Due to the discrete translational symmetry of the potential all y-coordinates (mod 2π) are identified and plotted within the unit cell $[0, 2\pi]$. For the total energy we can write $E = E_x + E_y$, where E_x and E_y are the instantaneous energies of each degree of freedom, e. g. $E_y = 1 + \cos y + v_y^2/2$. The outer boundary of the Poincaré surfaces of section is always given by the curve $E_y = E$. Consider first the zero-field case $\lambda = 0$. Here the equations of motion separate into two uncoupled pendulum equations, the energies E_x and E_y are each conserved, and the orbits appear as invariant curves of constant E_y. The latter can be constructed simply from the phase portrait of the pendulum. The two pendulum separatrices for $E_y = 2$ and $E_y = E - 2$ (i.e. $E_x = 2$) divide the plane into three regions as

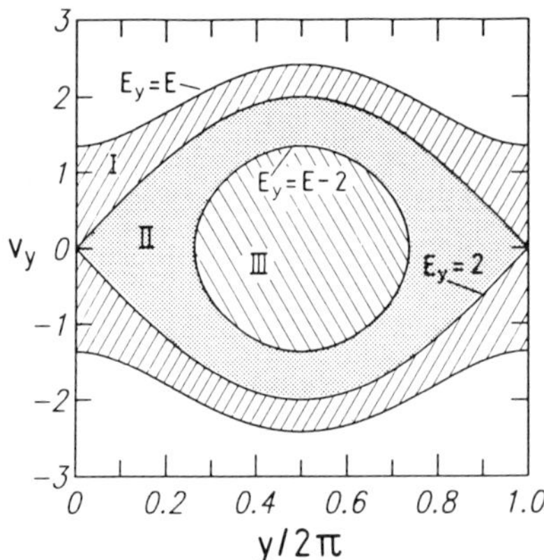

Figure 2: Poincaré surface of section at the potential minimum for the zero-field case ($\lambda = 0$). Two separatrices divide the phase space into regions of delocalized (drifting) motions (I, III) and localized motions (II).

depicted in Fig. 2: In region I, i.e. above the y-separatrix ($E_y > 2$) the orbits correspond to the running solutions of the y-pendulum. They are delocalized in y (drifting orbits) and thus must be localized in x (in the E-range considered here). In region III, for $E_y < E - 2$ we have $E_x > 2$ and thus the orbits correspond to running solutions of the x-pendulum and must be localized in y. Note that these two regions are interchanged under the symmetry operation $x \mapsto y$, $y \mapsto -x$. In the intermediate region II, we have $E_y < 2$ and $E_x < 2$ corresponding to swinging motion of both pendula and the particle performs local motions in both directions. The area of localized orbits between the two separatrices decreases with increasing total energy and disappears for $E \geq 4$, when the particle energy exceeds the potential maximum. In the section there is an elliptic fixed point at $y = \pi$, $v_y = 0$, corresponding to the minimum of the potential and a hyperbolic point at $y = 0$, $v_y = 0$, corresponding to the saddle point of V.

As we apply a magnetic field ($\lambda > 0$), in regions I and III regular drifting orbits continue to exist (Fig. 3). New elliptic and hyperbolic fixed points are created by non-linear resonances. Stochastic layers appear around the two separatrices as expected and grow with increasing magnetic field. The sharp boundaries between regions of localized (II) and delocalized orbits (I, III) are thus destroyed and a chaotic orbit in the layer may switch from a localized to a delocalized region and vice versa. This is the origin of diffusive motion. The layer around the outer separatrix generates diffusion in the y-direction. For symmetry reasons the inner layer plays the same role for the x-direction. The stochastic layers and thus the onset of $1\,\mathrm{D}$ diffusion are caused by a homoclinic intersection of the unstable and stable manifolds of the hyperbolic fixed point [9].

We now turn to the mechanism for the onset of $2\,\mathrm{D}$ diffusion. As is seen in Fig. 3 the inner and the outer stochastic layer are separated by invariant KAM-tori acting as barriers. An orbit in the outer layer thus cannot reach the inner layer and a particle diffusing in the y-direction cannot switch to free paths in the x-direction. With increasing perturbation the KAM-tori break up into cantori [10] and become partially penetrable thereby allowing $2\,\mathrm{D}$ diffusion. Fig. 4 shows the critical case $\lambda_c = 0.014$ where the two stochastic layers start being connected. The rate of switching between x-diffusion and y-

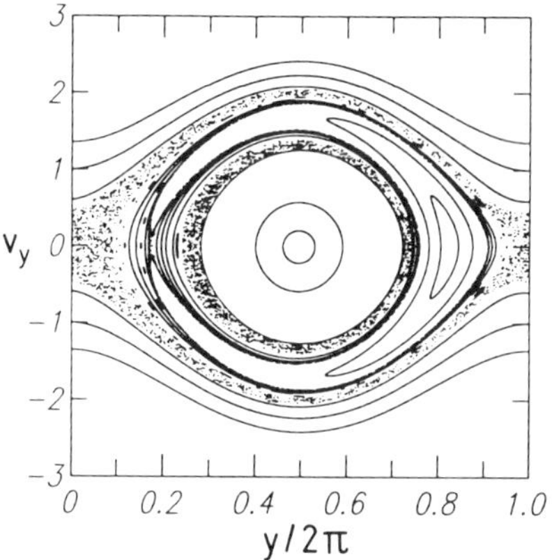

Figure 3: Same as Fig. 2 for $\lambda = 0.01$. The onset of 1 D diffusion consists in the creation of stochastic layers around the two separatrices and is caused by homoclinic intersections.

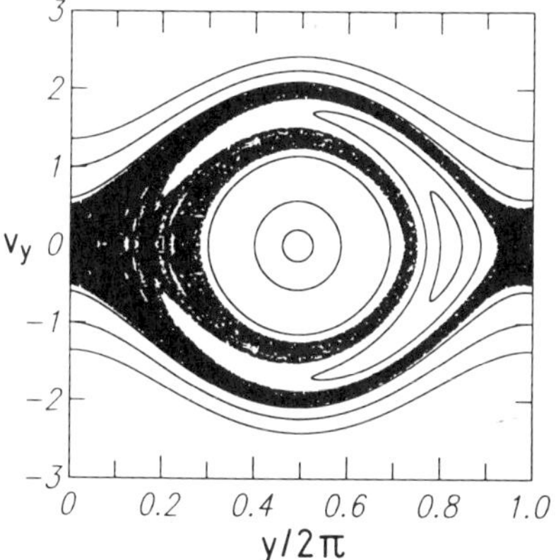

Figure 4: Same as Fig. 2 for $\lambda = 0.014$. 2 D diffusion sets in as the KAM-tori separating the two stochastic layers in Fig. 3 are destroyed. The connected stochastic region was generated by a single orbit.

diffusion is determined by the flux across two golden cantori. From work on the standard map [10] we may expect that it shows critical scaling behavior near λ_c with a power-law growth $(\lambda - \lambda_c)^{3.011722}$. The origin of the observed $1/f$-noise appears to be similar to the one which we analyzed in [4], more details are given in a forthcoming publication [11]. The above phenomena may show up experimentally in FIR-spectra measuring the frequency dependent conductivity (or the current power spectrum $S(\omega)$). A general conclusion to be drawn from our work is that the occurrence of chaotic diffusion limits the elastic mean free paths, even in a perfectly ordered medium as treated here.

References

[1] R.R. Gerhardts, D. Weiss, and K. v. Klitzing, Phys. Rev. Lett. **62**, 1173 (1989)

[2] R.W. Winkler, J.P. Kotthaus, and K. Ploog, Phys. Rev. Lett. **62**, 1177 (1989)

[3] K. Ismail, W. Chu, A. Yen, D.A. Antoniadis, and H.I. Smith, Appl. Phys. Lett. **54**, 460 (1989)

[4] T. Geisel, A. Zacherl, and G. Radons, Phys. Rev. Lett. **59**, 2503 (1987); T. Geisel, A. Zacherl, and G. Radons, Z. Phys. **B 71**, 117 (1988)

[5] T. Geisel, J. Wagenhuber, P. Niebauer, and G. Obermair, Phys. Rev. Lett. **64**, 1581 (1990)

[6] D.R. Hofstadter, Phys. Rev. **B 14**, 2239 (1976)

[7] C.W.J. Beenakker, Phys. Rev. Lett. **62**, 2020 (1989)

[8] A. Zacherl, T. Geisel, J. Nierwetberg, and G. Radons, Phys. Lett. **114 A**, 317 (1986)

[9] see e.g., M.V. Berry, in: Topics in Nonlinear Dynamics, S. Jorna, ed., AIP Conference Proc. No. 46, p. 16, American Institute of Physics, New York, (1978)

[10] R.S. MacKay, J.D. Meiss, and I.C. Percival, Physica **13 D**, 55 (1984)

[11] J. Wagenhuber, T. Geisel, P. Niebauer, G. Obermair, to be published

AUTHOR INDEX

SUBJECT INDEX